AF677566

Principles of Electricity and Electronics for the Automotive Technician

Principles of Electricity and Electronics for the Automotive Technician

Norm Chapman

South Puget Sound Community College
Olympia, WA

Delmar
Thomson Learning™

Africa • Australia • Canada • Denmark • Japan • Mexico • New Zealand • Phillippines •
Puerto Rico • Singapore • Spain • United Kingdom • United States

NOTICE TO THE READER

Publisher does not warrant or guarantee any of the products described herein or perform any independent analysis in connection with any of the product information contained herein. Publisher does not assume, and expressly disclaims, any obligation to obtain and include information other than that provided to it by the manufacturer.

The reader is expressly warned to consider and adopt all safety precautions that might be indicated by the activities herein and to avoid all potential hazards. By following the instructions contained herein, the reader willingly assumes all risks in connections with such instructions.

The publisher makes no representation or warranties of any kind, including but not limited to, the warranties of fitness for particular purpose or merchantability, nor are any such representations implied with respect to the material set forth herein, and the publisher takes no responsibility with respect to such material. The publisher shall not be liable for any special, consequential, or exemplary damages resulting, in whole or part, from the readers' use of, or reliance upon, this material.

Delmar Staff:
Business Unit Director: Alar Elken
Executive Editor: Sandy Clark
Acquisitions Editor: Vernon R. Anthony
Developmental Editor: Denise Denisoff
Editorial Assistant: Bridget Morrison
Executive Marketing Manager: Maura Theriault
Channel Manager: Mona Caron
Marketing Coordinator: Kasey Young
Executive Production Manager: Mary Ellen Black
Production Coordinator: Karen Smith
Project Editor: Barbara Diaz
Art/Design Coordinator: Cheri Plasse

Online Services

Delmar Online
To access a wide variety of Delmar products and services on the World Wide Web, point your browser to:
http://www.delmar.com
or email: info@delmar.com

Internet

COPYRIGHT © 2000
Delmar is a division of Thomson Learning. The Thomson Learning logo is a registered trademark used herein under license.

Printed in the United States of America

1 2 3 4 5 6 7 8 9 10 XXX 05 04 03 02 01 00

For more information contact Delmar at 3 Columbus Circle, PS Box 15015, Albany, New York 12212-5015; or find us on the Web at http://www.delmar.com

Asia
Thomson Learning
60 Albert Street, #15-01
Albert Complex
Singapore 189969

Australia/New Zealand
Nelson/Thomson Learning
102 Dodds Street
South Melbourne, Victoria 3205
Australia

Canada
Nelson/Thomson Learning
1120 Birchmont Road
Scarborough, Ontario
Canada M1K 5G4

International Headquarters
Thomson Learning
International Division
290 Harbor Drive, 2nd Floor
Stamford, CT 06902-7477
USA

Japan
Thomson Learning
Palaceside Building 5F
1-1-1 Hitotsubashi, Chiyoda-ku
Tokyo 100 0003
Japan

Latin America
Thomson Learning
Seneca, 53
Colonia Polanco
11560 Mexico D.F. Mexico

South Africa
Thomson Learning
Zonnebloem Building
Constantia Square
526 Sixteenth Road
P.O. Box 2459
Halfway House
South Africa

Spain
Thomson Learning
Calle Magallanes, 25
28015-Madrid
Espana

UK/Europe/Middle East
Thomson Learning
Berkshire House
168-173 High Holborn
London
WC1V 7AA United Kingdom

Thomson Nelson & Sons Ltd.
Nelson House
Mayfield Road
Walton-on-Thames
KT 12 5PL United Kingdom

All rights reserved Thomson Learning © 2000. No part of this work covered by the copyright hereon may be reproduced or used in any form or by any means—graphic, electronic, or mechanical, including photocopying, recording, taping, or information storage and retrieval systems—without the written permission of the publisher.

You can request permission to use material from this text through the following phone and fax numbers: Phone: 1-800-730-2214 Fax: 1-800-730-2215 or visit our web site at http://www.thomsonrights.com.

Library of Congress Cataloging-in-Publication Data
Chapman, Norman
Principles of electricity and electronics for the automotive technician / Norman Chapman.
p. cm.
Includes index.
ISBN 0-8273-8479-3 (alk. paper)
1. Automobiles-Electric equipment. 2. Automobiles—Electronic equipment. 3. Electronics 4. Electric engineering. I. Title.
TL272.C513 1999
629.25'4--dc21 99-14936
CIP

Contents

Chapter 11 Lighting Circuits

Chapter 12 Basics of Electronics and Computers

Chapter 13 Accessory Circuits and Basic Instrumentation

Chapter 14 Batteries and Testing

Chapter 15 Starting System Theory

Chapter 16
Charging System Theory

Chapter 17
Ignition System Theory

Preface

The *Principles of Electricity and Electronics for the Automotive Technician* has been written with the student in mind. Learning the basic principles of electricity and electronics should be fun and create a level of excitement for the subject, not intimidation. When compared to other textbooks on the same subject, most students will find the relaxed style of writing draws them to the subject. The need to learn the basics of electricity has never been more important than it is NOW!

The use of electronics in automobiles has grown drastically in the last twenty years. What used to be optional car equipment is now common in most cars, including rear window defoggers, climate control systems, intermittent wipers, air bags, digital instrument clusters, and memory seats. Somebody has to diagnose and fix these vehicle options when they go bad, and it might as well be you!

The automotive technician of the future must be solidly proficient in the basics of electricity. This electrical foundation must be a "sixth sense" for the technician. He can interpret the information in a wiring diagram and apply electrical concepts to the vehicle in order to accurately diagnose and repair the problem the first time.

This textbook is a "comprehensive exercise and resource guide" for the basics. It uses a comprehensive four-step approach for the acquisition of knowledge.

1. *Cognitive theory* To give students plenty of theory-based exercises, to prove *why* the circuit or device acts or reacts the way it does. Exercises are designed so repetition will bring mastery of the theoretical concepts.

2. *Case studies* By studying actual electrical case studies, the cognitive exercises can be proven as true. This expands the student's ability to see the entire operation of an electrical system before facing live troubleshooting on a vehicle.

3. *Hands-on vehicle tasks* Through the use of custom-designed tasks and the task list designed by the National Automotive Technicians Education Foundation (NATEF), a series of simple hands-on electrical tests common to most vehicles will complete the learning objectives.

4. *ASE style review questions* The most widely known national standard for the automotive technician is from the National Institute of Automotive Service Excellence or ASE. This textbook primarily uses the ASE multiple choice style review questions. They will be helpful to you in studying for all the ASE tests that have basic electrical questions: Electrical, Engine Repair, Engine Performance, Air conditioning, and Brakes.

We can all agree that times are changing for the automotive industry. Over the years the repair industry lost some good technicians because of their inability or unwillingness to upgrade their skills to modern technology. The vehicles of today cannot be repaired with skills from the sixties or seventies. Letting the "smoke" out of a solid state module because of your improper troubleshooting techniques can be expensive.

A technician with a solid background in electrical basics will be one of the most sought-after technicians in today's industry. Armed with wiring diagrams, various diagnostic charts, and the ability to use test equipment properly, a technician can solve the most complex electrical problems with accuracy.

If this textbook is your first experience with the basics of electricity and electronics, I hope it is a positive and enjoyable one.

The contributions of the following reviewers of the text are gratefully acknowledged.

Stephan Baldwin
Sullivan High School
Sullivan, IN

Walter Bertotti
Porter and Chester Institute
Wethersfield, CT

Thomas Broxholm
Skyline College
San Bruno, CA

William Caspole
Northwest Career Center
Dublin, OH

Earl Friedell
Dekalb Technical Institute
Clarkston, GA

John Gahrs
Ferris State University
Big Rapids, MI

Drew Goddard
Hennepin Technical College
Brooklyn Park, MN 55445

Robert Klauer
Metro Tech
Phoenix, AZ

James W. Manning
Santa Ana College
Santa Ana, CA

Paul Mueller
Brooklyn Center High School
Brooklyn Center, MN

Paul Pate
Community College of Southern Nevada
Boulder City, NV

Fred Raadsheer
British Columbia Institute of Technology
General Motors Group
Langley, British Columbia, Canada

Charles Statz
Temple Junior College
Temple, TX

Dr. Robert Wenzlaff
Fullerton College
Fullerton, CA

Dedication

This book is dedicated to several people who have been instrumental in my life over the last several years. To "Doc" Blevins who taught me decades ago a solid foundation of automotive technology that I continue to use today. To fellow instructors Jim Collins, Al Dent, and Steve McRavin who continue to have patience with all my "projects." To Denise Denisoff, my developmental editor at Delmar Publishers, for her positive attitude, encouragement, and tireless dedication to this project. To my children, Brandon, Brook, Bryan, Brittney, and the twins Jessica and Elizabeth, and my dear wife Barbara the "book widow." Without their fantastic support, patience, and encouragement for my writing efforts over the last two years, this project would not have been possible. And last, to my father, Warren, who instilled in me while growing up, all the personal traits and potential necessary for success in life.

1 Introduction to Electricity and Electronics

Introduction

The need for the automotive technician to understand the fundamentals of electricity has never been more important than *right now.* With complex electrical and electronic systems rapidly taking control over many mechanical aspects of the automobile, technicians who learn all that they can about this subject have open doors to a very rewarding career.

As consumers of electricity, we sometimes take for granted where it originates. Electricity can best be described as a natural form of energy. We cannot create energy, but we can successfully produce electricity and harness it by converting various other forms of natural energy. In this chapter we will look at natural forms of energy, and the people who have historically played a part in understanding and harnessing them for our use.

Objectives

When you complete this chapter, you should be able to:

- ❑ Explain the natural forms of energy used to produce electricity.
- ❑ Describe the people who were instrumental in developing modern electrical theory.

Electrical Sources of Power

Six sources of power will be discussed in this chapter. The first is **static electricity.** Static electricity is defined as electricity that is without motion or at rest, and will produce an electrical charge whenever two dissimilar nonmetallic materials are rubbed lightly together. Our knowledge of electricity began 2,500 years ago, when the Greeks discovered if they rubbed a piece of fossilized resin, called amber, with other materials it would attract lint, fibers, feathers, and other lightweight objects. The Greeks looked at this invisible force as a supernatural phenomenon. They didn't realize they discovered static electricity. Nearly everyone has experienced static electricity at some unexpected time. Examples are as simple as seeing the clothes stick together and crackle when they are removed from a hot dryer, running a comb through your hair and then picking up a piece of paper with the comb, or walking across a carpet on a dry day and then touching a metal object (Figure 1–1).

One way to see what is happening is to understand that static electricity is a surplus of electrons, or a lack of electrons, in an object. Electrons will jump from one object to another in order to return an atom to its neutral state of charge. In Figure 1–2, the electrons are transferred from the carpet to the rubber-soled shoe. The lower the humidity in the air, the greater the transfer of electrons.

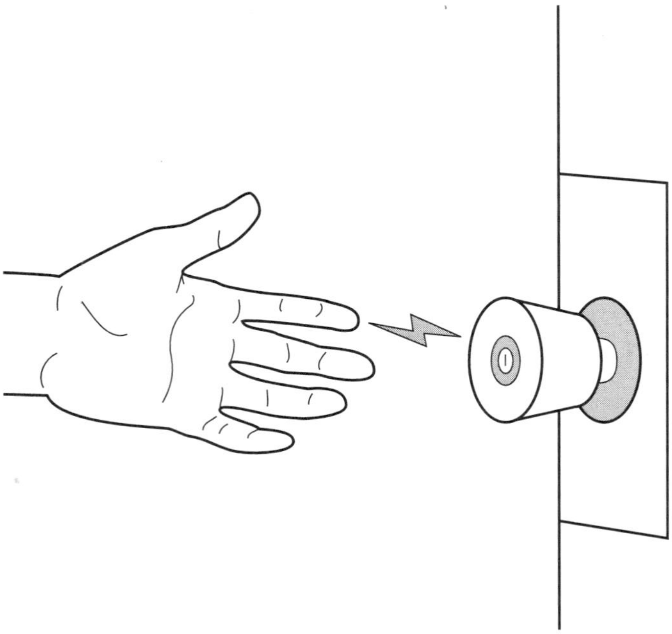

FIGURE 1–1 Example of static electricity.

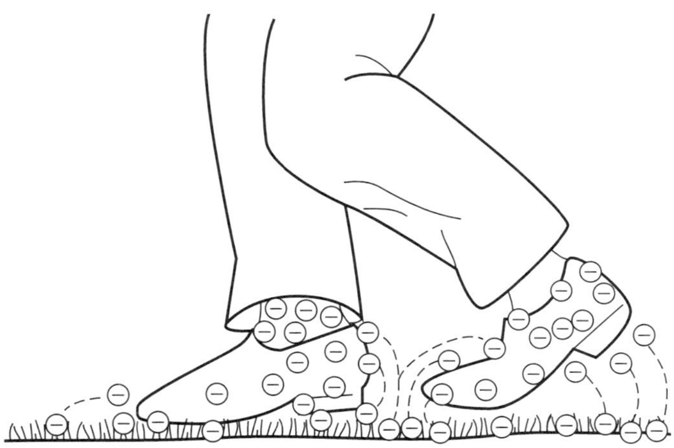

FIGURE 1–2 Excess electrons are transferred from the atoms in the carpet to the shoes.

These electrons will then transfer from the surface of your skin to any object that allows the transfer of the electrons (Figure 1–3).

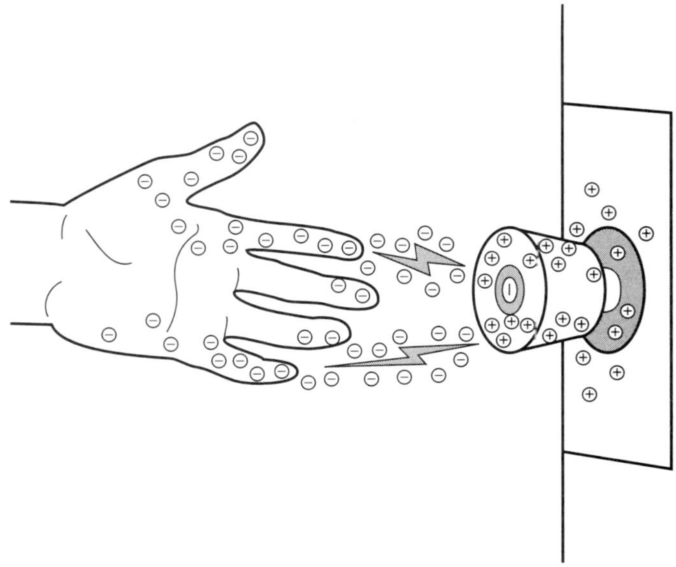

FIGURE 1–3 Example of electron transfer from a hand to a metal object.

One reason the characteristics of static electricity are important for the automotive technician to know, is **electrostatic discharge (ESD).** ESD is the process by which electrically charged objects transfer their charge to a nearby object. A common example of an ESD is a lightning storm. When Benjamin Franklin (inventor and scientist, 1706–1790) flew a kite during a thunderstorm (not a safe experiment to try yourself), he was trying to prove that the positive and negative distribution of electrons in the clouds produced the static electricity that causes lightning. Natural electron charge will produce lightning when it finds a suitable path to return to its neutral state of charge (Figure 1–4).

For the automotive technician, sliding across the vinyl seat of a car to service the solid state circuitry in the dash panel or to replace the vehicle computer can produce several thousand volts of static electricity. Vehicle components that are ESD sensitive are marked with ESD warning labels. A typical schematic diagram with an ESD warning label is shown in Figure 1–5.

To protect sensitive electronic components from ESD damage, the automotive technician should *always* use a grounding strap (Figure 1–6). These static straps are available from most electronic supply stores.

The second source of power is electricity generated from light energy. This is commonly known as **solar light energy,** which is light energy from the sun that is captured in a photovoltaic solar cell (Figure 1–7).

Light energy is composed of particles called **photons.** The photon is pure energy that contains no mass. When photon light energy strikes the surface of a photovoltaic cell, the energy in the photons is given to free electrons. This additional energy causes electrons to cross over the semiconductor material in a photovoltaic cell and produce a voltage. One of the most common

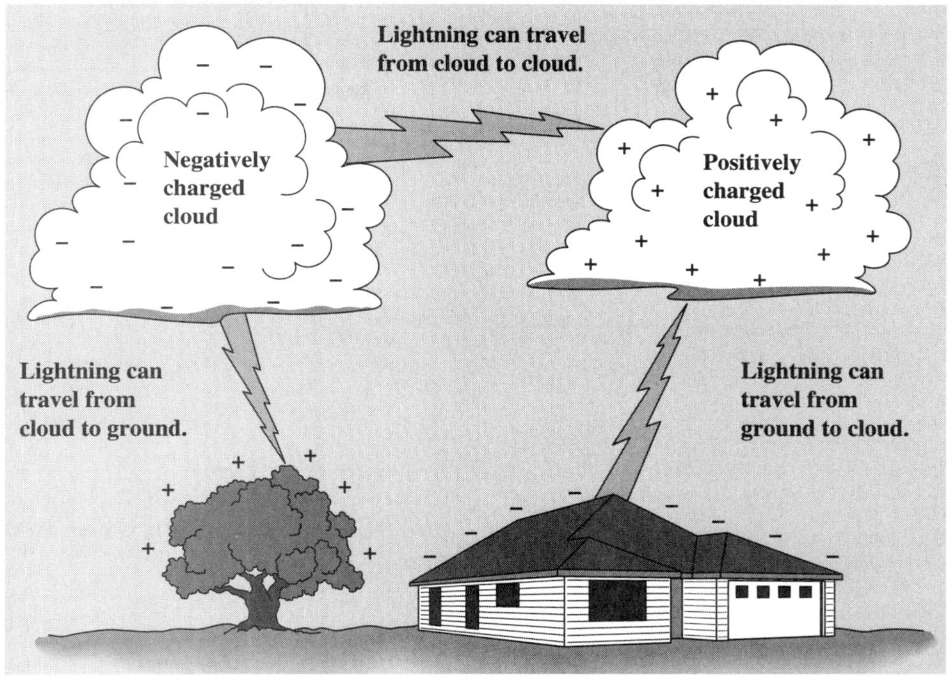

FIGURE 1–4 Electron induced charges in Earth's atmosphere.

(A)

(B)

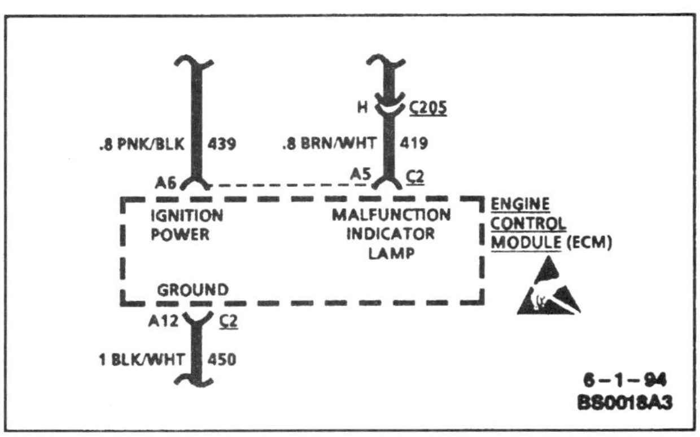

FIGURE 1–5 (A) Electrostatic Discharge (ESD) symbol and (B) sample ESD sensitive wiring circuit.

(Courtesy of General Motors Corporation, Service Technology Group)

applications of this solar technology is located on the face of a pocket calculator (Figure 1–8). The amount of voltage a photovoltaic cell can produce depends on the type of semiconductor material used, but is normally around 0.5 volts per cell in sunlight.

Another photovoltaic cell application is a photoelectric tachometer that uses light pulses to determine engine revolutions per minute (rpm) (Figure 1–9). To use this

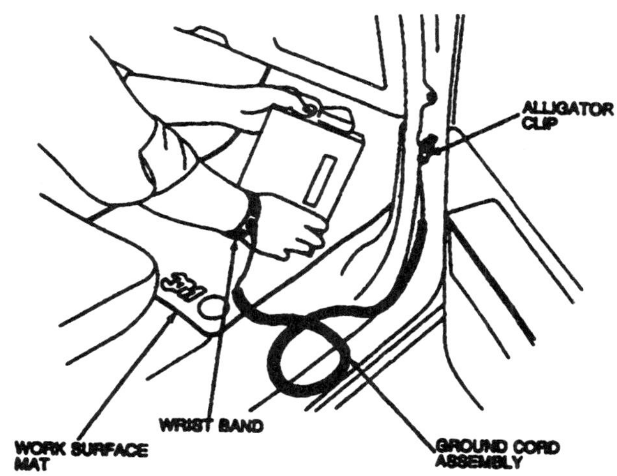

FIGURE 1–6 Using a grounding strap to protect sensitive electrical components from static discharge.

(Courtesy of Ford Motor Company)

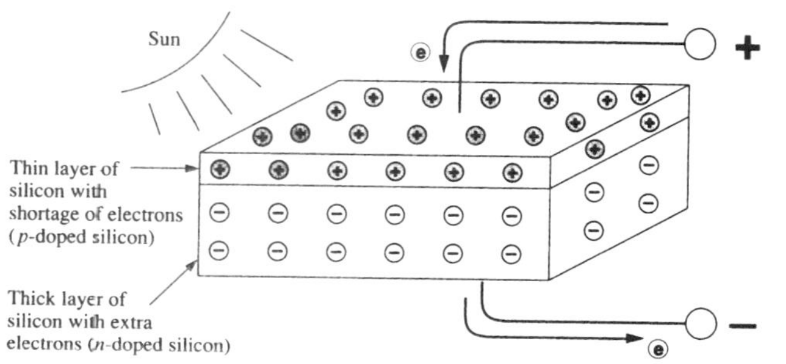

FIGURE 1–7 Simple photovoltaic solar cell.

FIGURE 1–9 Typical photoelectric tachometer uses reflective tape and light pulses to determine engine rpm.

(Courtesy of TIF Instruments, Inc.)

tachometer, a piece of photo reflective tape is placed on the front crankshaft pulley or harmonic balancer. With the engine running, the photoelectric tachometer is pointed towards the reflective tape, which reflects a series of light pulses back into the tachometer which then determines engine rpm.

In automotive applications, photovoltaic cells are used in the control circuits of several systems. Some of these are twilight sentinel headlamps, automatic adjustment control for a rearview mirror, and as part of a photoelectric distributor position sensor (Figure 1–10).

The schematic symbol for a photovoltaic cell is typically one of the three shown in Figure 1–11.

The third common source of power is **thermoelectricity,** which is electricity produced when two dissimilar metals are heated to generate an electrical voltage. In 1822, a German scientist named Seebeck found a direct connection between heating two dissimilar metals joined at one end, and a voltage produced at the open ends. This is also known as the **Seebeck Effect** (Figure 1–12).

One application of this source of power is a **thermocouple,** which is a small device that give off a voltage when the two dissimilar metals inside it are heated. These devices are used extensively in the diesel engine field, where they monitor the exhaust temperature of each

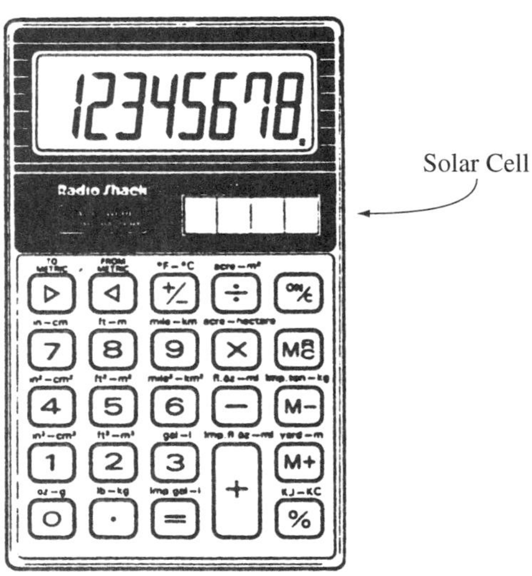

FIGURE 1–8 Pocket calculator powered by a photovoltaic solar cell.

(The EC-352 Metric Conversion Calculator is a trademark of the Radio Shack Division of Tandy Corporation)

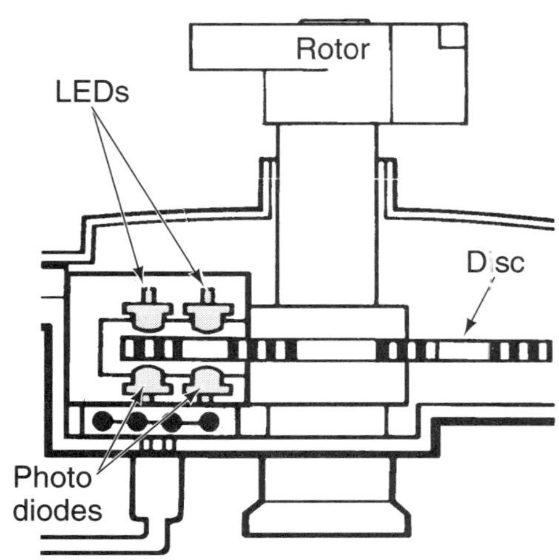

FIGURE 1–10 A photovoltaic crankshaft position sensor.

(Courtesy of DaimlerChrysler Corporation)

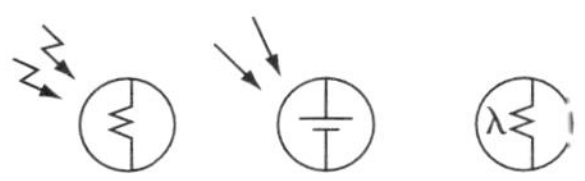

FIGURE 1–11 Common photovoltaic symbols used in wiring diagrams.

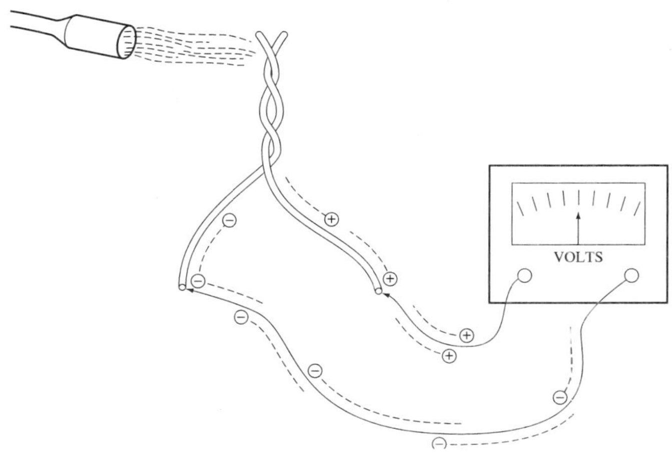

FIGURE 1–12 Thermoelectric production using two dissimilar metals and a heat source.

engine cylinder (Figure 1–13). Since exhaust temperatures usually exceed several hundred degrees, the use of a mercury glass thermometer would be impractical.

The fourth source of power is **piezoelectricity,** which is electricity produced from pressure. The word piezoelectric comes from the Greek word piezo which means pressure. When certain materials such as quartz or barium titanate undergo physical stress or vibration, a small oscillating voltage is produced (Figure 1–14).

The most common automotive application of a piezoelectric device is the knock sensor, or detonation sensor as it is sometimes called, which is used to control ignition spark knock (Figure 1–15).

The typical knock sensor produces about 300 millivolts (mV) and vibrates at a 6,000 hertz (cycles-per-second) frequency, which is about the same frequency of vibration as an ignition spark knock or detonation. When the engine detonates, either the fuel burn rate is too fast or the ignition advance timing is out of

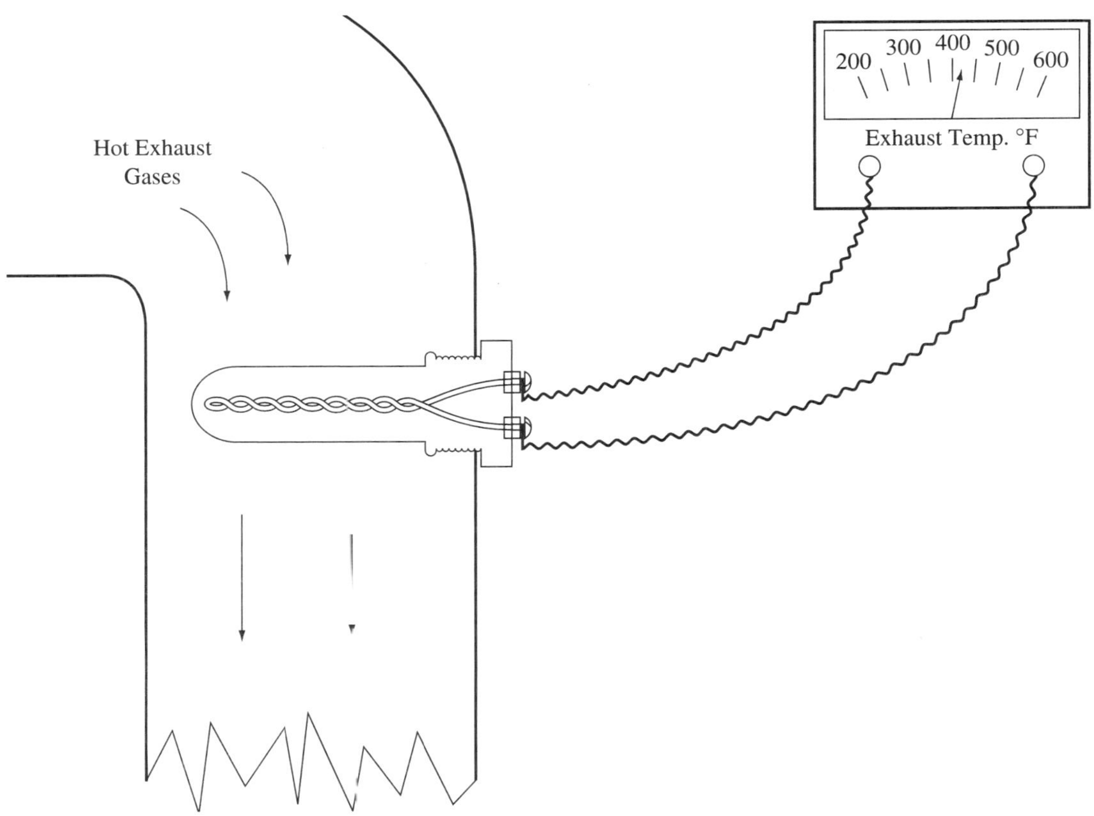

FIGURE 1–13 A simple thermocouple using the heat of an engine's exhaust to produce a voltage.

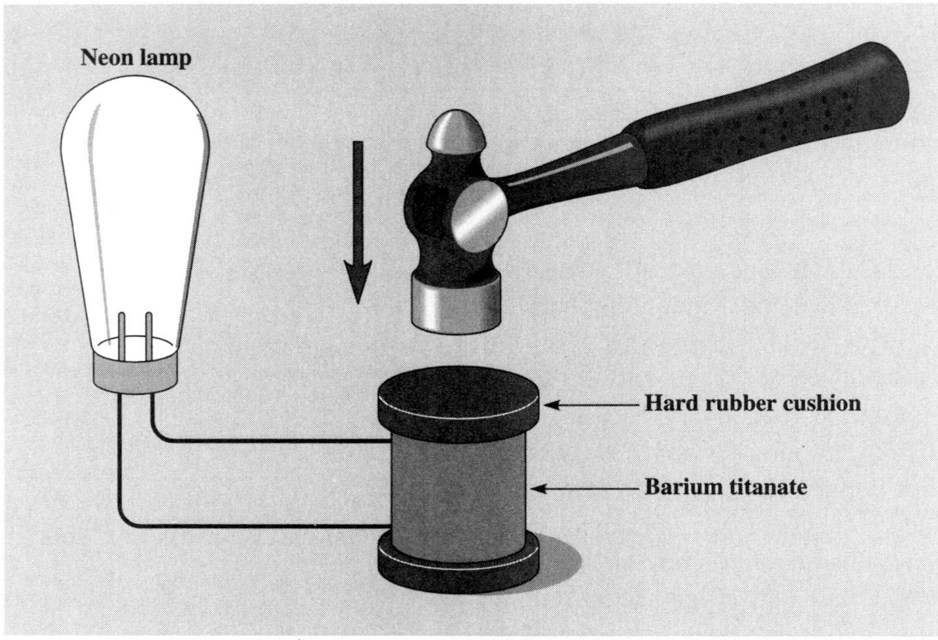

FIGURE 1–14 Producing a piezoelectric voltage.

specifications. The knock sensor will pick up this "mini explosion" and send the voltage signal to the engine control computer (Figure 1–16). The knock sensor is usually found on the lower part of the engine block to prevent an erroneous reading from the noise in the upper valve train.

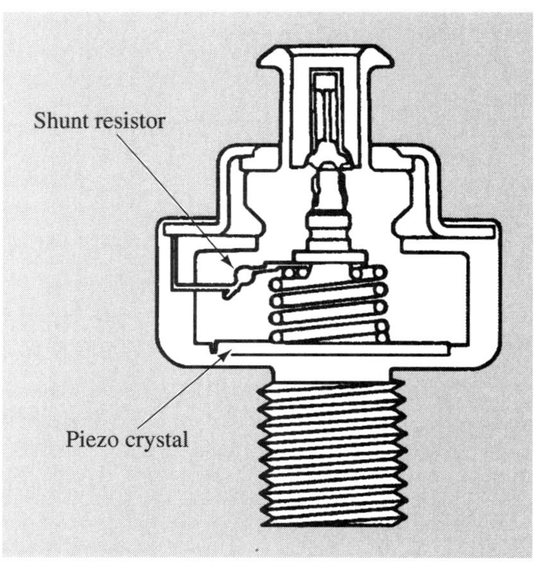

FIGURE 1–15 View of knock sensor with a piezoelectric sensing device. Shunt resistor prevents voltage spike (momentary high voltage) in the circuit.

(Courtesy of General Motors Corporation, Service Technology Group)

The fifth source of power is electricity produced from **chemical energy.** The most common form of this energy is the lead acid battery used in most automotive applications. One of the simplest forms of a battery is a lemon with two dissimilar metals like copper and zinc inserted as electrodes. The citric acid in the lemon acts as the electrolyte needed to produce a small voltage (Figure 1–17). Zinc will accept electrons better than copper, so zinc is considered the negative plate and copper is the positive plate.

A simple lead acid battery has a single cell with a positive electrode made from lead peroxide (PbO_2), and a negative electrode made from sponge lead (Pb) (Figure 1–18). The electrolyte is a mixture of sulfuric acid and distilled water (H_2SO_4). A chemical reaction takes place when the negative electrode releases electrons to the electrolyte and the positive electrode collects these electrons from the electrolyte.

Lead acid battery construction and testing will be discussed in Chapter 14.

The sixth and last source of electrical power is electricity produced from **electromagnetic induction (EMI).** When a current is carried through a conductor, a magnetic field will surround the wire and electromagnetism occurs (Figure 1–19).

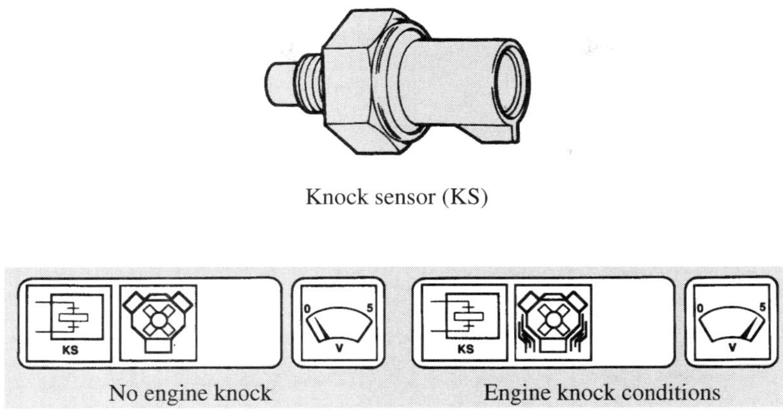

FIGURE 1–16 Knock sensor with related voltage signal.
(Courtesy of Ford Motor Company)

When additional loops are added to the conductor, the strength of the magnetic field increases accordingly (Figure 1–20).

A large-scale application of an electromagnet is where one pole of a magnet is formed in a shell, and the other is formed to the core inside the coil (Figure 1–21).

An automotive application of electromagnetism is a relay switch (Figure 1–22). Relays use an electric current

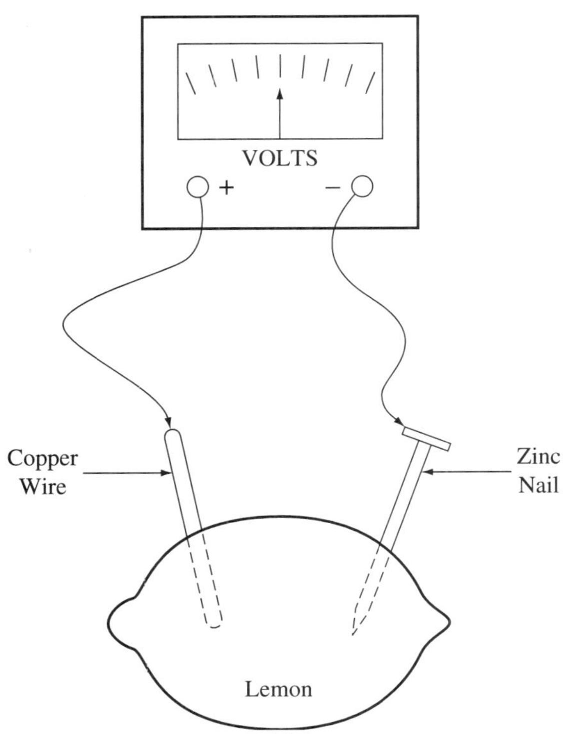

FIGURE 1–17 A simple lemon battery.

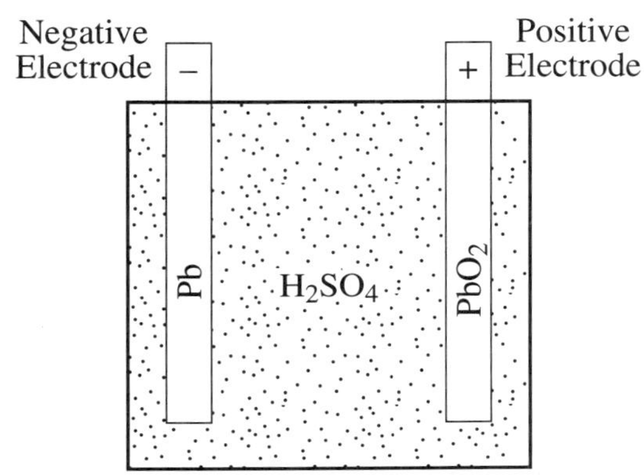

FIGURE 1–18 A simple lead acid battery.

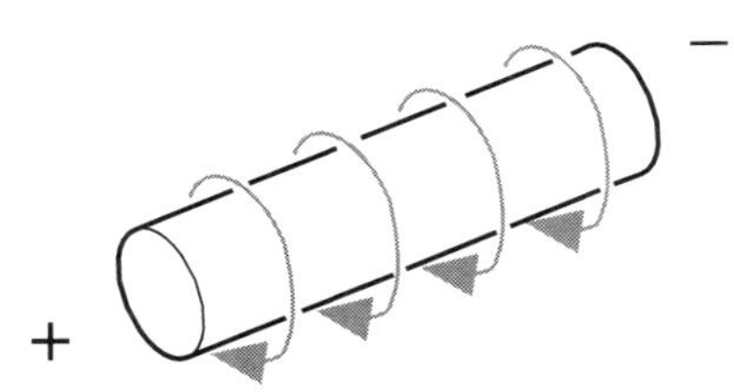

FIGURE 1–19 A magnetic field will surround a conductor when current flows through it.

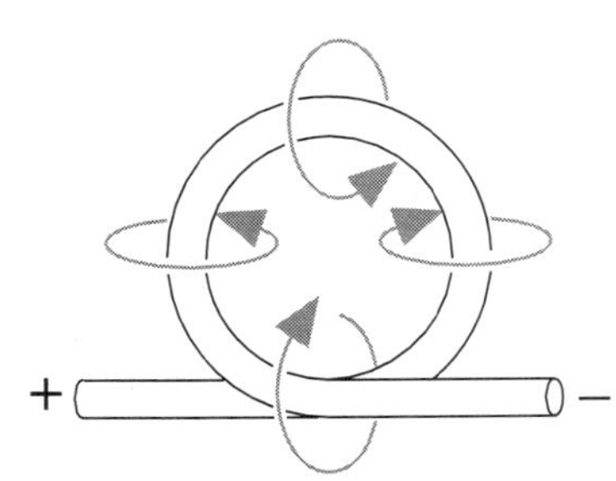

FIGURE 1–20 The magnetic field increases as the conductor is looped.

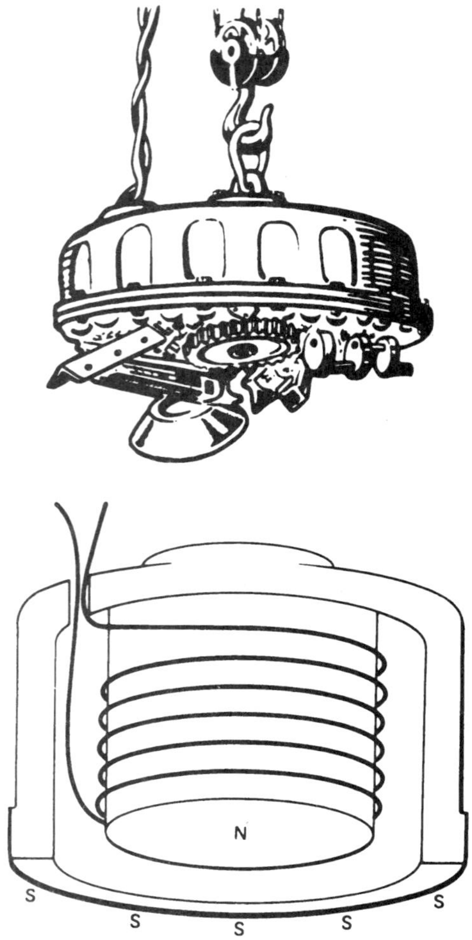

FIGURE 1–21 A simple lifting magnet.

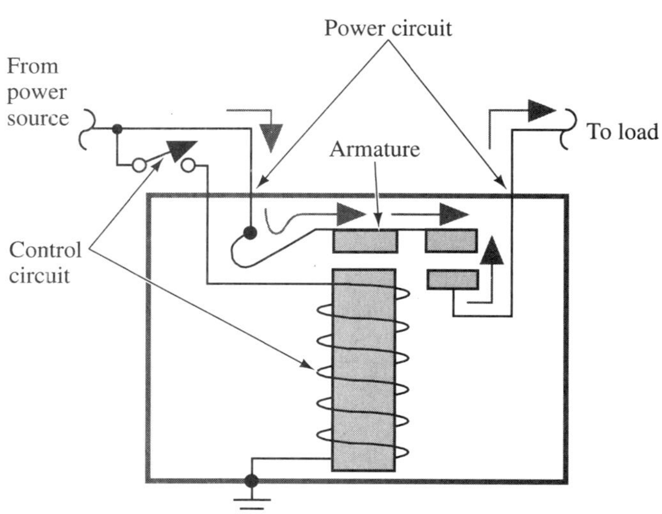

FIGURE 1–22 A simple relay using magnetism to energize the contact points.

to produce a magnetic field which pulls the contact points together to close the circuit.

In other chapters of this textbook we will discuss additional applications of producing electricity from electromagnetism, such as generators and alternators.

Historical Perspective

There are four people who have been instrumental in our modern day understanding and harnessing of electricity. The first is **André Marie Ampère,** (1775–1836). He was a French mathematician and scientist who established the importance of the relationship between electricity and magnetism. Ampére discovered that wires could behave like magnets by passing current through them, and that the polarity of their magnetism depended on the direction of the current. If the current in one wire is opposite of another wire, the wires will repel each other much the same as if you took two magnets of the same polarity and tried to touch them together (Figure 1–23).

Although it was not until sixty years after his death that his knowledge and accomplishments were given the respect they deserved, he did get the unit of electrical current, the ampere, or amp, named after him.

The second instrumental person was an Italian physicist named **Alassandro Volta,** (1745–1827). He was best known for his study with static electricity and current. His inventions ranged from the storage of static electricity, to the first device for producing electrical current, the battery. In 1775 he invented the electrophorus, which was a device used for storing an electrical charge. Today we call this device a condenser or capacitor. The invention that made him famous though, was the battery. He discovered that if two dissimilar metals were brought in contact with a salt solution, a current would be produced. This led to what is known as **voltaic piles,** which are a series of small round copper and zinc plates separated by cardboard soaked in saltwater (Figure 1–24).

Volta's most enduring reward was to have the unit of electrical potential, the volt, named after him.

The third person that influenced our modern day thinking about electricity was a German physicist named **George Simon Ohm** (1787–1854). In 1827 he showed there was a relationship between resistance, current, and voltage. As a result, the now famous **Ohm's Law** states that the current in an electrical circuit is inversely proportional to the resistance of the circuit, and directly

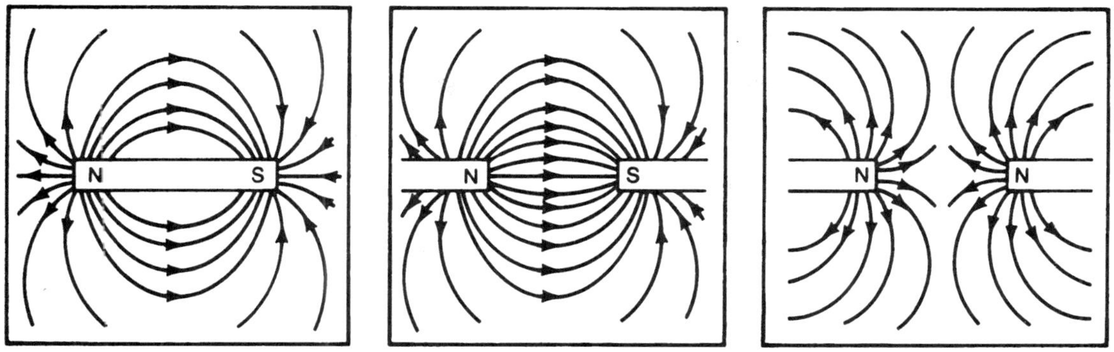

FIGURE 1–23 Invisible magnetic flux lines.

proportional to the voltage. This law has become the cornerstone in understanding the properties of electricity. With this law, scientists could accurately calculate the properties moving through an electrical circuit. We will be looking at Ohm's Law extensively in Chapter 2.

The last historical figure to influence electrical technology was a Scottish inventor named **James Watt** (1736–1819). Of all the inventors who made the Industrial Revolution possible, James Watt was probably the best known. In 1764 he realized the steam engine produced a considerable amount of wasted energy. Over the next several years he developed, patented, and refined several of his steam engines. To express the capacity of his engines, he developed a comparison between what his engine could do and the average work capacity of a normal size horse. He found a horse could produce 550 foot pounds of work per second, so he called this work one **horsepower** (Figure 1–25).

The results would be:

1 horsepower (hp) = 550 ft.-lb. per second or,
if you consider there are 60 seconds in a minute, then

1 horsepower = 33,000 ft.-lb. per minute

In electricity, the metric unit **joule** refers to the base unit of energy measurement. This unit is also called a watt, in honor of James Watt, where one watt is a rate of 1 joule per second. This relationship is best described in **Watt's Law,** which states that one watt is the amount of work done in one second by one volt moving a charge of one amp through a resistance of one ohm in a circuit. James Watt found horsepower could also be expressed as a unit of electrical power or watt. The simple conversion is:

1 horsepower = 746 watts

All four of these incredible inventors contributed substantial knowledge to the establishment of our current understanding of electricity. As a result, they have become standard household names associated with terms like a 60 watt light bulb, a 20 amp circuit breaker, a 120 volt wall outlet, and a 60 ohm electrical circuit.

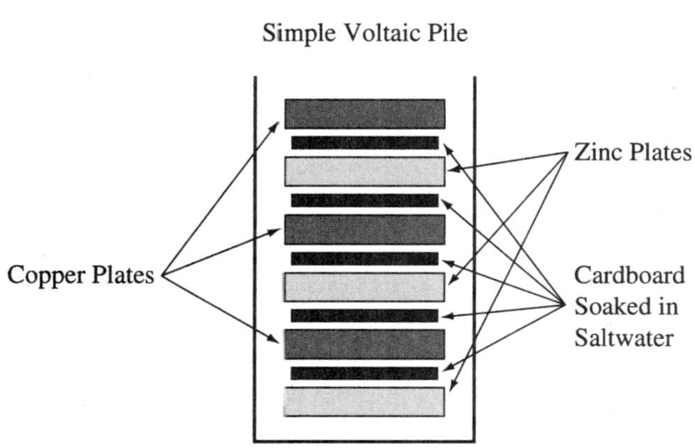

FIGURE 1–24 A simple voltaic pile.

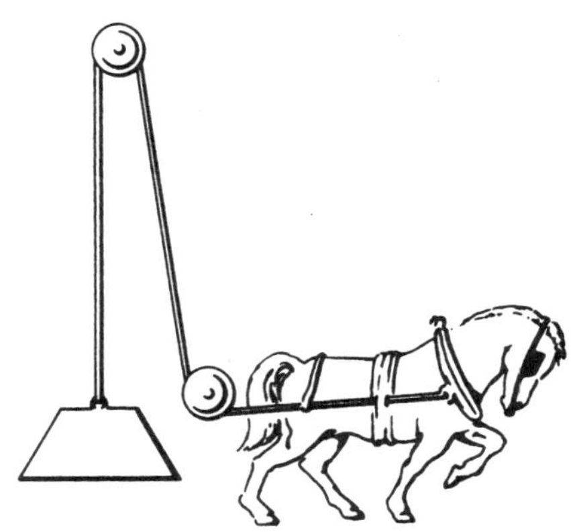

FIGURE 1–25 Horsepower.

Summary

- Static electricity is electricity without motion or at rest.
- Electrostatic discharge (ESD) is the process by which charged objects transfer their free electron charge to any neighboring object.
- An automotive technician should always use a static grounding strap when working with static-sensitive electronic devices.
- Solar energy is light energy from the sun that is gathered in a photovoltaic solar cell.
- A photon is pure energy that contains no mass.
- Thermoelectricity is electricity produced when two dissimilar metals are heated to generate an electrical voltage.
- A thermocouple is a small device that gives off a low voltage when two dissimilar metals are heated.
- Piezoelectricity is electricity produced when materials such as quartz or barium titanate are placed under pressure.
- The production of electricity from chemical energy is demonstrated in the lead acid battery.
- Electromagnetic induction is the production of electricity when a current is carried through a conductor and a magnetic field is produced.
- André Marie Ampére established the importance of the relationship between electricity and magnetism.
- Alassandro Volta discovered that if two dissimilar metals were brought in contact with a salt solution, a current would be produced. This invention is now known as the battery.
- George Simon Ohm showed a relationship between resistance, current, and voltage in an electrical circuit. He developed what is known as Ohm's Law.
- James Watt developed a method used to express a unit of electrical power known as Watt's Law.

Key Terms

André Marie Ampère
chemical energy
electromagnetic induction (EMI)
electrostatic discharge (ESD)
horsepower
joule
George Simon Ohm
Ohm's Law
photons
piezoelectricity
Seebeck effect
solar light energy
static electricity
thermocouple
thermoelectricity
Alassandro Volta
voltaic piles
James Watt
Watt's Law

Review Questions

Short Answer Essays

1. Describe how static electricity transfers from one object to another.
2. Name two sources of unwanted static electricity in automotive repair work, and what is used to prevent such occurrences.
3. Explain the use of photovoltaic in automotive applications.
4. Define what a thermocouple is and its most common use.
5. Describe the four people that have been instrumental in our modern day understanding and harnessing of electricity.

Fill-in-the-Blanks

1. Electricity that is without ______________ or at ______________, will produce an electrical charge whenever two dissimilar metallic materials are rubbed together lightly.
2. To protect sensitive electronic components from ______________ damage, the automotive technician should always use a ______________ ______________.
3. A photoelectric tachometer uses ______________ ______________ reflected back into the tachometer to determine engine revolutions per minute (rpm).
4. A simple lead acid battery consists of a single cell with a ______________electrode made from sponge lead, and a ______________ electrode made from lead peroxide.
5. In Ohm's Law, the ______________ in an electrical circuit is ______________ ______________ to the resistance of the circuit, and ______________ ______________ to the voltage.

ASE Style Review Questions

1. Technician A says static electricity can occur whenever the technician slides across a vinyl seat in a car. Technician B says one way to guard against static electricity is the use of a static wrist strap. Who is correct?

   A. A only

   B. B only

   C. Both A and B

   D. Neither A nor B

2. Technician A says static electricity is worse when there is high humidity in the air. Technician B says an electrostatic discharge (ESD) is the process by which a charged object transfers its charge to a neighboring object. Who is correct?
   A. A only
   B. B only
   C. Both A and B
   D. Neither A nor B

3. Technician A says the use of photovoltaic solar cells provides a way to capture energy from the sun. Technician B says a photon is the energy in sunlight that strikes the surface of a photovoltaic cell to create a voltage. Who is correct?
   A. A only
   B. B only
   C. Both A and B
   D. Neither A nor B

4. Technician A says a thermocouple is installed in the exhaust system of a diesel engine to measure the exhaust temperature. Technician B says a thermocouple operates the same way as a mercury glass thermometer. Who is correct?
   A. A only
   B. B only
   C. Both A and B
   D. Neither A nor B

5. Technician A says a knock sensor will produce a small voltage when there is an ignition spark knock. Technician B says the knock sensor is a piezoelectric device that will produce a voltage under physical stress or vibration. Who is correct?
   A. A only
   B. B only
   C. Both A and B
   D. Neither A nor B

6. Technician A says a lead acid battery uses positive and negative electrode plates that are the same type of material. Technician B says the chemical mixture of electrolyte in a lead acid battery is $PbSO_4$. Who is correct?
   A. A only
   B. B only
   C. Both A and B
   D. Neither A nor B

7. Technician A says the positive plate in a lead acid battery is made of lead peroxide (PbO_2). Technician B says the negative plate is made of a sponge lead material. Who is correct?
   A. A only
   B. B only
   C. Both A and B
   D. Neither A nor B

8. Two technicians are discussing the theory of electromagnetism. Technician A says any time an electrical current flows through a conductor, a magnetic field is created around the conductor. Technician B says electromagnetism is what causes a relay switch to operate. Who is correct?
   A. A only
   B. B only
   C. Both A and B
   D. Neither A nor B

9. Ohm's Law is being discussed: Technician A says the current is directly proportional to the resistance in a circuit, and inversely proportional to the voltage. Technician B says the current is inversely proportional to the voltage in a circuit, and directly proportional to the resistance. Who is correct?
   A. A only
   B. B only
   C. Both A and B
   D. Neither A nor B

10. Technician A says a *joule* is a metric unit of energy measurement. Technician B says 1 joule is the same as 100 Watts. Who is correct?
    A. A only
    B. B only
    C. Both A and B
    D. Neither A nor B

2 Basic Shop Safety

Introduction

Automotive shop safety is *everyone's* responsibility, and a shop is only as safe as the *least* safe person in it. Every technician, student, instructor, and boss must include safety as part of their daily routine. This begins with a positive attitude towards ensuring all precautions are taken to provide a safe working environment. This includes correct handling and use of hand tools and equipment, proper eye protection, personal attire, hazardous material safety and containment, location and type of fire extinguishers and first aid kits, and preventing electrical shock hazards. This chapter is intended to help you become aware of your working environment and develop the habits for a productive safe automotive career.

Objectives

When you complete this chapter you should be able to:

- ❑ Explain basic shop safety rules and protection.
- ❑ Identify hazardous waste products in the shop.
- ❑ Explain the use of material safety data sheets (MSDS).
- ❑ Define the purpose of OSHA and the EPA.
- ❑ Explain the different types of fires and fire extinguishing agents.
- ❑ Explain basic electrical safety.

Personal Safety

Personal Attire

Personal safety begins with the clothes you wear (Figure 2–1). Attire should display a professional image and be appropriate for the specific job environment. Loose fitting clothes with the shirt-tail or long sleeves dangling are accidents waiting to happen. Long hair should be tucked under a hat or tied back to prevent being caught in rotating machinery, fan belt, or drive axles. Rings or other jewelry items should not be worn. They may catch on moving parts, and pull fingers or hands into the moving part. Most jewelry is made of electrically conductive metal such as gold or silver that heat up instantly when contacted with electrical current. Shoes or boots should be intended for a professional working environment with slip-resistant soles, hardened leather or steel toe, and a grease/oil resistant material. Sports shoes, street shoes, or sandals are not recommended for the shop environment.

FIGURE 2–1 Working clothes should display a professional image.

Eye and Ear Protection

Eye Protection Eye protection should be worn at *all times* when in the automotive shop environment. Thousands of unnecessary and permanent eye injuries are reported every year from careless workers who do not wear them, or wear them *on top* of their head. There are several different types of eye protection available (Figure 2–2). Good eye protection should meet the following basic requirements:

- Comfortable to wear
- Unobstructed vision
- Safety plastic lens or tempered glass
- Have side shields
- Lenses will not pop-out under impact

Regular prescription eye wear does not meet safety impact standards and has no side shield protection. Some eye protection is designed to fit over prescription eye wear. For increased protection from splash or grinding hazards, a full face shield should be worn (Figure 2–3).

Ear Protection Extended exposure to high noise levels, called **decibels,** in the shop may lead to permanent hearing damage. Loud machinery, air tools, and vehicles running in a closed environment all produce harmful

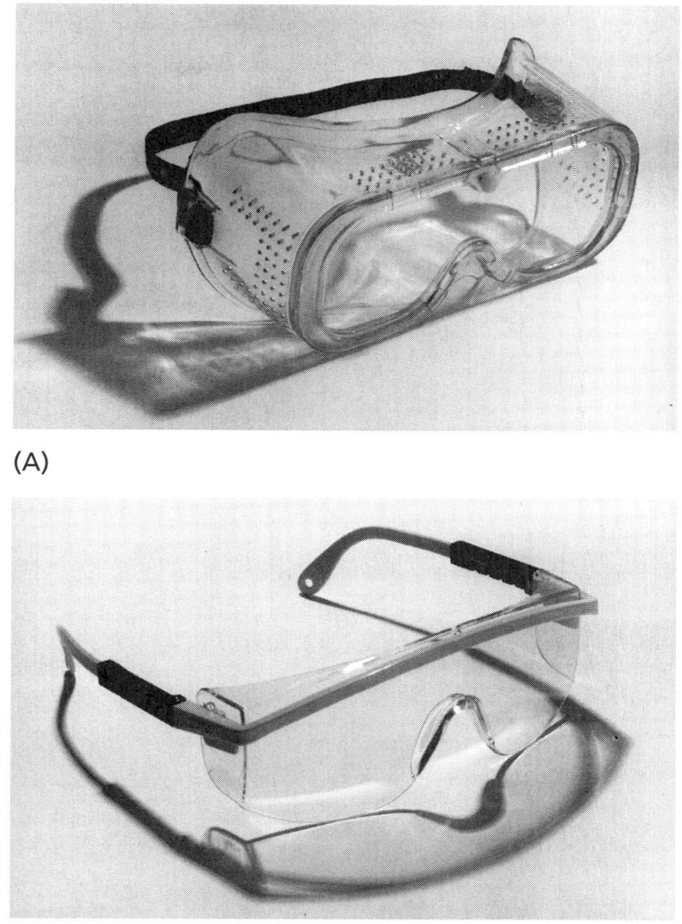

(A)

(B)

FIGURE 2–2 Typical styles of safety glasses with side shields (A)–(B).

(Courtesy of Sellstrom Mfg. Co.)

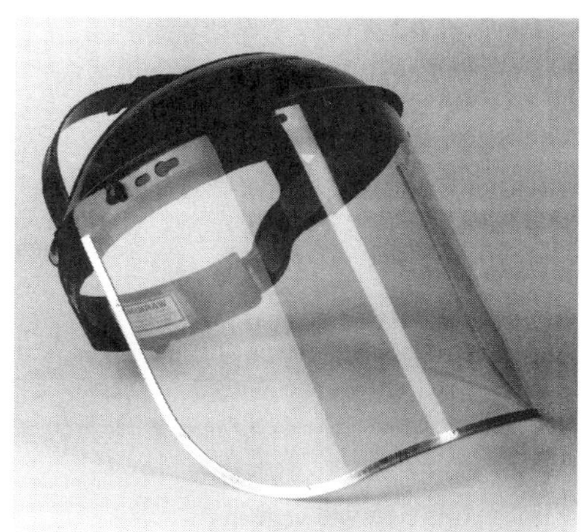

FIGURE 2–3 A full face shield provides increased protection.

(Courtesy of Thermadyne Industries, Inc.)

REPRESENTATIVE SOUND LEVELS		
SOUND LEVEL (dB)	**OPERATION OR EQUIPMENT**	
150	Jet engine test cell	**DANGER ZONE**
145	____________ Threshold of Pain	
130	Pneumatic press (close range) Pneumatic rock drill Riveting steel tank	
125	Pneumatic chipper Pneumatic riveter	
120	____________ Threshold of Discomfort Turbine generator	
112	Punch press Sandblasting	
110	Drills, shovels, operating trucks Drop hammer	
105	Circular saw Wire braiders, stranding machine Pin routers Riveting machines	
100	Can-manufacturing plant Portable grinders Ram turret lathes Automatic screw machine	**RISK ZONE**
90	Welding equipment Weaving mill Milling machine Pneumatic diesel compressor Engine lathes Portable sanders	
85	California freeway traffic (overpass)	
HEARING DAMAGE IF CONTINUED EXPOSURE ABOVE THIS LEVEL		
80	Tabulating machines, electric typewriters	
75	Stenographic room	
70	Electronics assembly plant	
65	Department store	**SAFE ZONE**
60	Conversation	
35	Quiet home forced air heating	
10	Whisper	

FIGURE 2–4 Typical decibel levels and zones of common noises.

levels of noise. Harmful levels consistently over 90 decibels can cause permanent hearing damage (Figure 2–4). To protect the worker, certified ear plugs or ear muffs should be worn (Figure 2–5).

Emergency Eyewash Station An emergency eyewash should be available in the immediate shop environment. When eyes are accidentally contaminated with foreign matter or dangerous liquids they should be

(A)

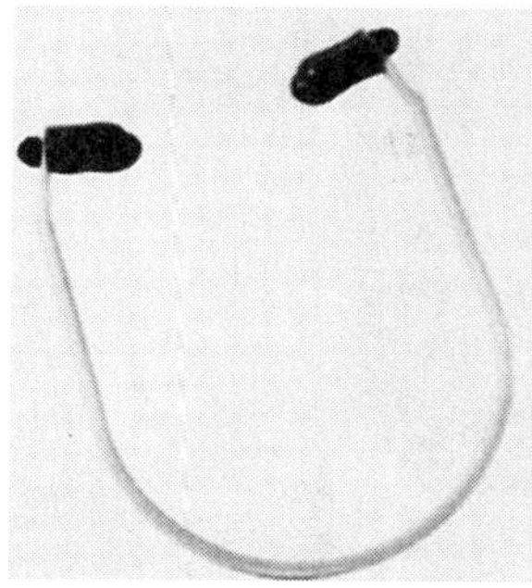

(B)

FIGURE 2–5 (A) Typical ear muffs, (B) and ear plugs. (Courtesy of Dalloz Safety)

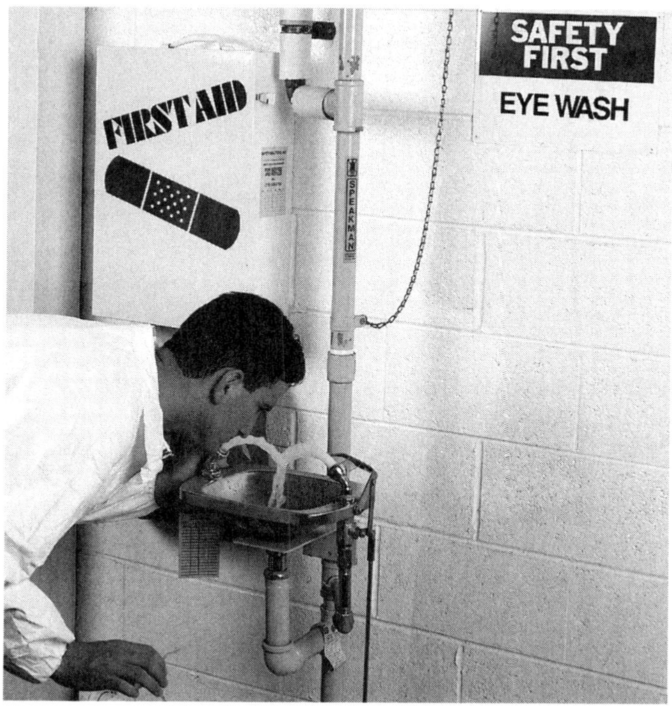

FIGURE 2–6 Typical emergency eye wash fountain. (Courtesy of Western Emergency Equipment)

flushed with water (Figure 2–6) and medical attention sought. It is important to know where the eyewash station is located in your shop, that it is in good working order, and the path is unobstructed.

Hazardous Materials

There are dozens of chemicals and materials considered hazardous to the worker. Some create injury or illness on contact. Other chemicals have long-term side effects that create a false sense of safety for the worker. Some of the more common hazardous shop chemicals are:

- Antifreeze/coolant (ethylene glycol)
- Air conditioning refrigerants (freon/R-12 and R134A)
- Cleaning solvents and chemicals
- Battery electrolyte acid
- Gasoline, diesel, and other fuels
- Asbestos from brakes and clutches
- Used motor oils
- Heavy metals

The regulation of the use of these chemicals and materials in the United States is through the **Occupational Safety and Health Administration (OSHA).** In 1970, the Federal government created OSHA to ensure consistent safety standards across the country that inform, teach, train, and enforce elements of a safe working environment.

Material Safety Data Sheet

Information about all chemicals used by the worker are listed on **material safety data sheets (MSDS)** (Figure 2–7). These sheets should be available to the worker, and detail the chemical composition and health, or safety, hazard precaution information of each product.

Hazardous Waste Management

Hazardous materials normally produce hazardous waste. The disposal of hazardous waste products is regulated by the **Environmental Protection Agency (EPA).** A product is considered a hazardous waste by the EPA if it has been identified specifically, or has one or more of the following characteristics:

- *Ignitability* a solid product that will spontaneously ignite, or a liquid with a flash point below 140° F.
- *Corrosivity* metals or other materials that will dissolve or melt, or will burn the skin on contact.
- *Reactivity* materials that generate toxic vapors, flammable gasses, or any material that reacts violently with water.

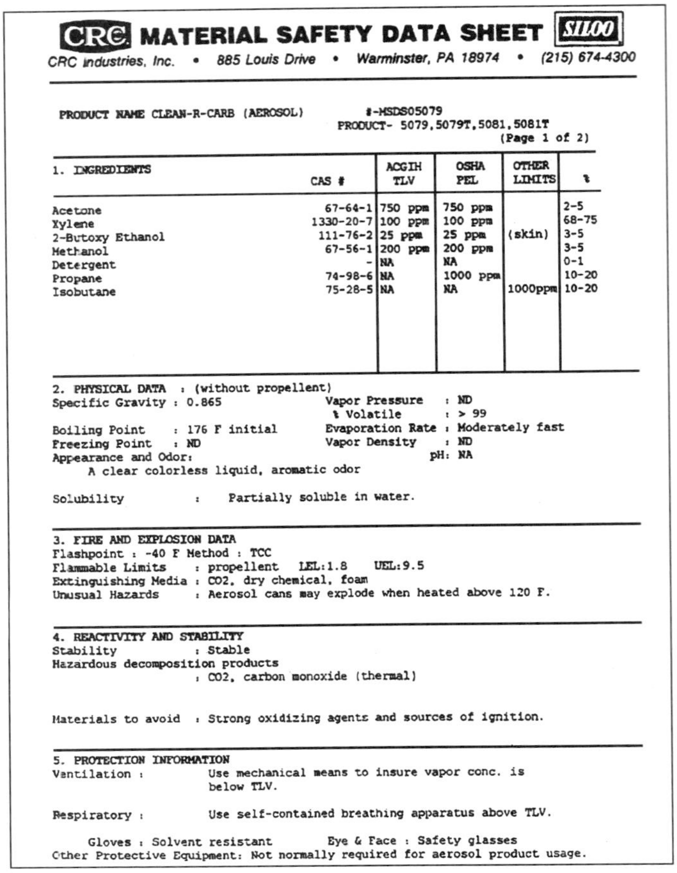

CRC MATERIAL SAFETY DATA SHEET SILOO

CRC Industries, Inc. • 885 Louis Drive • Warminster, PA 18974 • (215) 674-4300

PRODUCT NAME CLEAN-R-CARB (AEROSOL) #-MSDS05079
PRODUCT- 5079,5079T,5081,5081T
(Page 1 of 2)

1. INGREDIENTS	CAS #	ACGIH TLV	OSHA PEL	OTHER LIMITS	%
Acetone	67-64-1	750 ppm	750 ppm		2-5
Xylene	1330-20-7	100 ppm	100 ppm		68-75
2-Butoxy Ethanol	111-76-2	25 ppm	25 ppm	(skin)	3-5
Methanol	67-56-1	200 ppm	200 ppm		3-5
Detergent	-	NA	NA		0-1
Propane	74-98-6	NA	1000 ppm		10-20
Isobutane	75-28-5	NA	NA	1000ppm	10-20

2. PHYSICAL DATA : (without propellent)
Specific Gravity : 0.865 Vapor Pressure : ND
% Volatile : > 99
Boiling Point : 176 F initial Evaporation Rate : Moderately fast
Freezing Point : ND Vapor Density : ND
Appearance and Odor: pH: NA
A clear colorless liquid, aromatic odor

Solubility : Partially soluble in water.

3. FIRE AND EXPLOSION DATA
Flashpoint : -40 F Method : TCC
Flammable Limits : propellent LEL:1.8 UEL:9.5
Extinguishing Media : CO2, dry chemical, foam
Unusual Hazards : Aerosol cans may explode when heated above 120 F.

4. REACTIVITY AND STABILITY
Stability : Stable
Hazardous decomposition products
: CO2, carbon monoxide (thermal)

Materials to avoid : Strong oxidizing agents and sources of ignition.

5. PROTECTION INFORMATION
Ventilation : Use mechanical means to insure vapor conc. is below TLV.

Respiratory : Use self-contained breathing apparatus above TLV.

Gloves : Solvent resistant Eye & Face : Safety glasses
Other Protective Equipment: Not normally required for aerosol product usage.

FIGURE 2–7 Material safety data sheet (MSDS) should be available for every chemical used in the automotive shop.

(Courtesy of CRC Chemicals)

- *EP Toxicity* the leaching of heavy metals from materials in concentrations greater than 100 times drinking water standards.

All businesses that generate hazardous waste must be registered with the EPA and have a hazardous waste policy in force. When handling hazardous waste products, the use of proper equipment (Figure 2–8) will help contain the waste product.

Storage of hazardous waste products should be in approved containers (Figure 2–9) and cabinets (Figure 2–10).

FIGURE 2–8 Removing asbestos hazardous dust with a special vacuum system designed for this purpose.

(Courtesy of Nilfisk of America, Inc.)

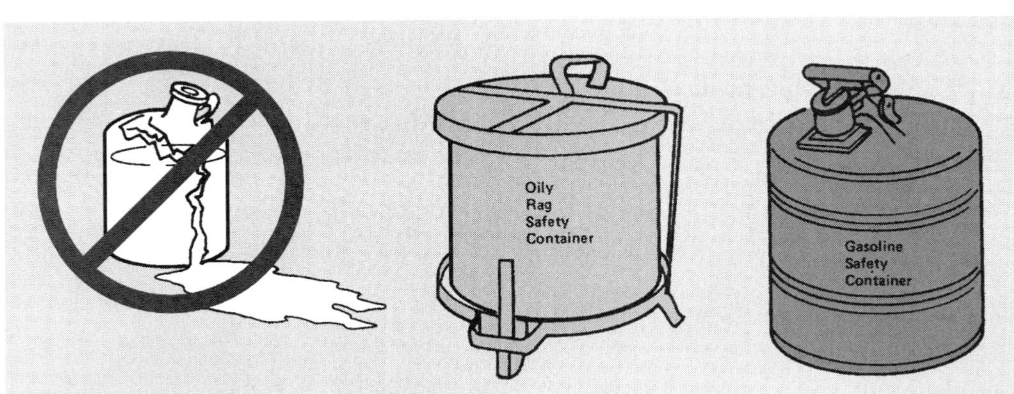

FIGURE 2–9 Typical combustibles safety containers.

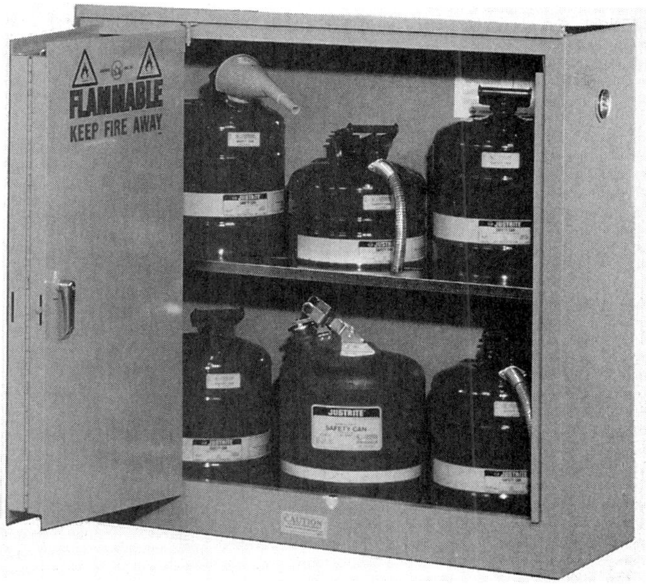

FIGURE 2–10 Typical self-closing flammables storage cabinet.

(Courtesy of Justrite Manufacturing Co.)

Fire Extinguishers and Agents

Fire extinguishers (Figure 2–11) are an important piece of shop equipment. Knowledge of the type of fire extinguisher and its location in the shop is extremely important when in a rush to extinguish a fire.

It is important to select the right fire extinguisher for each class of fire (Table 2-1, page 20).

FIGURE 2–11 Typical portable fire extinguishers.

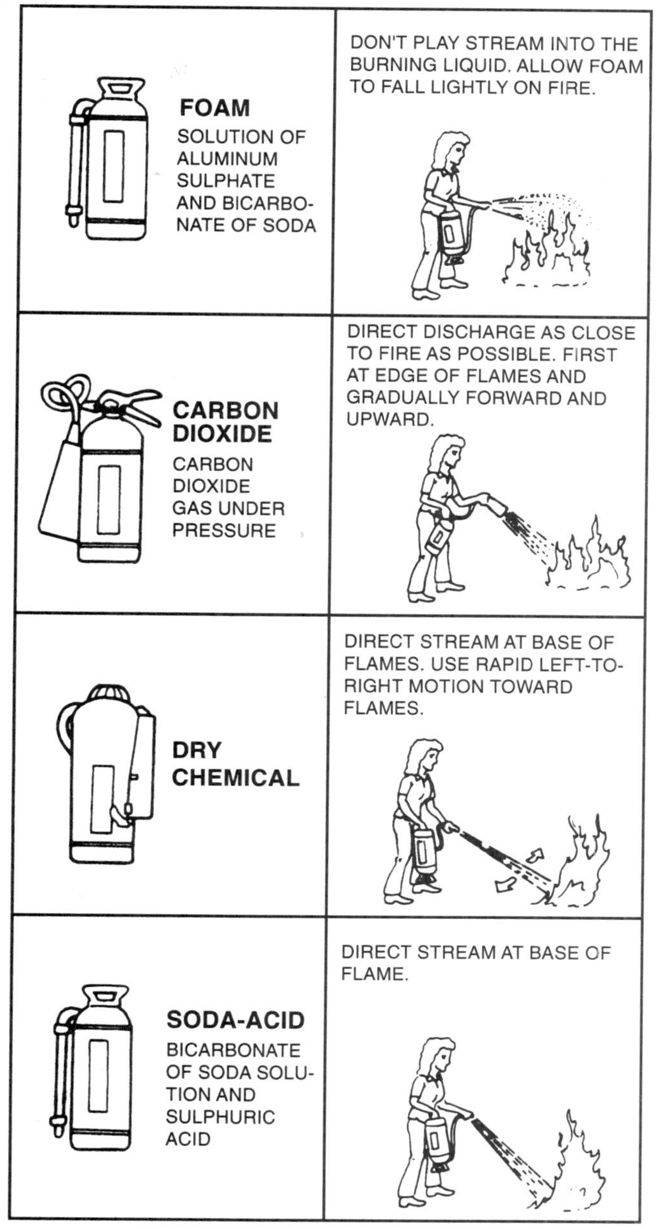

FIGURE 2–12 Extinguishing methods for each type of fire extinguisher.

The proper method of extinguishing a fire is partially determined by the type of fire extinguishing agent used (Figure 2–12). Always refer to label instructions printed on the fire extinguisher.

Electrical Safety

There are five basic safety precautions you should follow when working around shop or vehicle electrical systems:

TABLE 2–1 Guide to proper fire extinguisher selection for different classes of fires.

	Class of Fire	Typical Fuel Involved	Type of Extinguisher
Class A Fires (green)	**For Ordinary Combustibles** Put out a class A fire by lowering its temperature or by coating the burning combustibles.	Wood Paper Cloth Rubber Plastics Rubbish Upholstery	Water*[1] Foam* Multipurpose dry chemical[4]
Class B Fires (red)	**For Flammable Liquids** Put out a class B fire by smothering it. Use an extinguisher that gives a blanketing, flame-interrupting effect; cover whole flaming liquid surface.	Gasoline Oil Grease Paint Lighter fluid	Foam* Carbon dixoide[5] Halogenated agent[6] Standard dry chemical[2] Purple K dry chemical[3] Multipurpose dry chemical[4]
Class C Fires (blue)	**For Electrical Equipment** Put out a class C fire by shutting off power as quickly as possible and by always using a nonconducting extinguishing agent to prevent electric shock.	Motors Appliances Wiring Fuse boxes Switchboards	Carbon dioxide[5] Halogenated agent[6] Standard dry chemical[2] Purple K dry chemical[3] Multipurpose dry chemical[4]
Class D Fires (yellow)	**For Combustible Metals** Put out a class D fire of metal chips, turnings, or shavings by smothering or coating with a specially designed extinguishing agent.	Aluminum Magnesium Potassium Sodium Titanium Zirconium	Dry powder extinguishers and agents only

*Cartridge-operated water, foam, and soda-acid types of extinguishers are no longer manufactured. These extinguishers should be removed from service when they become due for their next hydrostatic pressure test.

Notes:

(1) Freezes in low temperatures unless treated with antifreeze solution, usually weighs over 20 pounds, and is heavier than any other extinguisher mentioned.

(2) Also called ordinary or regular dry chemical (sodium bicarbonate).

(3) Has the greatest initial fire-stopping power of the extinguishers mentioned for class B fires. Be sure to clean residue immediately after using the extinguisher so sprayed surfaces will not be damaged (potassium bicarbonate).

(4) The only extinguishers that fight A, B, and C classes of fires. However, they should not be used on fires in liquefied fat or oil of appreciable depth. Be sure to clean residue immediately after using the extinguisher so sprayed surfaces will not be damaged (ammonium phosphates).

(5) Use with caution in unventilated, confined spaces.

(6) May cause injury to the operator if the extinguishing agent (a gas) or the gases produced when the agent is applied to a fire is inhaled.

1. Wear proper attire. Electrical shock can be harmful and in some cases—fatal. Insulating your body through rubber-soled shoes or using approved electrically insulating mats or gloves will help prevent shock hazards. Voltage levels as low as 120 volts AC, or the effect of current flow as low as 0.15 amps through the body, can interrupt the natural electrical rhythm of the heart and cause fatal injury.
2. Never attempt to perform repairs on an electrical circuit with which you are unfamiliar.
3. Never wear metal objects such as watches, rings, or chains that will conduct electricity. If a metal ring came into contact with a live electrical current, it could instantaneously glow red hot, causing severe injury to the skin.

4. Treat every electrical circuit like a *live* circuit. Even though the safest electrical circuit to work on is an *inactive* or *dead* circuit, if you treat the circuit as if it is *active* or *live* the attention and cautions you exercise will be improved.

5. Use only a *Type C* fire extinguisher on electrical fires. Table 2-1 shows the proper precautions to fight a class C fire.

Shop Safety Exercise

Each repair shop has a layout and location of necessary safety equipment. Figure 2–13 is the layout of a typical automotive shop and important safety equipment.

❑ On a separate sheet of paper draw the layout of your automotive shop.

❑ List the location of the following shop safety items.

EQUIPMENT	LOCATION
1. first aid kit	
2. emergency eyewash station	
3. fire extinguishers	
4. exit doors	
5. hazardous waste containment	
6. flammable storage locker	
7. electrical circuit panel	
8. emergency phone numbers	
9. shop cleaning supplies	
10. MSDS sheets	
11. hazardous spill containment supplies	

❑ List the proper procedure for operating the fire extinguisher(s) in your shop.

____

____

❑ Exercise Summary. After performing the above shop safety exercise, what did you find most useful in performing this shop safety exercise.

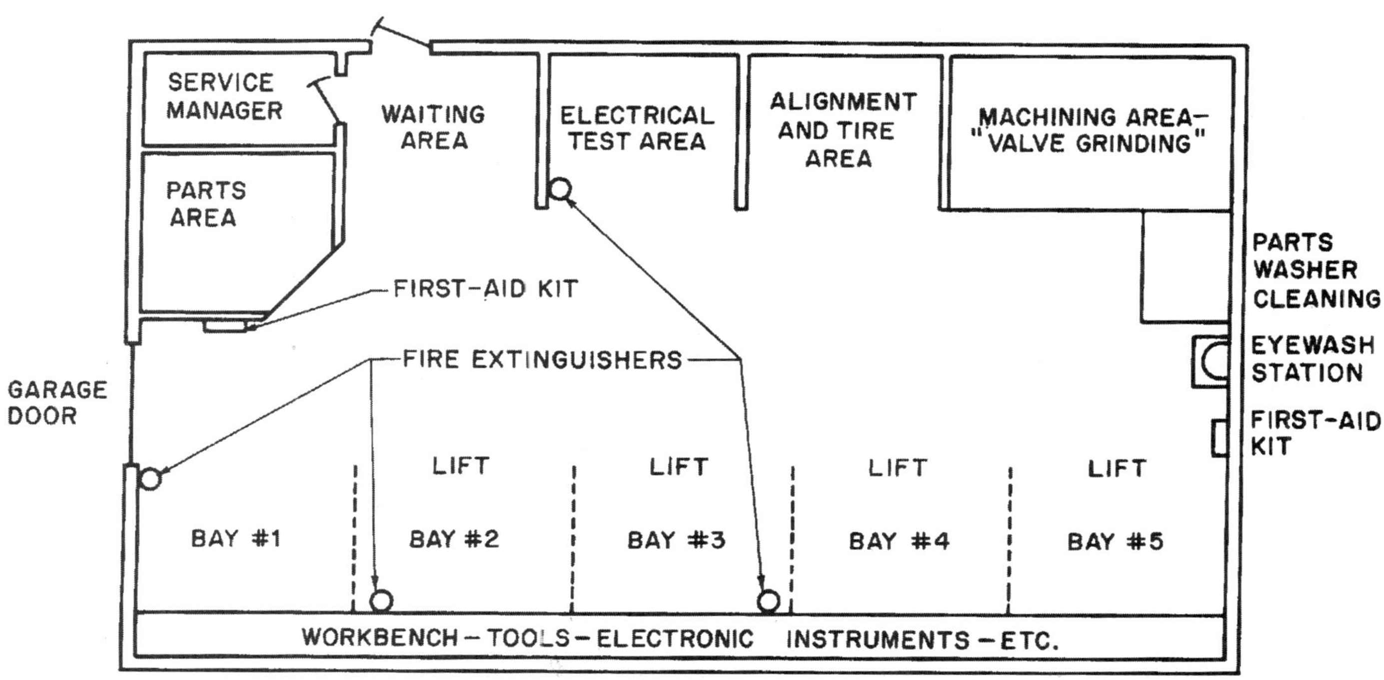

FIGURE 2–13 Typical layout of safety equipment in an automotive shop.

Summary

- ❑ Professional clothing should be worn that is appropriate for the specific job environment.
- ❑ Eye protection should be worn at all times when in the automotive shop environment.
- ❑ An emergency eyewash station should be available in the immediate shop environment.
- ❑ The regulation of hazardous chemicals is done through OSHA.
- ❑ An MSDS should be available for every chemical used in the automotive shop environment.
- ❑ The disposal of hazardous waste products is regulated by the EPA.
- ❑ A class A fire extinguisher is for ordinary combustibles, and extinguished with water, foam, or dry chemical powder.
- ❑ A class B fire extinguisher is for flammable liquids, and it extinguishes with foam, dry chemical powder, or carbon dioxide.
- ❑ A class C fire extinguisher is for electrical equipment, and it extinguishes with carbon dioxide or dry chemical powder.
- ❑ A class D fire extinguisher is for combustible metals, and it extinguishes with dry chemical powder only.
- ❑ Voltage levels as low as 120 volts or current flow as low as 0.15 amps can interrupt the natural electrical rhythm of the heart.

Key Terms

decibel

Environmental Protection Agency (EPA)

material safety data sheets (MSDS)

Occupational Safety and Health Administration (OSHA)

Review Questions

Short Answer Essays

1. Discuss some of the safety features of a well-designed shop.
2. Describe the four classes of fire and the proper extinguishing agents for each type.
3. Describe the basic safety precautions when working around electrical equipment.

Fill-in-the-Blanks

1. A technician with long hair should ____________________ or ____________________ when working in the automotive shop.
2. Eye protection should be worn ____________________ when working in the automotive shop.

3. Harmful sound levels over ______________ decibels can cause permanent hearing damage.

4. The regulation of chemicals and hazardous materials is through ______________.

5. Information about all chemicals used by the worker in the automotive shop are listed on the ______________.

ASE Style Review Questions

1. Technician A says safety glasses should be worn at all times while in the shop. Technician B says regular prescription eye wear may be substituted for safety glasses. Who is correct?

   A. A only

   B. B only

   C. Both A and B

   D. Neither A nor B

2. Technician A says if a hazardous substance dissolves metals and other materials, or burns the skin, it is said to have reactivity. Technician B says when a solid hazardous waste product spontaneously ignites it is considered to have ignitability. Who is correct?

   A. A only

   B. B only

   C. Both A and B

   D. Neither A nor B

3. Technician A ties his long hair behind his head while working in the shop. Technician B covers his long hair with a brimless cap. Who is correct?

   A. A only

   B. B only

   C. Both A and B

   D. Neither A nor B

4. Technician A uses carbon dioxide to extinguish a Class A fire. Technician B uses foam to extinguish a Class D fire. Who is correct?

   A. A only

   B. B only

   C. Both A and B

   D. Neither A nor B

5. Technician A says extended noise levels exceeding 90 decibels can damage the hearing. Technician B says all noises produced in the shop environment are harmful. Who is correct?
   A. A only
   B. B only
   C. Both A and B
   D. Neither A nor B

6. Technician A says good eye protection has side shields. Technician B says good eye protection has a plastic lens or tempered glass. Who is correct?
   A. A only
   B. B only
   C. Both A and B
   D. Neither A nor B

7. Technician A says used motor oil is considered a hazardous waste. Technician B says ethylene glycol coolant is only hazardous when in a used condition. Who is correct?
   A. A only
   B. B only
   C. Both A and B
   D. Neither A nor B

8. Technician A says OSHA is the federal regulating agency over the disposal of hazardous wastes. Technician B says the EPA is the federal regulating agency over the use of chemicals and hazardous materials. Who is correct?
   A. A only
   B. B only
   C. Both A and B
   D. Neither A nor B

9. Technician A says water is a suitable extinguishing agent on Class C electrical fires. Technician B says electrical shock hazards as low as 0.15 amps can be fatal. Who is correct?
   A. A only
   B. B only
   C. Both A and B
   D. Neither A nor B

10. Which of the following is/are important when working in the automotive shop?
    A. Using the proper tool for the job.
    B. Avoiding loose fitting clothes.
    C. Wearing steel tipped shoes.
    D. All of the above.

3 Basic Electrical Theory

Introduction

During the past several decades, automotive electrical systems have shown a steady growth in complexity. With the introduction of alternators in the 1960's, electronic ignition in the 1970's, computer-controlled systems in the 1980's, and advanced on-board diagnostics in the 1990's, automotive technicians need to prepare themselves for these and future advanced systems. By developing an in-depth understanding of electrical and electronic basics you can prepare yourself for the rapidly changing responsibilities that lie ahead. This chapter will provide a look at basic electrical theory and provide some practice using Ohm's Law and Watt's Law introduced in Chapter 1.

Objectives

When you complete this chapter you should be able to:

- ❑ Apply electrical laws and theories.
- ❑ Describe atomic structure, atoms, and electrons.
- ❑ Define conductors, semiconductors, and insulators.
- ❑ Define voltage, resistance, and current flow.
- ❑ Define the relationship between voltage, amperage, and resistance in Ohm's Law.
- ❑ Use the mathematical relationships associated with Ohm's Law to find the value of a missing factor.
- ❑ Define the relationship between watts, voltage, and amperage in Watt's Law.
- ❑ Use the mathematical relationships associated with Watt's Law to find the value of a missing factor.
- ❑ Identify commonly used electrical symbols.
- ❑ Define magnetism and electromagnetism.

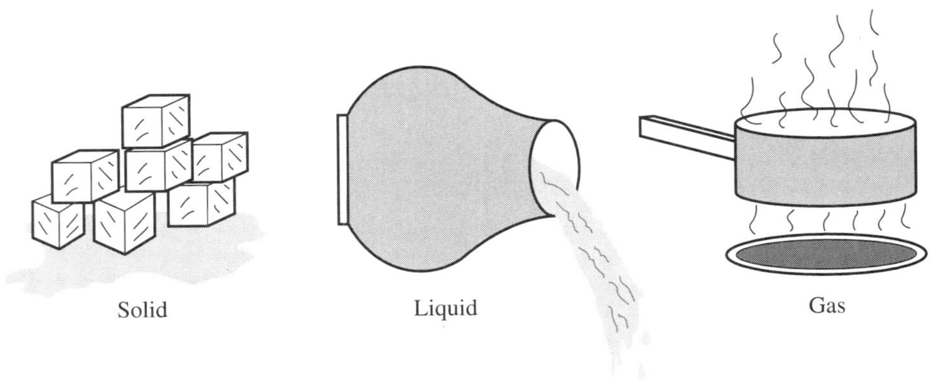

FIGURE 3–1 Three forms of matter.

Basics of Electricity

To understand electricity we will take a brief look at how atoms are structured, and then find out what makes some atoms better conductors of electricity and others good insulators. **Matter** can be defined as any substance that occupies a space and contains a mass. Using H_2O (water) as an example, mass may be present in three forms—a solid (ice), a liquid (water), or a gas (steam) (Figure 3–1).

This mass is made up of **molecules** which consist of a group of elements. An **element** is a cluster of one type of atom. We are all familiar with the well-known molecule of water and its combination of elements in H_2O (Figure 3–2).

There are over 100 different elements in nature. Table 3–1 lists the elements according to atomic order. This list is also referred to as "The Periodic Table of Elements."

TABLE 3–1 The Periodic Table of Elements.

ATOMIC NUMBER	NAME	VALENCE ELECTRONS	SYMBOL	ATOMIC NUMBER	NAME	VALENCE ELECTRONS	SYMBOL	ATOMIC NUMBER	NAME	VALENCE ELECTRONS	SYMBOL
1	Hydrogen	1	H	37	Rubidium	1	Rb	73	Tantalum	2	Ta
2	Helium	2	He	38	Strontium	2	Sr	74	Tungsten	2	W
3	Lithium	1	Li	39	Yttrium	2	Y	75	Rhenium	2	Re
4	Beryllum	2	Be	40	Zirconium	2	Zr	76	Osmium	2	Os
5	Boron	3	B	41	Niobium	1	Nb	77	Iridium	2	Ir
6	Carbon	4	C	42	Molybdenum	1	Mo	78	Platinum	1	Pt
7	Nitrogen	5	N	43	Technetium	2	Tc	79	Gold	1	Au
8	Oxygen	6	O	44	Ruthenium	1	Ru	80	Mercury	2	Hg
9	Fluorine	7	F	45	Rhodium	1	Rh	81	Thallium	3	Tl
10	Neon	8	Ne	46	Palladium	–	Pd	82	Lead	4	Pb
11	Sodium	1	Na	47	Silver	1	Ag	83	Bismuth	5	BI
12	Magnesium	2	Ma	48	Cadmium	2	Cd	84	Polonium	6	Po
13	Aluminum	3	Al	49	Indium	3	In	85	Astatine	7	At
14	Silicon	4	Si	50	Tin	4	Sn	86	Radon	8	Rd
15	Phosphorus	5	P	51	Antimony	5	Sb	87	Francium	1	Fr
16	Sulfur	6	S	52	Tellurium	6	Te	88	Radium	2	Ra
17	Chlorine	7	Cl	53	Iodine	7	I	89	Actinium	2	Ac
18	Argon	8	A	54	Xenon	8	Xe	90	Thorium	2	Th
19	Potassium	1	K	55	Cesium	1	Cs	91	Protactinium	2	Pa
20	Calcium	2	Ca	56	Barium	2	Ba	92	Uranium	2	U
21	Scandium	2	Sc	57	Lanthanum	2	La				
22	Titanium	2	Ti	58	Cerium	2	Ce		Artifical Elements		
23	Vanadium	2	V	59	Praseodymium	2	Pr				
24	Chromium	1	Cr	60	Neodymium	2	Nd	93	Neptunium	2	Np
25	Manganese	2	Mn	61	Promethium	2	Pm	94	Plutonium	2	Pu
26	Iron	2	Fe	62	Samarium	2	Sm	95	Americium	2	Am
27	Cobalt	2	Co	63	Europium	2	Eu	96	Curium	2	Cm
28	Nickel	2	Ni	64	Gadolinium	2	Gd	97	Berkelium	2	Bk
29	Copper	1	Cu	65	Terbium	2	Tb	98	Californium	2	Cf
30	Zinc	2	Zn	66	Dysprosium	2	Dy	99	Einsteinium	2	E
31	Gallium	3	Ga	67	Holmium	2	Ho	100	Fermium	2	Fm
32	Germanium	4	Ge	68	Erbium	2	Er	101	Mendelevium	2	Mv
33	Arsenic	5	As	69	Thulium	2	Tm	102	Nobelium	2	No
34	Selenium	6	Se	70	Ytterbium	2	Yb	103	Lawrencium	2	Lw
35	Bromine	7	Br	71	Lutetium	2	Lu				
36	Krypton	8	Kr	72	Hafnium	2	Hf				

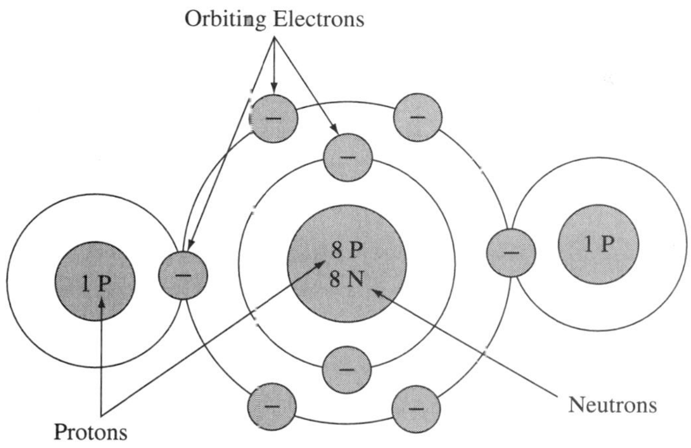

FIGURE 3–2 Two hydrogen atoms and one oxygen atom.

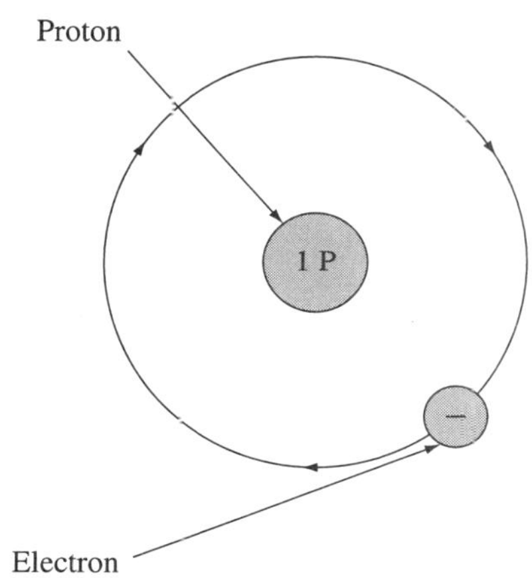

FIGURE 3–3 The hydrogen atom is the simplest atom known.

An **atom** is considered the smallest part of an element. It consists of *positively* charged **protons, neutrons** with *no* charge, and **electrons** that are *negatively* charged. Protons and neutrons combine to form the **nucleus** of the atom. The simplest atom is a hydrogen atom consisting of 1 proton and 1 orbiting electron, but no neutron (Figure 3–3). All other atoms in the atomic structure have neutrons.

The proton and neutron form the center of the nucleus, which is smaller than the electron, but over 1800 times heavier. The number of protons in an atom determines its atomic number. The neutron is a heavy particle and electrically neutral. Its weight and neutrality give the neutron an important property. The electron rotates in an orbit around the proton similar to Earth rotating around the sun. Even though a simple law of physics states that "like charges repel each other (Figure 3–4) and unlike charges attract each other" (Figure 3–5), the centrifugal force of the rotating, negatively charged electrons keeps them attracted to the positively charged protons.

To understand how atomic structure relates to electricity, look at a copper atom that has an atomic number of 29 and contains 29 protons, 34 neutrons, and 29 electrons (Figure 3–6). The number of electrons and protons in the copper atom's structure attempts to remain balanced or equal. If possible, the atom will give or take electrons from neighboring atoms to remain balanced.

An atom has several electron orbits (Figure 3–7). The electrons in the outer most orbit are called **valence electrons.** Because valence electrons become dislodged

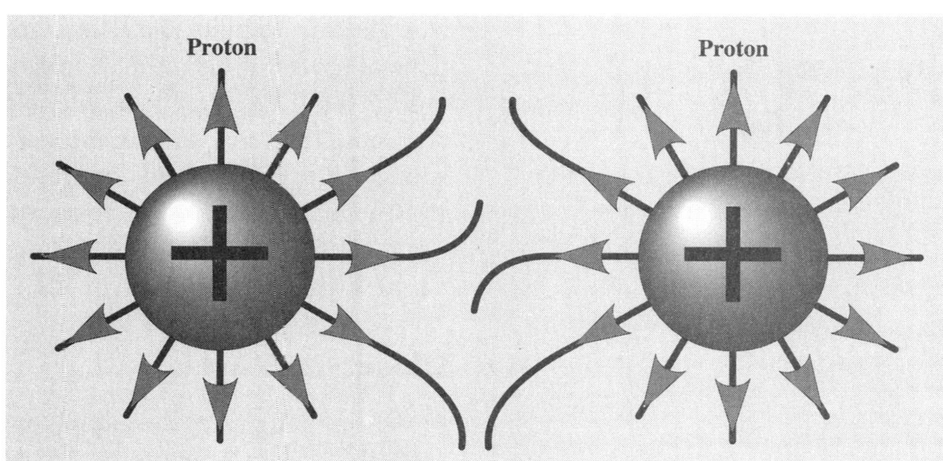

FIGURE 3–4 Like charges repel each other.

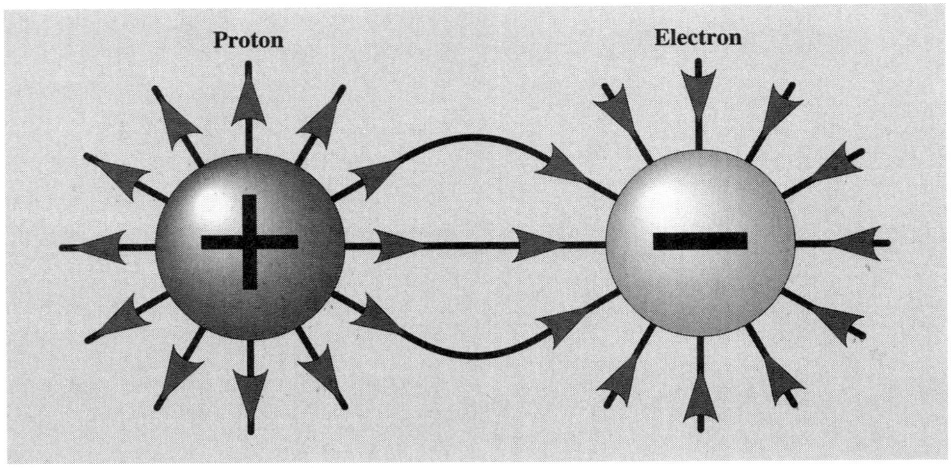

FIGURE 3–5 Unlike charges attract each other.

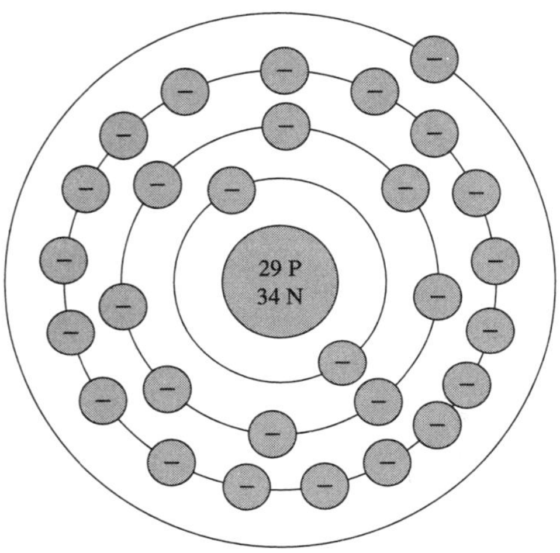

FIGURE 3–6 Atomic structure of a copper atom.

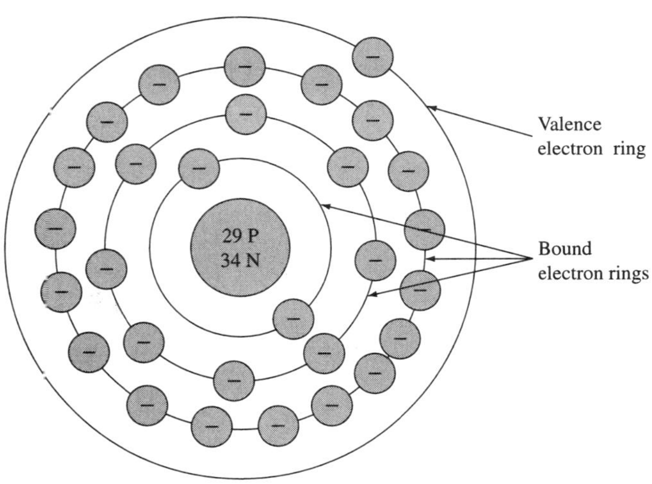

FIGURE 3–7 Bound and valence electron orbits of a copper atom.

from their orbit by some outside force, they are also called **free electrons.** Thus, they are commonly called either valence or free electrons.

The inner orbits have **bound electrons** which are free to move only within their own orbits at a fixed distance from the nucleus. The free, or valence, electrons is where our study of electron flow begins.

Note: Electricity is the movement of free electrons from atom to atom in a complete circuit.

Not all materials make good conductors of electricity. Figure 3–8 lists some of the most commonly used atoms in the automotive field. A good conductor must have three or less free electrons. The lower the number of free electrons, the better the conductor. A semiconductor contains exactly four free electrons and is used extensively in electronic circuits because of the material's electron stability. Insulators are usually combinations of several different molecules that have combined to form a tightly bound molecule with five or more free electrons. The force holding these free electrons is very strong, which makes them good insulators.

There are several other atoms we could list here, but our main focus is on the conductors and insulators we normally use in automotive circuits.

Direct Current

Most automotive electrical circuits operate on **direct current (DC)** that is produced by the battery or by the converted direct current from the charging system circuit. Direct current is a constant level of electrical potential

	ATOMIC NUMBER	NAME OF ATOM	# VALENCE ELECTRONS	SYMBOL USED
Good conductors 3 or less free electrons	47	Silver	1	Ag
	29	Copper	1	Cu
	79	Gold	1	Au
	13	Aluminum	3	Al
	26	Iron	2	Fe
Semi-conductors exactly 4 free electrons	32	Germanium	4	Ge
	14	Silicon	4	Si

The non metal materials below are considered tightly bound molecular combinations of different atoms that share free electrons.

	MATERIAL NAME	MOLECULE COMBINATION
Good insulators 5 or more free electrons	Plastics	Carbon, hydrogen, oxygen
	Dry wood	Carbon, hydrogen, oxygen & nitrogen
	Glass	Silicon, oxygen
	Diamond	Pure carbon
	Rubber	Carbon & hydrogen. Includes sulfur if vulcanized

FIGURE 3–8 Determining electrical conductivity of an atom.

that flows in one direction at a specific DC voltage as indicated on the scope screen (Figure 3–9). Some DC voltages may appear to have an up and down waveform as the electric component is operated. The DC voltage value may change from zero to a higher voltage, and then back to zero as the component is turned off. The waveform, though, will remain at only one side of the zero line.

Alternating Current

In electrical circuits, **alternating current (AC)** is produced in the charging system or any other device that has a conductor that rotates through a magnetic field, such as a wheel speed sensor in antilock brakes. With alternating current, the voltage begins at zero and rises

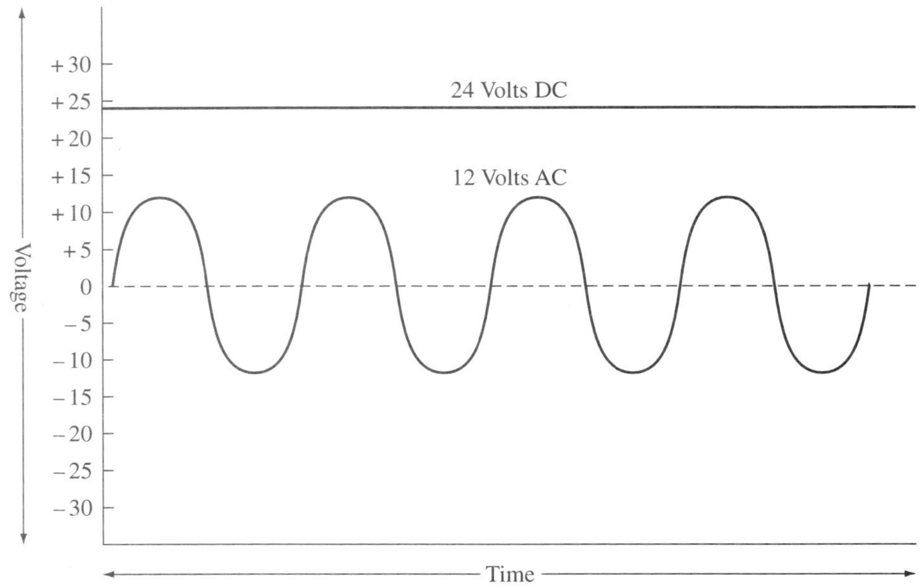

FIGURE 3–9 Comparison between direct current (DC) and alternating current (AC).

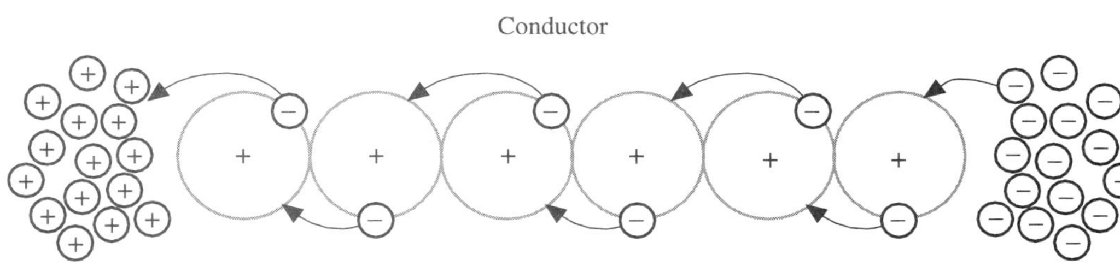

FIGURE 3–10 The flow of free electrons from atom to atom.

to a specified positive voltage, and then reverses to an equal negative value before returning to zero again (Figure 3–9). One compete rise and fall of the AC wave is called a cycle. A good example of an AC circuit is the typical lighting system found in houses of 110/120 volts at 60 cycles per second.

Current Flow

In Figure 3–10, the free electrons move to the left. The unbalanced positive charge on the left pulls a free electron from the right, causing a chain reaction. The flow of these free electrons from negative to positive is called **current flow.** It is measured in **Amperes** or Amps and flows at the speed of light. One ampere is equal to 6.28 billion billion free electrons flowing past a point in one second (Figure 3–11). That is a lot !

To represent amperage readings, the letter **I** is often used because it stands for "Intensity of current." The letter **A** is also used and it stands for amps. Both are used in electrical mathematical formulas. This textbook uses the letter I to represent current.

SAFETY TIP *The human body can conduct electricity. Because of this, anytime you perform inspection or service on electrical systems, beware that electrical hazards exist. Observe electrical safety rules at all times, and never wear jewelry while testing or servicing electrical systems.*

Another way to look at current flow is to compare it with the flow of water. Electron flow is measured by an ammeter, and water flow is measured by a meter that

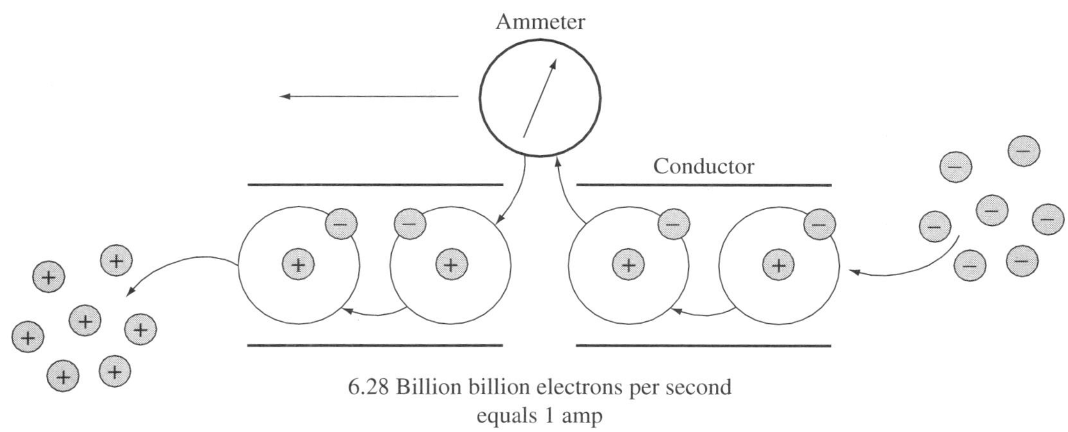

FIGURE 3–11 The measurement of electron flow.

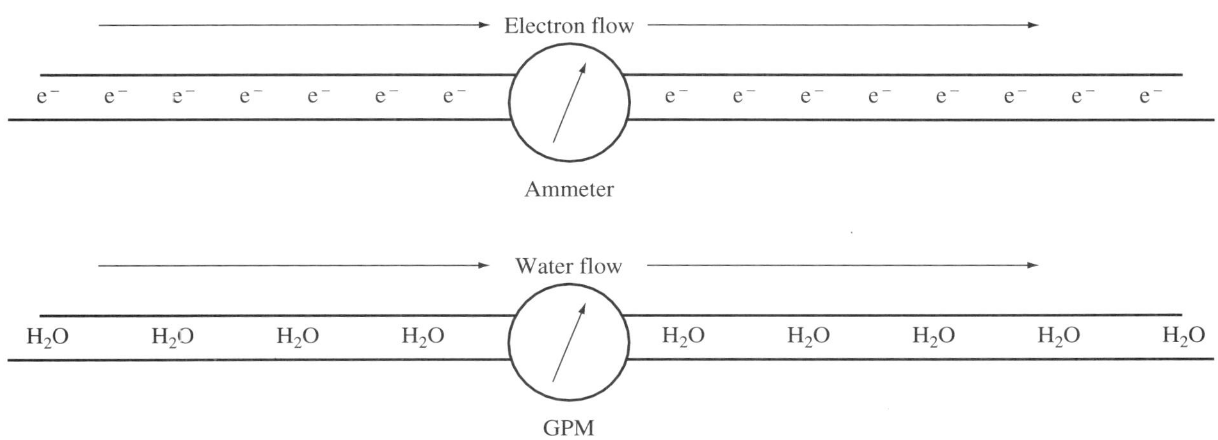

FIGURE 3–12 Comparison of the flow of electrons and the flow of water.

reads the flow of water in gallons per minute (gpm). (Figure 3–12).

Electron Flow Theory and Conventional Current Flow

Over the years, most automotive repair literature has described the flow of current using **Conventional Flow Theory.** This theory states that current flows from the most positive potential to the least positive, or negative, side of a circuit. In automotive applications the battery's negative terminal is the ground and its positive terminal is the "hot" or insulated side. Automotive wiring diagrams generally show the flow from positive to negative (Figure 3–13).

Electron Flow Theory suggests that since electrons are negatively charged, free electrons move from the negative, or surplus, side of the circuit towards an area having less free electrons, or the positive side of the circuit. Either theory applies because to have a current flow from a beginning point to an ending point there has to be a complete circuit.

Voltage

In order for the electrons to flow, a pressure must push the electrons through the circuit (Figure 3–14). This pressure is called an **Electromotive Force (EMF).** It is also known as **voltage** which is a measurement of pressure at a point of excess free electrons in a circuit to a point where there is a shortage of free electrons. The common symbols used for voltage are **E** for EMF, or **V** for voltage. The voltage pressure can be described much like the pressure in a water system (Figure 3–15) which

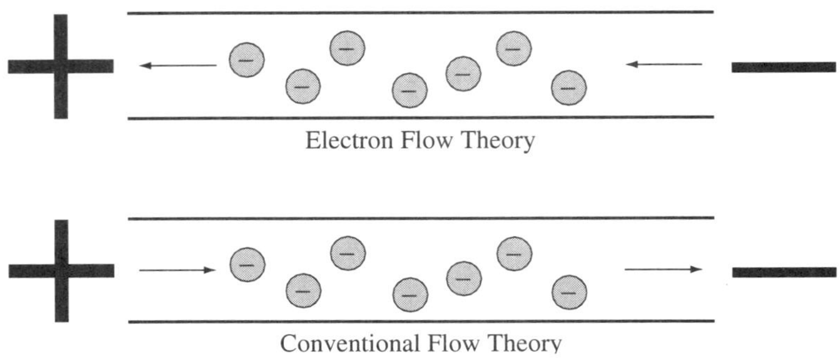

FIGURE 3–13 Comparison of Electron Flow Theory and Conventional Flow Theory.

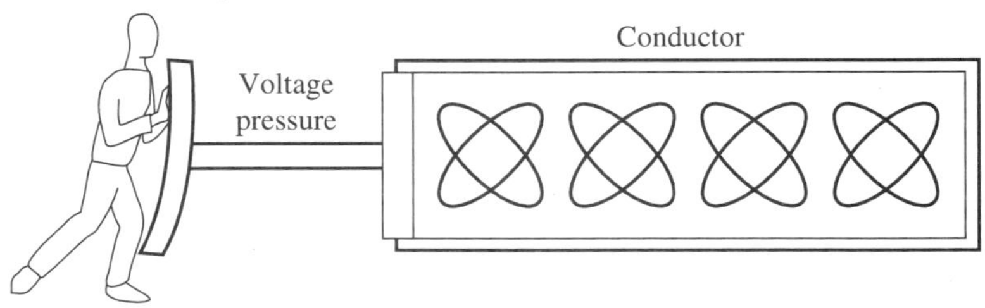

FIGURE 3–14 The pressure that causes electrons to flow is called voltage.

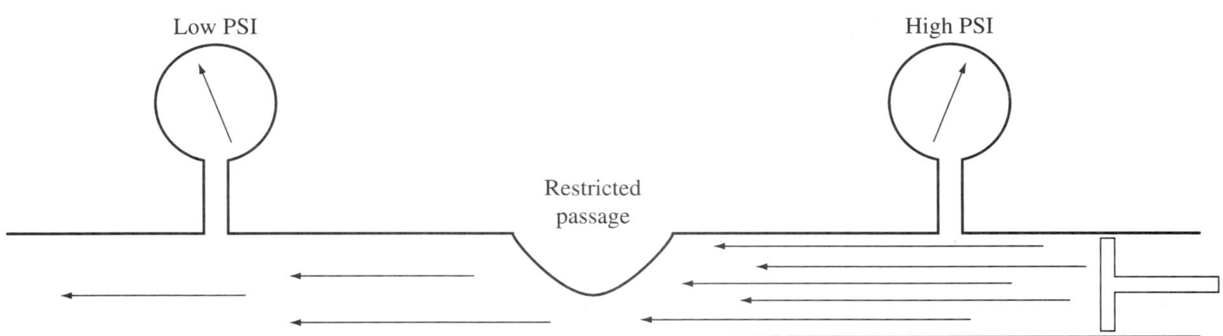

FIGURE 3–15 The psi in a water system operates similarly to voltage in an electrical circuit.

is measured in pounds per square inch (psi). When the pressure is equal on both sides of the circuit, then the circuit is considered near zero, or absolute (Figure 3–16).

If the psi at one end is higher than the psi at the other end, the water will flow. If the psi is equal on both ends, then there is no water flow except the flow due to gravity.

Resistance

Resistance is defined as any force or substance that restricts or opposes the flow of current in a circuit. The symbol most widely used for resistance is the Greek letter omega (Ω) which is used extensively in this text and in manufacturers' wiring diagrams. The letter **R** is also used in mathematical calculations to represent resistance.

Another way to describe resistance is the use of road blocks in some freeway lanes (Figure 3–17).

The more lanes that are blocked, the fewer cars will pass. This is similar to the restriction of water flow in a piping circuit. A nozzle on the end, or a small hose, would be a restriction (orifice) that would stop or slow down the flow of water (Figure 3–18).

The analogies we used to describe resistance can be summed up in Table 3–2.

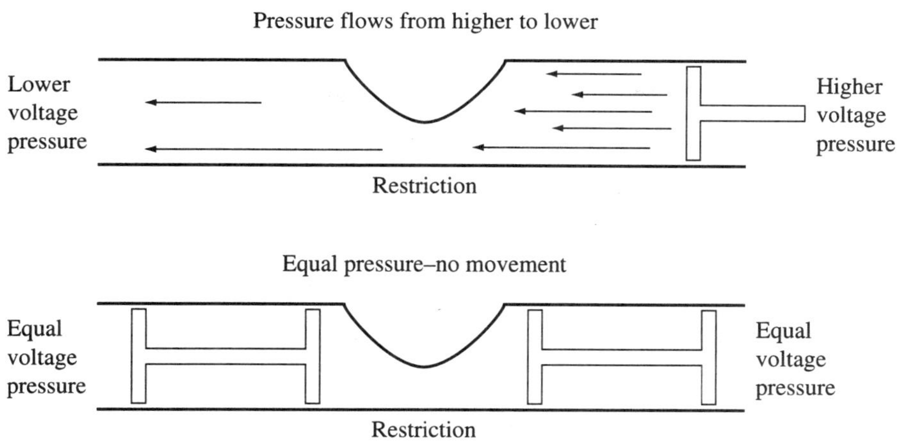

FIGURE 3–16 When the voltage pressure that causes electrons to flow is equal on both sides of a circuit, there is no electron flow.

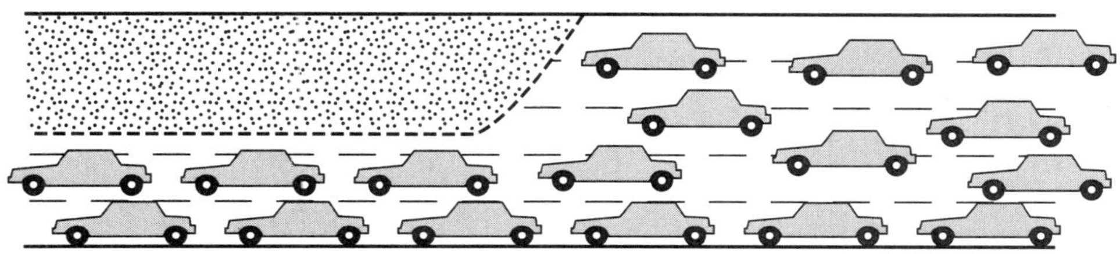

FIGURE 3–17 The more blocked lanes, the slower the flow of cars, and of course, angry drivers!

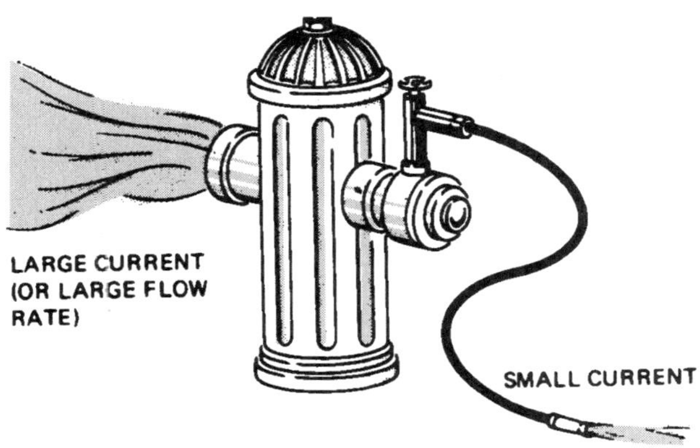

FIGURE 3–18 The higher the restrictive orifice, the smaller the flow.

TABLE 3–2 Summary of voltage analogies.

System	Pressure	Flow	Restriction
Electrical	EMF or volts	Electron flow or amps	Resistance or ohm's
Fluids	Pressure or psi	gpm	Restrictive orifice
Freeway		Movement of cars forward	Road blocks

Resistance is present in all circuits. Without resistance, a circuit would have no way to control current flow. There are several factors that aid in determining circuit resistance:

1. *Atomic structure* of the resistive material. Especially, the number of free electrons it has in its outer shell.

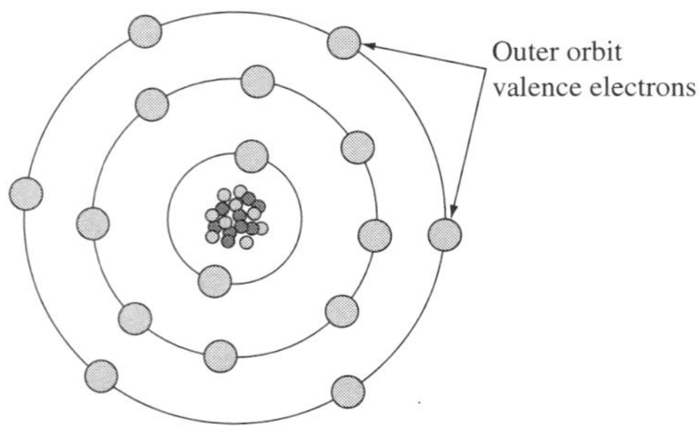

2. *Condition and quality of the circuit.* Anything that unintentionally restricts flow because of component failure or external damage.

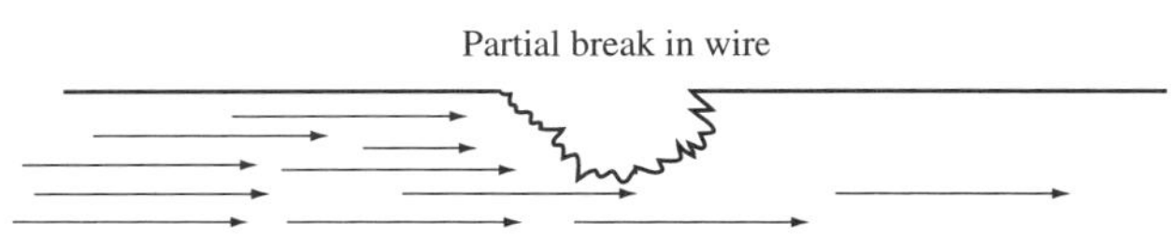

3. *Length of the wire or conductive device.* The resistance goes up as the wire gets longer.

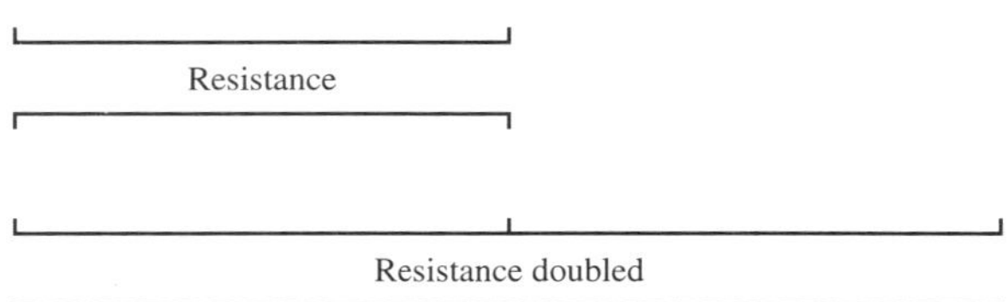

4. *Diameter of wire or conductive device.* The smaller the diameter, the higher the circuit resistance.

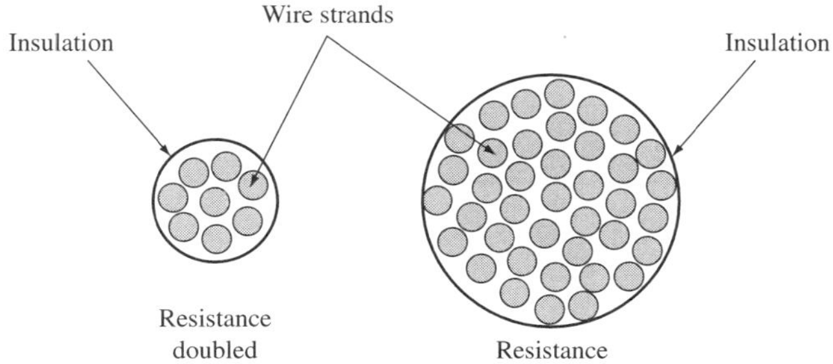

5. *Temperature of the circuit.* As the temperature of the circuit increases, the resistance also increases.

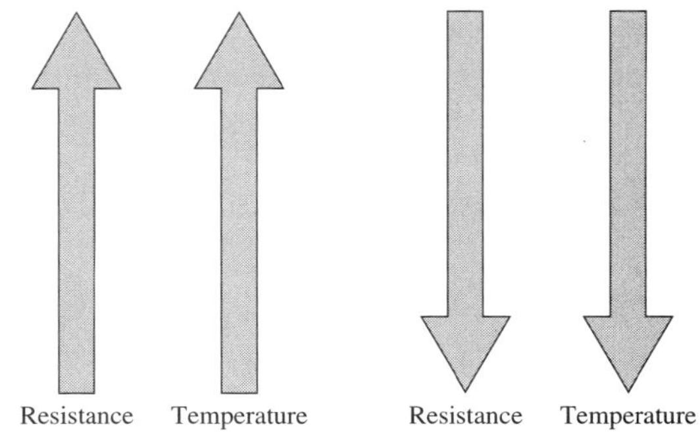

6. *Corrosion or foreign material at electrical connections.* Resistance will develop at any electrical connection that develops a loose or corroded connection, such as a battery cable-to-post connection.

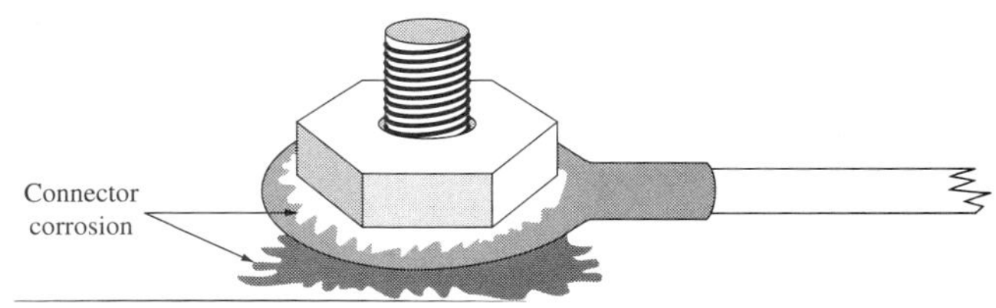

Quick Review of Voltage, Amperage, and Resistance

Definition	Symbol	
Voltage	Electromotive force or pressure	**E or V**
Amperage	Flow of electrons	**I or A**
Resistance	Opposition to electron flow	**R or Ω**

Ohm's Law

The relationship between voltage, current, and resistance in a circuit was proven in 1827 when German physicist George Simon Ohm (1787–1854) demonstrated a clearly defined relationship between voltage, current, and resistance in an electrical circuit. This relationship is known as **Ohm's Law,** which simply stated is:

The current in an electrical circuit is inversely proportional to the resistance of the circuit, and directly proportional to the voltage.

The amount of current flow in a circuit is determined by the voltage and resistance of the circuit. If the circuit resistance increases, the current flow will decrease, or vice versa, because voltage and resistance react *inversely*

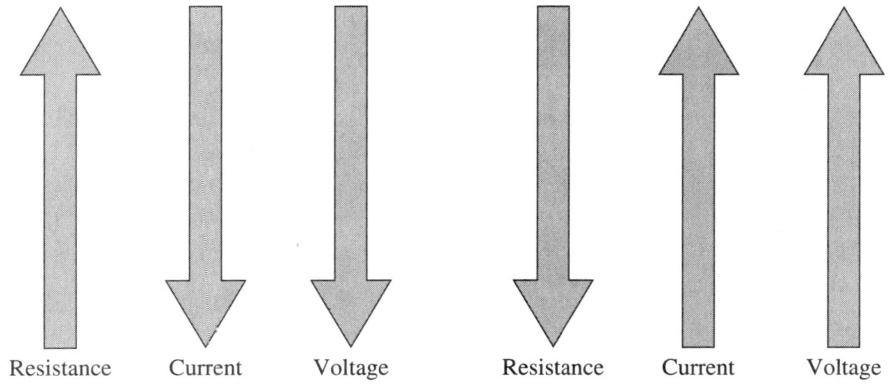

FIGURE 3–19 Relationship of resistance, current, and voltage in Ohm's Law.

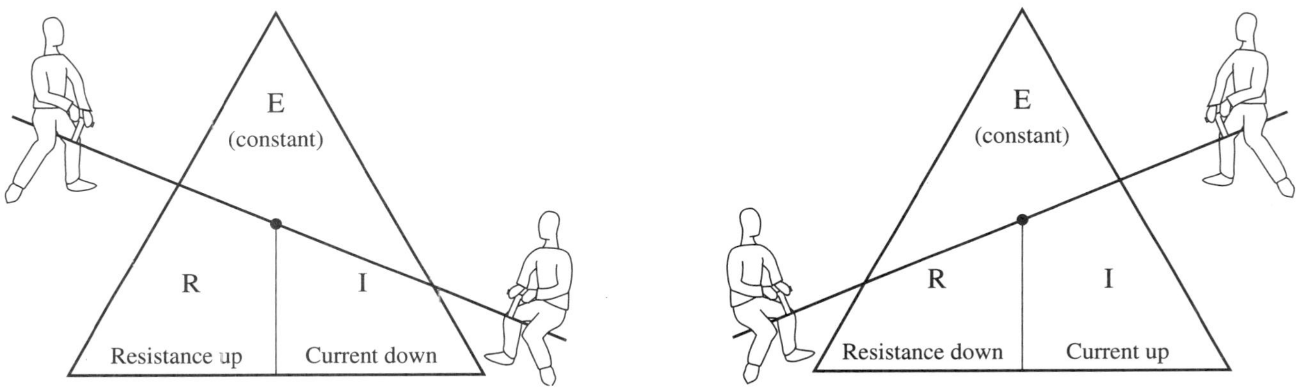

FIGURE 3–20 With constant voltage (E), current (I) versus resistance (R) relationship.

to each other. When the resistance remains constant, any change in the voltage will also change the current in proportion to the voltage (Figure 3–19).

In automotive circuits, system voltage remains relatively constant at 14.5 volts, or at 5 volts for some computer sensor circuits. A change in resistance will cause an opposite reaction to the current. A simple way to remember the principle of Ohm's Law whenever the voltage in a circuit is constant is to look at a teeter totter (Figure 3–20).

In any electrical circuit, the voltage, current, and resistance work together as a mathematical trio. One volt of electrical pressure (EMF), pushing one ampere of current, is the result of having one ohm (Ω) of resistance in the circuit. If you know any two parts of the trio, then you could mathematically figure out the third part. Figure 3–21 is an example of the **Ohm's Law Circle.** You may see this drawn as a triangle or square, but the equations inside the parameter remains the same.

In the circle, the E represents electromotive force or voltage, the I represents intensity or amperage, and the R represents resistance or ohms. Sometimes you may see a V for voltage, or an A for amperage, or a mixture of

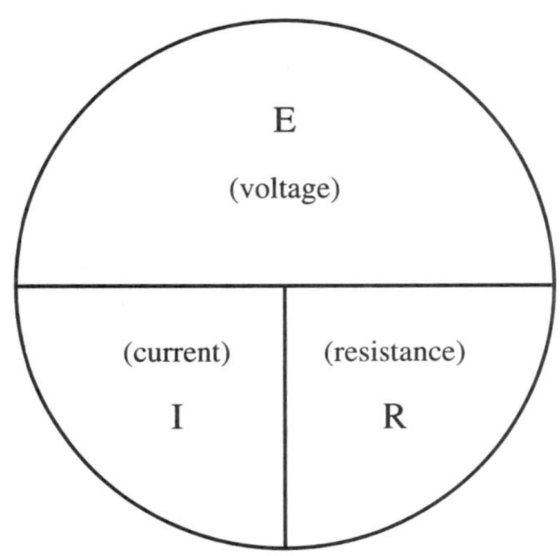

FIGURE 3–21 Ohm's Law Equation Circle.

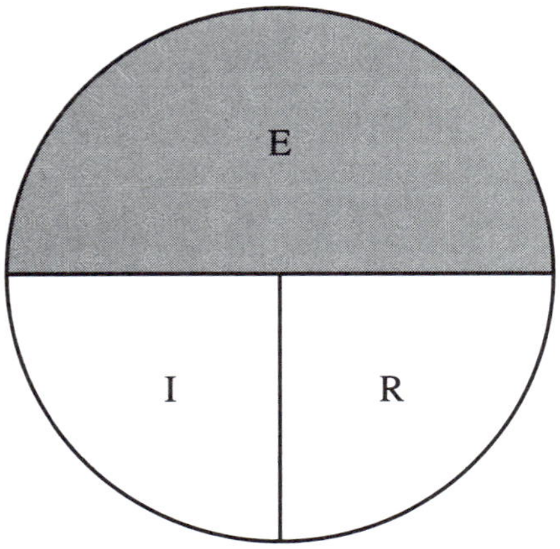

FIGURE 3–22 Use the exposed formula to find the unknown voltage (E).

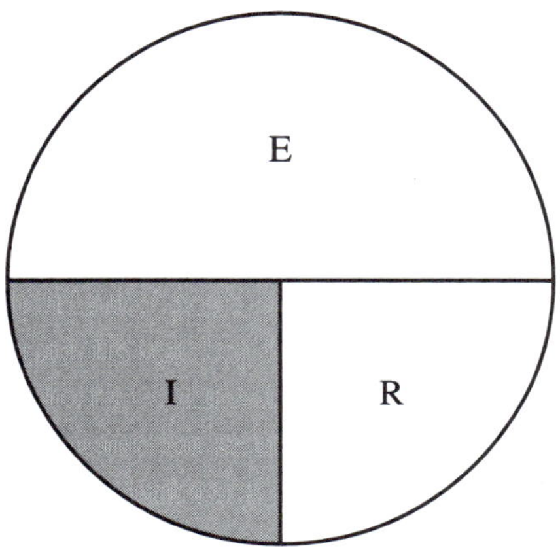

FIGURE 3–23 Use the exposed formula to find the unknown amperage (I).

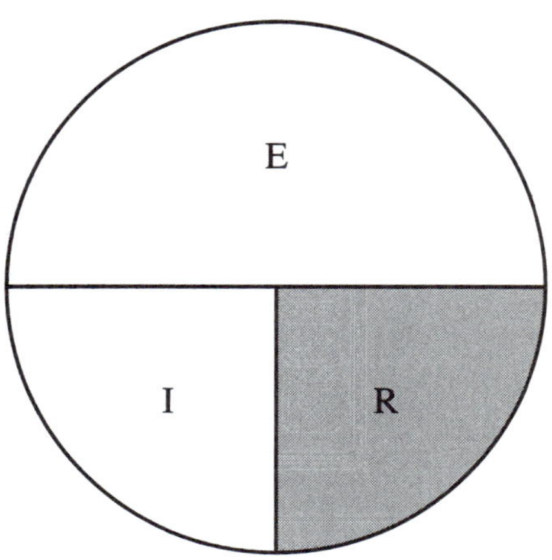

FIGURE 3–24 Use the exposed formula to find the unknown resistance (R).

these letters. As long as you know the basic Ohm's Law circle, the mathematical equations remain the same. When any two of the electrical factors are known, simply cover the unknown value, and calculate as follows.

To find *voltage* in a circuit ($E = I \times R$) (Figure 3–22).

To find *amperage* in a circuit ($I = E \div R$) (Figure 3–23).

To find *resistance* in a circuit ($R = E \div I$) (Figure 3–24).

An example of using Ohm's Law in a simple electrical circuit is shown in Figure 3–25. The source voltage is 12 volts, and the bulb resistance is 6 ohms. Using the Ohm's Law circle from Figure 3–23, calculate the circuit amperage.

The math equation for the above problem should look like the following.

$$I = \frac{E}{R} = \frac{12\text{ V}}{6\ \Omega} = 2\text{ A}$$

Automotive technicians rarely sit by a vehicle performing Ohm's Law calculations, but *knowing* these basic principles will help you determine how a circuit is supposed to operate. Figure 3–26 is the same circuit as Figure 3–25, except there is an added resistance in the wire or connection, between the source battery and the light bulb. The circuit resistance now totals 8 ohms. Using the same math equation as before, but first adding the series resistance together to total 8 Ω, perform Ohm's Law to find the total circuit amperage.

$$I = \frac{E}{R} = \frac{12\text{ V}}{8\ \Omega} = 1.5\text{ A}$$

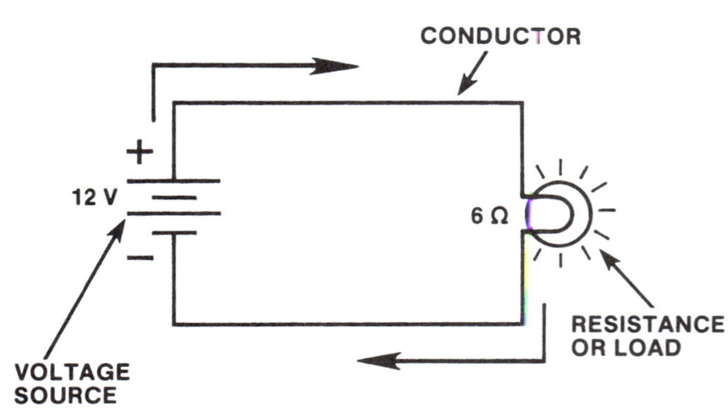

FIGURE 3–25 A simple electrical circuit to show the application of Ohm's Law.

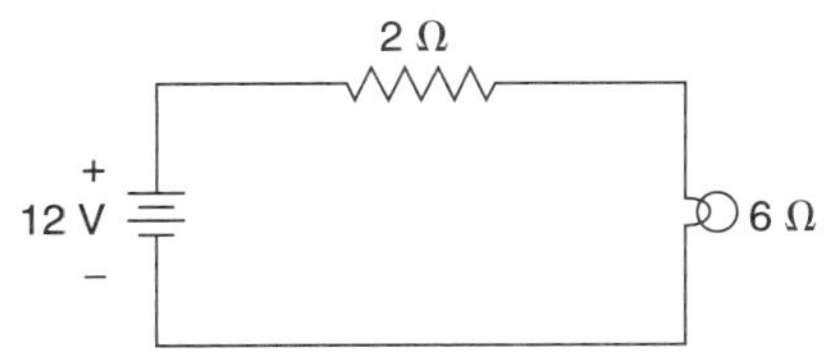

FIGURE 3–26 Same circuit as Figure 3–25, with the additional resistance caused by a corroded wire.

The result of increasing the resistance to 8 Ω caused the amperage to drop from 2 Amps to 1.5 Amps. Through an understanding of the principles of Ohm's Law, it is easy to see that as circuit resistance increases the current flow decreases.

Exercises—Ohm's Law

The following exercise is designed to help give you practice using the *Ohm's Law Circle* to determine the unknown value of an electrical circuit. Complete your answers to this exercise on a separate sheet of paper.

#	Voltage (E)	Resistance (R)	Amperage (I)
1.	12 V	24 Ω	
2.		.15 Ω	.015 A
3.	24,000 V	24,000 Ω	
4.	.057 V		.00114 A
5.	36 V	12,000 Ω	
6.	12 V	48 Ω	
7.		48,000 Ω	1 A
8.	24 V		6 A
9.	24 V	.024 Ω	
10.	12 V		8.0 A
11.	12 kV		6 m A
12.	12,000 V		6 A
13.	120 V	60 Ω	
14.		4.8 Ω	.012 A
15.	12 V	30 Ω	
16.		12 Ω	.0006 A
17.	24,000 V		.024 A
18.		7.5 Ω	15 A
19.	12 V	8 Ω	
20.	18.0 V	1.8 Ω	

Watt's Law

In the early 1800's, Scottish inventor James Watt was looking for a method to describe electrical activity as a measurement of power. The result is a relationship known as **Watt's Law,** which states:

One watt is the amount of work done in one second by one volt moving a charge of one amp in a circuit.

The standard unit of measurement for power is watts. The conversion factor between watts and horsepower is: *one horsepower equals 746 watts.* In any electrical circuit, power (watts), voltage, and current act as a mathematical trio. If you know any two parts of the trio, you can mathematically figure the third part. The **Watt's Law Circle,** is also known as the power triangle (Figure 3–27).

This circle is similar to the Ohm's Law Circle previously discussed. In the circle, the P represents power or watts of energy, the I represents intensity or amperage, and the E represents electromotive force or voltage. Sometime you may see a W to represent power, or an A for amperage, or a V for voltage, or a mixture of the above. As long as you know the basic Watt's Law circle, the mathematical equations remain the same. When any two of the basic factors are known, simply cover the unknown value, and calculate the same way you did when using Ohm's Law.

To find *power* in a circuit ($P = I \times E$) (Figure 3–28).

To find *amperage* in a circuit ($I = P \div E$) (Figure 3–29).

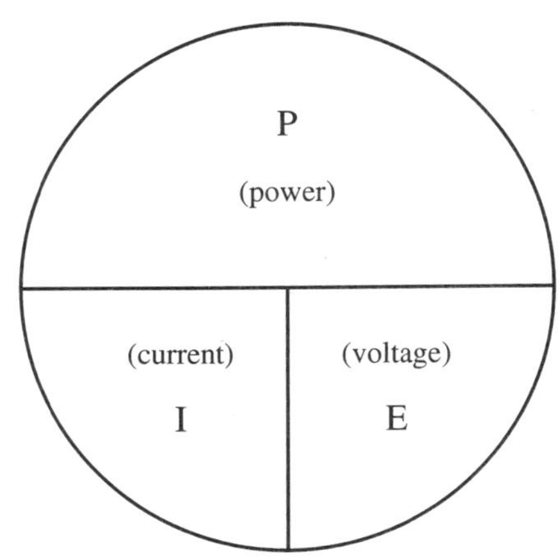

FIGURE 3–27 Watt's Law Equation Circle.

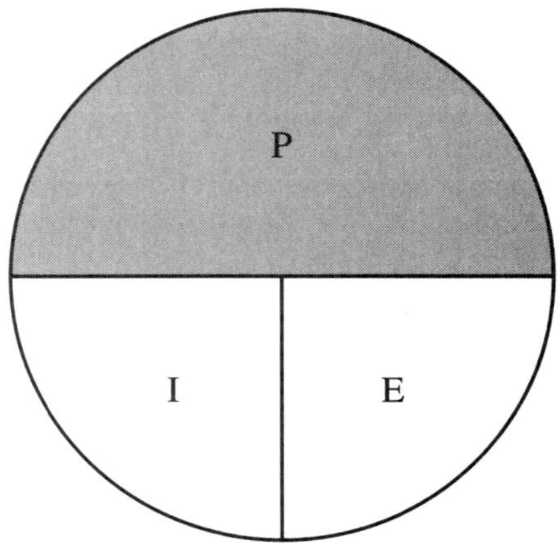

FIGURE 3–28 Use the exposed formula to find the unknown power (P).

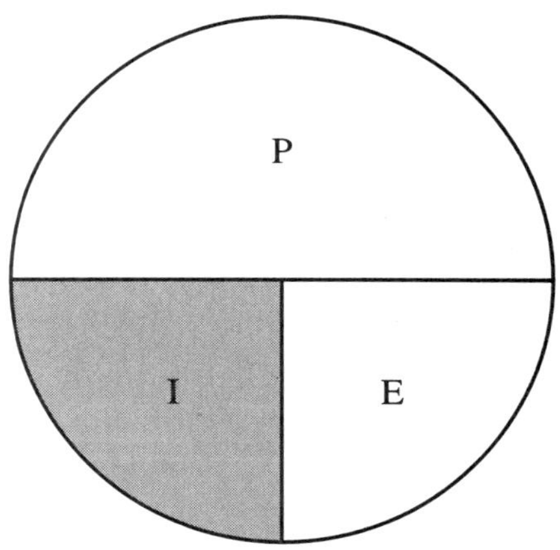

FIGURE 3–29 Use the exposed formula to find the unknown amperage (I).

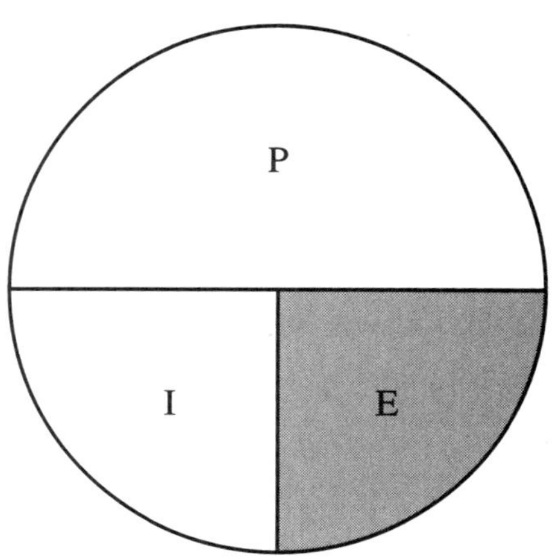

FIGURE 3–30 Use the exposed formula to find the unknown voltage (E).

To find *voltage* in a circuit (E = P ÷ I) (Figure 3–30).

The most well-known example of power measurement in electrical circuits is the wattage rating on household light bulbs. The higher the wattage rating, the brighter the bulb's intensity or output. Even though automotive technicians do not generally use Watt's Law to diagnose an electrical circuit, this law could aid you in a complete understanding of the circuit (Figure 3–31).

In Circuit A in Figure 3–31, the light bulb has 12 volts across its connections and 4 amps of current flow. Using the Watt's Law circle from Figure 3–28, calculate the circuit's power output.

$$P = E \times I, \quad P = 12\text{ V} \times 4\text{ A}, \quad P = 48\text{ W}$$

In Circuit B (Figure 3–31), the additional resistance in the circuit lowers the total current flow (remember the relationship between current and resistance in Ohm's Law), and the voltage across the bulb is now 9 volts. Using the same formula as above, calculate the power output.

$$P = E \times I, \quad P = 9\text{ V} \times 3\text{ A}, \quad P = 27\text{ W}$$

Circuit A produces 21 watts more power than Circuit B, so the bulb is almost twice as bright. Remember that the bulb wattage did not change, but only the characteristics of the circuit itself changed.

Although not as common to the automotive technician, there are other formulas in a **formula circle** that can be used to *find the missing value* of both Ohm's Law

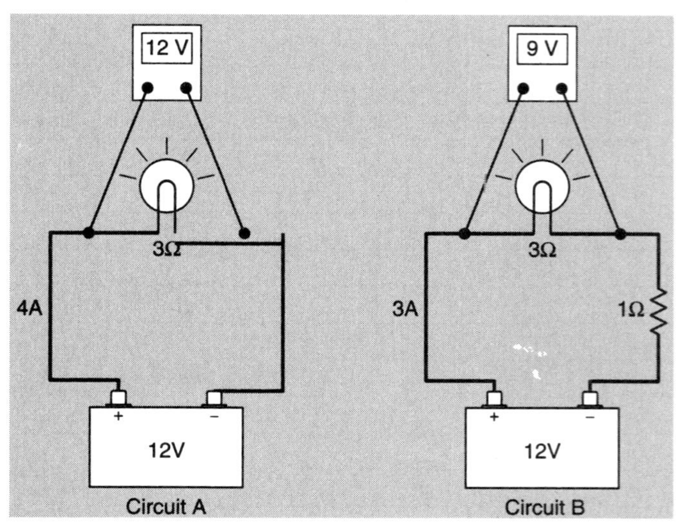

FIGURE 3–31 Simple circuits to demonstrate power measurement output.

and Watt's Law, depending on the known values (Figure 3–32). The formula circle is separated into four mathematical sections, with three formulas in each section. This formula circle can be used whenever the basic Ohm's Law or Watt's Law formulas have more than one of the factors missing, but enough factors to work with if both laws are combined.

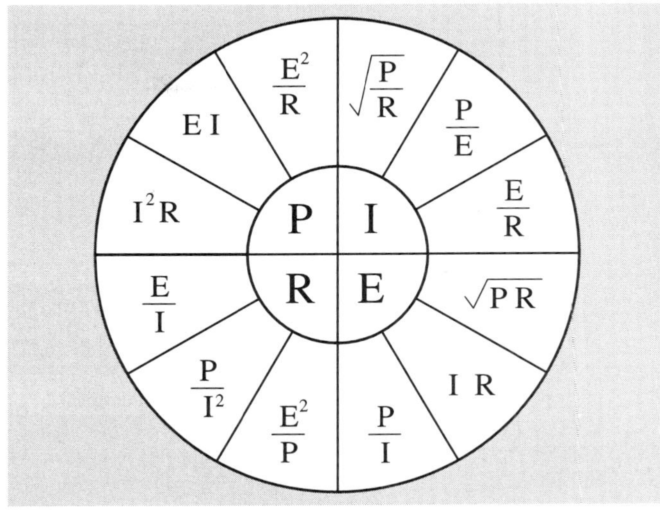

FIGURE 3–32 Formula circle for Ohm's Law and Watt's Law.

Exercise—Watt's Law

The following exercise is designed to help give you practice using the Watt's Law Circle to determine the unknown value of an electrical circuit. Complete your answers to this exercise on a separate sheet of paper.

#	Power (P)	Voltage (E)	Amperage (I)
1.	24 W	24 V	
2.		12.0 V	.025 A
3.	360 W	12.0 V	
4.	.072 kW		.00645 A
5.	18 W	12 V	
6.	12 W	.036 V	
7.		72 V	12 A
8.	24 W		6 A
9.	24 W	24 V	
10.	36 W		18.0 A
11.	36.0 W		1.8 A
12.	.036 W		18.0 A
13.	60 W	60 V	
14.		48 V	24 A
15.	24 W	.006 V	
16.		24 V	.0003 A
17.	2.4 W		.024 A
18.		12.5 V	15 A
19.	12 W	3 V	
20.	180 W	5.0 V	

Symbols

In automotive electrical repair we use a variety of **symbols** to abbreviate or describe a device or circuit. Depending on the manufacturer, you may find differences in the symbols used for the same device. Figure 3-33 is a sample list of common symbols used by a domestic manufacturer.

Figure 3–34 is a sample of common symbols used by an import manufacturer.

Be sure to use the symbol chart specific to your vehicle. Although there is some commonality of symbols, do not assume all symbols mean the same for each automotive manufacturer. We will be using electrical symbols extensively in future chapters of this text to study wiring diagrams and circuits.

Magnetism

In the study of electrical production in Chapter 1, we learned that electricity is a form of natural energy that is converted to a useable electrical source. One of the conversions that plays an important role in the production of electrical energy is **magnetism.** Magnetism is an invisible force that can repel or attract through an air space or solid item (Figure 3–35).

Earth itself is a large magnet with a North Pole and a South Pole. There are magnetic flux lines that wrap around Earth like a giant magnetic shell (Figure 3–36).

This phenomenon is based on the fact that a simple magnet has one north pole and one south pole. Magnetic poles that are alike will repel each other and that are unlike will attract each other (Figure 3–37).

Even though the lines of magnetic force are invisible, we can use iron filings surrounding a magnet to indicate where the lines of force are (Figure 3–38).

The closer to the magnet's core, the stronger the magnetic field. This will be discussed more in future

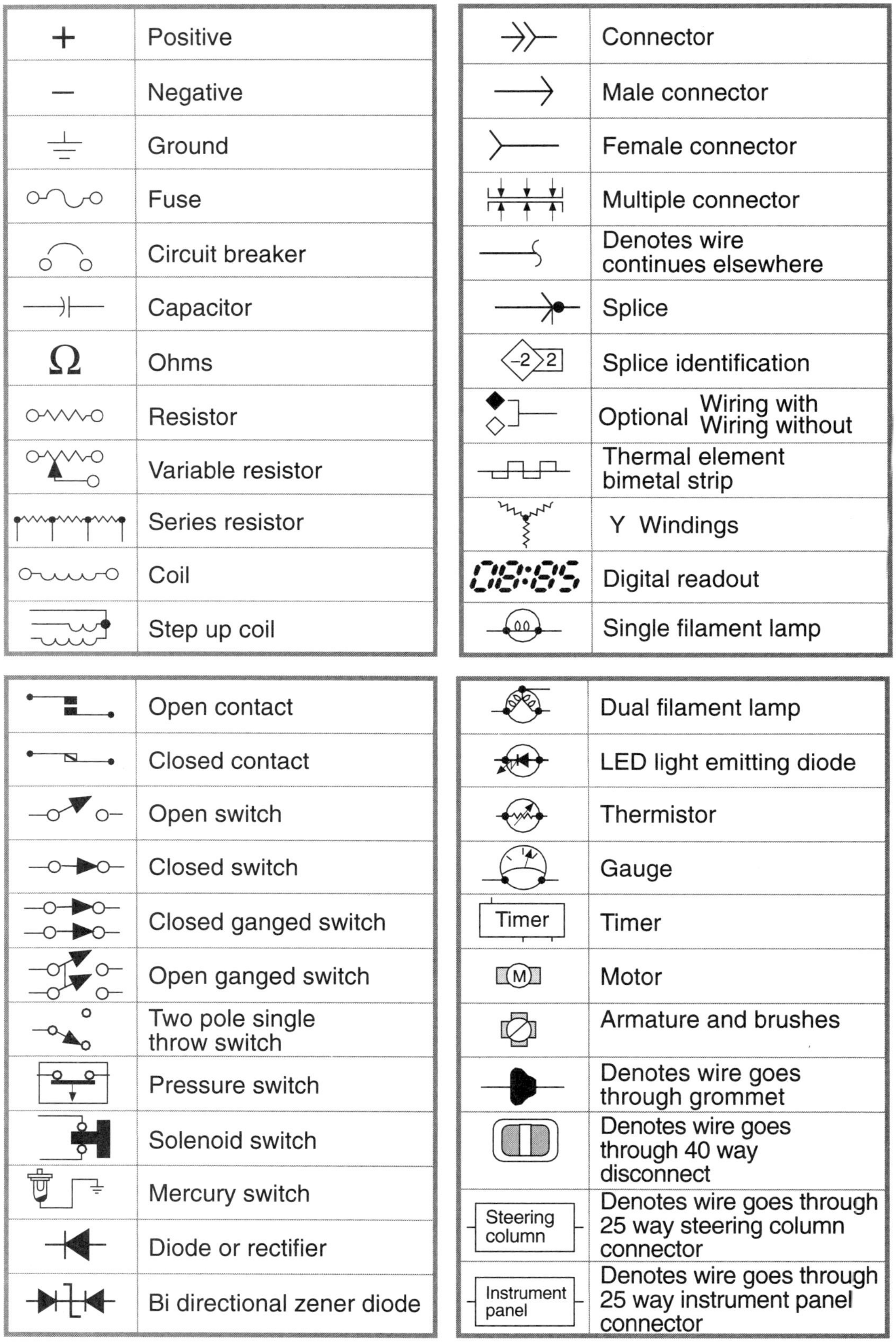

FIGURE 3–33 Some common electrical symbols used by domestic auto manufacturers.
(Courtesy of DaimlerChrysler Corporation)

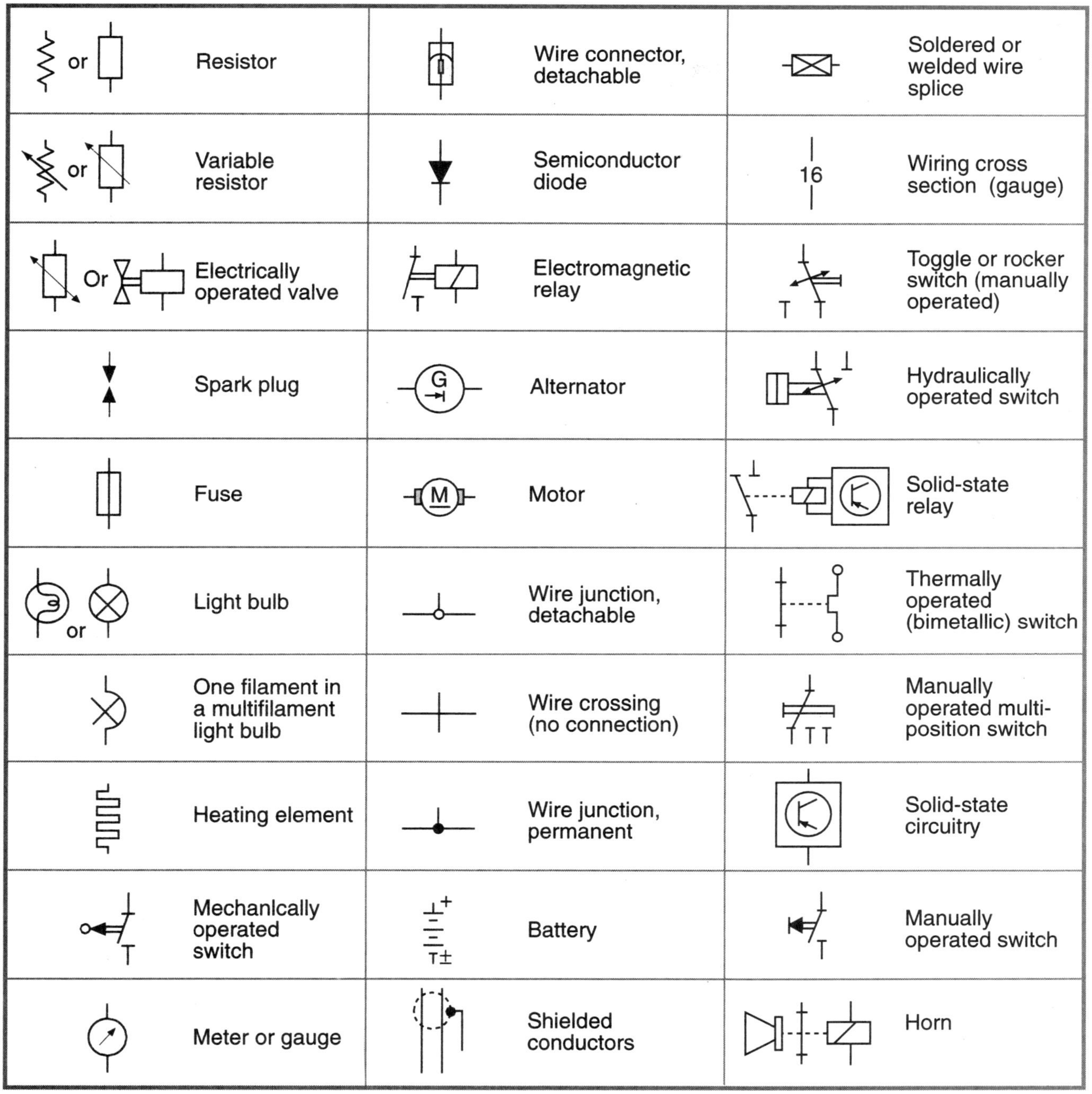

FIGURE 3–34 Some common electrical symbols used by import auto manufacturers.

chapters as we learn ways magnetism is harnessed for automotive electrical use.

Electromagnetism

In the early nineteenth century, André Marie Ampère discovered that holding a compass over a wire carrying current, the needle on a compass would point at right angles to the wire (Figure 3–39).

The movement of the compass needle is due to a circular magnetic field that forms around the wire and is perpendicular to the current flow through the wire. If the current increases, so does the strength of the magnetic field. The direction of the circular magnetic flux lines is determined by the **right hand rule,** which is based on conventional flow theory (positive-to-negative flow) (Figure 3–40).

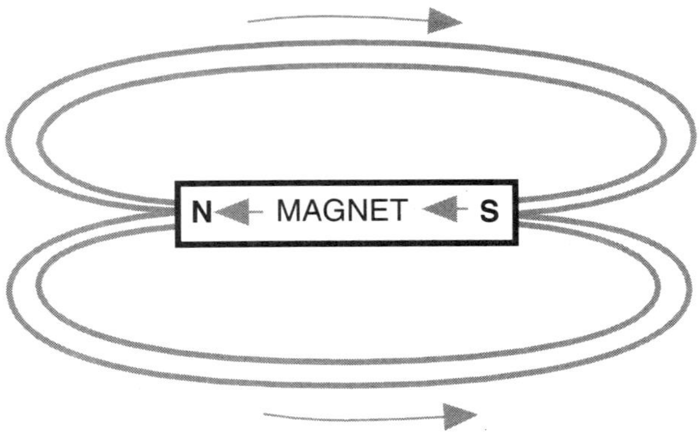

FIGURE 3–35 The invisible forces of magnetism.

If you place your right hand on a wire with your thumb pointing towards the direction of current flow, your fingers will point in the direction of the magnetic flux lines, and your thumb will point to the negative side of the circuit (Figure 3–41).

If you do not know which side of an electrical circuit is the positive and which is the negative, the following simple compass experiment will assist you (Figure 3–42).

Step # 1 Place a compass on a wire that has current flowing through it, and position the west and east compass headings parallel to the conductor, or wire.

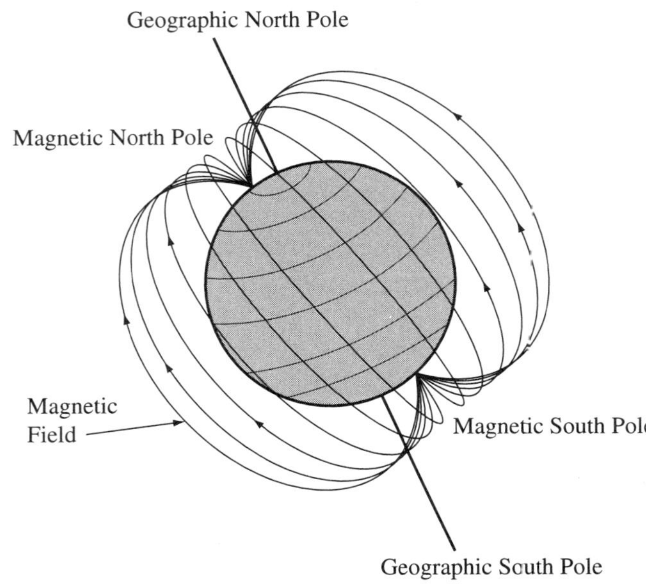

FIGURE 3–36 Earth's magnetic field.

Step # 2 Observe the compass needle direction. If the needle pointer points to the north heading, the positive end is to the left of the needle pointer, or west. If the needle pointer points to the south heading, then the positive end is also to the left of needle, or east.

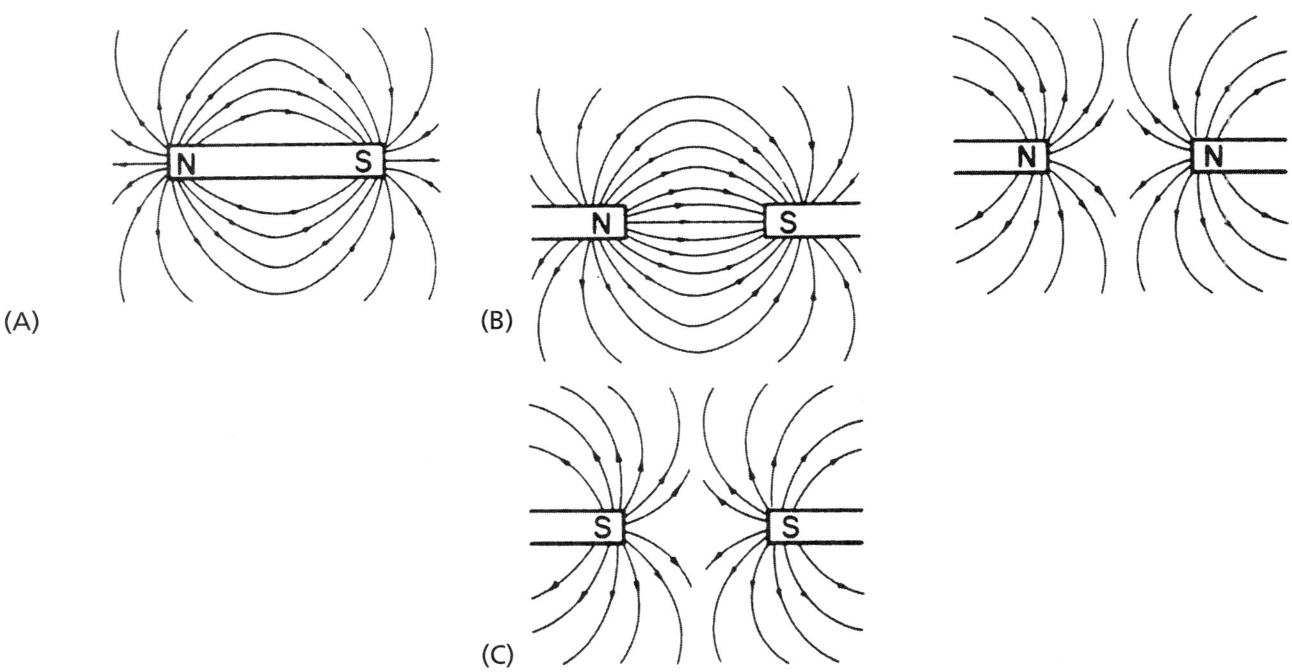

FIGURE 3–37 Principles of magnetism. (A) Magnetic north and south poles, (B) unlike poles attract each other, (C) like poles repel each other.

FIGURE 3–38 The invisible magnetic flux lines become visible.

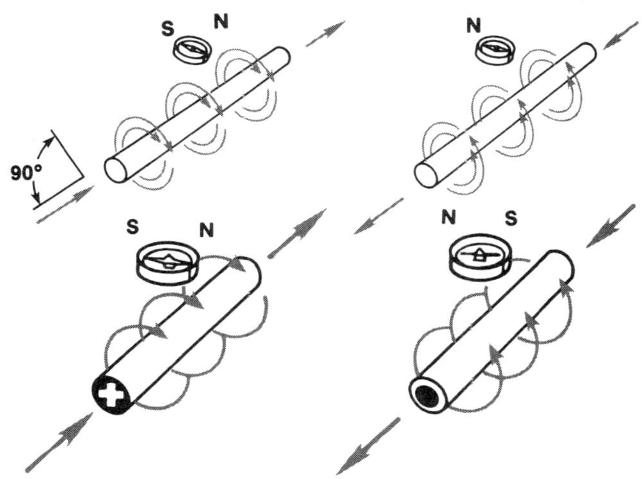

FIGURE 3–39 Magnetic flux lines cause the compass needle to deflect at a right angle to the conductor.

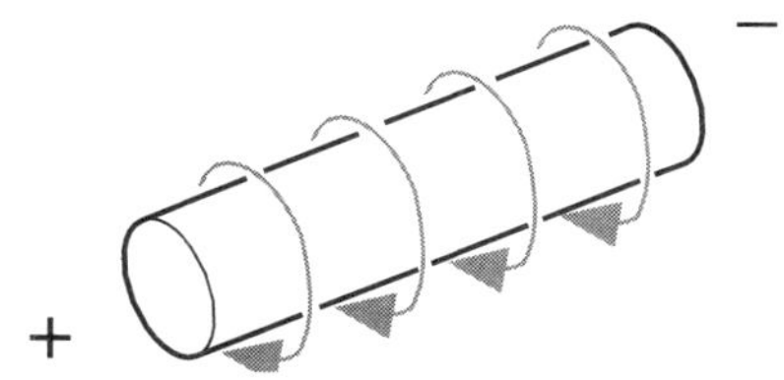

FIGURE 3–40 When current flows through a conductor, a magnetic field will surround it.

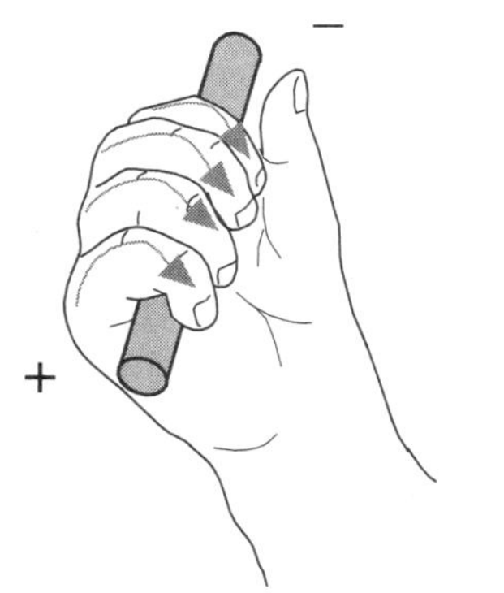

FIGURE 3–41 Right hand rule of magnetism.

Analysis When you place the west and east compass headings parallel to and over a wire, no matter which way in relation to polarity, the positive polarity will *always* be to the left of the compass pointer when the pointer is positioned upward, or pointing away from you.

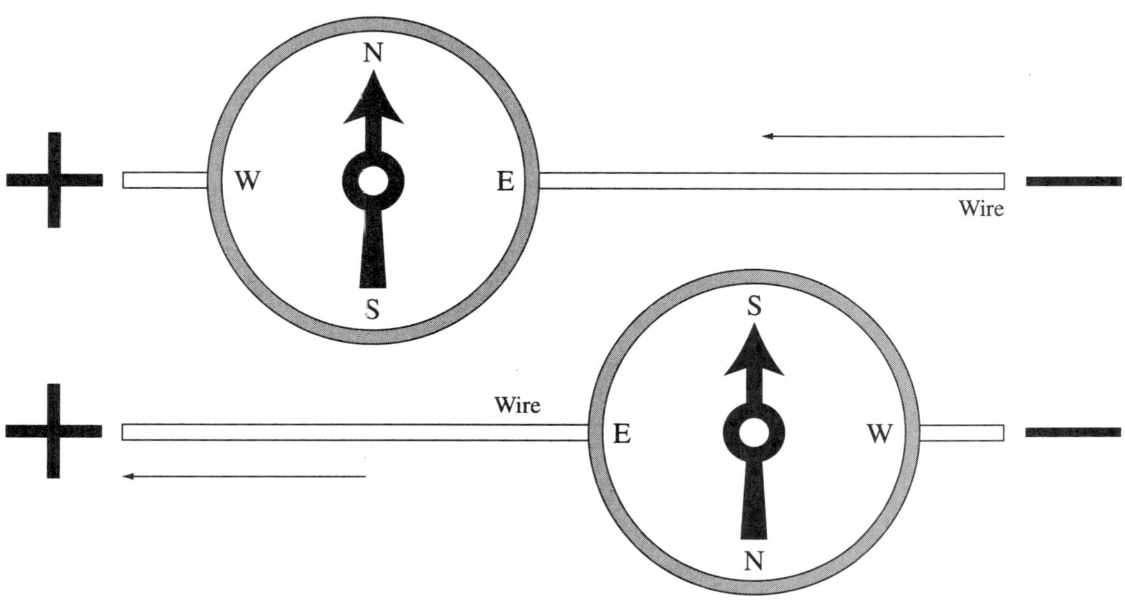

FIGURE 3–42 Compass experiment.

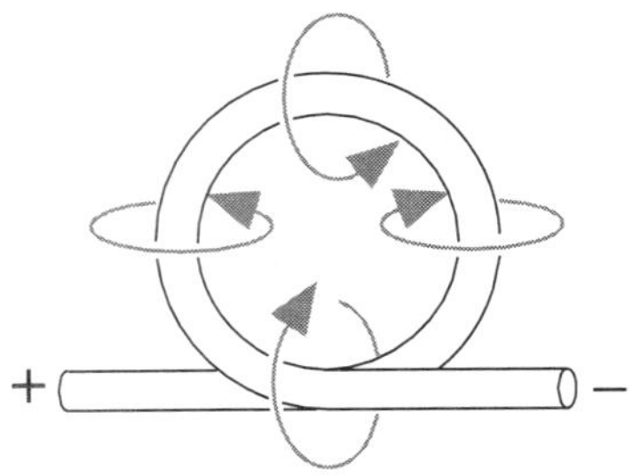

FIGURE 3–43 Magnetic field strengthens by looping the wire conductor into equal-size turns.

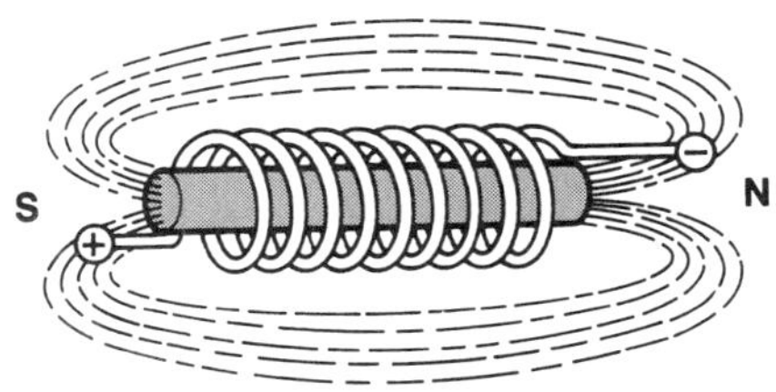

FIGURE 3–44 A simple electromagnet.

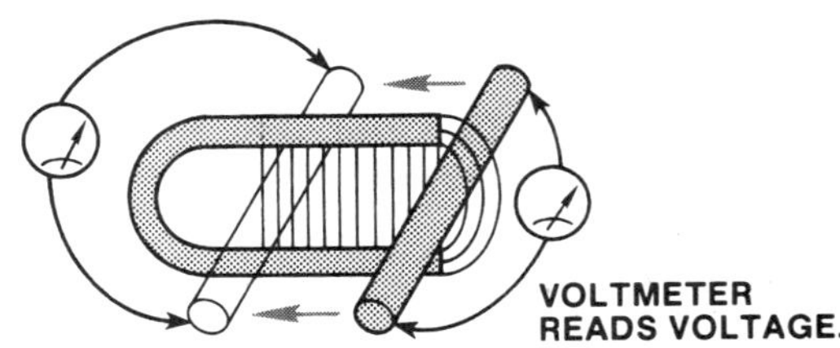

FIGURE 3–45 Moving a conductor through a magnetic field creates a voltage across the conductor.

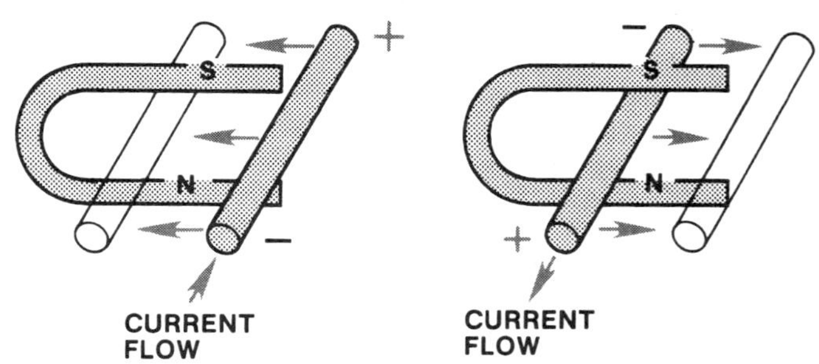

FIGURE 3–46 Voltage polarity is determined by the direction the conductor passes through in the magnetic field.

To increase the magnetic field strength around a wire, a wire can be is wound into equal size turns. By increasing the number of turns, the magnetic field is also increased (Figure 3–43). When current flows through this wound wire, the magnetic field that normally would surround a straight wire, instead combines to form a stronger magnetic field.

If a soft iron core is placed inside a coil of wire and a current is applied to the wire, the magnetic field will be further increased, and it becomes an **electromagnet** (Figure 3–44).

Electromagnetic Induction (EMI)

Magnetic fields can also be used to produce electricity. Instead of using current to generate a magnetic field, we can move a conductor within a magnetic field to generate a voltage. Or similarly, we can move a magnetic field across a conductor to produce a voltage (Figure 3–45).

As the conductor shown in Figure 3–46 moves rapidly through the magnetic field and into the magnet, a small voltage is induced in the conductor. This voltage has a specific polarity with a positive side and a negative side. When the conductor is brought back rapidly through the magnetic field from the inside, the polarity will switch (Figure 3–46).

The Electromagnetic Induction (EMI) induced in a wire or conductor depends upon five main factors:

1. The larger the conductor diameter, the lesser the internal resistance.
2. The higher the number of conductor turns, the higher the voltage induced.
3. If the intensity of the magnetic field is increased, then the voltage induced is increased.
4. As the gap between the magnetic field and the conductor decreases, the voltage induced increases.
5. As the speed of the conductors cutting through magnetic flux lines increases, the voltage induced increases.

The significance of learning magnetism and electromagnetism theory will be evident as you study future chapters that cover ignition systems, starting and charging systems, and electrical relay control circuits.

Summary

- ❑ Matter can be any substance that occupies a space and contains a mass.
- ❑ Molecules consist of a group of elements.
- ❑ Elements are known as a cluster of one type of atom.
- ❑ The Periodic Table of Elements contains over 100 different elements of nature.
- ❑ An atom is the smallest part of an element.
- ❑ Atoms contain protons, neutrons, and electrons.
- ❑ Protons are positively charged particles.
- ❑ Neutrons have no charge and combine with the protons to form the nucleus of the atom.
- ❑ Electrons are negatively charged particles.
- ❑ The electrons in the outer-most rotating orbit of an atom are called valence electrons.
- ❑ Free electrons are valence electrons that become dislodged from their orbit.
- ❑ A bound electron is free to move within its own perspective orbit only, and remain a fixed distance from the nucleus.
- ❑ Electricity is defined as the movement of free electrons from atom to atom.
- ❑ Good conductors of electricity have three or less valance electrons.
- ❑ Semiconductors have exactly four valance electrons.
- ❑ Good insulators of electricity have more then four valance electrons.
- ❑ Current flow is the mass movement of free electrons from atom to atom.
- ❑ One ampere of current flow is equal to 6.28 billion billion free electrons flowing past a point in one second.
- ❑ Conventional flow theory says current flows from positive to negative.
- ❑ Electron flow theory says current flows from negative to positive.
- ❑ Amperage is often recorded by using the letter I for "intensity of current" or the letter A for amps.
- ❑ EMF is Electromotive Force or voltage pressure in a circuit.
- ❑ Resistance is any force or substance that resists or opposes the flow of current in a circuit.
- ❑ The resistance symbol often used is the letter R or the Greek omega symbol Ω.
- ❑ Circuit resistance is determined by five factors: Atomic structure, condition and quality of the circuit, length of wire, diameter of wire, temperature of the circuit.
- ❑ Ohm's Law is a clearly defined relationship between voltage, current, and resistance.
- ❑ The Ohm's Law circle consists of voltage, current, and resistance working together as a mathematical trio.

- ❑ Watt's Law is a measurement of power in an electrical circuit.
- ❑ The Watt's Law circle consists of power, voltage, and current working together as a mathematical trio.
- ❑ The *formula circle* consists of all the combinations of mathematical formulas used in Ohm's Law and Watt's Law.
- ❑ Symbols are used by automotive manufacturers to abbreviate or describe a device or circuit.
- ❑ Magnetism is an invisible force where like poles repel each other, and unlike poles attract each other.
- ❑ The right hand rule of magnetism determines which way the magnetic flux lines point.
- ❑ The amount of EMI induced in a wire depends upon five factors: Diameter of the conductor, number of conductor turns, increasing the magnetic field, lowering the gap between the conductor and magnetic field, changing the speed of the conductor cutting through the magnetic field.

Key Terms

alternating current (AC)
amperes
atomic number
atoms
conductors
conventional flow theory
current flow
direct current (DC)
electromagnet
electromotive force (EMF)
electron flow theory
electrons
elements
formula circle
free electrons
insulators
magnetism
matter
molecules
neutrons
nucleus
Ohm's Law
Ohm's Law circle
protons
resistance
right hand rule
semiconductors
symbols
valence electrons
voltage
Watt's Law
Watt's Law circle

Review Questions

Short Answer Essays

1. What determines if an atom is a good conductor, semiconductor, or insulator of electricity?
2. Describe what current flow is in an electrical circuit.
3. Describe what voltage is in an electrical circuit.
4. Describe what resistance is in an electrical circuit.
5. Explain the five main factors that determine how much EMI is induced in a wire or conductor.

Fill-in-the-Blanks

1. Magnetism is an ____________________ force that can repel or attract through an ____________________ space or solid item.

2. The current in an electrical circuit is _______________ proportional to the resistance of the circuit, and _______________ proportional to the voltage.
3. Electron flow theory suggests that free electrons move from the _______________ side towards the _______________ side of the circuit.
4. The electrons in the outer most orbit are called _______________ electrons.
5. A typical atom consists of _______________ and _______________ that form the nucleus, and _______________ that rotate in orbits around the nucleus.

ASE Style Review Questions

1. Technician A says an atom is made up of protons, neutrons, and electrons. Technician B says the center of the atom is made up of protons and electrons only. Who is correct?
   A. A only
   B. B only
   C. Both A and B
   D. Neither A nor B

2. A simple law of physics states that "like charges repel each other, and unlike charges attract each other." Technician A says this law is what keeps the protons and electrons from being attracted to each other. Technician B says it is the centrifugal force of the rotating electrons in orbit that keeps them from being attracted to the protons. Who is correct?
   A. A only
   B. B only
   C. Both A and B
   D. Neither A nor B

3. A copper atom is an excellent conductor of electricity. Technician A says this is due to having one valance electron. Technician B says the neutrons determine the conductivity of the atom. Who is correct?
   A. A only
   B. B only
   C. Both A and B
   D. Neither A nor B

4. Technician A says the electrons in the outer-most orbit of an atom are the valence electrons. Technician B says the free electrons dislodge from their orbit and jump from atom to atom. Who is correct?
   A. A only
   B. B only
   C. Both A and B
   D. Neither A nor B

5. Technician A says the nucleus of the atom determines whether an atom is a good conductor or insulator of electricity. Technician B says the number of bound electron orbits determines if an atom is a good conductor or insulator of electricity. Who is correct?
   A. A only
   B. B only
   C. Both A and B
   D. Neither A nor B

6. Technician A says the flow of free electrons in a circuit is called current flow. Technician B says that 6.28 billion billion electrons flowing past a point in one second is one amp. Who is correct?
   A. A only
   B. B only
   C. Both A and B
   D. Neither A nor B

7. Technician A says that the letter I indicates amperage in an electrical circuit. Technician B says the letter A indicates amperes in an electrical circuit. Who is correct?
   A. A only
   B. B only
   C. Both A and B
   D. Neither A nor B

8. Technician A says conventional flow theory means current flows from the most positive potential to the least positive or negative. Technician B says conventional flow theory means the battery's negative terminal is the ground, and the positive terminal is the "hot" side. Who is correct?
   A. A only
   B. B only
   C. Both A and B
   D. Neither A nor B

9. Technician A says electron flow theory suggests that current flows from negative to positive. Technician B says that in electron flow theory the free electrons move from a surplus side to an area having less free electrons. Who is correct?
   A. A only
   B. B only
   C. Both A and B
   D. Neither A nor B

10. Technician A says the pressure in a circuit is called Electromotive Force or EMF. Technician B says the pressure in a circuit is called voltage. Who is correct?
    A. A only
    B. B only
    C. Both A and B
    D. Neither A nor B

11. Technician A says current will flow when the voltage is equal on both sides of a circuit. Technician B says voltage is a measurement of a pressure difference across a load. Who is correct?
    A. A only
    B. B only
    C. Both A and B
    D. Neither A nor B

12. Technician A says the greater the circuit resistance, the greater the current flow. Technician B says the lower the current flow, the lower the circuit resistance. Who is correct?
    A. A only
    B. B only
    C. Both A and B
    D. Neither A nor B

13. Technician A says the resistance of a wire increases as the length increases. Technician B says the resistance of a wire increases as the diameter increases. Who is correct ?
    A. A only
    B. B only
    C. Both A and B
    D. Neither A nor B

14. Technician A says Ohm's Law states that current is inversely proportional to the resistance in a circuit. Technician B says that as circuit resistance goes up, the current decreases. Who is correct?
    A. A only
    B. B only
    C. Both A and B
    D. Neither A nor B

15. Technician A says the current in an electrical circuit is directly proportional to the resistance. Technician B says current is inversely proportional to the voltage. Who is correct?
    A. A only
    B. B only
    C. Both A and B
    D. Neither A nor B

16. Technician A says to find the voltage of a circuit, the amperage is divided by the resistance. Technician B says the letter E represents electromotive force. Who is correct?

   A. A only
   B. B only
   C. Both A and B
   D. Neither A nor B

17. Technician A says a simple circuit with 12 volts and 8 ohms, should have an amperage total of 1.5 amps. Technician B says if the resistance increases on the same circuit from 8 ohms to 12 ohms, then the amperage total would also increase the same amount. Who is correct?

   A. A only
   B. B only
   C. Both A and B
   D. Neither A nor B

18. A simple circuit with 24 volts and 8 ohms is being discussed: Technician A says that to find the power of the circuit, you must first find the circuit current. Technician B says this circuit should have 72 watts of power. Who is correct?

   A. A only
   B. B only
   C. Both A and B
   D. Neither A nor B

19. Two technicians are discussing the formula circle that combines Ohm's Law and Watt's Law into four sections and twelve equations. Technician A says this formula circle could be used when only one factor is known for a circuit. Technician B says the known values of both laws can be combined to find a missing factor. Who is correct?

   A. A only
   B. B only
   C. Both A and B
   D. Neither A nor B

20. Technician A says all wiring diagram symbols are standard to all auto manufacturers. Technician B says there may be a difference in symbol usage between auto manufacturers. Who is correct?

   A. A only
   B. B only
   C. Both A and B
   D. Neither A nor B

21. Technician A says the following is the correct symbol for a circuit fuse.

Technician B says the following is the correct symbol for a circuit fuse.

Who is correct?

A. A only

B. B only

C. Both A and B

D. Neither A nor B

22. Technician A says magnetism is an invisible force that can be harnessed for electrical production. Technician B says the magnetic field is strongest at the magnet's core. Who is correct?

A. A only

B. B only

C. Both A and B

D. Neither A nor B

23. Technician A says the closer to the center of the magnetic field lines, the weaker the magnetism. Technician B says the right hand rule of magnetism determines the circular direction of the magnetic flux lines. Who is correct?

A. A only

B. B only

C. Both A and B

D. Neither A nor B

24. Technician A says a magnetic field can be used to produce electricity. Technician B says current flow in a conductor can produce a magnetic field. Who is correct?

A. A only

B. B only

C. Both A and B

D. Neither A nor B

25. Technician A says if a conductor is moved rapidly through a magnetic field a voltage is induced into the conductor. Technician B says that when the gap between the magnetic field and conductor increases the voltage induced in the conductor also increases. Who is correct?

A. A only

B. B only

C. Both A and B

D. Neither A nor B

4 Measurement System and Calculator Use

Introduction

During the past two hundred years, much of the world has utilized the International System (IS) of metric measurement. In contrast, the United States has used the U.S. Customary System (USC) of measurement, also known as the English System.

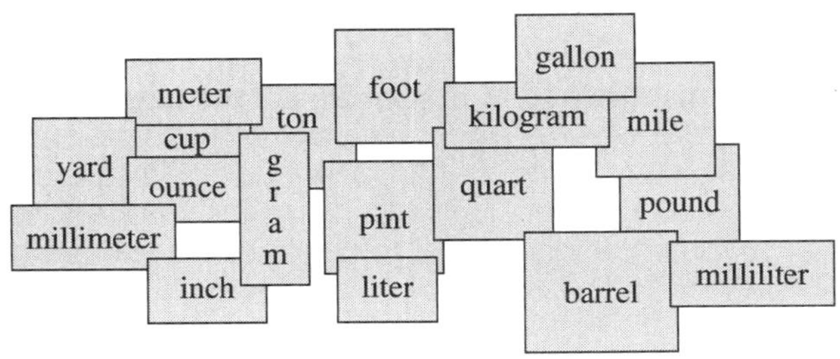

Since the early 1900's, the United States Congress has attempted to get the American public and the industrial community to convert to the metric system. In 1975, when Congress passed the Metric Conversion Act, the scientific community had been using the metric system for decades. The industrial community now started to move toward adopting the metric system, and especially the automotive manufacturers who began using more metrics in their products. This chapter provides a brief look at the metric system in preparation for a thorough understanding of the principles of electrical measurement.

Most modern digital multimeters (DMM) display only four digits, making it necessary to use measurement abbreviations. We will practice using the scientific notation features of a scientific calculator, which will be helpful in future chapters when we deal with electrical theory and associated mathematical formulas.

Objectives

When you complete this chapter you should be able to:

- ❑ Describe the metric system of measurement
- ❑ Apply basic metric system conversion principles
- ❑ Compare metric prefix terms to electrical prefix terms
- ❑ Demonstrate the basic features of a scientific calculator
- ❑ Use scientific notation in conversions

The Metric Measurement System

Today, all over the U.S. and the world, manufacturers, scientists, business people, government offices, the military, private homes, and even the auto industry use the metric system as the *primary* system of measurement. The metric system is very simple in design—an increase or decrease from the base unit of measurement always operates on the **Power of Ten** principle. Mastering the metric system will help when you learn electrical theory because you can directly apply its principles to the base unit identifiers used for electrical terms. More on this later, so stay tuned. The base units for the metric system are listed in Figure 4–1.

Meter for length	Liter for volume	Gram for weight

FIGURE 4–1 Basic metric units of measurements.

When each base unit is increased by a **power of ten,** it becomes a new number that is ten times the previous number. For example, if one gram is the base unit, increasing it by a power of ten (1×10) will make 10 grams. If we increase this new number by another power of ten, or 10^2 relative to the base unit, you will have 10×10 or 100 grams total.

Each power of ten has a name attached to its new unit, which is the **prefix.** Table 4–1 shows the most common units of measurements, prefixes, and symbols for the metric system.

For example, 1,000 grams can be referred to as 1 kilogram (1 kg), or 1 gram $\times 10^3$. Using this notation can help when dealing with awkwardly large or small numbers. When units of measurement are smaller then the base unit, you divide by powers of ten to make a smaller value number. For example, 1 gram divided by 10 (10^{-1}) is now *1/10* gram or 0.1 decigram (0.1 dg).

Because you will be working extensively with various units of measurements, you will need to convert between the different metric prefix values. An easy way to do this without using math formulas is to use a **linear scale** of metric conversions (Figure 4–2).

Unit values to the left of the base unit (0) are larger and unit values to the right of the base unit are smaller. For example, if you had 15,000 meters and wanted to change the number to kilometers (km) then you would need to move the decimal to the left three places to make 15 km (Figure 4–3).

If you had 84 grams and wanted to change the number to milligrams (mg), you would need to move the decimal to the right three places to make 84,000 mg (Figure 4–4).

Just remember the following when you change the unit value bigger or smaller.

LARGER UNIT ←——————→ SMALLER UNIT

Let's show a few more examples before giving you several to try yourself. Convert 275.73 µg (microgram)

TABLE 4–1 Metric prefix measurement table.

Prefix	Abbrev. Symbol	Name	Power of 10	Decimal Equivalent
tera	T	one trillion	10^{12}	1,000,000,000,000
giga	G	one billion	10^{9}	1,000,000,000
mega	M	one million	10^{6}	1,000,000
kilo	k	one thousand	10^{3}	1,000
hecto	h	one hundred	10^{2}	100
deka	da	ten	10^{1}	10
base unit		one unit	10^{0}	1
deci	d	one tenth	10^{-1}	0.1
centi	c	one hundredth	10^{-2}	0.01
milli	m	one thousandth	10^{-3}	0.001
micro	µ	one millionth	10^{-6}	0.000001
nano	n	one billionth	10^{-9}	0.000000001
pico	p	one trillionth	10^{-12}	0.000000000001

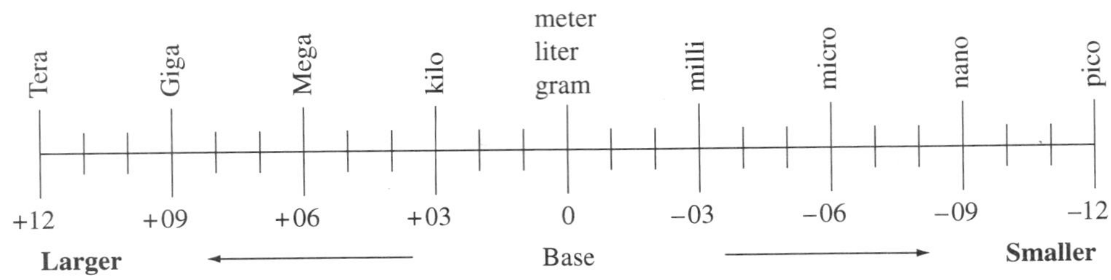

FIGURE 4–2 Linear scale of metric conversion.

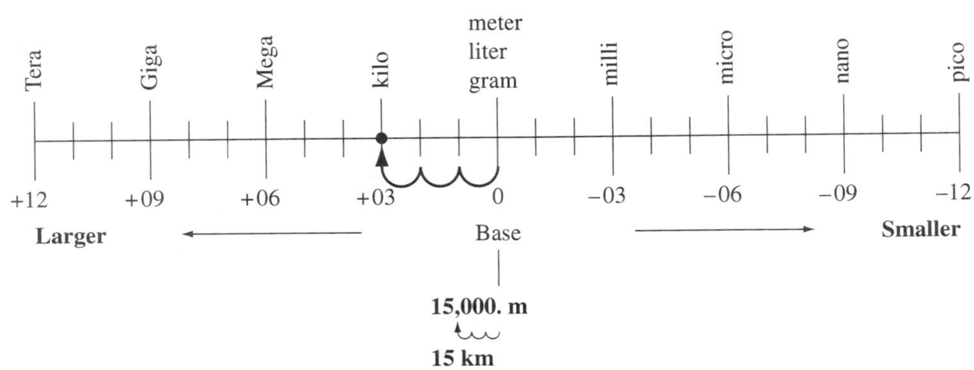

FIGURE 4–3 Conversion to a larger unit of measurement.

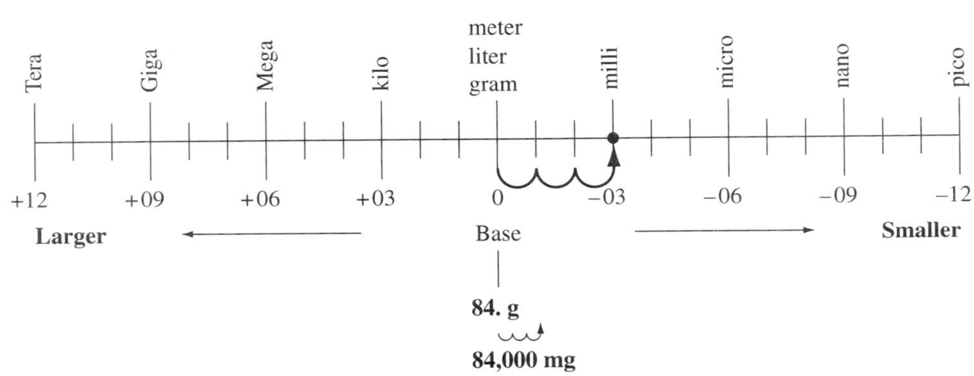

FIGURE 4–4 Conversion to a smaller unit of measurement.

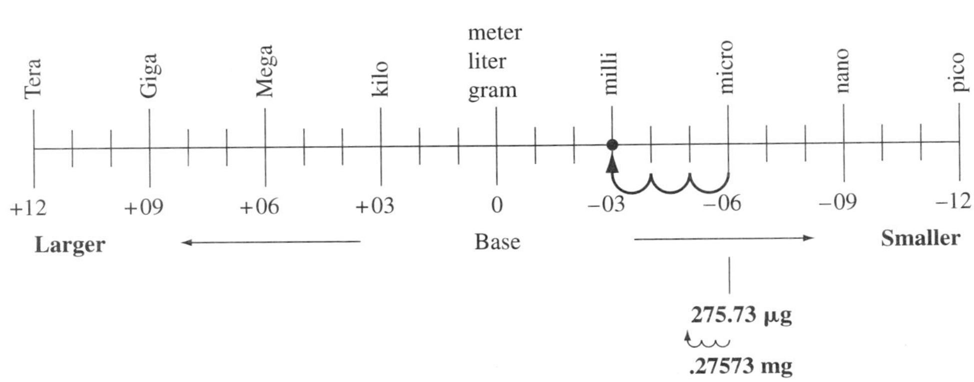

FIGURE 4–5 Conversion from a micro unit to a milli unit.

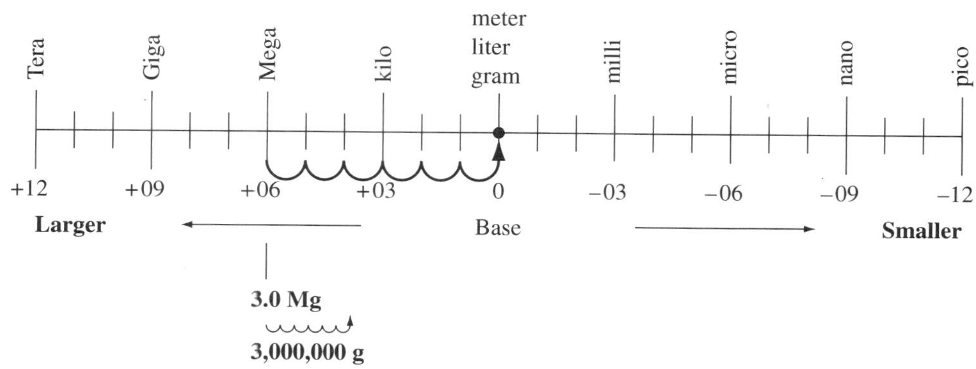

FIGURE 4–6 Conversion from a Mega unit to a whole unit.

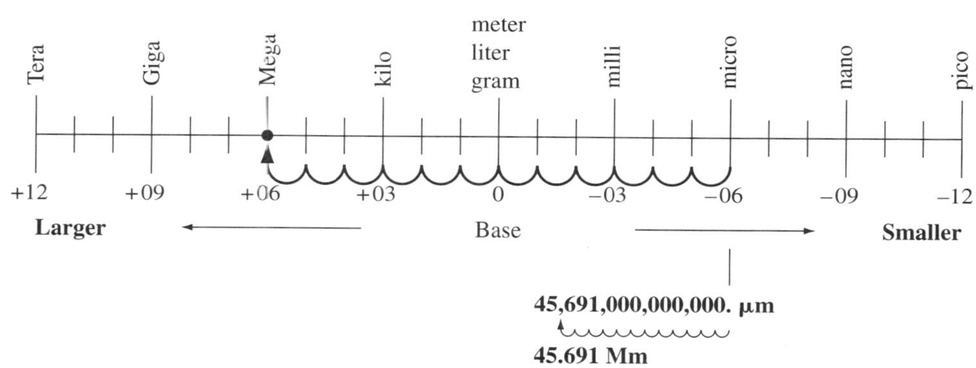

FIGURE 4–7 Conversion from a micro unit to a Mega unit.

to mg (milligram). This means the decimal must move 3 places to the left to reach the milli unit of measurement (Figure 4–5).

If you had 3.0 Mg (Megagram) and wanted to change to the whole number 3 million, you need to move the decimal to the right 6 places to get to grams (Figure 4–6).

Why not try changing an extra large number to a shorter number to make it easier to mathematically manipulate. If you had 45,691,000,000,000 µm and converted the number to Megameter (Mm), you would move the decimal to the left 12 places (Figure 4–7).

Try the practice conversions on the next page using the linear scale conversion method.

Write your answers on a *separate* sheet of paper.

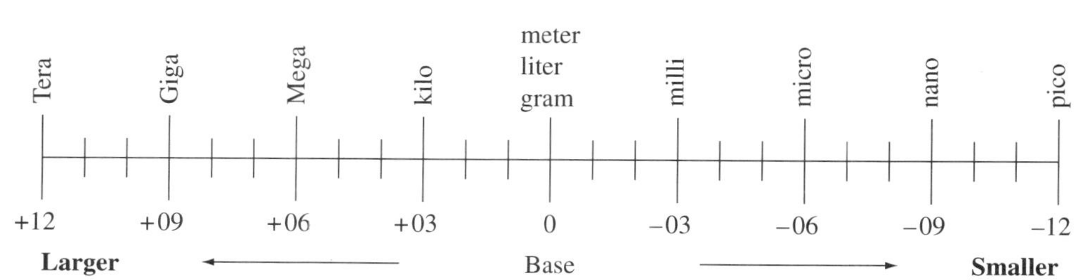

Original Number	Change To	Answer
1. 23.34 m (meters)	mm (millimeters)	23340 mm
2. .00267 Mm	m	
3. 51.9876 km	μm	
4. 20 μm	m	
5. 100 m	Mm	
6. 4888.269 mg	kg	
7. 56.9234 Mm	μm	
8. 22 mm	m	
9. 15 mg	g	
10. 100 g	mg	
11. 25789 mm	km	
12. 45.45 Mg	cm	
13. 1.56789 km	μm	
14. 39005.567 mg	Mg	
15. 1 m	μm	
16. 1 km	Mm	
17. 29.456 cm	mm	
18. 999999999.9 μm	Mm	
19. 999999999.9 Mg	μg	
20. 54.545454 m	cm	

Converting between metric units and English, or standard, units of measurements normally requires a metric conversion chart included in most automotive reference material (Appendix D). For example, if you were converting 55 miles to kilometers, look in Appendix D to see that miles is multiplied by 1.60935 to get kilometers. The answer is 88.51 kilometers. Another way to perform conversions is to use a metric conversion calculator. There are several brands available for this purpose, such as the one offered by Radio Shack (Cat No. 65-828) (Figure 4–8).

To perform a conversion in the example above, you enter 55 on the metric conversion calculator and push the upper left button labeled "*TO METRIC*." This takes you to the lower printed level on the calculator. To convert to kilometers, push the "mile/km" button. The answer should be the same as above—88.51. A metric calculator can be very useful for the automotive technician since you may encounter metric specifications when your tools or test equipment are in standard units of measurement.

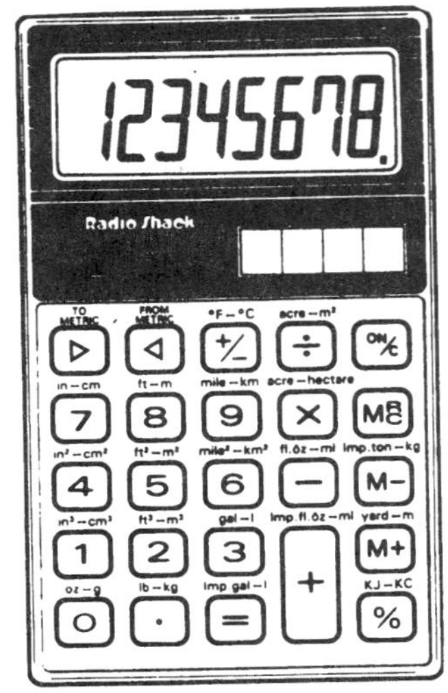

FIGURE 4–8 Metric conversion calculator.

(The EC-352 Metric Conversion Calculator is a trademark of the Radio Shack Division of Tandy Corporation.)

The three basic units of electrical measurements we discussed in Chapter 3 are:

Voltage for Pressure	Amperage for Flow	Resistance for Opposition

The linear scale we used for metric conversions can be modified for converting basic electrical units (Figure 4–9).

For example 52.753 A (amps) converted to mA (milliamps) would now be 52,753 mA. For more conversion practice, use the linear scale for electrical conversions (Figure 4–9) to solve the conversions on the next page. Write your answers on a *separate* sheet of paper.

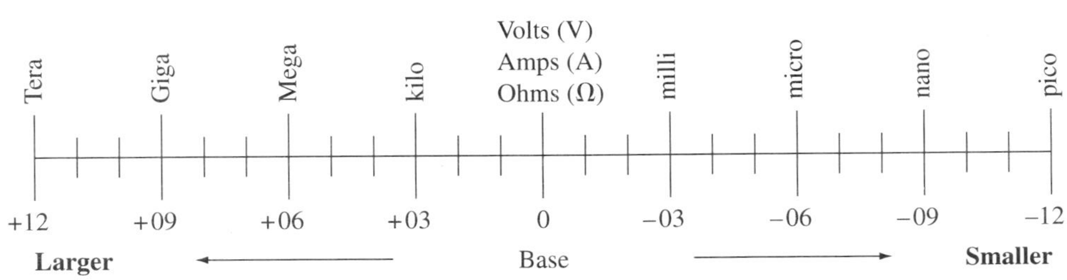

FIGURE 4–9 Linear scale of electrical conversion.

Original Number	Change To	Answer
1. 123.25 mV (millivolt)	V (volt)	.12325 V
2. .64 kΩ	Ω	
3. 50.244 μV	mV	
4. 12 Ω	μΩ	
5. .00078 A	mA	
6. .000294 MV	V	
7. .0000032473 mA	kA	
8. 3 V	μV	
9. 3 V	mV	
10. 3 V	kV	
11. 3 V	MV	
12. 6,000,000 μA	A	
13. 6 A	kA	
14. 99.9 kΩ	mΩ	
15. 99 MA	mA	
16. .034 μV	mV	
17. .000055 MA	A	
18. 55,000,000 mΩ	kΩ	
19. .000000439 MA	A	
20. 50.54 μV	mV	

Scientific Calculator Introduction

Automotive technicians working with electrical circuits of unknown values may need to perform mathematical calculations to determine the circuit operation. The use of a scientific calculator can help when performing these calculations for two important reasons. First, when you perform these calculations the number unit values *must* be the same. For example, you cannot multiply a kilo-ohm by a milli-amp without keeping track of the unit values. Second, sometimes these calculations contain extremely small or large numbers. Since most calculators contain only enough spaces for 8 digits, using the scientific notation features of a calculator permit you to deal with extra large or small numbers. Figure 4–10 is a typical scientific calculator.

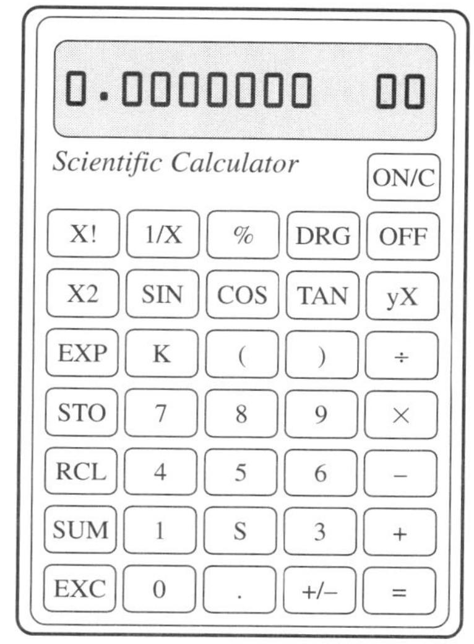

FIGURE 4–10 Typical scientific calculator.

There are two important features your scientific calculator must have. The first is an **EXP** feature, which is the power of ten we discussed earlier in this chapter and in Table 4–1. Some scientific calculators use an **EE** key for the power of 10 feature instead of the **EXP** key. The second feature is the **1/X** key. This is the reciprocal feature that is the inverse of a number. The location of these features is shown with shaded keys on the sample scientific calculator (Figure 4–11). If unsure of the actual location of these features on your calculator, refer to the calculator owner's manual. Some of the more expensive scientific calculators have several functions for each key so you may have to push a **2nd** function key to activate the **EXP** feature.

The **EXP** key is normally used for extra large or extra small numbers that do not fit on the standard 8-digit calculator display. When you push the **EXP** or **EE** feature on the calculator you will notice a separate set of zeros that appear offset on the right side of the display.

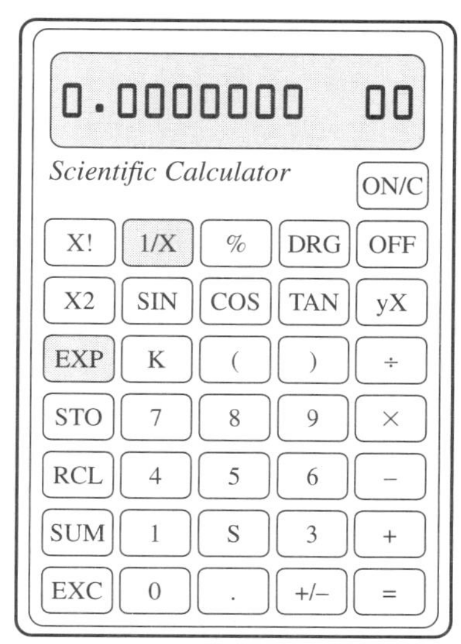

FIGURE 4–11 Important function keys for electrical calculations.

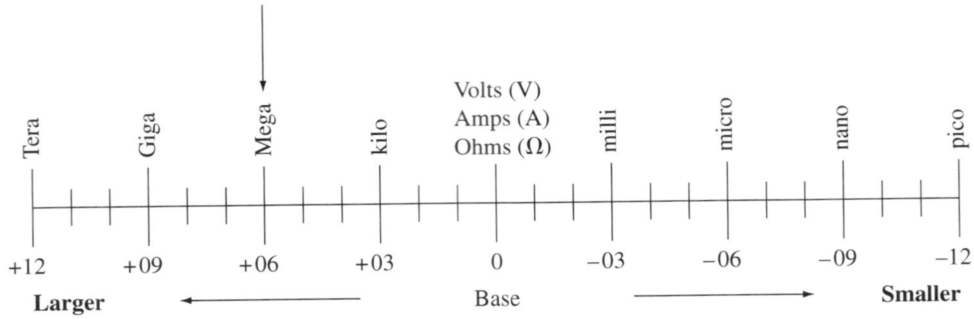

FIGURE 4–12 Conversion to a larger electrical unit value.

This is the scientific notation feature. As you work with complex problems, this feature can be a big help. The number 125 million would contain more then the 8 digits, so the **EXP** feature is useful here. With a number like 125,000,000 you could refer to the linear scale (Figure 4–12) and see that million is +06 power, or Mega.

On the calculator you would enter:

125 **EXP** 6

The calculator should display:

$125\ ^{06}$

When you need to enter a number with a unit value smaller than a whole number there is an additional calculator step necessary. Suppose you had the number 55 microamps (55 μA) in your calculation. On the linear scale (Figure 4–13) the unit value is −06 power, or micro. Without the scientific notation feature, a number would have to be converted to a whole unit value by moving the decimal 6 places to the left to get the value .000055 A.

On the calculator enter:

55 **EXP** 6 [+/−]

The calculator should display:

$55\ ^{-06}$

The additional key used was the **+/−** key which is located in the bottom left corner next to the = key in Figure 4–11. The **+/−** key allows you to toggle between the +06 power and −06 power each time you push the key.

Let us try adding a series of numbers together to gain experience using the **EXP** feature. Add the following numbers.

12 kΩ + 23 Ω + 34 mΩ + 45 MΩ + 560 μΩ

You would enter the following steps on the calculator.

12 **EXP** 3 **+** 23 **+** 34 **EXP** 3 **+/−** **+** 45 **EXP** 6 **+** 560 **EXP** 6 **+/−** **=**

The answer on the calculator may not be displayed in the unit value you are familiar. For example, the answer displayed on the author's calculator is:

$4.5012023\ ^{07}$

Moving the decimal one place to the right will make the number 45.012023^{06}, or 45.012023 Megaohm (MΩ).

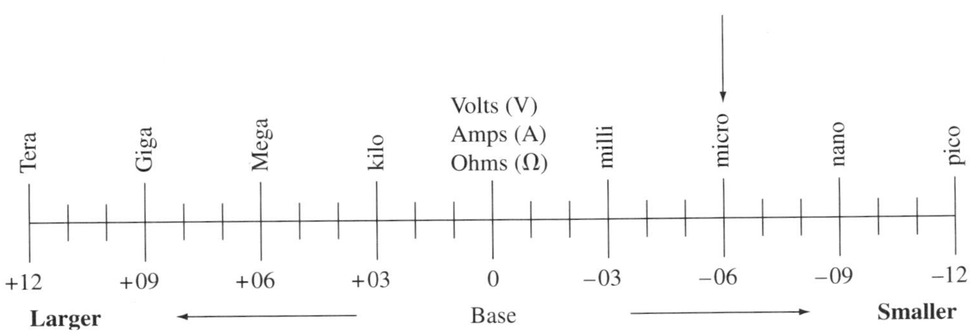

FIGURE 4–13 Conversion to a smaller electrical unit value.

Just remember to refer to the linear scale, and change the decimal to a value that is familiar.

The second important calculator feature that is helpful when performing mathematical calculations is the **1/X** key. This key automatically inverts a number, or creates the **reciprocal** of a number. As an example, the whole number 5 is really:

$$\frac{5}{1}$$

The reciprocal or inverse of 5 is the number .2.

$$\frac{1}{5} = .2$$

In *Chapter 8–Parallel Circuits,* we will see the reciprocal function is a vital step in finding the total resistance of a parallel circuit. Knowing how to use this feature on a scientific calculator will make electrical calculations easier for more complex circuits.

Summary

- ❑ The United States Congress passed the Metric Conversion Act in 1975.
- ❑ The metric system is based on the Power of Ten principle.
- ❑ Each power of ten has a prefix to identify it.
- ❑ The base units of metric measurement are the meter, liter, and gram.
- ❑ The base units of electrical measurement are the volt, ohm, and amp.
- ❑ The **EXP** feature on a scientific calculator is for converting extra large or extra small numbers to scientific notation.
- ❑ The **+/−** key on a scientific calculator allows the user to toggle the unit values of a number from positive to negative numbers.
- ❑ The **1/X** feature on a scientific calculator automatically inverts or reciprocates a number.

Key Terms

Giga	liter	milli	reciprocal
gram	Mega	pico	scientific notation
kilo	meter	Power of Ten	
linear scale	micro	prefix	

Review Questions

Short Answer Essays

1. Describe the Power of Ten principle in the metric measuring system.
2. Explain how to use the **1/X** reciprocal feature (**1/X** key) on a scientific calculator.
3. Explain how to convert between the various metric prefix values without using math formulas.

Fill-in-the-Blanks

1. In 1975, the United States Congress passed the ______________________ in an attempt to get the American public to convert to the ______________________ system.
2. When converting between metric prefix values, 250 grams would be the same as ______________________ kg.
3. The ______________________ feature key on a scientific calculator is used for extra large or extra small numbers that do not fit on a standard 8-digit display.

ASE Style Review Questions

1. Technician A says the base system of measurement for length is the kilometer. Technician B says the millimeter is the base unit of measurement for length. Who is correct?
   A. A only
   B. B only
   C. Both A and B
   D. Neither A nor B
2. Technician A says that 1000 grams is the same as 1 kilogram. Technician B says that 1000 grams is the same as 1,000,000 milligrams. Who is correct?
   A. A only
   B. B only
   C. Both A and B
   D. Neither A nor B
3. The linear scale of metric measurement is being discussed. Technician A says that to convert a number to a higher unit of value, the decimal point is moved to the left. Technician B says to move the decimal point to the right to make the unit a value smaller. Who is correct?
   A. A only
   B. B only
   C. Both A and B
   D. Neither A nor B
4. Two technicians are discussing the basic electrical system of measurement. Technician A says the base system of measurement for volts is the kilovolt. Technician B says the milliamp is the base unit of measurement for current. Who is correct?
   A. A only
   B. B only
   C. Both A and B
   D. Neither A nor B

5. The linear scale of electrical measurement is being discussed. Technician A says that to convert a number to a smaller unit of value, the decimal point is moved to the right. Technician B says move the decimal point to the left to make the unit of value larger. Who is correct?
   A. A only
   B. B only
   C. Both A and B
   D. Neither A nor B

6. The **EXP** feature is being discussed on a scientific calculator. Technician A says this is for numbers that are too large or too small for the calculator. Technician B says this feature is used to reciprocate a number. Who is correct?
   A. A only
   B. B only
   C. Both A and B
   D. Neither A nor B

7. The number 12.5^{03} is shown on the calculator display. The correct whole number is:
   A. 125,000
   B. .012500
   C. 1,250

8. Technician A says that the number 56.77^{-06} on the calculator display is really .00005677. Technician B says that the $^{-06}$ on the display means micro. Who is correct?
   A. A only
   B. B only
   C. Both A and B
   D. Neither A nor B

9. The **1/X** feature is being discussed on a scientific calculator. Technician A says this feature is used to invert a whole number to its reciprocal. Technician B says this feature is only for adding or subtracting extra large or extra small numbers. Who is correct?
   A. A only
   B. B only
   C. Both A and B
   D. Neither A nor B

10. Technician A says that 91 MΩ is the same as 91,000,000,000 mΩ. Technician B says that 91 MΩ is the same as 91,000 kΩ. Who is correct?
    A. A only
    B. B only
    C. Both A and B
    D. Neither A nor B

5 Components of an Electrical Circuit

Introduction

Today's automotive electrical system is a complex maze of subsystems that can be broken down into a collection of various components. A typical modern vehicle has 2,000 feet of wiring, dozens of connectors, and several hundred individual terminals. As a vehicle ages, electrical circuits and devices that have been exposed to deteriorating conditions like overheating, excessive moisture, road salts, vibration, and improper repairs by unqualified persons, can lead to early failure. Having the knowledge of the individual components of electrical circuits will help you perform not only a proper diagnosis, but effective and long-lasting repairs.

Objectives

When you complete this chapter you should be able to:

- ❑ Identify types of primary wire material, insulation properties, and proper wire size for an electrical circuit.
- ❑ Describe the proper procedures for repairing a wire splice.
- ❑ Identify various types of terminals, connectors and appropriate repair procedures.
- ❑ Identify resistor types and their resistive value from the resistor color bands.
- ❑ Explain the purpose of a circuit protection device, and identify the different types that are commonly used in automotive electrical circuits.
- ❑ Identify the various types of electrical switches and relays used to control electrical circuits.
- ❑ Describe the operation and construction of various types of lamps used in automotive lighting circuits.
- ❑ Explain capacitor applications, ratings, and construction in electrical circuits.

Automotive Wiring

Automotive **primary wire** is the term used for wiring circuits that carry *low* voltage. This is normally battery voltage or less. **Secondary wire** refers to wiring circuits that carry *high* voltages, like spark plug wires. Most primary wire is made of several strands of copper wire that are covered with a durable polyvinyl chloride (PVC) insulation (Figure 5–1).

Stranded wire is normally used in automotive applications instead of solid wire for two main reasons. First, the wire needs to be durable enough to withstand the vibrations and flexing that occur constantly. Second, a stranded wire has more surface area then a solid wire. Electrons flow on the surface of a wire, so higher current can be used for the same gauge wire (Figure 5–2).

For a load component to operate properly as designed, the wire size used in primary electrical circuits must not only handle the necessary current, but also any vibration and heat exposure. There are two common wire size standards used to designate a wire size—the **American Wire Gauge (AWG),** and the metric wire size currently used by most automotive manufacturers.

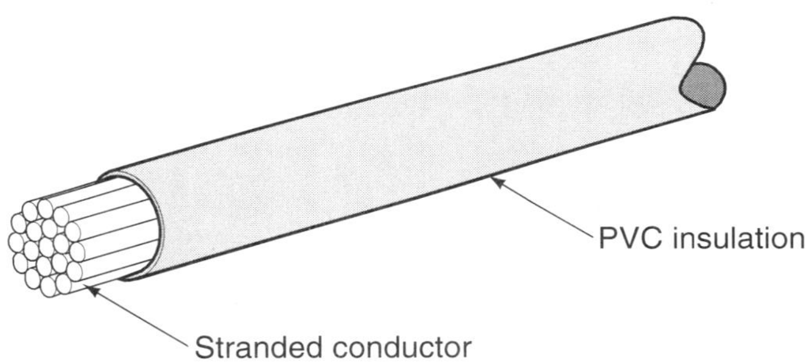

FIGURE 5–1 Stranded primary wire covered with PVC insulation.

The AWG standard assigns a number to a wire based on the diameter of the wire conductor, not including its insulation. The AWG wire rule states that the *higher* the AWG number, the *smaller* the wire conductor. For example, an 18 gauge wire is smaller in diameter than a 12 gauge wire. The majority of automotive primary electrical systems use 12 gauge to 18 gauge wires. The metric system of wire size is determined by the cross sectional area of the wire, which is expressed in square millimeters (mm^2). In the metric system the *smaller* the number is, the *smaller* the wire conductor is. Figure 5–3 is a cross reference between AWG and metric wire sizes. Many aftermarket parts suppliers only offer wire in AWG sizes, so it will be necessary to convert in most cases.

Another factor in determining the proper wire size for an electrical circuit is the length of the wire. As the wire gets longer, the internal resistance in the wire increases. The chart in Figure 5–4 indicates the proper wire gauge and length for a specific amperage load.

As an example, on the chart in Figure 5–4, a 16 gauge wire that is 10 feet long is capable of conducting 20 amperes. However, if the 20 amp current is to be carried for 15 feet, then 14 gauge wire is needed. Changing to 14 gauge wire prevents unnecessary voltage drops in the circuit due to inadequate wire size. Figure 5–4 can be a handy reference chart when installing aftermarket equipment on a vehicle.

Figure 5–5 is a summary of wire sizes and their effect on voltage drops and current carrying capacity.

Nearly all the wires in an automotive electrical circuit are covered with an insulation that has a **color code** or is identified with specific markings. Not all manufac-

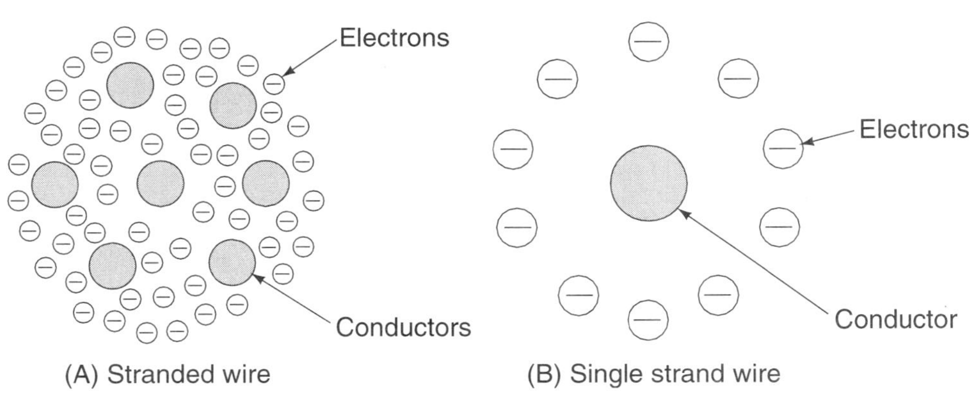

FIGURE 5–2 (A) Flexible stranded wire allows more electron flow than (B) solid wire.

Metric Size (mm^2)	AWG (Gauge) Size	Ampere Capacity
0.5	20	4
0.8	18	6
1.0	16	8
2.0	14	15
3.0	12	20
5.0	10	30
8.0	8	40
13.0	6	50
19.0	4	60

FIGURE 5–3 AWG-to-metric cross reference.

turers use a standard method to mark wires, so it may be necessary to refer to the manufacturer's color code index. Figure 5–6 is an example of the most common abbreviations used today.

Since wiring harnesses normally contain dozens of wires, it is necessary to double-up on some of the color abbreviations (Figure 5–7).

In wires with double color codes, the first letter(s) indicate the base color of the insulation, and the second group of letter(s) indicate the color of the tracer. A wire designated as BRN/Y would have a brown base color and a yellow tracer stripe.

Occasionally, a technician needs to replace a wire, and it is difficult to identify its proper gauge size. The chart and 5-step process (Figure 5–8) on page 67 explains the procedure used to determine the gauge size of unknown wires.

Total Approximate Circuit Amperes	Wire Gauge (for Length in Feet)								
12 V	3	5	7	10	15	20	25	30	40
1.0	18	18	18	18	18	18	18	18	18
1.5	18	18	18	18	18	18	18	18	18
2	18	18	18	18	18	18	18	18	18
3	18	18	18	18	18	18	18	18	18
4	18	18	18	18	18	18	18	16	16
5	18	18	18	18	18	18	18	16	16
6	18	18	18	18	18	18	16	16	16
7	18	18	18	18	18	18	16	16	14
8	18	18	18	18	18	16	16	16	14
9	18	18	18	18	16	16	16	14	12
10	18	18	18	18	16	16	16	14	12
11	18	18	18	18	16	16	14	14	12
12	18	18	18	18	16	16	14	14	12
15	18	18	18	18	14	14	12	12	12
18	18	18	16	16	14	14	12	12	12
20	18	18	16	16	14	14	12	12	10
22	18	18	16	16	12	12	10	10	10
24	18	18	16	16	12	12	10	10	10
30	18	16	16	14	10	10	10	10	10
40	18	16	14	12	10	10	8	8	6
50	16	14	12	12	10	10	8	8	6
100	12	12	10	10	6	6	4	4	4
150	10	10	8	8	4	4	2	2	2
200	10	8	8	6	4	4	2	2	1

Note: 18 AWG as indicated above this line could be 20 AWG electrically. 18 AWG is recommended for mechanical strength.

FIGURE 5–4 Application load amperage versus wire size and length.

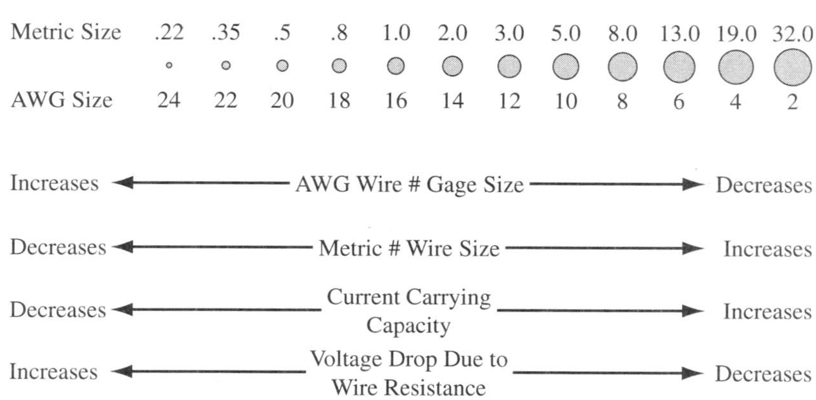

Note: Wire circle diameters not to exact scale

FIGURE 5–5 Summary of wire sizes and their effect on voltage drop and current flow.

Color	Abbreviations		
Aluminum	AL		
Black	BLK	BK	B
Blue (Dark)	BLU DK	DB	DK BLU
Blue	BLU	B	L
Blue (Light)	BLU LT	LB	LT BLU
Brown	BRN	BR	BN
Glazed	GLZ	GL	
Gray	GRA	GR	G
Green (Green)	GRN DK	DG	DK GRN
Green (Light)	GRN LT	LG	LT GRN
Maroon	MAR	M	
Natural	NAT	N	
Orange	ORN	O	ORG
Pink	PNK	PK	P
Purple	PPL	PR	
Red	RED	R	RD
Tan	TAN	T	TN
Violet	VLT	V	
White	WHT	W	WH
Yellow	YEL	Y	YL

FIGURE 5–6 Common color codes used in automotive applications.

Use the chart in Figure 5–8 to complete the following exercise by determining wire gauge size. Using the five steps in the chart as a guide, determine the gauge size of each of the exercises. Assume each wire example has 19 strands and put your answers on a *separate* sheet of paper.

	Wire Diameter	Gauge Size
1.	.0155	14 ga.
2.	.0385	________
3.	.0101	________
4.	.0244	________
5.	.0900	________
6.	.0429	________

Electrical Connectors

Today's complex wiring circuits use a variety of terminal connectors. Terminals must be able to perform with little or no resistance or opposition to current flow. A loose or corroded connection can cause unwanted voltage drop that will result in poor operation of the load component(s). The simplest terminal connector used in automotive circuits, especially older model vehicles, is the **solderless connector** (Figure 5–9).

Crimping a solderless connector is an acceptable method for splicing wires and connectors that are not subjected to dirt or excessive weather elements that can cause corrosion. It is necessary to use a specialized crimping tool (Figure 5–10) to obtain a solid connection. Other crimping tools like slip-joint pliers will not provide a tight enough connection.

The correct procedure for performing a solderless crimp is:

1. Use the correct size hole in the stripping area of the tool to remove enough insulation to allow the wire to completely penetrate the crimping end of the connector.

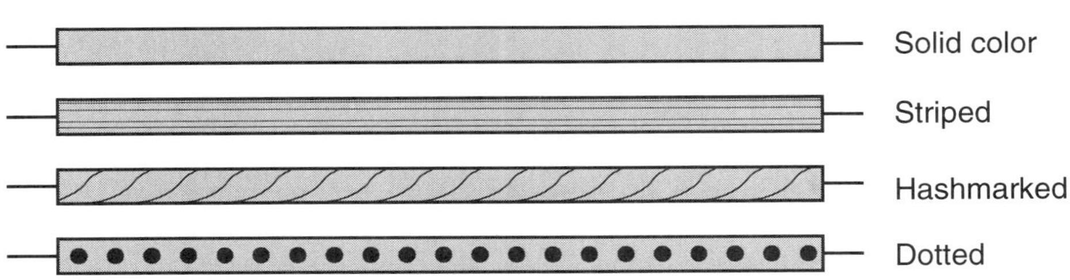

(D) Dot
(H) Hashmark
Stripe is understood

Examples:
BK–Solid black
BR–Y – Brown with yellow stripe
BK–Y H – Black with yellow hashmarks
O–BK D – Orange with black dots

FIGURE 5–7 Four methods the Ford Motor Company uses to color code wires.

(Courtesy of Ford Motor Company)

HOW TO MEASURE THE GAUGE OF CABLE

To determine the gauge of a cable, using the table, proceed as follows:

(1) Count the number of strands of wire.

(2) Measure the diameter of a single strand in thousandths of an inch, using a micrometer.

(3) In column A of the table, find the diameter of the wire you have measured, and on the same line, in column C, find its area.

(4) Multiply the area of a single wire by the number of strands, to get the total area.

(5) In column C, find the figure that is closest to the total area obtained by step 4, and on the same line, in column B, note the gauge number of a single wire having that area. This number is the gauge of the cable.

EXAMPLE: A cable is found to have 19 strands of wire, the individual strands (measured by the micrometer) being 0.0112 inch in diameter. The table (column C) shows the circular mil area of each strand to be 127. Multiplying this by the number of strands, 19, results in 2413 total circular mils. The closest figure in column C is 2583, and on the same line, in column B, we find that 16 is the nearest cable gauge.

SIZE AND AREA OF WIRE

(A) WIRE DIAMETER (INCHES)	(B) AMERICAN WIRE GAUGE	(C) CIRCULAR MIL AREA
.4600	0000	211600
.4096	000	167800
.3648	00	133100
.3249	0	105500
.2893	1	83690
.2576	2	66370
.2294	3	52640
.2043	4	41740
.2893	1	83690
.2576	2	66370
.2294	3	52640
.2043	4	41740
.1620	6	26250
.1285	8	16510
.1019	10	10380
.0808	12	6530
.0640	14	4107
.0508	16	2583
.0403	18	1624
.0319	20	1022
.0284	21	810.1
.0253	22	642.4
.0225	23	509.5
.0201	24	404.0
.0179	25	320.4
.0159	26	254.1
.0142	27	201.5
.0126	28	159.8
.0112	29	126.7
.0100	30	100.5
.0089	31	79.70
.0079	32	63.21
.0070	33	50.13
.0063	34	39.75
.0056	35	31.52
.0050	36	25.00

FIGURE 5–8 Determining the gauge size of unknown wire cable.
(Courtesy of General Motors Corporation, Service Technology Group)

2. Place the wire into the connector and use the crimping tool to perform a crimp (Figure 5–11). Place the open area of the connector towards the anvil of the crimping tool, and compress the tool.
3. Hold the connector and give the wire a gentle pull to verify the crimping is tight enough.

Another type of solderless connector is the **tap splice** (Scotch-lock) connector. This type of connector is can add an *additional* circuit to a feed circuit without needing to strip the insulation first. For example, a tap splice is used when a trailer plug is added to a vehicle's brake and tail light circuit. Figure 5–12 shows a tap

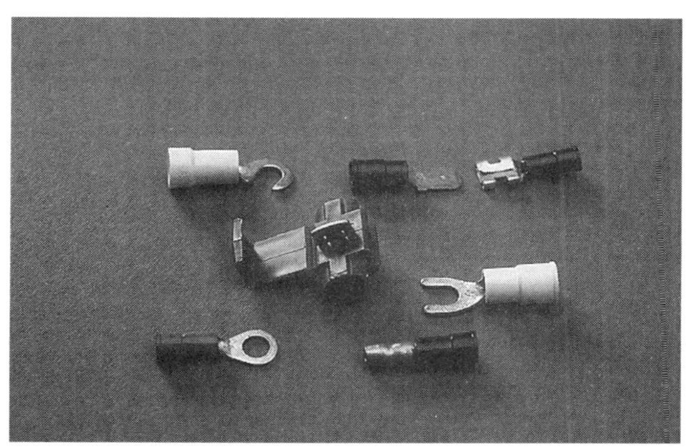

FIGURE 5–9 Common solderless crimp connectors.

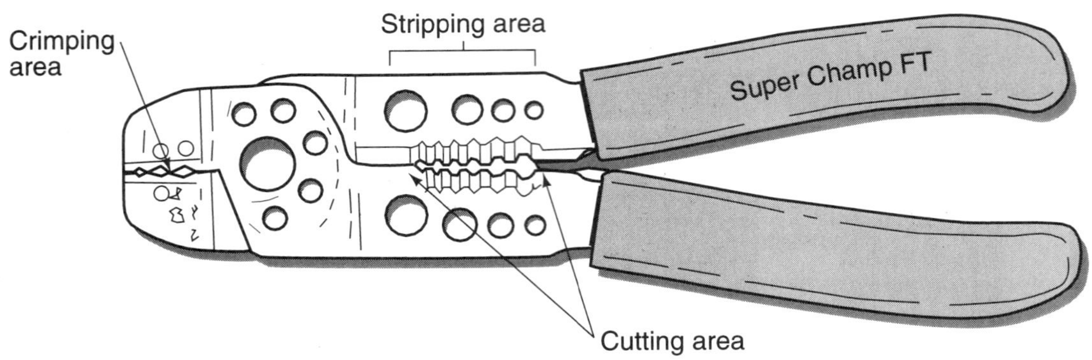

FIGURE 5–10 A crimping tool for solderless connectors.

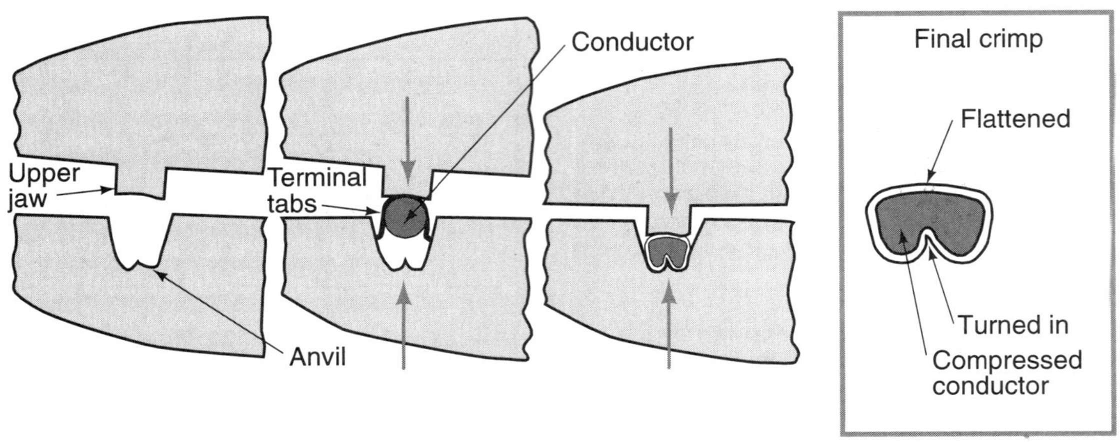

FIGURE 5–11 Using a crimping tool on a solderless connector.

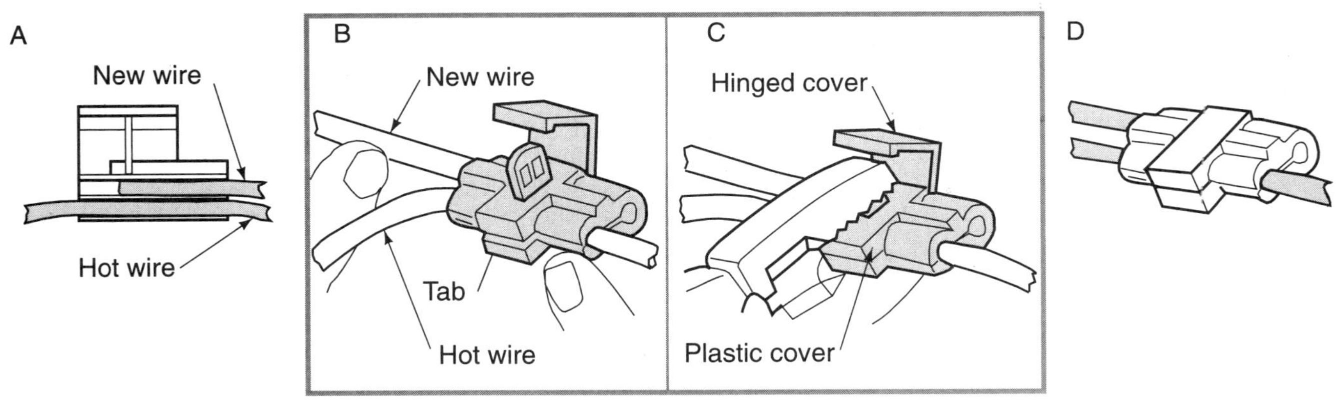

FIGURE 5–12 Using a tap connector to splice a wire into a circuit.

splice splicing a new wire to an existing hot wire. Extreme care should be exercised in using tap splice because wire strands may be cut during the crimping procedure, and the circuit being tapped may not handle the additional current load.

Note: Some automotive manufacturers do not recommend this method of wire repair.

Several other types of connectors have been used over the past several years to ensure weather-tight elec-

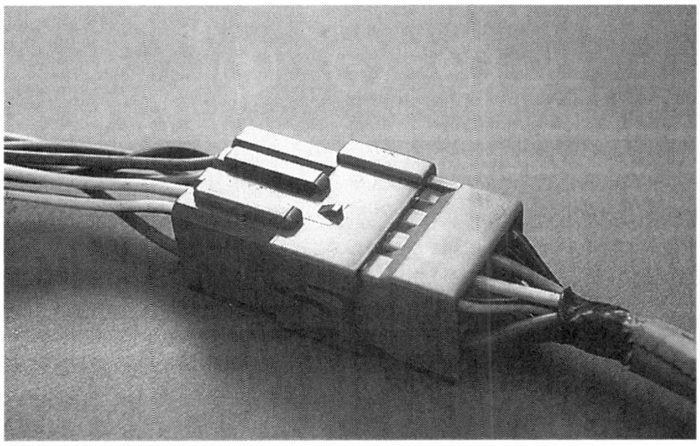

FIGURE 5–13 Hard shell connector with multiple wires.

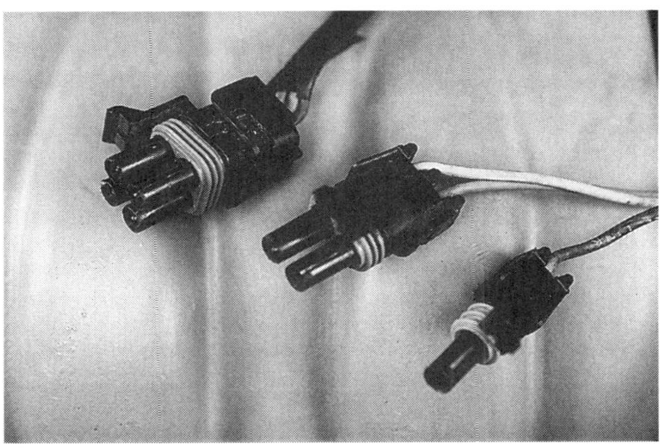

FIGURE 5–14 Weather-pack connector.

FIGURE 5–15 Metri-pack connector.

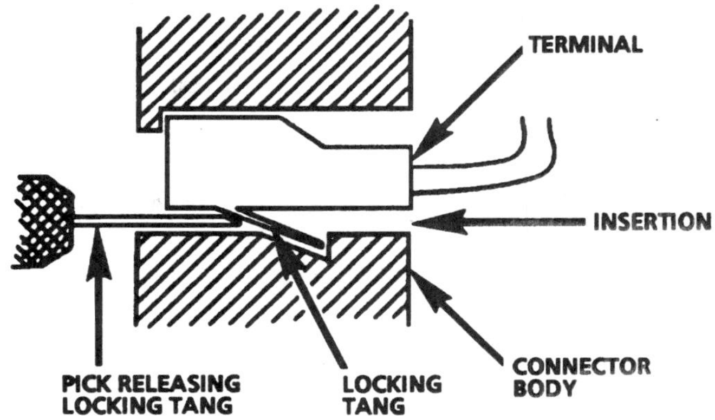

FIGURE 5–16 Typical push-to-seat connector.
(Courtesy of General Motors Corporation, Service Technology Group)

trical connections, which is especially important in modern vehicles having computer controls that need a more reliable connector. The first of these is a **hard shell connector** (Figure 5–13). This connector can have from one to a dozen wires molded into a connector component.

The next style is a **weather-pack connector** (Figure 5–14) that has a rubber seal on the terminal ends and on the covers of the connector halves, which provide for excellent corrosion protection.

The third style is a **metri-pack connector** (Figure 5–15). They are similar to weather-pack connectors, except they do not have the seal on the cover half of the connector and are smaller and more compact.

Metri-pack connectors use two different methods to fasten the terminal half and connector half together. The first is a **push-to-seat** connector and terminal (Figure 5–16).

To remove the terminal from the connector requires a special tool to unlock the male locking tang (Figure 5–17), or the female locking tang (Figure 5–18).

The second following method for a metri-pack connector is a **pull-to-seat** connector and terminal (Figure 5–19).

The pull-to-seat terminal is removed by inserting a pick tool into the connector face while prying up on the locking tang. The terminal can then be pushed through the front of the connector (Figure 5–20).

A special tool for unlocking the tang can be purchased at most tool supply stores. Figure 5–21 is an example of a universal wire terminal tool.

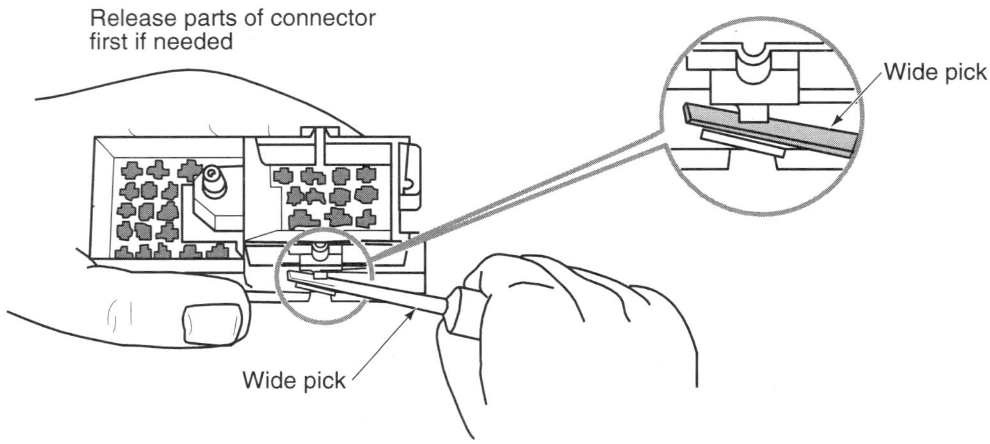

FIGURE 5–17 Using a special pick-tool to unlock the male terminal locking tang.
(Courtesy of General Motors Corporation, Service Technology Group)

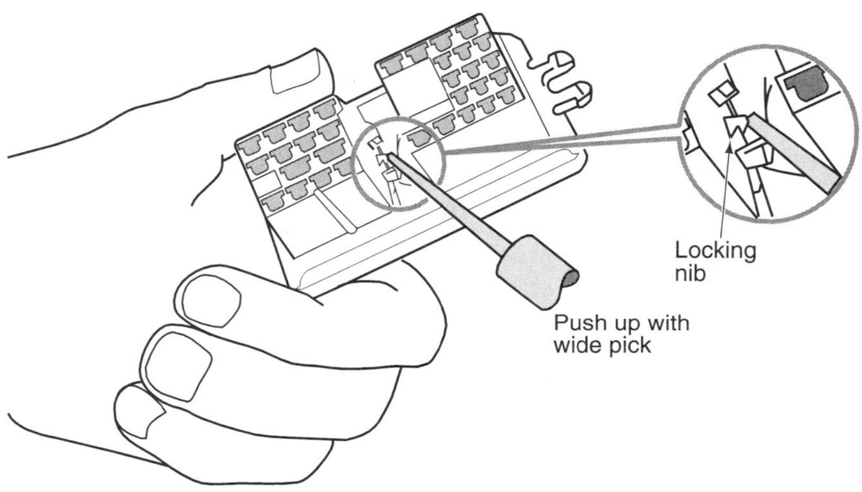

FIGURE 5–18 Using a special pick-tool to unlock the locking tab retainer for a female terminal.
(Courtesy of General Motors Corporation, Service Technology Group)

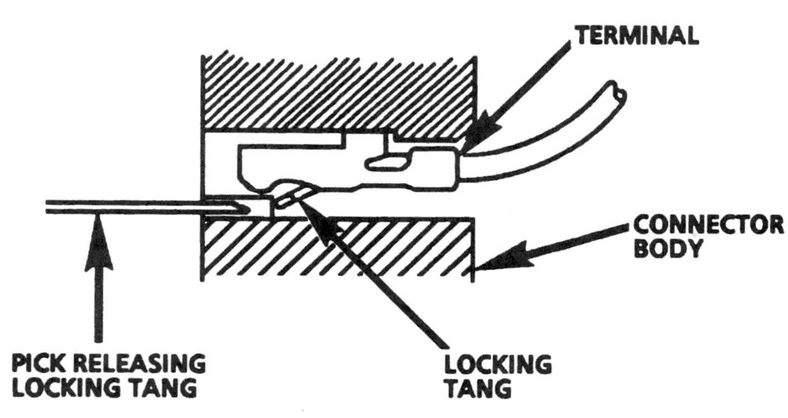

FIGURE 5–19 Typical pull-to-seat connector.
(Courtesy of General Motors Corporation, Service Technology Group)

A weather-pack connector is similar to a metri-pack in that a tang must be compressed to remove the terminal. Figure 5–22 and Figure 5–23 demonstrate the procedure to remove a weather-pack connector.

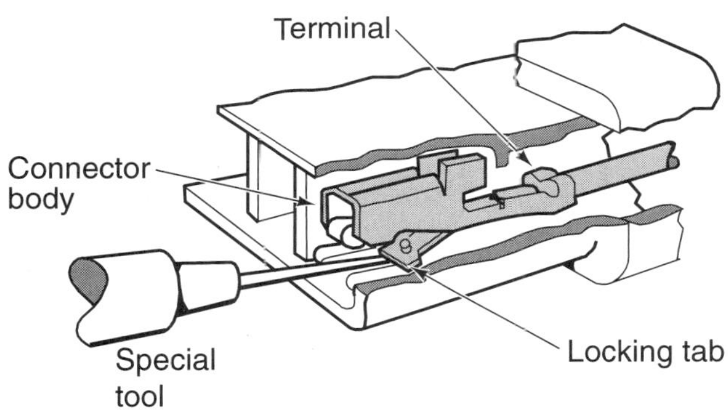

FIGURE 5–20 Pull up on the locking tang to release the terminal from the connector.

(Courtesy of General Motors Corporation, Service Technology Group)

FIGURE 5–21 Universal terminal tang tool.

(Courtesy of the Lisle Corporation)

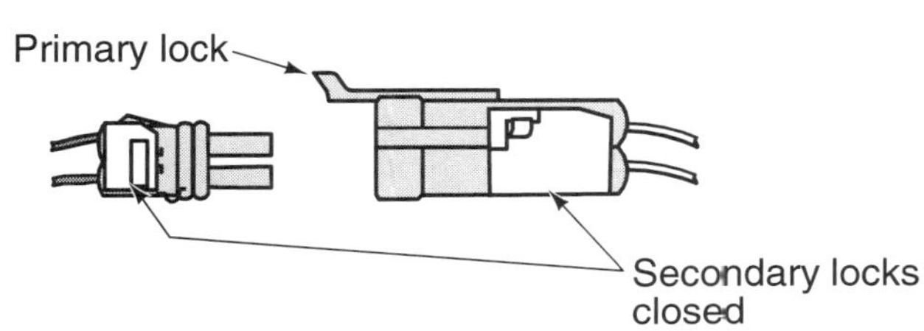

FIGURE 5–22 Primary lock for a weather-pack connector.

(Courtesy of General Motors Corporation, Service Technology Group)

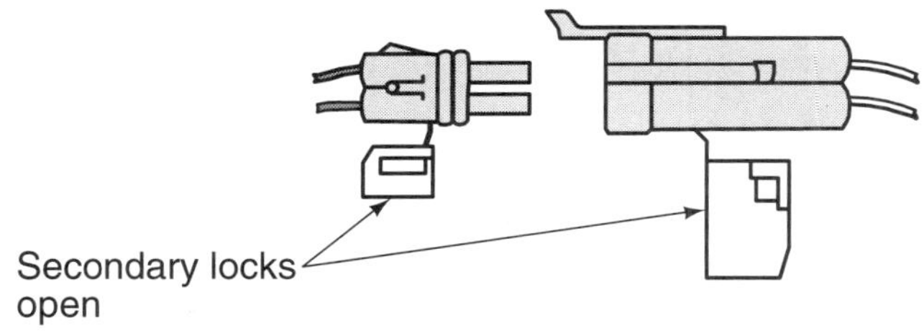

FIGURE 5–23 Secondary lock for a weather-pack connector.

(Courtesy of General Motors Corporation, Service Technology Group)

Soldering Procedures for Wire and Connectors

The best way to attach to a connector or to perform a wire splice is a solder connection. Electrical solder is an alloy of lead and tin along with a **rosin core** flux (cleaning agent). Solder usually comes in a 40/60 mixture (40% tin and 60% lead), but it also can come in a 50/50 or 60/40 mixture. A word of caution about choosing the correct mixture of solder, the higher the tin content is, the higher the heat will need to be to melt the solder. This high heat may damage sensitive electrical circuits. DO NOT USE acid core solder for wire or connector repairs. It can cause wire connections to corrode and lead to high resistance. Acid core solder is used for purposes *other* than electrical applications, such as radiators, joining copper pipes, and fusing light metals together. Photo Sequence 1 shows a proper method to for performing the soldering example.

Resistive Devices

Resistors are one of the most common components found in an electrical circuit. There are several different types of resistive devices, depending on the specific functions of the circuit, currently in use today. Resistive devices can control a circuit current, produce heat, light a bulb, or provide a designed in-voltage drop. For example, a resistor element made of nickel-chromium wire, or strip, is designed to produce heat in cigar lighters, and heated windows and mirrors. A light bulb

PHOTO SEQUENCE 1

Soldering Two Copper Wires Together

P5–1 Tools required to solder copper wire: 100-watt soldering iron, 60/40 rosin core solder, crimping tool, splice clip, heat shrink tube, heating gun, and safety glasses.

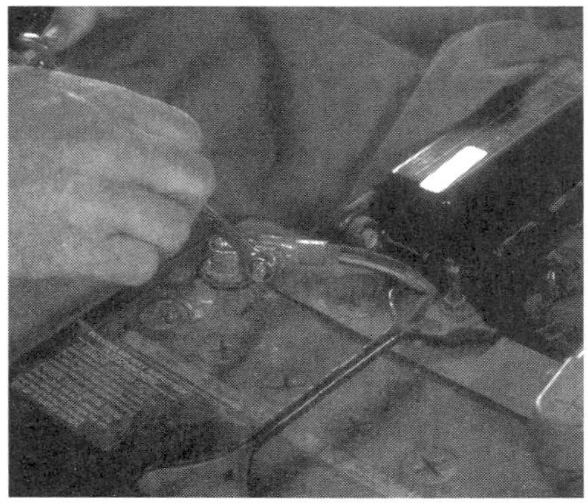

P5–2 Disconnect the fuse that powers the circuit being repaired. *Note:* If the circuit is not protected by a fuse, disconnect the ground lead of the battery.

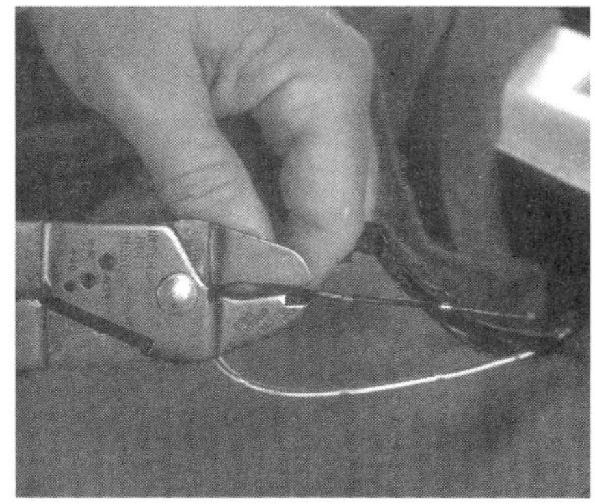

P5–3 Cut out the damaged wire.

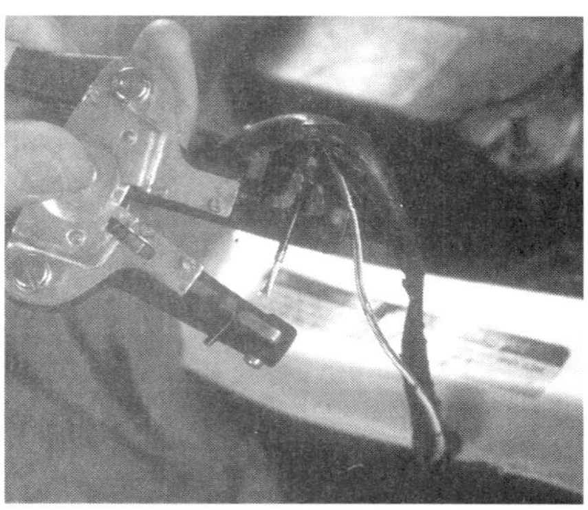

P5–4 Using the correct size stripper, remove about 1/2 inch of the insulation from both wires.

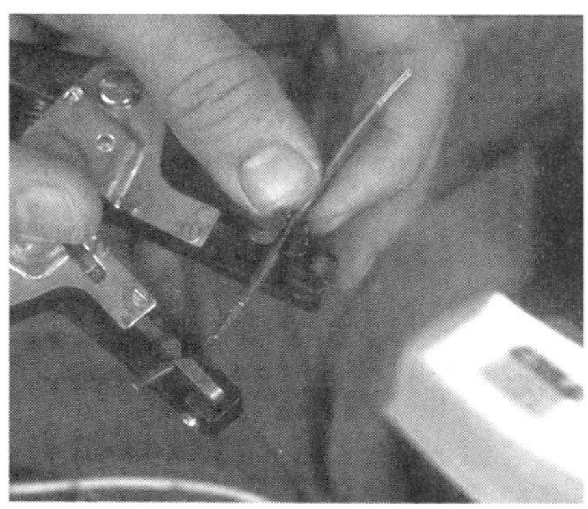

P5–5 Now remove about 1/2 inch of the insulation from both ends of the replacement wire. The length of the replacement wire should be slightly longer than the length of the wire removed.

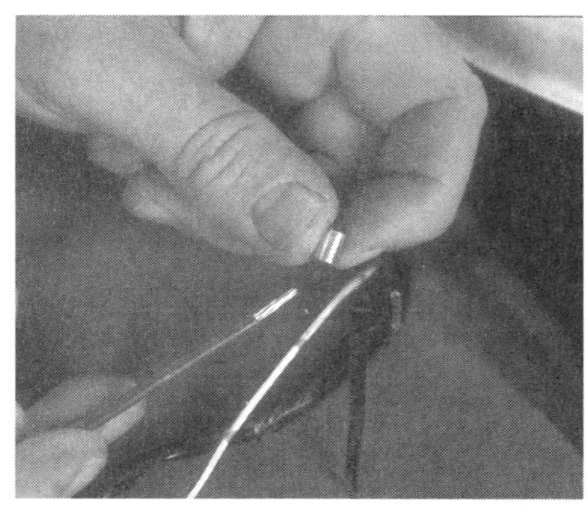

P5–6 Select the proper size splice clip to hold the splice.

PHOTO SEQUENCE 1

Soldering Two Copper Wires Together, continued

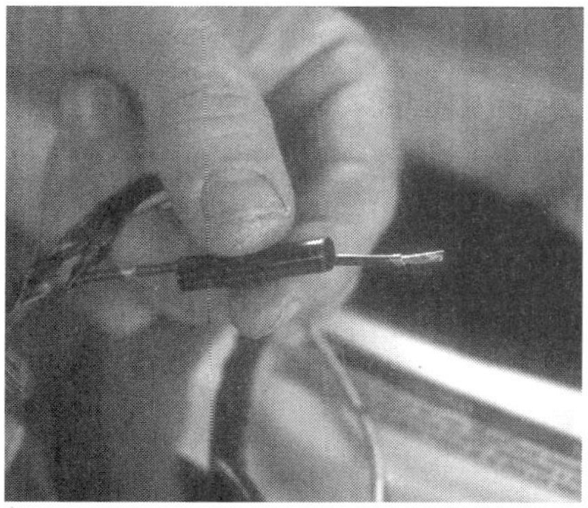

P5–7 Place the correct size and length of heat shrink tube over the two ends of the wire.

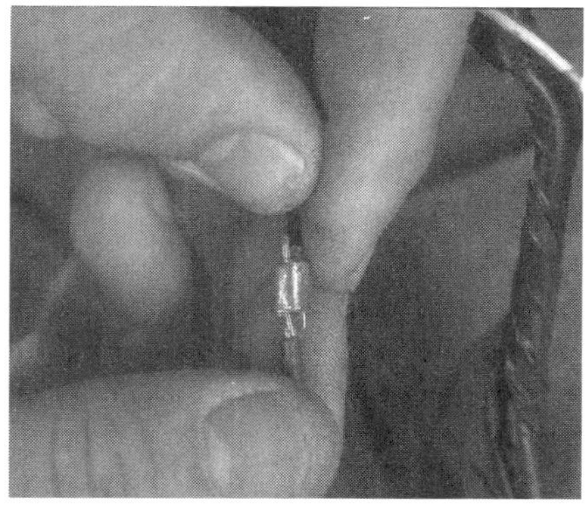

P5–8 Overlap the two splice ends and center the splice clip around the wires, making sure the wires extend beyond the splice clip in both directions.

P5–9 Crimp the splice clip firmly in place.

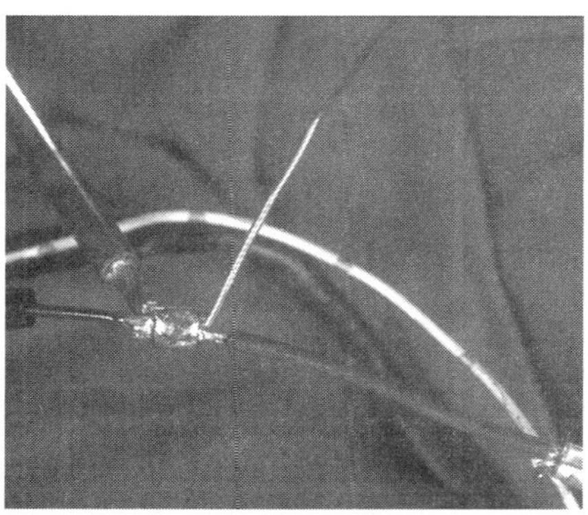

P5–10 Heat the splice clip with the soldering iron while applying solder to the opening of the clip. Do not apply solder to the iron. The iron should be 180 degrees away from the opening of the clip.

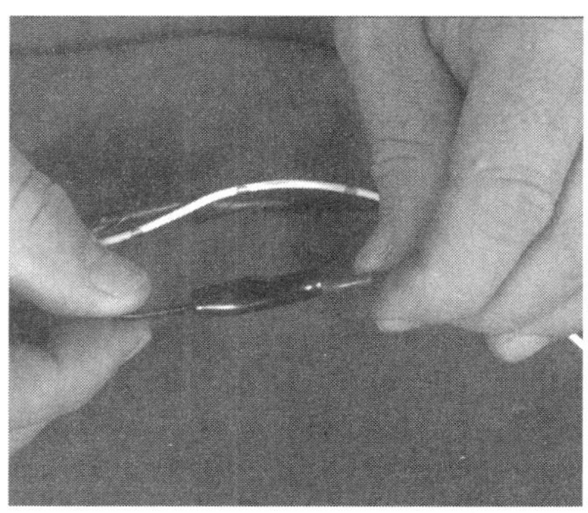

P5–11 After the solder cools, slide the heat shrink tube over the splice.

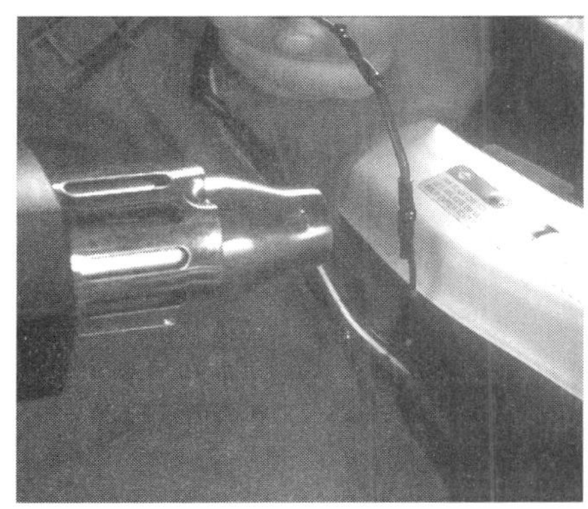

5–12 Heat the tube with the hot air gun until it shrinks around the splice. Do not overheat the heat shrink tube.

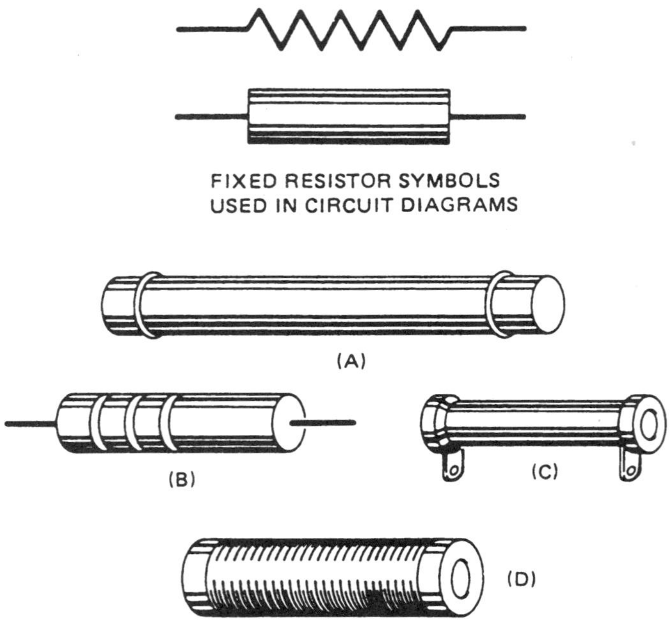

FIGURE 5–24 Several styles of a fixed resistor.

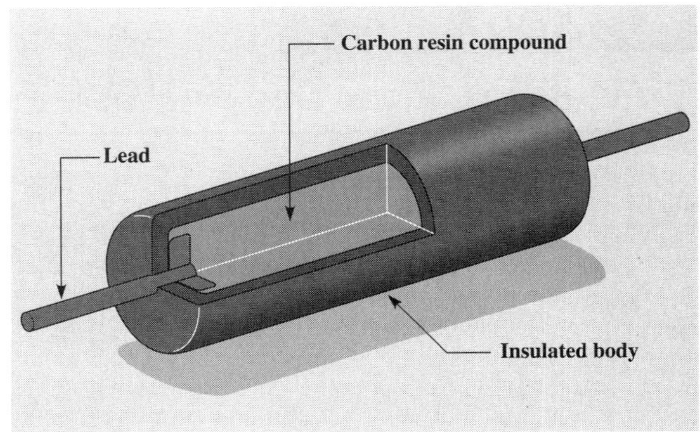

FIGURE 5–25 Typical carbon film fixed resistor.

produces light when the fine tungsten resistive element is heated.

The first style of resistor is a **fixed resistor.** It is probably the most common resistor used in electrical circuits today (Figure 5–24).

Fixed resistors are generally made of a carbon/graphite based material mixed with a resin bonding agent. The body is made from a variety of nonconductive enamel insulating materials. Figure 5–25 is a cutaway of a typical carbon film fixed resistor.

The size of a resistor does not determine its resistive value, but it does relate to the amount of heat it can dissipate without becoming damaged. This heat production is known as the power rating, in watts, of the resistor. Typically, the larger the resistor, the higher the wattage rating (Figure 5–26).

Fixed resistors with wattage ratings higher that 2 watts are normally wire wound resistors consisting of a hollow ceramic core to better dissipate excessive heat (Figure 5–27).

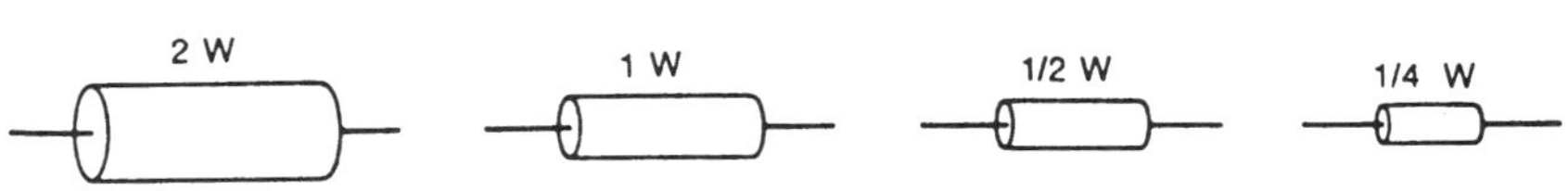

FIGURE 5–26 Watt rating increases with resistor size.

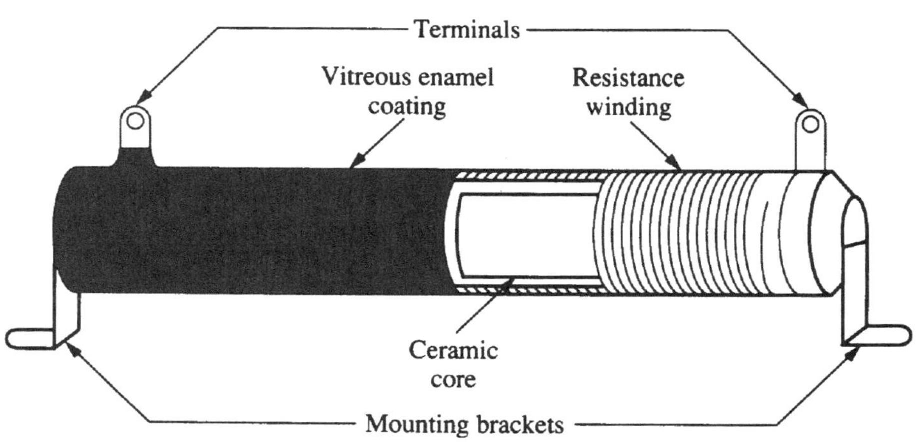

FIGURE 5–27 Wire wound resistor with hollow ceramic core.

The system of identifying the resistive value of a carbon film resistor is determined by a color code. The most common resistor color code system uses a four band system (Figure 5–28).

The first two color bands represent number values from the color code chart (Figure 5–29). The third color band represents the multiplying value, or the number of zeros to add to the right of the first two color band digits. The forth color band is the tolerance value of the resistor gold. A fourth band means the resistor has a five percent tolerance. For example, if a 1,200 Ω resistor had a gold tolerance band, the value limits of this resistor would be 1,200 × .05 = 60. This means the value of the 1,200 Ω resistor can be as high as 1,260 Ω, or as low as 1,140 Ω and still be considered a 1,200 Ω resistor. A fourth color band that is silver would mean a ten percent tolerance value. When there is *no* fourth color band, the tolerance value is twenty percent.

The two examples on the next page show how to determine the resistive value of color-banded fixed resistors.

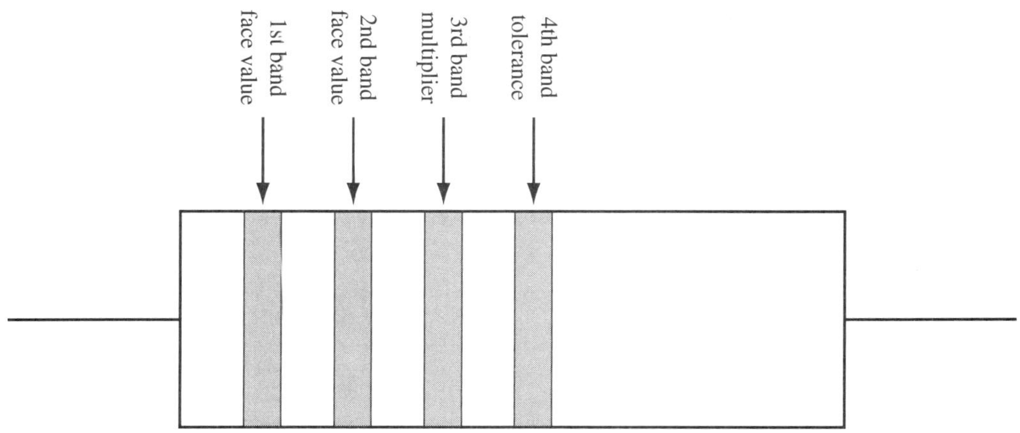

FIGURE 5–28 Four-color band fixed resistor.

OHMS RESISTANCE COLOR SYSTEM			
Color	Numerical Value of 1st and 2nd Figures	Multiplying Value	Percentage Tolerance (Fourth Band)
Black	0	1	
Brown	1	10	
Red	2	100	
Orange	3	1,000	
Yellow	4	10,000	
Green	5	100,000	
Blue	6	1,000,000	
Violet	7	10,000,000	
Gray	8	100,000,000	
White	9	1,000,000,000	
Gold	→	0.1	5%
Silver		0.01	10%
No color			20%

FIGURE 5–29 Resistor color code reference chart.

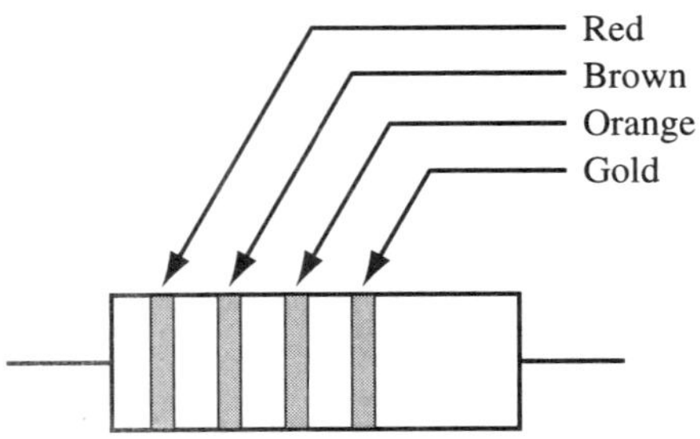

FIGURE 5–30 Resistive value = 21,000 Ω, 5%.

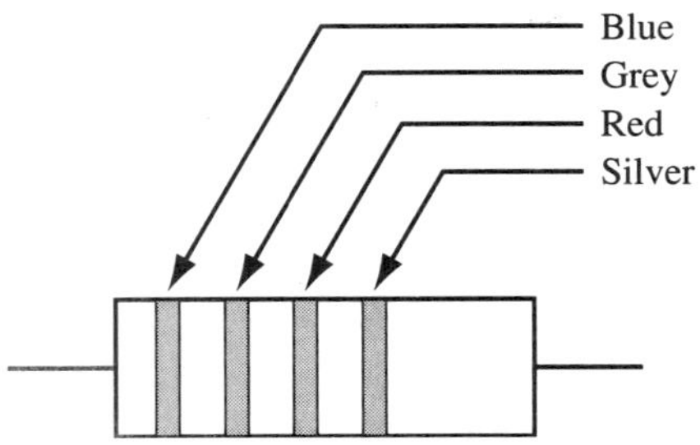

FIGURE 5–31 Resistive value = 6,800 Ω, 10%.

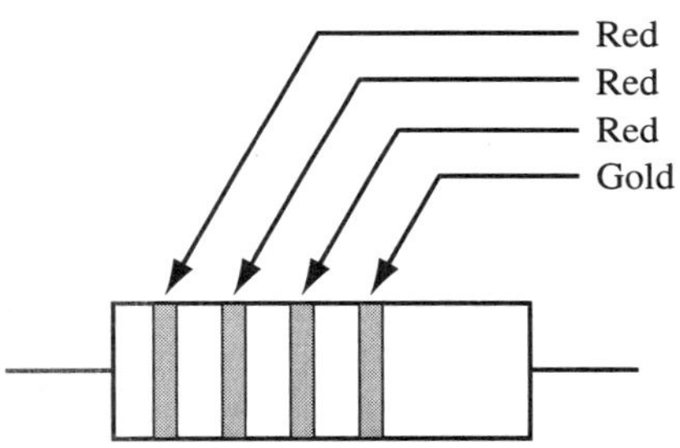

FIGURE 5–32 Resistive value unknown.

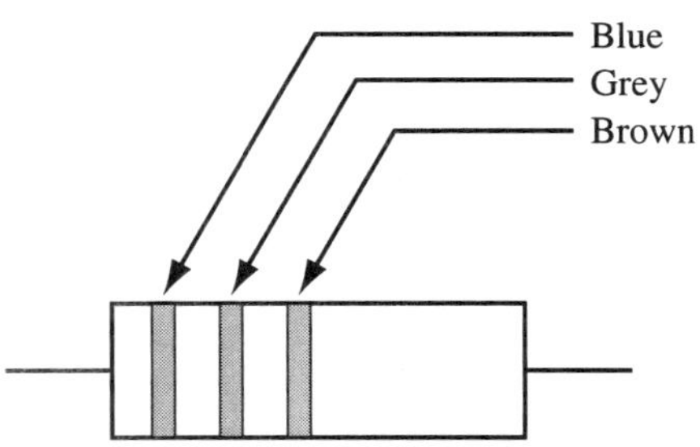

FIGURE 5–33 Resistive value unknown.

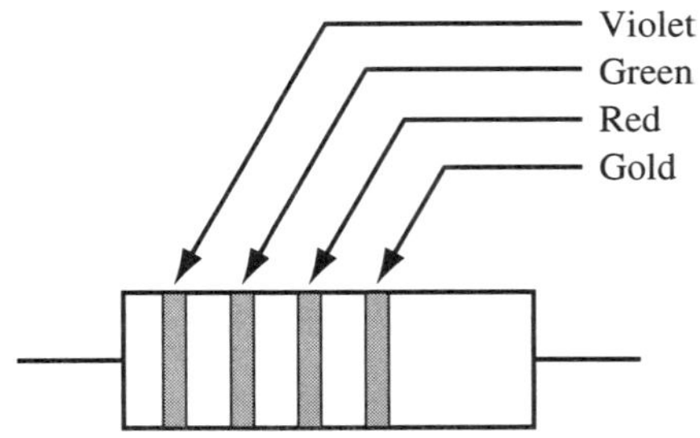

FIGURE 5–34 Resistive value unknown.

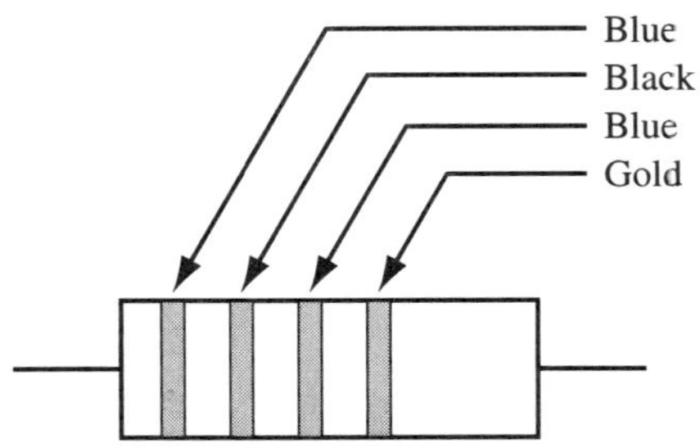

FIGURE 5–35 Resistive value unknown.

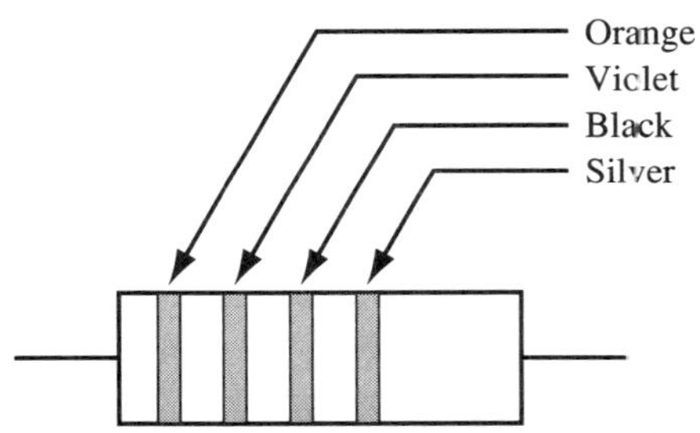

FIGURE 5–36 Resistive value unknown.

1. In Figure 5–30, the first color band is red, which corresponds to the numeric value of two as shown on the color code chart (Figure 5–29). The second color band is brown which is the numeric value one on the color chart. The third band is the multiplier, and is orange which corresponds to three. So, adding three zeros to the first two numbers yields 21,000. In other words, the number 21 multiplied by the multiplier 1,000 equals 21,000 Ω. The tolerance color is gold, so it is a 5% resistor. This means the resistor value can be as high as 21,000 + 1050 = 22,050, or as low as 21,000 − 1050 = 19,950.

2. In Figure 5–31, the first color band is blue which is six on the color code chart (Figure 5–29). The second color band is gray, which is eight on the color chart. The third band is the multiplier and is red, which is two. So, adding two zeros to the first two numbers yields 6,800. The tolerance color is silver,

so it is a 10% resistor. The resistor value can be as high as 6,800 + 680 = 7,480, or as low as 6,800 − 680 = 6,120.

Complete the remaining five resistance values by using the color code chart (Figure 5–29). Put your answers on a *separate* sheet of paper.

Resistance value of Figure 5–32 ____________
Resistance value of Figure 5–33 ____________
Resistance value of Figure 5–34 ____________
Resistance value of Figure 5–35 ____________
Resistance value of Figure 5–36 ____________

The second style of resistor is a **stepped resistor.** Automobiles commonly use a stepped resistor to control the fan motor's speed (Figure 5–37).

If the blower switch is in the low speed position, then the current in the circuit must travel through all three resistors to get to the blower motor. As the switch is moved to the other positions, there are fewer resistors (less total resistance) for the current to travel through (Figure 5–38).

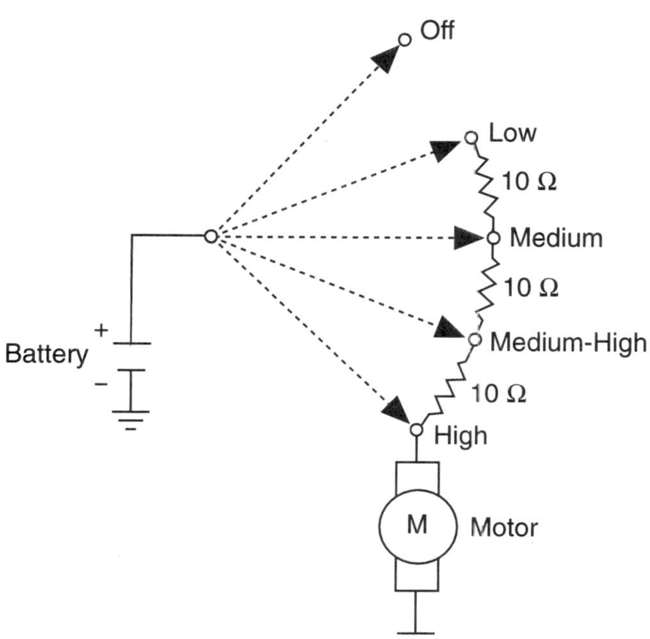

FIGURE 5–37 The total resistance is 30 ohms in the low-speed position, 20 ohms in the medium position, 10 ohms in the medium-high position, and zero ohms in the high position.

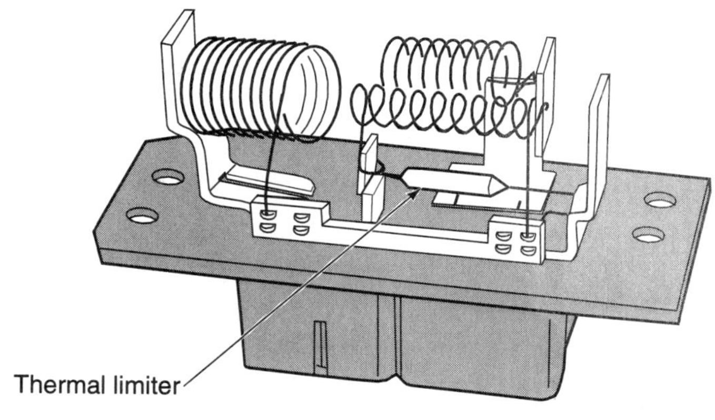

FIGURE 5–38 A stepped resistor used in a heater blower circuit.

(Courtesy of Ford Motor Company)

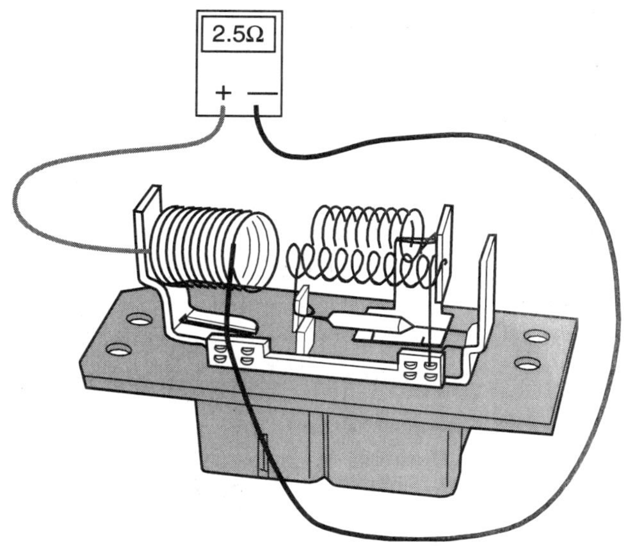

FIGURE 5–39 Testing a stepped resistor with an ohmmeter.

(Courtesy of Ford Motor Company)

Testing a stepped resistor is best done with the resistor unit removed from the circuit. Figure 5–39 shows each resistive coil being checked with an ohmmeter. Compare the results with the manufacturer's specifications.

The third style of resistor is a **variable resistor,** which can change the resistance potential of an electrical circuit. The most common types of variable resistors are **rheostats, potentiometers,** and **thermistors.** A rheostat has one terminal connected to a fixed end of a resistor, and a second terminal connected to a movable contact called a wiper (Figure 5–40).

The most common rheostat in the automobile is the headlight switch with a dimming knob feature for the

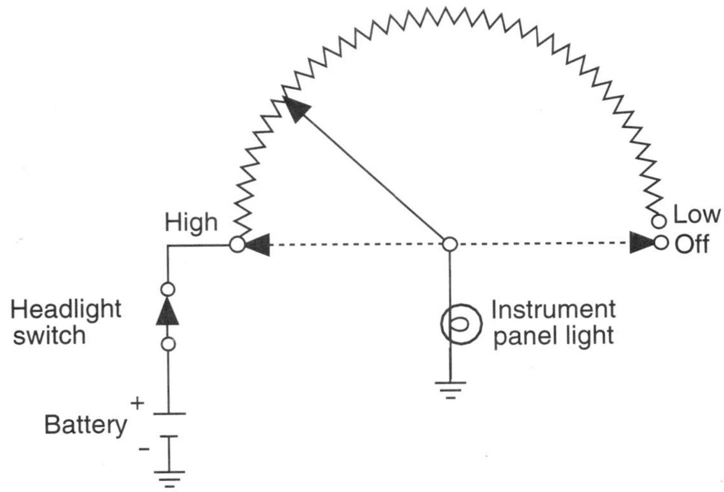

FIGURE 5–40 Using a rheostat to control the brightness of an instrument panel light.

instrument lights. By changing the position of the wiper on the resistor, the resistance can be increased (bulbs dim), or decreased (bulbs brighten) to control current (Figure 5–41).

A **potentiometer** (also called a pot) differs from a rheostat in that it has three terminals (or wires) and controls voltage. One terminal at the fixed end of the resistor is connected to a power source. The second terminal end of the resistor is connected to ground. The third terminal is connected to a movable wiper contact (Figure 5–42).

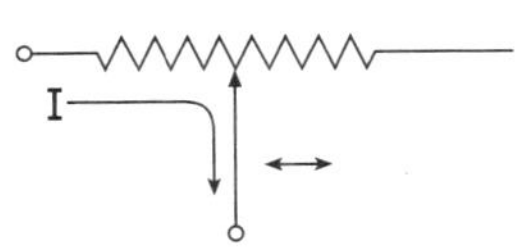

FIGURE 5–41 Wiring diagram for a rheostat.

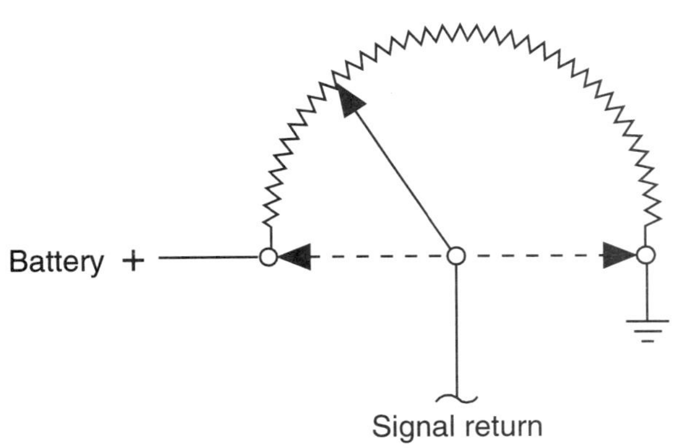

FIGURE 5–42 A typical potentiometer contains three wires.

The wiper, as it is moved over the resistor, creates a variable voltage drop. Because current is always flowing through the fixed portion of the resistor, the voltage drop generated by the potentiometer is very stable. For this reason, the potentiometer makes an excellent input sensor for a vehicle's computer system. An example of a potentiometer input sensor is a throttle position sensor (Figure 5–43).

The potentiometer sensor circuit in Figure 5–43 uses the amount of voltage drop to determine the throttle position. In Figure 5–44, a temperature mode door position switch using a three-wire potentiometer.

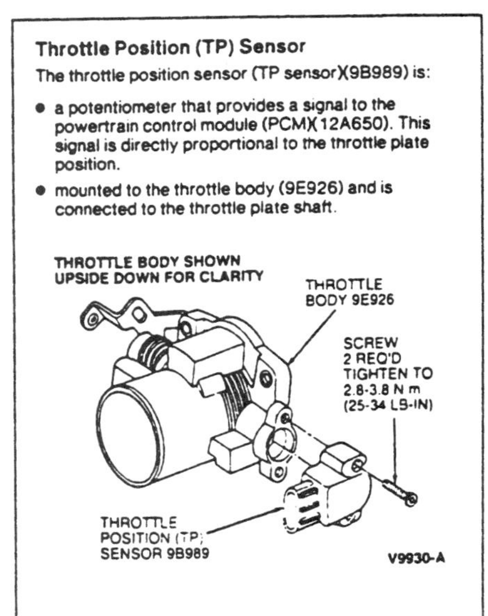

FIGURE 5–43 A throttle position (TP) sensor.

(Courtesy of Ford Motor Company)

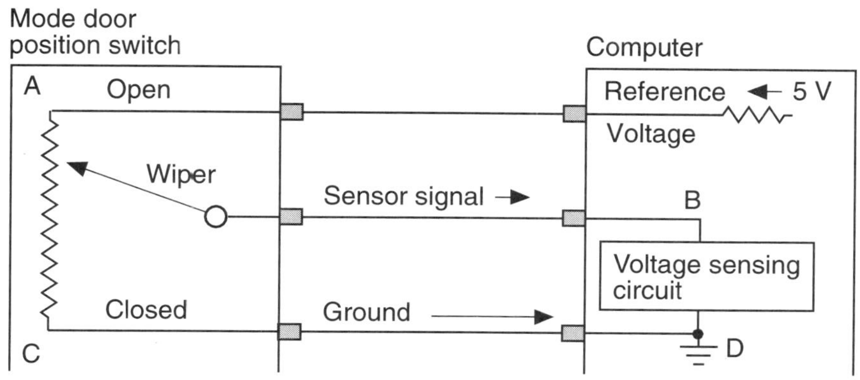

FIGURE 5–44 A temperature mode door position switch using a three-wire potentiometer.

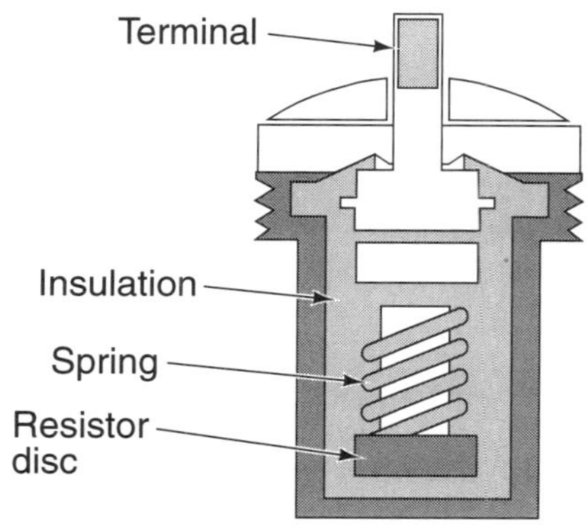

FIGURE 5–45 A thermistor sensor used to control engine temperature gauge.

The final style of variable resistor is a **thermistor.** The resistance value of a thermistor changes in proportion to its temperature and is typically used to measure coolant temperature. As coolant temperature rises, the thermistor resistance decreases and the current flow through a temperature gauge increases. When the coolant temperature lowers, the thermistor resistance increases, thus decreasing the current flow to the gauge (Figure 5–45).

Circuit Protection Devices

Most automotive electrical circuits are protected from damage from current levels that exceed the designed amperage capacity of the system. The most common **circuit protection device** is the fuse. There are three basic types of quick replacement fuses (Figure 5–46)—glass, ceramic, and the newer blade type fuse.

Glass and ceramic fuses are found mostly on older vehicles. Glass fuses are small glass cylinders with metal end caps. The metal strip that connects the two end caps is designed to blow quickly when the current indicated on the end cap code is exceeded. Codes such as AGA, AGW, and AGC indicate the length and diameter. Fuse lengths within each of these series codes are the same, but the current rating can vary. Glass fuses with an SFE code have the same diameter, but the length varies with the current rating. The ceramic fuse is used on many European imports and are available in two size codes, GBF (small) and the more common code GBC (large).

There are three different types of blade style fuses (Figure 5–47). These fuses consist of a plastic housing with a metal strip connected between two metal blade-type connectors.

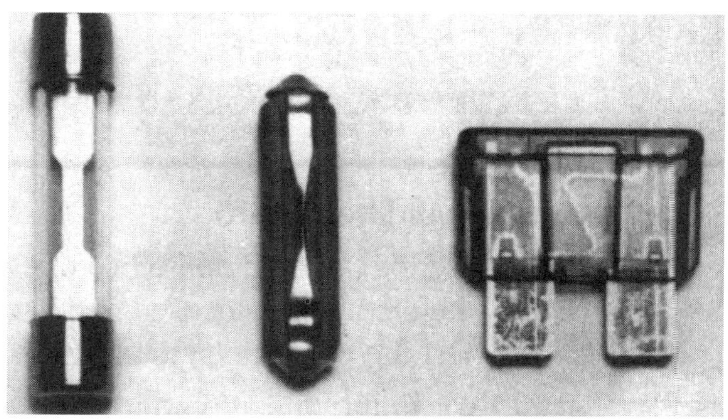

FIGURE 5–46 Three basic types of quick replacement automotive fuses—(A) glass cartridge, (B) ceramic, and (C) blade style.

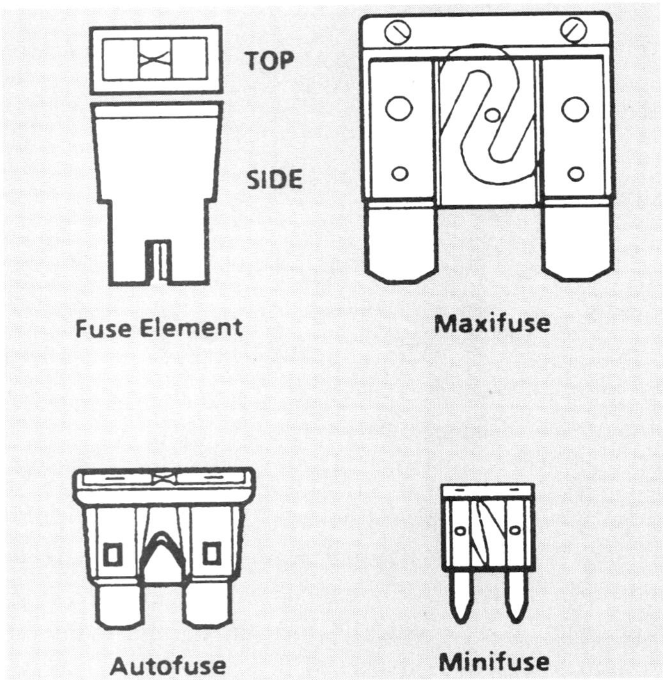

FIGURE 5–47 Pacific fuse element and three types of blade fuses.

(Courtesy of General Motors Corporation, Service Technology Group)

The fuse rating is indicated on top of the plastic housing, and the plastic is color coded for current rating (Figure 5–48). Blade fuses have code ratings of ATC or ATO.

The **autofuse** (3-30 amps) is the most common fuse used in today's vehicles to protect system components and wiring harnesses from damage. A **maxifuse** (20-80 amps) is larger than an autofuse and is a *slow blow* fuse. Common uses for this style of fuse is to place it between the battery and the main fuse box. The **minifuse** (5-30 amps) is similar in design and performance to an autofuse, but is smaller. Using this smaller fuse allows for more systems to have individual fuses within the same fuse box space. The last style of a quick replacement fuse is a **pacific fuse element** (30-60 amps) that was developed as a replacement for a **fusible link.** Both are designed to protect the electrical system from a direct short to ground and are normally located in the vicinity of the battery. Figure 5–49 is an example of a fusible link located near the battery. The current capacity of a fusible link is determined by its size. A fusible link is usually four wire sizes smaller than the circuit it protects. For example, a 14 gauge wire would require an 18 gauge fusible link. Remember the rule for wire size—the *smaller* the wire is, the *larger* its number is.

AUTOFUSE

CURRENT RATING	COLOR
3	VIOLET
5	TAN
7.5	BROWN
10	RED
15	BLUE
20	YELLOW
25	NATURAL
30	GREEN

MAXIFUSE

CURRENT RATING	COLOR
20	YELLOW
30	GREEN
40	AMBER
50	RED
60	BLUE
70	BROWN
80	NATURAL

MINIFUSE

CURRENT RATING	COLOR
5	TAN
7.5	BROWN
10	RED
15	BLUE
20	YELLOW
25	NATURAL
30	GREEN

PACIFIC FUSE ELEMENT

CURRENT RATING	COLOR
30	PINK
40	GREEN
50	RED
60	YELLOW

6-9-94
BB0019A5

FIGURE 5–48 Fuse current ratings and corresponding color codes.

A fuse is normally connected in series to an electrical circuit (Figure 5–50) ahead of all the loads in the circuit. However, there may be a fuse located at each individual circuit load device (Figure 5–51).

Normally, when fuses reach their current rating, the metal wire melts through causing the circuit to open (Figure 5–52). It is not always easy to see a blown fuse

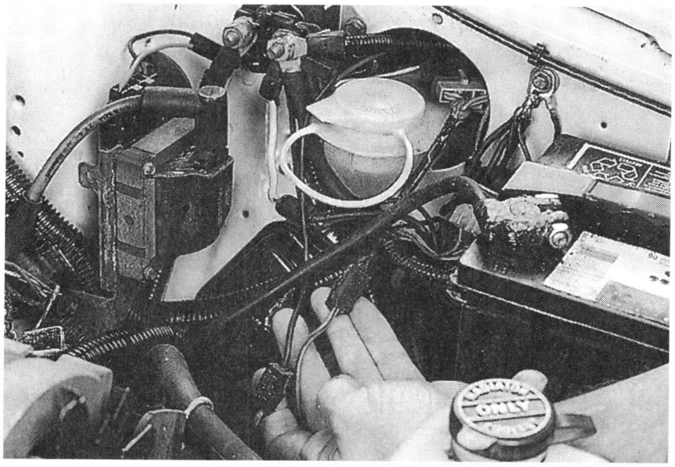

FIGURE 5–49 A fusible link located close to the battery.

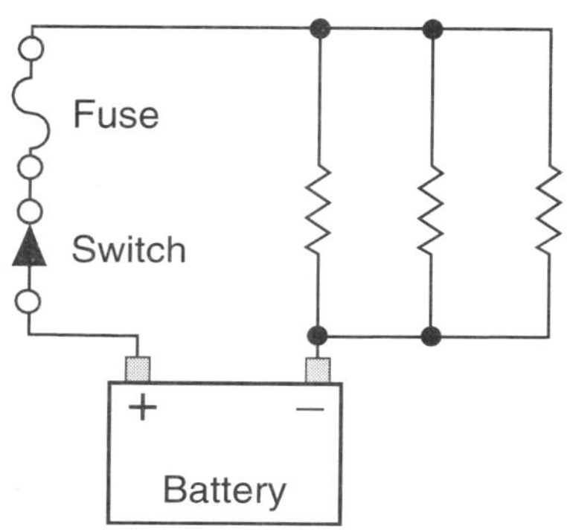

FIGURE 5–50 A fuse connected to protect an entire parallel circuit.

by visual inspection. Occasionally, technicians find hidden faults in fuses that look normal (Figure 5–53).

To prevent a visual misdiagnosis, a technician should use an ohmmeter, voltmeter, or test light to properly check a fuse or fusible link. When an ohmmeter is used to check a fuse or fusible link (Figure 5–54 and Figure 5–55), the fuse should first be removed or power

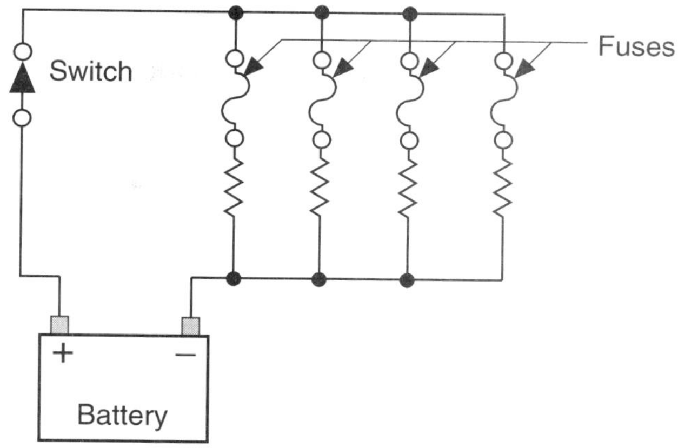

FIGURE 5–51 Individual fuses to protect each branch of a parallel circuit.

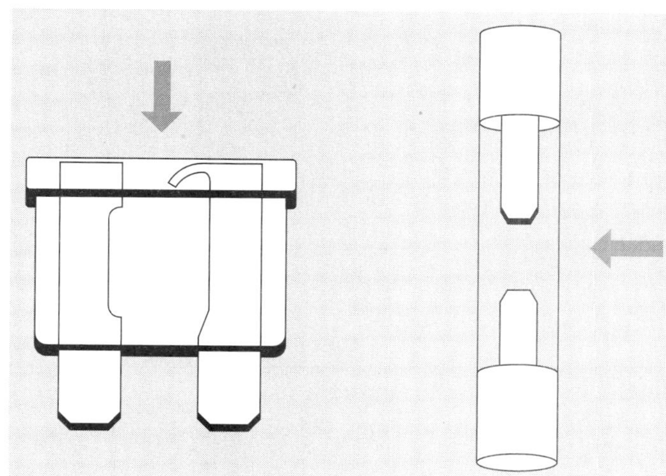

FIGURE 5–52 Typical view of a blown fuse.

removed from the fusible link. The ohmmeter should read 0 to 1 ohms if it is good, and infinite if it is blown.

To test a fuse or fusible link with a voltmeter or test light, check for available voltage at both terminals of the unit (Figure 5–56). If the device is good, there will be

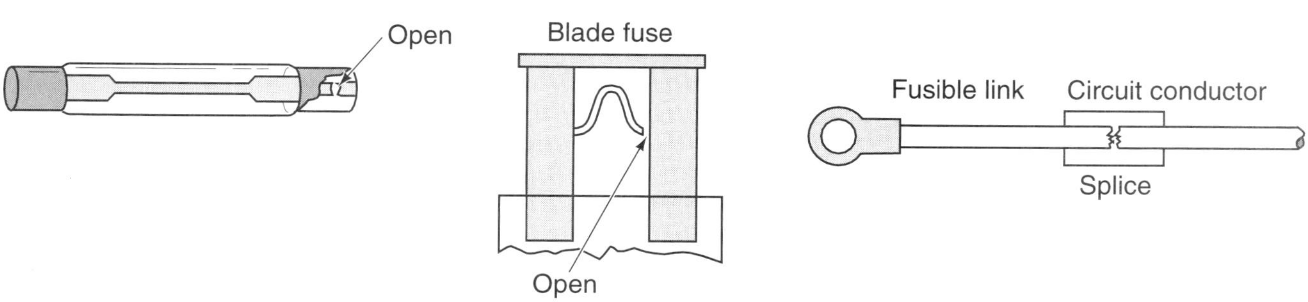

FIGURE 5–53 Hidden faults in a fuse can sometimes go undetected.

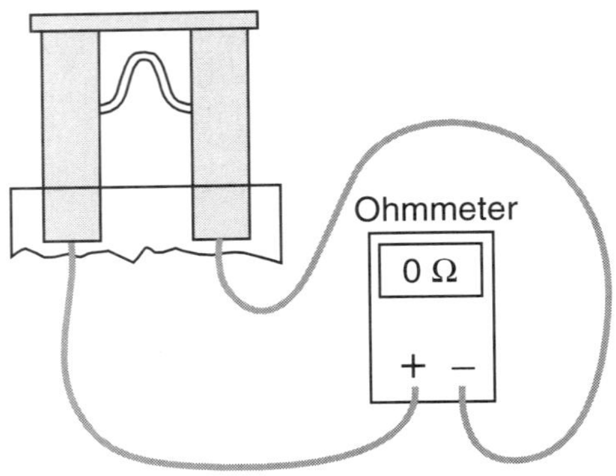

FIGURE 5–54 A good fuse will have zero resistance when tested with an ohmmeter.

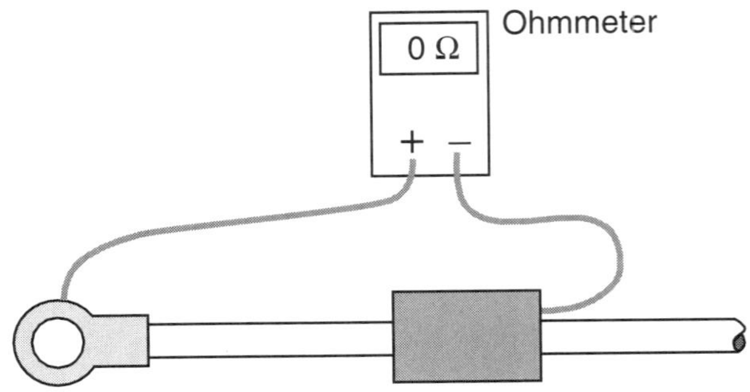

FIGURE 5–55 Testing a fusible link with an ohmmeter once it has been disconnected from the power source.

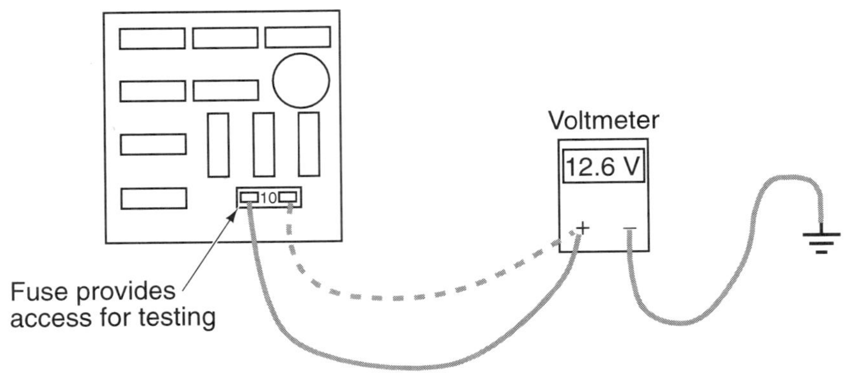

FIGURE 5–56 When testing a fuse with a voltmeter, there should be voltage present on both sides of the fuse.

voltage on both sides. A blown fuse will show no voltage after the fuse.

CAUTION *Fuses are rated by amperage for a specific circuit. Never bypass a fuse, or install a larger capacity circuit protection device other than what is designed by the manufacturer. Severe circuit damage or user injury could result. A typical fuse box or panel is normally marked with the appropriate fuse size for each circuit (Figure 5–57). On some older vehicles, the printing on the fuse box face may not be readable. In these cases, refer to manufacturers service manuals for proper fuse sizes.*

Circuit Breakers

Electrical circuits and load devices that are susceptible to overloads on a routine basis are usually protected by a **circuit breaker.** They can be mounted in the fuse panel or installed in-line with the electrical circuit. There are three common types of circuit breakers—the self-resetting circuit breaker (also called the cycling circuit breaker), the manual circuit breaker (also called the resettable circuit breaker), and the positive temperature coefficient (PTC) solid state circuit breaker.

Figure 5–58 shows an example of a self-resetting circuit breaker. It reacts to a higher-than-designed flow of current by heating a **bimetallic strip** to momentarily open a set of contact points. When the bimetallic strip cools, it will automatically close the contact points and let current flow in the circuit again.

A typical application for a self-resetting circuit breaker circuit is a power window circuit. Because the window is susceptible to overloading conditions, such as ice or snow buildup, a momentary current overload is possible. When this happens, the circuit breaker will open and close until the ice or snow buildup is removed.

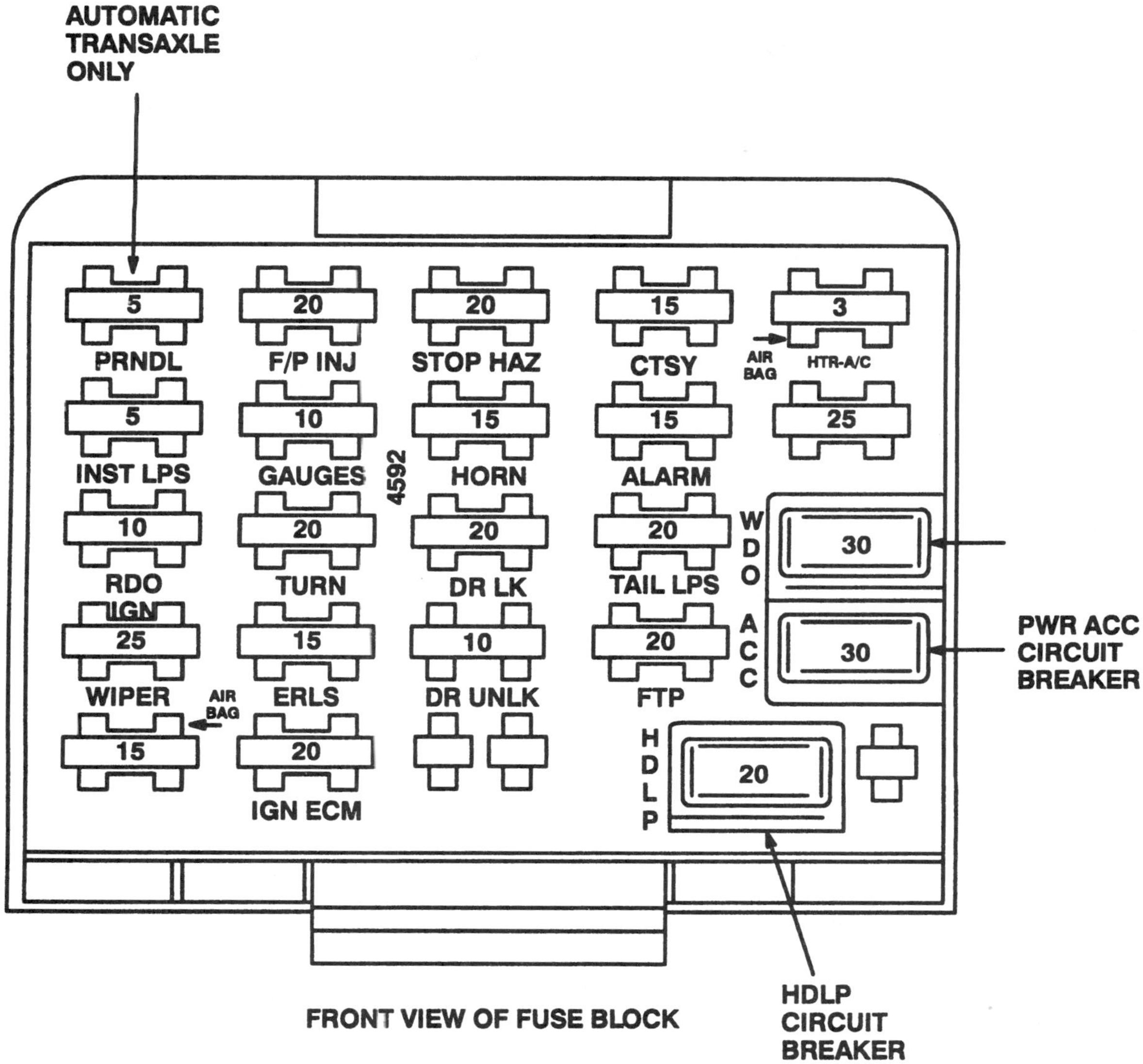

FIGURE 5–57 Typical fuse box or panel.

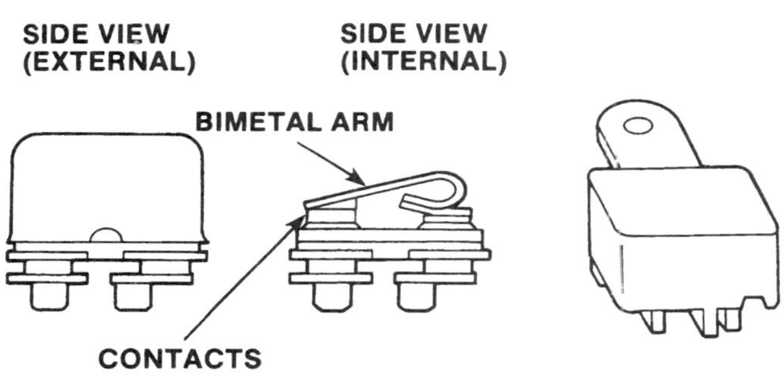

FIGURE 5–58 A self-resetting (cycling) circuit breaker.

The second type of circuit breaker is the manual resettable type, which is available in two styles. One style is reset by pushing a button on the circuit breaker, the other is reset by removing power from the circuit and letting the circuit breaker cool down (Figure 5–59).

The third type of circuit breaker is a PTC (positive temperature coefficient) solid state device (Figure 5–60). This type of circuit breaker is usually wired within a load component like a wiper motor or power window motor. As current exceeds its design limits, the resistance of the PTC circuit breaker increases until the circuit is opened. The PTC circuit breaker will reset only after the high current condition of the circuit no longer exists.

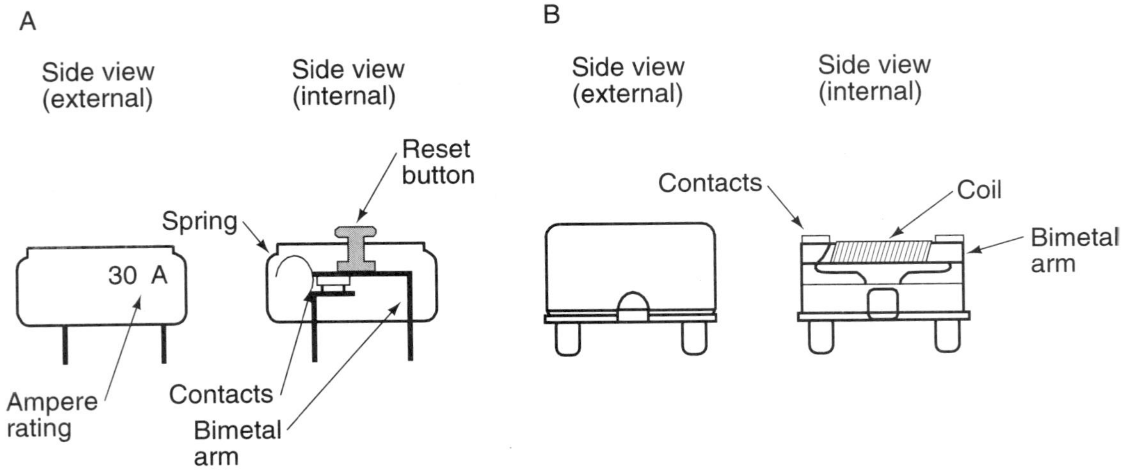

FIGURE 5–59 Typical manual reset circuit breakers. (A) Reset button style, (B) reset requires removal of power.

FIGURE 5–60 A PTC circuit breaker.

(Courtesy of General Motors Corporation, Service Technology Group)

Electrical Switches and Relays

An electrical switch is the most common way of controlling an on/off function, or of directing the flow of current in an electrical circuit. The simplest type of switch is the **single pole/single throw (SPST) switch** (Figure 5–61). The term *pole* refers to the number of input circuits, and *throw* refers to the number of output circuits.

Some SPST electrical switches are momentary contact switches. A horn switch (button) is an example of an SPST momentary switch. This switch has a spring loaded contact that keeps the contacts from closing unless external pressure is applied to the button (Figure 5–62).

Some electrical circuits require a **single pole/double throw (SPDT) switch.** A dimmer switch used in a headlight circuit is normally an SPDT switch (Figure 5–63). This switch has one input circuit and two output circuits. Either the high beam or low beam is turned on depending on the position of the switch.

The third type of switch is a **ganged switch.** This switch is also referred to as a multiple pole/multiple throw (MPMT) switch. An MPMT switch contains multiple wipers that are ganged (hooked) together so they all move together in unison. An ignition switch (Figure 5–64) is a common example of a ganged switch with five wipers that move together.

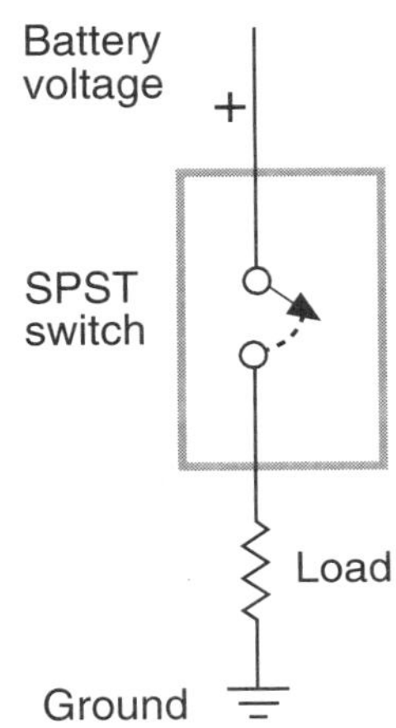

FIGURE 5–61 A simple single pole/single throw (SPST) switch.

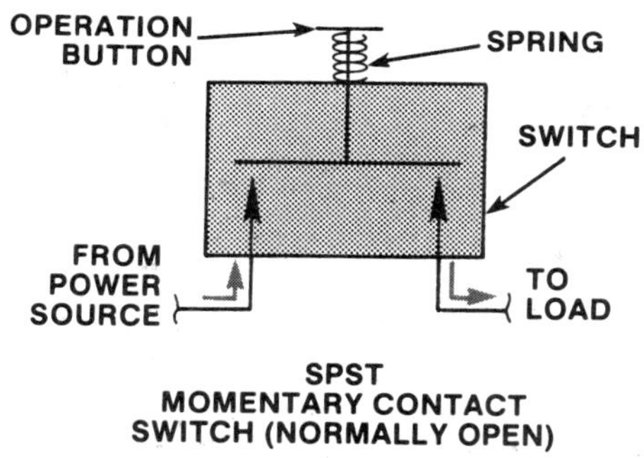

FIGURE 5–62 An SPST switch designed for momentary contact.

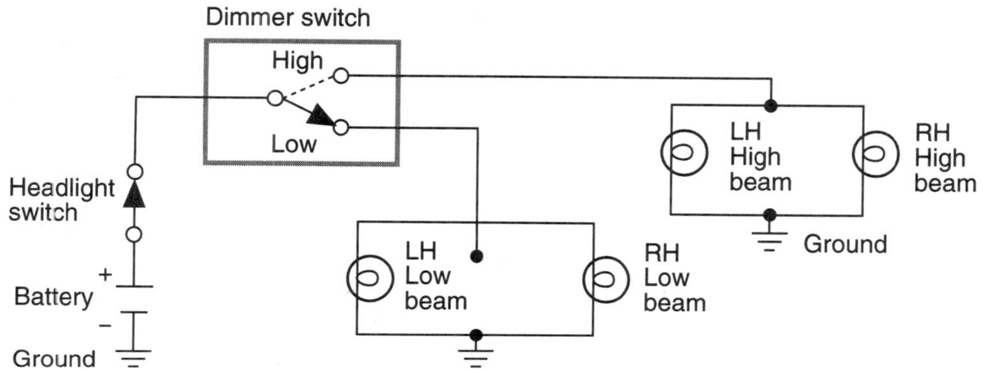

FIGURE 5–63 A simplified schematic of an SPDT dimmer switch in a headlight circuit.

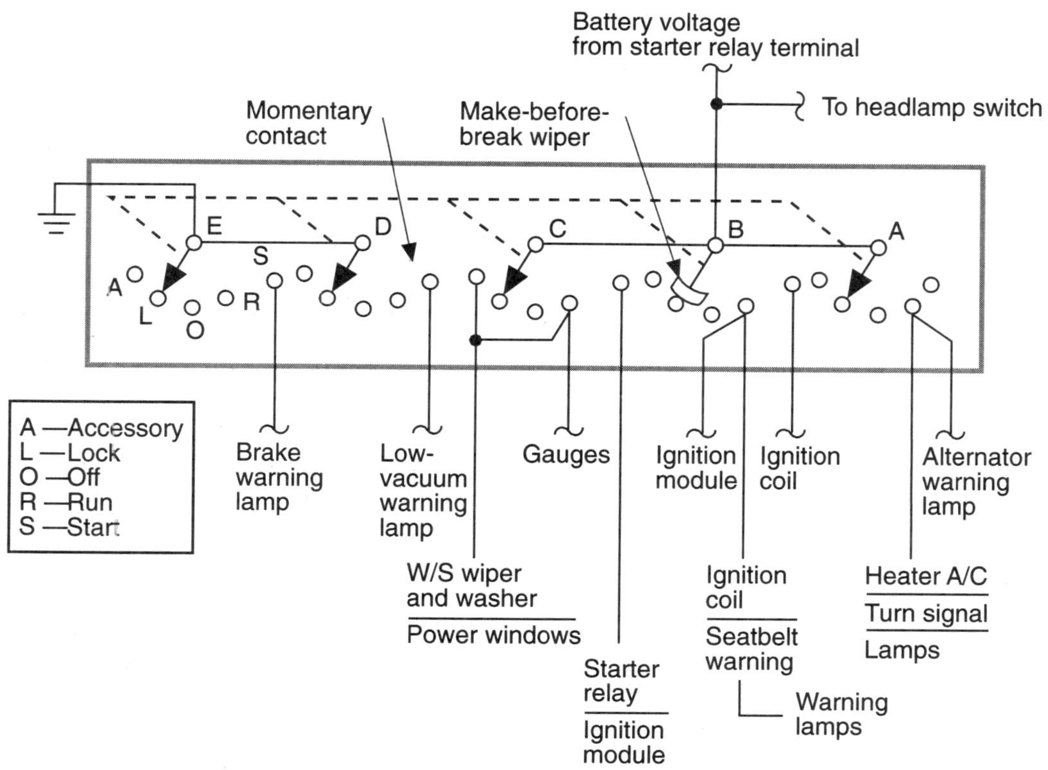

FIGURE 5–64 A typical MPMT ganged ignition switch.

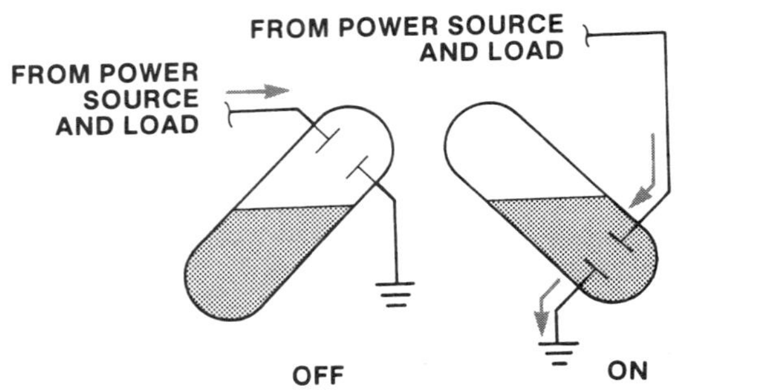

FIGURE 5–65 A typical mercury switch.

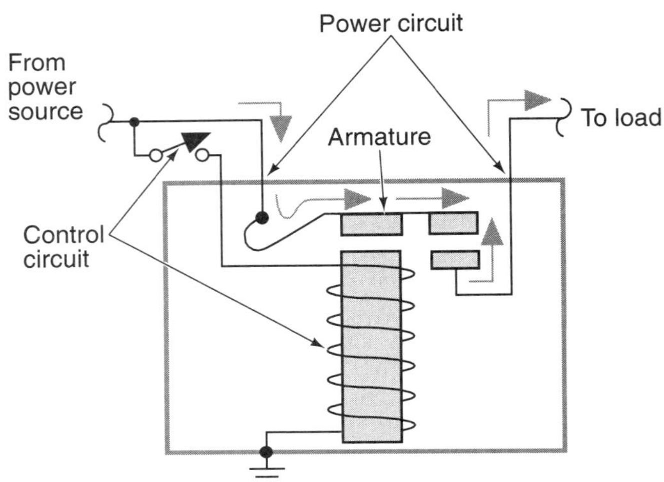

FIGURE 5–66 The relay contacts are closed by electro-magnetism created by the coil.

In some special electrical circuits, a **mercury switch** is sometimes used to control a light in a luggage compartment or engine compartment. Mercury is used in the switch (Figure 5–65) because mercury is a good conductor of electricity.

A mercury switch is usually mounted to a hood or trunk lid. When the lid is raised the mercury inside the switch capsule will move to the end to cover the electrical contacts and complete the circuit.

Another type of specialized switch used extensively in vehicles is an **electromagnetic switch.** This switch is also called a **relay** (Figure 5–66). It is designed to use a small amount of current to control a relay coil that will electro-magnetically close a set of contact points to complete a higher amperage circuit. A horn circuit is an example of a relay controlled application (Figure 5–67).

The relay coil in the horn circuit in Figure 5–67 will be energized when the horn switch button is closed. The coil develops a magnetic field which closes the contact points, and completes the circuit between the battery and the horn assembly. There are normally two tests to perform on a relay. On-vehicle testing of a relay will be covered in future chapters of this text. To bench test a relay when it is off the vehicle, first test the relay coil for continuity with an ohmmeter (Figure 5–68). If there is continuity in the coil, the relay winding is good. An open in the coil wiring would be indicated by an infinity reading on the ohmmeter.

Next, test the relay contacts (Figure 5–69). This is performed by connecting an ohmmeter to the relay contact terminals. When the relay coil is energized by a suitable battery, the ohmmeter should indicate continuity through the contact points. If the contact points are bad or burned, the resistance reading on the ohmmeter would be high.

There are several different brands and styles of relays used by vehicle manufacturers. The most common relay is a normally open (N/O) contact type with a four pin terminal. The terminal layout and diagram for a 4-pin N/O relay is shown in Figure 5–70.

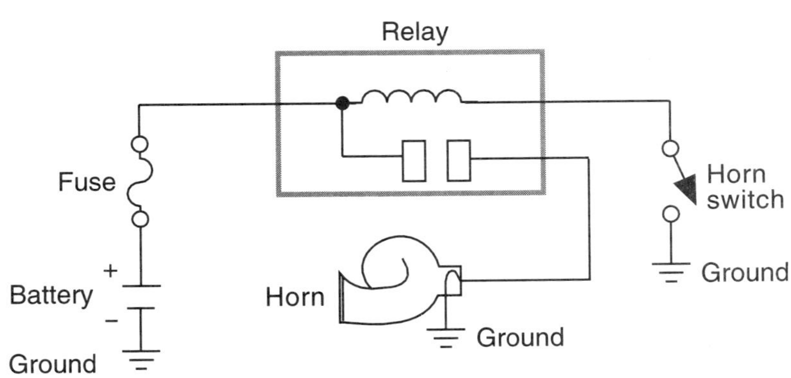

FIGURE 5–67 A relay used as a high current switch for a horn circuit.

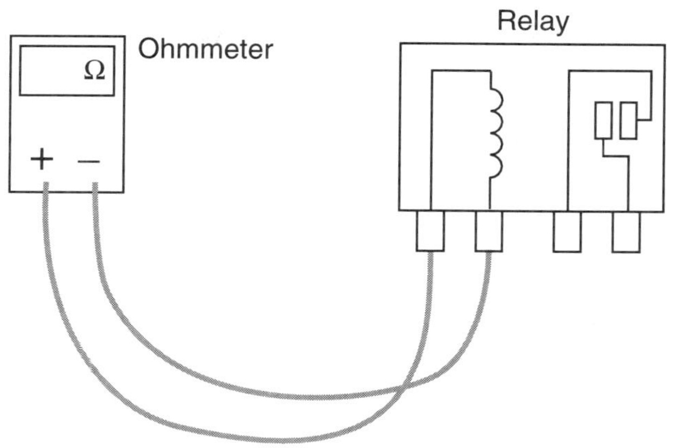

FIGURE 5–68 Testing the continuity of the coil in a relay.

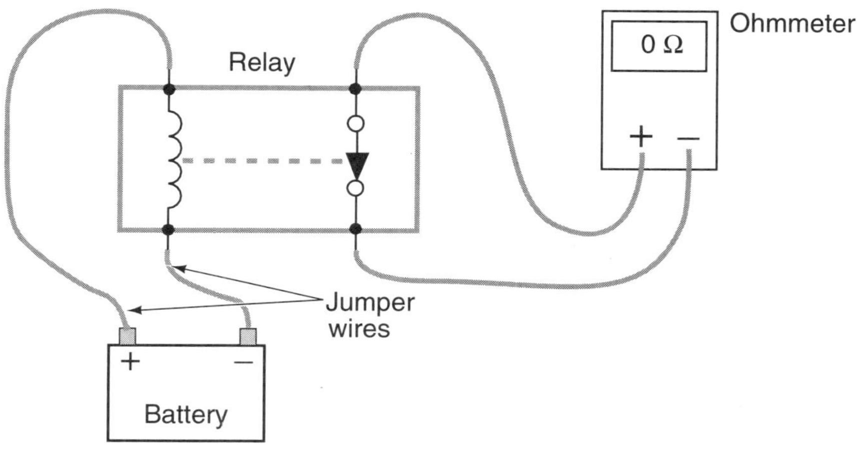

FIGURE 5–69 Checking the continuity of the relay contacts.

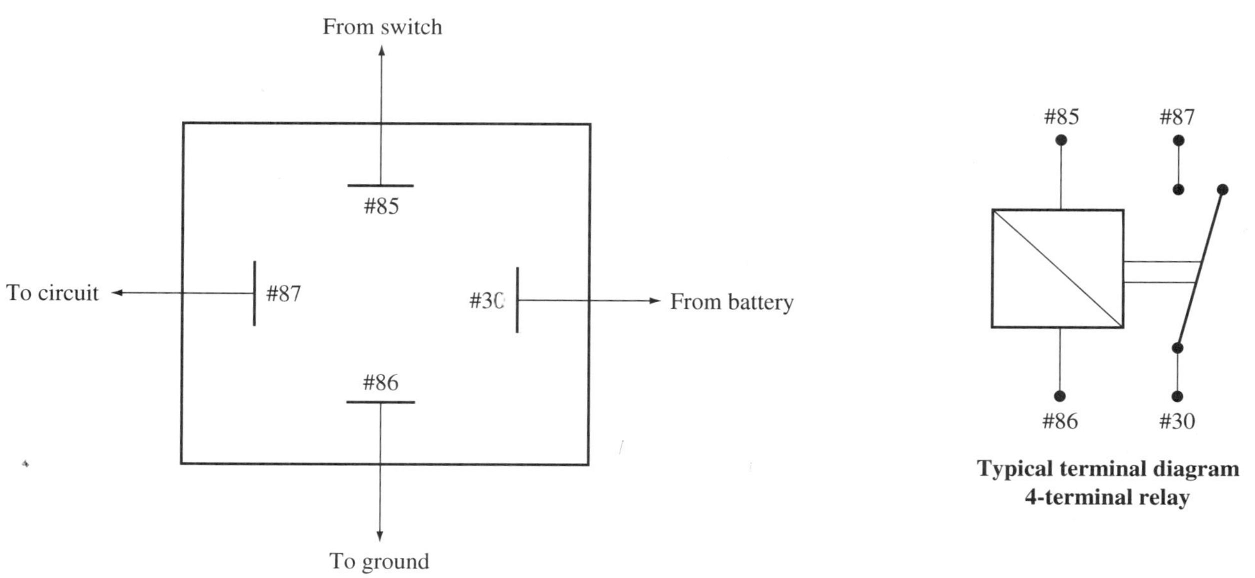

FIGURE 5–70 Terminal view and diagram of a 4-pin relay.

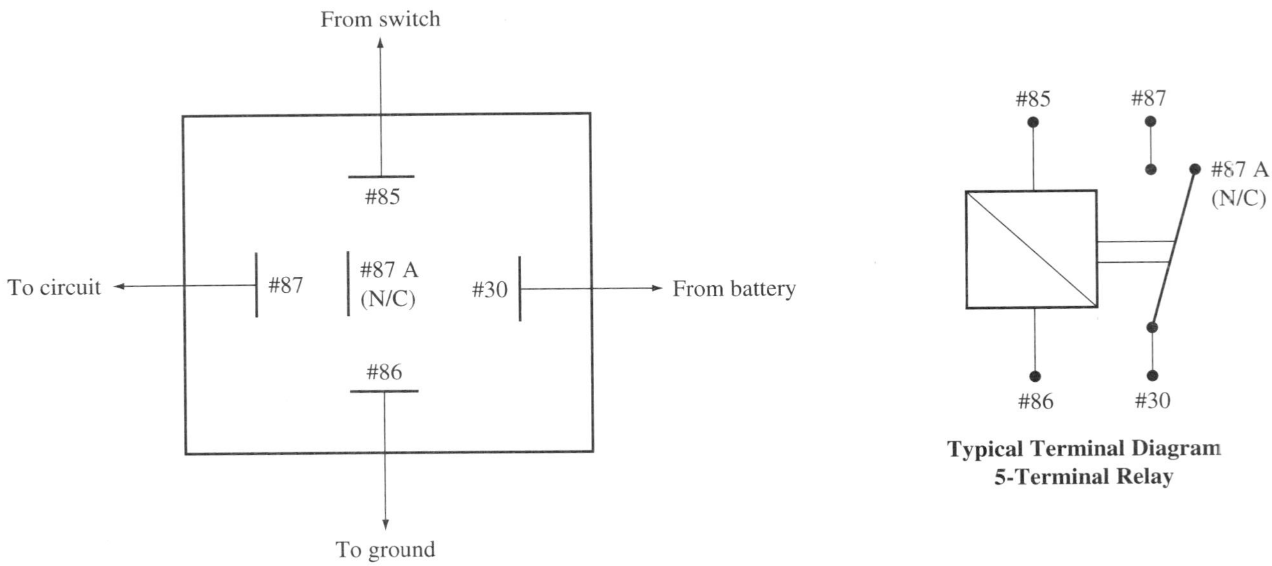

FIGURE 5–71 Terminal view and diagram of a 5-pin relay.

Whenever there is a five-pin relay, the fifth terminal of the relay is most often a normally closed (N/C) contact (Figure 5–71).

Capacitor Construction and Ratings

Some automotive electrical systems use a **capacitor,** or condenser as it is sometimes called, to store electrical charges from a DC circuit. Some of the most commonly used capacitors are shown in Figure 5–72.

A capacitor is a storage device for DC voltage. It is designed to absorb any voltage changes in the circuit to control voltage spikes. The majority of capacitors are connected in parallel with the load circuit(s) (Figure 5–73).

When charged, a simple capacitor (Figure 5–74) is constructed of a **positively charged ion** plate which has more protons than electrons, and a **negatively charged ion** plate which has more electrons than protons. The insulator between the two plates is made of a **dielectric** insulating material like plastic, glass, ceramic, or paper.

Capacitors operate on the basic principles that opposite charges attract each other, and that there is a potential voltage between any two points that are of opposite polarity. In Figure 5–75, when the system switch is closed, current will flow into the capacitor until the voltage charges are equal to the battery. The effect of electron movement toward the negative plate and away from the positive plate creates the effect of current flow.

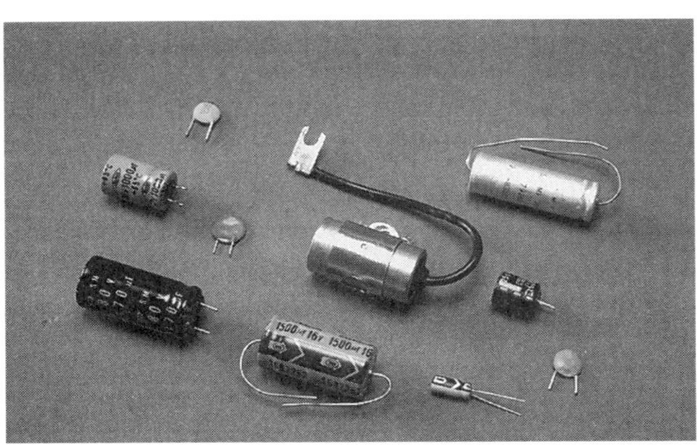

FIGURE 5–72 Common capacitors found in automotive electrical circuits.

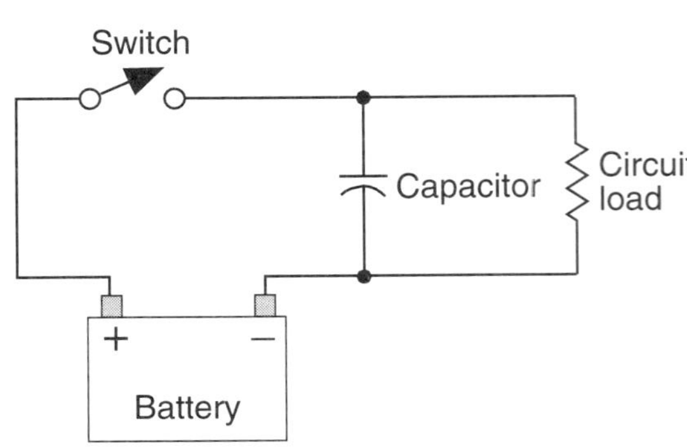

FIGURE 5–73 A capacitor connected in a simple circuit.

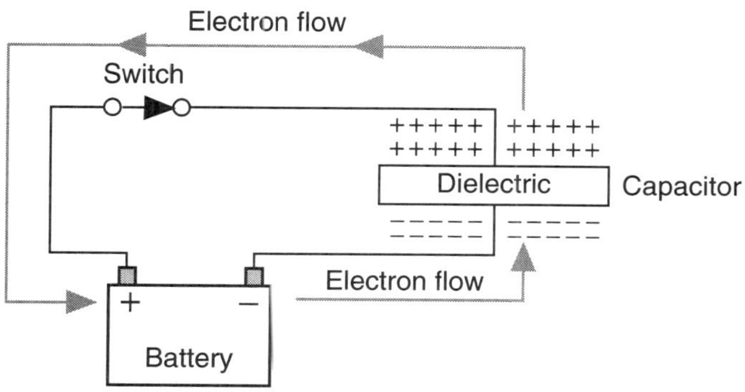

FIGURE 5–74 The flow of electrons in a capacitor.

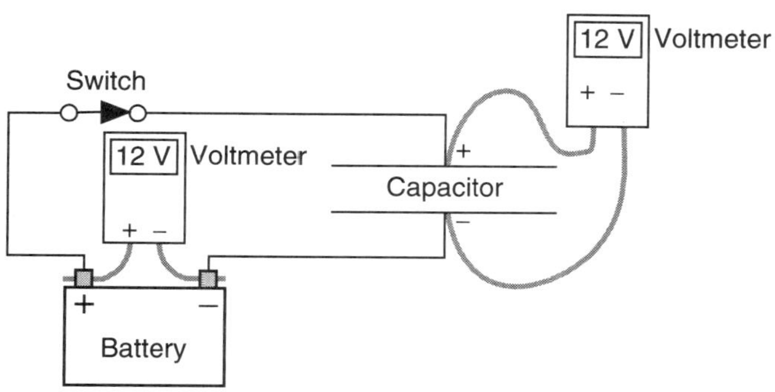

FIGURE 5–75 A capacitor that is fully charged.

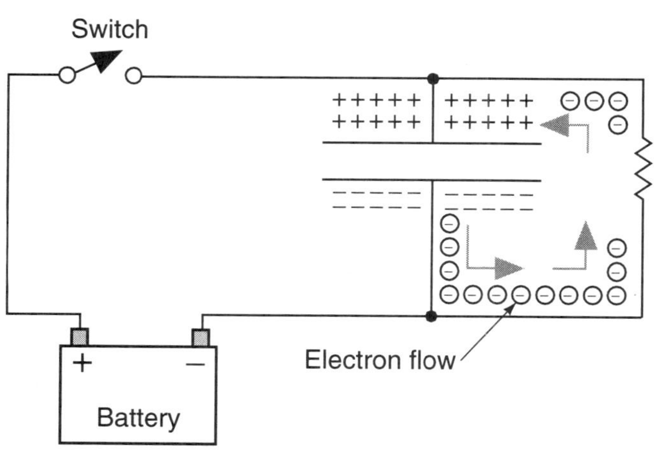

FIGURE 5–76 Current flow with the switch open and the capacitor discharging.

Electrons do not actually pass through the capacitor's **electrostatic field,** which is between the two opposite charge plates, but are stored on the plates as static electricity.

When a DC circuit switch is opened, the capacitor will discharge the electrons from the surface of the negative capacitor plate, through the resistive load, and to the positive capacitor plate until a neutral state is reached (Figure 5–76).

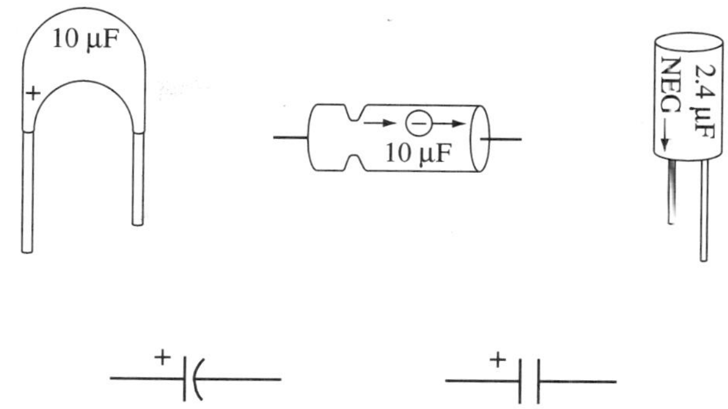

FIGURE 5–77 Common capacitor markings and schematic symbols.

Capacitors are normally rated by their maximum voltage rating and capacitance value. The standard capacitance unit value is the **farad.** Common automotive capacitors are usually rated in microfarads (μF) which are one millionth of a farad. This value may be marked on the capacitor. Common capacitors are made of plastic, ceramic, or electrolytic material. Figure 5–77 is an example of capacitor markings and common schematic symbols.

Automotive Lamps

The lighting circuits of today's vehicles can consist of more than 50 bulbs of different designs and sizes. These bulbs are for an array of applications that include headlights, taillights, parking lights, stop lights, side marker lights, dash lights, courtesy lights, and so on. Knowing the proper size bulb for a specific application is important for proper circuit operation (Figure 5–78).

The simplest lamp design is an **incandescent lamp.** This lamp produces a light as a result of current flowing through a coiled tungsten wire filament inside a glass bulb envelope. Inside the glass bulb is an inert gas that replaces the oxygen and lengthens the life span of the filament. Lamps are normally one of two types—**a single-filament bulb** (Figure 5–79) or a **double-filament bulb** (Figure 5–80). The double-filament bulb is designed to serve more than one electrical function, such as stoplamp and taillamp functions. The stoplamp function uses the lower resistance filament because it needs to be brighter, and the taillamp function uses the higher resistance filament.

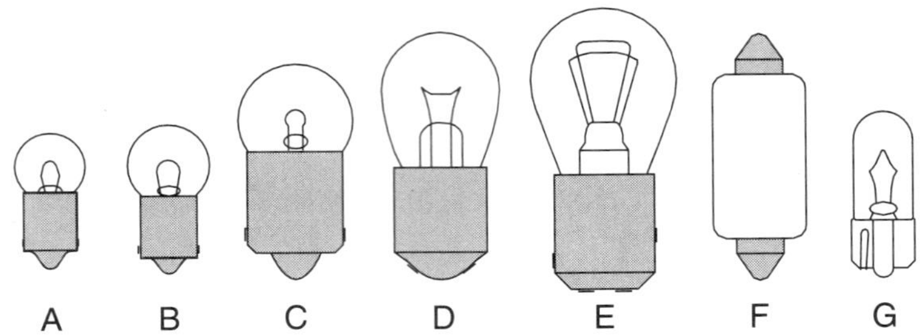

FIGURE 5–78 Common automotive bulb designs.

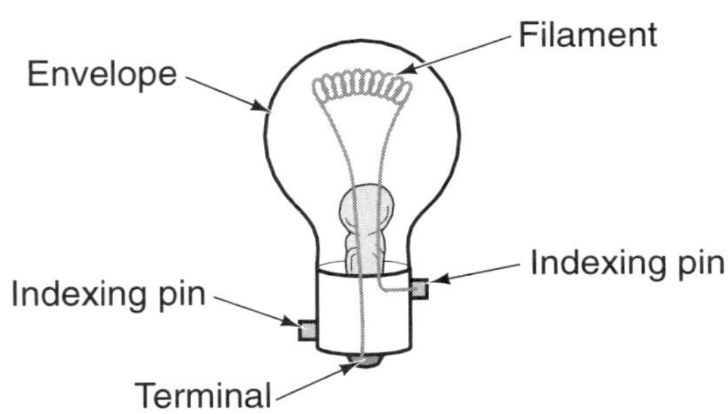

FIGURE 5–79 A single-filament incandescent bulb.

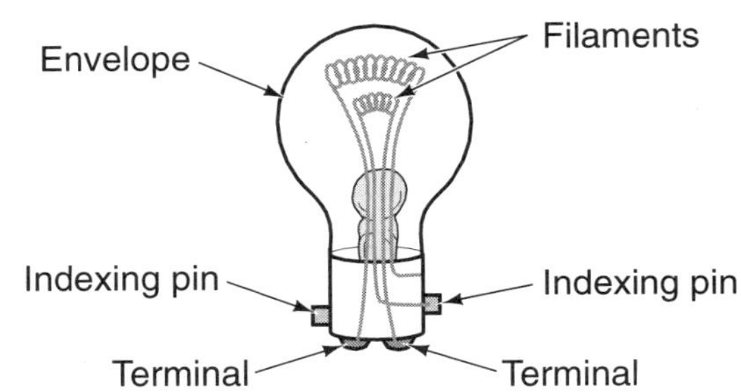

FIGURE 5–80 A double-filament incandescent bulb.

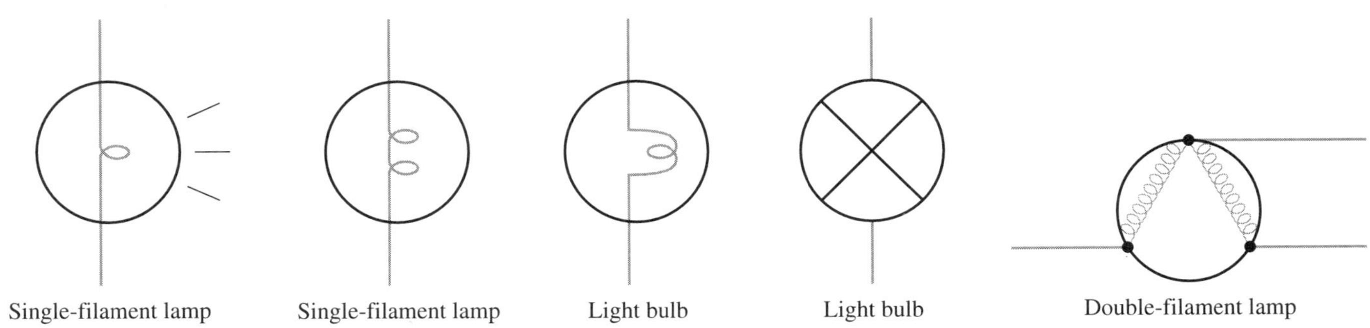

FIGURE 5–81 Common lamp identification symbols.

When replacing a lamp, it is important to use the correct lamp for the application. The *Lamp Standard Trade Number* chart in Appendix B is a handy reference to determine the design voltages, amperes, and watts of a specific bulb circuit.

Lamp identification symbols on wiring schematics can vary with manufacture of a vehicle (Figure 5–81).

Automotive Headlamps

There are four types of headlamps used in vehicles today: (1) sealed beam, (2) halogen sealed beam, (3) composite, and (4) High Intensity Discharge (HID).

The first type is the basic round **sealed beam headlamp** that was used extensively between 1939 and 1975. The introduction of the rectangle headlamp in 1975 allowed vehicle manufacturers to streamline the front-end design of their vehicles. Both the round and rectangle sealed beam headlamps use similar construction (Figure 5–82).

In a sealed beam headlamp, the glass lens acts as a prism to direct the light beam from the reflector into a specific pattern (Figure 5–83). Inside the lamp, argon gas prevents the filament from becoming oxidized.

The direction of the light beam is controlled by the location of the filament in relation to the reflector (Figure 5–84). In a double-filament lamp, the lower filament is used for the high beam, and the upper filament is used for the low beam.

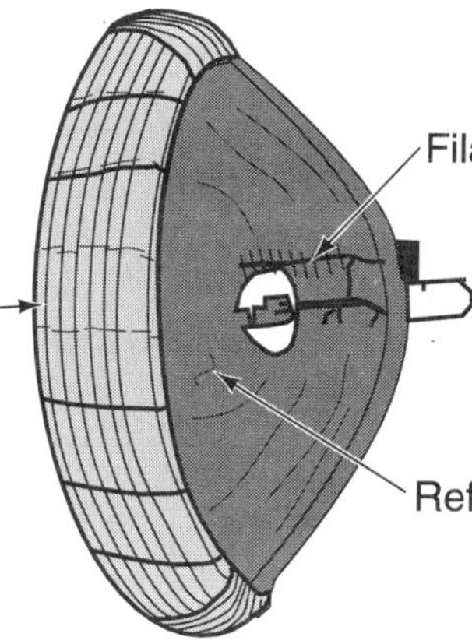

FIGURE 5–82 Sealed beam headlight construction.

The second style of headlamp currently used is the **halogen headlamp** (Figure 5–85). This type of headlamp uses a halogen filled inner bulb positioned in a glass housing. Its tungsten filament is able to burn approximately 25% brighter than a conventional sealed beam headlamp. If the outer lens is cracked or broken,

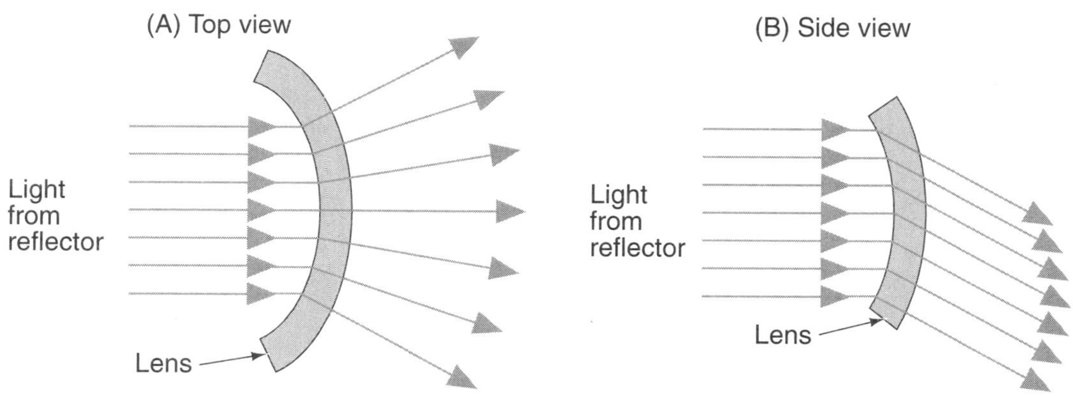

FIGURE 5–83 The lens directs the beam into (A) a flat horizontal pattern and (B) downward.

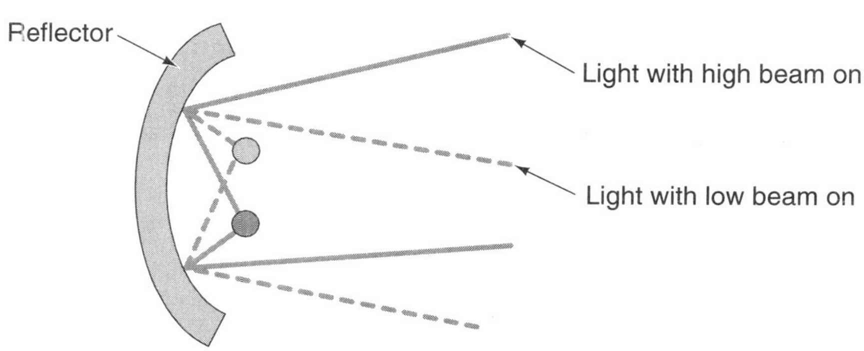

FIGURE 5–84 Filament placement controls projection of the light beam.

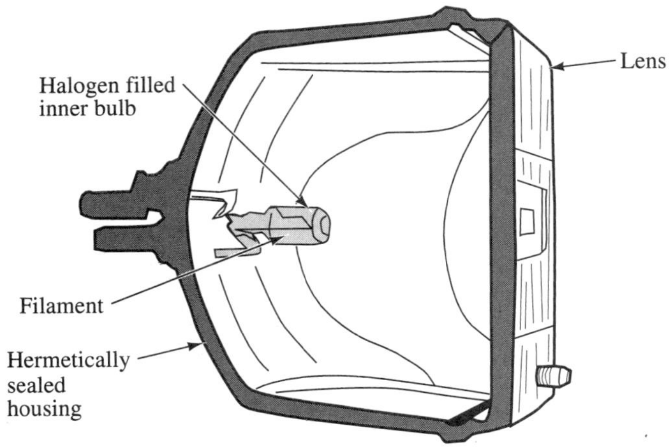

FIGURE 5–85 A sealed beam headlight with a halogen-filled inner bulb.

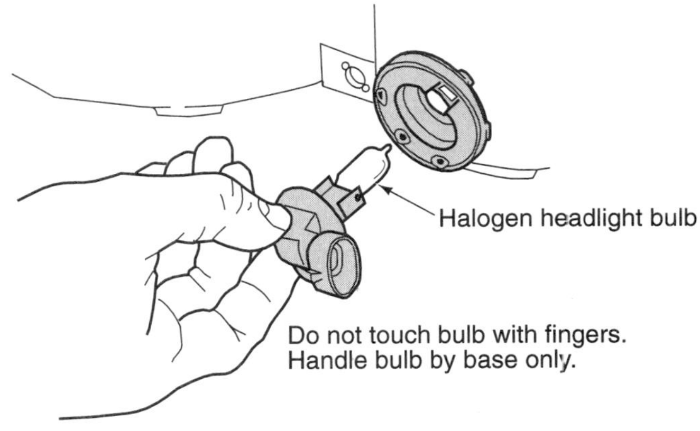

FIGURE 5–87 The correct method to handle a composite bulb during replacement.
(Courtesy of Ford Motor Company)

the filament will continue to operate, but the light quality will be reduced.

The third type of headlamp is the **composite headlamp** (Figure 5–86). This style of halogen headlamp allows the vehicle manufacturer to produce any shape of lens needed for vehicle styling and aerodynamics. Some of these composite headlamps are vented, so condensation that may naturally develop on the inside of the lens assembly does not affect headlight operation. Refer to a vehicle's service manual if you are unsure if a particular composite headlight is vented or non-vented. A non-vented composite lamp should be replaced if it collects moisture.

Be Careful, and do not touch halogen bulbs when replacing them (Figure 5–87). The oily residue on the skin will shorten the life of the bulb.

The remaining type of headlamp used in vehicles today is the **high-intensity discharge (HID)** or arc-discharge, light source (Figure 5–88). Instead of using a tungsten filament found in conventional headlamps, HID headlamps use a high-voltage arc discharge tube that jumps a gap inside a capsule filled with xenon gas. The technology is similar to the electronic flash assembly on a camera. The arc produces a bluish-white light that is three to five times brighter than a glowing filament, and lasts up to 2,000 hours (life of a vehicle). HID lighting systems are not available with a double-filament bulb because of the limits of the single-arc discharge tube. An HID system requires a ballast that transforms 12 volts DC battery voltage up to 20,000 volts AC within 0.3 seconds for initial start-up. After a short warm-up period, the steady-state light system voltage is

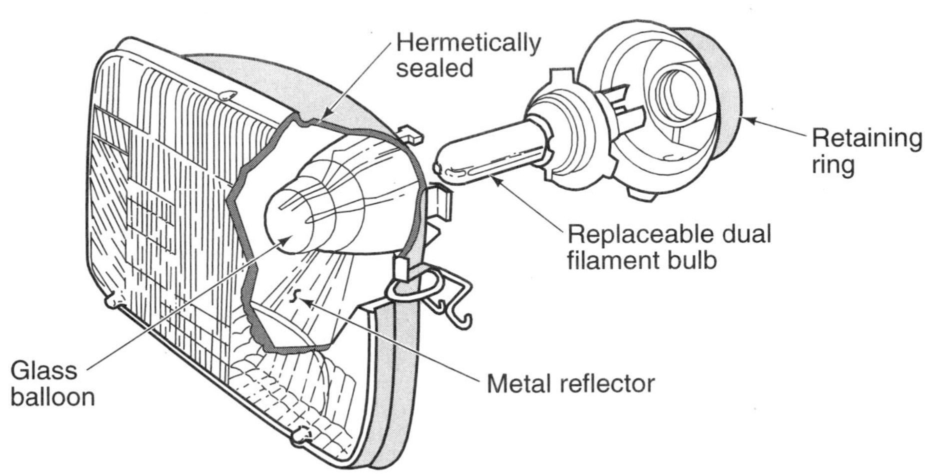

FIGURE 5–86 A replacement halogen bulb in a composite headlight.

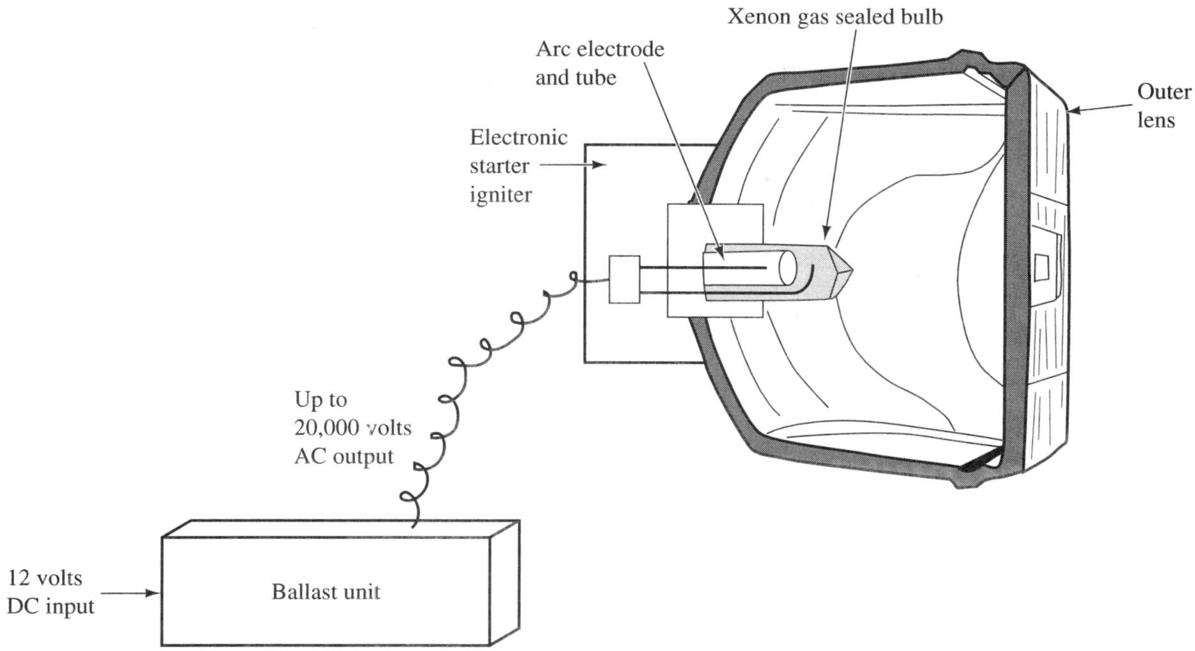

FIGURE 5–88 Typical HID light source and components.

around 90 volts AC at about 3 amps. This makes HID lighting more efficient than typical automotive head-lamp systems. This style of headlamp first appeared on select-model vehicles from BMW in 1993, Ford in 1995, and Porsche in 1996.

Hands-On Vehicle Tests

The following two hands-on vehicle checks are included in the **1996 NATEF** (National Automotive Technician Education Foundation) **Task List.** The task reference numbers are the same as the NATEF reference numbers. Complete your answers to the following questions on a *separate* sheet of paper.

Performance Task #1 (NATEF Task # VI-A-8)

Task Description

Inspect and test fusible links, circuit breakers, and fuses; replace as needed.

Task Objectives

❑ Obtain a vehicle that can be used for this task. What model and year of vehicle are you using for this task? ______________________________

❑ Identify the following components and list their location on the vehicle.

1. Fusible link(s)______________________________
2. Fuse panel(s)______________________________
3. Circuit breaker(s) ______________________________

4. List the fuses in the fuse panel and their current rating.

a. ______________ e. ______________ i. ______________

b. ______________ f. ______________ j. ______________

c. ______________ g. ______________ k. ______________

d. ______________ h. ______________

Circuit Checks

- ❑ Check continuity of a fusible link with an ohmmeter. Disconnect power to the fusible link before checking continuity. What is the result? ______________
- ❑ Describe the procedure for replacing a fusible link. ______________
- ❑ What is the general rule in determining what size fusible link to use for a circuit? ______________
- ❑ Check a fusible link with a voltmeter. What is the result?______________
- ❑ If you measured a voltage drop across the fusible link what would this indicate? ______________
- ❑ Check all fuses in the fuse panel with a voltmeter or test light. What are the results if the fuse is good?______________.

  What would be the results on a blown fuse? ______________
- ❑ Check a fuse that is out of the vehicle with an ohmmeter. What are the results if the fuse is good?______________.

  What would be the results if the fuse is blown?______________
- ❑ Check a circuit breaker for proper operation with an ohmmeter or voltmeter. Determine if the circuit breaker is a self-resetting or a resettable type. Record the results of your findings. ______________

Task Summary

After performing the above NATEF task, what can you determine that will be helpful in future inspection of fusible links, fuses, and circuit breakers? ______________

Performance Task # 2 (NATEF Task # VI-A-9)

Task Description

Inspect and test switches, connectors, relays, and wires of electrical/electronic circuits. Repair or replace as needed.

Task Objectives

❑ Obtain a vehicle that can be used for this task. What model and year of vehicle are you using for this task? ______

❑ Identify at least six of each of the following components and list their location on the vehicle.

1. Switches

   a. ______ c. ______ e. ______

   b. ______ d. ______ f. ______

2. Connectors

   a. ______ c. ______ e. ______

   b. ______ d. ______ f. ______

3. Relays

   a. ______ c. ______ e. ______

   b. ______ d. ______ f. ______

4. Wiring (harnesses)

   a. ______ c. ______ e. ______

   b. ______ d. ______ f. ______

Circuit Checks

❑ Check the continuity of a switch with an ohmmeter. Disconnect power to the switch before checking continuity. What is the result? ______

❑ Describe the procedure for replacing a switch. ______

______

❑ Check a switch with a voltmeter. What is the result? ______

❑ If you measured a voltage drop across the switch what would this indicate? ______

______

❑ Perform a visual inspection of all accessible connectors on the vehicles. Check for loose or corroded connections. What are the results?______

❑ Describe how to check an electrical connection for a voltage drop?______

______

❑ Perform a bench test of a relay's coil and contact points. Using an ohmmeter, what is the reading on the relay coil? ______.

❑ Using a battery to energize the relay coil, what is the ohmmeter reading on the relay contact points? ______

❑ Using a voltmeter, test the proper operation of a relay on the vehicle. What voltage reading do you get on each side of the relay coil?

Input ______________________. Output ______________________.

- ❑ What voltage reading do you get on each side of the relay contact points?

Input ______________________. Output ______________________.

- ❑ Inspect all accessible wiring harnesses in the vehicle. Check for frayed wires, deteriorated insulation, missing straps or tie downs. Record results here. ______________________________________

____

Task Summary

After performing the above NATEF task, what can you determine will be helpful in future inspections of fusible links, fuses, and circuit breakers? ______________________________________

____

Summary

- ❑ Automotive primary wire is used for low-voltage circuits.
- ❑ Automotive secondary wire is used for high-voltage circuits, like spark plug wires.
- ❑ Stranded copper wire is used for most automotive wiring applications.
- ❑ Wire size is measured by the AWG standard or by metric sizes.
- ❑ Metric wire size is measured by determining the cross sectional area of the wire.
- ❑ A solderless connector uses a special crimping tool to connect it to the wire.
- ❑ Rosin core flux solder is used when soldering electrical connections.
- ❑ A hard-shell connector contains multiple wires in one connector component.
- ❑ A weather-pack connector uses a rubber seal on the terminal and connector ends.
- ❑ A metri-pack connector is similar to a weather-pack connector except there is no seal on the cover half of the connector.
- ❑ Pull-to-seat and push-to-seat are forms of metri-pack connectors.
- ❑ Resistors are one of the most common components found in electrical circuits.
- ❑ Fixed resistors are generally made of carbon/graphite-based material.
- ❑ A stepped resistor is sometimes used to control blower motor speeds in electrical circuits.
- ❑ Variable resistors have the ability to change the resistance potential of an electrical circuit.
- ❑ A rheostat has two terminals—one to a fixed end and the other to a movable wiper contact.
- ❑ A potentiometer has three terminals—one on each fixed end and the third to a movable wiper.
- ❑ A thermistor's resistive value changes with temperature.
- ❑ Fuses are the most common circuit protection device.

- ❑ A glass fuse uses end cap codes of AGA, AGW, or AGC to indicate length and diameter.
- ❑ A glass fuse with the end cap code of SFE have the same diameter, but the length varies with the current rating.
- ❑ Blade fuses have code ratings of ATC or ATO.
- ❑ A maxifuse and pacific fuse element were developed to replace the fusible link.
- ❑ Two types of circuit breakers are the self-resetting(cycling) and the resettable.
- ❑ A SPST switch has one input pole circuit and one output throw circuit.
- ❑ A SPDT switch has one input pole circuit and two output throw circuits.
- ❑ A MPMT switch is also known as a ganged switch and contains multiple wiper contacts that are all ganged (hooked) together.
- ❑ Mercury switches contain a small vial of mercury to connect two contacts when moved to a specified position.
- ❑ A relay is a specialized switch that electro-magnetically closes a set of contact points.
- ❑ An incandescent lamp uses a tungsten wire filament in a glass bulb envelope.
- ❑ There are four types of headlamps used in vehicles today—sealed beam, halogen sealed beam, composite halogen, and high-intensity discharge.
- ❑ A capacitor, or condenser, is used to store electrical charges in a circuit.

Key Terms

american wire gauge (AWG)
autofuse
bimetallic strip
capacitance
circuit breaker
circuit-protective device
color code
composite lamp
dielectric
double-filament bulb
electromagnetic switch
electrostatic field
farad
fixed resistor
fusible link
ganged switch
halogen lamp
hard-shell connector
high-intensity discharge (HID)
incandescent lamp
maxifuse
mercury switch
metri-pack connector
minifuse
negatively charged ion
pacific fuse element
positively charged ion
potentiometer
primary wire
pull-to-seat
push-to-seat
relay
rheostat
rosin core
sealed beam lamp
secondary wire
single-filament bulb
solderless connector
single pole/double throw (SPDT) switch
single pole/single throw (SPST) switch
stepped resistor
tap splice
thermistor
variable resistor
weather-pack connector

Review Questions

Short Answer Essays

1. Describe the difference between automotive primary wire and secondary wire.

2. Define the difference and use of acid core solder and rosin core solder.
3. Describe how to identify the resistive value of a carbon film resistor.
4. Describe the purpose of a capacitor in an automotive electrical circuit.
5. Describe the difference between a sealed beam headlamp and a halogen headlamp.

Fill-in-the-Blanks

1. The standard rule with AWG wire is the ________________________ the gauge number, the ________________________ the wire conductor.
2. A tap splice connector is sometimes used when a(n) ________________________ circuit is added to a(n) ________________________ circuit.
3. A pacific fuse element was developed as a replacement for a ________________________ ________________________.
4. A ________________________ is used in an electrical circuit to store charges from the DC circuit.
5. A ________________________ ________________________ bulb is designed to serve more than one electrical function.

ASE Style Review Questions

1. Technician A says that automotive primary wire is used to carry low voltages. Technician B says that spark plug wires are secondary wires that carry high voltages. Who is correct?
   A. A only
   B. B only
   C. Both A and B
   D. Neither A nor B

2. Technician A says that most primary wire is made of several strands of copper wire wound together and covered with an insulation. Technician B says there is more resistance in a stranded wire that in the same gauge solid wire. Who is correct?
   A. A only
   B. B only
   C. Both A and B
   D. Neither A nor B

3. Wire size identification standards are being discussed. Technician A says the rule with AWG standard wire size is the higher the number, the larger the wire conductor. Technician B says in the metric system of determining wire size, the smaller the number, the smaller the wire conductor. Who is correct?
   A. A only
   B. B only
   C. Both A and B
   D. Neither A nor B

4. Wires with double color codes are being discussed. Technician A says the first letters indicate the base color of the insulation. Technician B says the second set of letters indicate the color of the tracer. Who is correct?
   A. A only
   B. B only
   C. Both A and B
   D. Neither A nor B

5. Technician A says solderless crimp connectors are used extensibly in automotive circuits because they provide excellent protection against corrosion and dirt. Technician B says solderless crimp connectors are normally crimped to the wire with slip-joint pliers.
   A. A only
   B. B only
   C. Both A and B
   D. Neither A nor B

6. Technician A says solder used for electrical use has a rosin core flux. Technician B says solder with a 60/40 mixture means 60% lead and 40% tin. Who is correct?
   A. A only
   B. B only
   C. Both A and B
   D. Neither A nor B

7. Technician A says a hard-shell connector has 4 wires molded into a connector component. Technician B says a hard-shell connector has 6 wires molded into a connector component. Who is correct?
   A. A only
   B. B only
   C. Both A and B
   D. Neither A nor B

8. Technician A says a weather-pack connector provides for excellent corrosion protection. Technician B says a weather-pack connector has a rubber seal on the terminal end and on the connector half. Who is correct?
   A. A only
   B. B only
   C. Both A and B
   D. Neither A nor B

9. Technician A says metri-pack connectors have a push-to-seat connector. Technician B says metri-pack connectors have a pull-to-seat connector. Who is correct?
   A. A only
   B. B only
   C. Both A and B
   D. Neither A nor B

10. Technician A says a stepped resistor is used to control temperatures in a circuit. Technician B says a rheostat has one fixed end and one movable wiper contact. Who is correct?
    A. A only
    B. B only
    C. Both A and B
    D. Neither A nor B

11. Technician A says a potentiometer uses three wires—one movable wiper and two fixed ends. Technician B says a potentiometer can sense a variable voltage drop. Who is correct?
    A. A only
    B. B only
    C. Both A and B
    D. Neither A nor B

12. Technician A says glass fuses have end cap codes of AGA, AGW, or AGC. Technician B says glass fuses with the end code SFE have the same length but their diameter depends on the current rating. Who is correct?
    A. A only
    B. B only
    C. Both A and B
    D. Neither A nor B

13. Technician A says a maxifuse is a replacement fuse designed as a fusible link in a modern vehicle. Technician B says a pacific fuse element is a replacement fuse designed as a fusible link in a modern vehicle. Who is correct?

 A. A only

 B. B only

 C. Both A and B

 D. Neither A nor B

14. Technician A says all circuit breakers in a vehicle's electrical system are self-resetting. Technician B says all circuit breakers in a vehicle's electrical system are cycling. Who is correct?

 A. A only

 B. B only

 C. Both A and B

 D. Neither A nor B

15. Technician A says an SPST switch has a single input and two outputs. Technician B says an SPST switch can be a momentary switch. Who is correct?

 A. A only

 B. B only

 C. Both A and B

 D. Neither A nor B

16. Technician A says a ganged switch uses multiple wipers hooked together. Technician B says a ganged switch has one input and multiple outputs. Who is correct?

 A. A only

 B. B only

 C. Both A and B

 D. Neither A nor B

17. Technician A says a relay's contact points are energized by a small internal coil. Technician B says a relay's contact points are energized electro-magnetically. Who is correct?

 A. A only

 B. B only

 C. Both A and B

 D. Neither A nor B

18. Technician A says a sealed beam headlamp can have a single internal filament. Technician B says a sealed beam headlamp can have two internal filaments. Who is correct?
    A. A only
    B. B only
    C. Both A and B
    D. Neither A nor B

19. Technician A says a halogen headlamp uses a halogen filled inner bulb sealed in a glass housing. Technician B says in a composite halogen headlamp, the entire headlamp assembly needs to be replaced when burned out. Who is correct?
    A. A only
    B. B only
    C. Both A and B
    D. Neither A nor B

20. Technician A says automotive capacitors generate voltage spikes in a DC electrical circuit. Technician B says a standard capacitor unit is a farad. Who is correct?
    A. A only
    B. B only
    C. Both A and B
    D. Neither A nor B

6 Meters and Measuring Devices

Introduction

An automotive technician working on electrical and electronic circuits must be familiar with the proper use and limitations of instruments that measure electrical properties. When a test instrument is hooked up correctly, it provides the "eyes" for the technician for what's inside the circuit. In this chapter, we will study several types of electrical meters, their common names, and a brief look at how each is used.

Objectives

When you complete this chapter you should be able to:

- ❑ Describe the design characteristics and limitations of an analog multimeter.
- ❑ Describe the design characteristics and limitations of a digital multimeter.
- ❑ Demonstrate how to properly connect a multimeter to a circuit to measure system voltage.
- ❑ Demonstrate how to properly connect a multimeter to a circuit to measure system amperage.
- ❑ Demonstrate how to properly connect a multimeter to a circuit to measure system resistance.
- ❑ Demonstrate the proper use of a test light to measure for voltage in a circuit.
- ❑ Describe the design characteristics of a digital oscilloscope.
- ❑ Explain the proper use and limitations of a digital oscilloscope.
- ❑ Describe the main characteristics found on most computer scan testers.

Analog Meters

For many years, one of the more traditional meters found in automotive shops for basic electrical diagnosis has been the **analog meter.** This type of meter uses a pointer needle that moves across the face of a printed scale (Figure 6–1), much the same as a speedometer or standard dash gauges.

Analog meters operate by a **D'Arsonval movement** (Figure 6–2). This movement is also called a **moving coil meter,** where a horseshoe style permanent magnet surrounds a small coil of wire attached to a pointer. When current flows through the coil, it interacts with the magnetic field of the magnet and causes the pointer needle to move. The direction of pointer movement is determined by the direction of current flow through the coil.

Analog meters have a **rotary switch** that allows the meter to be used for multiple ranges of volts, ohms, and amps. This type of meter is also known as a **multimeter** or **VOM,** which is the abbreviation for volt-ohm-

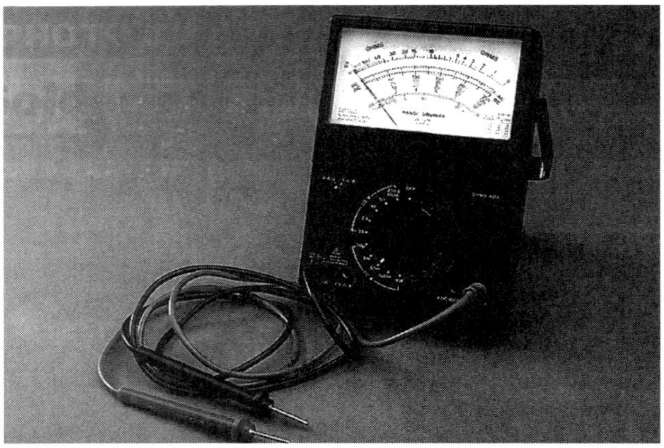

FIGURE 6–1 A simple analog meter.

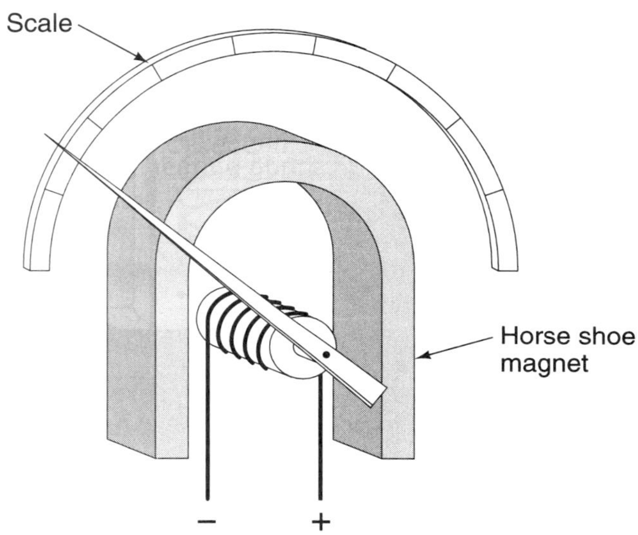

FIGURE 6–2 A D'Arsonval movement used in an analog meter.

milliamp meter. Figure 6–3 shows an example of an analog multimeter's scale face. Reading this face can sometimes be confusing for a technician, and can lead to a misdiagnosis. The top scale of this meter represents ohms of resistance. When the rotary switch is set on the Rx1 setting, the resistance reading of 1.6 Ω is read directly from the meter scale. If the rotary switch is set on the Rx1000 scale, the reading must be multiplied by 1,000, or 1.6 × 1,000 = 1600 Ω. Anytime the ohms value is changed with the rotary switch, the meter must be re-calibrated by touching the meter leads together and re-setting the pointer to zero on the ohms scale with the adjustment knob.

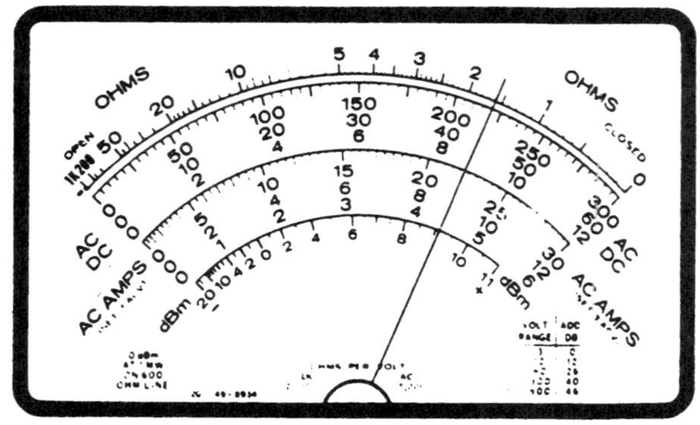

FIGURE 6–3 A typical analog meter face with multiple functions.

The second scale on the meter in Figure 6–3 represents **AC (alternating current)** or **DC (direct current)** voltage. The scale has three sets of numbers depending on the location of the rotary switch. The meter reading would be 220 volts on the 0-300 scale, 44 volts on the 0-60 scale, and 8.8 volts on the 0-12 scale. The third scale represents AC amps, and the last scale is for measuring in decibels. Both the AC amp and decibel scales require the use of special attachments.

Analog meters have three service limitations that the technician needs to keep in mind. The first is choosing the proper range before taking a reading with the meter. If the range selection is too *low,* the meter pointer will peg-over to the fullest extent and possibly damage the pointer mechanism. The safest method is to choose a *higher* range initially, and then move to a lower scale as needed to get the most accurate reading.

The second service limitation of an analog meter is the **polarity** of the test leads in relation to the circuit polarity. The analog meter can measure voltage or amperage in one direction only. If the meter leads are accidentally reversed, the pointer will deflect to the left and possibly damage the pointer mechanism. To help prevent this, meter leads are normally labeled and color coded black for negative, and red for positive (Figure 6–4).

The last service limitation of an analog meter is its internal resistance (**impedance**). This is the meter's internal ability to limit the current increase of the electrical circuit being measured. Most automotive manufacturers specify that sensitive solid state circuits like digital dash clusters, and engine control sensors, should be tested only with a meter that has 10 megaohm (million), or higher, internal impedance. The high impedance means the meter will draw very little current from the

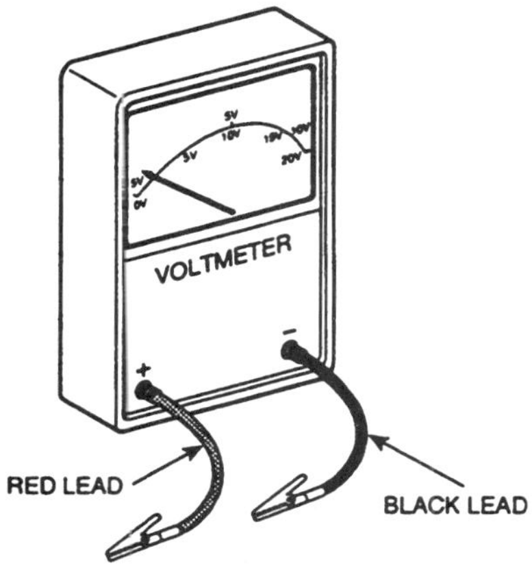

FIGURE 6–4 Analog meters are polarity sensitive.

(Courtesy of Ford Motor Company)

circuit or component being tested. Analog meters *do not* have this high level of internal impedance and should not be used for current sensitive electronic circuits.

Digital Multimeters

For diagnosing today's automotive electrical and electronic systems, the **digital multimeter (DMM)** has become a necessary tool for the automotive technician. Another name for a digital multimeter is a **digital volt-ohm meter (DVOM).** A quality digital meter (Figure

FIGURE 6–5 This digital multimeter has eighteen functions.

(Courtesy of TIF Instruments, Inc.)

6–5) has a 10 megaohm internal impedance, and uses electronic circuitry and a liquid crystal display to provide more precise readings than analog meters.

Some lower-cost digital multimeters require scale ranges to be set manually (Figure 6–5). If an out-of-range reading is taken with this type of meter, an **overload symbol (OL)** is displayed on the digital face, or the display will flash digital numbers. The most popular DMM types used by technicians have an **auto-ranging** feature which automatically adjusts to the best range for the circuit being measured. Unlike an analog meter, an auto-ranging digital meter will zero itself when performing resistance readings. Figure 6–6 is an example of an auto-ranging DMM.

Some of the key features of a high-quality DMM are shown on the meter in Figure 6–6. When the button in the center of the rotary dial is held for 2 seconds, the meter is in the **touch-hold** setting. While in this setting, anytime a reading is taken on a circuit, the reading is displayed on the meter face and held there after the test leads are removed. This can be a handy feature when taking an electrical reading in a tight area such as under the dash panel. A high-quality DMM will also have current jacks that are circuit protected by a fuse if overloaded. Overloading a current measurement is a common error made by even the most experienced technician when testing circuits where the amperage is unknown.

The display face of the meter in Figure 6–7 indicates several important features. Touch-hold is indicated by a small circle with a center dot. If the meter leads are reversed the negative polarity sign will show to the left of the reading. If you recall on an analog meter, negative polarity could damage the meter. This is not the case with a digital meter. For example, if the reading is normally 12 volts on the meter face, with the leads reversed it would be −12 volts. The low battery indicator lets the technician know when there is eight hours or less of meter battery life remaining. The **analog bar graph** along the bottom of the face, acts as a simulated analog meter. This is helpful because digital numbers react too rapidly to changes in the circuit being checked, and the sweep of the analog bar graph react more slowly.

The rotary switch on a DMM controls its features similar to an analog meter. The symbols on the face of the meter are called **icons** (Figure 6–8).

The description of the icons shown on Figure 6–8 are:

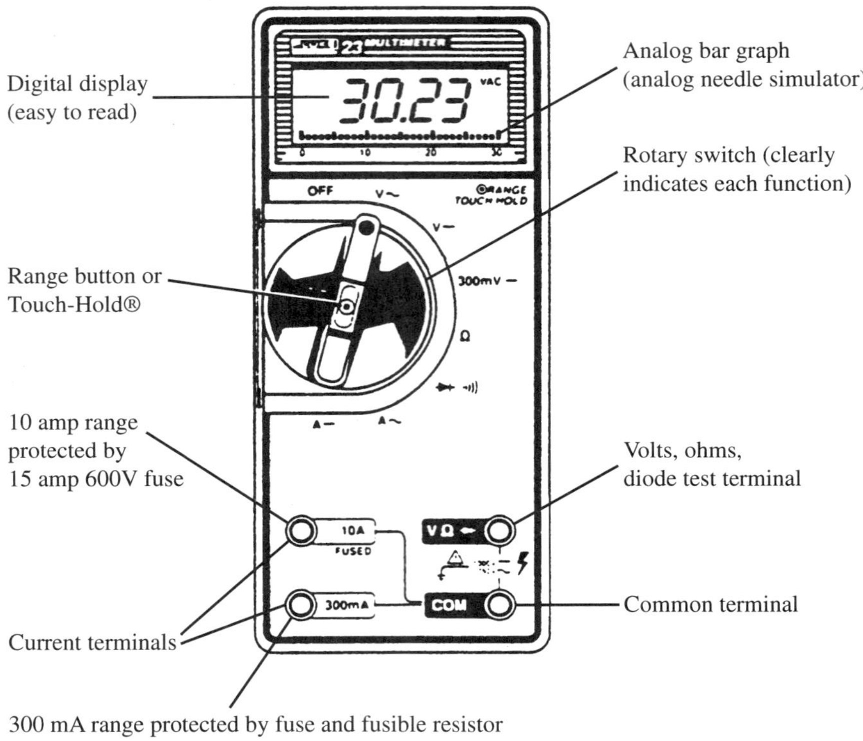

FIGURE 6–6 Typical functions on an auto-ranging digital multimeter.

(Courtesy of The Fluke Corporation)

FIGURE 6–7 A digital multimeter liquid crystal display (LCD).

(Courtesy of The Fluke Corporation)

Volts AC—volts alternating current. Used in household power, generator units, ABS brake wheel speed sensors, and vehicle speed sensors.

Volts DC—volts direct current. Used in most automotive electrical circuits and chemically produced in batteries.

Millivolts DC—volts direct current of 1/1000 of a volt scale.

Ohms (resistance)—normally shown by the Greek omega symbol Ω for resistance. An internal battery in the meter supplies the power needed to operate in this mode.

Continuity—continuously beeps when attached to a circuit of 150 Ω or less.

Diode test—supplies a small voltage to check the continuity of a diode, which is an electronic check valve.

Amps AC—amps alternating current. Used in household circuits and some automotive computer sensors.

Amps DC—amps direct current. Used in most automotive electrical circuits to measure current flow.

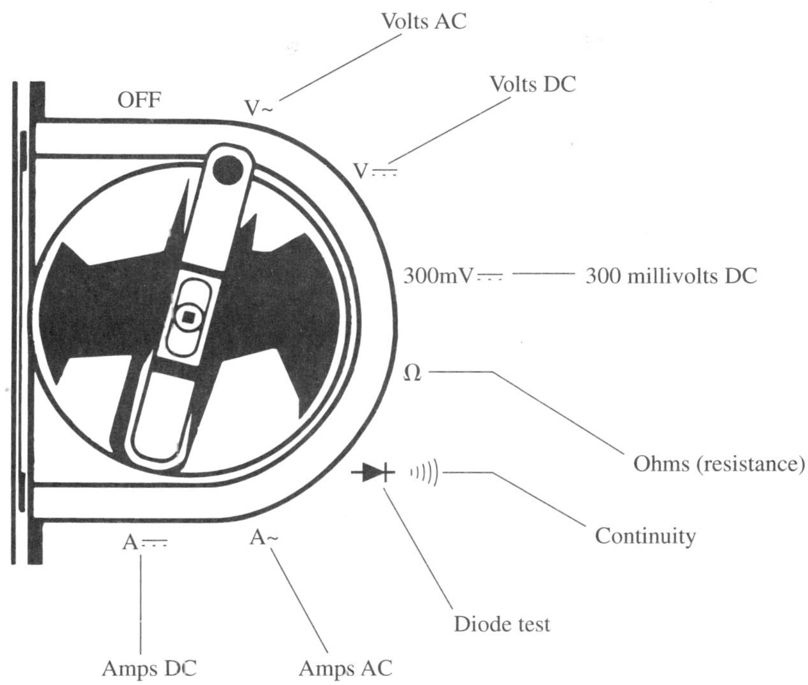

FIGURE 6–8 Digital multimeter rotary switch and icons.
(Courtesy of The Fluke Corporation)

Measuring Voltage

A voltmeter (analog or digital) is used to measure voltage potential across an electrical circuit's connections, terminals, switches, wiring, and components. There are three rules that MUST be followed when using a voltmeter to measure voltage potential.

Rule # 1 An electrical circuit *must* be energized to measure voltage potential with a voltmeter.

Rule # 2 A voltmeter is always hooked in *parallel* across the component being measured (Figure 6–9).

Rule # 3 A voltmeter is NEVER intentionally placed in *series* (in-line) with the component being measured (Figure 6–10).

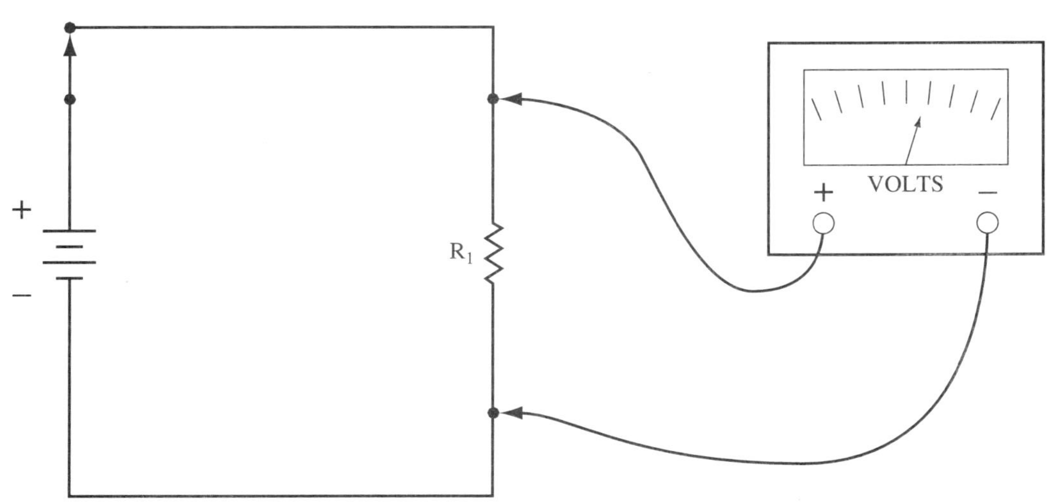

FIGURE 6–9 Meter leads are placed in parallel across the load component of a live circuit.

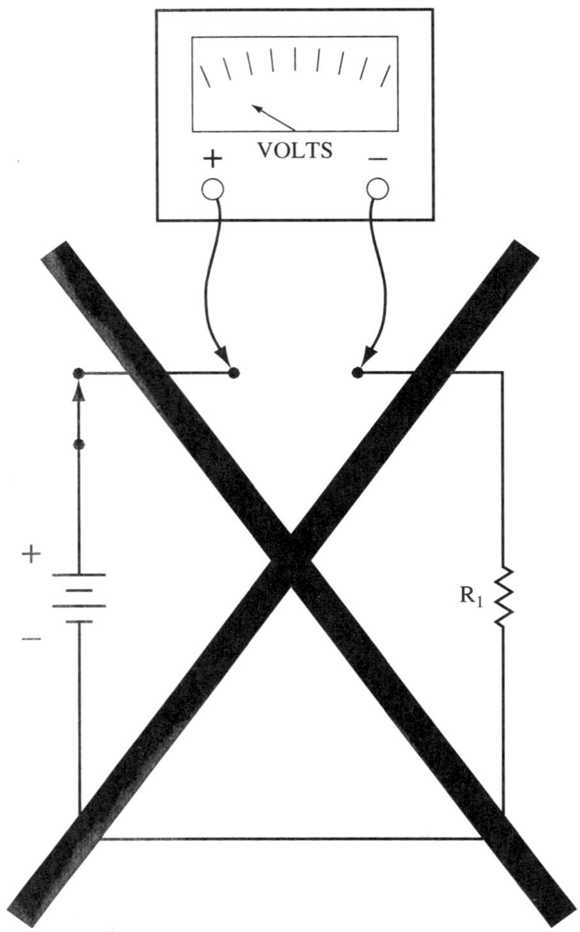

FIGURE 6–10 Never place a voltmeter in series with the load component(s) of a live circuit.

In an electrical circuit, voltage is the pressure that pushes the current through the circuit. Some of the source voltage intended for the circuit components is lost due to resistance through wiring and connections. This loss is called **voltage drop,** and can be measured with a voltmeter. In a circuit, the sum of any voltage drop (Figure 6–11) through the connections, wiring and load components *always equals* the source voltage.

The longer and more complex a circuit is, the more chance there is for resistance to develop in connections, this may drop the voltage to a level that will affect the proper operation of the load component(s). The following general specifications are used when checking a circuit for voltage drops.

Connection Point	Allowed Voltage Drop
Ground Connections	0.1 volts
Switch Contacts	0.2 to 0.4 volts
System Wiring	0.2 volts
Terminal Connections	0.0 volts
Computer Sensor Connections	0.0 to 0.05 volts

Figure 6–12 is an example of a starter circuit being tested for a voltage drop. This circuit is designed to operate the starter motor on the voltage supplied by the battery. In this starter circuit, there are ten areas to check for a voltage drop—three battery cables, six connections, and the solenoid switch. Excessive voltage drop in any one connection or series of connections, could effect the proper operation of the starter motor.

Measuring Amperage

An ammeter (analog or digital) is used to measure the flow of electrons (current flow) in an electrical circuit. There are three rules that MUST be followed when measuring current.

Rule # 1 An electrical circuit MUST be energized to measure the current flow.

Rule # 2 An ammeter is *always* hooked up in *series* (in-line) with the component or circuit being measured (Figure 6–13).

Rule # 3 An ammeter is NEVER placed directly in *parallel* with the circuit or component (Figure 6–14). An ammeter connected in parallel will be damaged or blow a fuse.

Because the ammeter must be placed in series with the circuit, this normally requires disconnecting a wire leading up to, or from the component. This may not be easily performed in some electrical circuits. In cases where the ammeter cannot be placed in series, or the meter's designed current capacity is exceeded, an **inductive pick-up** adapter is available (Figure 6–15). This adapter senses the amount of current flow by measuring the strength of the magnetic field surrounding the conductor. The adapter is not accurate for measuring low current flow.

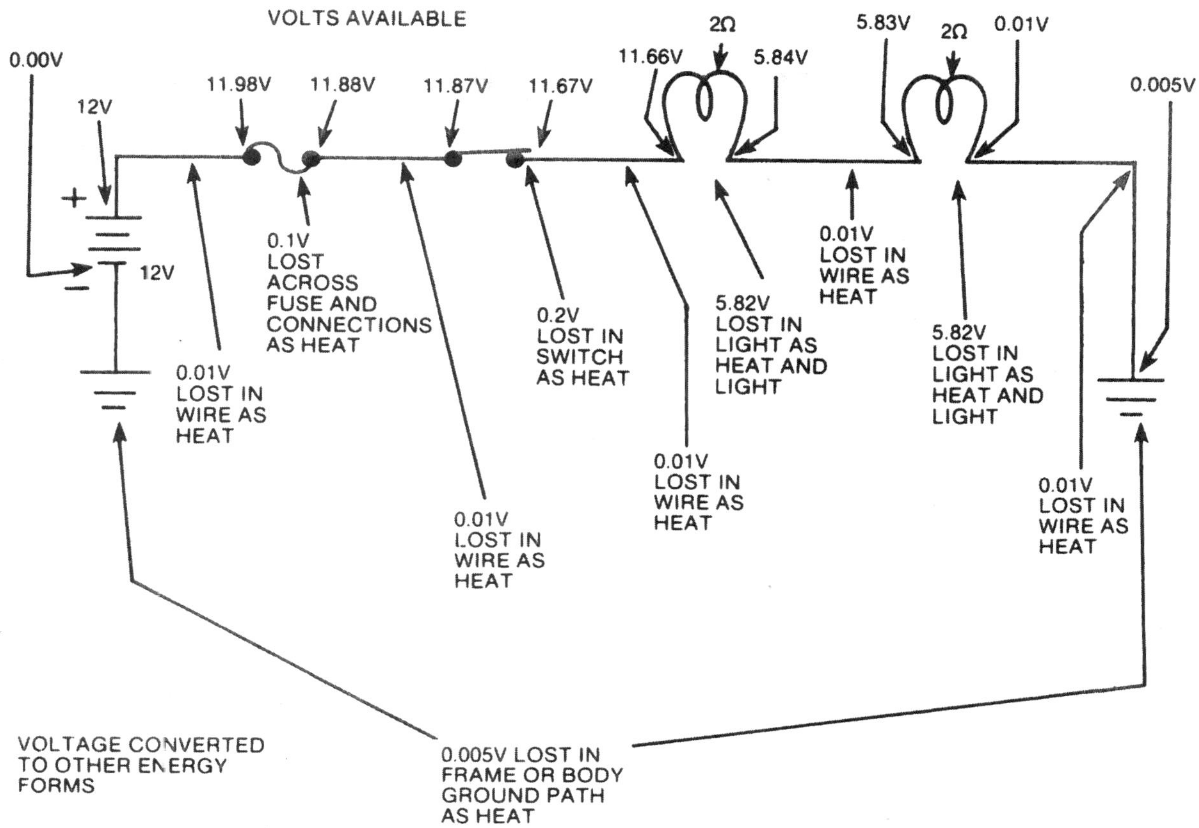

FIGURE 6–11 Measuring voltage drops of connections and components of a circuit.

Most hand-held analog or digital multimeters have a current limit of 2 to 20 amps maximum, so an inductive current clamp would be necessary when measuring several hundred amps such as the starter circuit (Figure 6–16).

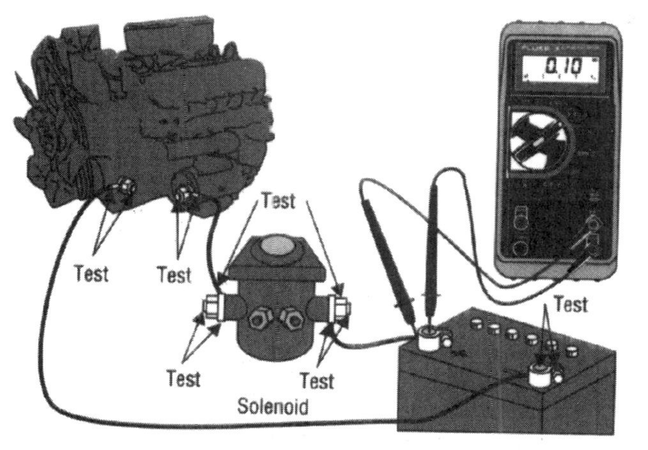

FIGURE 6–12 Testing for excessive voltage drop in a starter circuit.

(Courtesy of The Fluke Corporation)

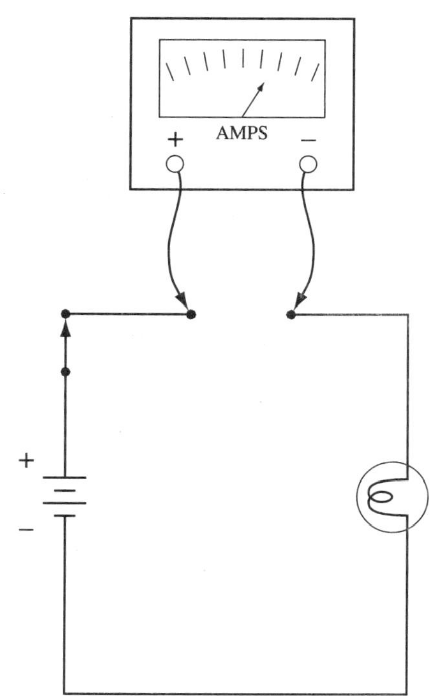

FIGURE 6–13 Ammeter leads are placed in series with the load components of a live circuit.

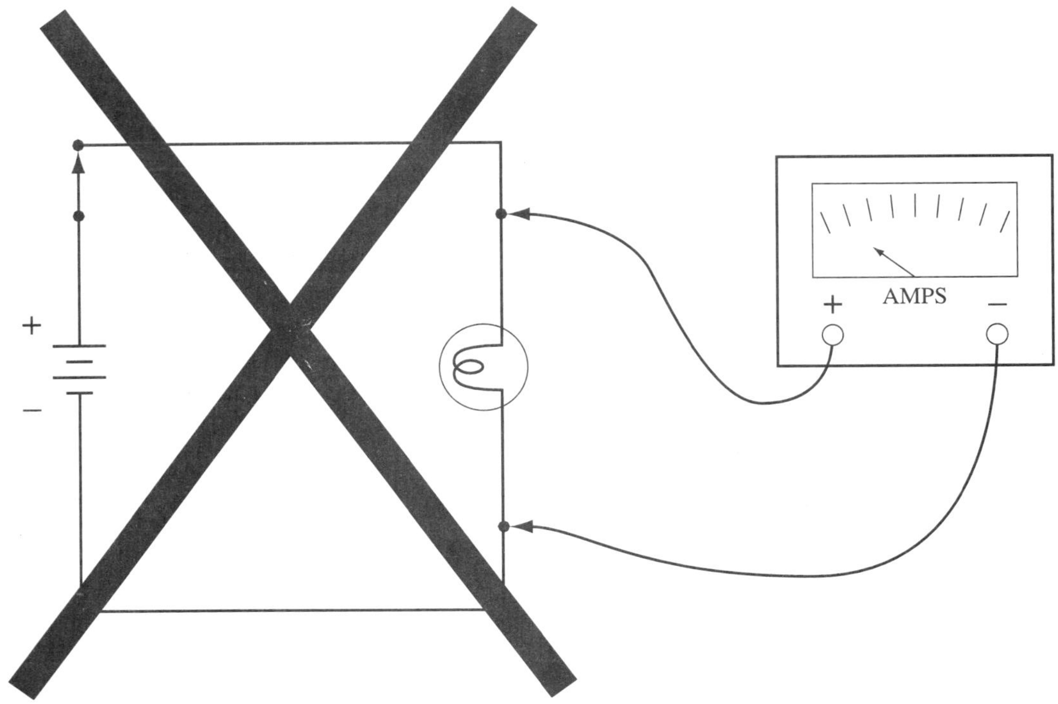

FIGURE 6–14 Never place an ammeter in parallel with the load component(s) of a live circuit.

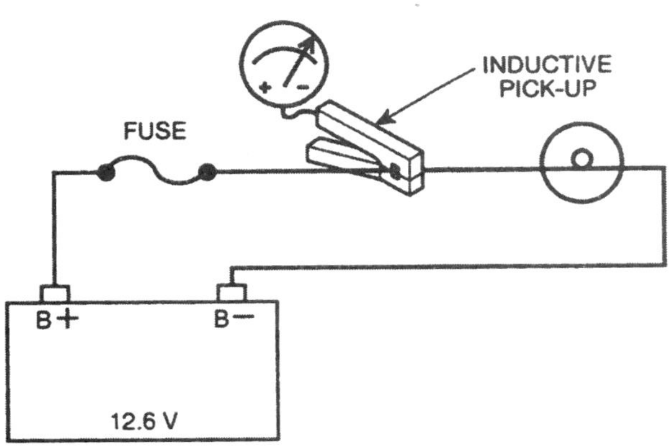

FIGURE 6–15 Using an inductive pick-up to measure current flow.

(Courtesy of Ford Motor Company)

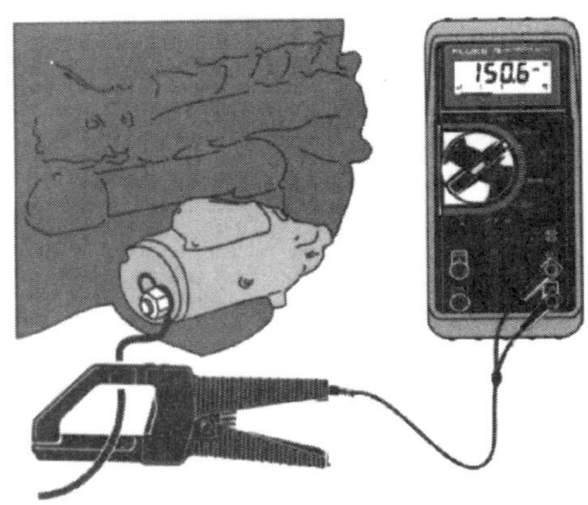

FIGURE 6–16 Measuring starter current draw with an inductive pick-up clamp.

(Courtesy of The Fluke Corporation)

Measuring Resistance

An ohmmeter (analog or digital) is used to measure the resistance, in ohms (Ω), in a electrical circuit. There are three rules that MUST be followed when measuring resistance.

Rule # 1 An electrical circuit MUST be *disconnected* from its power source in order to measure the circuit or component resistance. The meter uses its own battery as a power source for measuring resistance.

Rule # 2 An ohmmeter is hooked up in *parallel* across the component or circuit being measured (Figure 6–17).

Rule # 3 An ohmmeter is NEVER placed in *series* (in-line) with the component being measured (Figure 6–18).

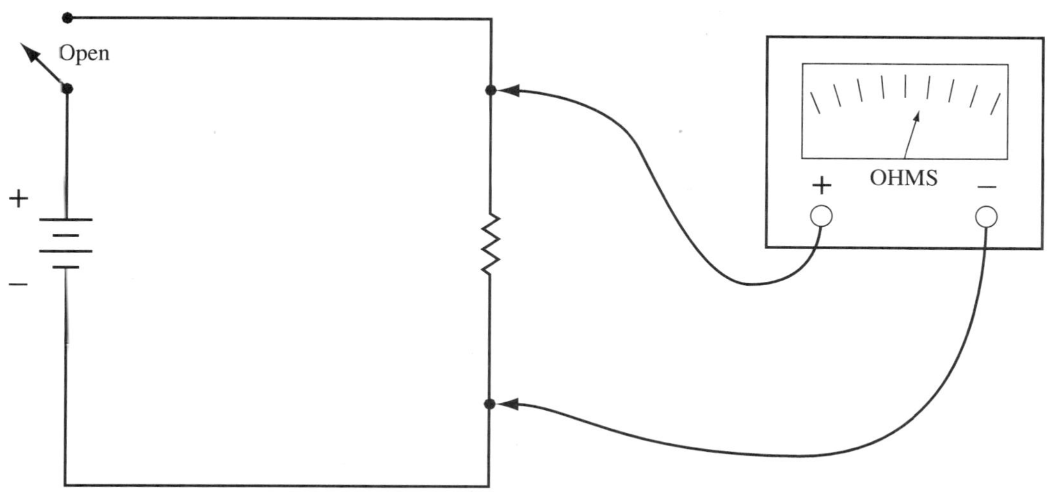

FIGURE 6–17 Ohmmeter leads are placed in parallel across the load component of a dead circuit.

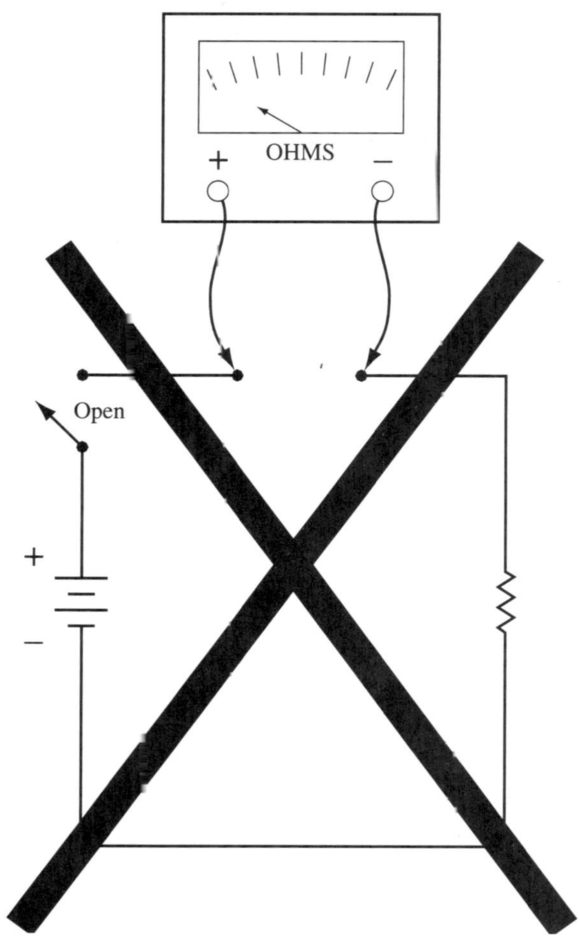

FIGURE 6–18 Never place an ohmmeter in series with the load components of a dead circuit.

When there are multiple load components in a circuit that share common connections, each individual component needs to be isolated from the rest of the circuit to get an accurate resistance reading (Figure 6–19). In this circuit, if the component R_3 is not isolated, the resistance reading will be for all three of the load components.

Summary of Meter Use

Meter	Connection	Do Not	Power
Voltmeter	Place in parallel	Place in series	ON
Ammeter	Place in series	Place in parallel	ON
Ohmmeter	Place in parallel to isolated component	Place in series	OFF

Test Lights

There are two conventional types of bulb test lights used by automotive technicians to diagnose simple electrical circuits—a non-powered test light and a self-powered test light.

Non-powered Test Light

A **non-powered test light** is one of the simplest tools for checking for voltage in an electrical circuit. It consists of a pointed probe, a 12 volt light bulb, and a flexible ground lead (Figure 6–20). It is used merely to check for

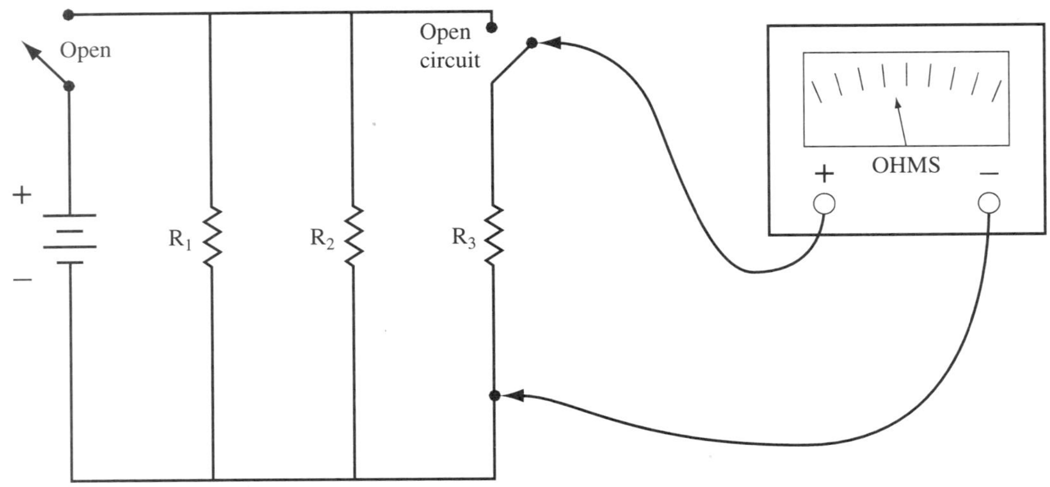

FIGURE 6–19 Always isolate the component from the circuit to obtain a correct resistance reading.

the presence of voltage in a circuit. When the bulb is lit, it indicates a voltage is present (Figure 6–21).

CAUTION *A test light should not be used to test for power in a solid state electronic circuit. Voltage in a sensitive circuit should only be checked with a voltmeter with a 10 megaohm (10 million ohm), or higher internal impedance.*

Self-powered Test Light

A **self-powered test light,** also known as a **continuity light,** is used to check the continuity of *open* circuits. It looks similar to a regular test light, except it contains a small internal battery attached between a light bulb and two leads (Figure 6–22).

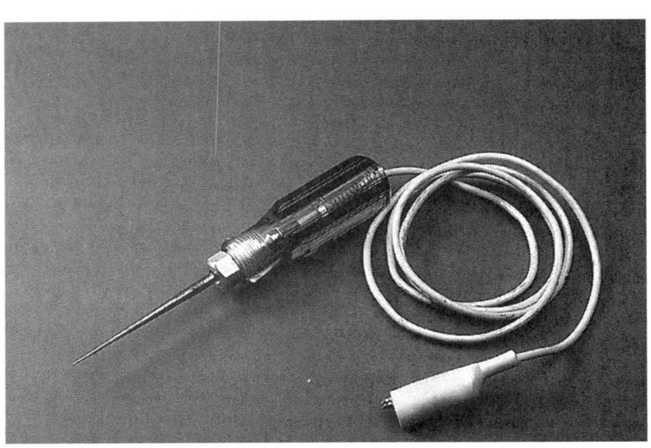

FIGURE 6–20 A typical non-powered test light.

With a self-powered test light, if the two leads are touched together or connected through a low resistance circuit (Figure 6–23), the bulb will light because doing so completes the circuit between the bulb and the test light's internal battery. The same caution that applies to a non-powered test light also applies to a self-powered light.

Automotive Oscilloscopes

Over the past several years, oscilloscopes have become a major piece of diagnostic equipment in repair shops for diagnosing ignition or charging system problems. The

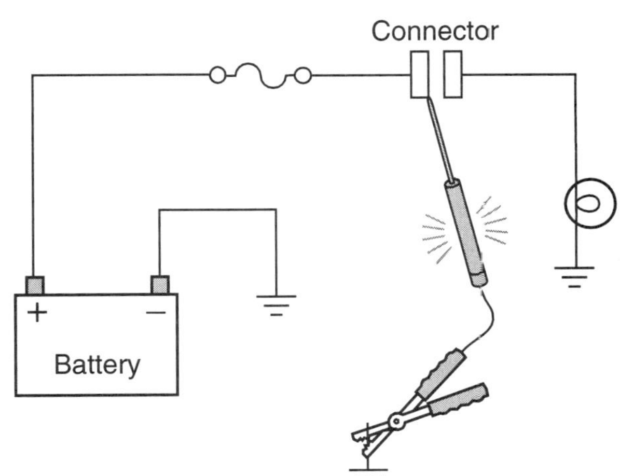

FIGURE 6–21 Non-powered test light will glow when voltage is present.

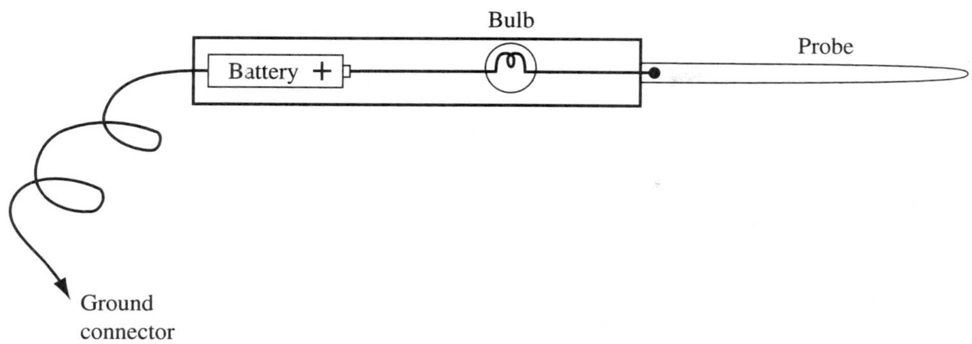

FIGURE 6–22 A typical self-powered test light.

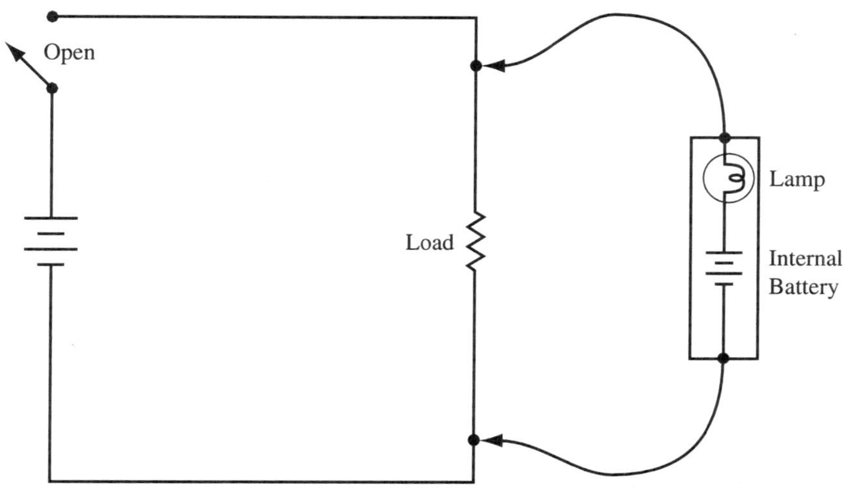

FIGURE 6–23 Using a self-powered test light to check continuity in an open circuit.

scope is basically a visual image voltmeter with an extremely fast reaction time to changes in voltage over a period of time. Two types of oscilloscopes are used by technicians to perform electrical diagnosis.

Analog Oscilloscope

The first type is an **analog oscilloscope.** This scope is normally too heavy to be hand carried. It reads and displays electrical signals as they occur in the circuit. They are also known as *real-time scopes* where you see on the screen what is presently occurring at the point being measured. This is helpful, and updates are continuous though somewhat slow. Figure 6–24 is an example of a small semi-portable analog scope. Some of these scopes are encased in large roll-around cabinets.

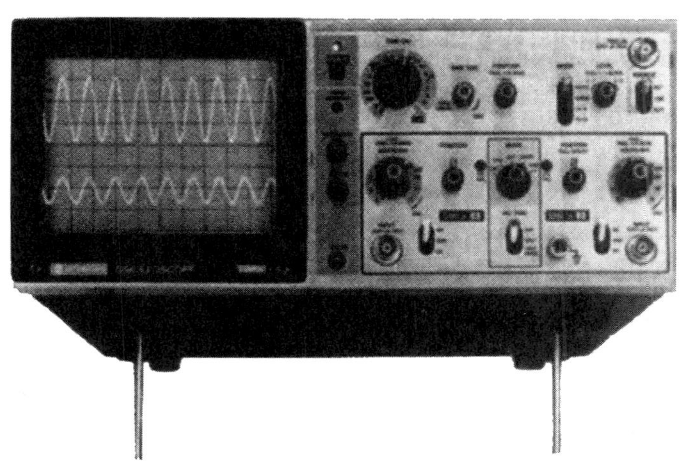

FIGURE 6–24 A typical semi-portable analog oscilloscope.

Digital Storage Oscilloscope

A **digital storage oscilloscope (DSO)** (Figure 6–25) is highly portable, making it popular with technicians because it can be taken in the vehicle on test drives. A DSO is also known as a scope-meter or lab-scope because it has selections for digital multimeter functions. A DSO continuously converts the input voltage signal into digital information, and then stores it into short term memory for screen display or to transfer to other diagnostic equipment. Some DSO's sample the input electrical signal at up to 40 billionths of a second which is 47,000 times faster than analog scopes. The fast capture rate allows the technician to *freeze* the captured signal of input and output sensors or actuators over a period of time for a step-by-step playback and analysis at a later time. The freeze feature combined with the sampling speed allows the DSO to display glitches that cannot be seen by analog scopes or multimeters.

The screen on a analog or digital scope is divided into grid squares that are used for measurement purposes. Voltage, or **amplitude,** measurement is represented by the vertical Y axis, and time measurement is represented by the horizontal X axis (Figure 6–26).

The range parameter setting on the scope determines the voltage reading for each grid line from the bottom edge to the top, and the time division reading per grid line from the left to the right. For example, in Figure 6–27 the scope is set on 2 volts per grid division, and 30 mS (milliseconds) per time division. The result would be a maximum of 8 volts above the zero line and 8 volts below the zero line, for a total of 16 volts that can be displayed on the screen. Since each of the ten time divisions equal 30 mS, then there would be a total of 300 mS displayed. The ideal range setting depends on how precise the reading need to be.

The picture produced on a scope screen is called a **waveform** or **trace.** Because a scope displays a voltage measurement over a period of time, the waveform moves from the left to the right on the grid screen. A typical AC

FIGURE 6–25 A portable digital storage oscilloscope (DSO).

(Courtesy of The Fluke Corporation)

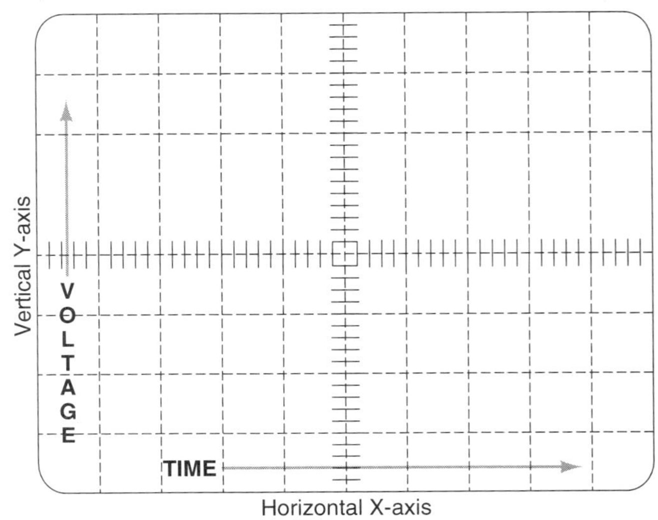

FIGURE 6–26 Grids on a scope screen which serve as a time and voltage reference.

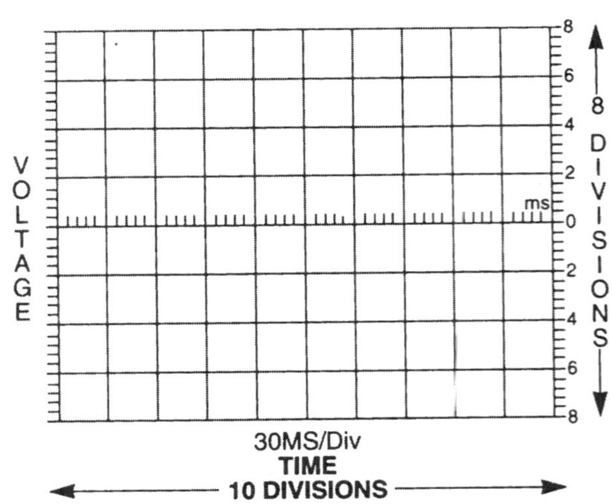

FIGURE 6–27 Setting the appropriate range selection for precise measurement.

(Courtesy of Ford Motor Company)

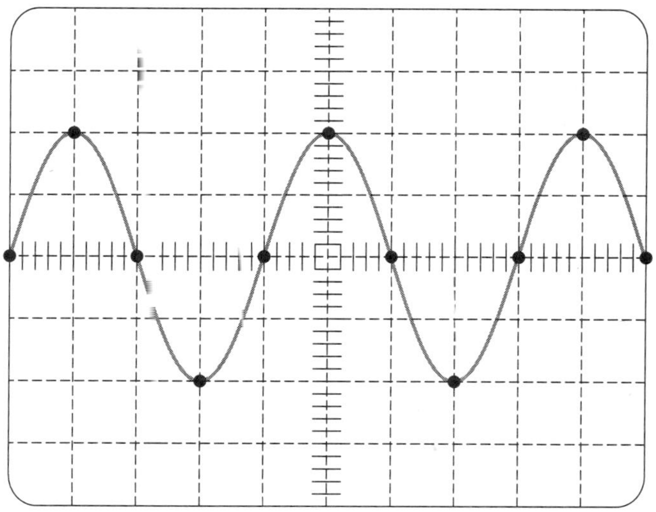

FIGURE 6–28 A typical AC (alternating current) sine wave.

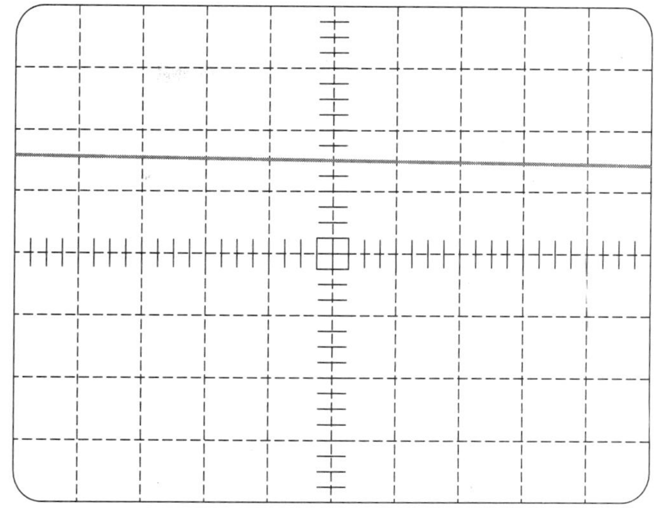

FIGURE 6–29 A typical DC (direct current) constant waveform pattern.

(alternating current) **sine wave** is shown in Figure 6–28. One complete sine wave cycle shows the voltage moving from the zero line to its positive peak, and then down through zero to its negative peak and back to zero again. If the rise and fall from positive to negative are the same, then the wave cycle is considered to be **sinusoidal.** The number of cycles that occur per second is the frequency of the signal.

A typical DC (direct current) waveform pattern is shown in Figure 6–29. The DC voltage waveform will appear as a constant straight line across the screen at a specific voltage. Any changes in the voltage would raise or lower the straight line.

Analog or digital oscilloscopes that have the ability to display two waveforms at a time are known as **dual-trace** oscilloscopes (Figure 6–30). Observing an abnormal waveform above or below a good waveform can be helpful for efficient diagnosis.

Some of the modern hand-held DSO's have technician-friendly controls with preset programs. The DSO in Figure 6–31 can display four voltage waveforms or up to six multimeter functions at a time.

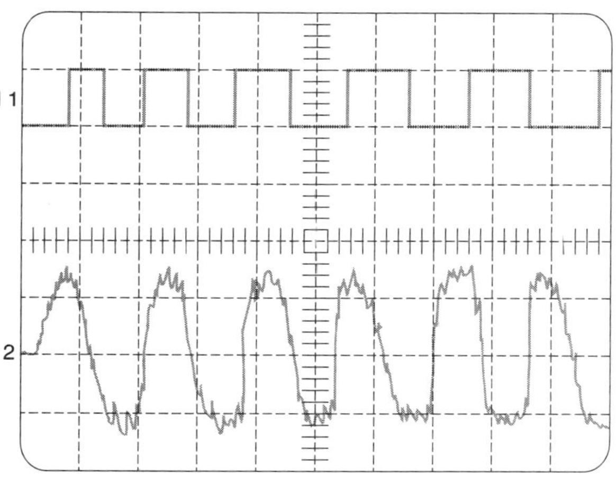

FIGURE 6–30 The benefit of a dual-trace scope is showing two waveforms for comparison.

Analog Voltage Signal

An analog voltage signal produced on a DSO screen shows a variable voltage within a specific range and time period (Figure 6–32). A common vehicle sensor producing an analog signal is an anti-lock brake (ABS) wheel speed sensor. As the wheel speed increases or decreases the analog sine wave will vary in frequency and amplitude (height) (Figure 6–33).

Digital Voltage Signal

A typical digital voltage signal is either ON (high) or OFF (low). When a signal in an electronic circuit is turned ON and OFF, the wave signal is known as a **square wave** (Figure 6–34). As the voltage is turned on (leading edge), the waveform line increases to a specified ON voltage and remains constant until the circuit is shut off (trailing edge), and the voltage decreases (Figure 6–35) to its OFF voltage.

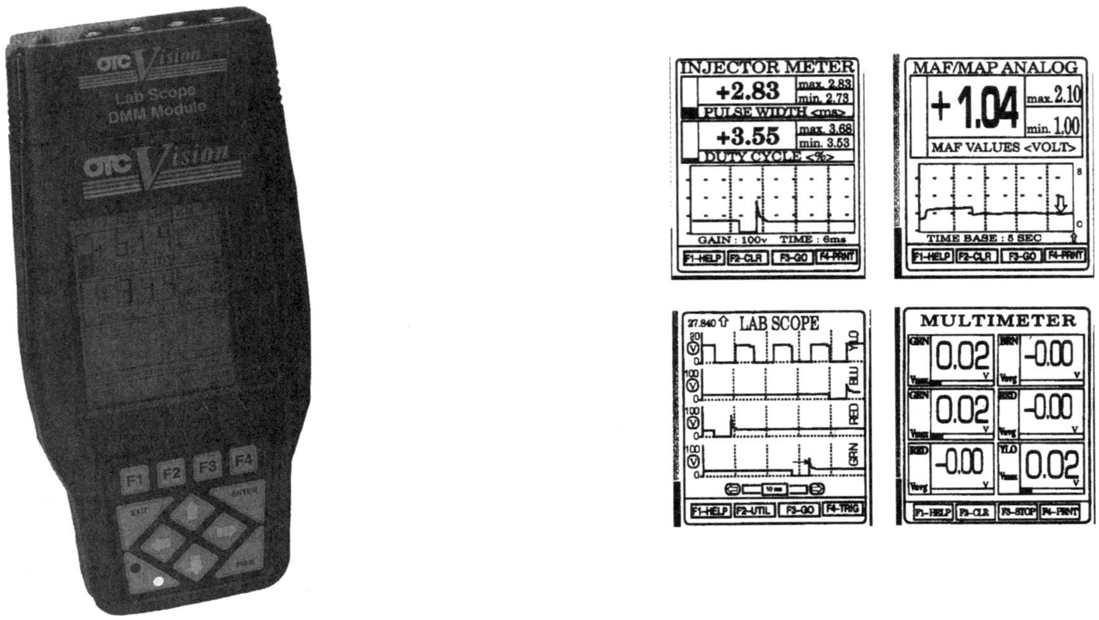

FIGURE 6–31 A DSO with display of four voltage waveforms or six multimeter functions.

(Courtesy of SPX Corporation)

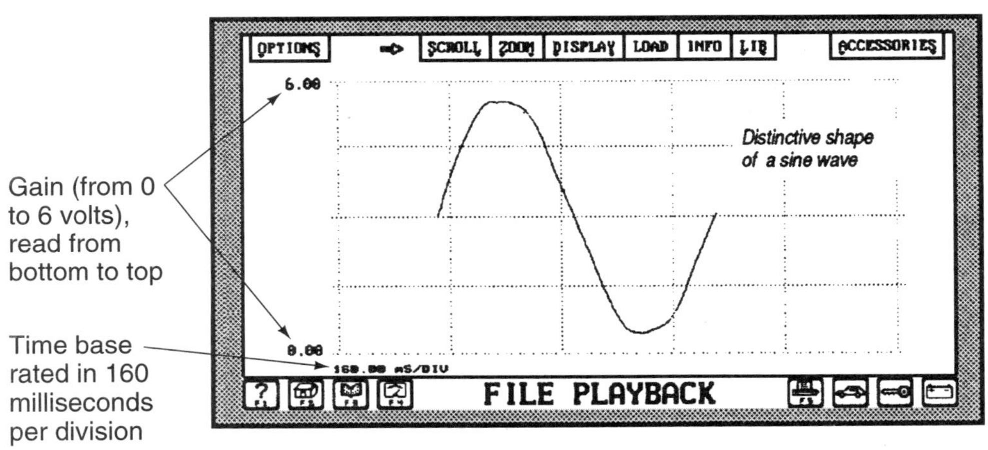

FIGURE 6–32 Analog voltage signal on a DSO screen.

(Courtesy of Snap-On Diagnostics)

The amount of time the voltage signal in a digital circuit is ON (or OFF) in a digital circuit is known as the **pulse width.** The distance between the leading edge of one digital signal and the leading edge of the next signal is one cycle. The number of cycles within a defined period of time (usually one second) is known as the **frequency** of the circuit. The amount of ON time versus the OFF time in a digital signal is called its **duty cycle.** If the component's ON and OFF time are equal, the duty cycle is 50% (Figure 6–36). If a component has a 75% duty cycle, it is ON for 75% of the time, and OFF for 25% of the time. Some computer control outputs will vary the pulse width while keeping the frequency constant. This is known as **pulse width modulation.**

Computer Scan Testers

The introduction of computer controls for the various vehicle systems listed in Figure 6–37, brought with it the need for diagnostic equipment capable of troubleshooting these complex systems. The best tool to communi-

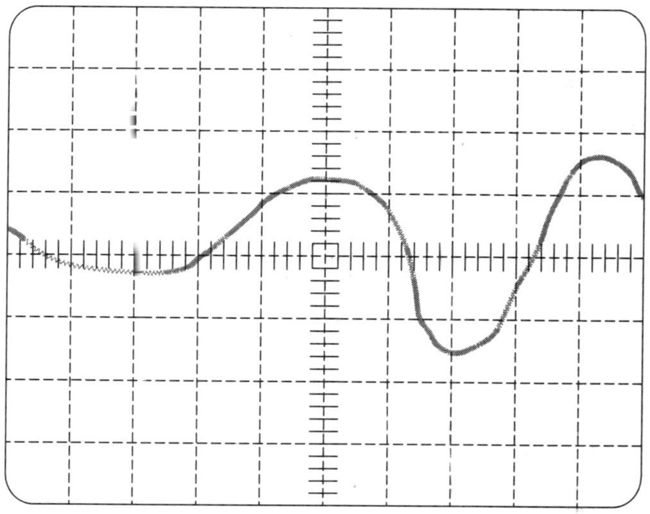

FIGURE 6–33 An AC signal from an ABS wheel spin sensor.

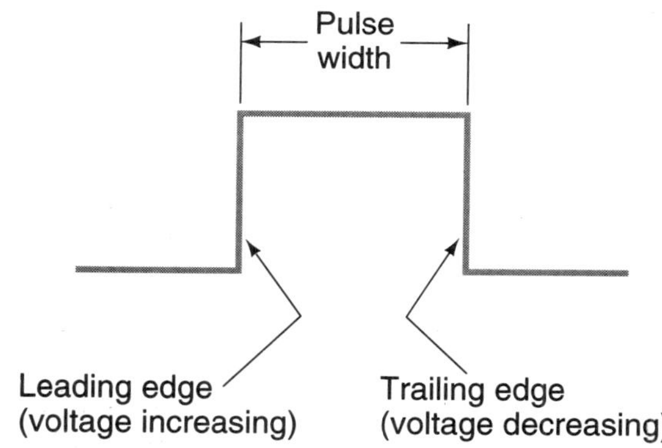

FIGURE 6–35 The component of time in a digital signal is called pulse width.
(Courtesy of Snap-on Diagnostics)

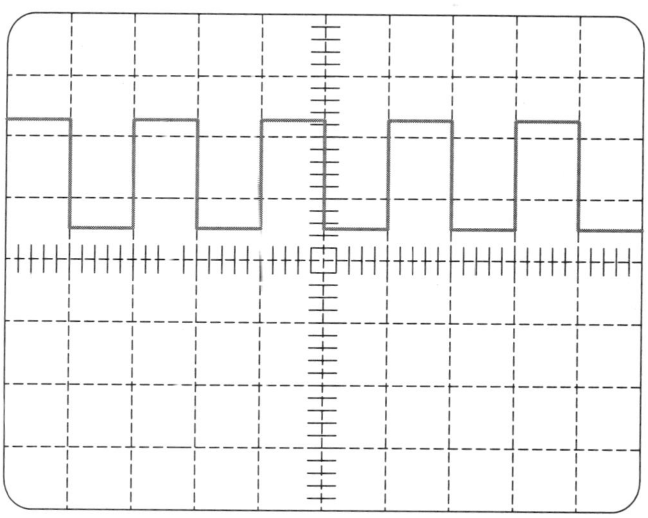

FIGURE 6–34 A typical square wave voltage signal.

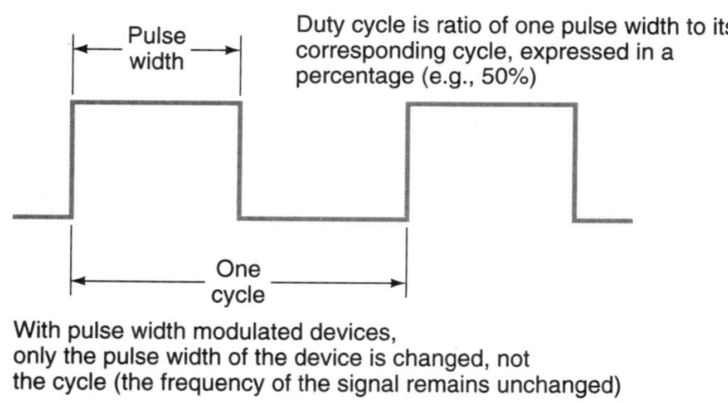

FIGURE 6–36 The relationship between ON/OFF time in a digital signal is called the duty cycle.
(Courtesy of Snap-on Diagnostics)

cate with the vehicle's computer and access stored diagnostic trouble codes is the **scan tool** (Figure 6–38).

A scan tool is a portable microprocessor-based instrument designed to connect to the vehicle's computer, through a diagnostic link connector (DLC) (Figure 6–39), to monitor the system's operation. Typical scan tools use a plug-in **programmable read-only memory (PROM)** chip that is specific to the vehicle being tested (Figure 6–40).

The benefit of having a diagnostic scan tool that is portable is being able to take it on a road test with the vehicle. This allows the technician to record the serial data from the on-board computer at the exact time that any fault is occurring. After the test drive, the information can be viewed on the scan tool's display screen (Figure 6–41), or interfaced with a printer or personal computer's (PC) software (Figure 6–42 and Figure 6–43).

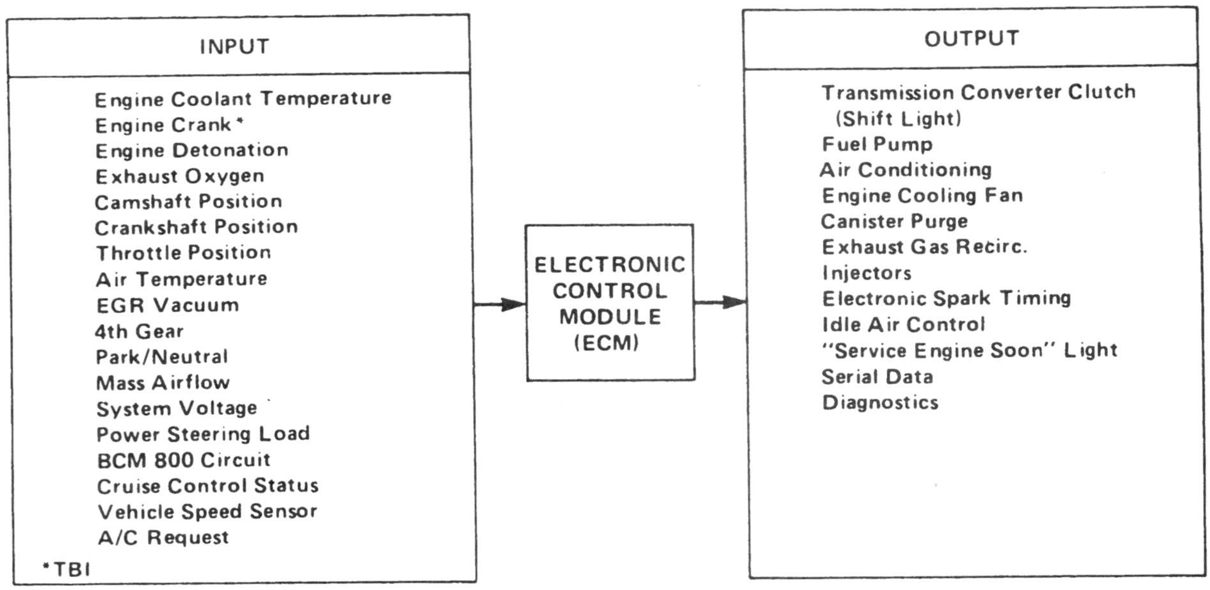

FIGURE 6–37 Computer-controlled inputs and outputs for a GM advanced fuel injection system.

(Courtesy of General Motors Corporation, Service Technology Group)

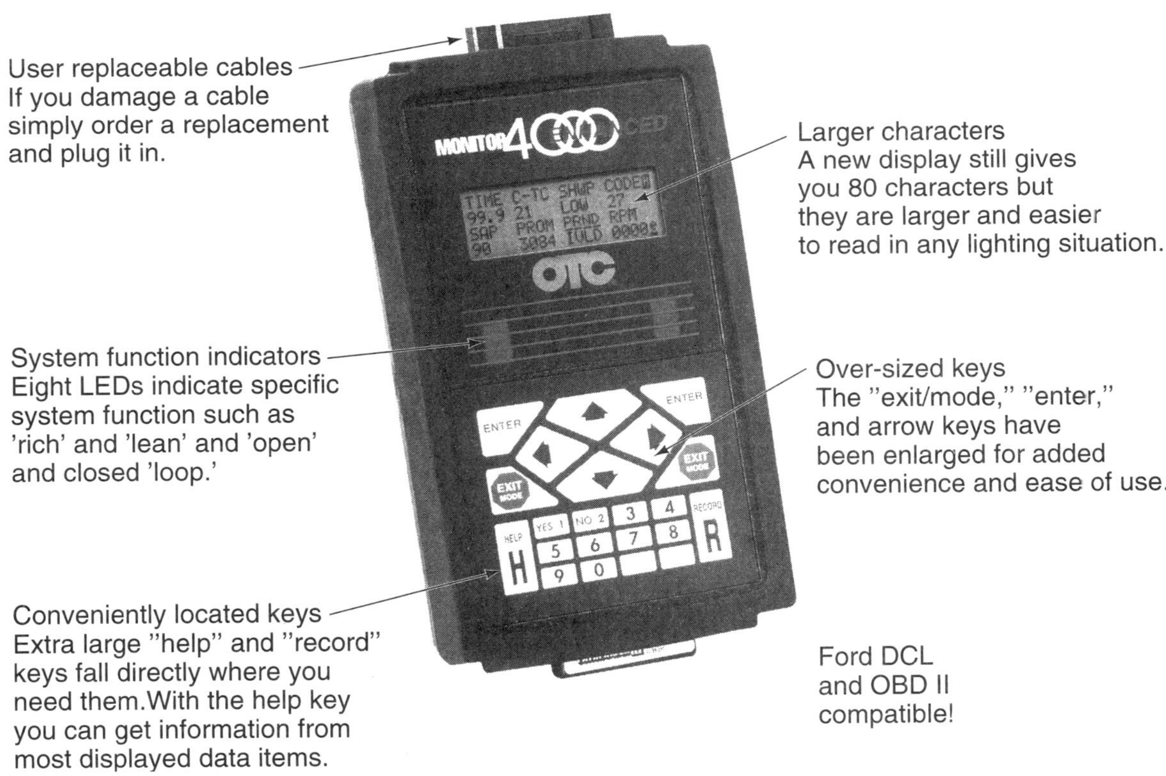

FIGURE 6–38 Hand-held scan tester's multiple features.

(Courtesy of SPX Corporation)

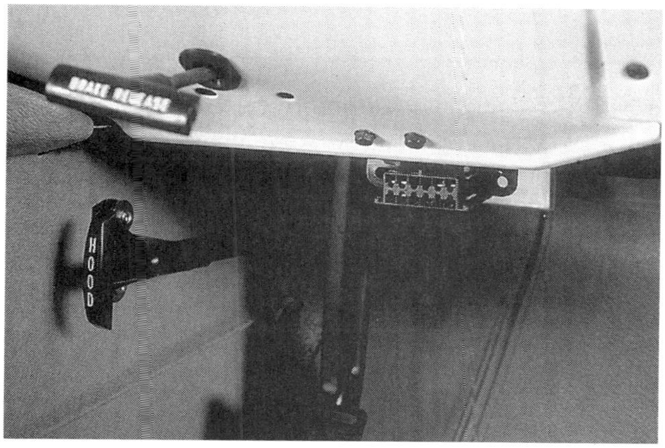

FIGURE 6–39 Typical data link connector under the dash on a GM vehicle.

FIGURE 6–40 Inserting a memory cartridge into a portable scan tester.

(Courtesy of SPX Corporation)

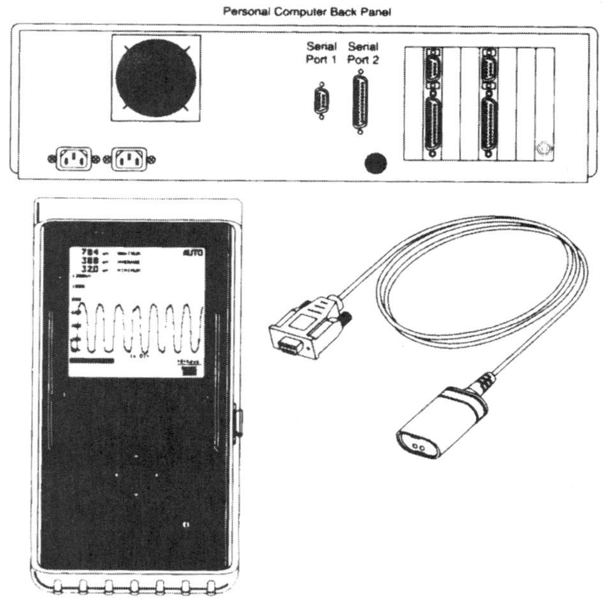

FIGURE 6–42 Connecting a portable scan tool to a personal computer for extended view.

(Courtesy of The Fluke Corporation)

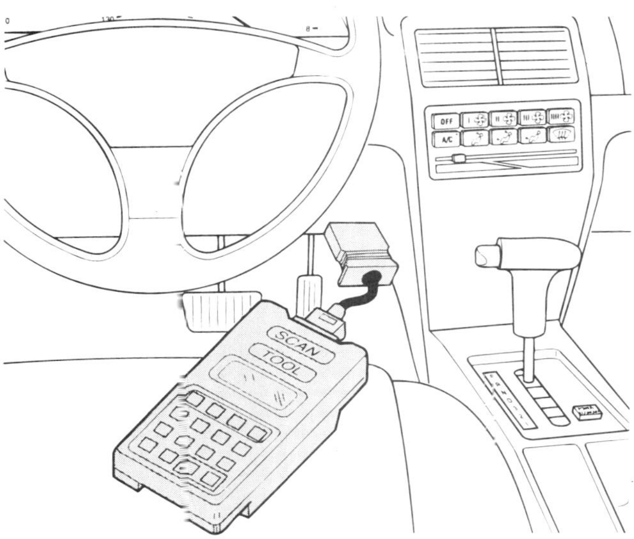

FIGURE 6–41 Using a portable scan tool on a road test of a vehicle.

(Courtesy of General Motors Corporation, Service Technology Group)

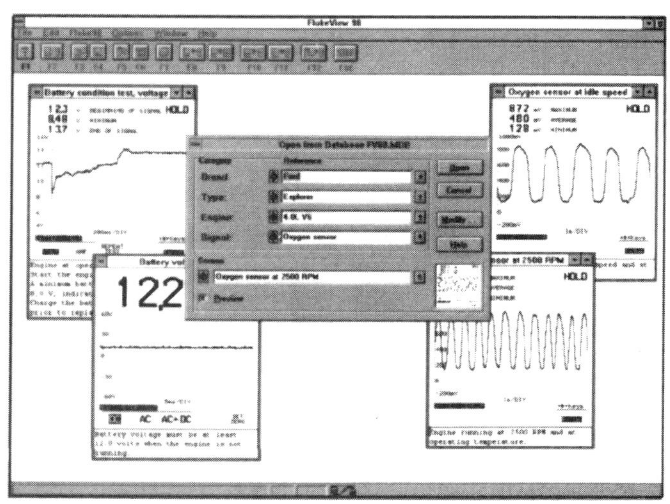

FIGURE 6–43 Comparing scan tool images with known good images from a database.

(Courtesy of The Fluke Corporation)

Hands-On Vehicle Tests

The following three hands-on vehicle checks are included in the **NATEF** (National Automotive Technician Education Foundation) **Task List.** The task reference numbers are the same as the NATEF reference numbers. Complete your answers to the below questions on a *separate* sheet of paper for all three tasks.

Performance Task # 1 (NATEF Task # VI-A-2)

Task Description

Check electrical circuits with a test light; determine needed repairs.

Task Objectives

- ❑ Obtain a vehicle that can be used for this task. What model and year of vehicle are you using for this task? ______
- ❑ Using a wiring diagram for the vehicle, determine electrical circuits that are safe to check with a test light. Verify circuits to test with your instructor.
- ❑ List the circuit(s) to be tested with the test light.

  1. ______ 4. ______
  2. ______ 5. ______
  3. ______ 6. ______

- ❑ List the procedure(s) used to safely hook up a test light to a vehicle's electrical circuit.

  ______
  ______
  ______

- ❑ Check the operation of the circuits listed above with the test light. What can you determine about using a test light on each of the circuits.

  1. ______ 4. ______
  2. ______ 5. ______
  3. ______ 6. ______

- ❑ What would be the benefit(s) of using a test light on electrical circuits compared to other electrical testing equipment?

  ______

Task Summary

After performing the above NATEF task, what can you determine that will be helpful in using test lights to perform simple electrical circuit troubleshooting? ______

______

Performance Task # 2 (NATEF Task # VI-A-3)

Task Description

Check voltage and voltage drop in electrical/electronic circuits using a digital multimeter (DMM); determine needed repairs.

Task Objectives

❑ Obtain a vehicle that can be used for this task. What model and year of vehicle are you using for this task? ____________________

❑ Using a wiring diagram for the vehicle, identify an electrical circuit that is suitable for checking voltages and voltage drops. Verify circuits to test with your instructor.

❑ List the circuit(s) to be tested for voltage and voltage drops:

1. ____________________ 3. ____________________

2. ____________________ 4. ____________________

❑ List the procedure(s) used to test for system voltage and voltage drops in an electrical circuit.

____

____

____

____

❑ What is the system voltage of each circuit tested and any voltage drop(s) measured?

System voltage ________. Voltage drops ________, ________, ________, ________.

System voltage ________. Voltage drops ________, ________, ________, ________.

System voltage ________. Voltage drops ________, ________, ________, ________.

System voltage ________. Voltage drops ________, ________, ________, ________.

❑ Do the voltage drops in each circuit tested equal the system total voltage?

____

____

Task Summary

After performing the above NATEF task, what can you determine that will be helpful in knowing how to check for voltage drops in an electrical circuit in reference to system voltage? ____________________

____

____

____

____

Performance Task # 3 (NATEF Task # VI-A-4)

Task Description

Check current flow in electrical/electronic circuits and components using an ammeter; determine needed repairs.

Task Objectives

- ❑ Obtain a vehicle that can be used for this task. What model and year of vehicle are you using for this task? ______________________

- ❑ Using a wiring diagram for the vehicle, determine electrical circuits that are safe to check the current flow with an ammeter. Verify circuits to test with your instructor.

- ❑ List the circuit(s) to be tested for current flow.

  1. ______________ 3. ______________

  2. ______________ 4. ______________

- ❑ List the procedure(s) used to safety hook up an ammeter to a vehicle's electrical circuit.

  ____

  ____

  ____

- ❑ Check the operation of the circuits listed above with an ammeter. What can you determine about testing for current flow on each of the circuits.

  1. ______________ 3. ______________

  2. ______________ 4. ______________

- ❑ What would be the benefit(s) of using an ammeter on electrical circuits compared to other electrical testing equipment? ______________________

  ____

  ____

Task Summary

After performing the above NATEF task, what can you determine that will be helpful in testing a circuit with an ammeter to troubleshoot an electrical circuit? ______________________

____

Summary

- ❑ An analog meter uses a pointer needle that moves across the face of a printed scale.
- ❑ A D'Arsonval movement is the moving coil movement in a analog meter.
- ❑ Analog meters operate between various range scales with a rotary switch.

- ❑ A multimeter is also known as a VOM.
- ❑ Alternating current is also known as AC current.
- ❑ Direct current is also known as DC current.
- ❑ Analog meters are polarity sensitive. The test leads must be hooked to the circuit according to proper polarity.
- ❑ The internal impedance of a meter is its ability to internally resist any current flow.
- ❑ A digital multimeter is also known as a DMM or DVOM.
- ❑ When a DMM is overloaded the meter will display OL, for overloaded on some brands.
- ❑ Some high quality DMM's have a touch-hold setting that captures a reading for display at a later time.
- ❑ The Greek symbol for resistance is the omega symbol Ω.
- ❑ An electrical circuit must be energized to measure voltage potential.
- ❑ A voltmeter is hooked in parallel across the component being measured.
- ❑ A voltage drop in a circuit is the intended or un-intended drop in voltage through a component, wiring, or connection point.
- ❑ An ammeter is used in an electrical circuit to measure the flow of electrons or current.
- ❑ An ammeter is hooked in series to the component or circuit.
- ❑ An inductive pick-up may be used to measure current flow of a circuit when the current exceeds the normal capacity of the meter.
- ❑ An ohmmeter is used to measure resistance in an electrical component.
- ❑ An ohmmeter is hooked in parallel across the component or circuit being measured, with the circuit power disconnected.
- ❑ A non-powered test light is used to check for voltage in a simple electrical circuit.
- ❑ A self-powered test light is used to check for continuity in an open circuit.
- ❑ Non-powered and self-powered test lights should not be used for voltage sensitive circuits.
- ❑ Analog oscilloscopes read and display electrical signals as they occur in a circuit.
- ❑ Digital storage oscilloscopes convert an input voltage signal into digital information.
- ❑ Voltage or amplitude measurement is represented by the vertical (Y) axis on a scope screen.
- ❑ Time measurement is represented by the horizontal (X) axis on a scope screen.
- ❑ The picture produced on a scope screen is called a waveform or trace.
- ❑ A typical AC sine wave with equal rise and fall is called sinusoidal.
- ❑ A typical DC waveform pattern is a straight horizontal line on the scope screen.
- ❑ A pulse width is the amount of time the voltage is ON per cycle.
- ❑ A number of cycles within a period of time is known as the frequency.

- ❑ When the pulse width varies in a circuit, but the frequency remains constant is known as pulse width modulation.
- ❑ A computer scan tool accesses the serial data from the vehicle computer to determine diagnostic trouble codes and system data.
- ❑ A scan tool uses a PROM chip to read specific vehicle information.

Key Terms

alternating current (AC)
amplitude
analog bar graph
analog meter
analog oscilloscope
auto ranging
continuity light
D'Arsonval movement
digital
digital storage oscilloscope (DSO)
direct current (DC)
digital multimeter (DMM)
dual-trace
duty cycle
digital volt ohm meter (DVOM)
frequency
icons
impedance
inductive pick-up
moving coil meter
multimeter
non-powered test light
polarity
programmable read-only memory (PROM)
pulse width
pulse width modulation
rotary switch
scan tool
self-powered test light
sine wave
sinusoidal
square wave
touch-hold
trace
waveform
voltage drop
VOM

Review Questions

Short Answer Essays

1. Describe the three service limitations technicians need to keep in mind when using an analog meter.
2. What are the three rules that must be followed when using a voltmeter to measure voltage potential?
3. What are the three rules that must be followed when using an ammeter to measure current in a circuit?
4. Describe how a typical AC sine wave would look like on a scope screen.
5. Describe the purpose of a computer scan tester in diagnosing today's vehicles.

Fill-in-the-Blanks

1. A voltmeter is always hooked in ____________________ across the component being measured.
2. An ammeter is ____________________ placed directly in parallel with the circuit or component.
3. The picture produced on a scope screen is called a ____________________.
4. A number of cycles within a period of time is known as the ____________________ of the circuit.
5. A typical scan tool uses a ____________________ chip that is specific to the vehicle being tested.

ASE Style Review Questions

1. Technician A says that analog meters use a pointer that moves across the face of a printed scale. Technician B says that analog meters are not polarity sensitive to the circuit being measured. Who is correct?
   A. A only
   B. B only
   C. Both A and B
   D. Neither A nor B

2. Technician A says a rotary switch is used on an analog meter to select the desired meter range. Technician B says if a rotary switch is set on the Rx1000 scale, the resistance reading must be multiplied by 1000. Who is correct?
   A. A only
   B. B only
   C. Both A and B
   D. Neither A nor B

3. Technician A says a quality DMM has an internal resistance of at least 10,000 ohms. Technician B says some DMM's have an auto-ranging feature to automatically adjust the measured range. Who is correct?
   A. A only
   B. B only
   C. Both A and B
   D. Neither A nor B

4. Technician A says the symbol for resistance on a DMM is the omega symbol. Technician B says the symbol for resistance on a DMM is Ω. Who is correct?
   A. A only
   B. B only
   C. Both A and B
   D. Neither A nor B

5. Technician A says a voltmeter must be hooked in series with the circuit being measured. Technician B says an electrical circuit must be open to measure the voltage potential of a circuit. Who is correct?
   A. A only
   B. B only
   C. Both A and B
   D. Neither A nor B

6. Technician A says that voltage is the pressure that pushes the current through a circuit. Technician B says resistance at a bad connection in a circuit results in a voltage drop. Who is correct?
   A. A only
   B. B only
   C. Both A and B
   D. Neither A nor B

7. Technician A says that typical terminal connections allow a maximum drop of 0.0 volts per connection. Technician B says that typical ground connections allow a maximum drop of 0.5 volts per connection. Who is correct?
   A. A only
   B. B only
   C. Both A and B
   D. Neither A nor B

8. Technician A says an electrical circuit must be energized to measure the current flow through the circuit. Technician B says the ammeter is hooked in series with the component or circuit being measured. Who is correct?
   A. A only
   B. B only
   C. Both A and B
   D. Neither A nor B

9. Technician A says an inductive pick-up is used by a DMM to measure current that exceeds the normal capacity of the meter. Technician B says an inductive pick-up is used to measure very small current flow in a circuit. Who is correct?
   A. A only
   B. B only
   C. Both A and B
   D. Neither A nor B

10. Technician A says an electrical circuit must be energized to get an accurate resistance reading with an ohmmeter. Technician B says an ohmmeter is hooked up in parallel across the component or circuit being measured. Who is correct?
    A. A only
    B. B only
    C. Both A and B
    D. Neither A nor B

11. Technician A says a non-powered test light can be used to check simple solid-state circuits. Technician B says a self-powered test light is used to check an energized circuit. Who is correct?
    A. A only
    B. B only
    C. Both A and B
    D. Neither A nor B

12. Technician A says the analog scope reads and displays electrical signals as they occur in the circuit. Technician B says analog scopes are known as "real time" scopes. Who is correct?
    A. A only
    B. B only
    C. Both A and B
    D. Neither A nor B

13. Technician A says a digital storage oscilloscope can sample input electrical signals at up to 40 billionth of a second. Technician B says DSO's have the ability to freeze the captured input or output signals. Who is correct?
    A. A only
    B. B only
    C. Both A and B
    D. Neither A nor B

14. Technician A says the scope waveform for a voltage measurement is represented by the vertical (Y) axis. Technician B says the scope waveform for amplitude is represented by the horizontal (X) axis. Who is correct?
    A. A only
    B. B only
    C. Both A and B
    D. Neither A nor B

15. Technician A says the equal rise and fall of an AC sine wave is considered to be sinusoidal. Technician B says one complete sine wave is known as a cycle. Who is correct?
    A. A only
    B. B only
    C. Both A and B
    D. Neither A nor B

16. Technician A says some scopes can display a pre-programmed pattern next to a live pattern as a reference. Technician B says some scopes can show two live patterns of different circuit components at the same time. Who is correct?
    A. A only
    B. B only
    C. Both A and B
    D. Neither A nor B

17. Technician A says a typical square wave digital voltage signal is either ON or OFF. Technician B says one cycle of time in a digital voltage signal is from the leading edge of one signal to the leading edge of the next signal. Who is correct?
    A. A only
    B. B only
    C. Both A and B
    D. Neither A nor B

18. Technician A says the amount of time a voltage is ON or OFF in a digital circuit is the pulse width. Technician B says the number of cycles with a period of time is known as the duty cycle. Who is correct?
    A. A only
    B. B only
    C. Both A and B
    D. Neither A nor B

19. Technician A says a computer scan tool hooks to the vehicle's computer through a data link connector. Technician B says a computer scan tool uses the parallel data from the on-board vehicle computer to determine an intermittent fault. Who is correct?
    A. A only
    B. B only
    C. Both A and B
    D. Neither A nor B

20. Technician A says a computer scan tool can interface with a personal computer to display the findings. Technician B says a computer scan tool uses a plug-in PROM chip that is specific to the vehicle being tested. Who is correct?
    A. A only
    B. B only
    C. Both A and B
    D. Neither A nor B

7 Series Circuits

Introduction

In this chapter, you will study the application of voltage, current, and resistance in *series circuits* that are commonly used in the automotive field. You will be introduced to some electrical circuit laws and formulas that pertain *only* to series circuits. As you apply these laws and associated formulas to electrical circuits, you will be guided through each new concept step by step and learn the particular behavior of a series circuit. The repetition of the exercises in this chapter are designed to help you remember the *series circuit laws* and to demonstrate that all series circuits, even complex series circuits, use the same circuit laws and formulas.

Objectives

When you complete this chapter you should be able to:

- ❑ Define a series circuit.
- ❑ Apply series circuit laws for voltage, current, and resistance.
- ❑ Define Kirchoff's Voltage Law.
- ❑ Apply Ohm's Law to series circuit calculations.
- ❑ Apply troubleshooting and testing techniques to series circuits.
- ❑ Perform basic hands-on series circuit exercises on the vehicle.

Series Circuit Explanation

A circuit is a **series circuit** where there is only a *single* path for current to flow through one or more resistive loads in a complete circuit. Because there is only one path for the current to flow, the current is the same at any point in the circuit (Figure 7–1).

There are several series circuits used by vehicle manufacturers as primary automotive electrical circuits. Some of the most common are heater control circuits, car burglar alarms, and trunk/hatch release solenoids. They are also used as part of more complex automotive electrical circuits that will be discussed later. The dependence of individual circuit devices on each other is the main reason for using series circuits for car burglar alarms. Any interruption of the single path of current flow would trigger the alarm circuit. It is for this same reason that a series circuit is not used in a taillight circuit. It would be inconvenient to have all the taillight bulbs go out just because one of them burns out. Figure 7–2 is an example of a simple series circuit for a trunk lid release, with a voltage source, fuse protection, control release switch, load device (solenoid), and wiring.

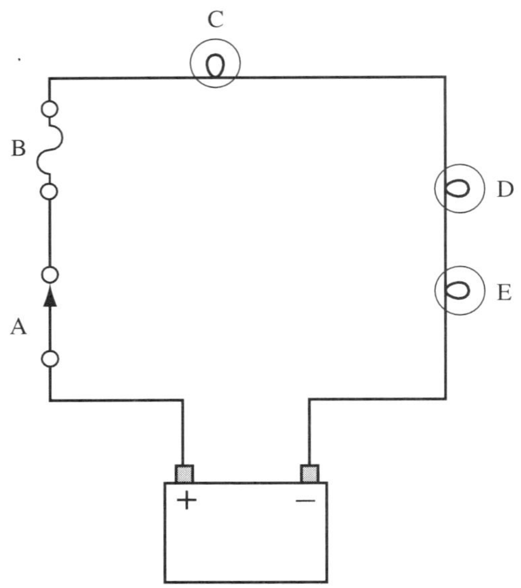

FIGURE 7–1 A basic series circuit including (A) a switch, (B) a fuse, (C) lamp #1, (D) lamp #2, (E) lamp #3.

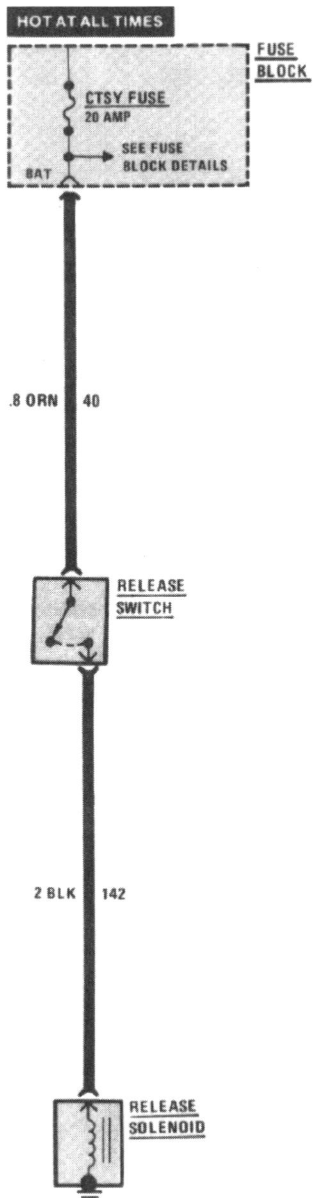

FIGURE 7–2 Luggage compartment lid release. (Courtesy of General Motors Corporation, Service Technology Group)

Circuit Laws

There are three circuit laws that pertain *only* to series circuits.

Series Law #1

In a series circuit the current is the same at any point in the circuit. Because there is only a single path for the current to flow, the current must pass though each resistive device in the circuit. Each resistive load will pass the same current through it. Figure 7–3 is an example of the amperage readings you would get on a simple series circuit with three resistors (R_1, R_2, R_3) wired in series.

In simple math terms, the circuit in Figure 7–3 can be expressed with I_T representing current total, and I_1, I_2, and I_3 representing the three resistors as follows.

$$I_T = I_1 = I_2 = I_3 \qquad \text{(same or equal)}$$

If you knew the current at any point in the circuit, then you would know the current anywhere else in the circuit. This is important because if your access to the whole circuit is limited, you might find a convenient point to measure the current.

Series Law #2

The total resistance in a series circuit is equal to the *sum* of the individual resistance loads in the circuit. Remembering Ohm's Law—as the resistance goes up the current flow goes down—then the more resistive loads there are, the lower the current flow. In other words, current and resistance are inversely proportional to each other. Figure 7–4 is an example of a series circuit with several resistors.

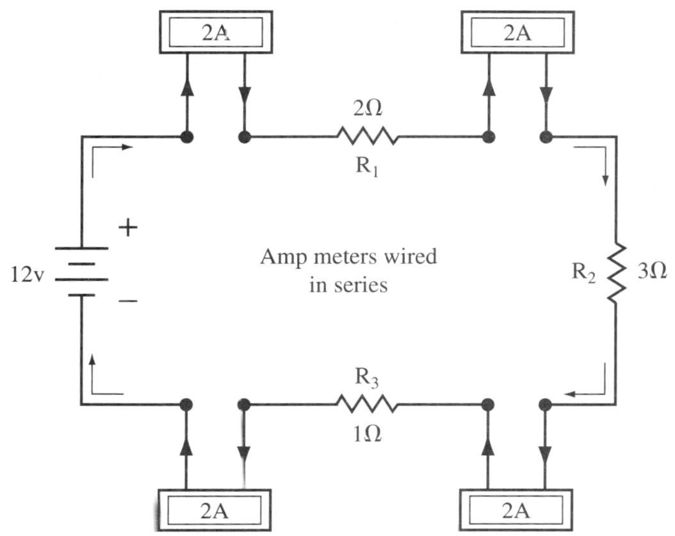

FIGURE 7–3 Amperage readings in a series circuit.

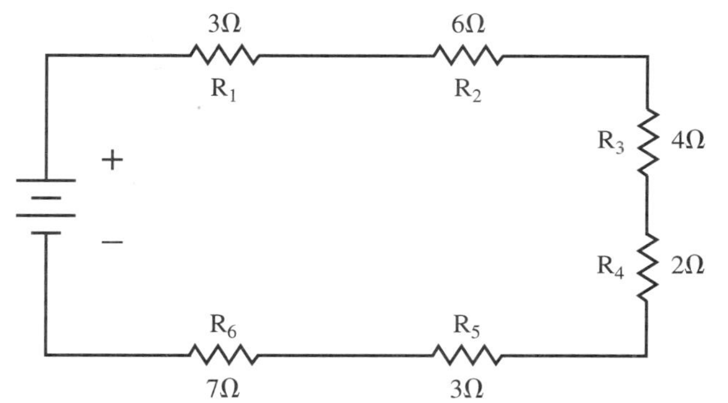

R_1 = Resistor #1 of 3Ω
R_2 = Resistor #2 of 6Ω
R_3 = Resistor #3 of 4Ω
R_4 = Resistor #4 of 2Ω
R_5 = Resistor #5 of 3Ω
R_6 = Resistor #6 of 7Ω

FIGURE 7–4 Resistance in a series circuit.

In math terms, Figure 7–4 can be expressed with R_T representing total resistance, and R_1 through R_6 representing the six resistive loads as follows.

$$R_T = R_1 + R_2 + R_3 + R_4 + R_5 + R_6 \text{ (additive)}$$

Using the numbers in Figure 7–4, the total resistance is:

$$R_T = 3\,\Omega + 6\,\Omega + 4\,\Omega + 2\,\Omega + 3\,\Omega + 7\,\Omega$$

$$R_T = 25\,\Omega$$

Series Law #3

The *total* voltage in a series circuit is equal to the *sum* of the individual voltage drops of each resistive load in the circuit. This is known as **Kirchoff's Voltage Law.** In simple math terms, V_T represents total voltage, and V_1 through V_3 (or more), represents the voltage drop of each resistive device, as follows

$$V_T = V + V_2 + V_3 + \text{etc.} \quad \text{(additive)}$$

This principle is also known as the **Voltage Drop** of a series circuit. Figure 7–5 is an example of a series circuit showing what voltmeters would measure for the voltage drop of each resistive device in the circuit.

When resistive loads differ from one another, the voltage drop for each load will will also differ. More precisely, voltage drops are proportional to the values of each resistive load in the circuit. If one resistive load is twice as large as another, then the voltage drop will also be twice as large. You can see this in Figure 7–5, where the 2 Ω resistive load is twice as large as the 1 Ω value, the voltage drop for the 2 Ω is also twice as large as the 1 Ω resistive load.

There are three primary ways to find the voltage drops in a series circuit. The first is to take the resistance values and use Ohm's Law. To do this using the values given in Figure 7–5, you first need to find the resistance total.

$$R_T = R_1 + R_2 + R_3\text{, or}$$
$$R_T = 1\,\Omega + 2\,\Omega + 3\,\Omega\text{, or}$$
$$R_T = 6\,\Omega$$

Having the resistance total, you next use Ohm's Law to find the amperage total.

12 volts total ÷ 6 Ω resistance total =
2 amps current total

Since you know the series circuit law for current, you can then use Ohm's Law to apply the current total, or 2 amps, to each resistive device in the circuit to find the voltage drop for each device as follows.

2 amps × 1 Ω = 2 volts
2 amps × 2 Ω = 4 volts
2 amps × 3 Ω = 6 volts

The sum of the individual voltage drops should equal the voltage source or total.

2 volts + 4 volts + 6 volts = 12 volts

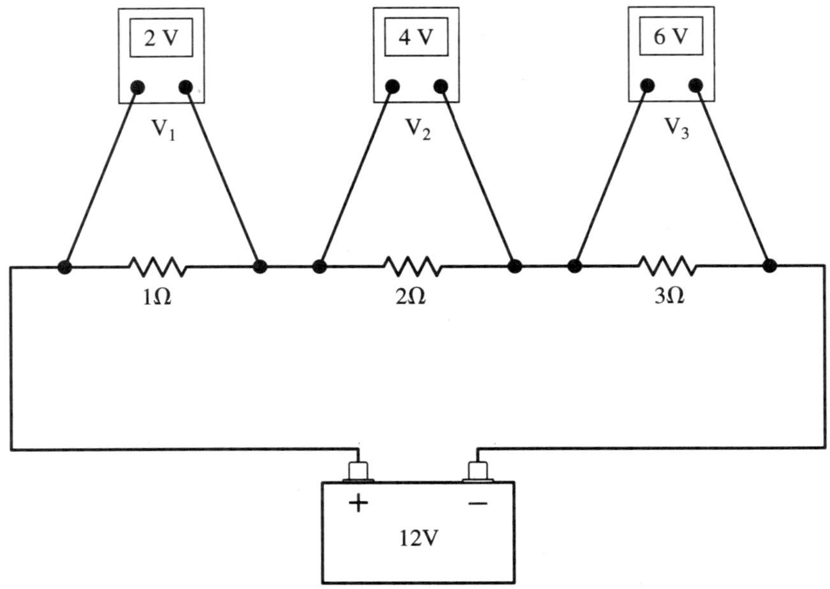

FIGURE 7–5 Measuring voltage drops in a series circuit.

The second way to find voltage drops is to use Kirchoff's Voltage Law. This method will only work if you know the voltage source total and two of the three voltage drops, then you can take what you know and subtract it from the voltage total to get what you don't know. Using the same values as above, figure the voltage drop of the 3 Ω value.

$$V_T = V_1 + V_2 + V_3, \text{ or } 12\text{ V} = 2\text{ V} + 4\text{ V} + \text{(the unknown volt drop), or } 12\text{ V} - 6\text{ V} = 6\text{ V}$$

The third way to find voltage drops in a series circuit is to use Watt's Law. This is not the most efficient way to find voltage drops, but will work if you do not know individual circuit resistance's, but do know their perspective amperage and wattage.

$$W_T = W_1 + W_2 + W_3, \text{ or}$$
$$24\text{ W} = 4\text{ W} + 8\text{ W} + 12\text{ W}$$

$$4\text{ W} \div 2\text{ A} = 2\text{ V},$$
$$8\text{ W} \div 2\text{ A} = 4\text{ V},$$
$$12\text{ W} \div 2\text{ A} = 6\text{ V}$$

The individual voltage drops add up to equal the source voltage.

Because the voltage and resistance are both additive in series circuits, the amount of each individual voltage or resistance value is also proportional to its own total. This is known as **Voltage and Resistance Percentage.** Figure 7–6 is an example of a simple series circuit with individual resistive loads that add up to 40 Ω total.

To find the percentage of the resistance total that each individual resistor load takes, use the following calculation where R_n is R_1, R_2, etc.

$$R_n \div R_T \times 100 \text{ (for percent)} =$$

As an example, what percent of the resistance total is from resistive load #3 (R_3)?

$$20\ \Omega \div 40\ \Omega = 0.50$$
$$0.50 \times 100 = 50\%$$

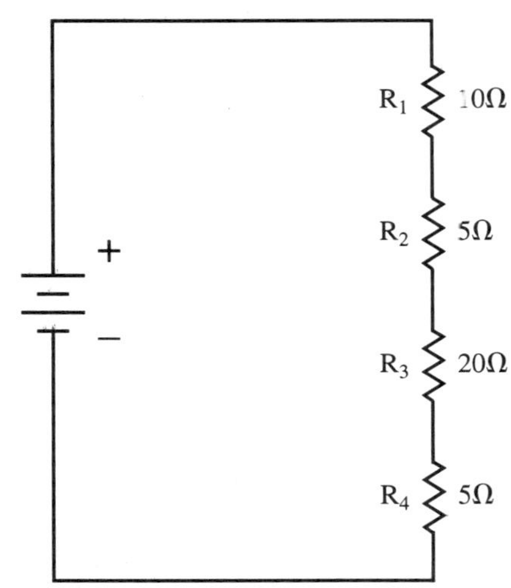

FIGURE 7–6 Simple series circuit with individual resistive loads.

If you perform this procedure for all the remaining resistive values, they should add up to 100 percent. The same procedure can be used to find the voltage percentage of individual resistive loads in a series circuit.

$$V_n \div V_T \times 100 \text{ (for percent)} =$$
Voltage percentage of individual loads

As an example, if the voltage total of Figure 7–6 is 12 V, then you would know that 12 V ÷ 40 Ω would produce a current total of .3 A. This would make a voltage drop of R_3 equal 6 V.

$$6 \text{ V} \div 12 \text{ V} = 0.50$$

$$0.50 \times 100 = 50\%$$

As you can see, the percentage numbers using the resistive values and using the voltage drops produce the same values of percentage numbers. This means you can use either set of numbers to find the percentage that an individual resistive load takes in a series circuit.

Before we continue, here is a brief overview of the series circuit laws to use as a quick reference guide.

Basic Laws of Series Circuit Electricity

	SERIES
CURRENT	Equal or same $I_T = I_1 = I_2 = I_3 =$ etc.
VOLTAGE	Additive $E_T = E_1 + E_2 + E_3 +$ etc.
RESISTANCE	Additive $R_T = R_1 + R_2 + R_3 +$ etc.

Exercises—Series Circuits

The following cognitive exercises are designed to help give you practice using the series laws and methods explained previously. The exercises will increase in complexity in an effort to test your understanding of the basic series laws, and to prepare you for the actual hands-on series circuit lab exercises that will follow.

Exercise #1

Instructions Identify the type of circuit in Figure 7–7. On a *separate* sheet of paper make an answer table like

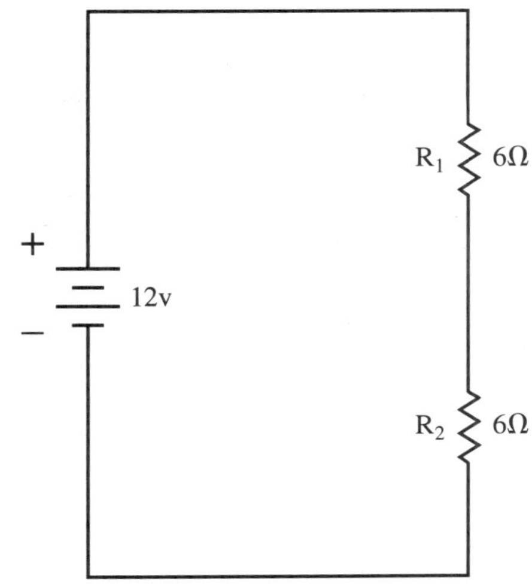

FIGURE 7–7 Basic series circuit.

Table 7–1. Using the table, fill in the known circuit values from the series diagram. This is a helpful first step by giving you an idea of where to begin and what answers you need. Calculate the answers in the remaining spaces using one of the methods previously learned.

R_T = Resistance total, I_T = Amperage total,
E_T = Voltage total, W_T = Watts total

R_1 = Resistor #1, I_1 = Amperage #1,
E_1 = Voltage #1, W_1 = Watts #1

R_2 = Resistor #2, I_2 = Amperage #2,
E_2 = Voltage #2, W_2 = Watts #2

Table 7–1

R_T ______	I_T ______	E_T ______	W_T ______
R_1 ______	I_1 ______	E_1 ______	W_1 ______
R_2 ______	I_2 ______	E_2 ______	W_2 ______

Circuit Analysis

To ensure that you are familiar with the series laws, the correct steps and answers for the previous problem are shown below. If your answers are different, refer to the discussion section at the front of this chapter for further study.

Step #1 Known values: E_T = 12 V, R_1 = 6 Ω, R_2 = 6 Ω. Fill them in on a *separate* sheet of paper like Table 7–1.

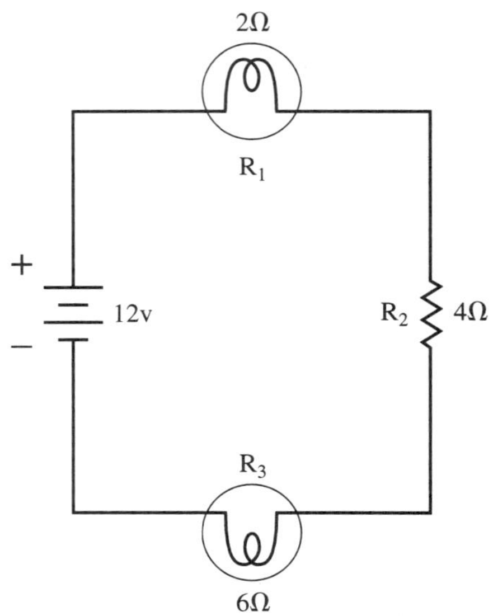

FIGURE 7–8 Identify basic circuit.

Step #2 Remembering that the Resistance Law is additive in series circuits, add them up and fill in the resistance total of 12 Ω.

$R_1 + R_2 = R_T$

$6\ \Omega + 6\ \Omega = 12\ \Omega$

Step #3 $E \div R = I$

Take the voltage total 12 V and divide by the resistance total 12 Ω to get the amps total of 1 A. Remembering that the Current Law states that the current is the same in series circuits, fill in the other current values with the same 1 A.

Step #4 $I \times R = E$

Next, multiply the individual amperages by the individual resistance values to get the voltage drops of E_1 and E_2.

Step #5 $I \times E = W$

To find the power consumption, use Watt's Law. Volts total times Amps total will equal Watts total, which is 12 W. (If unsure, refer back to Chapter 3). Do the same procedure to find W_1 and W_2.

Table 7-2 is a summary of the right answers.

Table 7–2

R_T 12 Ω	I_T 1 A	E_T 12 V	W_T 12 W
R_1 12 Ω	I_1 1 A	E_1 12 V	W_1 12 W
R_2 12 Ω	I_2 1 A	E_2 12 V	W_2 12 W

Instructions Follow the instructions used on the previous problem for the remaining series circuit exercises. Remember to put your answers on a *separate* sheet of paper.

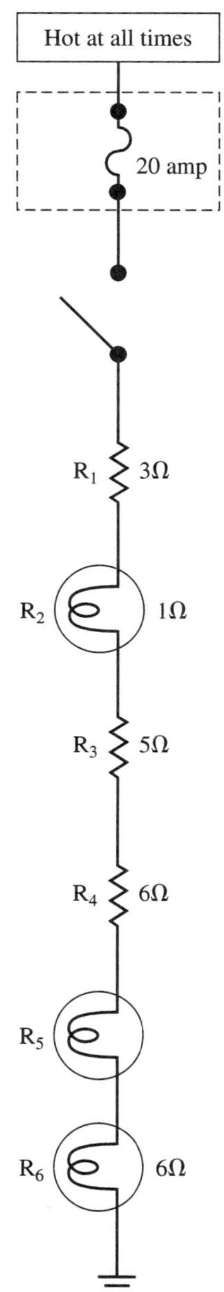

FIGURE 7–9 Basic series circuit.

Exercise #2

Note: After you have identified the type of circuit in Figure 7–8, fill in the known values on a table like Table 7–3 on a *separate* sheet of paper.

Table 7–3

R_T	I_T	E_T	W_T
R_1	I_1	E_1	W_1
R_2	I_2	E_2	W_2
R_3	I_3	E_3	W_3

Exercise #3

Note: Figure 7–9 is a series circuit that is drawn a little differently. You may find some actual vehicle wiring diagrams drawn similar to this. Just because the schematic is not drawn from left to right, or from top to bottom, does not mean the theory of series circuits is different. Identify the circuit by following the wiring from the positive side of the battery, through each resistive load of the circuit, and then to the negative side.

Complete the answers to the above exercise by filling in a table like Table 7–4 on a *separate* sheet of paper.

Table 7–4

R_T	I_T .5 A	E_T	W_T
R_1 3 Ω	I_1	E_1	W_1
R_2 1 Ω	I_2	E_2	W_2
R_3 5 Ω	I_3	E_3	W_3
R_4 6 Ω	I_4	E_4	W_4
R_5	I_5	E_5 1.5 V	W_5
R_6 6 Ω	I_6	E_6	W_6

Begin by filling in the known circuit values and then look for the best Series Law(s) to apply first.

HINT: Remember the series circuit law for current.

Exercise #4

Figure 7–10 is a series circuit with a missing resistive value. Complete your answers to Exercise #4 on a *separate* sheet of paper similar to Table 7-5.

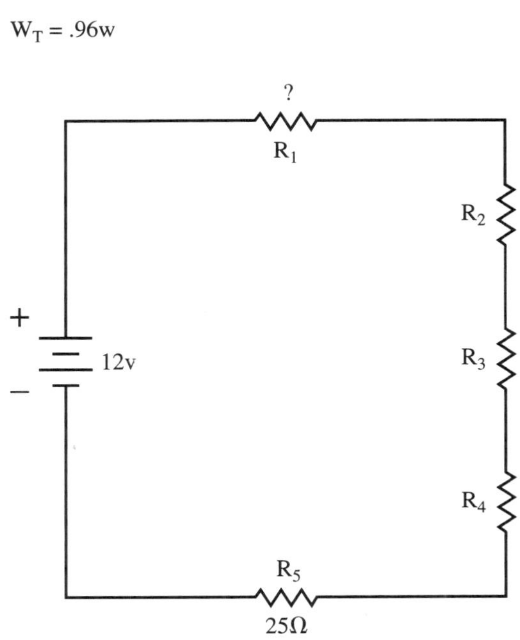

FIGURE 7–10 Basic skills circuit.

Table 7–5

R_T	I_T	E_T	W_T
R_1	I_1	E_1	W_1
R_2	I_2	E_2	W_2
R_3	I_3	E_3	W_3
R_4	I_4	E_4	W_4
R_5	I_5	E_5	W_5

Now that you have practiced with some basic series circuit examples, let's try a more difficult circuit problem to see how well you know the series circuit laws.

HINT: No matter how complex the circuit looks it always uses the same basic series circuit laws. The better you understand the laws applying to series circuits, the easier electrical problem solving will become.

Exercise #5

Using Figure 7–11 as a reference, use a *separate* sheet of paper and complete Table 7–6.

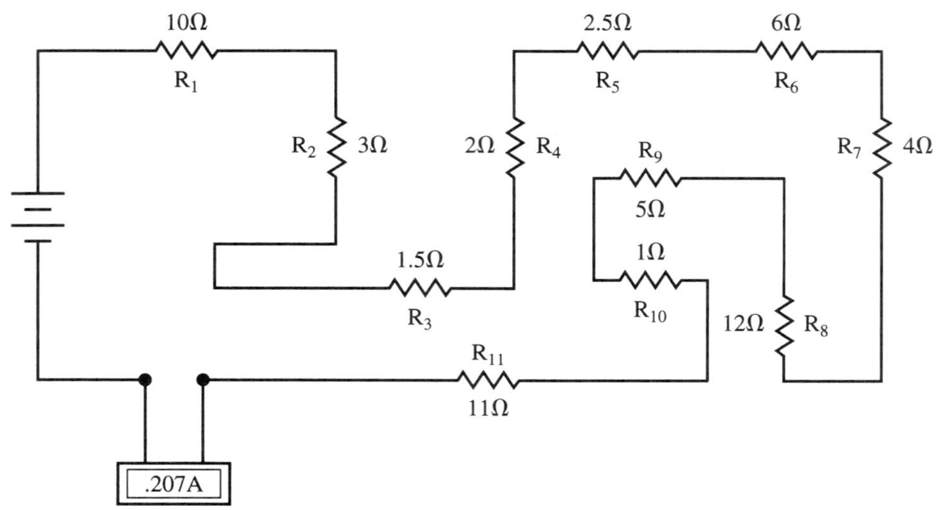

FIGURE 7–11 Complex series circuit.

TABLE 7–6

R_T ________	I_T ________	E_T ________	W_T ________
R_1 ________	I_1 ________	E_1 ________	W_1 ________
R_2 ________	I_2 ________	E_2 ________	W_2 ________
R_3 ________	I_3 ________	E_3 ________	W_3 ________
R_4 ________	I_4 ________	E_4 ________	W_4 ________
R_5 ________	I_5 ________	E_5 ________	W_5 ________
R_6 ________	I_6 ________	E_6 ________	W_6 ________
R_7 ________	I_7 ________	E_7 ________	W_7 ________
R_8 ________	I_8 ________	E_8 ________	W_8 ________
R_9 ________	I_9 ________	E_9 ________	W_9 ________
R_{10} ________	I_{10} ________	E_{10} ________	W_{10} ________
R_{11} ________	I_{11} ________	E_{11} ________	W_{11} ________

You've spent some time now learning series circuit theory and laws, and developing the knowledge on how series circuits perform in relation to voltage, amperage, and resistance. So let's apply theory to some real-life case studies and prove the theories you have learned.

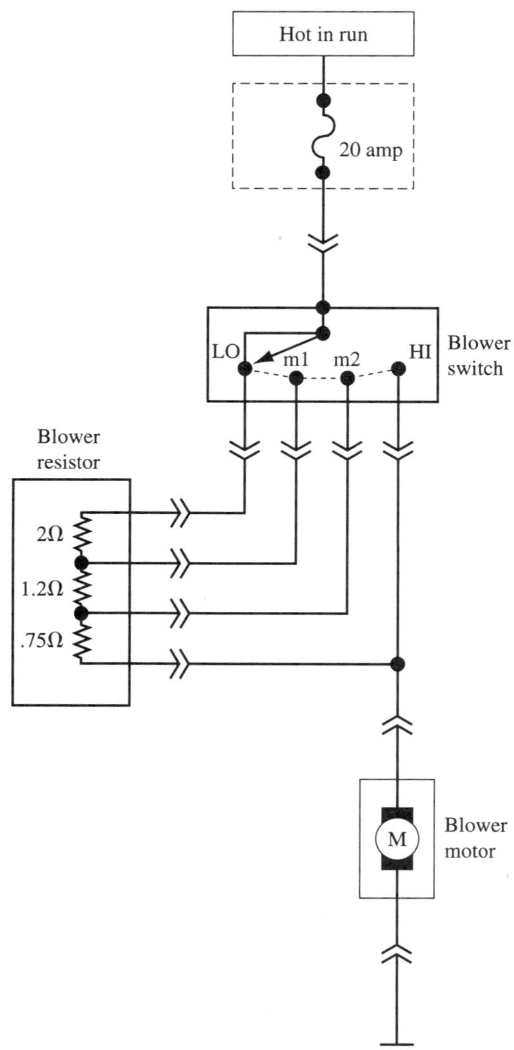

FIGURE 7–12 Typical series heater motor circuit.

CASE STUDIES

CASE #1

The following case study is typical of a live series circuit problem that technicians often face. Knowledge of series circuits can help you perform the diagnosis quickly and efficiently. So let's give it a try.

Customer Complaint

Vehicle heater circuit does not work when the blower switch is in the *LO* and *M1* positions, but does work properly in the *M2* and *HI* positions. Figure 7–12 shows the layout of the blower motor circuit. The customer replaced the blower switch, thinking this was the cause. The results were the same as before.

Known Information

- ❑ Vehicle operating voltage = 14 volts
- ❑ Blower switch has been replaced
- ❑ System wiring and connectors between blower switch and resistor assembly have been checked, and are satisfactory
- ❑ The system fault with *LO* and *M1* happened suddenly
- ❑ Blower motor has windings with a resistance of 1.5 Ω
- ❑ All other resistance information located on wiring diagram Figure 7–12

Circuit Analysis

Answer the following questions on a *separate* sheet of paper:

1. What is the most likely cause of the inoperative blower circuit in the *LO* and *M1* positions only?

   ______________________________________.

2. What is the total circuit resistance of the blower circuit assuming that the blower switch, connections, and wiring have no measurable resistance? ______________________.

3. What is the total amperage of the circuit at the following positions if an amp meter was placed in series with the circuit at the fuse.
   - Amperage reading with blower switch on *LO* position? ______________________
   - Amperage reading with blower switch on *M1* position? ______________________.
   - Amperage reading with blower switch on *M2* position? ______________________.
   - Amperage reading with blower switch on *HI* position? ______________________.

4. What is the best way to repair this system? ______________________________.

5. What series circuit law(s) were helpful when troubleshooting this system or explaining it to the customer? ______________________________

   ______________________________________.

6. What is your analysis of the above case study? ______________________________

   ______________________________________.

CASE #2

Using the same circuit diagram, Figure 7–12, solve the customer complaint below.

Customer Complaint

Vehicle heater does not work in any blower switch position. Customer replaced the fuse, but the results were the same.

Known Information

- ❑ Vehicle operating voltage = 14 volts
- ❑ 20 amp fuse has been replaced
- ❑ System fault happened suddenly during heater blower operation

Circuit Analysis

Answer the following questions on a *separate* sheet of paper.

1. If you knew the fuse was good, what system components, and in what priority, would you check in the circuit? List the most important component first, and the procedure you would use to check the it.

   Component #1 ____________________. Troubleshooting Procedure ____________________.

   Component #2 ____________________. Troubleshooting Procedure ____________________.

   Component #3 ____________________. Troubleshooting Procedure ____________________.

   Component #4 ____________________. Troubleshooting Procedure ____________________.

   Component #5 ____________________. Troubleshooting Procedure ____________________.

2. What series circuit law(s) were helpful when troubleshooting this system or explaining it to the customer? ____

   ____

   ____.

3. What is your analysis of the above case study? ____________________________________

   ____.

Figure 7–13 is an example of a series heater circuit that includes a blower relay which needs to be energized by the ignition switch in order to provide voltage to the blower circuit.

The blower motor in this circuit is located before the blower resistor instead of after as in Figure 7–12. Remember, in a series circuit there is only one path for the current to flow, so the blower motor can be located before or after the blower controls and still operate. The primary difference between the two blower circuits is that the positive side of the blower motor contains the source voltage at all times in Figure 7–12, and it is also on the negative side of the circuit where the resistive speed control block is located. This changes the way you would perform your troubleshooting techniques. If you were to do a quick analysis of the circuit in Figure 7–13, you would see the following.

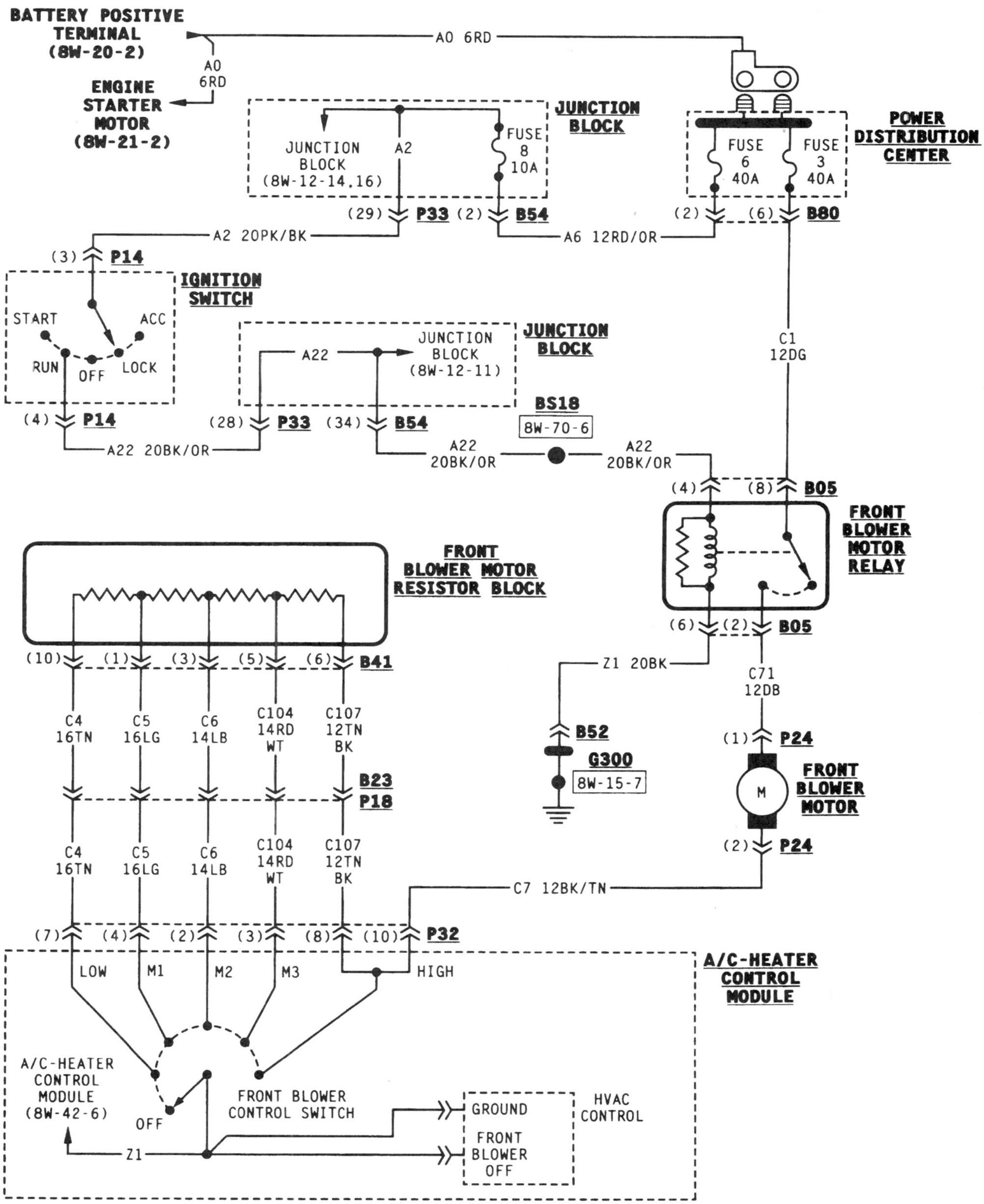

FIGURE 7–13 Heater blower circuit.

(Courtesy of DaimlerChrysler Corporation)

1. Power is *hot* at all times to the front blower motor relay through the 40 amp fuse in the power distribution center.
2. The front blower motor relay is energized every time the key is in the *run* position.
3. Ground for the coil side of the relay is supplied on circuit wire *Z1*.
4. Ground for the blower motor is controlled through the blower motor resistor and switch.
5. With the blower motor in the *low* position, the circuit's current must go through the blower motor, through all four resistors in the resistor block, and then through the LOW position on the blower switch to ground.

Hands-On Vehicle Tests

The following two hands-on vehicle checks are included in the **NATEF** (National Automotive Technicians Education Foundation) **Task List.** The task reference numbers are the same as the NATEF reference numbers. Complete your answers to the below questions on a *separate* sheet of paper.

Performance Task #1 (NATEF #V-E-5)

Task Description
Check the operation of the parking brake indicator light system.

Task Objectives

- ❑ Obtain a vehicle that can be used for this task. What model and year of vehicle are you using for this task? ________________
- ❑ Find a wiring diagram that shows the parking brake indicator circuit for this vehicle. Does the parking brake circuit follow the characteristics of a series circuit? ________________.
- ❑ Identify the following components and list their location on the vehicle.
  1. Instrument panel cluster ________________.
  2. Parking brake indicator bulb(s) ________________.
  3. Parking brake switch ________________.
  4. Circuit fuse and fuse # ________________.
  5. Other electrical circuits that share the same fuse ________________.
  6. Circuit ground(s) ________________.
  7. Other circuit components ________________.

     ________________.

Circuit Checks

- ❑ Check operation of parking brake switch. Disconnect wire connections at switch. Set a DVOM to continuity and hook it to the switch terminals and activate the parking brake. What happens as a result? ______________________________.
- ❑ Reconnect the wires to parking brake switch.
- ❑ Remove the parking brake fuse and hook the DVOM in series in its place. What is the total amperage of the circuit? ______________________________.
- ❑ With the parking brake off, is there still current flowing through the DVOM? ______________.
- ❑ If the answer is *yes* to the above question, where is the amperage coming from? ______________.
- ❑ What is the total circuit voltage? ______________________________.
- ❑ Perform a voltage drop test through the circuit components, connectors, and wires. Does the measurable voltage drop(s) add up to the total circuit voltage? ______________________.
- ❑ Does this parking brake circuit follow the circuit laws for a series circuit? ______________.

Task Summary

After performing the above NATEF task, what can you determine that will be helpful in future parking brake circuit diagnosis? ______________________________

______________________________.

Performance Task #2 (NATEF #VII-D1-2)

Task Description

Inspect and test the A/C heater blower, motors, resistors, switches, relays, wiring, and protection devices; repair or replace as needed.

Task Objectives

- ❑ Obtain a vehicle that can be used for this task. What model and year of vehicle are you using for this task? ______________________________.
- ❑ Find a wiring diagram that shows the heater control and motor circuit for this vehicle. Does the heater control circuit follow the characteristics of a series circuit? ______________________.
- ❑ Identify the following components and list their location on the vehicle.

  1. Heater motor control switch ______________________________.
  2. Blower motor assembly ______________________________.
  3. Blower resistor block assembly ______________________________.
  4. Blower motor relay (if applicable) ______________________________.
  5. Blower circuit fuse and fuse # ______________________________.
  6. Blower motor relay fuse #(if applicable) ______________________________.
  7. Other electrical circuits that share the same fuse(s) ______________________________.
  8. Circuit ground(s) ______________________________.

9. Other circuit components ______________________________.

______________________________.

Circuit Checks

- ❑ To ensure that all DVOM readings of the circuit are characteristic of the actual circuit operation, the vehicle should be running in a well ventilated area (keeps the battery from discharging).
- ❑ With the blower switch in the *LOW* position, measure the voltage available to the circuit. This voltage reading is most accurately taken from the fuse block on the positive side to the ground connection at the end of the circuit. Voltage total of the blower circuit is ______________.
- ❑ Measure the voltage drop across the points listed below. Actual points to record the voltage drops may vary slightly depending on the circuit design.

1. Across circuit fuse ______________________________.
2. From fuse box to blower switch ______________________________.
3. From blower switch LOW to resistor block ______________________________.
4. Across each resistor in the blower resistor block assembly. Normally there are at least three resistors in this block assembly. There are vehicles with more or less, so complete as many as necessary.

   resistor #1 voltage drop ______________________________.

   resistor #2 voltage drop ______________________________.

   resistor #3 voltage drop ______________________________.

   resistor #4 voltage drop ______________________________.
5. Across blower motor ______________________________.
6. From blower motor to actual circuit ground connection in vehicle ______________.

- ❑ Do the measured voltage drops add up to equal the measured voltage total? ______________.
- ❑ With the blower switch in the *M1* position, measure the voltage available to the circuit in the same way as you did above. Voltage total of the blower circuit ______________.
- ❑ Measure the voltage drop across the points listed below. Actual points to record the voltage drops may vary slightly depending on the circuit design.

1. Across circuit fuse ______________________________.
2. From fuse box to blower switch ______________________________.
3. From blower switch *M1* to resistor block ______________________________.
4. Across each resistor in the blower resistor block assembly. There are vehicles with more or less, so complete as many as necessary.

   resistor #1 voltage drop ______________________________.

   resistor #2 voltage drop ______________________________.

   resistor #3 voltage drop ______________________________.

resistor #4 voltage drop ______________________________.

5. Across blower motor ______________________________.

6. From blower motor to actual circuit ground connection in vehicle______________________________.

❑ Are the voltage drop readings in *M1* the same as in the LOW position? ______________________________.

Why? ______________________________.

❑ Do the measured voltage drops add up to equal the measured voltage total? ______________________________

❑ With the blower switch in the *M2* position, measure the voltage available to the circuit in the same way as you did above. Voltage total of the blower circuit ______________________________.

❑ Measure the voltage drop across the points listed below. Actual points to record the voltage drops may vary slightly depending on the circuit design.

1. Across circuit fuse ______________________________.
2. From fuse box to blower switch ______________________________.
3. From blower switch *M2* to resistor block ______________________________.
4. Across each resistor in the blower resistor block assembly. There are vehicles with more or less, so complete as many as necessary.

   resistor #1 voltage drop ______________________________.

   resistor #2 voltage drop ______________________________.

   resistor #3 voltage drop ______________________________.

   resistor #4 voltage drop ______________________________.

5. Across blower motor ______________________________.
6. From blower motor to actual circuit ground connection in vehicle ______________________________.

❑ Are the voltage drop readings in *M2* the same as in the *LOW* and *M1* positions? ______________________________.

Why? ______________________________.

❑ Do the measured voltage drops add up to equal the measured voltage total? ______________________________.

❑ With the blower switch in the *HI* position, measure the voltage available to the circuit in the way same as you did above. Voltage total of the blower circuit ______________________________.

❑ Measure the voltage drop across the points listed below. Actual points to record the voltage drops may vary slightly depending on the circuit design.

1. Across circuit fuse ______________________________.
2. From fuse box to blower switch ______________________________.
3. From blower switch HI to resistor block ______________________________.
4. Across blower relay (if applicable) ______________________________.
5. Across each resistor in the blower resistor block assembly. There are vehicles with more or less, so complete as many as necessary.

resistor #1 voltage drop ______.

resistor #2 voltage drop ______.

resistor #3 voltage drop ______.

resistor #4 voltage drop ______.

6. Across blower motor ______.
7. From blower motor to actual circuit ground connection in vehicle______.

- ❑ Are the voltage drop readings in the *HI* position the same as in the *LOW, M1* and *M2* positions?

  Why? ______
- ❑ Does the measured voltage drops add up to equal the measured voltage total? ______.
- ❑ Turn blower circuit off and install a DVOM in series to measure circuit amperage. Normally, the easiest place to perform this procedure is at the fuse or after the blower motor assembly. What is the circuit amperage with the blower switch in the following positions:
  1. *LOW* ______
  2. *M1* ______
  3. *M2* ______
  4. *HI* ______
- ❑ What can you conclude about voltage drops in a series circuit?______.

  ______.
- ❑ What effect do resistors wired in a series line have on the circuit current?______

  ______.

Task Summary

After performing the above NATEF task, what can you determine will be helpful in future parking brake circuit diagnosis? ______

______.

Summary

- ❑ In a series circuit there is only one path of flow for the electrons to travel.
- ❑ In a series circuit the current is the same at any point in the circuit.
- ❑ In a series circuit the total resistance is the sum of the individual resistance loads in the circuit.
- ❑ In a series circuit the voltage drops in each system component add up to equal the source voltage. This is known as Kirchoff's Law.
- ❑ Voltage and resistance are both additive in a series circuit, so the Voltage and Resistance Percentage calculation can be used to find individual percentage loads.

Key Terms

Kirchoff's voltage law
series circuit
series circuit laws
voltage and resistance percentage
voltage drop

Review Questions

Short Answer Essays

1. Describe the circuit law for voltage in a series circuit.
2. Describe the circuit law for amperage in a series circuit.
3. Describe the circuit law for resistance in a series circuit.

Fill-in-the-Blanks

1. The amperage of a load component in a series circuit is the ____________________ as the amperage ____________________.
2. In a series circuit there is only ____________________ path for ____________________ to travel.
3. ____________________ and ____________________ are both additive in a series circuit.

ASE Style Review Questions

1. Technician A says that in a series circuit the voltage of each circuit load device is the same as the total circuit voltage. Technician B says the amperage of each circuit load is the same throughout the circuit. Who is correct?
   A. A only
   B. B only
   C. Both A and B
   D. Neither A nor B

2. In the following series light circuit the bulbs do not light when the switch is closed. Technician A says the switch can be bad. Technician B says one of the light bulbs could be burned out. Who is correct?
   A. A only
   B. B only
   C. Both A and B
   D. Neither A nor B

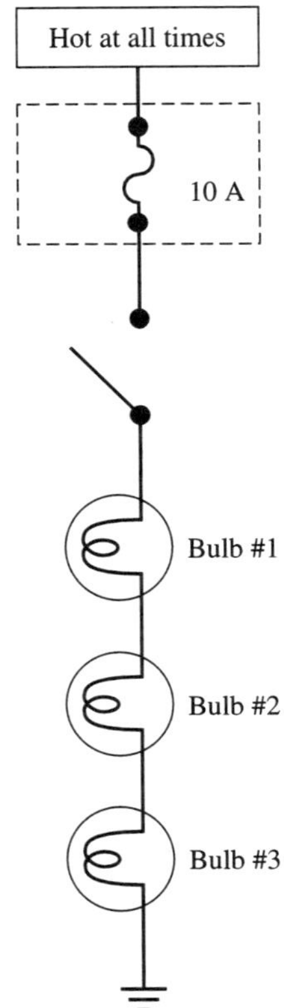

3. A circuit has 4 resistors wired in series; 10 Ω, 5 Ω, 4 Ω, 2.5 Ω. What is the resistance total? The correct answer is:
   A. 21.5 Ω
   B. 1.05 Ω
   C. 20.5 Ω
   D. None of the above

4. Technician A says the resistance total in a series circuit is always the same as the amperage total. Technician B says the resistance total in a series circuit is always one half of the voltage total. Who is correct?
   A. A only
   B. B only
   C. Both A and B
   D. Neither A nor B

5. Technician A says that adding more series loads to a series circuit increases the total circuit amperage. Technician B says that if a series circuit is opened, the entire circuit will not operate. Who is correct?
   A. A only
   B. B only
   C. Both A and B
   D. Neither A nor B

6. In a typical heater blower circuit, the circuit does not work in the low position, but works fine in all other switch positions. Technician A says the switch could be at fault. Technician B says the blower resistor block could be at fault. Who is correct?
   A. A only
   B. B only
   C. Both A and B
   D. Neither A nor B

7. A series circuit has three resistors of 20 Ω each. The voltage drop of each resistor is 4 volts. Technician A says the voltage total is 12 volts. Technician B says the resistance total is 60 Ω. Who is correct?
   A. A only
   B. B only
   C. Both A and B
   D. Neither A nor B

8. Technician A says the individual resistive loads in a series circuit will add up to equal the resistance total. Technician B says voltage drops in a series circuit will add up to equal the voltage total. Who is correct?
   A. A only
   B. B only
   C. Both A and B
   D. Neither A nor B

9. A vehicle blows a fuse in a series blower motor circuit only when the switch is in the *HI* position. Technician A says the blower motor could be drawing too much amperage in this position for the size of the fuse. Technician B says a wrong amperage fuse could be the problem. Who is correct?
   A. A only
   B. B only
   C. Both A and B
   D. Neither A nor B

10. 150 milliamps total is flowing through a circuit that consists of three resistors in series, 30 Ω, 40 Ω, and 10 Ω. Technician A says the source voltage for the circuit is 120 volts. Technician B says the voltage is 1.2 volts. Who is correct?
    A. A only
    B. B only
    C. Both A and B
    D. Neither A nor B

8 Parallel Circuits

Introduction

In Chapter 7 you were introduced to series circuit laws pertaining to voltage, current, and resistance. In this chapter you will be exposed to some new electrical circuit laws and formulas pertaining to parallel circuits *only.* Rest assured though, as you apply these electrical laws and formulas to voltage, current, and resistance in parallel circuits you will be guided through each new concept step-by-step. When you have completed this chapter your confidence in working with parallel circuits should be greatly enhanced.

Objectives

When you complete this chapter you should be able to:

- ❑ Define a parallel circuit.
- ❑ Apply parallel circuit laws for voltage, current, and resistance.
- ❑ Define Kirchoff's Current Law.
- ❑ Perform parallel circuit calculations.
- ❑ Apply troubleshooting and testing techniques to parallel circuits.
- ❑ Perform basic hands-on parallel circuit exercises on a vehicle.

Parallel Circuit Explanation

Parallel circuits contain two or more resistive loads with all the *positive terminals* connected to a *common* junction and all *negative terminals* connected to a second *common* junction. The current will flow to each branch independent of any other branch.

Note: In Figure 8–1, after the current leaves the positive side of the battery it will split three ways through the three resistors. The amount of current flowing through each resistive load depends on the value of the individual resistor. An open circuit in one resistive load does not effect the operation of the other loads of the circuit.

Remember that Ohm's Law states the higher the resistance, the lower the current flow. Current and resistance are inversely proportional to each other.

In the automobile there are a lot of simple parallel circuits. Some of these are headlights, back-up lights, courtesy lights, rear window defroster grids, and taillight circuits. Figure 8–2 is an excellent example of a simple exterior lamp parallel circuit. All of the circuit bulbs share a common positive at splice S408, and eventually share a common ground at connection G400.

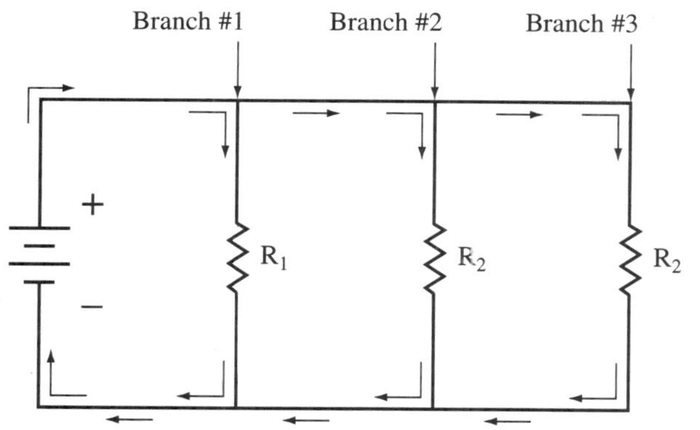

FIGURE 8–1 Simple parallel circuit.

Circuit Laws

Now that we know what a basic parallel circuit is, let's look at the three parallel circuit laws that pertain *only* to parallel circuits.

Parallel Law #1

In a parallel circuit, total current flow is equal to the *sum* of the current flowing through the individual load circuits (**Kirchoff's Current Law**). In simplified terms, $I_T = I_1 + I_2 + I_3 + \text{etc....}$ (additive). No matter how

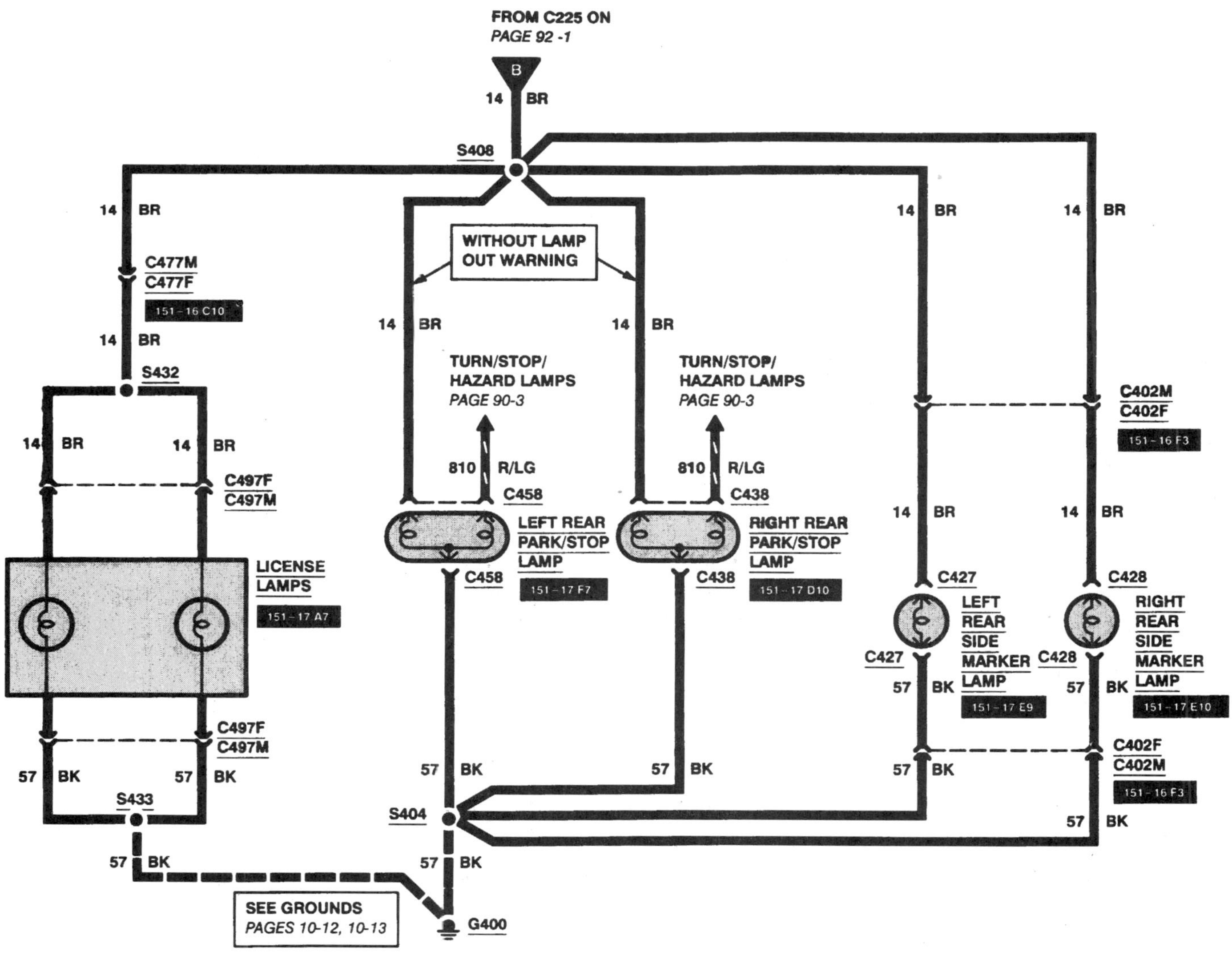

FIGURE 8–2 Simple parallel lighting circuit.
(Courtesy of Ford Motor Company)

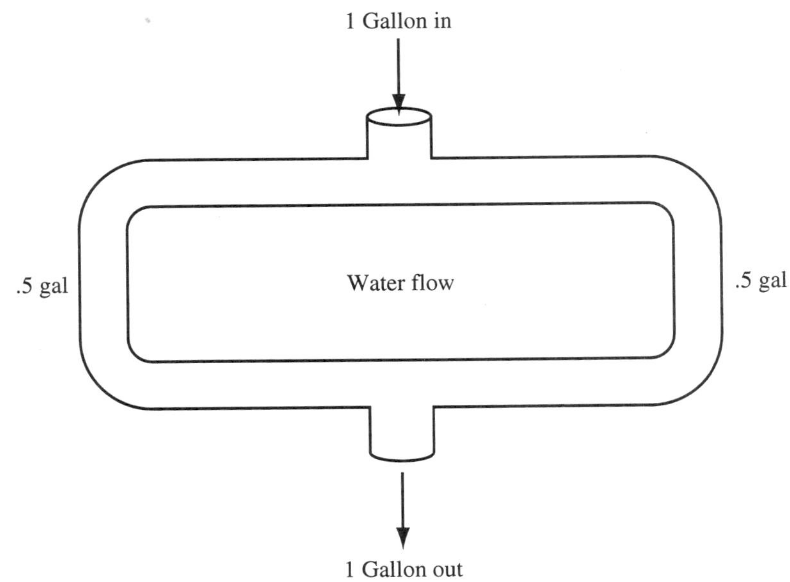

FIGURE 8–3 Water flow example.

many resistive loads are connected together at a common junction, Kirchoff's Law still applies. If one circuit load develops an open circuit, the others continue to operate as designed.

If you were to compare this with the Series current law in Chapter 7, you will note the difference: $I_T = I_1 = I_2 = I_3 =$ etc. . . . Keep in mind that current flow in series circuits is equal throughout the circuit. If one part of the circuit quits, the whole circuit quits.

Another way of describing the flow of current through a parallel circuit is to use the flow of water through a piping system. If 1 gallon of water was poured into the system (Figure 8–3) and each leg allowed 1/2 gallon of flow, then the result must be 1 gallon of total output. What goes *into* it, must go *out* of it!

Simple Experiment

You can try an experiment yourself to demonstrate this theory by taking an empty milk jug and cutting two square half inch holes close together in the bottom corner. Cover both holes with tape and fill the jug with water. Using a stop watch that reads in seconds, let the water out of the jug by uncovering *one* of the holes. Record the results in seconds ____________________. Refill the jug with water. While using the stop watch, let the water out again by uncovering both holes. The results should be approximately one half of the first reading.

The theory of fluid behind this experiment is directly related to an electrical circuit diagram (Figure 8–4).

The circuit in Figure 8–4 has 1 amp of total current flow from the 12 volt battery that it divides into two loads having equal resistance of 24 Ω each. This means the amperage will divide evenly between the two loads, or .5 amps per branch load as indicated by each ammeter wired in series. As you learn the other two parallel circuit laws in the next few pages, you will gain a better understanding of how to calculate current flow in the various loads of a parallel circuit.

For example, Figure 8–5 is a parallel circuit with four resistive loads. If you knew the total current of the circuit and the current of three of the four branches, you could subtract what you know about the three branches from the total to get what you don't know—the fourth branch.

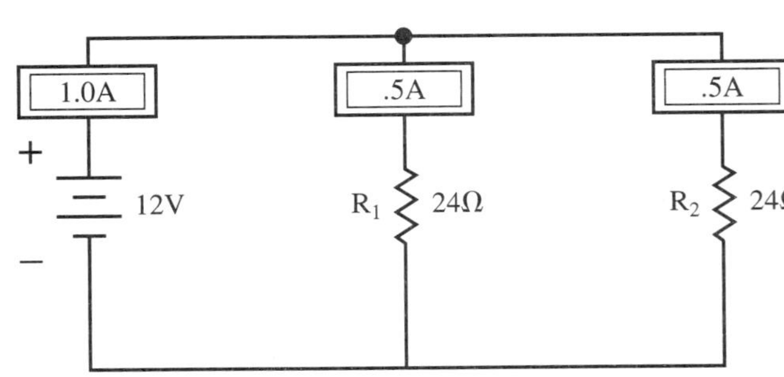

FIGURE 8–4 Parallel circuit flow.

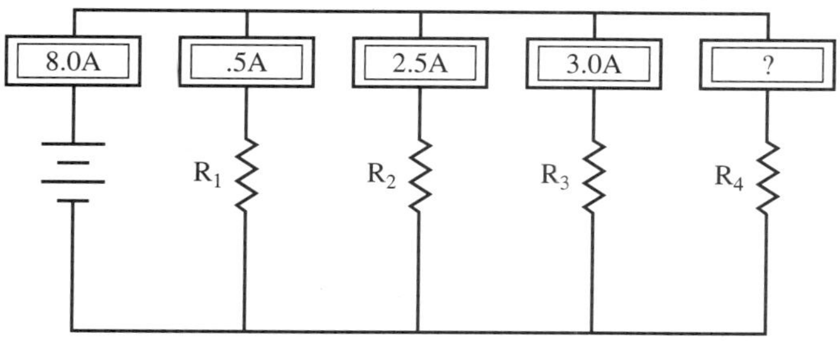

FIGURE 8–5 Parallel circuit with four resistive loads.

$$I_T = I_1 + I_2 + I_3 + I_4, \text{ or}$$
$$8A = .5A + 2.5A + 3A + (\text{the unknown amps of } I_4)$$

If you know I_1, I_2, and I_3, just subtract them from I_T to get I_4, which is 2 amps.

Keep this principle in mind because it could assist you in solving future exercises, and also solving actual electrical problems.

Parallel Law #2

The source voltage (total) applied to a parallel circuit is the same as the voltage applied to each branch load in the circuit. This is also stated in **Kirchoff's Voltage Law,** where the sum of voltages around any closed circuit is equal to the sum of the applied voltage.

$$E_T = E_1 = E_2 = E_3 = \text{etc.} \ldots \text{(same or equal)}$$

This law is different than the voltage law you learned for the series circuit in Chapter 7: $E_T = E_1 + E_2 + E_3 +$ etc. . . . (additive). Make sure to keep these two laws separate, or you are sure to make some big diagnosis errors!

In Figure 8–6, you will see that all the parallel loads connect to a common positive. The other end of the loads connect to a common negative. The voltage meters show the voltage throughout the circuit is the same as the source voltage of 12 volts. Knowing this principle makes it easier to perform calculations and electrical diagnosis on parallel circuits. No matter how many circuit loads there are, as long as there is no circuit resistance between the source voltage and each individual load, and back to the source voltage, then the voltage is THE SAME as the source voltage. *Don't forget this!*

Parallel Law #3

In a parallel circuit, the total circuit resistance (R_T) is *always* less than the value of the lowest resistive branch in the circuit.

$$R_T = \text{less than } (<) \text{ the lowest resistive branch}$$

The series resistance law $R_T = R_1 + R_2 + R_3 +$ etc. . . . (additive) is quite different from a parallel circuit, so remember to keep them separated.

Computing resistance in a parallel circuit can be simplified depending on the method that you use to get your answers. There are five different methods that you can use to find resistance in a parallel circuit. Since your objective is finding the answer using the quickest and easiest way possible, choose a method comfortable with your math abilities. And remember to use your calculator whenever possible!

Method #1: Product Over Sum Method You can use this method for a parallel circuit with *two* resistive loads only (see Figure 8–7). If there are more than two resistors, then this method can be used in *groups of two* to get the answer.

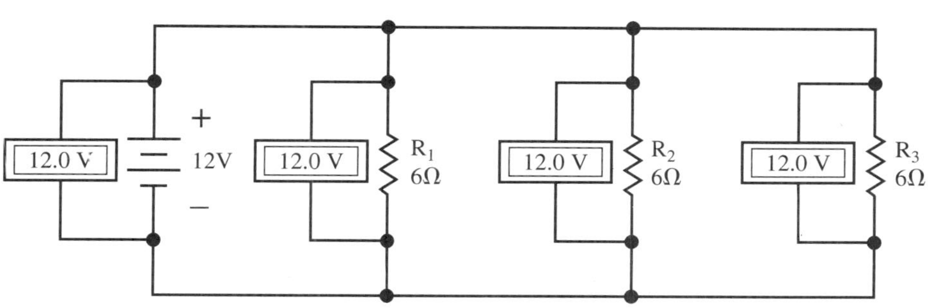

FIGURE 8–6 Voltages of parallel circuit branches.

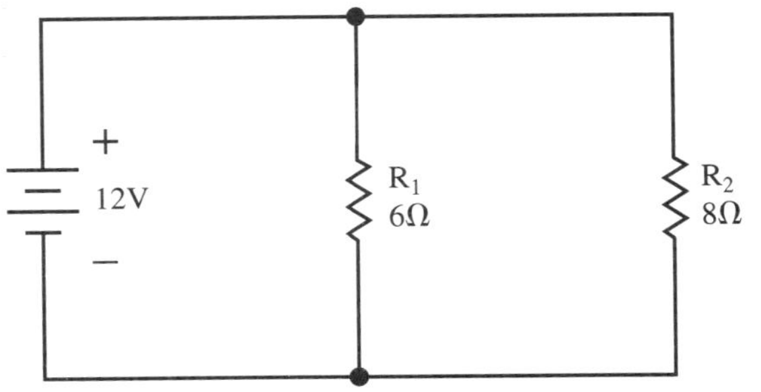

FIGURE 8–7 Parallel circuit with two resistive loads.

HINT: If there are more then two resistive loads, this method is too time consuming. There is an easier way we will see later in this chapter.

Calculation: $R_T = \dfrac{R_1 \times R_2}{R_1 + R_2}$

To perform the calculation:

Step 1 $R_T = \dfrac{6\Omega \times 8\Omega}{6\,\Omega + 8\,\Omega}$

Step 2 $R_T = \dfrac{48\,\Omega}{14\,\Omega}$

Step 3 $R_T = 3.42857\,\Omega$, or $3.43\,\Omega$

If there are more than two resistive loads in a parallel circuit (Figure 8–8), and you wanted to use Method #1, use the following process to calculate the resistance of the circuit. (Remember you can only do two resistive loads at a time.)

Calculation: $R_T = \dfrac{R_1 \times R_2}{R_1 + R_2}$

Step 1 $R_A = \dfrac{4\,\Omega \times 3\,\Omega}{4\,\Omega + 3\,\Omega}$

$R_A = \dfrac{12\,\Omega}{7\,\Omega}$ $R_A = 1.714\,\Omega$

Next, use R_A (1.714 Ω) as the first of your two numbers and 6 Ω (R_3) as the second.

Step 2 $R_B = \dfrac{1.714\,\Omega \times 6\,\Omega}{1.714\,\Omega + 6\,\Omega}$

$R_B = \dfrac{10.284\,\Omega}{7.714\,\Omega}$ $R_B = 1.333\,\Omega$

Next, use 1.333 Ω as the first number and 8 Ω (R_4)as the second number.

Step 3 $R_T = \dfrac{1.333\,\Omega \times 8\,\Omega}{1.333\,\Omega + 8\,\Omega}$

$R_T = \dfrac{10.664\,\Omega}{9.333\,\Omega}$

$R_T = 1.143\,\Omega$

Even though Method #1 eventually gets the right answer, there are other methods that work better when there are more than two resistive branches in a parallel circuit.

Method #2: Same Value Method This method can be used if all the resistive loads in the parallel circuit are of the same value. Simply take the value of the resistors and divide by the number of resistors in the circuit (Figure 8–9).

There are nine resistive branches with a 30 Ω load in each branch. Divide 30 by 9 to get R_T, which is 3.33 Ω.

Remember, this method applies only when all resistive loads have the same value. If you had an electrical circuit where some of the loads were the same and some

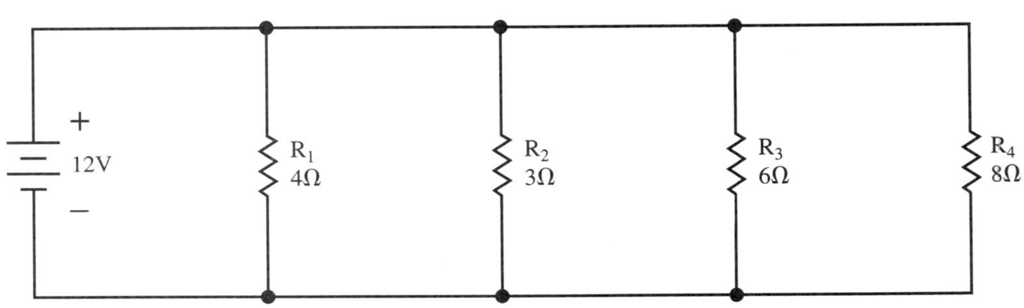

FIGURE 8–8 Parallel circuit with more than two resistive loads.

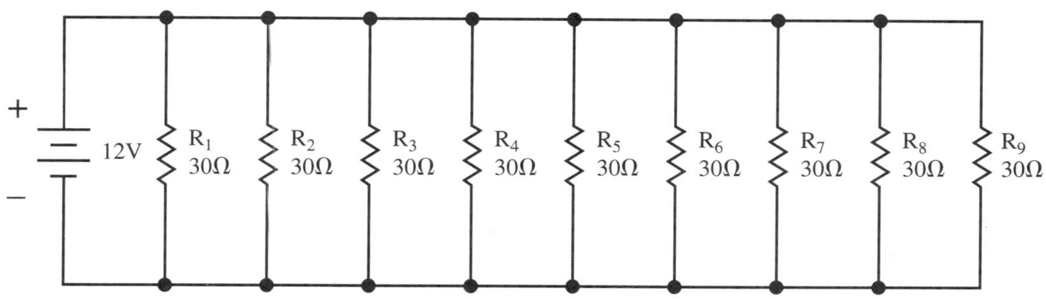

FIGURE 8–9 Parallel circuit with all resistance values the same.

were different, you could save time by using Method #2 to find the common value of those resistive loads with the same value. Then you could use Method #1 to find the total circuit resistance. This approach takes a considerable amount of time, so we will look at other methods more suited to this type of problem.

Method #3: Assumed Voltage Method Because the applied voltage is constant in a parallel circuit, we can use Ohm's Law to find the resistance. No matter what voltage you use for the calculation, the amperage will be proportional to the voltage. The resistance total can be calculated accurately using this method.

For example, use the same four parallel resistances used in Figure 8–8.

$$4\ \Omega, 3\ \Omega, 6\ \Omega, 8\ \Omega$$

And, assume a constant voltage of 12 volts.
Using Ohms's Law: $E \div R = I$, then:

$$\frac{12\text{ V}}{4\ \Omega} = 3\text{ A} \qquad \frac{12\text{ V}}{3\ \Omega} = 4\text{ A}$$

$$\frac{12\text{ V}}{6\ \Omega} = 2\text{ A} \qquad \frac{12\text{ V}}{8\ \Omega} = 1.5\text{ A}$$

Remember that the circuit law for current in a parallel circuit is additive.

$$I_T = I_1 + I_2 + I_3 + \text{etc.} \ldots$$

Add the answers from above.

$$3\text{ A} + 4\text{ A} + 2\text{ A} + 1.5\text{ A} = 10.5\text{ A} = I_T$$

Take the assumed voltage total and divide by the amps total to get resistance total.

$$\frac{12.0\text{ V}}{10.5\text{ A}} = 1.143\ \Omega = R_T$$

R_T equals 1.143 Ω and is the same answer as when using the four resistive values of Method #1. Let's try Method #3 again, assuming a source voltage of 38 volts instead to show that this method always works.

$$\frac{38\text{ V}}{4\ \Omega} = 9.5\text{ A} \qquad \frac{38\text{ V}}{3\ \Omega} = 12.67\text{ A}$$

$$\frac{38\text{ V}}{6\ \Omega} = 6.33\text{ A} \qquad \frac{38\text{ V}}{8\ \Omega} = 4.75\text{ A}$$

Add the amps as before.

$$9.5\text{ A} + 12.67\text{ A} + 6.33 + 4.75\text{ A} = 33.25\text{ A} = I_T$$

$$\frac{38.0\text{ V}}{33.25\text{ A}} = 1.143\ \Omega = R_T$$

Note: As you can see, an R_T of 1.143 Ω is the same for both problems. If you used any other value for the voltage, the answer would still come out the same. Remember that voltage and amperage are proportional to each other. Try a few voltage examples of your own to show the answer will be the same (1.143 Ω).

Method #3 works, but can be very time consuming in finding a resistance total. There are other methods that are better suited for this type of circuit.

Method #4 Reciprocal Formula Method This is the method most often used in textbooks and taught by electrical educators. We will explore this method before moving on to another method. For consistency, let's use the same values as in the previous parallel circuit example:

$$4\ \Omega, \quad 3\ \Omega, \quad 6\Omega, \quad 8\ \Omega$$

Formula for Method #4:

$$R_T = \frac{1}{\frac{1}{R_1} + \frac{1}{R_2} + \frac{1}{R_3} + \frac{1}{R_4}}$$

Step 1 $$R_T = \frac{1}{\frac{1}{4\,\Omega} + \frac{1}{3\,\Omega} + \frac{1}{6\,\Omega} + \frac{1}{8\,\Omega}}$$

At this point, you can either calculate the fractions by finding their lowest common denominator, or by converting the reciprocal numbers to their decimal equivalents.

Let's do it both ways!!

Step 2 Fraction style. The lowest common denominator in this problem is 24.

$1/4 = 6/24$, $1/3 = 8/24$,
$1/6 = 4/24$, $1/8 = 3/24$

Step 3 $$R_T = \frac{1}{\frac{6}{24} + \frac{8}{24} + \frac{4}{24} + \frac{3}{24}}$$

Step 4 $$R_T = \frac{1}{\frac{21}{24}}$$

Step 5 $$R_T = 1 \times \frac{24}{21}$$

Step 6: $24 \div 21 = R_T = 1.143\ \Omega$

Great! This is the same answer as in the previous two methods!!

Now let's try using Method #4 again, but this time converting the same resistance values to decimals.

Step 1 $$R_T = \frac{1}{\frac{1}{4\,\Omega} + \frac{1}{3\,\Omega} + \frac{1}{6\,\Omega} + \frac{1}{8\,\Omega}}$$

Step 2 Divide each number into 100 to get decimal equivalents.

$$\frac{100}{4} = 25 \qquad \frac{100}{3} = 33.3$$

$$\frac{100}{6} = 16.67 \qquad \frac{100}{8} = 12.5$$

Step 3 Add up the decimal answers of each to find the total decimal value.

$25 + 33.3 + 16.67 + 12.5 = 87.47$

Step 4 Because of the numerator 1 over the original calculation (in Step 1) we need to divide 87.47 into 100.

$$\frac{100.00}{87.47} = 1.143\ \Omega$$

This is the same answer as before!!

Note: As you recall, the *primary* goal when finding the resistance total of a parallel circuit is to get a correct answer the fastest and easiest way possible. Let's do that now.

Method #5: Scientific Calculator Method This method really performs all the calculations of Method #4, except you are letting the calculator do all the work. If a calculator is available, use it!

Step 1 Identify the **1/X** key on your scientific calculator. On some calculators you need to push the **2nd** function key first because the **1/X** key is on the lower printed portion of the calculator face. Consult your calculator reference guide if you are unsure of location of the **1/X** key.

Step 2 Use the same four parallel circuit load values as before.

4 Ω, 3 Ω, 6 Ω, 8 Ω

Step 3 Press the keys exactly as shown:

4 **1/X** + 3 **1/X** + 6 **1/X** + 8 **1/X** **=** **1/X**

The answer on the calculator should be 1.1428571. Round off to 1.143 Ω

Note: This is the *same* answer as using the previous methods.

By entering this series of numbers and functions, the calculator performed the reciprocal calculations for you. A common mistake is to forget to push the **1/X** key one final time after you push the = key. No matter how complex the parallel resistance numbers are, the process in Method #5 is still the same. If a circuit has a mixture of ohms (Ω), kiloohms (kΩ), and megaohms (MΩ), then you would need to apply the **EXP** function key to the process.

For example, suppose the parallel circuit has resistive loads of 12 kΩ, 2.5 MΩ, 640 Ω, and 275 kΩ. You would enter the keys as follows:

12 **EXP** 3 **1/X** + 2.5 **EXP** 6 **1/X** +
640 1/X + 275 **EXP** 3 **1/X** = **1/X**

Your answer should be 606.11 Ω.

That last example made the process look simple. Using the calculator properly makes Method #5 the simplest method to use when finding the resistance total in a parallel circuit.

Because of this, all future discussions and practice examples in this text dealing with parallel circuit resistance calculations will primarily use Method #5. Having an understanding of parallel circuit laws and methods will enable you to focus more on circuit and electrical diagnosis than on performing complex math calculations.

Let us start with a few easy exercises in parallel circuits to verify that you know the circuit laws and calculations, and then move on to more complex circuits with real electrical diagrams. Through repetition of these parallel exercises you will build a solid foundation of parallel circuit laws that will aid you in diagnosing complex circuits.

Before we continue, though, here is a brief overview of series and parallel circuit laws. Mixing up the different circuit laws could be frustrating, so use this overview as a quick reference guide.

Review of Basic Laws of Electricity

	Series	Parallel
CURRENT	Equal or Same $I_T = I_1 = I_2 = I_3$ = etc.	Additive $I_T = I_1 + I_2 + I_3$ + etc.
VOLTAGE	Additive $E_T = E_1 + E_2 + E_3$ + etc.	Equal or Same $E_T = E_1 = E_2 = E_3$ = etc.
RESISTANCE	Additive $R_T = R_1 + R_2 + R_3$ + etc.	Less Than R_T = < lowest # resistor

Exercises—Parallel Circuits

The following exercises are designed to help give you practice using the parallel laws and methods previously explained. The exercises increase in complexity in an effort to test your understanding of the basic parallel laws, and to prepare you for the actual hands-on parallel circuit lab exercises that follow.

Exercise #1

Instructions Identify the type of circuit in Figure 8–10. On a *separate* sheet of paper make an answer table like Table 8-1. Using the table, fill in the known circuit values from the parallel circuit diagram. This is a helpful first step that gives you an idea of where to begin and what answers you need. Then, calculate the answers and write them in the remaining spaces using one of the methods previously learned.

R_T = Resistance total, I_T = Amperage total,
E_T = Voltage total, W_T = Watts total

R_1 = Resistor #1, I_1 = Amperage #1,
E_1 = Voltage #1, W_1 = Watts #1

R_2 = Resistor #2, I_2 = Amperage #2,
E_2 = Voltage #2, W_2 = Watts #2

Table 8–1

R_T____	I_T____	E_T____	W_T____
R_1____	I_1____	E_1____	W_1____
R_2____	I_2____	E_2____	W_2____

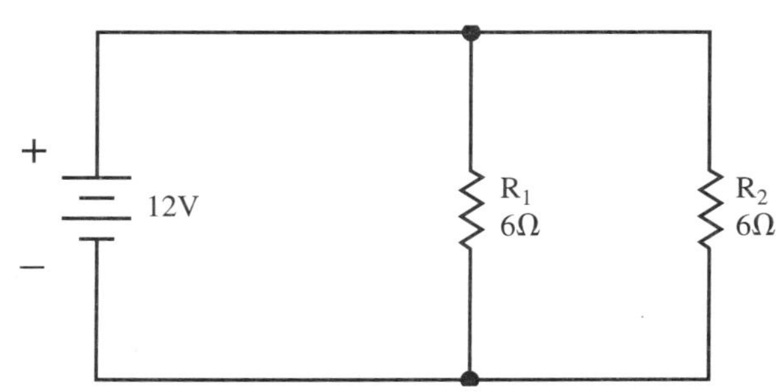

FIGURE 8–10 Parallel circuit.

Circuit Analysis Here are the correct steps and answers for Exercise #1. If your answers are different, refer to the discussion section at the front of this chapter for further study.

Step #1 Known values: $E_T = 12$ V, $R_1 = 6\ \Omega$, and $R_2 = 6\ \Omega$. Remember that the Voltage Law states that the voltage is constant in parallel circuits, so fill in the voltage for E_1 and E_2 with 12 volts.

Step #2 To find total resistance use Method 1, 2, or 5. We will use Method 5 here because it will give you practice using the calculator's **1/X** key. Begin by entering the following into the calculator: 6 **1/X + 6 1/X = 1/X.** The resistance total should be 3 Ω Add this to your table.

Step #3 Take the voltage total of 12V and divide it by the resistance total of 3 Ω to get the amps total of 4 A. Divide E_1 (12 V) by R_1 (6 Ω) to get I_1 which equals 2 A. Do the same with E_2 and R_2 to get I_2. To verify the correct answer make sure the amps add up to equal the total amps, or $I_T = I_1 + I_2$.

Step #4 To find the Watts total, use Watts Law. Voltage total times amperage total equals the Watts total, or 48 W (if unsure, refer back to Chapter 3). Use the same procedure to find W_1 and W_2.

Table 8–2 is a summary of the correct answers.

Table 8–2

R_T 3 Ω	I_T 4 A	E_T 12 V	W_T 48 W
R_1 6 Ω	I_1 2 A	E_1 12 V	W_1 24 W
R_2 6 Ω	I_2 2 A	E_2 12 V	W_2 24 W

Note: Follow the steps used in Exercise #1 for the remaining parallel circuit exercises. Remember to put your answers on a *separate* sheet of paper.

Exercise #2

Note: Make sure you always look at the known values! In Figure 8–11 the voltage was changed from 12 volts to 24 volts. Complete your answers on a table like Table 8–3.

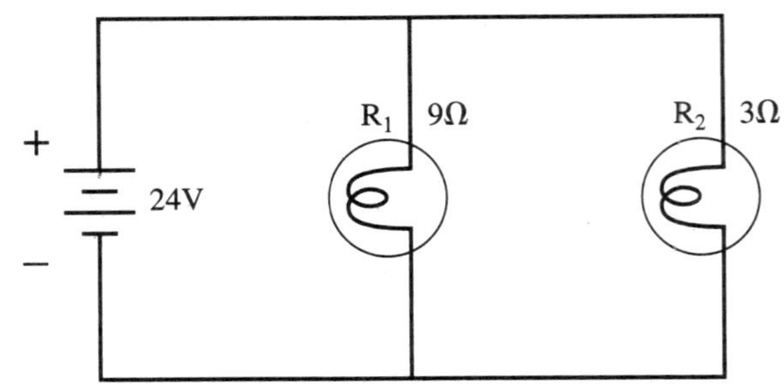

FIGURE 8–11 Simple parallel light circuit.

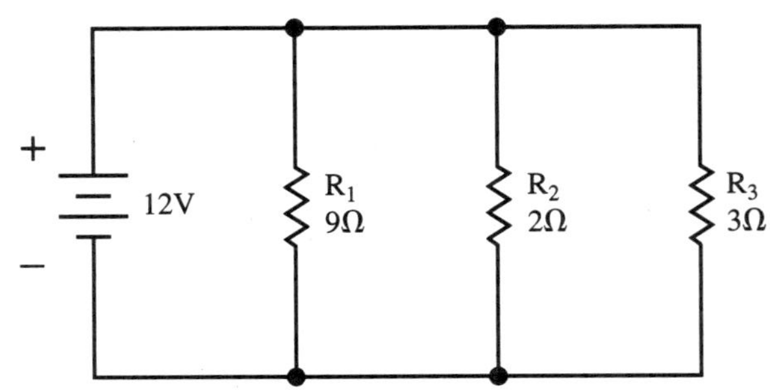

FIGURE 8–12 Simple parallel circuit.

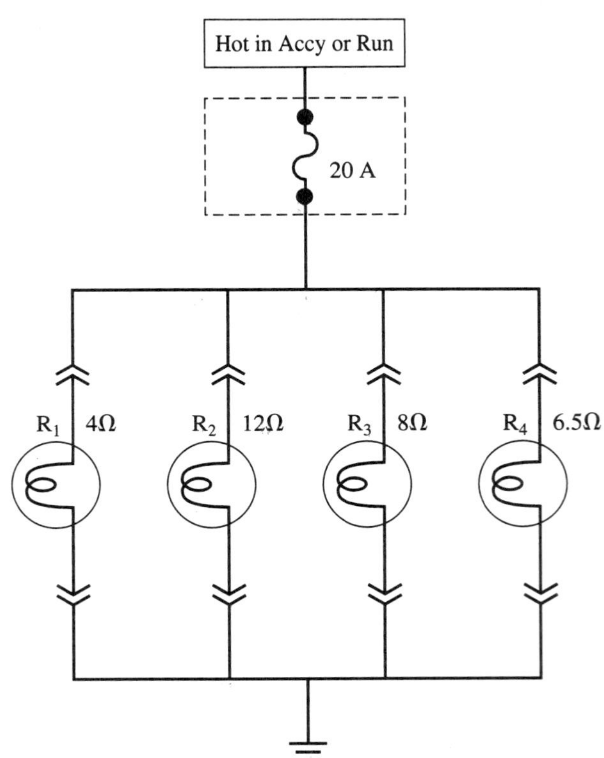

FIGURE 8–13 Typical parallel light circuit.

Table 8–3

R_T		I_T		E_T	24 V	W_T	
R_1	9 Ω	I_1		E_1		W_1	
R_2	3 Ω	I_2		E_2		W_2	

Exercise #3

Note: After you have identified the type of circuit in Figure 8–12, fill in the known values on a table like Table 8–4.

Table 8–4

R_T		I_T		E_T		W_T	
R_1		I_1		E_1		W_1	
R_2		I_2		E_2		W_2	
R_3		I_3		E_3		W_3	

Exercise #4

Note: The circuit in Figure 8–13 has been drawn differently than in the previous exercises. If you look closely, you can see that the circuit starts at the top, and then splits three ways depending on the resistive value of each light bulb. The circuit then recombines to form a common ground at the bottom of the page. Complete the answers to this parallel circuit on a table like Table 8–5.

Table 8–5

R_T		I_T		E_T		W_T	
R_1		I_1		E_1		W_1	
R_2		I_2		E_2		W_2	
R_3		I_3		E_3		W_3	
R_4		I_4		E_4		W_4	

Exercise #5

Note: For this exercise the voltage is the unknown property (see Figure 8–14). After you find R_T you can then use I_T and Ohm's Law to find the voltages. Once you have all the voltages, you can get the remaining individual amperages. Complete your answers on a table like Table 8–6.

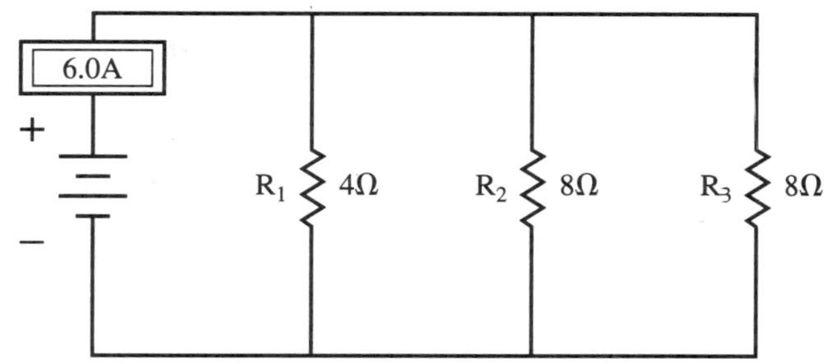

FIGURE 8–14 Parallel circuit with an unknown voltage.

Table 8–6

R_T		I_T		E_T		W_T	
R_1		I_1		E_1		W_1	
R_2		I_2		E_2		W_2	
R_3		I_3		E_3		W_3	

Exercise #6

Note: Figure 8–15 shows a parallel circuit drawn a little differently. Just because the schematic is not drawn from left to right, or from top to bottom, does not mean the theory of parallel circuits is different. Identify the circuit by following the wiring from the positive side of the battery to each resistive branch independently of each other. This can also be done from the negative side of each branch to the battery negative. This circuit can be redrawn in more simplified forms like the three examples in Figure 8–16.

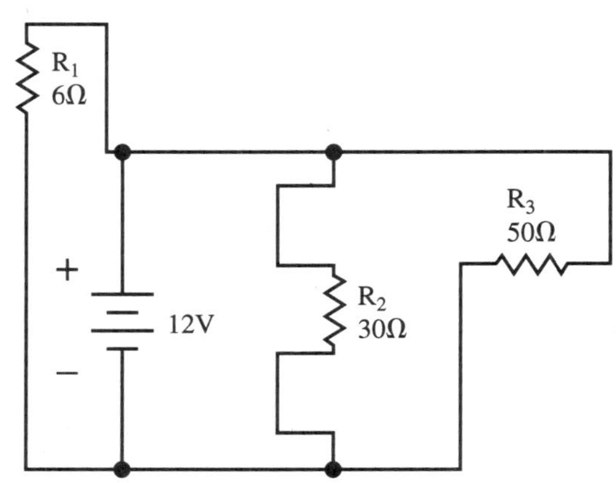

FIGURE 8–15 Simple parallel circuit.

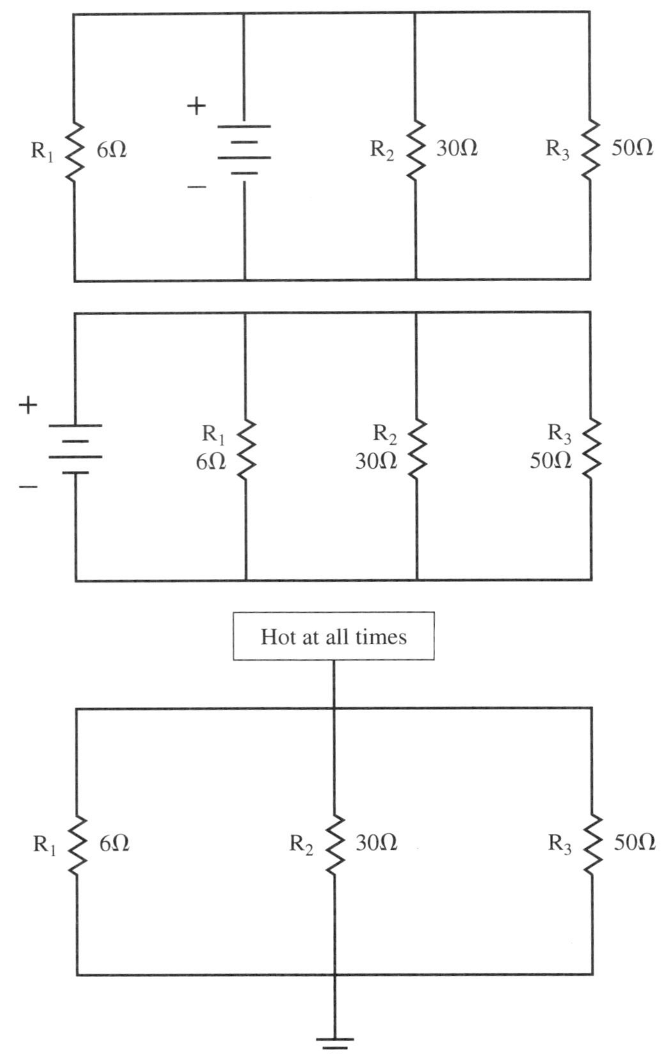

FIGURE 8–16 Three different drawings of the same parallel circuit.

Complete the answers to the above Exercise #6 by filling in a table like Table 8–7.

Table 8–7

R_T ________	I_T ________	E_T ________	W_T ________
R_1 ________	I_1 ________	E_1 ________	W_1 ________
R_2 ________	I_2 ________	E_2 ________	W_2 ________
R_3 ________	I_3 ________	E_3 ________	W_3 ________

Now that you have practiced with examples of basic parallel circuits, it is time to increase the difficulty of the parallel circuit exercises. Always remember, no matter how complex the circuit looks it *still* uses the same basic parallel circuit laws. The better you understand the laws applying to parallel circuits, the easier electrical problem solving will become.

Exercise #7

Identify the type of circuit in Figure 8–17. Using the known values for this circuit from the Table 8–8, complete the answers for this exercise on a *separate* sheet of paper.

Table 8–8

R_T ________	I_T 3.75 A	E_T ________	W_T ________
R_1 ________	I_1 1.50 A	E_1 ________	W_1 ________
R_2 ________	I_2 1.00 A	E_2 ________	W_2 ________
R_3 ________	I_3 ________	E_3 18 V	W_3 ________
R_4 ________	I_4 .75 A	E_4 ________	W_4 ________

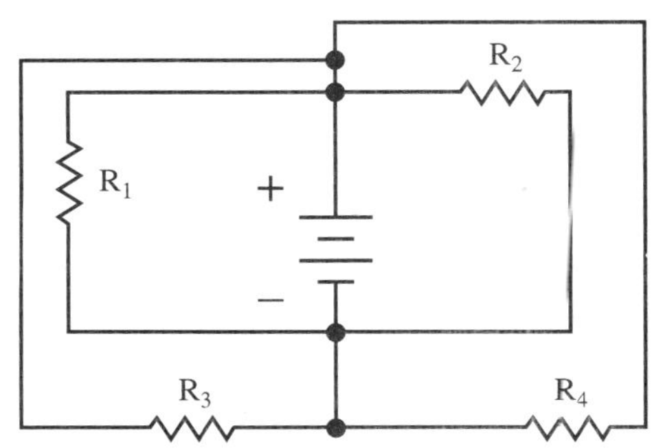

FIGURE 8–17 It looks different, but it is still a parallel circuit.

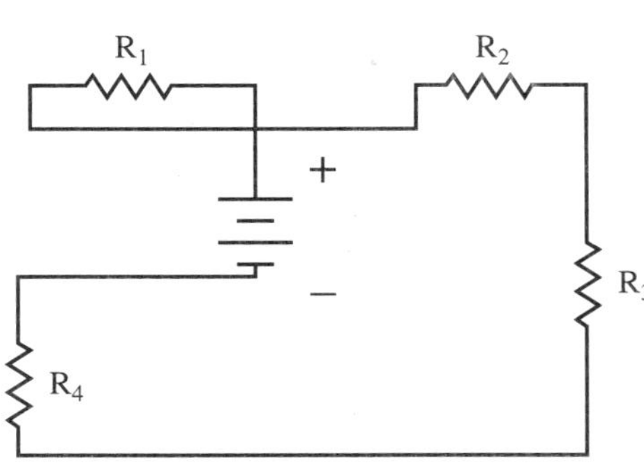

FIGURE 8–18 Identify unknown electrical circuit.

Exercise #8

Using the circuit in Figure 8–18 as a reference, use a *separate* sheet of paper and complete Table 8–9.

Table 8–9

R_T		I_T		E_T		W_T	
R_1	2 Ω	I_1		E_1		W_1	
R_2	4 Ω	I_2		E_2	12 V	W_2	
R_3	3 Ω	I_3		E_3		W_3	
R_4	6 Ω	I_4		E_4		W_4	

Stop If you completed Exercise #8 using the parallel circuit laws, you made an *error!* If you follow the circuit from the battery positive, then back to the battery negative you will see it is a *series* circuit. Make sure you always know what type of circuit you are working on. If you noticed this right away, congratulations, you are very observant!

Exercise #9

The parallel circuit shown in Figure 8–19 has an operating voltage of 14 volts. Because resistance in a bulb cannot be measured directly (hot bulb resistance), refer to

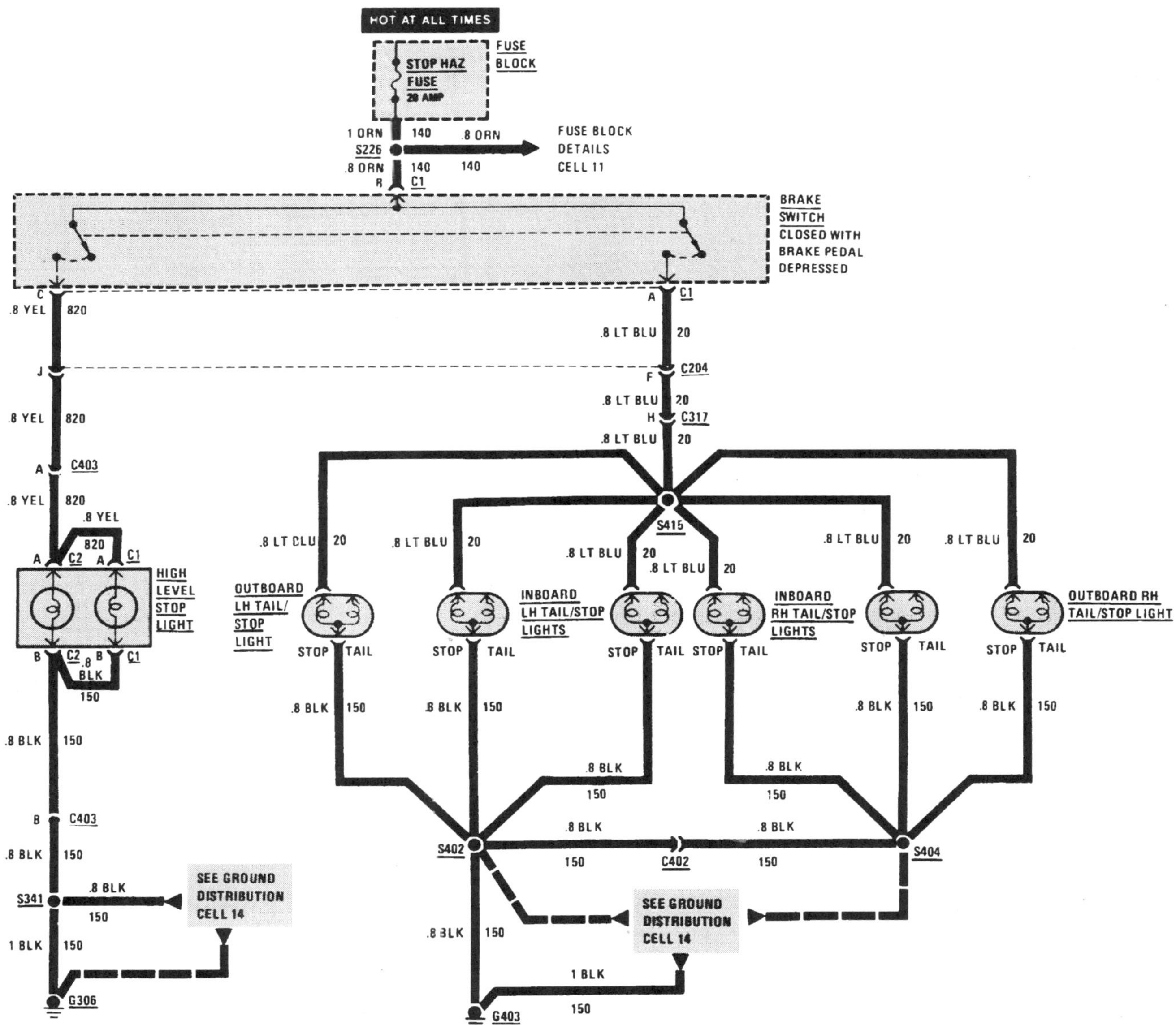

FIGURE 8–19 Exterior stoplight circuit.

(Courtesy of General Motors Corporation, Service Technology Group)

the operating amperages listed in the Lamp Standard Trade Number Chart in Appendix B. The two high-level stoplight bulbs are GE-194 bulbs which use .27 amps each at 14 volts. The six stoplights are GE-2057 bulbs which use 2.297 amps each at 14 volts (2.10 amps each at 12.8 V). On a *separate* sheet of paper complete the calculations to complete Table 8–10.

Table 8–10

R_T ______	I_T ______	E_T ______	W_T ______
R_1 ______	I_1 ______	E_1 ______	W_1 ______
R_2 ______	I_2 ______	E_2 ______	W_2 ______
R_3 ______	I_3 ______	E_3 ______	W_3 ______
R_4 ______	I_4 ______	E_4 ______	W_4 ______
R_5 ______	I_5 ______	E_5 ______	W_5 ______
R_6 ______	I_6 ______	E_6 ______	W_6 ______
R_7 ______	I_7 ______	E_7 ______	W_7 ______
R_8 ______	I_8 ______	E_8 ______	W_8 ______

Exercise #10

Complete your answers to Exercise #10 on a *separate* sheet of paper using a table similar to Table 8–11. Use 12 volts (as shown in Figure 8–20) as the circuit operating voltage.

Table 8–11

R_T ______	I_T ______	E_T ______	W_T ______
R_1 ______	I_1 ______	E_1 ______	W_1 ______
R_2 ______	I_2 ______	E_2 ______	W_2 ______
R_3 ______	I_3 ______	E_3 ______	W_3 ______
R_4 ______	I_4 ______	E_4 ______	W_4 ______
R_5 ______	I_5 ______	E_5 ______	W_5 ______
R_6 ______	I_6 ______	E_6 ______	W_6 ______
R_7 ______	I_7 ______	E_7 ______	W_7 ______

Using Method #5 will come in handy for finding the resistance total in this exercise. Because of the mixture of kiloohms and ohms, enter the following on your calculator to find the resistance total.

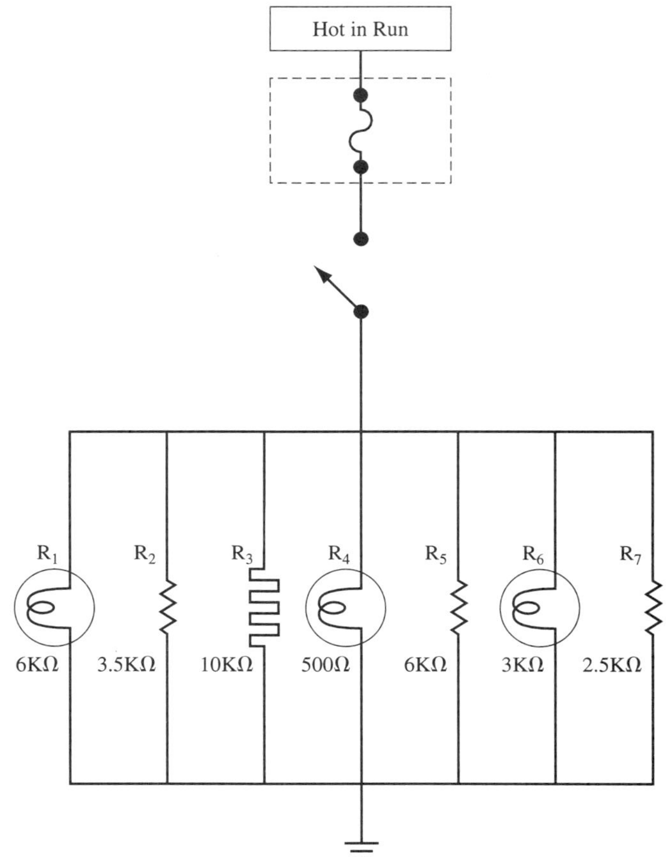

FIGURE 8–20 Parallel circuit with operating voltage of 12 volts.

6 **EXP** 3 **1/X** + 3.5 **EXP**
3 **1/X** + 10 **EXP** 3 **1/X** + 500

1/X + 6 **EXP** 3 **1/X** +
3 **EXP** 3 **1/X** 2.5 **EXP** 3 **1/X** = **1/X**

The answer is 289.66 Ω

In Exercise #10, what percentage of the current flows through I_2? To answer this, you need to start with the following calculation: I_A divided by I_T. You then multiply this result by 100 to get a percentage.

$$\frac{I_2}{I_T} = \frac{.0034\ \text{A}}{.0414\ \text{A}} = .082, \qquad .082 \times 100 = 8.2\,\%$$

You can determine the remaining six circuit loads by using the above calculations.

Total percent of branch I_1 ________.
Total percent of branch I_2 ________.
Total percent of branch I_3 ________.

Total percent of branch I_4 ________.
Total percent of branch I_5 ________.
Total percent of branch I_6 ________.
Total percent of branch I_7 ________.

The total of all the branch percentages should equal 100%. Do they? ________.

In Exercise #10, what percentage of the resistance total is R_2? To answer this, you need to first use the following calculation: R_T divided by R_n. You then multiply this result by 100 to get a percentage.

$$\frac{R_T}{R_2} = \frac{289.66\ \Omega}{3500\ \Omega} = .082, \qquad .082 \times 100 = 8.2\,\%$$

You can determine the remaining six resistive loads by using the above calculations.

Total percent of branch R_1 ________.
Total percent of branch R_2 ________.
Total percent of branch R_3 ________.

Total percent of branch R_4 ________.
Total percent of branch R_5 ________.
Total percent of branch R_6 ________.

Total percent of branch R_7 ________.

The total of all the branch percentages should equal 100%. Do they? ________.

If you look at the answers to the current and resistance percentage of Exercise #10, you will notice that they are the SAME. The reason for this is resistance and current in parallel circuits are not constant, but are divided proportionally into their own totals. Because of this, you could use either current or resistance to determine the same percentage answers. Remember this for future problem solving.

Exercise #11

HINT: The parallel circuit in Figure 8–21 appears more complicated than it really is. Redraw it, if necessary, to see that each load shares a common positive, and that each negative shares its own common. Just remember Ohm's Law, Watt's Law, and the Parallel Circuit Laws.

Complete your answers to Exercise #11 on a separate sheet of paper similar to Table 8–12.

Table 8–12

R_T ____	I_T ____	E_T ____	W_T ____
R_1 25 Ω	I_1 ____	E_1 ____	W_1 ____
R_2 25 Ω	I_2 ____	E_2 ____	W_2 ____
R_3 10 Ω	I_3 ____	E_3 ____	W_3 ____
R_4 6 Ω	I_4 ____	E_4 ____	W_4 ____
R_5 ____	I_5 4.83 A	E_5 ____	W_5 70.04 W
R_6 7.5 Ω	I_6 ____	E_6 ____	W_6 ____

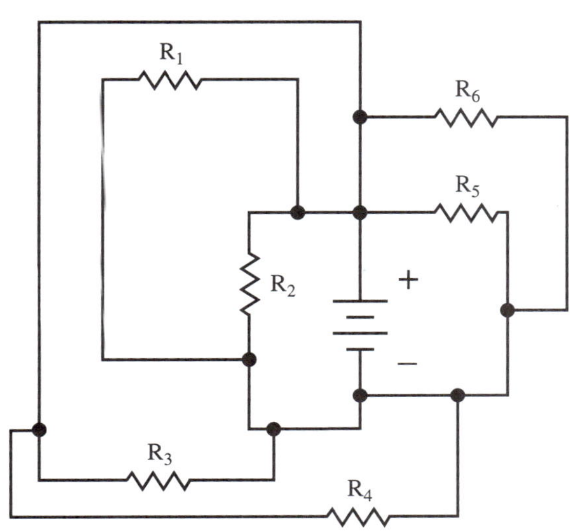

FIGURE 8–21 Complex parallel circuit.

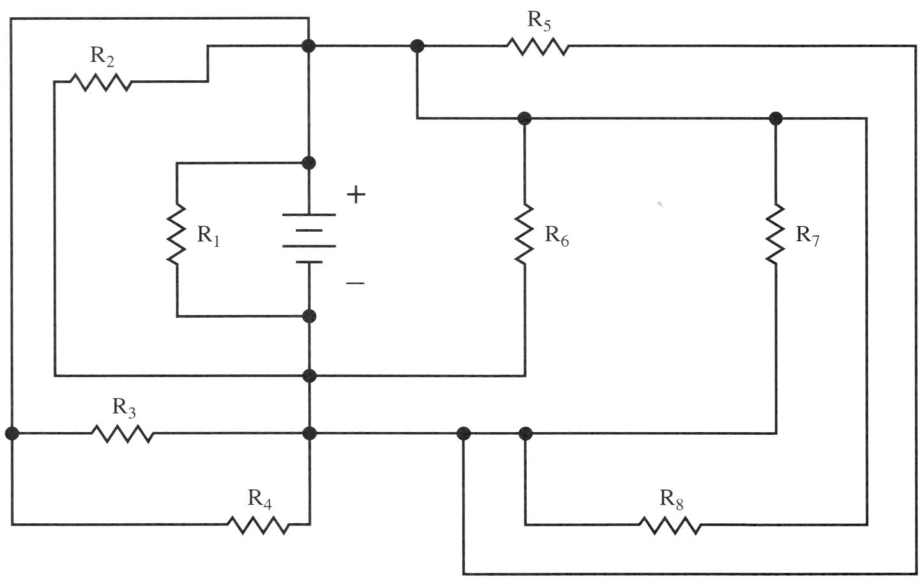

FIGURE 8–22 Complex parallel circuit.

Exercise #12

HINT: When you look at the number values given in Figure 8–22, and at which ones are similar, you should keep in mind the circuit laws you have learned and that amps and resistance are inversely proportional to each other. Complete your answers to this exercise on a *separate* sheet of paper using a table similar to Table 8–13.

You have spent some time learning the parallel circuit theory laws. Even though you most likely will not be standing next to a vehicle electrical problem with a calculator in hand, by performing the past parallel exercises, you should have the basic knowledge to know how a parallel circuit is supposed to perform. Just remember to keep the series and parallel circuit laws separate.

Table 8–13

R_T		I_T	76 A	E_T		W_T	
R_1	4 Ω	I_1		E_1		W_1	
R_2		I_2	8 A	E_2		W_2	
R_3	4 Ω	I_3		E_3		W_3	
R_4		I_4	16 A	E_4		W_4	
R_5		I_5	6 A	E_5		W_5	
R_6		I_6	4 A	E_6		W_6	
R_7	4 Ω	I_7		E_7		W_7	
R_8		I_8	6 A	E_8		W_8	

CASE STUDIES

CASE #1

The following case study is typical of a live parallel circuit problem that technicians often face. Kncwledge of parallel circuits can help you perform the diagnosis quickly and efficiently.

Customer Complaint

The vehicle's parking lamps and taillamps work fine when not hooked up to a camping trailer. As soon as the owner hooks up a trailer, the 10 amp taillamp fuse blows. The customer thinks there must be a short circuit in the trailer lights, and wants it fixed. Figure 8–23 indicates the location of the lights on the vehicle and trailer.

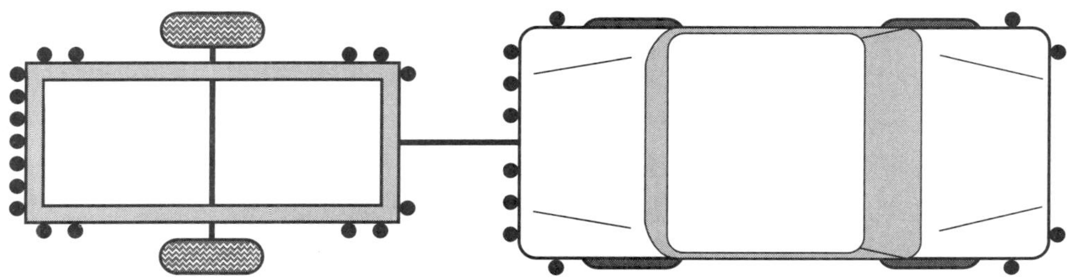

FIGURE 8–23 Vehicle and trailer lighting layout.

Unless you have had considerable experience with exactly the same circuits before, first assumptions about the wiring layout could lead to a misdiagnosis. Figure 8–24 provides a clearer picture for the technician to determine what type of circuit it is, and what circuit laws apply.

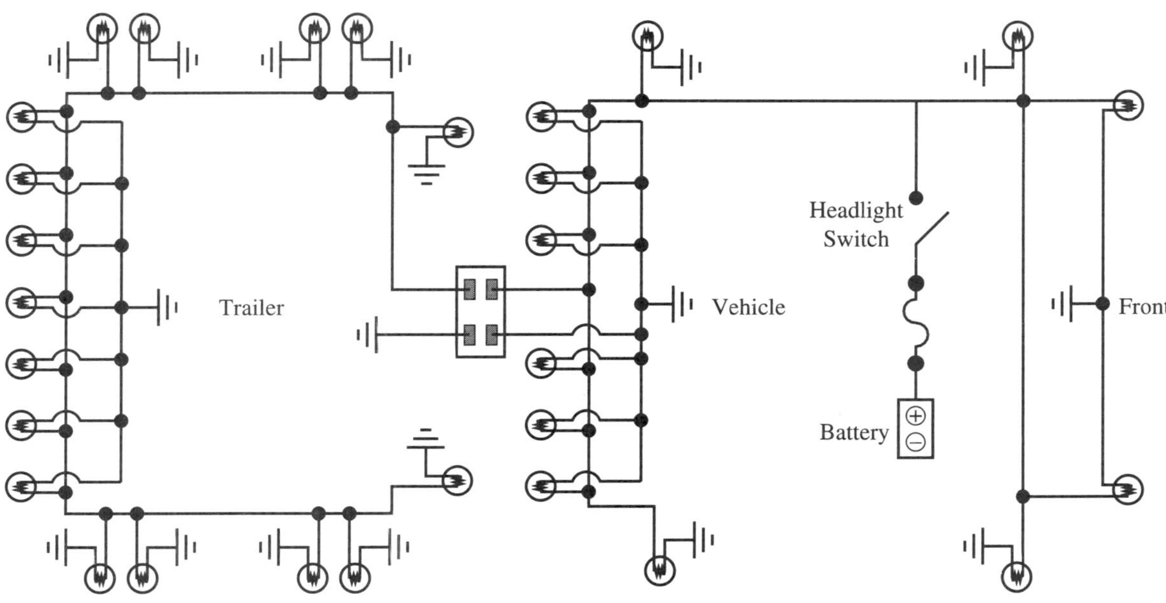

FIGURE 8–24 Vehicle and trailer schematic diagram.

Known Information

- ❑ Vehicle operating voltage equals 14 volts
- ❑ Vehicle's front parking lamps (#1157 bulb) have a design amperage of .59 A each, and a hot bulb resistance of 23.73 Ω each
- ❑ Vehicle's side marker lamps (#194 bulb) have a design amperage of .27 A each, and a hot-bulb resistance of 51.85 Ω each
- ❑ Vehicle's rear taillamps (#1157 bulb) have a design amperage of .59 A each, and a hot-bulb resistance of 23.73 Ω each
- ❑ Trailer's side and front marker lamps (#57 bulb) have a design amperage of .24 A each, and a hot-bulb resistance of 58.33 Ω each
- ❑ Trailer's rear taillamps (#1157 bulb) have a design amperage of .59 A each, and a hot-bulb resistance of 23.73 Ω each

Circuit Analysis

Answer the following questions on a *separate* sheet of paper:

1. What is the total amperage of the vehicle's parking lamp/taillamp circuit? ____________________.
2. What is the total hot-bulb resistance of the vehicle's parking lamp/taillamp circuit? ______________.
3. Is the 10 amp fuse designed for the circuit suitable for the vehicle's amperage load only? ________.
4. What is the hot-bulb resistance of the trailer parking lamp/taillamp circuit? ____________________.
5. With the trailer hooked to the vehicle, what is the total resistance of the parking lamp/taillamp circuit? ____.
6. What is the amperage total of the circuit with the trailer hooked to the vehicle? ________________.
7. Is the rated 10 amp circuit fuse satisfactory for the vehicle and trailer combined?________________.
8. What are at least three ways to repair this problem?
   1. ____.
   2. ____.
   3. ____.
9. What circuit laws were helpful when troubleshooting this system, or explaining it to the customer?

   ____.
10. What is your analysis of Case Study #1? ____

    ____

    ____.

CASE #2

The back-up lights in a modern sedan do not work (Figure 8–25).

Because this is a sedan, the station wagon part of the diagram should be disregarded. If you were to do a quick analysis of the diagram you would see the following:

1. The turn signal flasher shares the same fuse as the back-up lights.
2. The back-up lights should work only in the reverse position.
3. All the back-up lights share a common ground at G 403.
4. The voltage to the back-up bulbs should be the source voltage of 12 volts.
5. There are other circuits that share the G 403 common ground.
6. A further study of the ground distribution chart in Figure 8–26 shows all common links at G 403 ground.

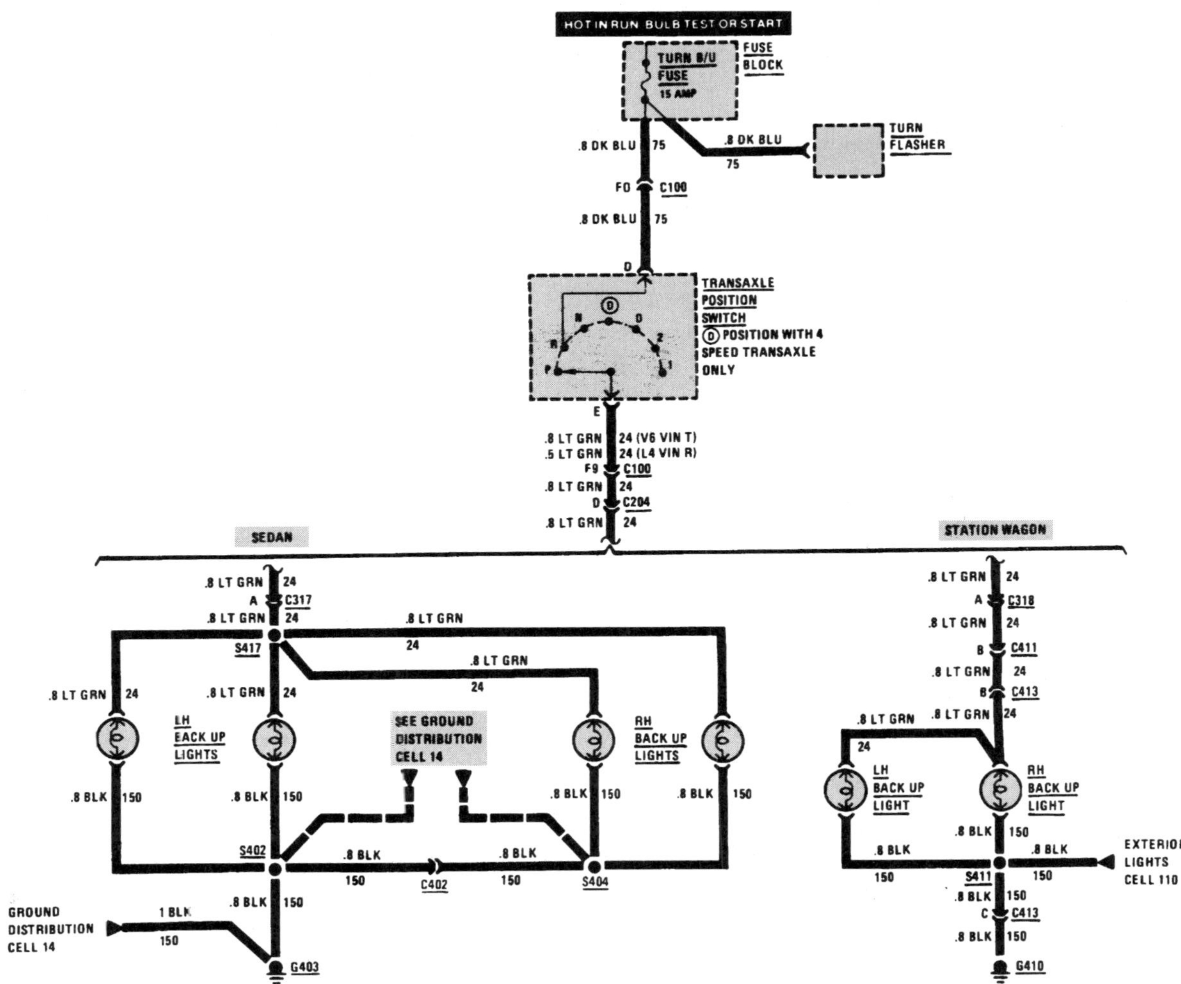

FIGURE 8–25 Back-up light circuit.

(Courtesy of General Motors Corporation, Service Technology Group)

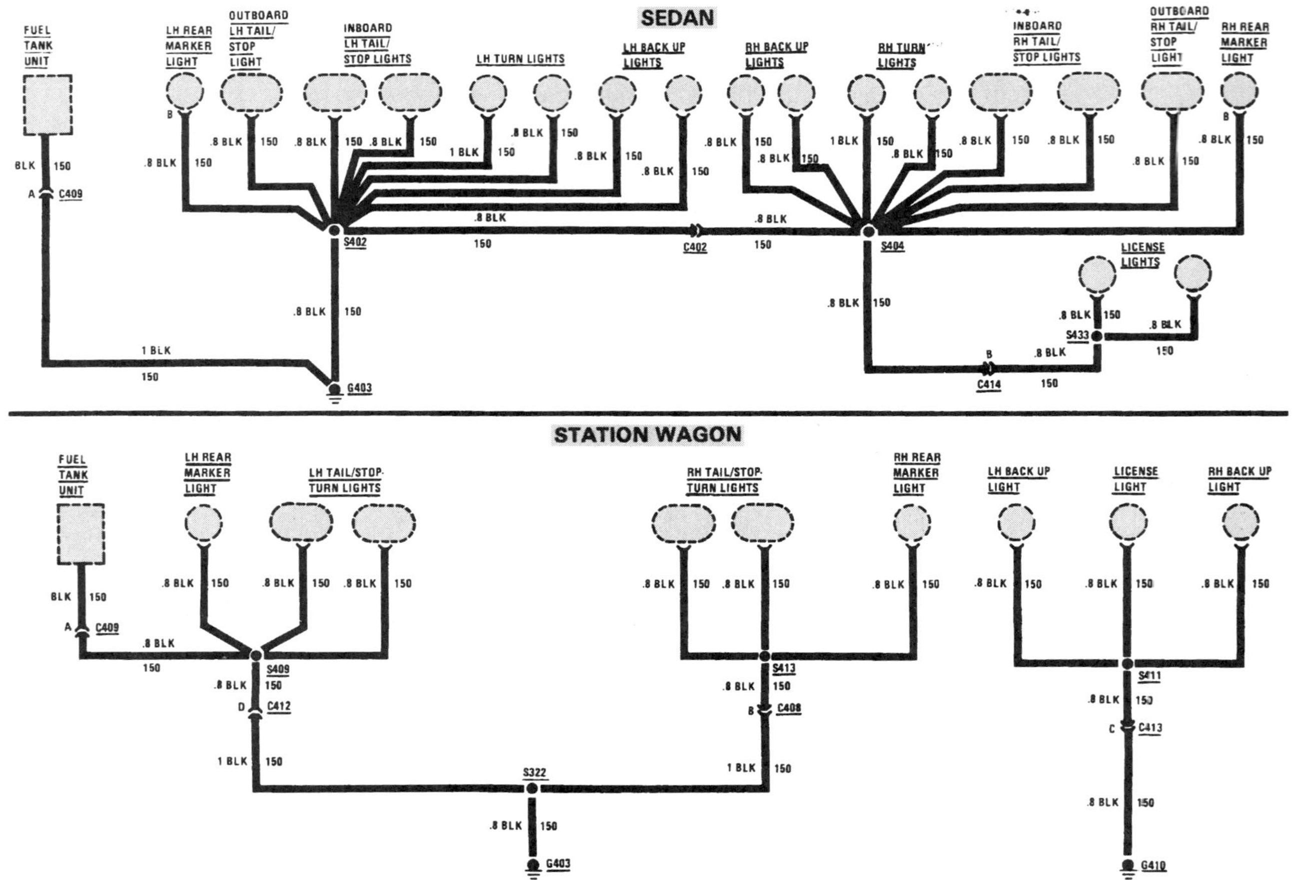

FIGURE 8–26 Typical ground distribution system.
(Courtesy of General Motors Corporation, Service Technology Group)

To quickly isolate the problem you could refer to the wiring diagrams in Figure 8–25 and Figure 8–26, and do the following.

1. Operate the turn signals. If they work, the fuse is OK.
2. If the taillights and fuel gauge work fine, the G 403 ground is OK.
3. Put transmission selector in reverse, there should be 12 volts at each back-up light.

More extensive checks would consist of the following.

1. Check for voltage at the transaxle position switch (terminals D and E). There should be source voltage at both connection points. If there is, the switch is OK. If there is voltage only at the inlet side (D), then suspect a bad switch. The switch can be double checked with an ohmmeter by disconnecting both leads to the switch, and hooking the ohmmeter across the switch contacts. There should be continuity through the switch.
2. If there is source voltage at the back-up light bulbs, check the bulbs to see if all are burned out. It could happen!
3. If the bulbs are OK, check the circuit ground wiring leading up to the G 403 connection.
4. If the bulbs were a #1156 bulb with a designed amperage of 2.10 amps, what would the circuit amperage be if all four bulbs were operating as designed?

5. Knowing the theoretical amperage total would be helpful in case there is suspected resistance at a connection point that may cause a voltage drop. By removing the circuit fuse and installing an ammeter in its place, could tell you the total circuit amperage.

6. What are at least three ways to quickly isolate a back-up light circuit problem? ________________

____

____.

7. What is your analysis of the above case study? ________________

____

____.

CASE #3

The right rear and left rear side marker lamps do not work on a 1996 Ford Taurus sedan (Figure 8–27). The circuit begins at the top of Figure 8–27 at point C which is a continuation of another circuit on "PAGE 92-1" (Figure 8–28).

If you were to do a quick analysis of the diagram you would see the following.

1. The source power starts at the 15 amp park lamp fuse and goes through the headlamp switch. The circuit then splits in the I/P fuse panel.

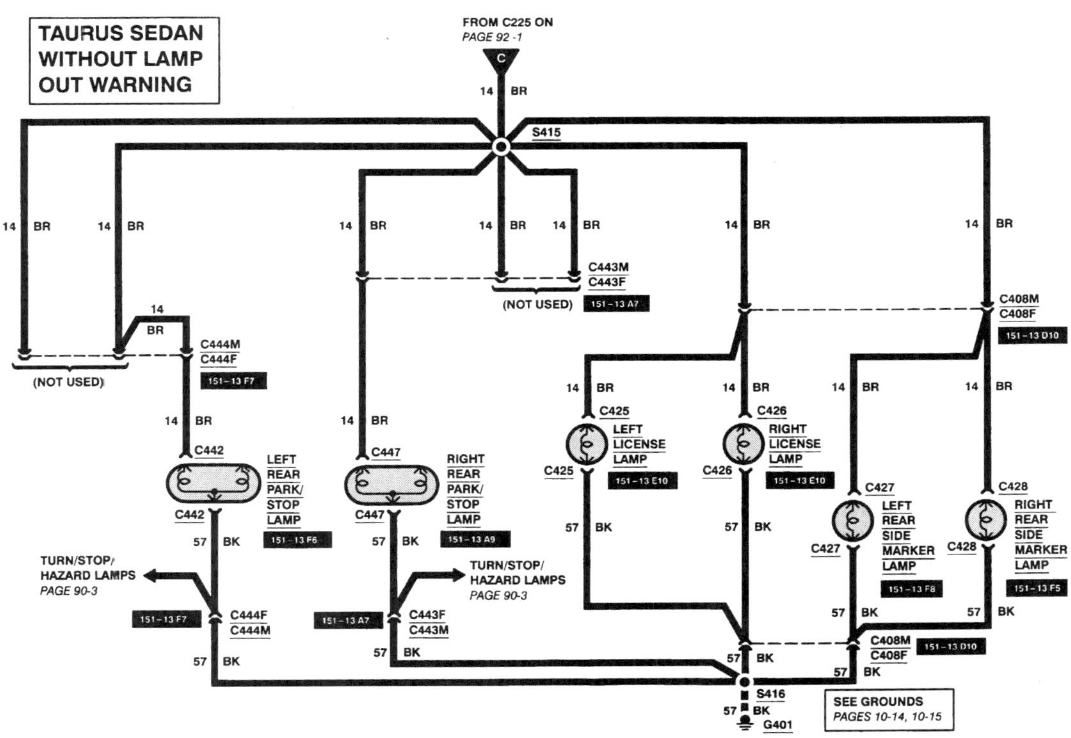

FIGURE 8–27 Exterior lamp circuit.

(Courtesy of Ford Motor Company)

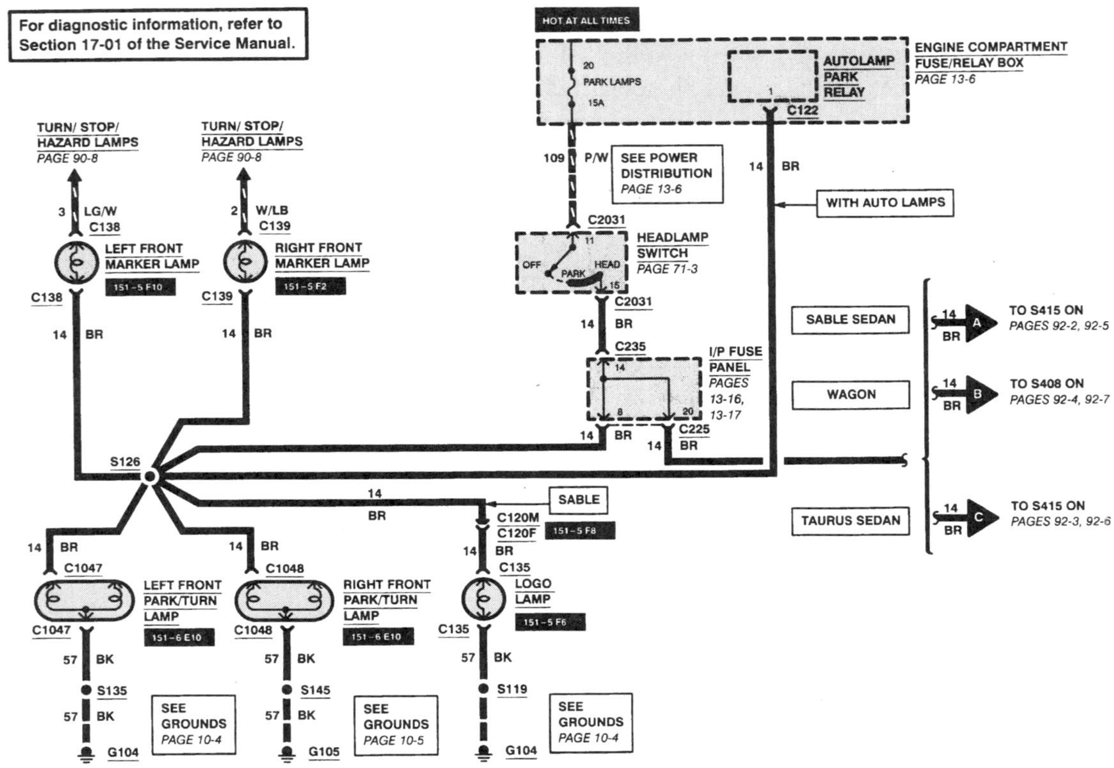

FIGURE 8–28 Exterior lamp circuit.
(Courtesy of Ford Motor Company)

2. One part of the circuit goes to the front lights, the other part goes to point C where the circuit we need begins.
3. The circuit divides into seven parallel sub-sections at the S 415 connection. Three of these connections are not used in this model vehicle. The remaining sub-sections go to lamps.
4. Two of the sub-sections further divide into parallel branch circuits.
5. All the lamps recombine at S 416 and then share a common ground at G 401.

To quickly isolate the problem, you could do the following.

1. Make sure all the remaining lamps work in the circuit. This would tell you that there is power coming in at point C and the grounds at G 401 are good.
2. Since the left rear and right rear side marker lamps both work off the same parallel branch, check the connection at point C 408 coming into the bulbs.
3. If there is power at C 408, check the C 408 ground side connection.
4. If both of these are OK, then the problem is in the bulbs or the bulb sockets.

What is your analysis of the above case study? ____

____

____.

Hands-On Vehicle Tests

The following two hands-on vehicle checks are included in the **NATEF** (National Automotive Technician Education Foundation) **Task List.** The task reference numbers are the same as the NATEF reference numbers. Complete your answers to the following questions on a *separate* sheet of paper.

Performance Task #1 (NATEF #V-E-6)

Task Description

Check the operation of the brake stoplight system. Adjust and service as needed.

Task Objectives

- ❑ Obtain a vehicle that can be used for this task. What model and year of vehicle are you using for this task? ____________________
- ❑ Find a wiring diagram that shows the stoplight circuit for this vehicle.
- ❑ Does the stoplight circuit follow the characteristics of a parallel circuit? ____________________
- ❑ Identify the following components and list their location on the vehicle.

1. Left stoplight bulb(s). ____________________.
2. Right stoplight bulb(s). ____________________.
3. Stoplight switch. ____________________.
4. Circuit fuse and fuse #. ____________________.
5. Circuit ground(s). ____________________.
6. Other circuit components. ____________________

____________________.

Circuit Checks

- ❑ Check the operation of the stoplight switch. Disconnect wire connections at the switch. Connect a DVOM, set on continuity, to the switch terminals and then depress the brake pedal. What happens as a result? ____________________
- ❑ Remove the stoplight fuse and connect the DVOM, in series, in place of the fuse. What is the total amperage of the circuit? ____________________
- ❑ Although not a normal diagnostic procedure, leave the DVOM connected up to read amps so you can verify that Kirchoff's Law, which you used in the previous parallel exercises, is true. Remove each stoplight bulb and record the amperage difference between the old amps total and each new total amps. Replace each bulb before removing the next.

Bulb #1 __________ Bulb #2 __________ Bulb #3 __________ Bulb #4 __________

Bulb #5 __________ Bulb #6 __________ Bulb #7 __________ Bulb #8 __________

- ❑ Do the totals of all branch amperages measured above equal the same as the circuit amperage total?

____

- [ ] If amperages do not add up to the same as the total, where are the remainder amps going?

____

Task Summary

After performing the above NATEF task, what can you determine will be helpful in future stoplight circuit diagnosis? ____

____

____.

Performance Task #2 (NATEF #VI-H-2)

Task Description

Diagnose the cause of poor, intermittent, or no heated glass operation. Repair as needed.

Task Objectives

- [ ] Obtain a vehicle that can be used for this task. What model and year of vehicle are you using for this task? ____.
- [ ] Find a wiring diagram that shows the rear window heated glass circuit for this vehicle.
- [ ] Does this heated glass circuit follow the characteristics of a parallel circuit? ____________________.
- [ ] Identify the following components and list their location on the vehicle.
    1. Rear window grid. ____.
    2. Control switch. ____.
    3. Relay (if applicable). ____.
    4. Circuit fuse and fuse #. ____.
    5. Circuit ground(s). ____.
    6. Circuit indicator light. ____.
    7. Circuit timer assembly (if applicable) . ____.
    8. Other circuit components. ____.

Circuit Checks

- [ ] Check the operation of the indicator light. If operating, proceed to rear window heating grid. If the indicator is not operating, check to see if the rear window grid feels warm. If not, is the circuit fuse OK? ____.
- [ ] Check for circuit source voltage at the rear window heater grid. DVOM voltage reading __________.
- [ ] What is the voltage from the positive end of the heating grid to the negative end? ________________.
- [ ] If you leave the DVOM connected to the negative side and slowly move the positive lead along a grid line, what happens to the voltage readings? __________________________. Why? ________________.
- [ ] Why are rear window grid lines wired in parallel? ____.

- ❑ Remove the circuit fuse and install the DVOM, in series, in its place. What is the amperage of the circuit?____________________.
- ❑ What happens to the total circuit amps as the rear heating grid starts to heat up? ________________.

Task Summary

After performing the above NATEF task, what can you determine will be helpful in future rear window heating grid circuit diagnosis? ____________________

____

____.

Summary

- ❑ In a parallel circuit there is more than one path for the flow of electrons to travel.
- ❑ In a parallel circuit, the voltage is the same at any point in the circuit.
- ❑ In a parallel circuit, the branch amperages are added to equal the total amperage.
- ❑ In a parallel circuit, the resistance total is always less than the lowest-value resistive device in the circuit.
- ❑ The "product over sum" method can be used to compute parallel resistance total for two resistance's at a time.
- ❑ The "same value" method can be used to compute parallel resistance total for multiple resistors with the same value.
- ❑ The "assumed voltage" method can be used to compute parallel resistance total for circuits where the voltage is unknown.
- ❑ The "reciprocal formula" method is the mathematical formula for computing parallel resistance total when there are multiple resistors with different values.
- ❑ The "scientific calculator" method can be used to compute parallel resistance total using the **1/X** key on a scientific calculator.

Key Terms

Assumed Voltage Method
Kirchoff's Current Law
Kirchoff's Voltage Law
parallel circuits
parallel circuit laws
Product Over Sum Method
Reciprocal Formula Method
Same Value Method
Scientific Calculator Method

Review Questions

Short Answer Essays

1. Describe the circuit law for voltage in a parallel circuit.
2. Describe the parallel circuit law for amperage in a parallel circuit.
3. Describe the parallel circuit law for resistance in a parallel circuit.
4. Explain how to use the **1/X** feature on a scientific calculator when computing resistance total in a parallel circuit.

Fill-in-the-Blanks

1. The amperage in a load component in a parallel circuit is ________________ to ________________ the total amperage.
2. In a parallel circuit, the voltage ________________ is the ________________ as the voltage ________________ each branch load in the circuit.
3. When current ________________ in a parallel circuit increases, the resistance ________________.

ASE Style Review Questions

1. Technician A says that in a parallel circuit the voltage across each circuit load adds up to the total circuit voltage. Technician B says the amperages of each parallel circuit load are equal because they share a common positive and a common negative. Who is correct?
   A. A only
   B. B only
   C. Both A and B
   D. Neither A nor B

2. In the following parallel light circuit the bulbs do not light when the switch is closed. Technician A says the switch can be bad. Technician B says both light bulbs could be burned out. Who is correct?
   A. A only
   B. B only
   C. Both A and B
   D. Neither A nor B

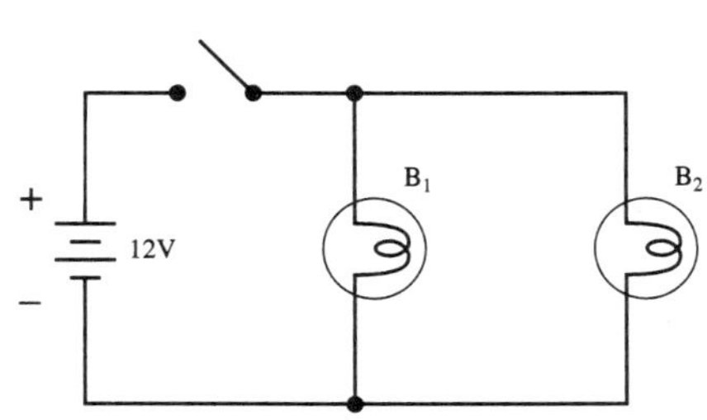

3. A circuit has the following four fixed resistors wired in parallel—10 Ω, 5 Ω, 10 Ω, and 2.5 Ω. What is the resistance total of the circuit?
   A. 27.5 Ω
   B. 1.25 Ω
   C. 125 Ω
   D. None of the above

4. Technician A says the resistance total in a parallel circuit is always less then the smallest resistive branch in the circuit. Technician B says the parallel circuit laws are the same as series circuit laws. Who is correct?
   A. A only
   B. B only
   C. Both A and B
   D. Neither A nor B

5. Technician A says that adding resistive branches to a parallel circuit increases the total circuit amperage. Technician B says if a parallel branch is opened, the circuit source voltage is lowered. Who is correct?
   A. A only
   B. B only
   C. Both A and B
   D. Neither A nor B

6. In a rear window heating grid, some of the grid lines are not working. Technician A says the grid lines have a break in them creating an open. Technician B says some of the grid line will continue to work because they are wired in parallel. Who is correct?
   A. A only
   B. B only
   C. Both A and B
   D. Neither A nor B

7. In the circuit below, the lights do not operate, but the relay can be heard making a clicking sound when the switch is energized. Technician A says the relay could be bad. Technician B says the bulb ground could be bad. Who is correct?
   A. A only
   B. B only
   C. Both A and B
   D. Neither A nor B

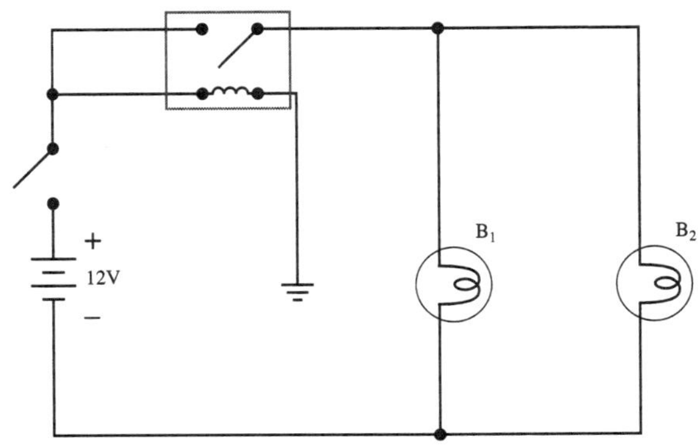

8. Technician A says the proper way to find the resistance of a "hot bulb" parallel circuit is to measure the amperage and voltage of the circuit, and then calculate the circuit resistance total. Technician B says that the "hot bulb" resistance total of a parallel circuit can be determined by taking the "hot bulb" resistance of each circuit bulb and finding the reciprocal of the sum of each parallel branch. Who is correct?
   A. A only
   B. B only
   C. Both A and B
   D. Neither A nor B

9. A vehicle has 6 taillights wired in parallel with .75 amps per bulb. When a trailer is hooked up to the vehicle, an additional four lights with .50 amps per bulb is added to the circuit. What is the amperage total of this parallel circuit?
   A. 4.5 amps total
   B. 6.0 amps total
   C. 1.25 amps total
   D. None of the above

10. 2.4 milliamps total are flowing through a circuit that consists of three resistors in parallel—20 kΩ, 20 kΩ, and 10 kΩ. Technician A says the source voltage of the circuit is 1.2 volts DC. Technician B says the voltage is 120 volts DC. Who is correct?
    A. A only
    B. B only
    C. Both A and B
    D. Neither A nor B

9 Series/Parallel Circuits

Introduction

In this chapter you will apply the individual series and parallel circuit laws learned in previous chapters to a combination circuit consisting of some components connected in series and some in parallel. Series/parallel electrical circuits seem complicated, but are in fact fairly simple to understand if you remember which circuit laws apply to each circuit load component. Most vehicle electrical circuits used today contain several series/parallel circuits, or portions of a series/parallel circuit.

Objectives

When you complete this chapter you should be able to:

- ❑ Define a series/parallel electrical circuit.
- ❑ Identify the individual series and parallel circuit loads of a compound series/parallel circuit.
- ❑ Perform series/parallel circuit calculations for voltage, current, and resistance.
- ❑ Apply troubleshooting and testing techniques to series/parallel circuits.
- ❑ Perform basic "hands-on" series/parallel exercises on the vehicle.

Series/Parallel Circuit Explanation

A **series/parallel circuit** is defined as a circuit that contains both series circuits and parallel circuits. This type of circuit is also known as a **combination circuit** (Figure 9–1). The simple circuit in Figure 9–1 has resistor #1 in series from the battery, then splits into two parallel branches (resistor #1 and resistor #2) before recombining and returning to the battery.

There is no specific law or formula that pertains to the whole series/parallel circuit for voltage, amperage, and resistance. Instead, it is a matter of determining which branch loads of the circuit are in series and which are in parallel, simplifying the circuit where possible, and using the circuit laws that apply to each of these branches to find the value totals. Figure 9–2 is an

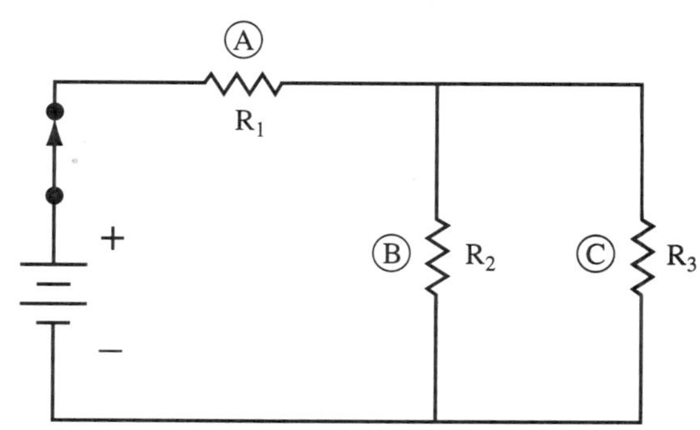

FIGURE 9–1 A simple series/parallel circuit.

example of a series/parallel circuit with resister #1 (6 Ω) in series, and resistors #2 and #3 (12 Ω each) in parallel.

To find the **series/parallel resistance** total for the circuit in Figure 9–2, perform the following steps.

1. Resistance total equals resistor #1 + the reciprocal sum of resistor #2 and resistor #3.
2. $R_T = R_1 +$ (reciprocal sum of R_2 and R_3)
3. $R_1 = 6\,\Omega$
4. R_2 and R_3 = 12 **1/X** + 12 **1/X** = **1/X**
5. $R_T = 6\,\Omega$(series) + $6\,\Omega$(sum parallel)
6. $R_T = 12\,\Omega$

As you can see by the steps performed, it was necessary to simplify the series/parallel circuit to find the resistance total. Figure 9–3 shows how the circuit is eventually converted to a series-only circuit to calculate the resistance total. Remember that the series circuit law for resistance, $R_T = R_1 + R_2 + R_3 +$ etc. is additive, so once the sum of the parallel branch loads is calculated it is reduced to a single value.

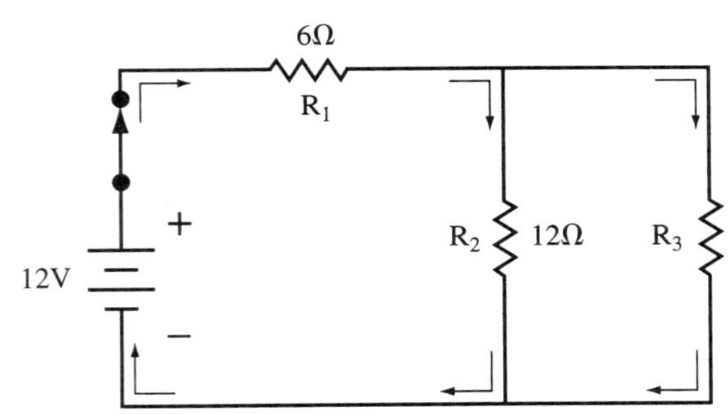

FIGURE 9–2 Simple series/parallel circuit.

To find the **series/parallel current** total of Figure 9–2, perform the following step.

1. Use Ohm's Law to find current total.

   $I = E \div R$

   $12\ V_T \div 12\ \Omega_T = 1$ amp total

With the above circuit totals, it is now possible to find the **series/parallel voltage** drops by performing the following steps.

1. Because resistor #1 (6 Ω) is in series, the current is the same as the current total of 1 amp.

   $V = I \times R$

   $1\ A \times 6\ \Omega = 6\ V$

   The voltage drop across resistor #1 is 6 V. If you subtract the voltage drop from the voltage total of 12 V, the remaining voltage available for the parallel circuit is 6 volts.

2. Because the voltage is constant in a parallel circuit, then resistor #2 and resistor #3 share the same 6 volts.

3. Find the current of R_2 using Ohm's Law.

   $I = E \div R$

   $6\ V \div 12\ \Omega = .5\ A$

4. Find the current of R_3 using Ohm's Law.

   $I = E \div R$

   $6\ V \div 12\Omega = .5\ A$

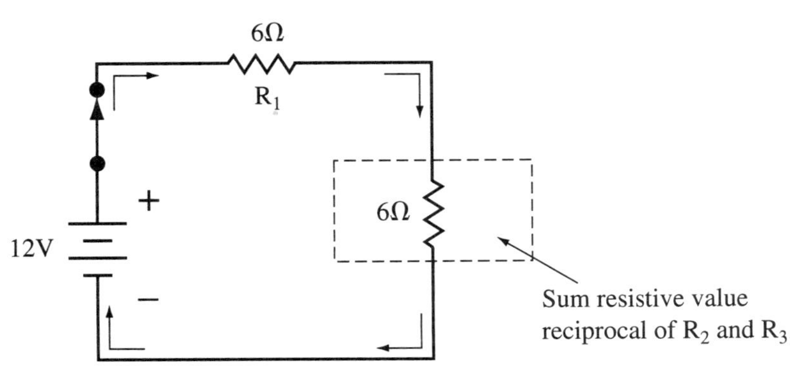

FIGURE 9–3 Simplify series/parallel circuit.

Note: Even though the parallel resistance values in the above calculations were identical, and the current total could have been divided by two, it was important to perform this step because the values are not always going to be the same.

The procedure for finding circuit values is the same even if the circuit's series and parallel branches are located in different places such as in Figure 9–4.

In Figure 9–4, resistor #1 and resistor #2 are in parallel, and resistor #3 is in series. The values will still end up being the same as in the previous problem because the voltage drop at the series load device determines the remaining voltage for the parallel circuit.

Figure 9–5 is a series/parallel circuit which consists of **multiple circuits** that in turn consist of more than one circuit that uses the same circuit laws.

In Figure 9–5, resistor #1 and resistor #4 are in series with the 12 volt source, and resistor #2 and resistor #3 are in parallel. To find the resistive values of this circuit, perform the following steps.

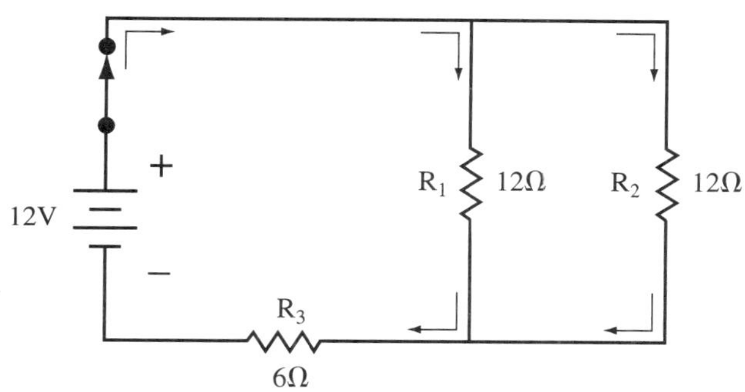

FIGURE 9–4 A simple series/parallel circuit.

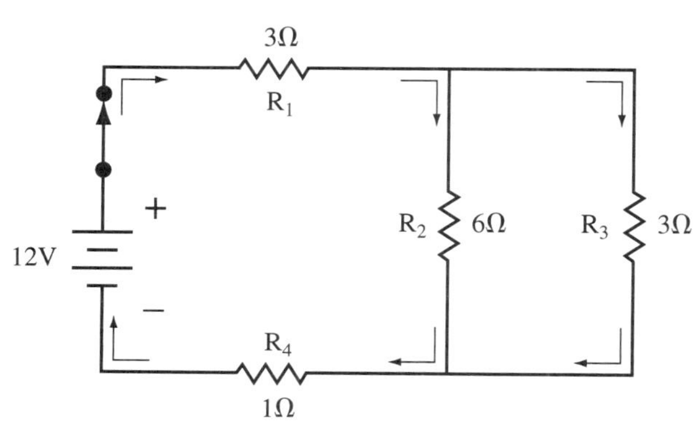

FIGURE 9–5 Series/parallel circuit with multiple series loads.

1. Resistance total equals resistor #1 + resistor #4 + the sum reciprocal of resistors #2 and #3.
2. $R_T = R_1 + R_4 +$ (sum reciprocal of R_2 and R_3)
3. $R_1 + R_4 = 3\ \Omega + 1\ \Omega$
4. Sum parallel of R_2 and R_3 = 6 **1/X** + 3 **1/X** = 2 Ω
5. $R_T = 3\ \Omega + 1\ \Omega$(series) + 2 Ω(sum parallel)
6. $R_T = 6\ \Omega$

   To find the current total of Figure 9–5, perform the following step.

   12 V ÷ 6 Ω = 2 amps total

With the above circuit totals of 12 V, 6 Ω, and 2 A, it is now possible to find the voltage drops of the series values (R_1 and R_4) and the remaining parallel voltage by performing the following steps.

1. V = I × R, resistor #1 voltage = 3 Ω × 2 A = 6 V
2. V = I × R, resistor #4 voltage = 1 Ω × 2 A = 2 V
3. Add the voltage drops of R_1 and R_4= 6 V + 2 V = 8 V
4. Subtract the series voltage drops from the voltage total to get the parallel voltage.

   12 V − 8 V = 4 V
5. Because the voltage is constant in a parallel circuit, both resistor #2 and resistor #3 share the same 4 volts.
6. Find the current of R_2 using Ohm's Law.

   I = E ÷ R

   4 V ÷ 6 Ω = .667 A
7. Find the current of R_3 using Ohm's Law.

   4 V ÷ 3 Ω = 1.333 A
8. Add the current of R_2 and R_3 which should then be the same as each series load and the total circuit current.
9. R_2 (.667) + R_3 (1.333) = 2 amps

Review of Basic Circuit Laws

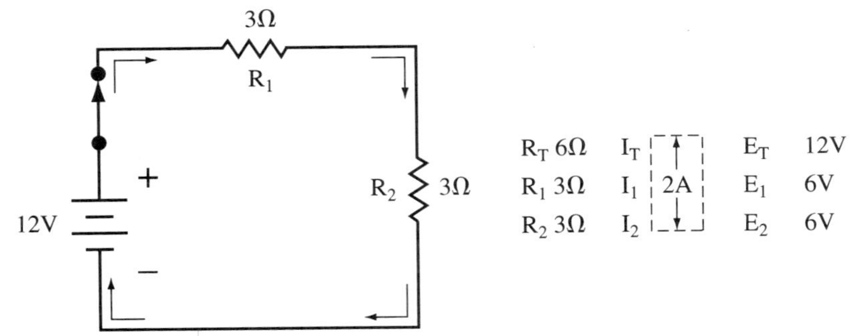

FIGURE 9–6 Basic series circuit.

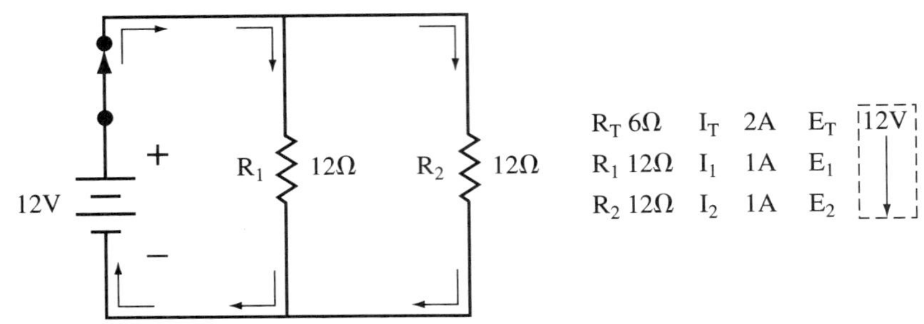

FIGURE 9–7 Basic parallel circuit.

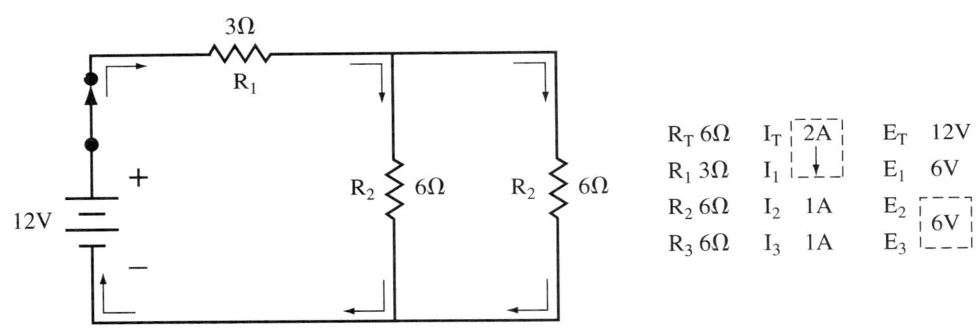

FIGURE 9–8 Basic series/parallel circuit.

Exercises—Series/Parallel Circuits

The following exercises are designed to help give you practice using the examples previously explained. The exercises will increase in complexity in an effort to test your understanding of the mixture of series and parallel circuit laws, and prepare you for the actual hands-on series/parallel electrical exercises that follow later in this chapter.

Exercise #1

Instructions Identify the type of circuit in Figure 9–9. On a *separate* sheet of paper, make an answer table like Table 9–1. Using the answer table, fill in the known circuit values from Figure 9–9. This is a helpful first step to give you an idea of where to begin, and what answers you need. Calculate the answers in the remaining spaces using one of the methods previously learned.

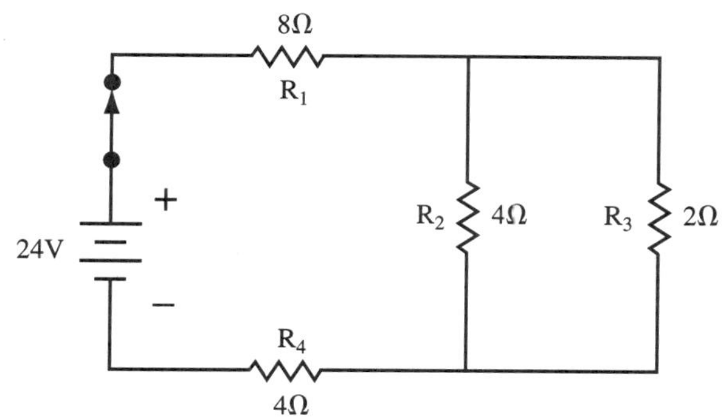

FIGURE 9–9 Simple series/parallel circuit.

R_T = Resistance total, I_T = Amperage total,
E_T = Voltage total, W_T = Watts total

R_1 = Resistor #1, I_1 = Amperage #1,
E_1 = Voltage #1, W_1 = Watts #1

R_2 = Resistor #2, I_2 = Amperage #2,
E_2 = Voltage #2, W_2 = Watts #2

R_3 = Resistor #3, I_3 = Amperage #3,
E_3 = Voltage #3, W_3 = Watts #3

R_4 = Resistor #4, I_4 = Amperage #4,
E_4 = Voltage #4, W_4 = Watts #4

Table 9–1

R_T ______	I_T ______	E_T ______	W_T ______
R_1 ______	I_1 ______	E_1 ______	W_1 ______
R_2 ______	I_2 ______	E_2 ______	W_2 ______
R_3 ______	I_3 ______	E_3 ______	W_3 ______
R_4 ______	I_4 ______	E_4 ______	W_4 ______

To verify that you are on the right track for finding the circuit values of series/parallel circuits, look at Table 9–2. It is a summary of the correct answers.

Table 9–2

R_T 1.33 Ω	I_T 1.80 A	E_T 24 V	W_T 43.2 W
R_1 8 Ω	I_1 1.80 A	E_1 14.4 V	W_1 25.92 W
R_2 4 Ω	I_2 .6 A	E_2 2.4 V	W_2 1.44 W
R_3 2 Ω	I_3 1.2 A	E_3 2.4 V	W_3 2.88 W
R_4 4 Ω	I_4 1.80 A	E_4 7.2 V	W_4 12.96 W

Instructions Follow the instructions used on the previous problem for the remaining series/parallel exercises. Remember to put your answers on a *separate* sheet of paper.

Exercise #2

Note: Use Figure 9–10 to fill in Table 9–3. The voltage is not listed on the circuit diagram, but the amperage total is shown on the ammeter mounted in series with the battery.

Table 9–3

R_T ______	I_T ______	E_T ______	W_T ______
R_1 ______	I_1 ______	E_1 ______	W_1 ______
R_2 ______	I_2 ______	E_2 ______	W_2 ______
R_3 ______	I_3 ______	E_3 ______	W_3 ______
R_4 ______	I_4 ______	E_4 ______	W_4 ______
R_5 ______	I_5 ______	E_5 ______	W_5 ______

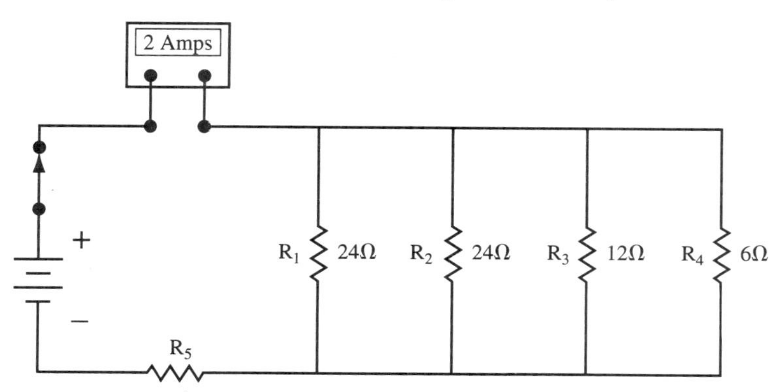

FIGURE 9–10 Measuring current total on a series/parallel circuit.

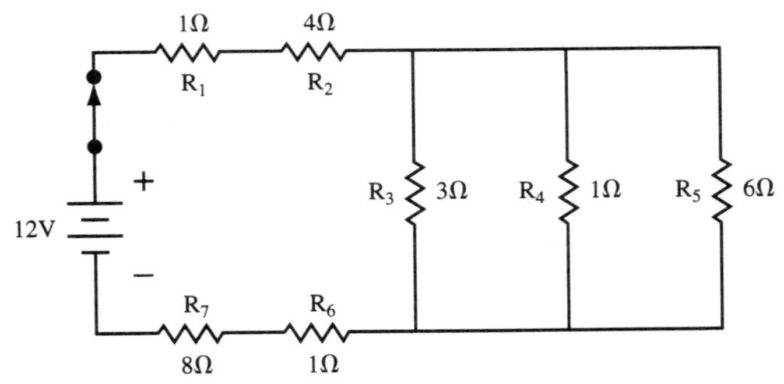

FIGURE 9–11 Expanded series/parallel circuit.

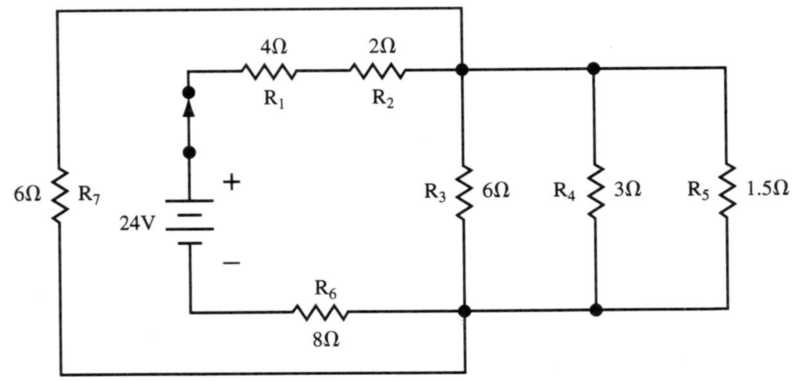

FIGURE 9–12 Complex series/parallel circuit

Exercise #3

Table 9–4

R_T ______	I_T ______	E_T ______	W_T ______
R_1 ______	I_1 ______	E_1 ______	W_1 ______
R_2 ______	I_2 ______	E_2 ______	W_2 ______
R_3 ______	I_3 ______	E_3 ______	W_3 ______
R_4 ______	I_4 ______	E_4 ______	W_4 ______
R_5 ______	I_5 ______	E_5 ______	W_5 ______
R_6 ______	I_6 ______	E_6 ______	W_6 ______
R_7 ______	I_7 ______	E_7 ______	W_7 ______

Note: The series/parallel circuit in Figure 9–12 is drawn a little differently than in the previous circuits. Identify the circuit by following the wire from the positive side of the battery through resistor #1 and resistor #2 (in series), and then as it splits four ways to resistors #3, #4, #5, and #6 (in parallel) before going through resistor #8 (in series). Write your answers in a chart like Table 9–4.

Table 9–5

R_T ______	I_T ______	E_T ______	W_T ______
R_1 ______	I_1 ______	E_1 ______	W_1 ______
R_2 ______	I_2 ______	E_2 ______	W_2 ______
R_3 ______	I_3 ______	E_3 ______	W_3 ______
R_4 ______	I_4 ______	E_4 ______	W_4 ______
R_5 ______	I_5 ______	E_5 ______	W_5 ______
R_6 ______	I_6 ______	E_6 ______	W_6 ______
R_7 ______	I_7 ______	E_7 ______	W_7 ______

Now that you have learned how to find the values for the combined series and parallel circuits, it is time to test your abilities on some real-life case studies. Just remember to trace out the circuit to determine what part of the circuit is in series and what part is in parallel.

CASE STUDIES

CASE #1

The case study in Figure 9–13 and Figure 9–14 are typical of a live series/parallel interior dash light circuit which technicians often encounter. Knowledge of series/parallel circuits can help you verify the values of circuit components quickly and efficiently, and perform an accurate diagnosis.

Known Information

- ❑ System operating volts = 14.0 volts
- ❑ Light switch dimmer rheostat resistance on maximum (HI) or DIM position = 15 Ω
- ❑ Light switch dimmer rheostat resistance on maximum (LOW) or BRIGHT position = 5 Ω
- ❑ Interior dash light bulb specifications = #194 bulb, design amperes .27 amps, hot bulb resistance = 51.85 Ω (see Appendix A)
- ❑ There are five bulbs not specifically shown on Figure 9–13, located in the Instrument Cluster Printed Circuit

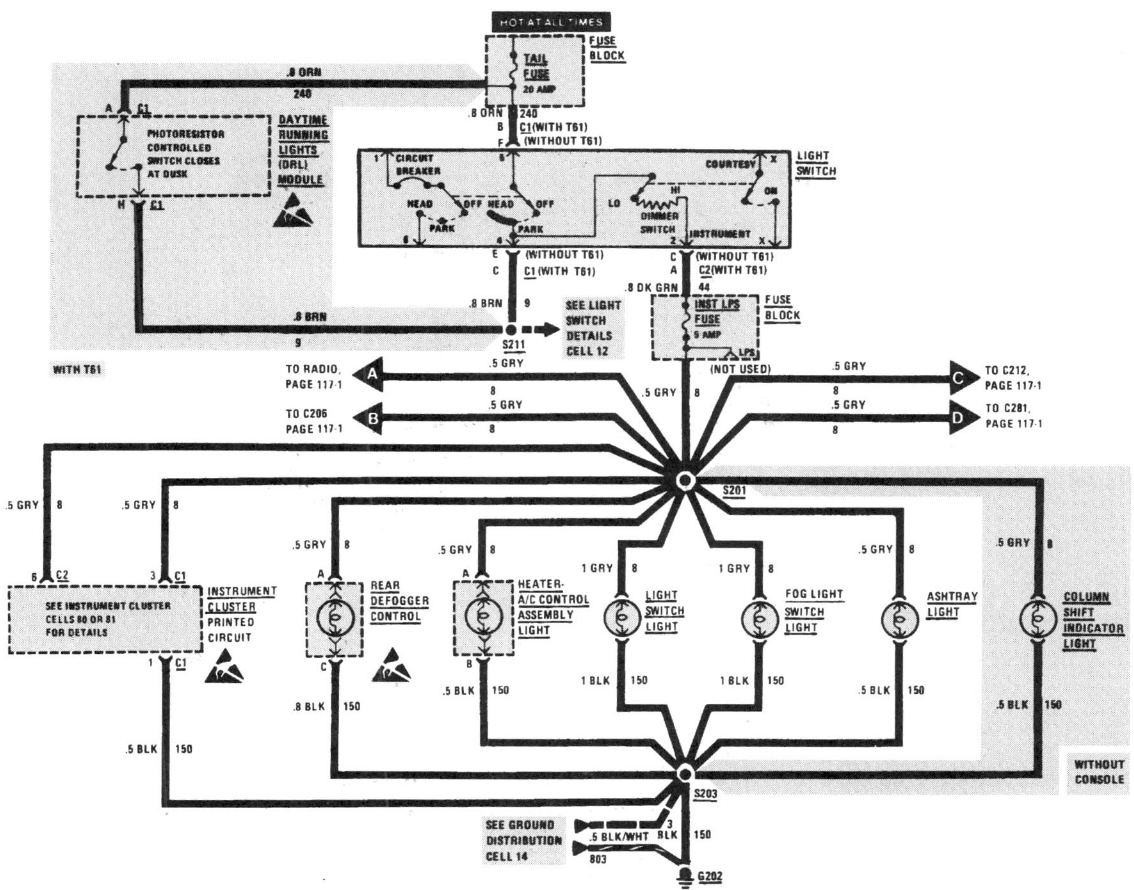

FIGURE 9–13 Interior dash light circuit.

(Courtesy of General Motors Corporation, Service Technology Group)

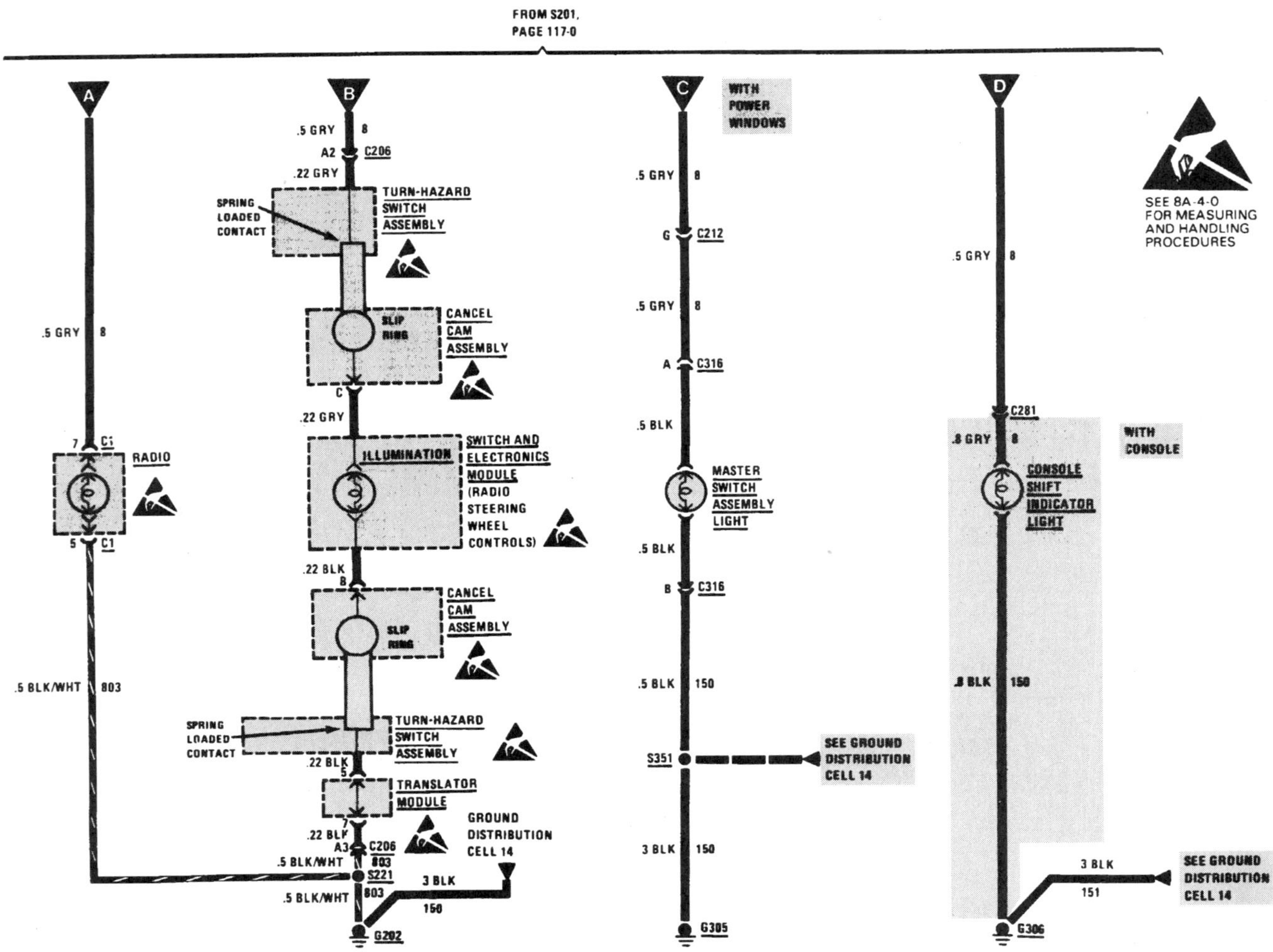

FIGURE 9–14 Interior dash light circuit.
(Courtesy of General Motors Corporation, Service Technology Group)

Circuit Analysis

Answer the following on a *separate* sheet of paper.

1. What component(s) of the circuit are in series?__________.
2. What component(s) of the circuit are in parallel? __________.
3. What is the total number of #194 bulbs in this circuit?__________.
4. What is the resistance total of bulbs *only?*__________.
5. What is the resistance total of the bulbs and dimmer switch in the LOW position? __________.
6. What is the resistance total of the bulbs and dimmer switch in the HI position? __________.

7. What is the amperage total of the circuit with the dimmer switch in the LOW position?__________.

8. What is the amperage total of the circuit with the dimmer switch in the HI position? __________.

9. What is your analysis of the above case study? ______________________________

____.

CASE #2

Using the same circuit diagrams as in Case #1, Figure 9–13 and Figure 9–14, diagnose the customer complaint below.

Customer Complaint

Interior dash lights do not work in any position. Customer replaced the 20 amp tail fuse and the 5 amp Instrument lamp fuse, but the results were the same.

Known Information

- ❑ Vehicle operating voltage = 14 volts
- ❑ Both circuit fuses have been replaced
- ❑ Bulbs are not burned out

Circuit Analysis

Answer the following on a *separate* sheet of paper.

1. If you knew both fuses were good, which system components would you check in the circuit and in what priority order? List the most important component first, and the procedure you would use to perform the check.

   Component #1 ____________________. Troubleshooting Procedure ____________________.

   Component #2 ____________________. Troubleshooting Procedure ____________________.

   Component #3 ____________________. Troubleshooting Procedure ____________________.

   Component #4 ____________________. Troubleshooting Procedure ____________________.

2. What circuit law(s) were helpful when troubleshooting this system or explaining it to the customer?

   ____.

3. What is your analysis of the above case study? ______________________________

   ____.

CASE #3

The case study in Figure 9–15 is typical of a back-up light circuit that operates parallel lights from a series fuse and switch.

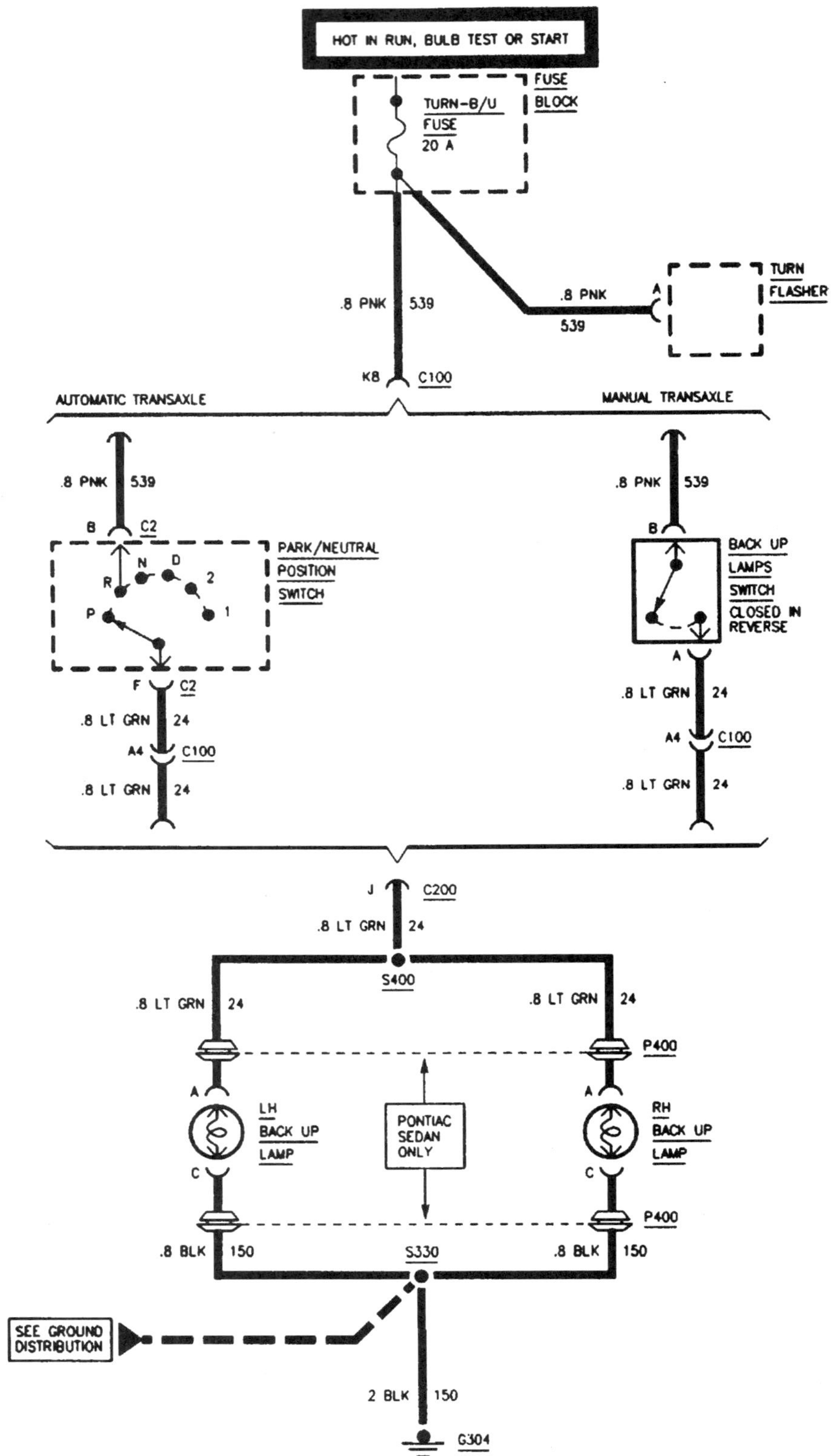

FIGURE 9–15 Back-up light circuit.

(Courtesy of General Motors Corporation, Service Technology Group)

Customer Complaint

Back-up lights do not operate when position switch is put in the reverse (R) position.

Known Information

- ❑ Vehicle operating voltage = 14 volts
- ❑ Turn signals operate normally
- ❑ 20 amp Turn-B/U fuse is OK
- ❑ Automatic transmission vehicle
- ❑ Back-up lamps, #1156 bulb, design amperes 2.10, hot bulb resistance = 6.10 Ω each

Circuit Analysis

Answer the following on a *separate* sheet of paper.

1. If you knew the fuse was good, which system components would you check in the circuit and in what priority order? List the most important component first, and the procedure you would use to perform the check.

   Component #1 ____________________. Troubleshooting Procedure ____________________.

   Component #2 ____________________. Troubleshooting Procedure ____________________.

   Component #3 ____________________. Troubleshooting Procedure ____________________.

   Component #4 ____________________. Troubleshooting Procedure ____________________.

2. What is the resistance total of this circuit? ________________________________.
3. What is the amperage total of the circuit? ________________________________.
4. What is the amperage of each bulb in the circuit? ________________________________.
5. What is your analysis of the above case study? ________________________________

   ____.

CASE #4

The circuit in Figure 9–16 operates correctly, except for the RH Seat Roof Vanity Mirror.

Known Information

- ❑ Vehicle operating voltage available at points A and B at the top of circuit diagram = 12 volts
- ❑ RH Seat Roof Vanity Mirror lamps are OK
- ❑ Power available at the RH Seat Roof Vanity Mirror = 12 volts
- ❑ RH Seat Roof Vanity Mirror dimmer rheostat has a resistance of 5 Ω in the maximum bright position, and 20 Ω in the maximum dim position
- ❑ All other vanity mirrors work satisfactorily

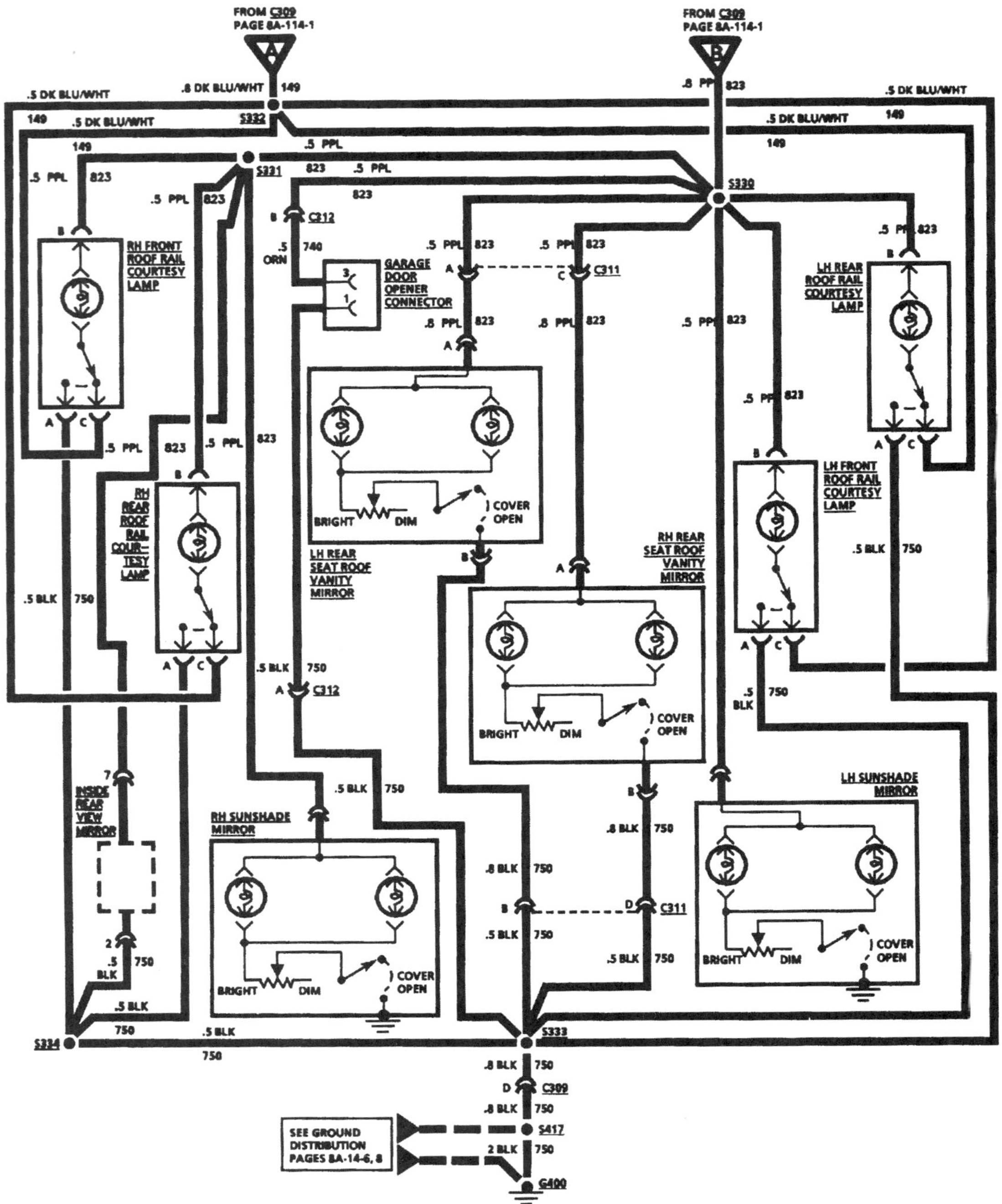

FIGURE 9–16 Interior vanity light circuit.

(Courtesy of General Motors Corporation, Service Technology Group)

Circuit Analysis

Answer the following on a *separate* sheet of paper.

1. If you knew there was 12 V at connection A going to the RH vanity mirror, what would you check in the circuit and in what priority order? List the most important component check first, and the procedure you would use to check the item.

Component #1 ____________________. Troubleshooting Procedure ____________________.

Component #2 ____________________. Troubleshooting Procedure ____________________.

Component #3 ____________________. Troubleshooting Procedure ____________________.

Component #4 ____________________. Troubleshooting Procedure ____________________.

2. If the lamps for the RH rear seat roof vanity mirror were 51 Ω each, and the dimmer rheostat was at the maximum bright position, what is the resistance total of this portion of the Interior Lamp circuit?
____.

3. What is the amperage going to the RH rear seat roof vanity mirror at connector point A?
____.

4. What is the amperage coming from the RH rear seat roof vanity mirror at connector point B?
____.

5. What is your analysis of the above case study? ____________________
____.

Hands-On Vehicle Tests

The following hands-on vehicle check is included in the **NATEF** (National Automotive Technician Education Foundation) **Task List.** The task reference number is the same as the NATEF reference number. Complete your answers to the questions below on a *separate* sheet of paper.

Performance Task #1 (NATEF #VI-E-1)

Task Description

Diagnose brighter than normal, intermittent, dim or no light operation.

Task Objectives

❑ Obtain a vehicle that can be used for this task. What model and year of vehicle are you using for this task? ______________________________.

❑ Find a wiring diagram that shows the interior dash light circuit for this vehicle.

❑ Does the dash light circuit follow the characteristics of a series/parallel circuit? ________________.

❑ Identify the following components and list their location on the vehicle.

1. Light switch assembly ______________________________.
2. Circuit fuse(s) and fuse numbers ______________________________.
3. Other electrical circuits that share the same fuse ______________________________.
4. Number of instrument panel dash lights ______________________________.
5. Circuit ground(s) ______________________________.
6. Other circuit components ______________________________.

Circuit Checks

- ❑ Check operation of the headlight switch. Rotate dimmer switch throughout full range from maximum DIM to maximum BRIGHT position. Does it operate as designed? ___________________.
- ❑ Identify inlet and outlet wires to the dimmer switch. Connect a DVOM, set on voltage, in parallel with the inlet and outlet wires to the dimmer switch. Measure the voltage drop at maximum DIM ________________________, and maximum BRIGHT ________________________.
- ❑ What is the available voltage remaining for the parallel dash lights? ________________________.
- ❑ Remove the circuit fuse and install an ammeter (in series) in place of fuse. Measure the amperage total with the dimmer switch in the fully DIM position ________________________.
- ❑ Measure the amperage total with the dimmer switch in the maximum BRIGHT position _________.
- ❑ What would happen to the dash light circuit if the dimmer switch became inoperative?

  ____.
- ❑ What would happen to the dash light if one of the lights burned out? ________________________.

Task Summary

After performing the above NATEF task, what can you determine will be helpful in future light circuits that operate other than designed? ____

____.

Summary

- ❑ In a series/parallel circuit, there are individual series load components and parallel load components.
- ❑ In a series/parallel circuit, the resistance total is computed by adding the series resistance to the sum reciprocal of the parallel branches.
- ❑ In a series/parallel circuit, voltage drops exist at the series loads, with the remaining voltage available for the parallel load branch(s).
- ❑ A series/parallel circuit is also called a combination circuit.
- ❑ Series/parallel circuits can have several series load branches and several parallel load branches.
- ❑ An interior dash light circuit uses a series dimmer rheostat to adjust the parallel lamp circuit brightness.
- ❑ In a series/parallel circuit, failure of a main series load device will disable the entire circuit.

Key Terms

combination circuit
multiple circuits
series/parallel circuit
series/parallel current
series/parallel resistance
series/parallel voltage

Review Questions

Short Answer Essays

1. Describe how to find the current total in a compound circuit.
2. Describe how to find the resistance total in a compound circuit.
3. Describe how to find the voltage total in a compound circuit.

Fill-in-the-Blanks

1. The ________________________ of a series load component in a compound circuit is the same as ________________________ total.
2. In a series/parallel circuit the resistance total is computed by ________________________ the series load components with the sum of the ________________________ load components.
3. A compound electrical circuit with a series load of 10 Ω and two parallel loads of 10 Ω each, has a resistance total of ________________________.

ASE Style Review Questions

1. Series/parallel theory is being discussed. Technician A says that a compound circuit contains both series and parallel load components. Technician B says a series parallel circuit is also known as a combination circuit. Who is correct?
   A. A only
   B. B only
   C. Both A and B
   D. Neither A nor B

2. In the following simple series/parallel circuit, the bulbs do not light when the switch is closed. Technician A says the switch could be bad. Technician B says both bulbs could be burned out. Who is correct?
   A. A only
   B. B only
   C. Both A and B
   D. Neither A nor B

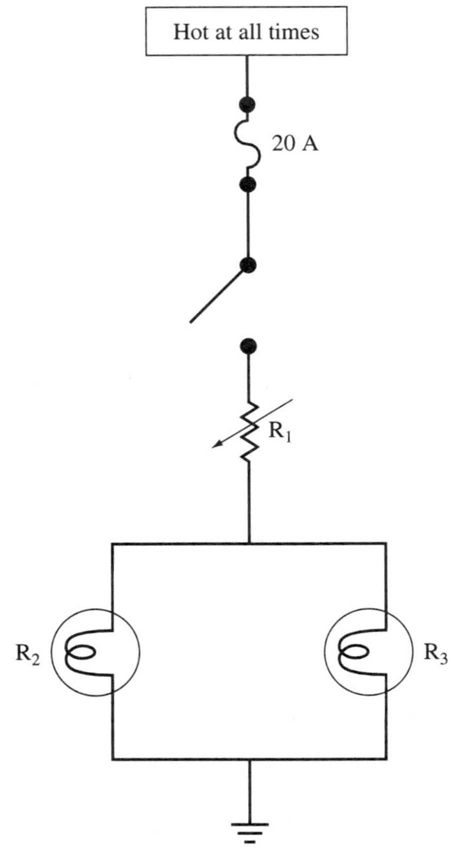

3. Two technicians are discussing circuit laws that pertain to a series/parallel circuit. Technician A says that in a compound circuit the resistance total is computed by adding the series load devices with each individual parallel load device. Technician B says that in a compound circuit the circuit voltage is the same throughout the whole circuit. Who is correct?
   A. A only
   B. B only
   C. Both A and B
   D. Neither A nor B

4. Technician A says the amperage of each series load in a compound circuit is the same as the amperage total. Technician B says the amperage of each parallel load in a compound circuit is the same as the amperage total. Who is correct?
   A. A only
   B. B only
   C. Both A and B
   D. Neither A nor B

5. A series/parallel circuit has two series resistors, 3 Ω and 4 Ω, and four parallel resistors, 6 Ω, 6 Ω, 3 Ω, and 1.5 Ω. What is the resistance total?
   A. 23.5 Ω
   B. 7.75 Ω
   C. .5217 Ω
   D. None of the above

6. Two technicians are discussing a simple series/parallel circuit. Technician A says if a series load component fails in a compound circuit the entire circuit will quit working. Technician B says if a parallel load component fails in a compound circuit the rest of the circuit will continue to operate as designed. Who is correct?
   A. A only
   B. B only
   C. Both A and B
   D. Neither A nor B

7. The amperage total is being measured in a series/parallel circuit. Technician A says to remove a series load device and install an ammeter in its place in order to read the amperage total. Technician B says the ammeter is installed in series in place of the circuit fuse to read amperage total. Who is correct?
   A. A only
   B. B only
   C. Both A and B
   D. Neither A nor B

8. A series/parallel circuit has a series load device with an amperage reading of 8 amps, and three parallel load devices with readings of 4.5 amps, 1.5 amps, and 2.0 amps. What is the amperage total of this circuit?
   A. 8 amps
   B. 16 amps
   C. 12.5 amps
   D. None of the above

9. Voltage total of a simple series/parallel circuit is being discussed. There are two series load devices with voltage drops of 6 volts each, and four load devices in parallel with a voltage reading across each device of 12 volts. What is the voltage total of this circuit?
   A. 12 volts
   B. 24 volts
   C. 60 volts
   D. None of the above

10. The simple series/parallel lamp circuit (below) does not work when the switch is closed. Technician A says the circuit will not work if only lamp C is burned out. Technician B says the circuit will not work if only lamp B is burned out. Who is correct?
    A. A only
    B. B only
    C. Both A and B
    D. Neither A nor B

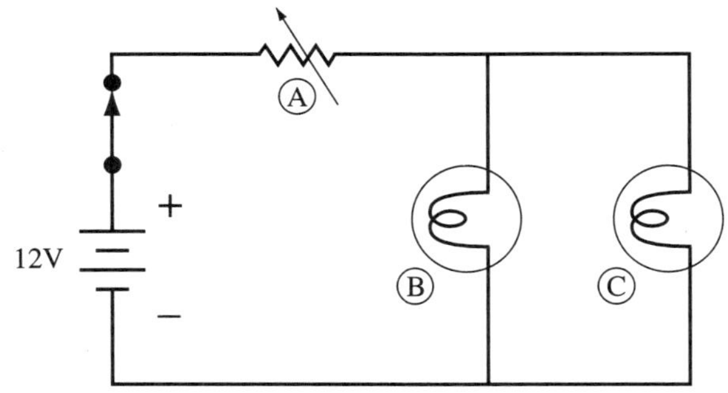

10 Basic Troubleshooting Techniques and Tips

Introduction

Automotive technicians working on electrical circuits must be familiar with commonly used troubleshooting techniques for locating unwanted circuit faults. Successful troubleshooting of a circuit begins with the knowledge of electrical theory and circuit laws you learned in previous chapters. In this chapter you will apply this knowledge to find faults in an electrical circuit.

Objectives

When you complete this chapter you should be able to:

- ❑ Define an open circuit.
- ❑ Troubleshoot an electrical circuit to find an unintentional open.
- ❑ Define a short-to-voltage.
- ❑ Troubleshoot an electrical circuit to find a short-to-voltage.
- ❑ Define a short-to-ground.
- ❑ Troubleshoot an electrical circuit to find a short-to-ground.
- ❑ Define excessive resistance in an electrical circuit.
- ❑ Troubleshoot an electrical circuit to find excessive resistance.
- ❑ Measure unwanted voltage drops in electrical circuits.
- ❑ Diagnose the cause of a key-off parasitic battery drain.
- ❑ Define a sneak circuit and its characteristics.

Circuit Faults

Open Circuits

An **open circuit** is defined as a break in the current flow of a complete circuit. When an electrical circuit is open, the current does not flow and the load component will not work. There are two types of open circuits that effect current flow in a circuit. The first type is an **intentional open circuit** (Figure 10–1). The "open" is designed as a circuit control device like a switch or relay that turns on the circuit load when needed by the driver. Almost all electrical circuits have a switching device for on/off control.

The second type of open circuit is an **unintentional open circuit.** This type of open circuit can be created

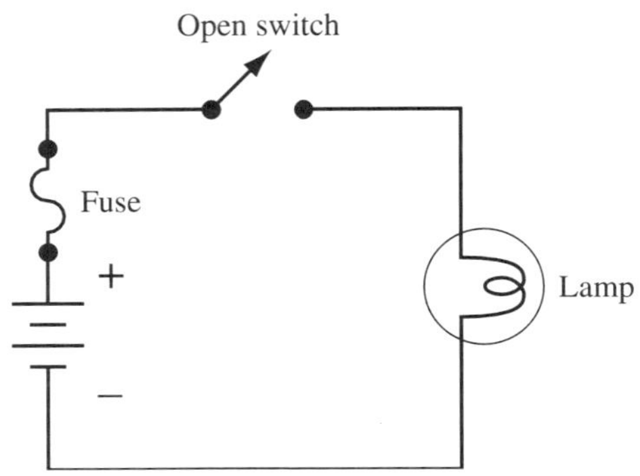

FIGURE 10–1 A switch is an intentional open circuit.

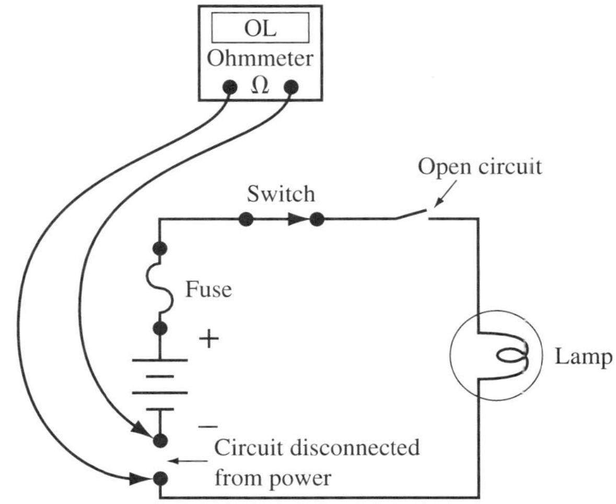

FIGURE 10–3 Testing for an open circuit using an ohmmeter.

when there is an electrical problem in the circuit (Figure 10–2). The current flow will stop when a bulb (or other load device) burns out, a wire is broken or disconnected, or a circuit protection device (fuse or circuit breaker) blows. Poor electrical connection are normally the most common causes of open circuits because they are most often hidden from view within a connector. Open circuits caused by a blown fuse are normally associated with other circuit problems that create enough excessive current to blow the fuse.

Depending on the electrical circuit being tested and the accessibility of the components that need testing, several acceptable devices that can be used to troubleshoot are an ohmmeter, voltmeter, test light, self-powered test light, and a fused jumper wire. To test a circuit, the technician must know the correct operation of the circuit. This is where your knowledge of series and parallel circuit laws, and Ohm's Law will come in handy.

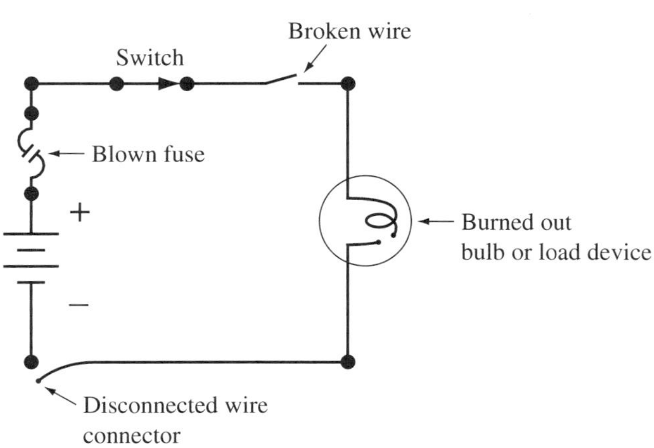

FIGURE 10–2 Several faults could cause an unintentional open circuit—a blown fuse, a broken disconnected wire, or a burned out bulb or load device.

To test for an open circuit using an ohmmeter (Figure 10–3) the circuit MUST be disconnected from the power source before installing the ohmmeter in parallel. A circuit with an open will read over limit (OL) or Infinite. A complete (not open) circuit would read the designed circuit resistance (usually less than 100 Ω).

To test for an open circuit using a voltmeter (Figure 10–4), first hook up the meter in parallel as close to the circuit load as possible, and then work backwards through the circuit until the open circuit is found. The voltage reading in the circuit will be zero *after* the open circuit, and will be the source voltage *before* the open circuit, as long as the rest of the circuit has no other problems.

The more complex a circuit is, the harder it may be to diagnose an open circuit. This is why it is important to remember the circuit laws for series and parallel circuits, so you know how the circuit is supposed to operate. In Figure 10–5, lamp #1 and lamp #3 are parallel to each other with 12 volts going to each bulb, and with a common ground. When bulb #1 develops an open circuit (Figure 10–6) on the ground side, bulb #2 and bulb #4 become wired in series with bulb #1, causing them to glow dimly. Bulb #3 continues to glow brightly, as designed, because the ground is unaffected by the circuit's open.

To test for an open circuit using a test light (Figure 10–7), hook up the ground lead of the test light to a

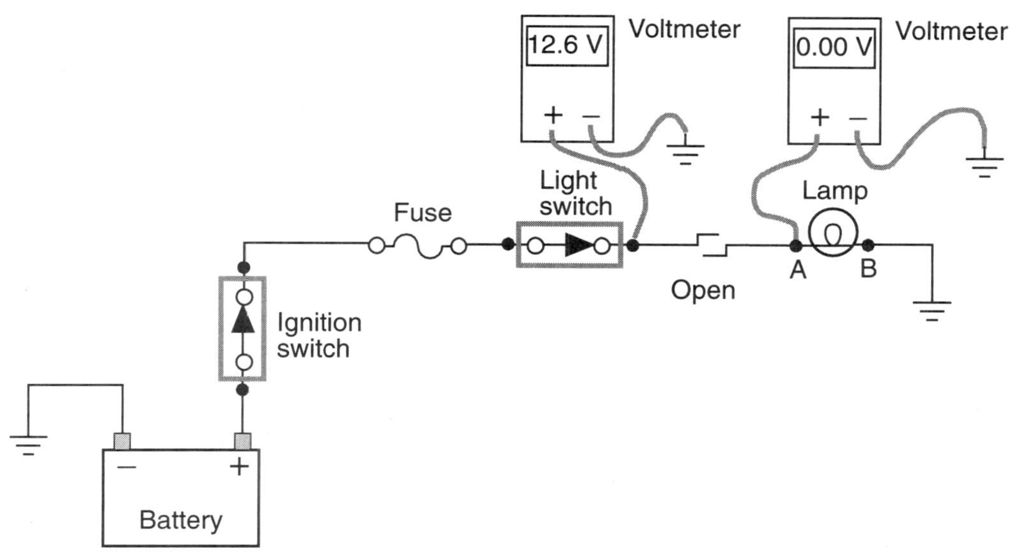

FIGURE 10–4 An open circuit is located between the point where a voltage is measured and where no voltage is found.

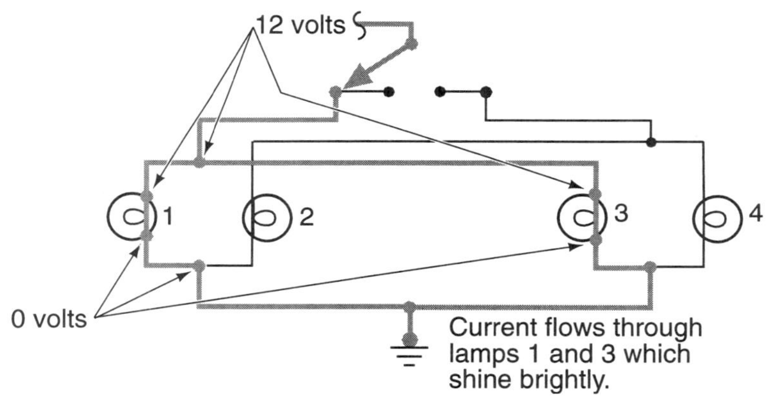

FIGURE 10–5 A properly operating parallel circuit.

(Courtesy of DaimlerChrysler Corporation)

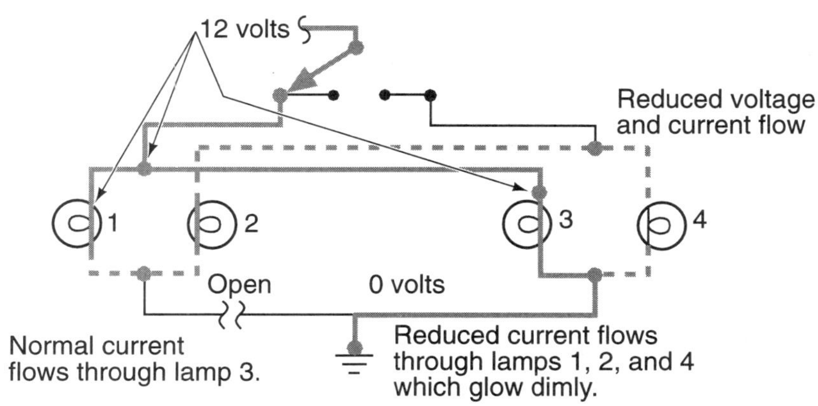

FIGURE 10–6 An open in the combined ground path of a parallel circuit may change the circuit into a series circuit.

(Courtesy of DaimlerChrysler Corporation)

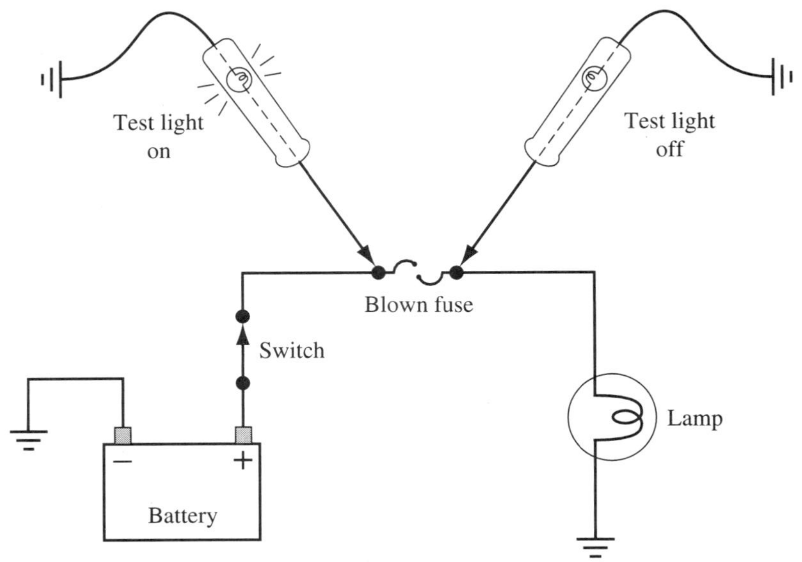

FIGURE 10–7 Using a test light to check for an open circuit.

known good ground to prevent any misdiagnosis of the circuit. Ensure that the lamp is working properly by first touching a known good power connection, and then probe along the circuit until the test lamp goes out. This is the point that an open circuit exists.

CAUTION *Test lights and self-powered test lights should not be used around electronic or computer control circuits. Damage may result to the sensitive circuit due to excessive current flow from the test light.*

Using a self-powered test light to test for an open circuit can be a useful tool for checking for circuit continuity. Because the test light uses its own internal battery, all it needs is a completed path of current flow through a circuit to energize the test light (Figure 10–8). The test lamp will light as long as there is continuity through the circuit being tested. When an open is found with the test lamp probe, the test lamp will not light.

Individual circuit components that are removed from the circuit can be checked for continuity, or an open, with a self-powered test light. Figure 10–9 shows a

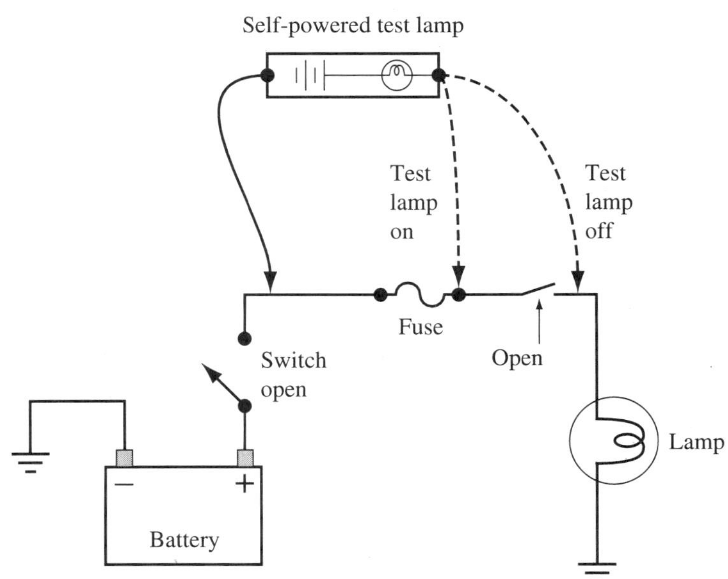

FIGURE 10–8 Using a self-powered test lamp to check for an open circuit.

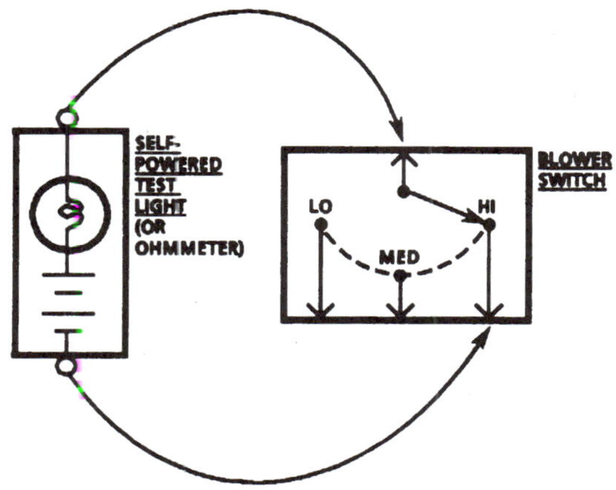

FIGURE 10–9 Checking a blower switch for continuity using a self-powered test light.

(Courtesy of General Motors Corporation, Service Technology Group)

blower switch HI contact being checked for continuity. If an open exists, the lamp will not light.

To test for an open circuit using a fused jumper wire can be a simple way of diagnosing circuits where the load control switch is suspected to be faulty. Using an in-line fuse along with the jumper wire ensures circuit safety if the jumper wire is connected incorrectly. In Figure 10–10, a fused jumper wire is being used to bypass a suspected bad light switch. Jumper wires can also be used on the ground side of the circuit load to diagnose opens in the ground.

CAUTION *Never use a fuse in the jumper wire with a rating higher than the fuse that protects the circuit that is being tested.*

Short Circuits

A **short circuit** is defined as an unwanted or accidental bypass of the current load in a circuit. There are two types of short circuits that can effect the current flow in a circuit. The first is a **short-to-voltage** (Figure 10–11). In this type of short, there is an unwanted copper-to-copper connection between two separate circuits causing one or more circuits to operate when not needed. In the case of Figure 10–11, every time the brake light switch is pushed, the horn will honk, or if the horn button is pushed, the brake lights will go on. Sometimes, with this type of short the circuit fuse may blow if the designed amperage flow exceeds the fuse limits.

Locating a short-to-voltage can sometimes be very difficult to find. If a visual check of the wiring, insulation, connectors, and circuit components does not show the cause of the short, then isolating the circuit will be necessary. An excellent, home-built diagnostic tool that can be used to help find a short-to-voltage is a 12 volt buzzer with an in-line fuse (Figure 10–12).

If the affected circuits share a common fuse (see Figure 10–11), then remove the fuse and install a buzzer that has been fitted with terminals to fit across the fuse

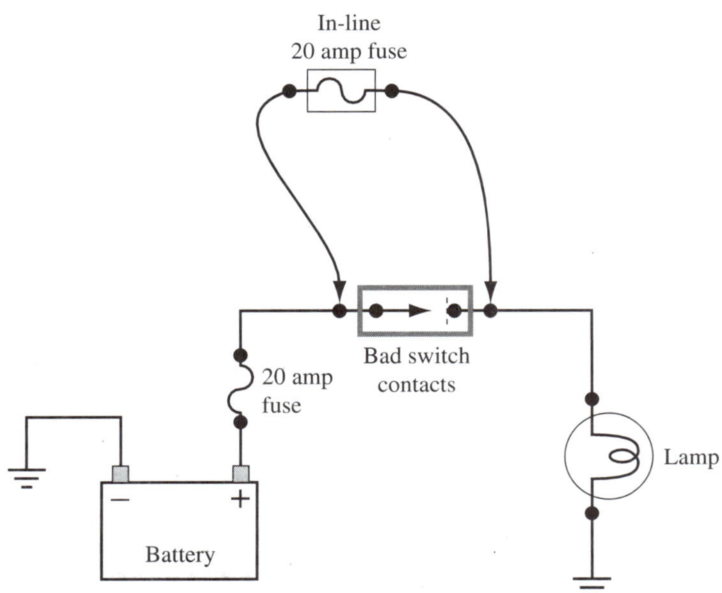

FIGURE 10–10 Using a fused jumper wire to bypass a suspected bad switch.

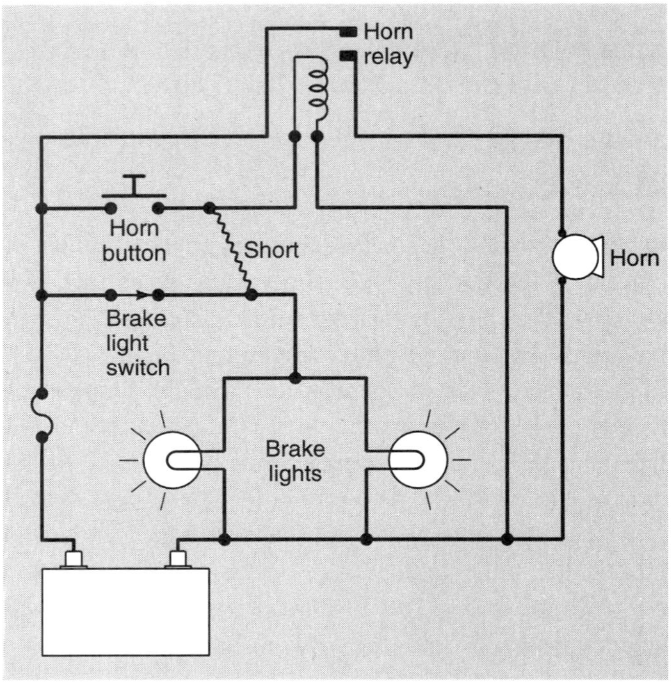

FIGURE 10–11 An unwanted copper-to-copper wire short circuit.

holder terminals (Figure 10–13). Using a buzzer is helpful when you isolate circuit components through a process of elimination by disconnecting them. When the affected spot is reached the buzzer will stop.

Similarly, activate the circuit that the buzzer is connected to and disconnect all of the loads that are supposed to be activated by each switch. Then, disconnect the wire connectors at the circuit that go from the load(s) back to the switch(es). If the buzzer stops when a connector is disconnected, the short-to-voltage is in that portion of the circuit.

The second type of short circuit, is a **short-to-ground** (Figure 10–14), also known as a **grounded circuit.** In this type of short, there is an unwanted path for current to flow directly to ground. This normally occurs when wire insulation breaks down from vibration or unwanted rubbing, and the wire touches a ground. This problem provides a low resistance path for the current to flow, which will most likely blow the circuit protection fuse or overheat wiring insulation and connectors.

Where a short-to-ground is located within a circuit can affect the circuit differently depending on where the control switch and circuit load device(s) are located. In Figure 10–15, if the short-to-ground occurs before the circuit switch and load, the fuse protection device will blow even if the switch is open.

In Figure 10–16, the short-to-ground is located after the circuit switch, but before the load. The fuse will blow only after the switch is closed.

If the short-to-ground is located after the load, but before the control switch, then the load will remain on all the time (Figure 10–17). This type of short will not blow the fuse but may run the circuit battery down.

There are several methods that can be used to diagnose a short-to-ground. If the circuit fuse blows, a test light or voltmeter would be difficult to use. To prevent this from occurring, connect a cycling (self-resetting) circuit breaker in place of the fuse (Figure 10–18). The circuit breaker will cycle open and closed to let the technician test for voltage.

With a cycling circuit breaker installed (Figure 10–19), current initially flows through the circuit creating a magnetic field surrounding the wire while the breaker is closed. When the short-to-ground causes the circuit breaker to open, the magnetic field collapses.

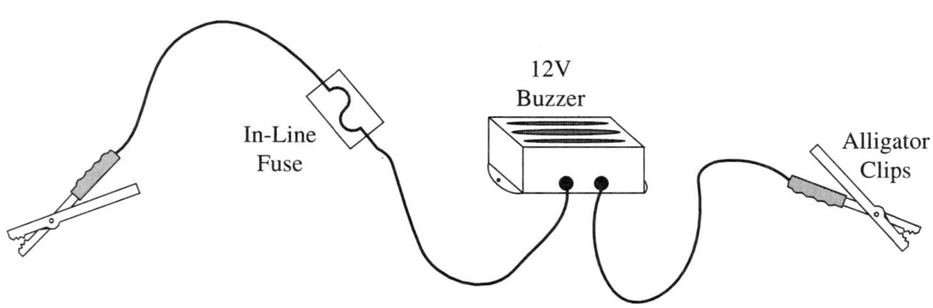

FIGURE 10–12 Buzzer project: Old seat belt buzzer, or Radio Shack part #273-055, or equivalent; in-line glass fuse holder, Radio Shack part #270-1217 or equivalent, or blade fuse holder, Radio Shack part #270-1213 or equivalent; alligator clips, Radio Shack part #270-375 or equivalent.

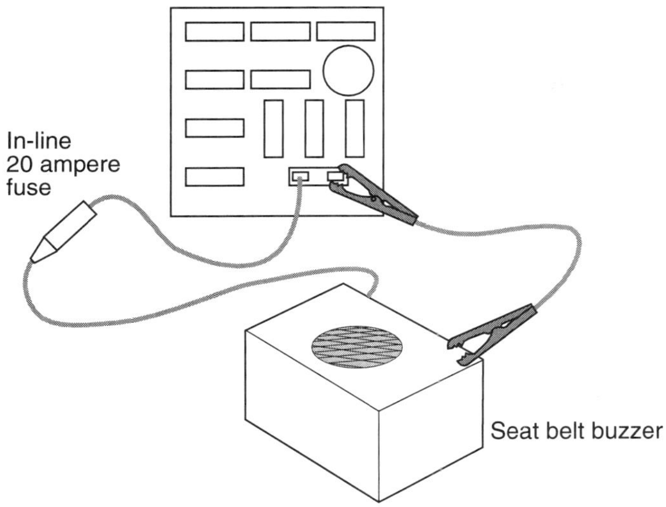

FIGURE 10–13 Using a fused 12 volt buzzer in series to help isolate a short-to-voltage.

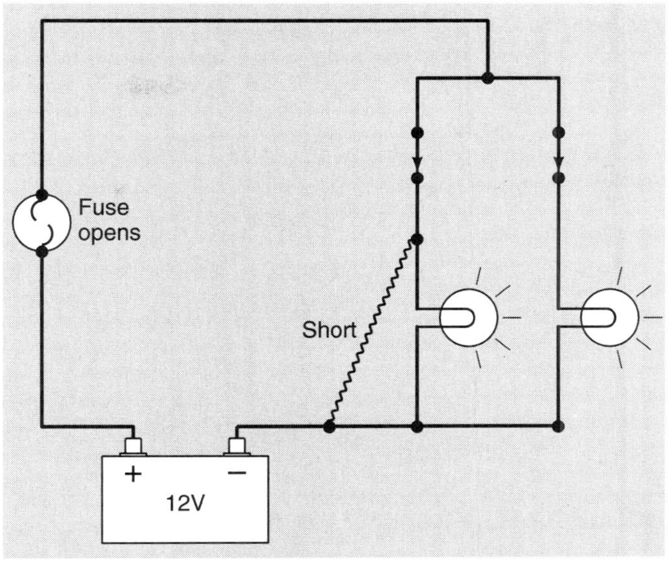

FIGURE 10–14 A typical short-to-ground in a parallel circuit.

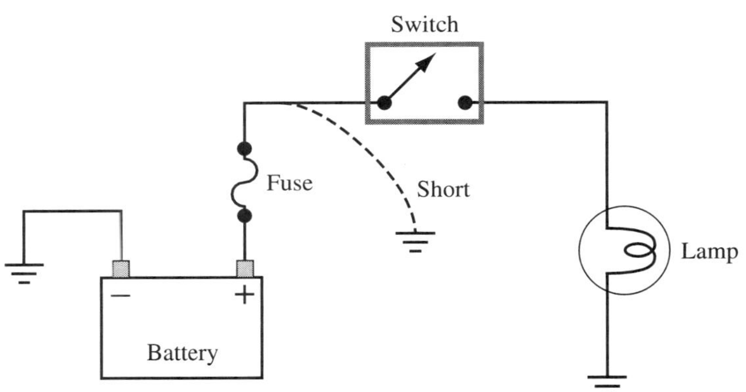

FIGURE 10–15 A short-to-ground located before the circuit control switch.

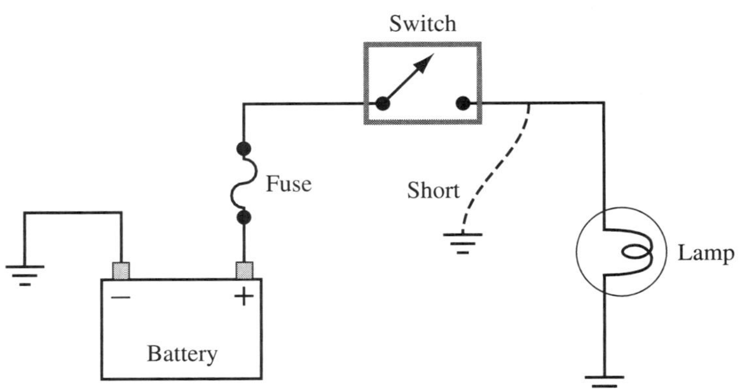

FIGURE 10–16 A short-to-ground located after the circuit control switch.

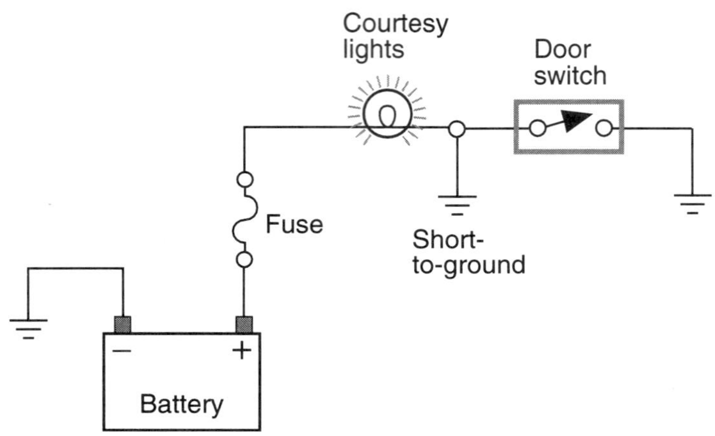

FIGURE 10–17 A short-to-ground located after the load, but before the circuit control switch.

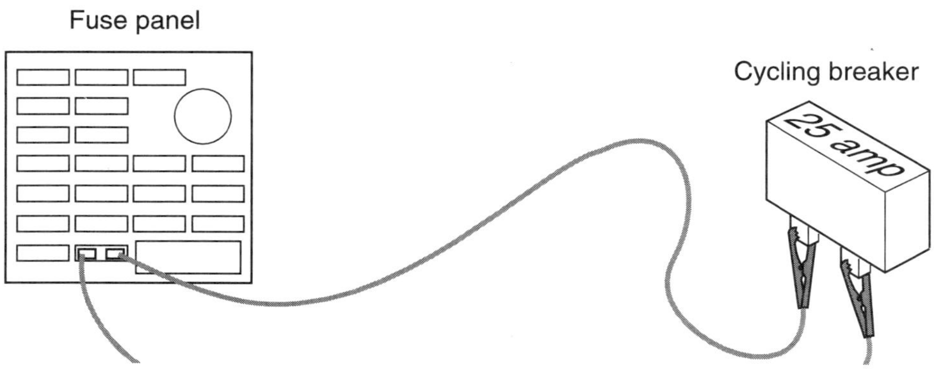

FIGURE 10–18 To protect the circuit while checking for a short-to-ground, install a cycling circuit breaker in place of the circuit fuse.

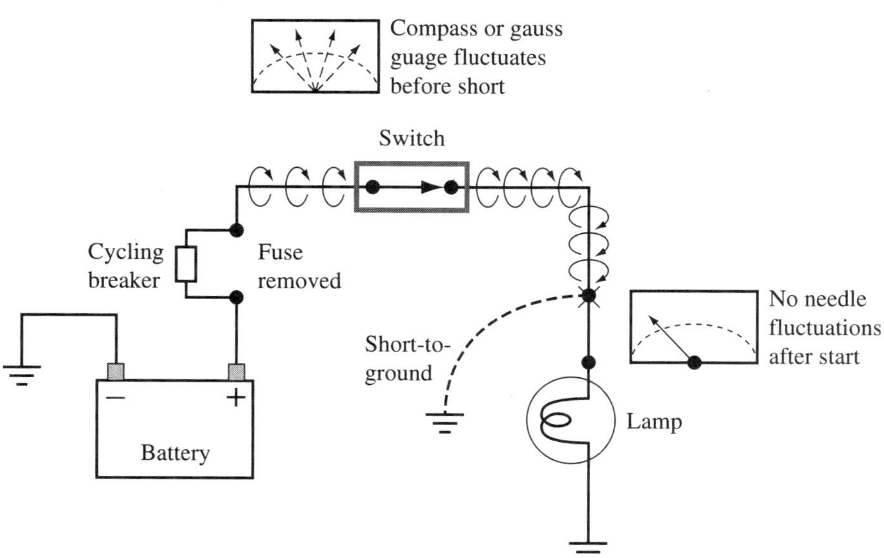

FIGURE 10–19 Finding a short using the fluctuations of a compass or gauss meter.

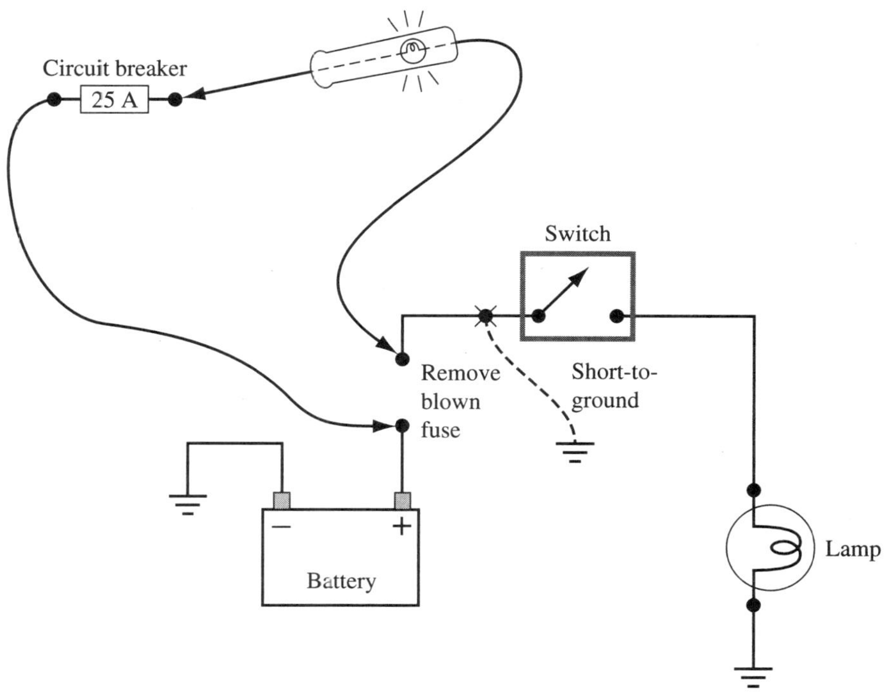

FIGURE 10–20 Using a test light to find a short-to-ground.

When the breaker resets itself, this cycling repeats continuously. By holding a compass or gauss gauge over the circuit wiring, the needle fluctuates rapidly from the pulsing magnetic field *up to* the location of the short. *After* the short, there is no current flow and the compass or gauge needle will not fluctuate. This method works well even when the wire is covered by the vehicle's trim.

You can test for a short-to-ground with a test light or buzzer (Figure 10–20), which should be used in series with a cycling circuit breaker for protection. The test light or buzzer provides resistance in the circuit and allows you to isolate the short by disconnecting the circuit components and connectors until the problem is found. If the short-to-ground is intermittent, this method is helpful because you can wiggle the harness every few inches until the test light glows or the buzzer sounds. This would be the location of the short.

A self-powered test light or ohmmeter can also be useful in finding a short-to-ground (Figure 10–21). Remove the blown fuse from the shorted circuit, disconnect the battery and the circuit load component(s) before hooking up the self-powered test light. The test lamp will glow, or the ohmmeter will record a reading if there is a short-to-ground. To locate the short, wiggle the harness every few inches until the test lamp goes out.

CAUTION *Be careful when using test lights and self-powered test lights around electronic circuits like air bags or computer controlled circuits. Damage may occur to sensitive circuits from the current flow of the test light.*

Voltage Drops

When an electrical circuit develops problems that cause, for example, dim lights or a slowly operating starting motor, it is normally a result of unwanted excessive resistance in the circuit. **Excessive resistance** is a higher than normal resistance that causes the current flow to be lower. Remember Ohm's Law, which we learned in previous chapters, concerning resistance and current being inversely proportional to each other.

The best way to test for excessive resistance in a complete circuit is an **unwanted voltage drop test** which is defined as the portion of the applied circuit voltage that is used up in points of the circuit other than the designed load component(s). It is the resistance in the circuit that reduces the available electrical pressure beyond the excessive resistance point(s). Voltage drops may occur on the positive side or the ground (negative) side of the circuit. Whenever a voltage drop is suspected, both sides of a circuit should be checked for excessive

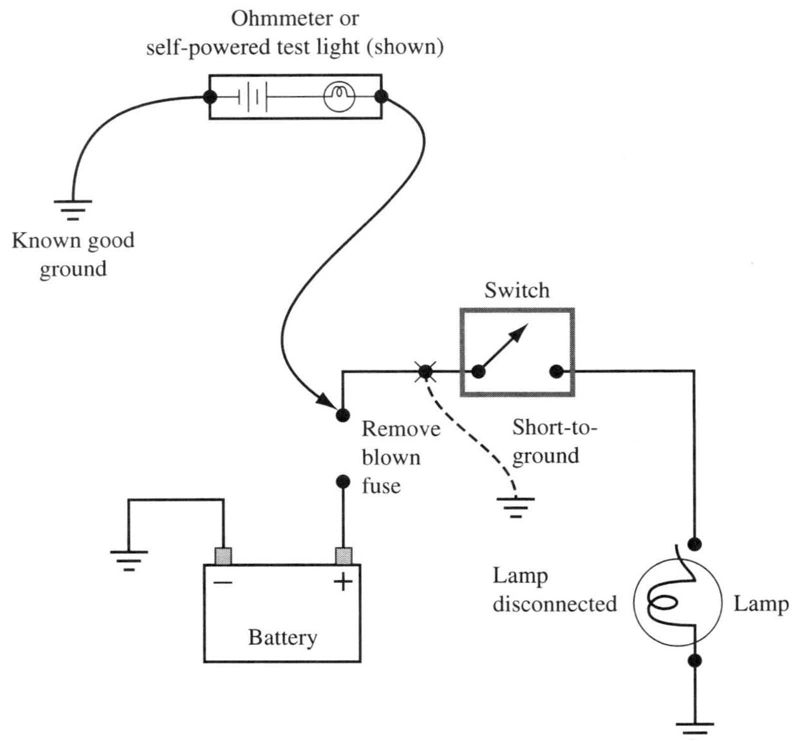

FIGURE 10–21 Using a self-powered test light to find a short-to-ground.

resistance. In Figure 10–22, a voltmeter is being used to check for the cause of the dim light in the load of a simple series circuit.

If you recall that the series circuit law for voltage states that $V_T = V_1 + V_2 + V_3 +$ etc., then you would know that the voltage across the lamp in Figure 10–22 should be 12 volts. In this case it reads 6 volts, so unwanted resistance exists somewhere in the circuit. Figure 10–23 identifies where corrosion has caused the voltage loss in the circuit causing the lamp to operate dimly.

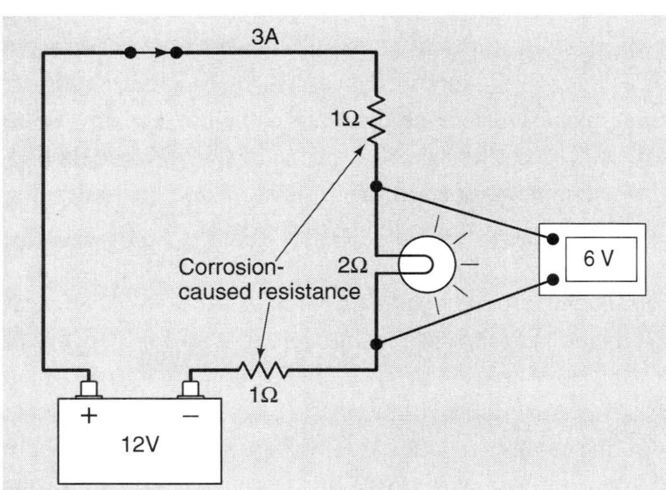

FIGURE 10–22 Checking a simple series circuit for an unwanted voltage.

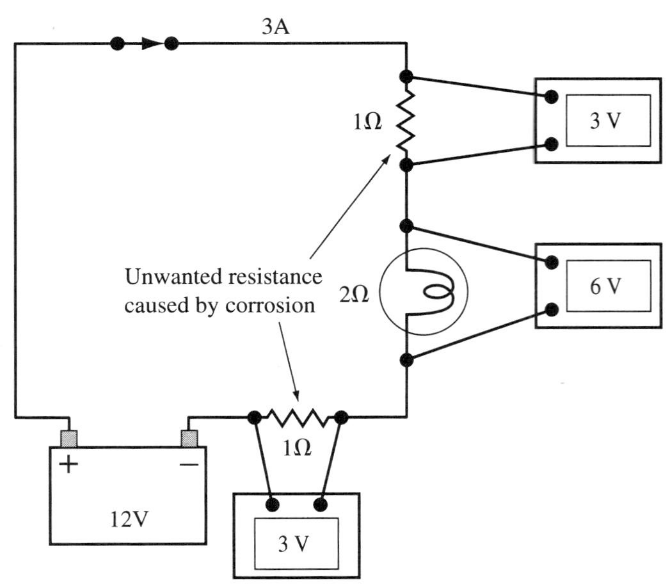

FIGURE 10–23 Using a voltmeter to find unwanted voltage drops in a circuit.

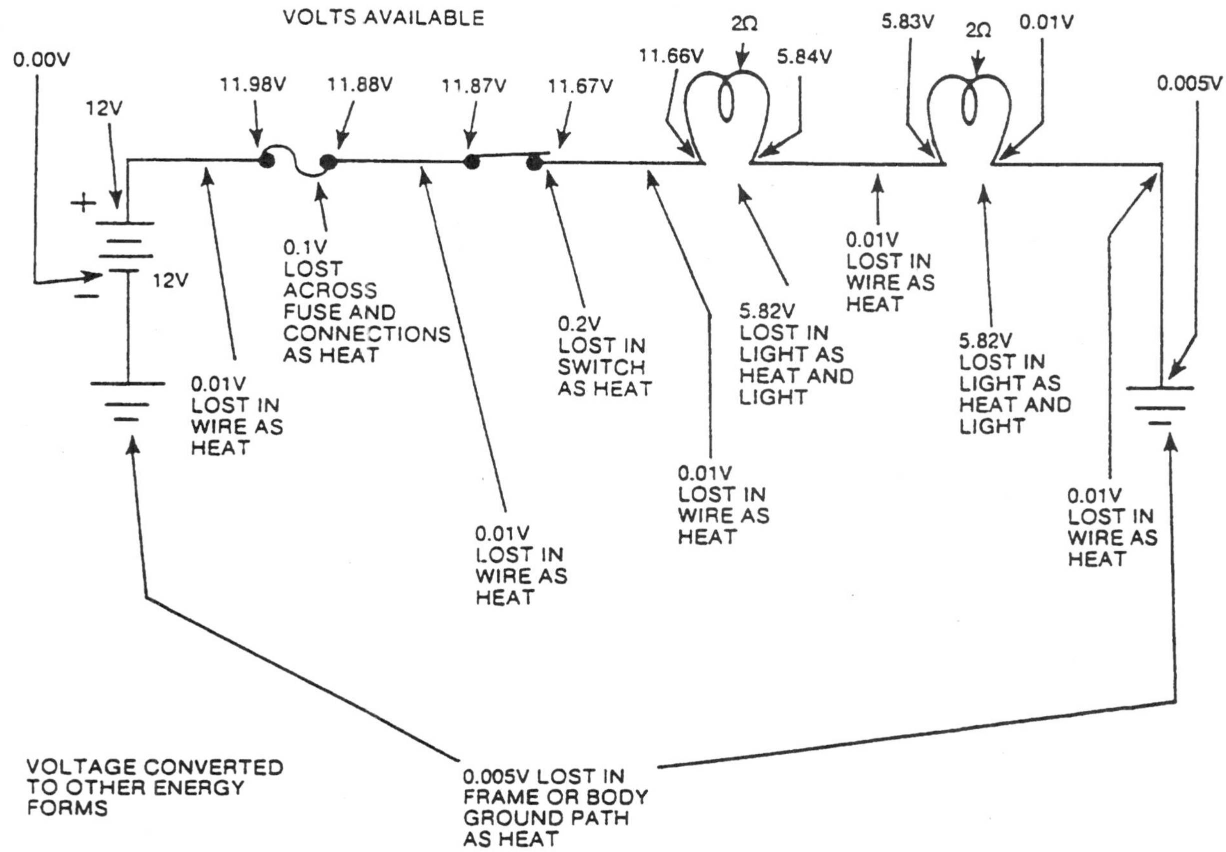

FIGURE 10–24 Adding up all the unwanted voltage drops in a circuit.

The following general specifications are general rules for permitted voltage drops in circuits when manufacturers' specifications are not available.

Connection Point	Allowed Voltage Drop
Ground connections	0.1 volts
Switch Contacts	0.2 to 0.4 volts
System Wiring	0.2 volts
Terminal Connections	0.0 volts
Computer Sensor Connections	0.0 to 0.05 volts

The more connections and wiring a complex circuit contains, the greater chance there is for unwanted resistance to eventually develop that will cause a voltage drop. In Figure 10–24, there are several small voltage drops that add up to over .3 volts. Even though this is a minimum amount overall, the voltage drops could eventually increase and cause the lamps to operate dimly.

Figure 10–25 is an example of a starter circuit being tested for a voltage drop. This circuit is designed to operate the starter motor on the entire 12 volts supplied by the battery. In this starter circuit, there are ten areas to check for unwanted resistance that would cause a voltage drop—three battery cables, six connections, and the solenoid switch. A voltage drop in any one of these connections or a series of connections, could effect the

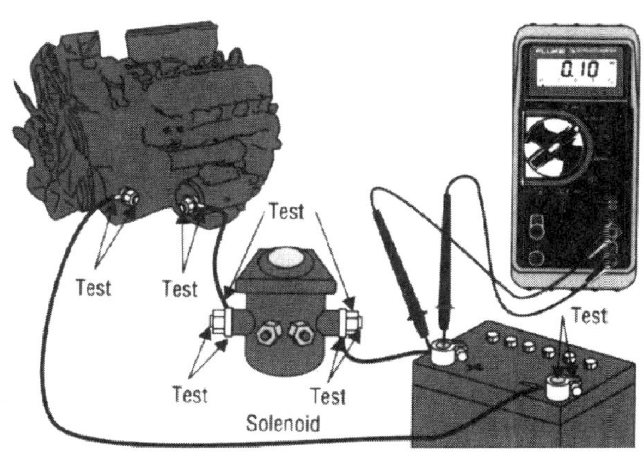

FIGURE 10–25 Checking a starter circuit for a voltage drop.

(Courtesy of The Fluke Corporation)

proper operation of the starter motor. Large current circuits like a starter motor must have very low resistance (voltage drop) in order for the 12 volts source to push as much as 200 amps through the circuit.

Parasitic Drains

When customers complain about the vehicle battery going dead if the vehicle has been sitting for a few days or more, and if the battery has been tested and ruled out as the cause, an unwanted Key Off current drain can be suspected. A **parasitic drain** is defined as a current drain present in the electrical system after the ignition switch is in the Key Off position. There are several possible causes of an excessive parasitic drain, but the most common cause is a light not being turned off, such as a trunk light, engine compartment light, or glove box light.

Acceptable limits of parasitic loads will depend on the vehicle manufacturer and the amount of electrical accessories on the vehicle, but most consider the maximum limit at 35 mA-50 mA (.035 amps-.050 amps). Vehicles with several on-board computers and electrical accessories may have as much as 100 mA (.1 amp) parasitic drain designed into the system that may take several minutes to shut off or reach the "sleep mode." Some of the more common acceptable parasitic drains are listed in Table 10–1.

TABLE 10–1

Load Device	Parasitic Drain
ECM	3 mA-10 mA
PCM	5 mA - 3 mA
ABS Computer	1 mA-3 mA
Digital Clock	2 mA-5 mA
Electronic Radio	3 mA-8 mA
CD Changer	3 mA-5 mA
Memory Seats	2 mA-4 mA
Alternator	.5 mA-2 mA
Door lamp/Chime module	.5 mA-1 mA
Horn module	.25 mA-.5 mA
Twilight Sentinel module	3 mA-8 mA

Normal Electrical Parasitic Drains

The procedure for performing a parasitic drain test may vary with each manufacturer due to specific vehicle applications. There are five common methods that have been used over the years depending on the complexity of the vehicle electrical system. The first three methods are NOT considered reliable on newer computerized vehicles, and are included here to give you a complete understanding of common troubleshooting practices from the past.

Method #1—Voltmeter in Series This method is also known as the Voltage Draw Test, which is NOT a reliable test on newer computerized vehicles, or when used with a high impedance digital voltmeter. An *analog* voltmeter is connected in series with the negative battery cable (Figure 10–26), and a reading higher than 6 volts would indicate a parasitic drain. A digital voltmeter would read the source voltage of 12 volts even if there is not a parasitic drain, which makes this method unreliable with a digital voltmeter.

Method #2—Test Light in Series This method uses the test light connected in series in the same way as the voltmeter is used in method #1. This method is also NOT a reliable test. The brightness of the test light is used to determine if a parasitic drain exists. However, it is hard to determine how much parasitic drain exists by looking at how bright the test lamp is.

Method #3—Ohmmeter in Parallel This method uses an ohmmeter connected in parallel to the vehicle's electrical circuit and in place of the battery (Figure 10–27). This method is NOT a reliable test because some computer circuits have to be energized before going into a standby or sleep mode, and the ohmmeter battery will not operate computer switching circuits.

Method #4—Ammeter in Series This method uses an ammeter in series with the vehicle electrical circuit and placed between the negative battery post and the negative battery cable (Figure 10–28). This method is a RELIABLE test that is used widely by technicians because it shows the exact current draw on the electrical system. However, there is a precaution that should be observed when using this method. The ammeter has a current capacity limit that should never be exceeded

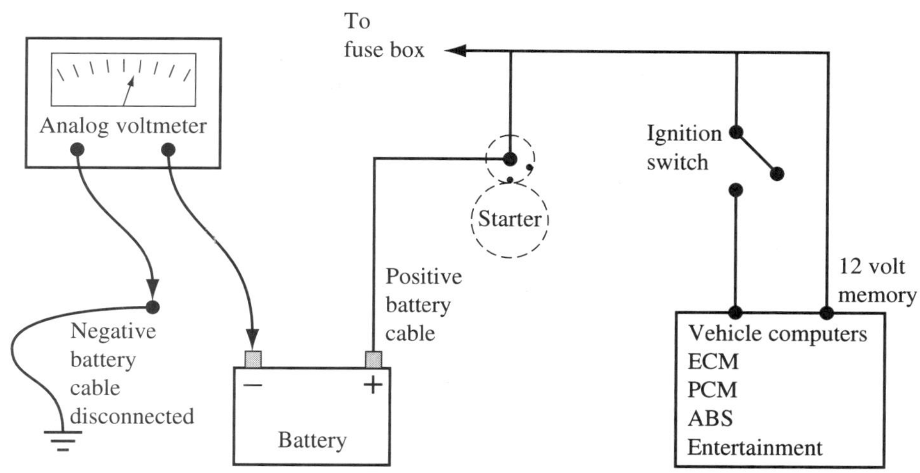

FIGURE 10–26 An analog voltmeter connected in series to the negative cable to find a parasitic drain.

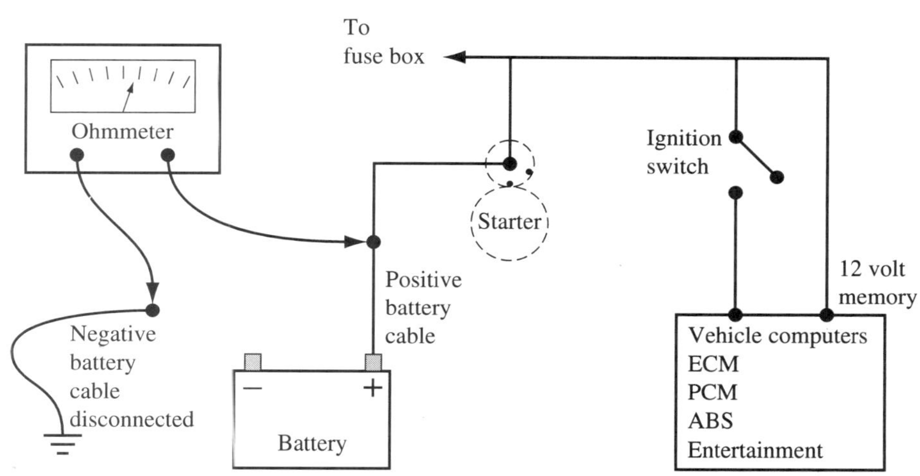

FIGURE 10–27 With the battery disconnected, an ohmmeter is connected in parallel to the circuit to check for a parasitic drain.

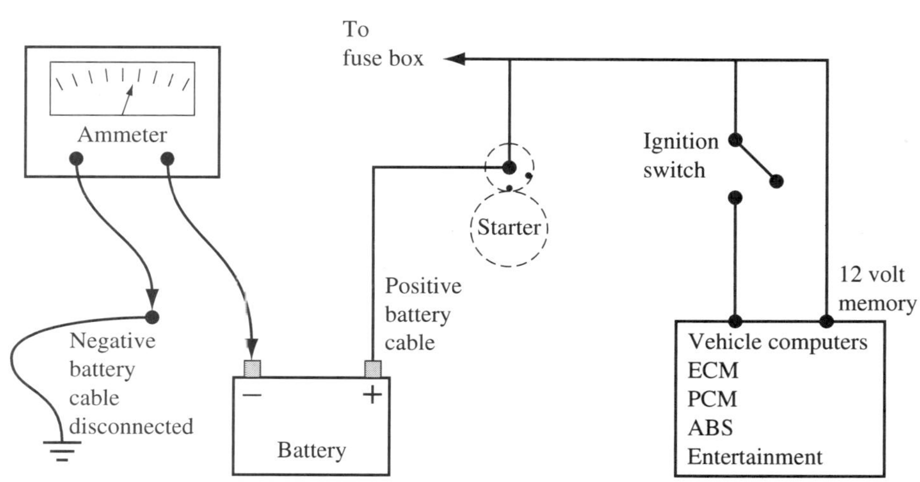

FIGURE 10–28 An ammeter is connected in series to the negative cable to find a parasitic drain.

when checking for parasitic drains. Most quality digital ammeters are fuse protected to prevent overloading the meter and can handle up to 10 amps satisfactorily. Typical parasitic drains are less then 5 amps, unless major circuit problems exist.

Method #5—Voltage Drop This method is very effective in testing for parasitic drains in a circuit under actual voltage conditions. Some computer circuit faults may be voltage-sensitive related and only show up when stressed under normal working conditions. In this method, a fused 1 ohm, 10 watt resistor is put in series between the negative battery post and the negative battery cable (Figure 10–29). A digital voltmeter is connected in parallel across the 1 ohm resistor. If there is a parasitic drain present in the electrical system, the current will have to pass through the 1 ohm resistor and develop a voltage drop across it. Using Ohm's Law (E = I x R), the amount of voltage drop across the resistor is directly related to the amount of current flowing in the circuit.

For example, if the voltage reading across the 1 ohm resistor is. 050 volts, this would be the same as .050 amps (.050 volts ÷ 1 ohm = .050 amps). If you remember the conversion method learned in Chapter 3, then .050 amps is the same as 50 mA. Because this method is growing in popularity, making a "Parasitic Drain, 1 ohm Test Tool" for under $10.00 can be a handy diagnostic addition to have. Figure 10–30 is a sample project that can be made from simple parts from your local electronic parts supply store. Radio Shack part numbers are shown for reference only.

After you have selected a diagnostic method to measure the unwanted parasitic drain, the Fuse Pull technique is the most common way to isolate the electrical circuit that is causing the problem. In Figure 10–31, the fuses are being removed one at a time while viewing the ammeter for any change in its reading.

When the vehicle on-board computer circuits take several minutes to shut off or go into their sleep mode, a parasitic draw test switch (Figure 10–32) may be needed to accurately test for parasitic drains. This test switch is connected in series between the negative battery post and the negative battery cable, and allows the vehicle to operate as normal when the switch is in the ON position. With the ammeter connected across the test switch terminals, wait for all delay circuits to shut off (this can be up to 5 minutes depending on manufacturer). Change the control on the parasitic test switch to the *off* position. The ammeter is now in series with the electrical circuit. The amount of parasitic drain can now be recorded. *Remember* to close the switch before attempting to operate any of the vehicle's electrical systems. This switch is for accurately testing for parasitic drains only.

Sneak Circuits

A **sneak circuit** is defined as an unwanted back-feed of current from one circuit through another that shares the

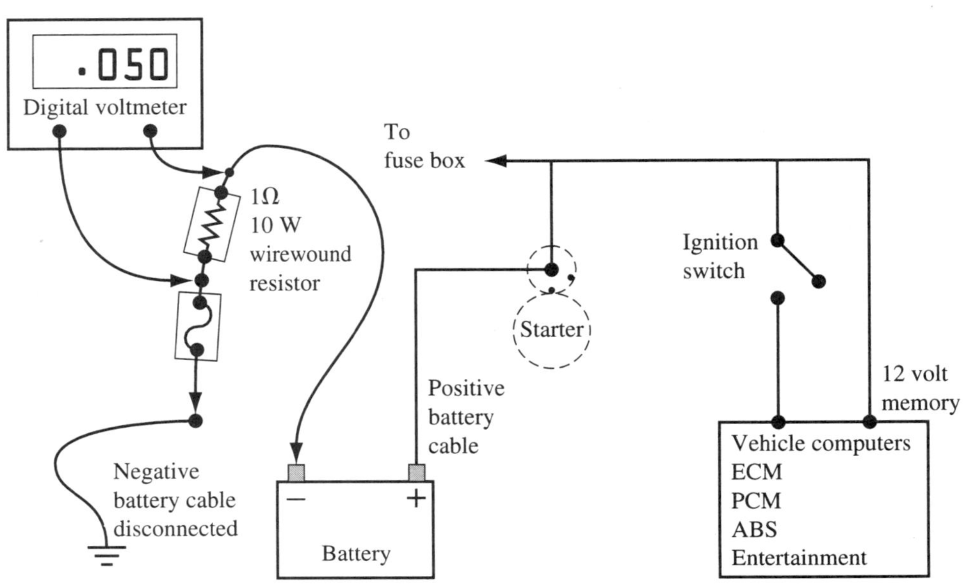

FIGURE 10–29 Checking for a parasitic drain with the dynamic voltage drop method.

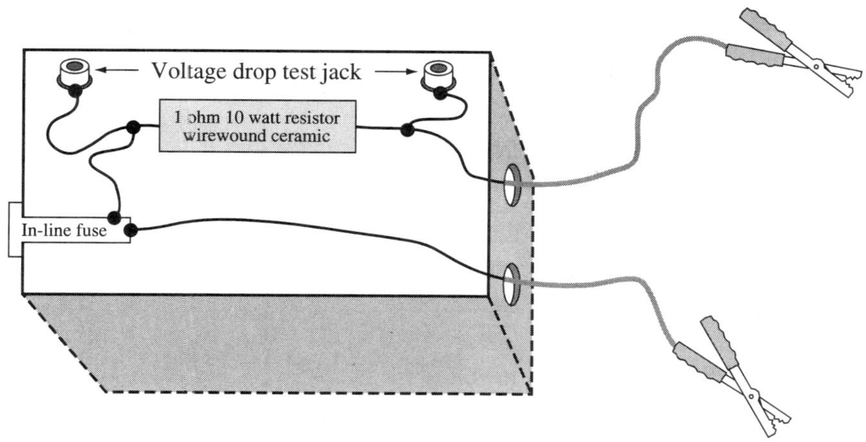

FIGURE 10–30 A parasitic drain test tool project.

same power feed or ground path. This can happen when a shared fuse is blown or a shared ground is disconnected. Figure 10–33 is an example of a sneak circuit where there is a 10 amp fuse for the left low and high sealed beam headlamps and a separate 10 amp fuse for the right low and high sealed beam headlamp circuit. On the ground side, the left and right high beams share a ground, and the left and right low beams share a separate ground. Having the grounds combined allows the driver to operate between low and high beams with one switch. When either circuit fuse blows, a series circuit is created and the headlamps on the side with the blown fuse will dimly glow because voltage from the good headlamp fuse will *sneak* through the bulbs to the headlamp switch and then to ground. Customers may not realize a sneak circuit is taking place, and replace the headlamp bulbs, only to find the same problem.

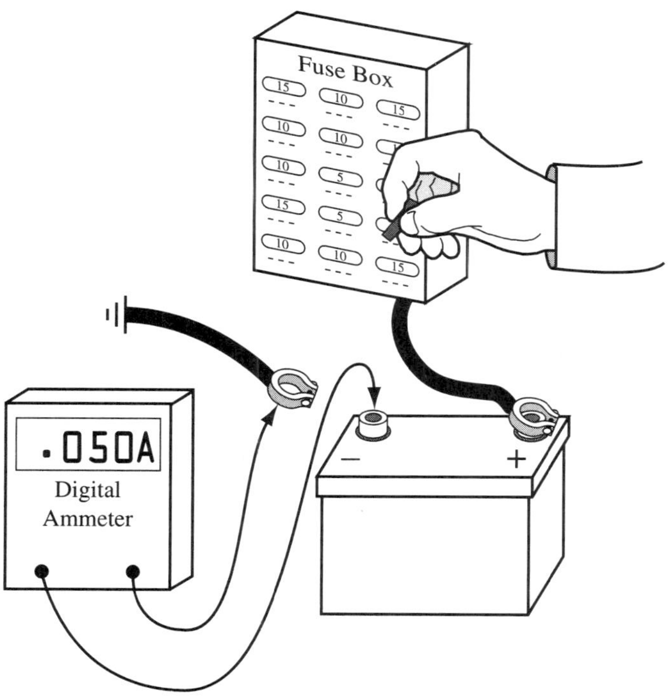

FIGURE 10–31 Removing circuit fuses to isolate the parasitic drain.

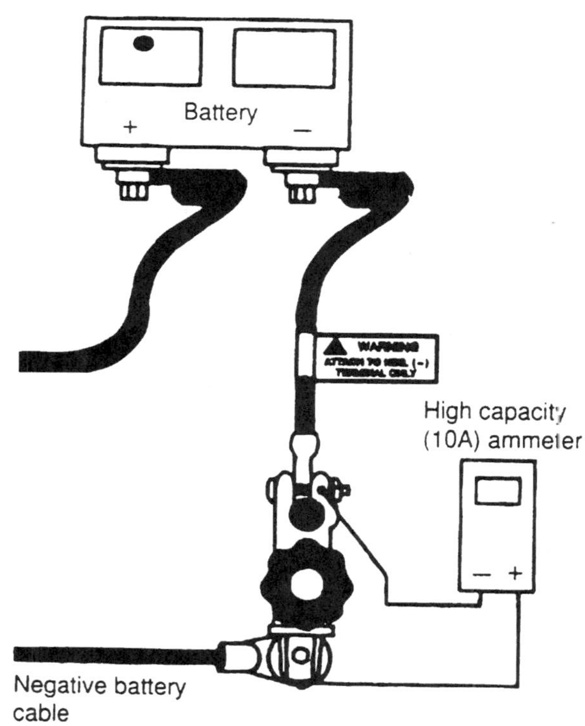

FIGURE 10–32 Checking for a parasitic battery drain using a parasitic draw test switch.

(Courtesy of General Motors Corporation, Service Technology Group)

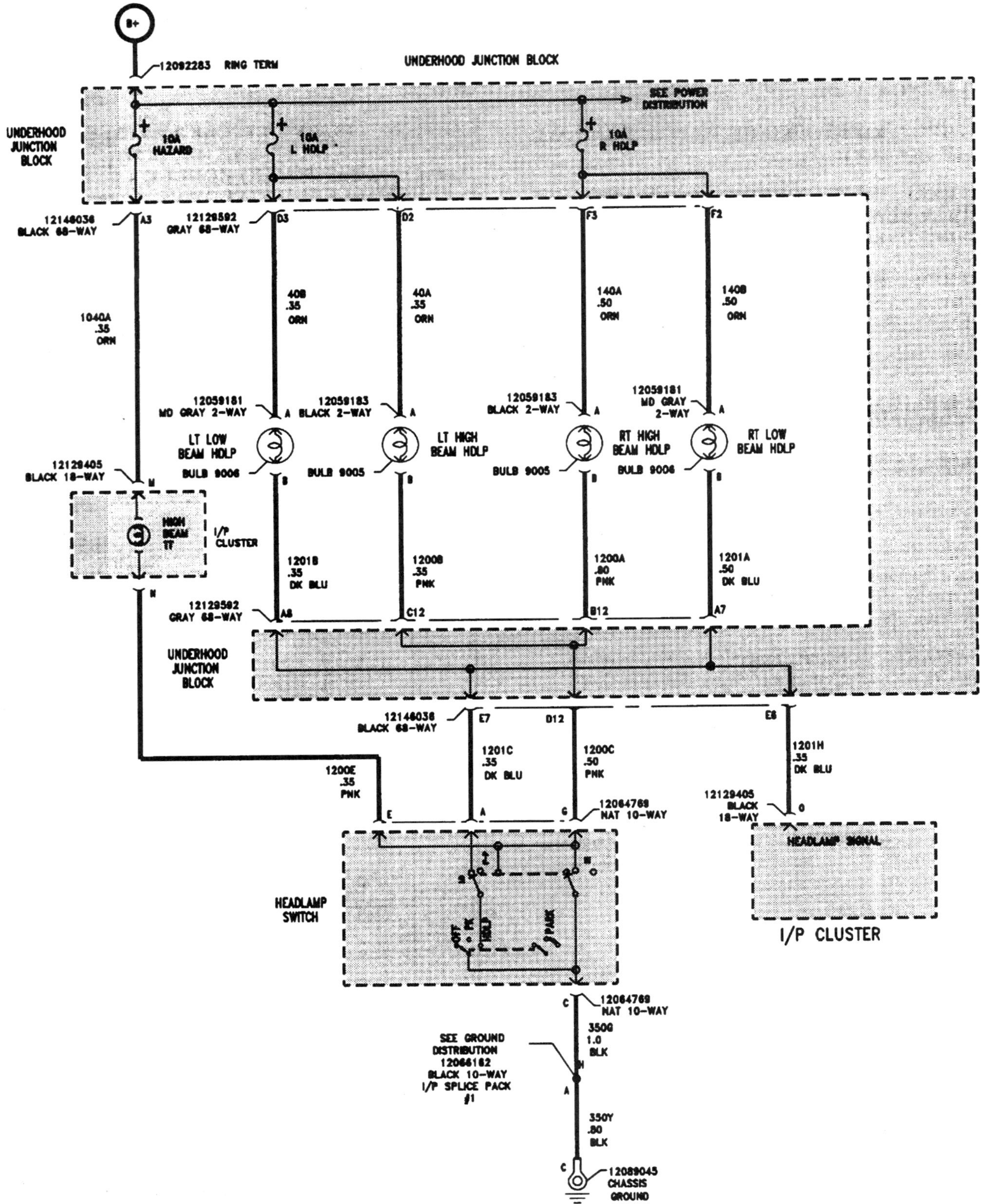

FIGURE 10–33 Locating a sneak circuit in a headlamp schematic.

(Courtesy of The Saturn Corporation)

CASE STUDIES

CASE #1

The following case study is typical of an open circuit problem that technicians often encounter. Knowledge of the circuit and the simple test equipment needed can help you perform the diagnosis quickly and efficiently.

Customer Complaint

A customer brought his car into the shop because the left rear taillamp does not work. The customer replaced the left rear taillamp, but the lamp still does not light. The rest of the rear lamps appear to operate properly (Figure 10–34).

Known Information

- ❑ Vehicle operating voltage = 14 volts
- ❑ Circuit fuse is OK
- ❑ All taillamps, license lamp, and marker lamp bulbs are OK
- ❑ Headlamp switch is OK

Circuit Analysis

Answer the following questions on a *separate* sheet of paper.

1. What is the most likely cause of the left rear taillamp to not illuminate?__________________________________.
2. What should the voltage be at connector A of the .8 BRN wire at the LH taillamp?____________________________.
3. Can the .8 BLK wire from the LH taillamp to the S 317 connector be open?________________.

   Why?____________________________________.
4. How could you verify the circuit problem using a jumper wire? ________________________________.
5. What is the best way to repair this system? ______________________________.
6. What diagnostic steps were helpful when troubleshooting this system or explaining it to the customer? ________________________________.
7. What is your analysis of the above case study? ________________________________.

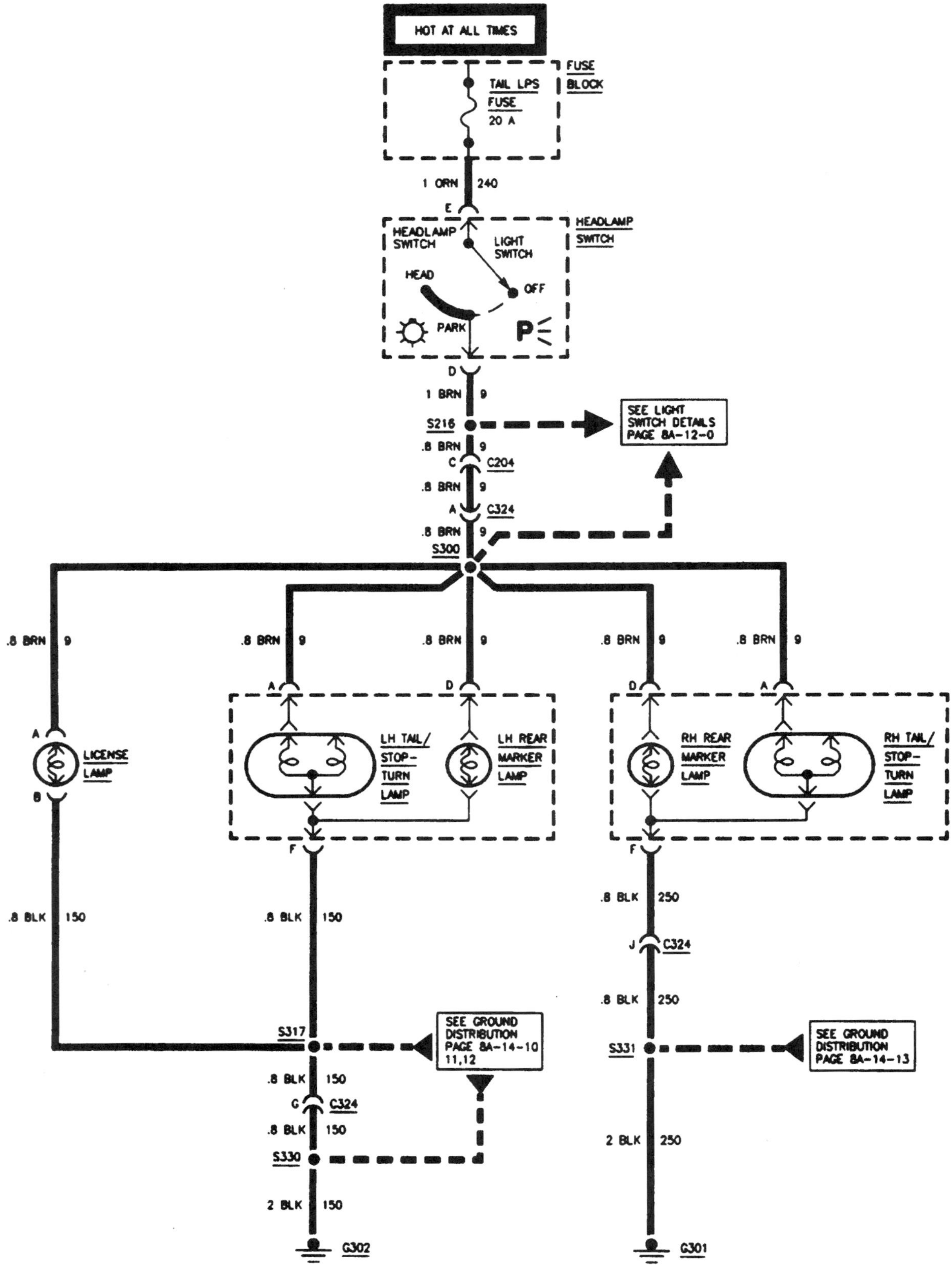

FIGURE 10–34 Exterior rear taillight schematic.

(Courtesy of General Motors Corporation, Service Technology Group)

CASE #2

Using the same circuit diagram, Figure 10–34, complete the customer complaint below.

Customer Complaint

The 20 amp taillamp fuse will blow when the headlamp switch is turned on to the Park or Headlamp position. The customer replaced the fuse four times, but it blows instantly.

Known Information

- ❑ Vehicle operating voltage = 14 volts
- ❑ 20 amp fuse has been replaced by the customer four times
- ❑ Headlamp switch is OK

Circuit Analysis

Answer the following questions on a *separate* sheet of paper.

1. What is the most likely cause of the taillamp fuse to blow when the headlight switch is turned on?

   ____.

2. What diagnostic steps and tools would use to find the suspected problem in this taillamp circuit?

   ____

   ____.

3. If you disconnected the .8 BRN wire at C 324 and the fuse still blows, but does not blow when the .8 BRN wire is disconnected at C 204, what would be the suspected problem in the circuit? __________

   ____.

4. How could you verify the circuit problem using a jumper wire? ______________________.
5. What is the best way to repair this system? ______________________

   ____.

6. What diagnostic steps were helpful when troubleshooting this system or explaining it to the customer? ______________________

   ____

7. What is your analysis of the above case study? ______________________

   ____

   ____.

CASE #3

The following case study shown in Figure 10–35 is typical of a short circuit problem that technicians often encounter. Knowledge of the circuit and use of simple diagnostic test tools can help you perform the diagnosis quickly and efficiently.

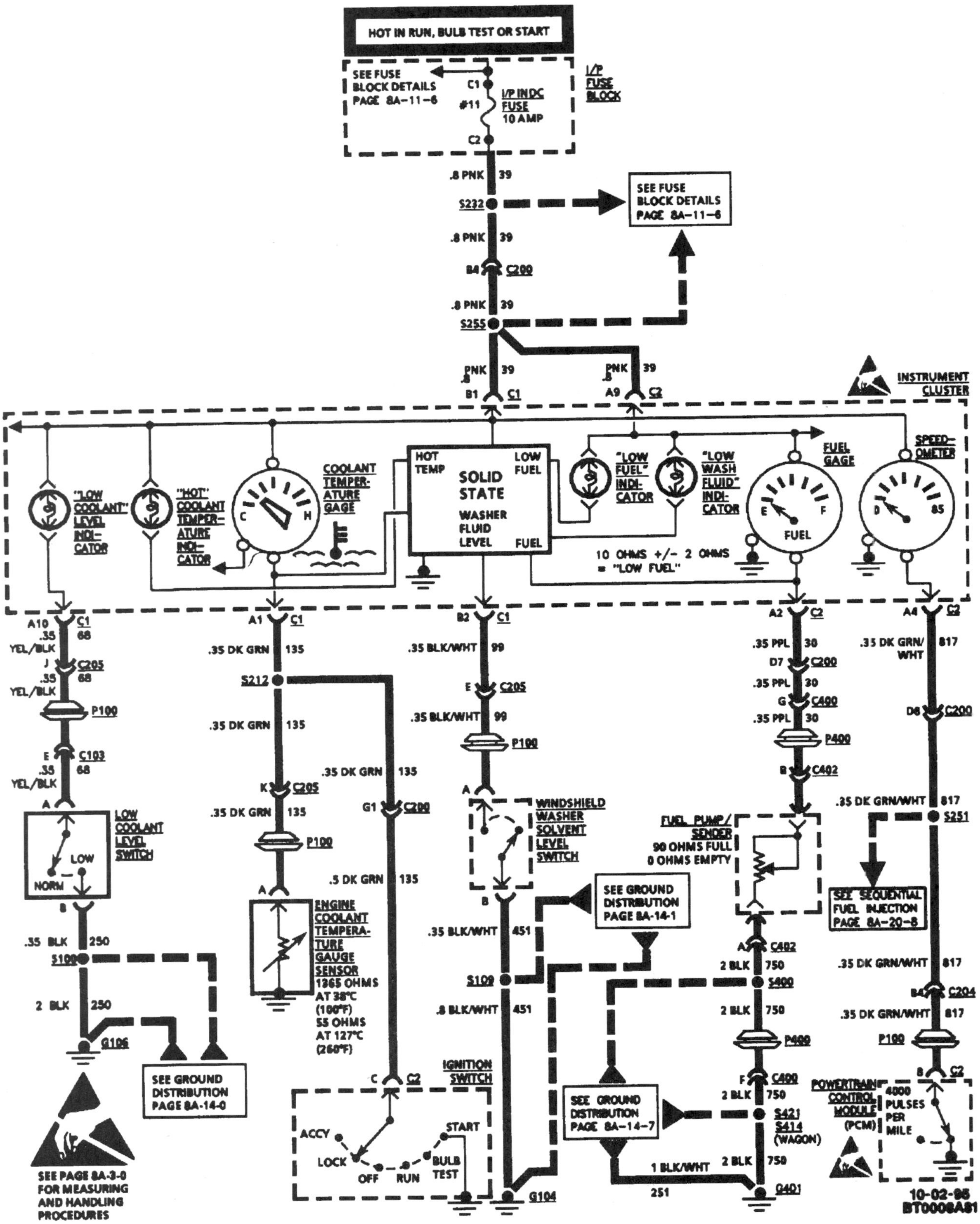

FIGURE 10–35 Instrument cluster and control circuit schematic.

(Courtesy of General Motors Corporation, Service Technology Group)

Customer Complaint

A customer complains the coolant temperature gauge reads hot anytime the ignition switch is on or the vehicle is running. The customer replaced the temperature gauge and the temperature gauge sensor, but the problem still exists.

Known Information

- ❑ Vehicle operating voltage = 14 volts
- ❑ All other instrument cluster gauges and indicator lights operate properly
- ❑ Ignition switch is OK

Circuit Analysis

Answer the following questions on a *separate* sheet of paper.

1. What is the most likely cause for the temperature gauge to read hot if the gauge and sending unit have both been replaced? ____
____.
2. What should happen to the temperature gauge reading if the coolant temperature gauge sensor was disconnected while the engine was running?____.
3. If you disconnected the temperature gauge sensor, and the gauge still showed hot, but moved back to cold when the .35 DK GRN wire was disconnected at the C 205 connector, what would be the most likely cause of this circuit problem? ____
____.
4. What is the best way to repair this system? ____.
5. What diagnostic steps were helpful when troubleshooting this system or explaining it to the customer? ____
____.
6. What is your analysis of the above case study? ____
____
____.

CASE #4

The following case study in Figure 10–36 is typical of a parasitic drain problem that technicians often are required to diagnose. Knowledge of the circuit, test procedures, and use of simple diagnostic test tools can help you diagnose the circuit safely and efficiently.

Customer Complaint

A customer complains that the vehicle battery runs down after the vehicle sits for a few days. The battery has been replaced twice, and the alternator tested satisfactorily.

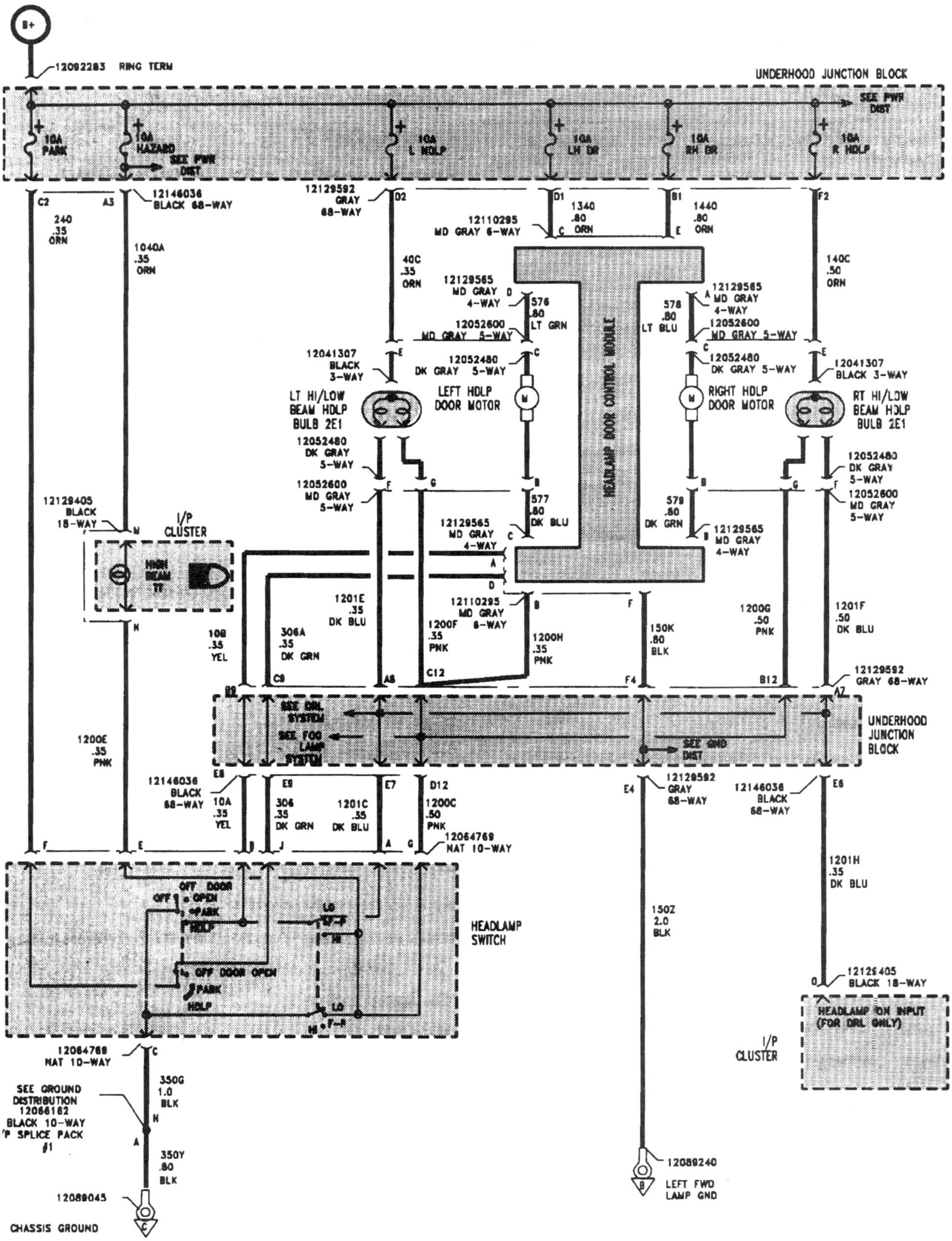

FIGURE 10–36 High/low beam headlamp schematic.

(Courtesy of Saturn Corporation)

Known Information

- ❑ Vehicle operating voltage = 14 volts
- ❑ Battery and alternator both test satisfactory
- ❑ A parasitic drain is suspected

Circuit Analysis

Answer the following questions on a *separate* sheet of paper.

1. What diagnostic steps and test tools would you use to check this vehicle for a parasitic drain? ______ ______.
2. The vehicle was drawing 2 amps of parasitic current until the 10 amp RH DR (right hand door) fuse was pulled out. The parasitic drain then went down to 35 mA. What would be the MOST likely cause for the excessive parasitic drain in the system? ______.
3. How could you verify that the circuit problem existed in the identified problem component? ______ ______.
4. What is the best way to repair this system? ______ ______.
5. What diagnostic steps were helpful when troubleshooting this system or explaining it to the customer? ______ ______.
6. What is your analysis of the above case study? ______ ______.

Hands-On Vehicle Tests

The following seven hands-on vehicle checks are included in the **NATEF** (National Automotive Technicians Education Foundation) **Task List.** The task reference numbers are the same as the NATEF reference numbers. Complete your answers to the below questions on a *separate* sheet of paper.

Performance Task #1 (NATEF #VI-A-1)

Task Description

Use wiring diagrams during diagnosis of electrical circuit problems.

Task Objectives

- ❑ Obtain a vehicle that can be used for this task. What model and year of vehicle are you using for this task? ______.

- ❑ What electrical circuit problem(s) exist with this vehicle? ______________________________
____
____.
- ❑ Locate a wiring diagram(s) that will show the electrical circuit identified.
- ❑ Using the wiring diagram(s), identify areas in the circuit such as connectors, wires, fuses, and components, that can cause the problem.

Identified Area	Symptom	Cause	Test Procedure

- ❑ Perform the diagnostic steps and procedures identified in the previous question. Does this help identify the circuit problem(s)?____.
- ❑ What diagnostic steps were helpful when troubleshooting this system? ____________________.
- ❑ What is your analysis of using wiring diagrams to help diagnose electrical circuit problems? ______
____.

Performance Task #2 (NATEF Task #VI-A-2)

Task Description
Check electrical circuits with a test light. Determine needed repairs.

Task Objectives

- ❑ Obtain a vehicle that can be used for this task. What model and year of vehicle are you using for this task? ____.
- ❑ Using a wiring diagram for the vehicle, determine electrical circuits that are safe to check with a test light. Verify circuits to test with your instructor.
- ❑ List the circuit(s) to be tested with the test light.

  1. ____________________ 4. ____________________
  2. ____________________ 5. ____________________
  3. ____________________ 6. ____________________

- ❑ List the procedure(s) used to safety connect a test light to a vehicle's electrical circuit.
____
____
____

❑ Check the operation of the circuits you listed with the test light. What can you determine about using a test light on each of the circuits.

1. ____________________ 4. ____________________

2. ____________________ 5. ____________________

3. ____________________ 6. ____________________

❑ What would be the benefit(s) of using a test light on electrical circuits compared to other electrical testing equipment? ____

____.

Task Summary

After performing the above NATEF task, what can you determine will be helpful in using test lights to perform simple electrical circuit troubleshooting? ____

____.

Performance Task #3 (NATEF Task #VI-A-3)

Task Description

Check voltage and voltage drop in electrical/electronic circuits using a digital multimeter (DMM). Determine needed repairs.

Task Objectives

❑ Obtain a vehicle that can be used for this task. What model and year of vehicle are you using for this task? ____.

❑ Using a wiring diagram for the vehicle, identify an electrical circuit that is suitable for checking voltages and voltage drops. Verify circuits to test with your instructor.

❑ List the circuit(s) to be tested for voltage and voltage drops.

1. ____________________ 4. ____________________

2. ____________________ 5. ____________________

3. ____________________ 6. ____________________

❑ List the procedure(s) used to test for system voltage and voltage drops in an electrical circuit.

____

____

____

❑ What is the system voltage of each circuit tested, and any voltage drop(s) measured.

System voltage __________. Voltage drops __________, __________, __________, __________.

System voltage __________. Voltage drops __________, __________, __________, __________.

System voltage __________. Voltage drops __________, __________, __________, __________.

System voltage __________. Voltage drops __________, __________, __________, __________.

❑ Do the voltage drops in each circuit tested equal the system total voltage? ______________________

____.

Task Summary

After performing the above NATEF task, what can you determine will be helpful in knowing how to check for voltage drops in an electrical circuit in reference to system voltage? ______________________

____.

Performance Task #4 (NATEF Task #VI-A-4)

Task Description

Check current flow in electrical/electronic circuits and components using an ammeter. Determine needed repairs.

Task Objectives

❑ Obtain a vehicle that can be used for this task. What model and year of vehicle are you using for this task? ____.

❑ Using a wiring diagram for the vehicle, determine electrical circuits that are safe to check the current flow with an ammeter. Verify circuits to test with your instructor.

❑ List the circuit(s) to be tested for current flow.

1. ______________________ 3. ______________________

2. ______________________ 4. ______________________

❑ List the procedure(s) used to safely connect an ammeter to a vehicle's electrical circuit.

____

____

____

❑ Check the operation of the circuits listed above with an ammeter. What can you determine about testing for current flow on each of the circuits.

1. ______________________ 3. ______________________

2. ______________________ 4. ______________________

❑ What would be the benefit(s) of using an ammeter on electrical circuits compared to other electrical testing equipment? ____

____

____.

Task Summary

After performing the above NATEF task, what can you determine will be helpful in testing a circuit with an ammeter to troubleshoot an electrical circuit? ______________________________________

____.

Performance Task #5 (NATEF Task #VI-A-5)

Task Description

Check electrical circuits using jumper wires. Determine needed repairs.

Task Objectives

- ❑ Obtain a vehicle that can be used for this task. What model and year of vehicle are you using for this task? ______________________________.
- ❑ Obtain a wiring diagram of the horn circuit for the vehicle. Determine that the circuit is safe to use with a jumper wire to bypass circuit components. Verify circuit components to bypass with your instructor.
- ❑ List the circuit components to be bypassed with the jumper wire.

  1. ______________ 3. ______________

  2. ______________ 4. ______________
- ❑ List the procedure(s) used to safely connect the jumper wire to bypass circuit components.

  ____

Task Summary

After performing the above NATEF task, what can you determine will be helpful in knowing how to use a jumper wire as an effective tool in diagnosing a horn circuit problem or other electrical circuit problems? ______________________________

______________________________.

Performance Task #6 (NATEF Task #VI-A-6)

Task Description

Find shorts, grounds, opens, and resistance problems in electrical/electronic circuits. Determine needed repairs.

Task Objectives

Short-to-Voltage

- ❑ Obtain a vehicle that can be used for finding a short-to-voltage in an electrical system. What model and year of vehicle are you using for this task? ______________________________.
- ❑ Determine what electrical circuit has a short-to-voltage. ______________________________.
- ❑ What symptoms are present in this short-to-voltage?______________________________.
- ❑ Obtain a wiring diagram of the circuit that has the short-to-voltage. List the components, connectors, and wiring that can be isolated to determine the cause of the short. ______________________________

  ______________________________.
- ❑ List the diagnostic procedure(s) and test equipment used to safely locate the short-to-voltage.

  ______________________________.

- ❑ Where is the short-to-voltage in the circuit, and what caused it to happen? ______________________________.
- ❑ What diagnostic steps were helpful when troubleshooting for a short-to-voltage? ______________________________.
- ❑ What is your analysis of finding a short-to-voltage in an electrical circuit? ______________________________.

Short-to-Ground

- ❑ Obtain a vehicle that can be used for finding a short-to-ground in an electrical system. What model and year of vehicle are you using for this task? ______________________________.
- ❑ Determine what electrical circuit has a short-to-ground. ______________________________.
- ❑ What symptoms are present in this short-to-ground? ______________________________.
- ❑ Obtain a wiring diagram of the circuit that has the short-to-ground. List the components, connectors, and wiring that can be isolated to determine cause of the short. ______________________________.
- ❑ List the diagnostic procedure(s) and test equipment used to safely locate the short-to-ground. ______________________________.
- ❑ Where is the short-to-ground in the circuit, and what caused it to happen? ______________________________.
- ❑ What diagnostic steps were helpful when troubleshooting for a short-to-ground? ______________________________.
- ❑ What is your analysis of finding a short-to-ground in an electrical circuit? ______________________________.

Open Circuit

- ❑ Obtain a vehicle that can be used for finding an open in an electrical system. What model and year of vehicle are you using for this task? ______________________________.
- ❑ Determine what electrical circuit has an open. ______________________________.
- ❑ What symptoms are present in this open? ______________________________.
- ❑ Obtain a wiring diagram of the circuit with the open. List the components, connectors, and wiring that can be isolated to determine cause of the open. ______________________________.
- ❑ List the diagnostic procedure(s) and test equipment used to safely locate the open. ______________________________.

- ❑ Where is the open in the circuit, and what caused it to happen? ______________________________

  ______________________________.
- ❑ What diagnostic steps were helpful when troubleshooting for an open? ______________________________

  ______________________________.
- ❑ What is your analysis of finding an open in an electrical circuit? ______________________________

  ______________________________.

High Resistance Circuit

- ❑ Obtain a vehicle that can be used for finding high resistance in an electrical system. What model and year of vehicle are you using for this task? ______________________________.
- ❑ Determine what electrical circuit has a high resistance. ______________________________.
- ❑ What symptoms are present in this high resistance? ______________________________.
- ❑ Obtain a wiring diagram of the circuit with the high resistance. List the components, connectors, and wiring that can be isolated to determine the cause of the high resistance. ______________________________.
- ❑ List the diagnostic procedure(s) and test equipment used to safely locate the high resistance. ______

  ______________________________.
- ❑ Where is the high resistance in the circuit, and what caused it to happen? ______________________________

  ______________________________.
- ❑ What diagnostic steps were helpful when troubleshooting for high resistance? ______________________________

  ______________________________.
- ❑ What is your analysis of finding a high resistance in a electrical circuit? ______________________________

  ______________________________.

Task Summary

After performing the above NATEF task, what can you determine will be helpful in testing a circuit with an ammeter to troubleshoot an electrical circuit? ______________________________

______________________________.

Performance Task #7 (NATEF Task #VI-A-7)

Task Description

Measure and diagnose the cause(s) of abnormal key-off battery drain. Determine needed repairs.

Task Objectives

- ❑ Obtain a vehicle that can be used for finding a parasitic drain (key-off battery drain). What model and year of vehicle are you using for this task? ______________________________.
- ❑ Describe the procedure you would use to find an unwanted current drain on the electrical system?

  ____

- ❑ What electrical circuit did you find with the unwanted parasitic drain? ________________________.
- ❑ How much current is flowing through this parasitic drain? ________________________________.
- ❑ What is the maximum amount of parasitic drain allowed by this vehicle manufacturer? __________.
- ❑ Using a wiring diagram for the vehicle, determine what electrical circuits are part of the circuit that you isolated as causing the parasitic drain? ____

  ____.
- ❑ Check the operation of each circuit that you identified by the wiring diagram for proper operation. What circuit can be isolated as the cause of the parasitic drain? ____________________________.
- ❑ List the procedures that you would use to repair this electrical circuit to manufacturer standards.

  ____

  ____.

Task Summary

After performing the above NATEF task, what can you determine will be helpful in knowing how to check for parasitic key-off drains in an electrical circuit? ____________________________________

____.

Summary

- ❑ An open circuit is defined as a break in the current flow of a complete circuit.
- ❑ An intentional open circuit is designed into the circuit, such as a switch or relay for on/off control.
- ❑ An unintentional open circuit is created when there is a problem in the circuit—such as a broken wire, tripped circuit breaker or blown fuse, or loose connection.
- ❑ A poor electrical connection is the most common cause of open circuits.
- ❑ An ohmmeter can be used to test for an open circuit. A circuit with an open will read OL or infinite.
- ❑ A voltmeter can be used to test for an open circuit. Voltage readings should be zero after the open, and at the circuit voltage before the open.
- ❑ A test light can be used to test for an open circuit. The test light will light before the open, and go out after the open.
- ❑ A self-powered test light can be used to test for an open circuit. The test light will remain lit until the test lamp probe goes beyond the open.
- ❑ Test lights and self-powered test lights should not be used around electronic or sensitive computer-controlled circuit.
- ❑ Testing for an open circuit with a jumper wire is an excellent way to bypass bad circuit switches.
- ❑ A short circuit is defined as an accidental bypass of the current flow in a circuit.

- ❑ A short-to-voltage is an unwanted copper-to-copper connection between two separate circuits.
- ❑ A short-to-ground is an unwanted part for current to flow directly to ground.
- ❑ A short-to-ground will blow the circuit protection fuse.
- ❑ A voltage drop is defined as the portion of the applied circuit voltage that is lost at points of the circuit other than the designed load component(s).
- ❑ A parasitic drain is defined as an unwanted current drain still present in the electrical system after the ignition switch is turned to the Key Off position.
- ❑ Acceptable parasitic drains vary with manufacturer, but most allow up to 35-50 mA of current drain.
- ❑ The most reliable methods for testing for parasitic drains are to use an ammeter or perform a dynamic voltage drop test.
- ❑ Some electrical circuits take several minutes to "shut off" or go into the "sleep mode," which can possibly give the technician erroneous information.
- ❑ A sneak circuit is defined as an unwanted back-feed of current from one circuit through another that shares the same power feed or ground path.

Key Terms

excessive resistance
grounded circuit
intentional open circuit
open circuit
parasitic drain
short circuit
short to ground
short to voltage
sneak circuit
unintentional open circuit
unwanted voltage drop

Review Questions

Short Answer Essays

1. Describe how you would test for an unintentional open circuit using a voltmeter.
2. Describe what a short circuit is in an electrical circuit.
3. What are the general specifications allowed for voltage drops in an electrical circuit.
4. Describe how to test an electrical circuit for a parasitic drain.
5. Describe how an electrical circuit can develop a sneak circuit.

Fill-in-the-Blanks

1. An ________________ circuit is defined as a ________________ in the current flow of a complete circuit.
2. An ohmmeter will read ________________ or ________________ when connected in parallel to an open circuit.

3. Current flow through an unwanted path to ground is called a ________________.
4. Dim lights or a slow turning motor could be caused by ________________ resistance.
5. Maximum parasitic loads normally allowed for a vehicle are ________________.

ASE Style Review Questions

1. Technician A says an open circuit could be caused by a switch. Technician B says a bad relay could be the cause of an open circuit. Who is correct?
   A. A only
   B. B only
   C. Both A and B
   D. Neither A nor B

2. A simple series 12 volt light circuit has a short-to-ground (Figure 10–37). Technician A says the current flow will be higher than normal through the lamp. Technician B says the switch will not turn off the lamp. Who is correct?
   A. A only
   B. B only
   C. Both A and B
   D. Neither A nor B

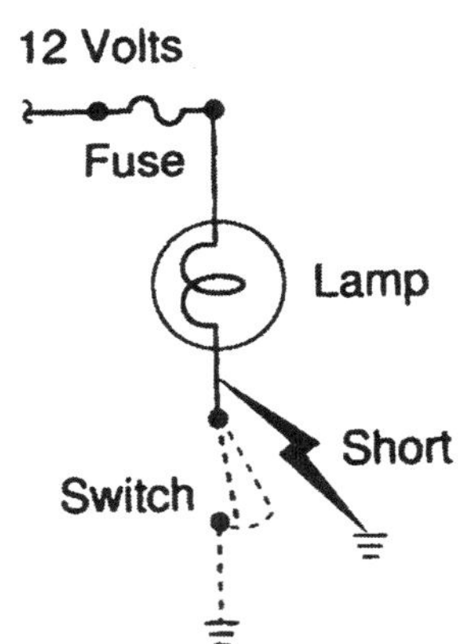

(Courtesy of DaimlerChrysler Corporation)

3. Technician A says an intentional open circuit is caused by a switch or relay. Technician B says an unintentional open circuit could be caused by a blown fuse. Who is correct?
   A. A only
   B. B only
   C. Both A and B
   D. Neither A nor B

4. Two technicians are discussing the use of a test light to check for an open circuit (Figure 10–38). Technician A says the test lamp will light when probing the circuit from the battery positive connector up to the open, and not after the open. Technician B says the test light will only light after the open connection. Who is correct?

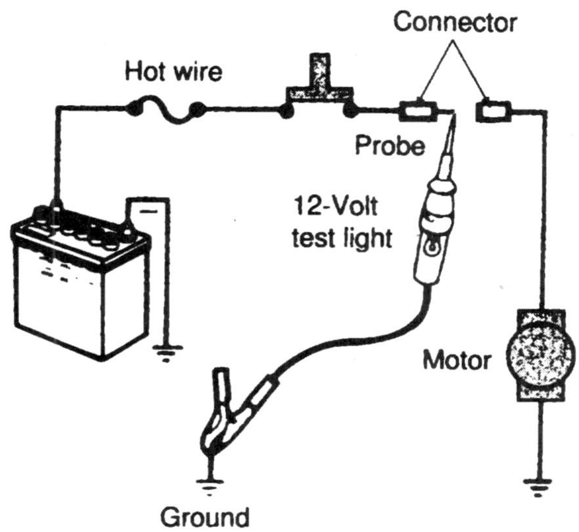

A. A only
B. B only
C. Both A and B
D. Neither A nor B

5. Technician A says a fused jumper wire can be used to bypass a switch to determine if a circuit is working. Technician B says a self-powered test light can be a useful tool for checking for circuit continuity. Who is correct?
A. A only
B. B only
C. Both A and B
D. Neither A nor B

6. Technician A says a short-to-voltage is an unwanted copper-to-copper connection between two circuits. Technician B says a short-to-voltage may not blow the circuit fuse. Who is correct?
A. A only
B. B only
C. Both A and B
D. Neither A nor B

7. A test light is being used to test for a short (Figure 10–39). Technician A says the test light will not light with the circuit fuse removed as shown. Technician B says the test light should not be used to check this circuit. Who is correct?

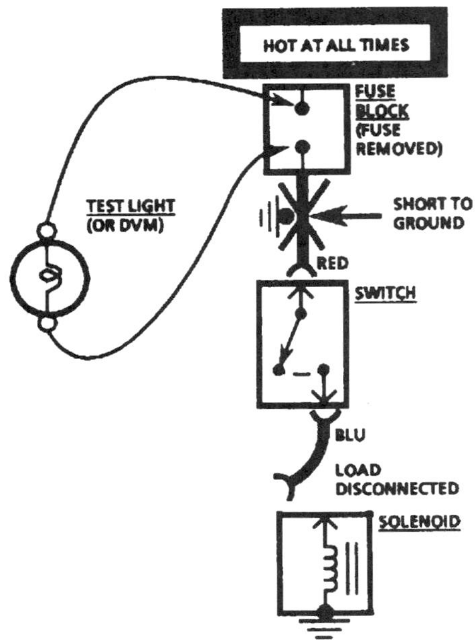

(Courtesy of General Motors Corporation, Service Technology Group)

A. A only
B. B only
C. Both A and B
D. Neither A nor B

8. Technician A says a short-to-ground is always located after the circuit load device(s). Technician B says a short-to-ground can occur when wire insulation breaks down from vibration. Who is correct?

A. A only
B. B only
C. Both A and B
D. Neither A nor B

9. Technician A says a cycling circuit breaker can be used in place of the circuit fuse when diagnosing the cause of a short-to-ground. Technician B says a short-to-ground will always cause to circuit fuse to blow. Who is correct?

A. A only
B. B only
C. Both A and B
D. Neither A nor B

10. Technician A says a test light can be connected in series with a cycling circuit breaker to test for a short-to-ground. Technician B says a buzzer can be connected in series with a cycling circuit breaker to test for a short-to-ground. Who is correct?

    A. A only

    B. B only

    C. Both A and B

    D. Neither A nor B

11. Technician A says that test lights and self-powered test lights should never be used around sensitive electronic circuits. Technician B says computer control circuits have fail-safe controls that make the use of test lights safe to use. Who is correct?

    A. A only

    B. B only

    C. Both A and B

    D. Neither A nor B

12. Technician A says excessive circuit resistance in connectors could be the cause of an unwanted voltage drop. Technician B says bad wiring could be the cause of an unwanted voltage drop. Who is correct?

    A. A only

    B. B only

    C. Both A and B

    D. Neither A nor B

13. Technician A says that voltage drops of .5 volts at ground connections are satisfactory. Technician B says that voltage drops at terminal connections of .2 volts are satisfactory. Who is correct?

    A. A only

    B. B only

    C. Both A and B

    D. Neither A nor B

14. Technician A says a parasitic drain of 1 amp is the maximum allowed by most manufacturers. Technician B says some parasitic drains may take several minutes to shut off or reach the sleep mode after the key is turned off. Who is correct?

    A. A only

    B. B only

    C. Both A and B

    D. Neither A nor B

15. Technician A says a sneak circuit is when current can back-feed through a circuit when two circuits share a common power feed. Technician B says a sneak circuit is when current can back-feed through a circuit when two circuits share a common ground path. Who is correct?
    A. A only
    B. B only
    C. Both A and B
    D. Neither A nor B

11 Lighting Circuits

Introduction

In the last several chapters you learned basic electrical theory, concepts, circuit laws, and troubleshooting techniques. In this chapter, you can apply what you learned to a variety of exterior and interior lighting circuits. Electrical lighting circuits of modern vehicles can consist of complex circuit boards, hidden wiring looms and connectors, several lamps, circuit protectors, relays, and control switches. In addition, there are after-market accessories that may have been installed over the car's life by an unqualified technician. All of this can make the job of diagnosing circuits that do not work challenging for the technician.

Objectives

When you complete this chapter you should be able to:

- ❑ Explain the operation and basic diagnosis of the following exterior light systems—parking and side marker lamps, headlights and dimmer switches, flash-to-pass, taillights, brake and collision avoidance lights, turn signal/hazard lights and flashers, fog lights, and back-up lights.
- ❑ Explain the operation and basic diagnosis of the various interior light systems, including courtesy lights, dome lights, and reading/map/vanity lights.

Exterior Lighting Circuits

Exterior lighting circuits of today's vehicles are largely regulated by federal laws. Even though there are differences between manufacturers in lighting circuit designs and style, the result is a vehicle that meets or exceeds all vehicle safety requirements and laws.

Headlight Systems

Although we discussed headlamp types and design features in Chapter 5, we will now apply them to various headlight circuits common in today's vehicles. A basic headlight circuit normally consists of a headlight switch, dimmer switch, low and high beam headlights, high beam indicator, and circuit wiring and connectors. Figure 11–1 is an example of a headlight circuit with the dimmer switch directing the battery voltage *only* to the low beam circuit.

When the dimmer switch is changed to the high beam position (Figure 11–2), the battery voltage is directed *only* to the high beams and the high beam indicator lamp.

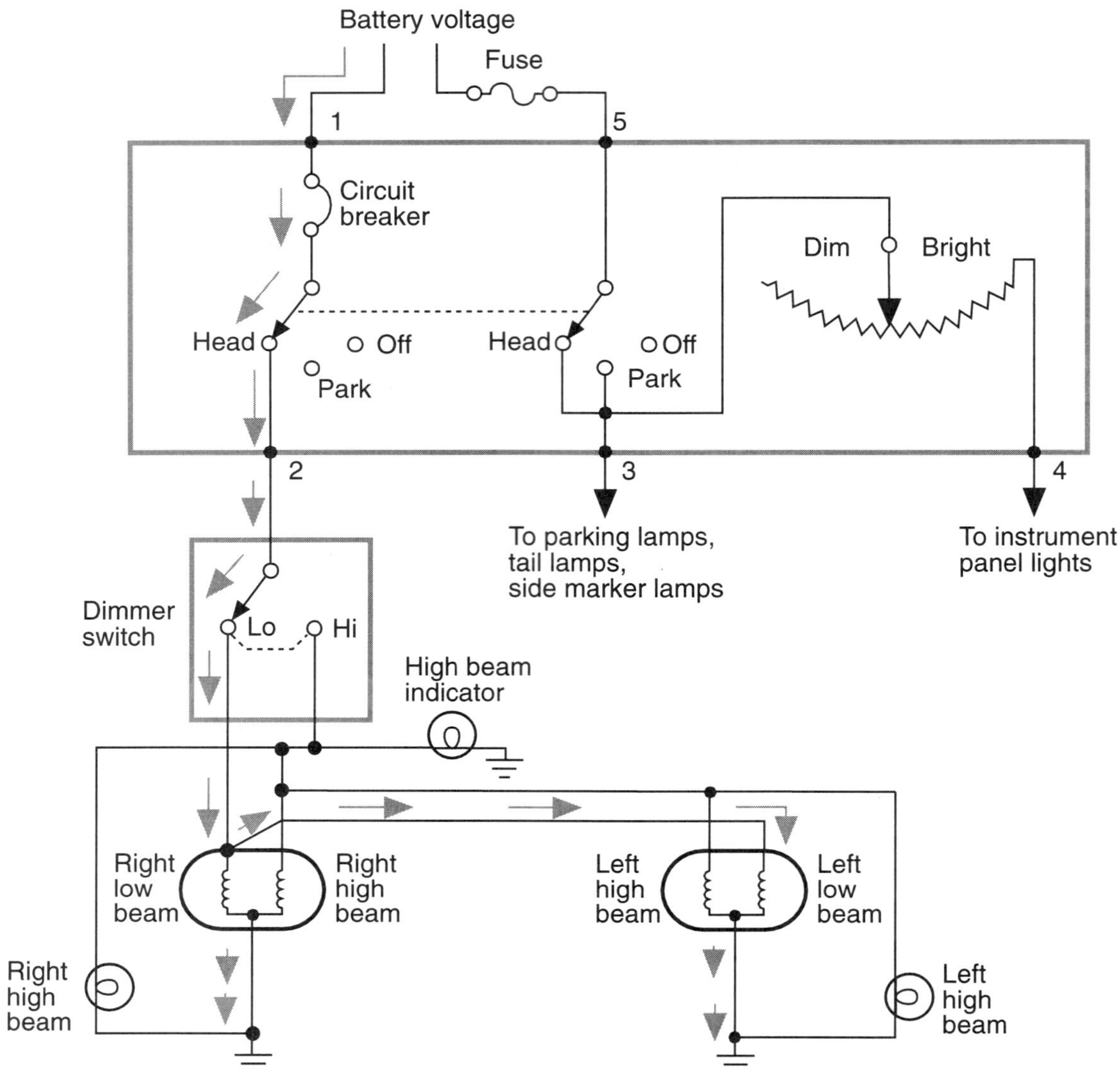

FIGURE 11–1 Current flow through the dimmer switch to the low beams.

In the headlight circuit just discussed, the headlight switch and dimmer switch are both on the positive side of the circuit and the headlamps complete the path to ground. In many import vehicles the headlamps are insulated on the positive side of the circuit and the headlight switch and dimmer switch complete the path to ground (Figure 11–3).

Headlight Switch Regardless of the type of headlight system, a headlight switch controls the operation of the circuit. The typical headlight switch could be mounted either in the dash or instrument panel (Figure 11–4), or in the steering column (Figure 11–5).

The most common dash panel headlight switch is a three position style with Off, Park, and Headlight positions. The switch is normally combined with the ceramic rheostat used to control dimming of the instrument panel lights (Figure 11–6). When this switch has to be removed for troubleshooting or replacement, the control knob can be removed from the switch by pushing on the release button located on the bottom of the switch (Figure 11–7). It is important to know about the location of the release button. Some switches are hard to see and are not easily accessible, so the technician has to rely on feel to locate the release button and remove the control knob.

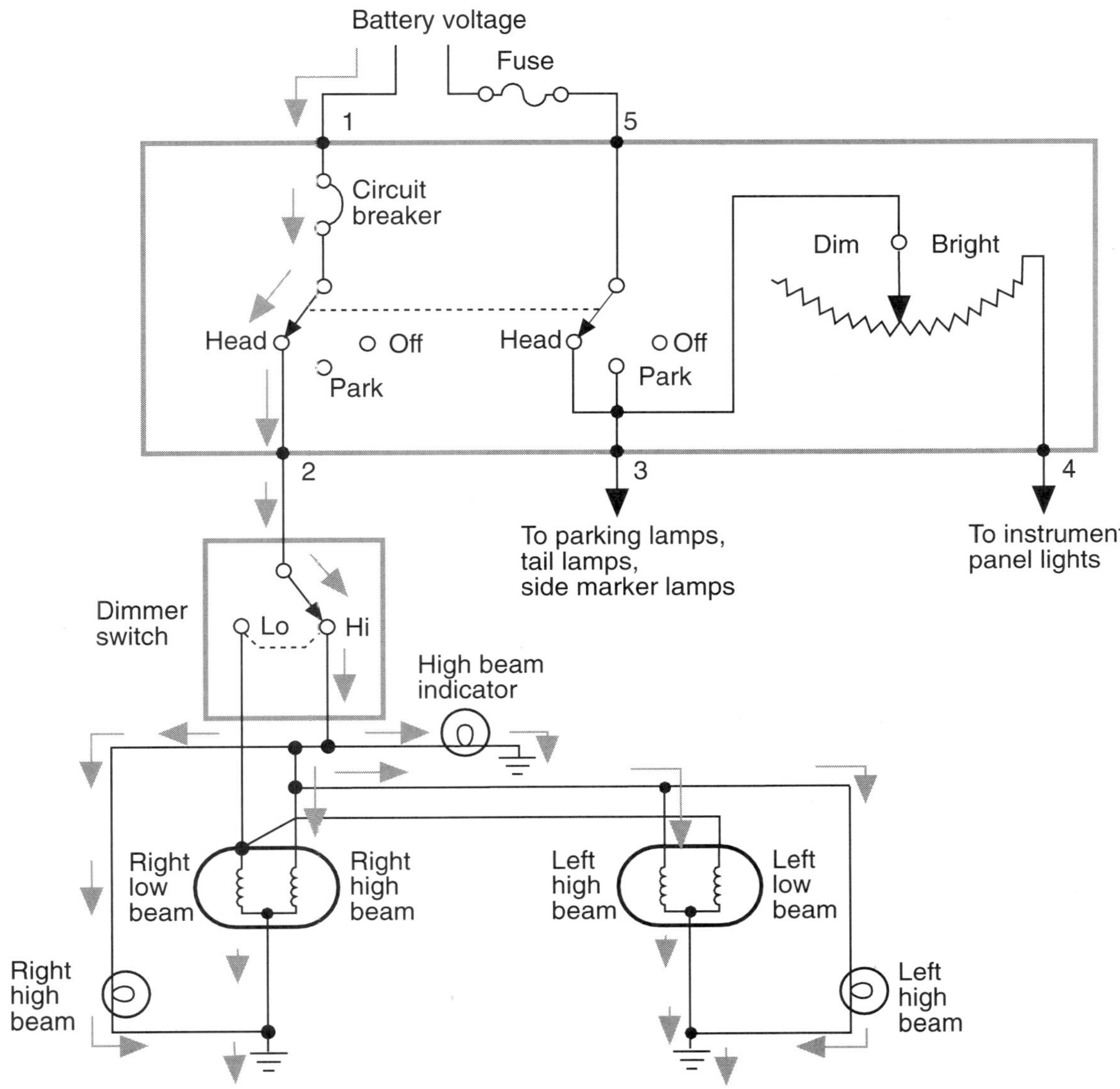

FIGURE 11–2 Current flow through the dimmer switch to the high beam.

In a typical three position headlight switch there are two terminals that normally have battery voltage. This allows the headlight switch to operate the lights when the ignition switch is in the Off position (Figure 11–8).

When the headlight switch is pulled to the Park position (Figure 11–9), the battery voltage present at terminal #5 is directed to the parking lamps, taillamps, side marker lamps, license plate lamp, and through a ceramic dimming rheostat to the instrument cluster lamps, but *not* the headlights.

When the switch is pulled to the Headlight position (Figure 11–10), the battery voltage present at terminal #1 is directed through the circuit breaker to the headlamps. The circuit breaker is designed so it does not totally disable the headlights when temporary overloads exist in the system. The voltage at terminal #5 continues to feed battery voltage to the park circuit.

Dimmer Switches In older vehicles, the dimmer switch was located next to the left kick panel on the floor board (Figure 11–11). This location subjected the switch to water, dirt, and other contaminates that shortened the service life of the switch. Newer vehicles locate the dimmer switch on the steering column as a separate switch (Figure 11–11), or as part of a **multifunction switch**

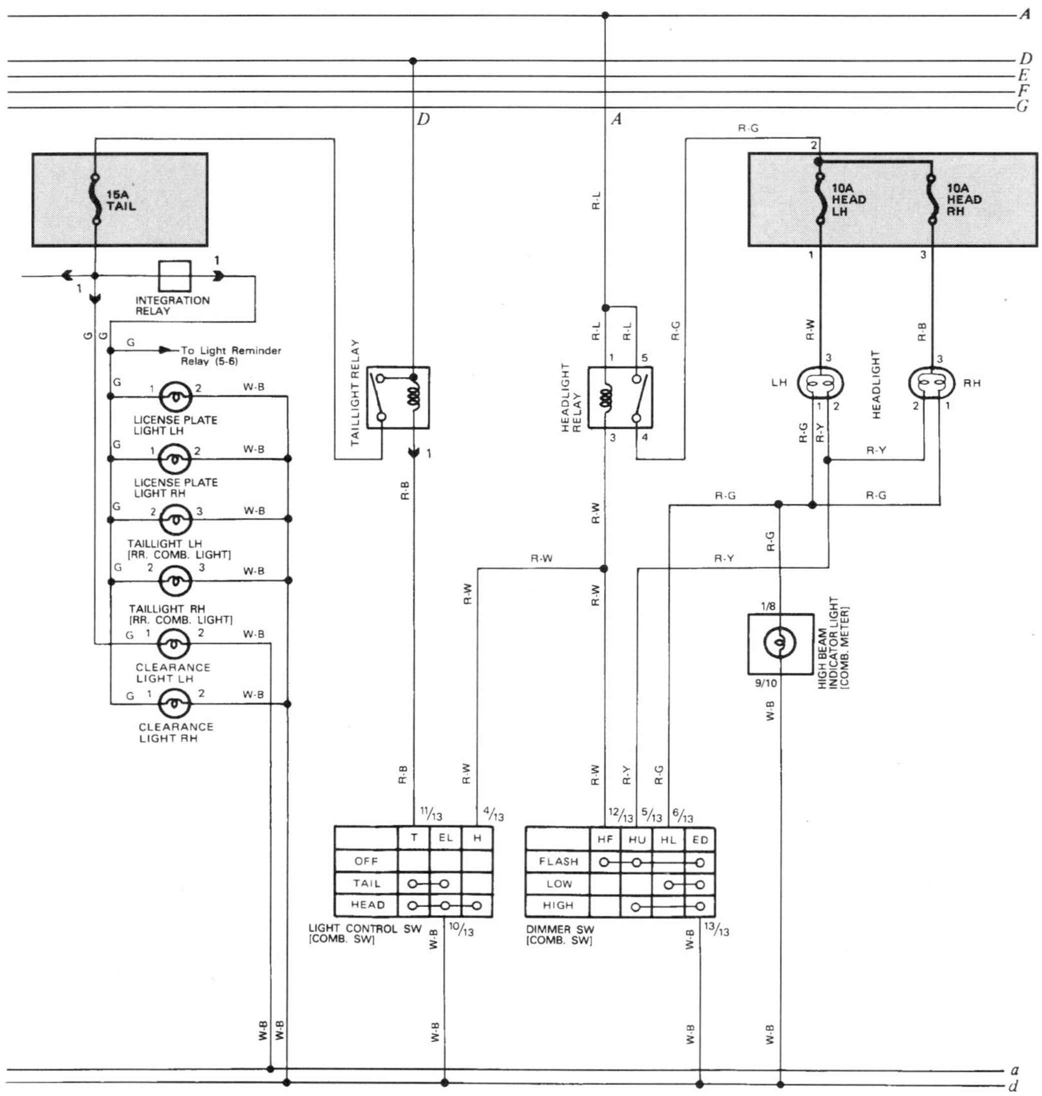

FIGURE 11–3 Headlight circuit with the control switches on the ground side.

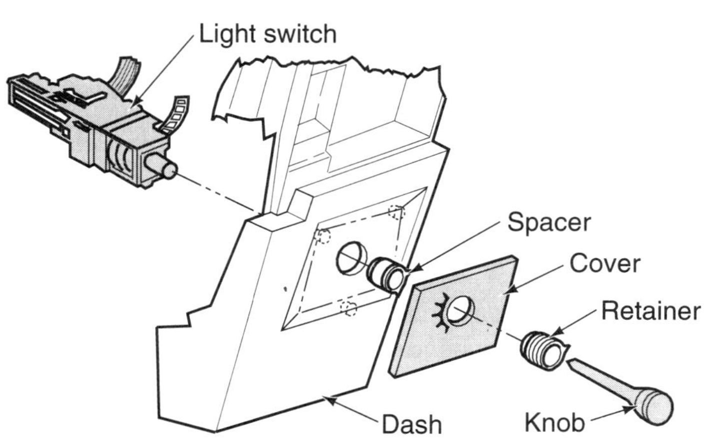

FIGURE 11–4 Typical dash-mounted headlight switch.

(Courtesy of Ford Motor Company)

FIGURE 11–5 Typical headlight switch mounted on the steering column, with a separate control on the dash for dimming the instrument panel lights.

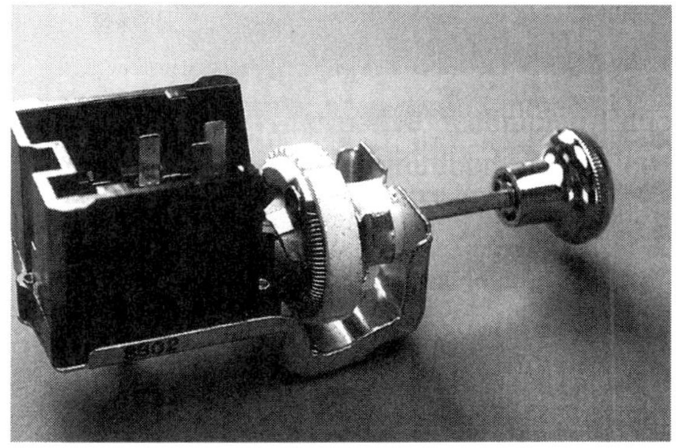

FIGURE 11–6 Typical dash-mounted headlight switch.

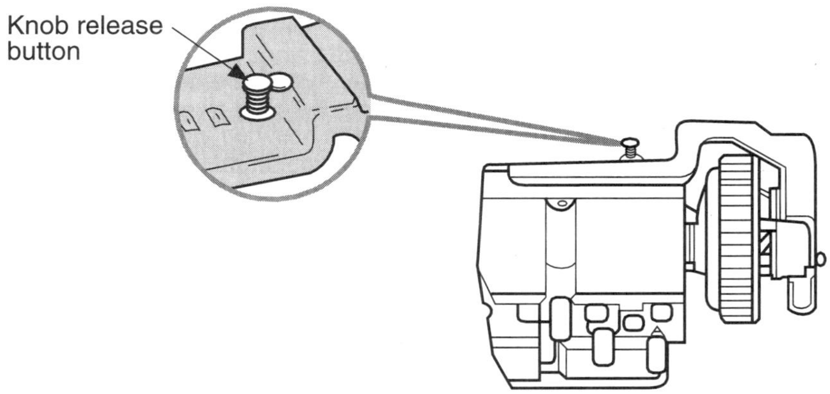

FIGURE 11–7 Typical location of the headlight switch knob release button.

(Courtesy of Ford Motor Company)

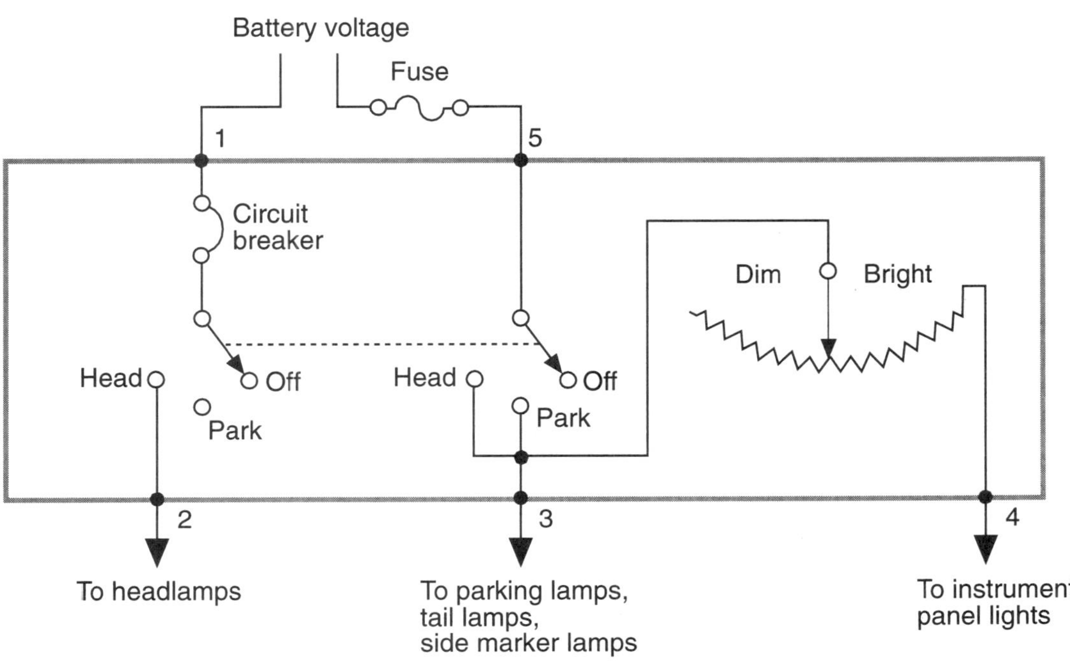

FIGURE 11–8 A headlight switch in the Off position.

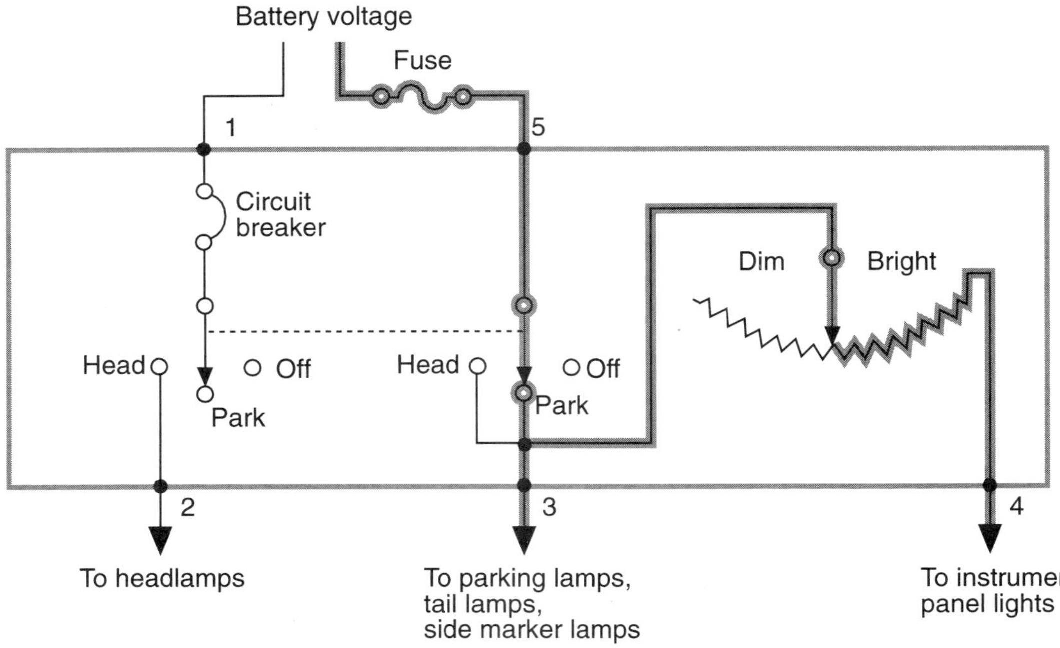

FIGURE 11–9 A headlight switch in the Park position.

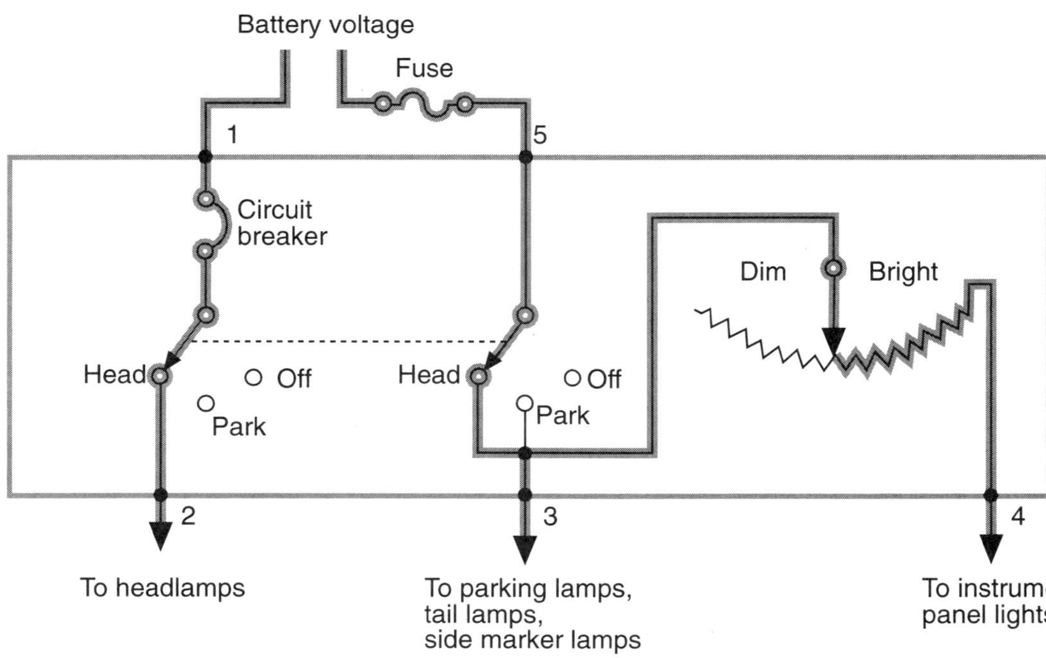

FIGURE 11–10 A headlight switch in the Headlight position.

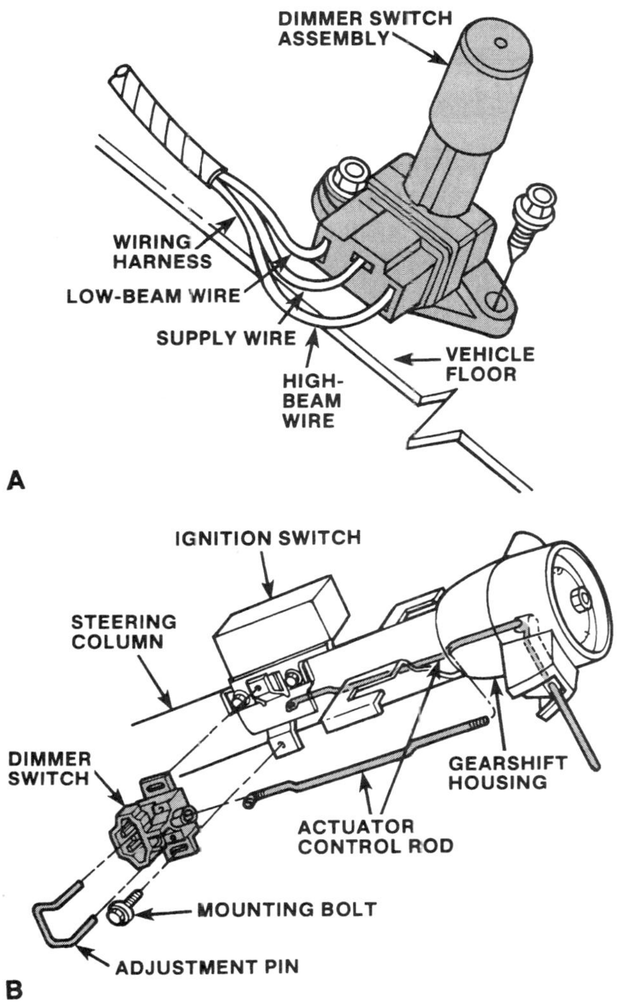

FIGURE 11–11 (A) Floor mounted dimmer switch, and (B) steering column mounted dimmer switch. (Courtesy of Ford Motor Company)

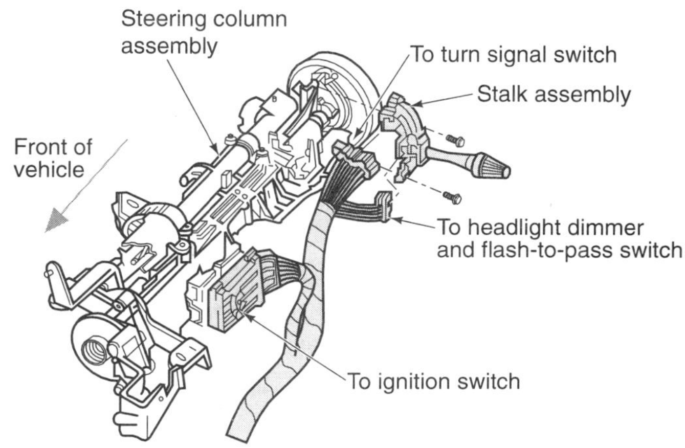

FIGURE 11–12 A multifunction switch with dimmer switch incorporated into it.

(Figure 11–12) which combines the dimmer switch with one or more of the following switches—turn signal, headlight, hazard, horn, and flash-to-pass.

To diagnose a suspected malfunctioning dimmer switch, a fused jumper wire can be used to bypass the switch to either low or high beams (Figure 11–13). The dimmer switch is typically connected in series with the headlight circuit, so when the switch is bypassed, the low or high beams will go on unless something else is wrong in the circuit.

On dimmer switches that are part of the multifunction switch (Figure 11–14), it may be possible to test the switch without removing it from the steering column. Normally there is a multi-connector at the base of the column that can be used for testing the switch. If not, it may be neces-

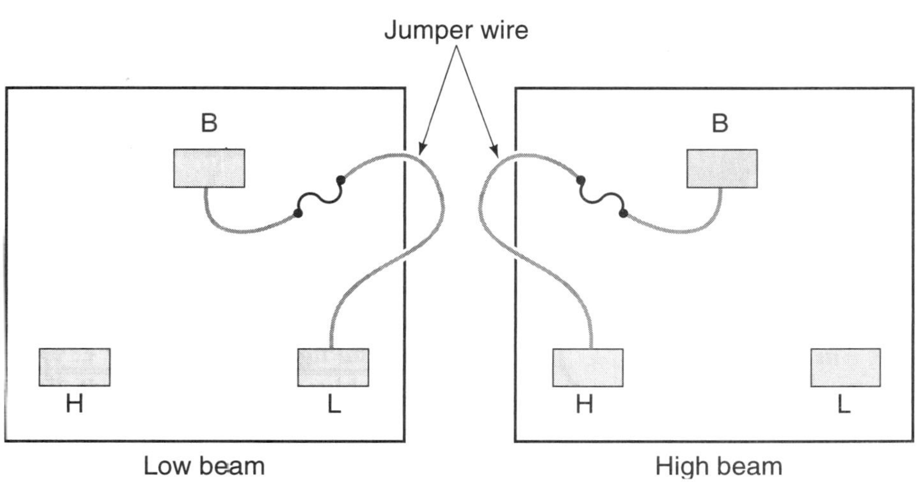

FIGURE 11–13 Using a jumper wire to bypass a suspected malfunctioning dimmer switch.

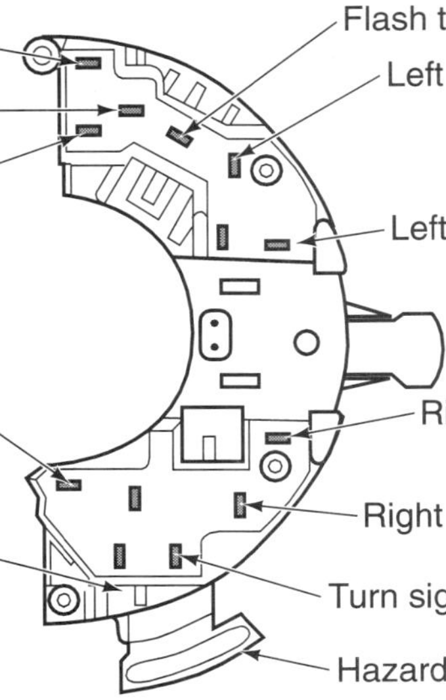

FIGURE 11–14 Multifunction steering column switch connection identification.
(Courtesy of Ford Motor Company)

sary to remove the column covers to gain access. Refer to specific vehicle manufacturer guidelines for this procedure to prevent damaging access covers or connectors.

Some dimmer switches have an additional feature incorporated into them—(**flash-to-pass**)—which switches on the high beams, and in some vehicles—the low beams, to light even if the headlight switch is off The circuit in Figure 11–15 supplies battery voltage from the B1 terminal at the headlight switch through the 14LG wire to the dimmer switch at connector L2. This allows battery voltage at the dimmer switch even if the headlight switch is in the Off or Park position. When the driver activates the flash-to-pass feature, the dimmer switch completes the circuit to the high beams and, in some vehicles, the low beams as well.

On many luxury vehicles the dimmer switch circuit contains a photoelectric cell and dimmer module (Figure 11–16) that automatically dim the headlights when the shine of on-coming headlights approach, and then return to high beams after the vehicle passes. The photoelectric cell is normally located in the front grill area, the front-part of the dash board, or an area that will pick-up the shine of on-coming lights.

Concealed Headlight Systems In an effort to improve vehicle styling and aerodynamics, some headlights are concealed behind movable doors (Figure 11–17). When the headlight switch is off the headlight doors are closed, but automatically open when the headlight switch is turned to the Headlight position, but not the Park position.

The headlight doors are controlled by either a vacuum circuit with vacuum motors or an electric circuit with electric motors. In Figure 11–18, a vacuum distribution valve is attached to the back of the normal headlight switch to vent the vacuum from the motors that hold the doors closed (Figure 11–19). If vacuum is used to close the headlight doors and springs are used to open the doors as the vacuum is vented, this assures the doors will be open if vacuum to the system is lost.

In electrically operated headlight doors, a door motor is normally located at each headlight door (Figure 11–20); or, when only one motor is used, a torsion bar connects both doors together (Figure 11–21).

In both vacuum controlled motors and electric motor systems, a means of manually opening the doors is provided in the event the system fails. In Figure 11–22, a bypass valve supplies vacuum directly to the doors from the engine. In Figure 11–20, the manual knob allows the doors to be turned up manually. Remember to check appropriate service information before attempting manual procedures.

CAUTION *Remember to keep hands clear of door mechanisms when performing testing or repair. Doors may suddenly open with a snap due to a bind or malfunction. Play it safe! Pay attention!*

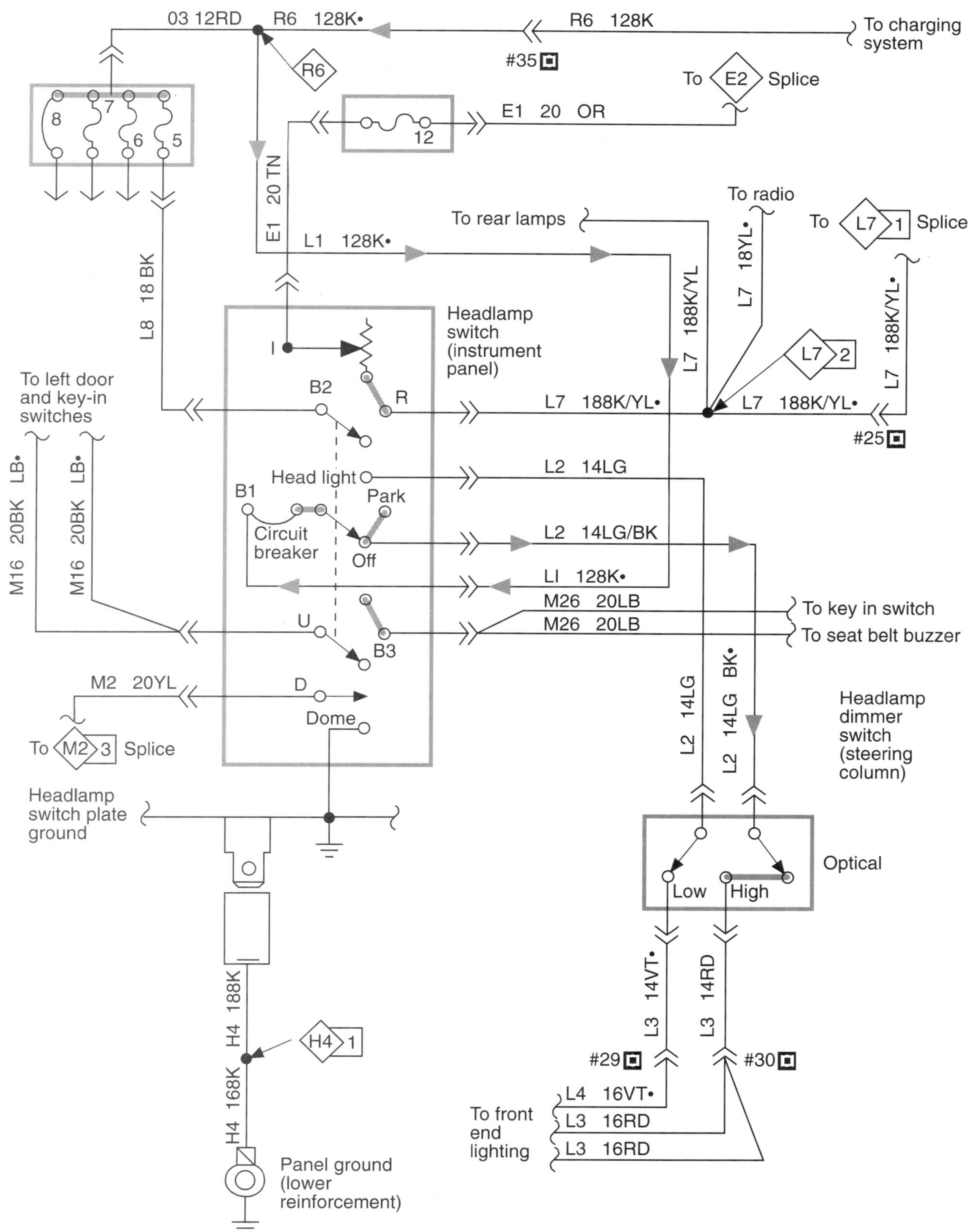

FIGURE 11–15 A typical flash-to-pass headlight circuit.

(Courtesy of DaimlerChrysler Corporation)

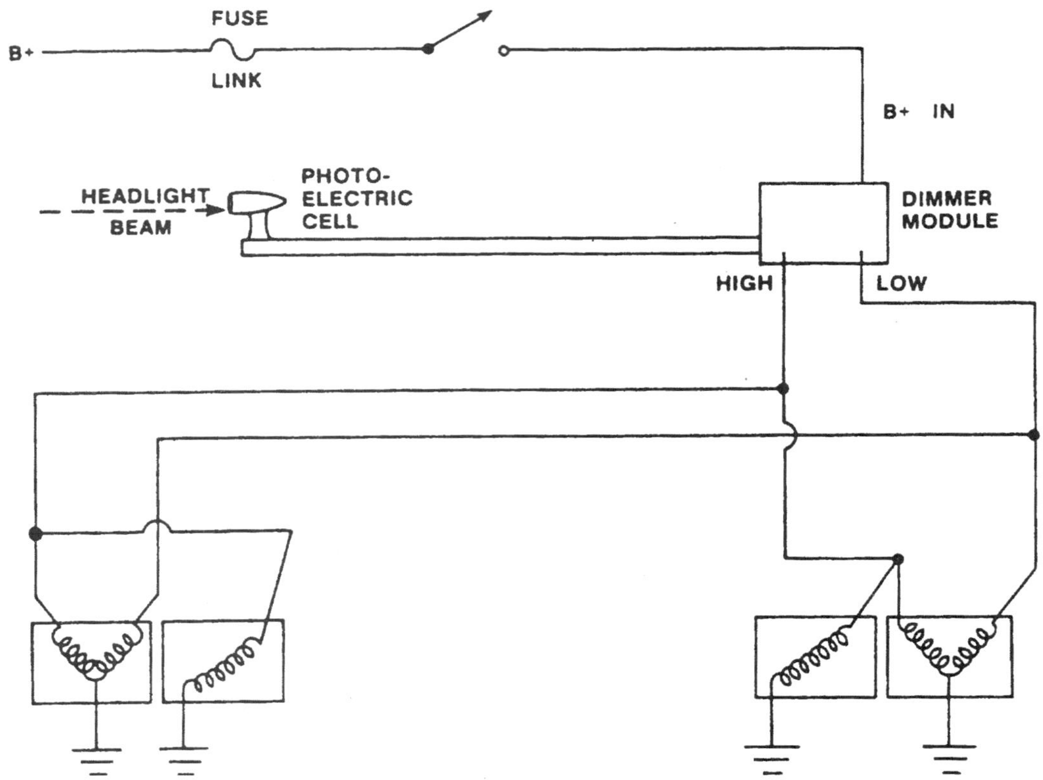

FIGURE 11–16 Typical photoelectric headlight dimming circuit.

FIGURE 11–17 Concealed headlight designed for styling and aerodynamics.

FIGURE 11–18 A headlight switch combined with a vacuum distribution valve.

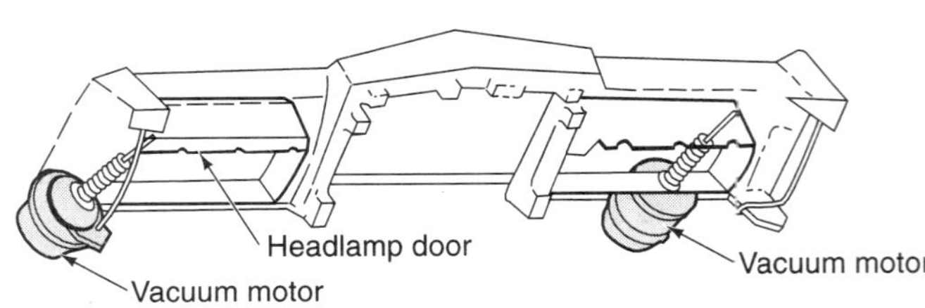

FIGURE 11–19 Typical headlight doors connected to vacuum control motors.

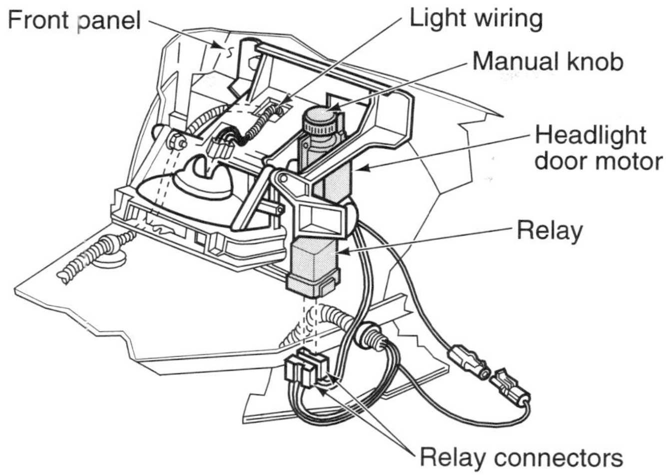

FIGURE 11–20 Typical electrically operated headlight doors.
(Courtesy of General Motors Corporation, Service Technology Group)

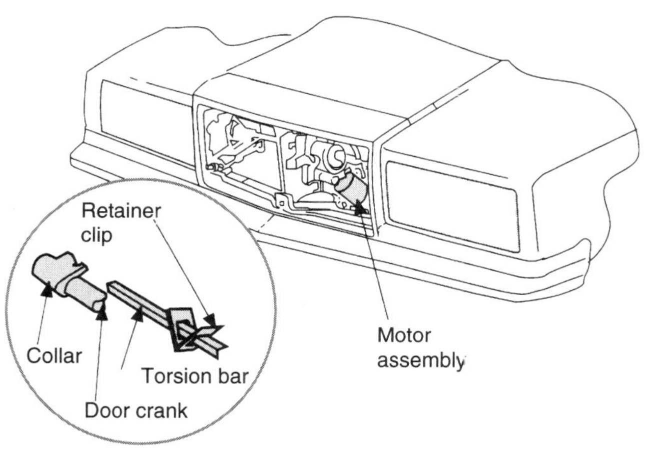

FIGURE 11–21 Torsion bar operated headlight door system with single motor.

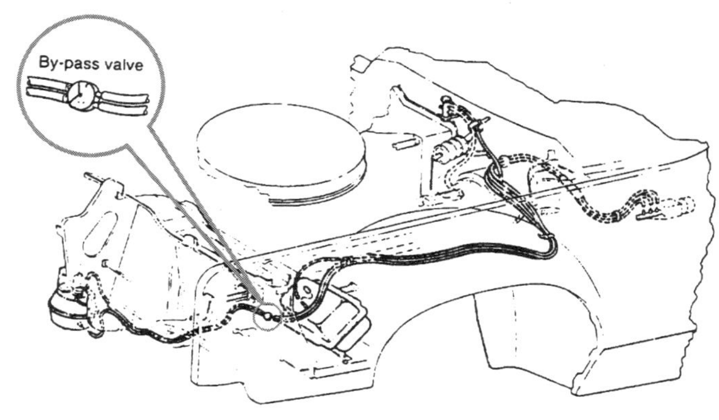

FIGURE 11–22 A by-pass valve used to open headlight doors in the event of a system failure.
(Courtesy of DaimlerChrysler Corporation)

Daylight Running Lights (DRL) In an effort to increase vehicle safety, Canadian and US federal laws require that vehicles be equipped with **daylight running lights (DRL),** for daylight safety. Depending on the system, the low beam or high beam lights operate anytime the vehicle is running, but usually at a reduced voltage and intensity to extend headlamp life. In Figure 11–23, the DRL relay is connected directly to battery voltage. When the vehicle is running during daylight hours and the parking brake is off, the DRL control module in the instrument cluster microprocessor will ground the DRL relay coil, which will then energize the low beam lamps in *series* with each other. This drops the voltage to approximately 6 volts per low beam bulb. When the headlight switch is activated for night or poor light conditions, the DRL system is deactivated and the lamps operate at regular brightness.

Headlight Aiming When headlights are replaced, vehicle ride height is altered by suspension changes, or a vehicle is driven with a load condition that changes the headlight aim, headlights should be adjusted to meet all laws and regulations. Some federal, state, and local laws specify requirements for headlamp aim that *must* be fol-

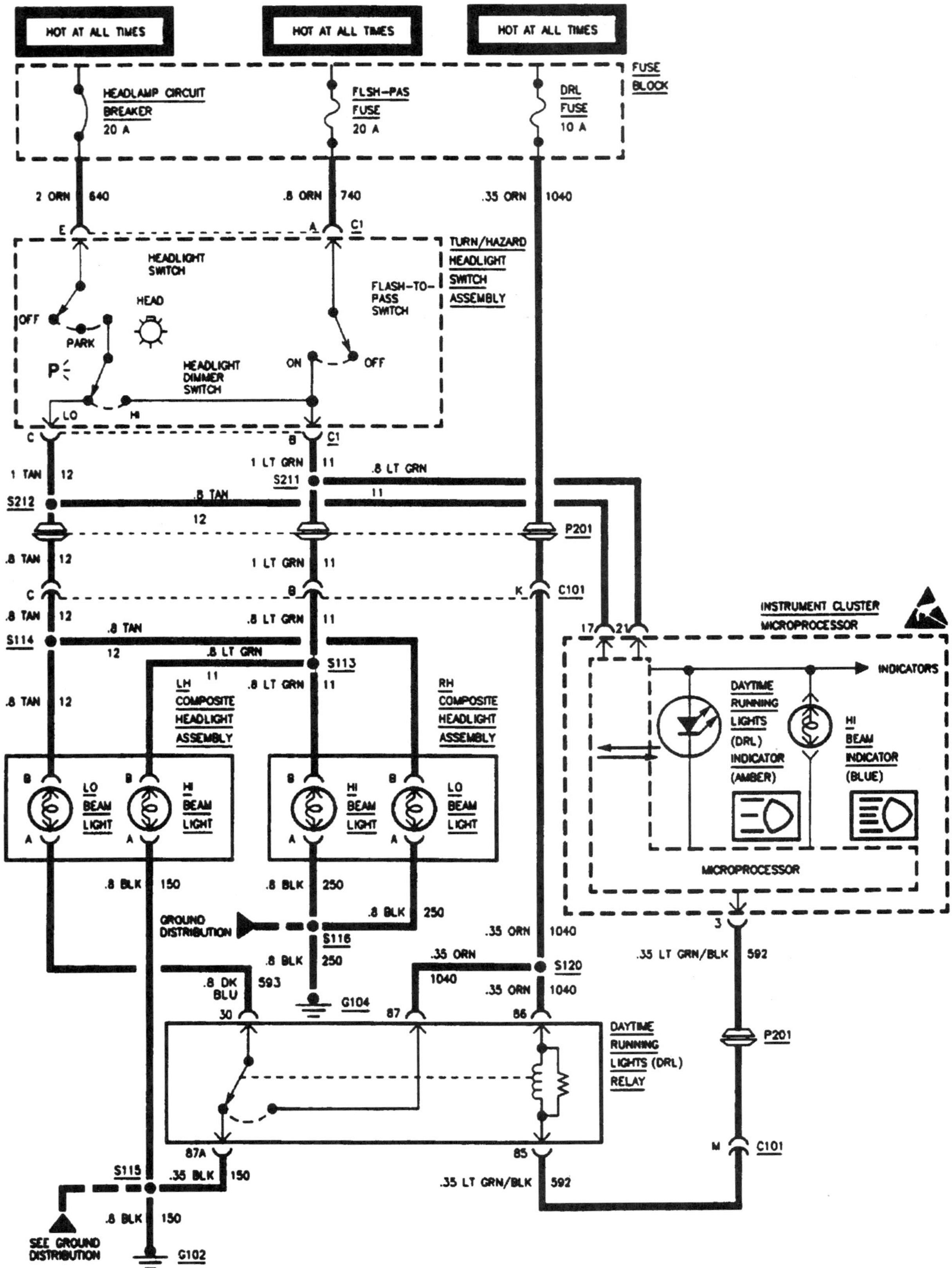

FIGURE 11–23 Headlight system with daylight running light (DRL) circuit.

(Courtesy of General Motors Corporation, Service Technology Group)

lowed. Before headlights are aimed, the following preparation should be done to the vehicle.

- Remove any excess dirt, mud, snow, etc., from the vehicle that will affect the ride height.
- Consult with the customer about load conditions on the vehicle. A vehicle continually loaded with heavy material in the luggage compartment could cause the light to shine into on-coming traffic.
- Place vehicle on a flat level surface. If this is not available, the floor slope must be compensated for with the aiming tools.
- Jounce vehicle suspension so it rests in its normal driving plane.
- Make sure fuel tank is as least half full. Some manufacturers recommend the fuel tank be filled completely for the most accurate headlight adjustment.
- Adjust tire pressures to recommended settings.
- Determine weight distribution of vehicle occupants and cargo. This could be important on the ride height of smaller vehicles with weaker suspension systems.

For accuracy, most shops use portable headlight aiming equipment and adapters (Figure 11–24). The aiming units adapt to the headlight aiming pads (Figure 11–25) by adapters that fit each style of headlight. Composite headlights have additional adapters that adjust to the aerodynamic contour of the headlight lens (Figure 11–26).

The vertical adjustment on the headlight is accomplished by adjusting the vertical adjustment screw located above the headlight in the center of the retainer trim (Figure 11–27). The vertical indicator on the headlight aiming unit is a *spirit bubble* that is centered when adjustment is complete (Figure 11–28).

The horizontal adjustment on the headlight is accomplished by adjusting the horizontal adjustment screw located on the inside edge of the retainer trim (Figure 11–27). The horizontal indicator on the headlight aiming unit is a *split image* that is aligned when the adjustment is complete (Figure 11–29).

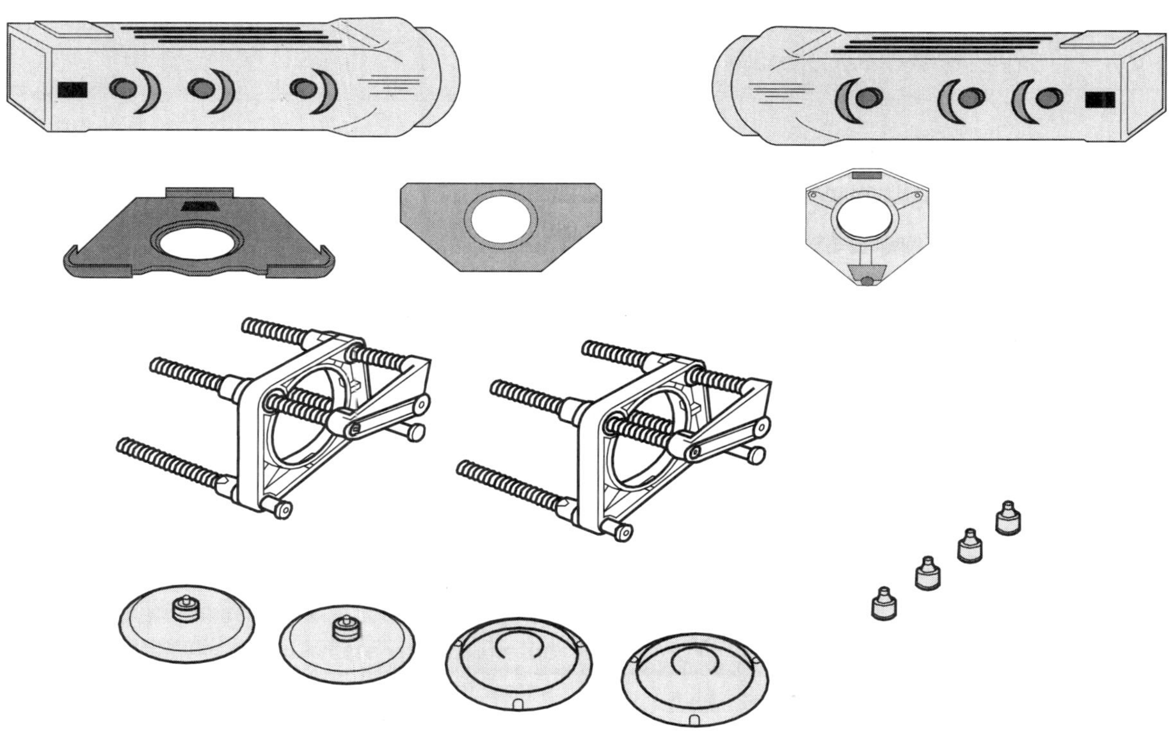

FIGURE 11–24 Mechanical headlight aiming equipment and adapters.
(Courtesy of General Motors Corporation, Service Technology Group)

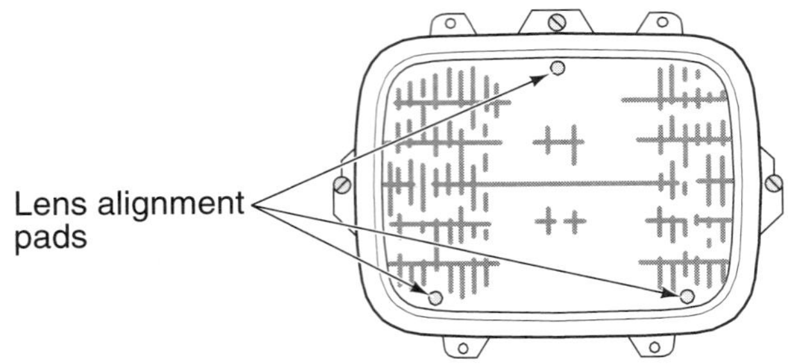

FIGURE 11–25 Lens alignment pad location on a headlight.

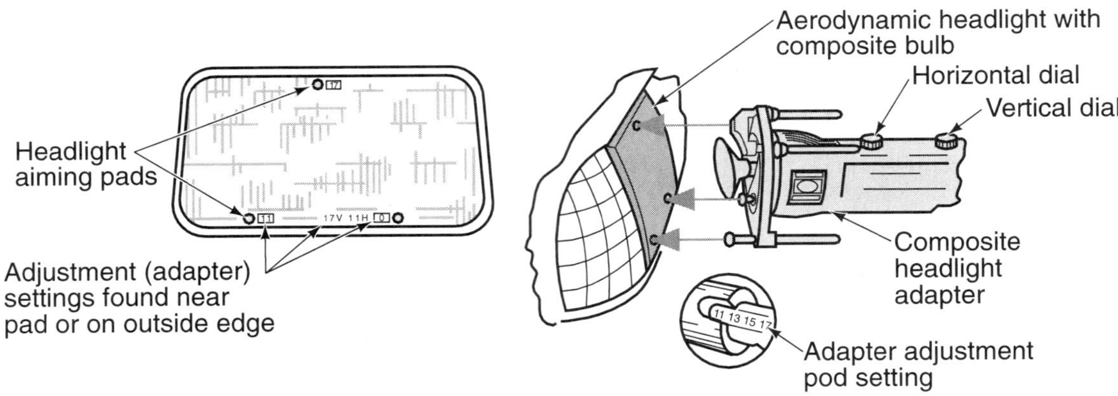

FIGURE 11–26 Special composite headlight aiming adapters and adjustment settings.

(Courtesy of General Motors Corporation, Service Technology Group)

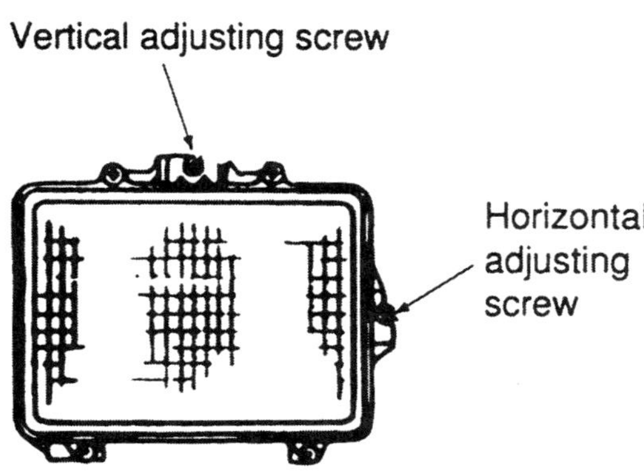

FIGURE 11–27 Headlight vertical and horizontal adjusting screw locations.

(Courtesy of General Motors Corporation, Service Technology Group)

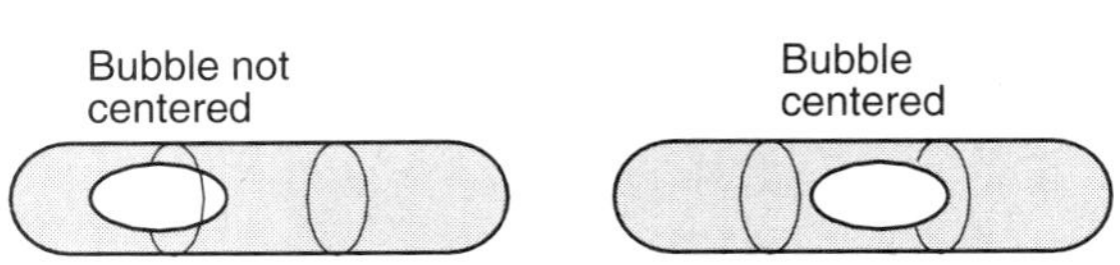

FIGURE 11–28 Adjusting a headlight vertical setting with a spirit bubble.

(Courtesy of DaimlerChrysler Corporation)

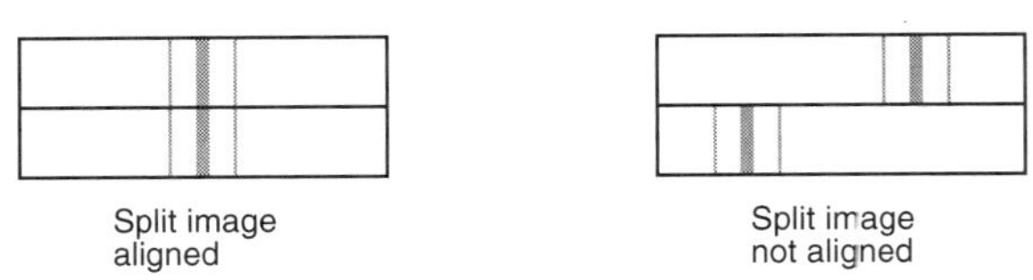

FIGURE 11–29 Adjusting a headlight horizontal setting with a split image target.

(Courtesy of DiamlerChrysler Corporation)

Composite headlights typically have their adjustment screws located on the back of the composite unit or close by on vehicle trim components(Figure 11–30).

Headlight Voltage Drop Test Dimmer than normal headlights are most often the result of a voltage drop caused by excessive resistance in the system due to a bad connection. Other common causes of dim headlights could be low alternator output, wrong type of lamp, or excessive electrical loads on the system. Headlights do not wear out and get dimmer over time unless the light has been damaged or chipped. To test the headlight system for voltage drop from excessive resistance, determine the number and location of connections, switches, and components from a wiring diagram and component locator diagram. Using a voltmeter, measure each point in the system that has been identified as a possible cause of the voltage drop (Figure 11–31). Follow the manufacturer specification for voltage drop, or use the general table listed in Chapter 10.

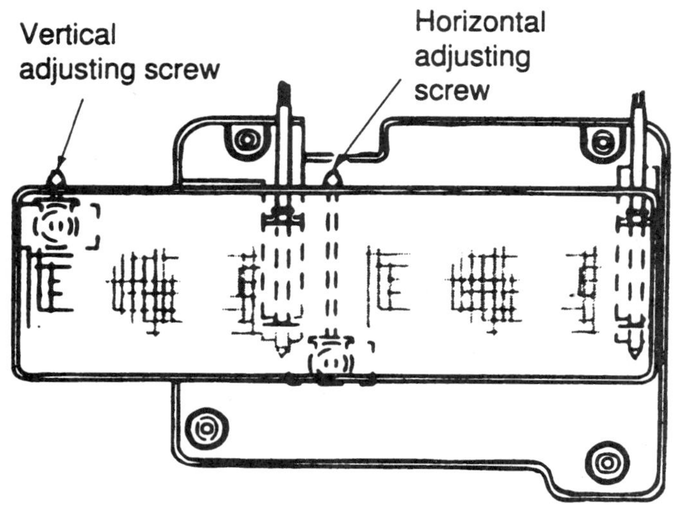

FIGURE 11–30 Typical composite headlight vertical and horizontal adjusting screws.
(Courtesy of General Motors Corporation, Service Technology Group)

Parking and Taillight Circuits

Parking and taillight circuits are normally controlled by the headlight switch. When the headlight switch is pulled to the first detent position, the parking lamps, license lamp, marker lamps, and taillamps will turn on, but the headlights will remain off. Figure 11–32 is an example of a two bulb taillight circuit. In this circuit a dual-filament bulb is used on each side of the circuit for

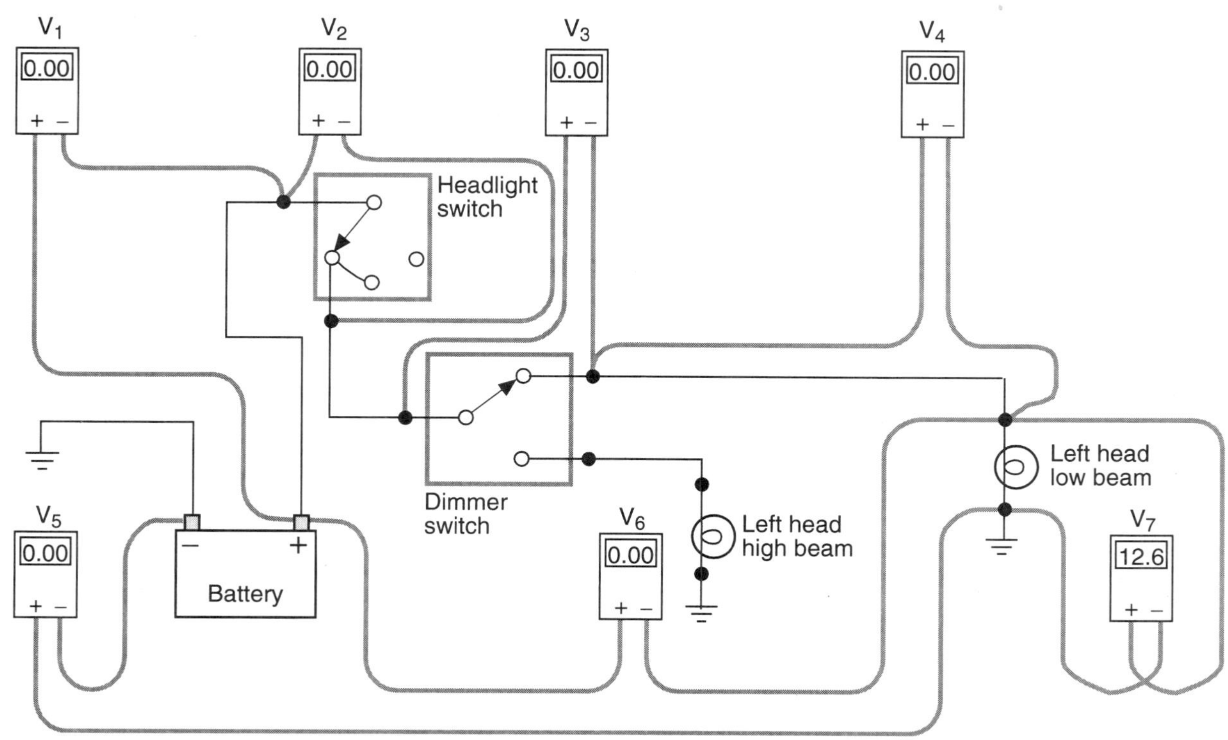

FIGURE 11–31 Checking for voltage drops in a headlight circuit.

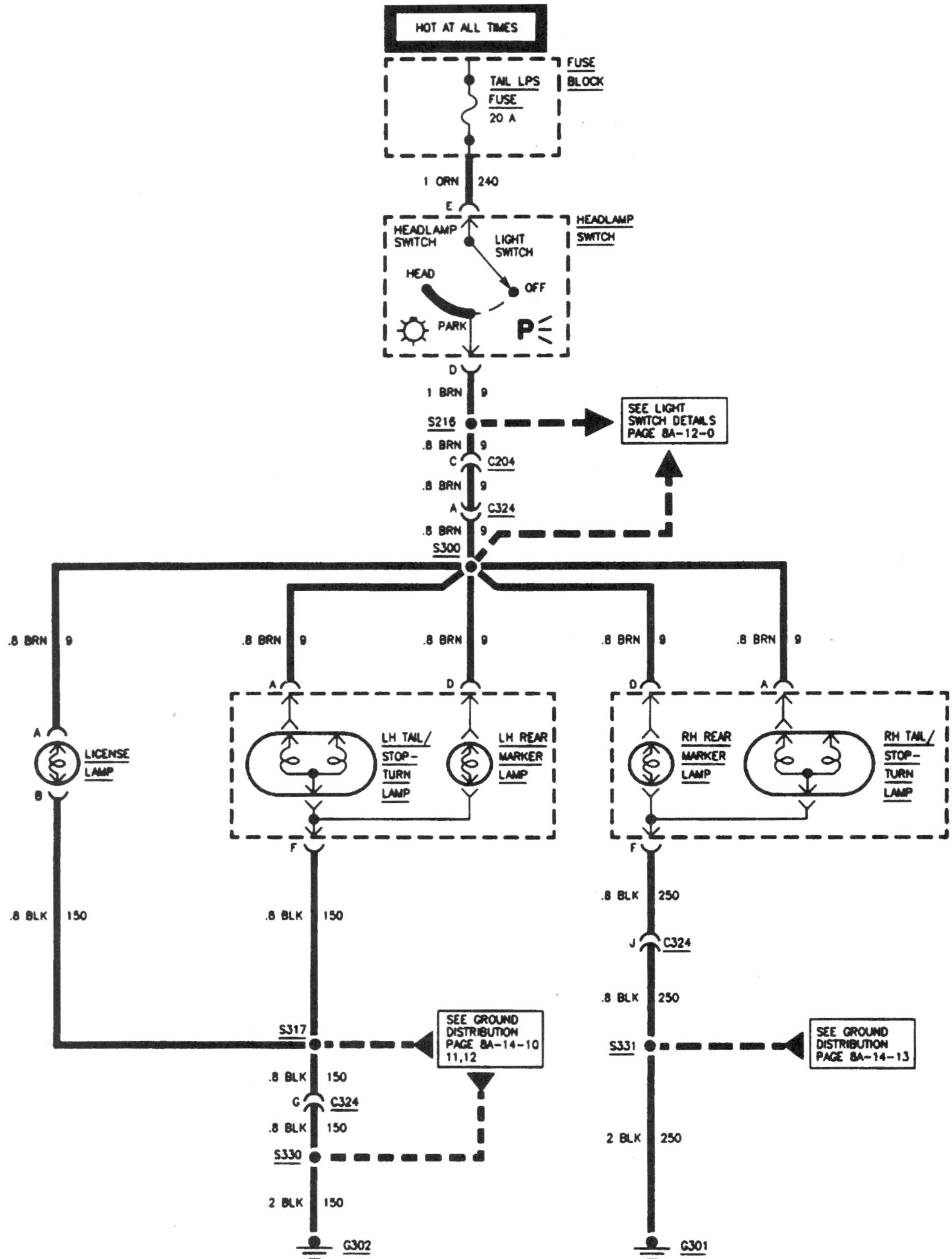

FIGURE 11–32 Typical two bulb taillight circuit.

(Courtesy of General Motors Corporation, Service Technology Group)

taillights and brake lights. If a single-filament bulb is mistakenly installed into a socket designed for a dual-filament bulb, the single contact may short across the dual filament contacts in the socket causing problems in the operation of the attached circuits (Figure 11–33).

Figure 11–34 is an example of a three bulb taillight circuit with individual control for each bulb circuit. Most import vehicles and several newer domestic vehicles use this type of system to prevent lighting problems that might occur when joining two separate systems together.

Brake Lights and Turn Signal Circuits

Brake Light Switch The brake lights in a three bulb taillight circuit are controlled by a mechanical brake light switch attached to the brake pedal arm (Figure 11–35). When the brakes are applied, the switch contacts will close and complete the circuit to the brake lights.

Center High Mounted Stop Light Beginning in 1986, all vehicles must have a **center high-mounted stop light (CHMSL).** This light is also referred to as a **collision avoidance light.** In a three bulb system, the CHMSL is wired in parallel to the brake light circuit because this circuit is separate from the turn signal circuit (Figure 11–36).

In a two bulb brake light system, the CHMSL can be wired in two common methods. The first method is to wire the CHMSL to the brake light circuit after the brake light switch but before the turn signal switch (Figure 11–37). This increases the amount of wiring needed in the circuit. The second method is to install diodes in the system (Figure 11–38) which isolate the CHMSL from the turn signal functions.

Turn Signal Switch The turn signal switch is an important safety feature in any vehicle. The turn signal switch on a two bulb system determines which brake

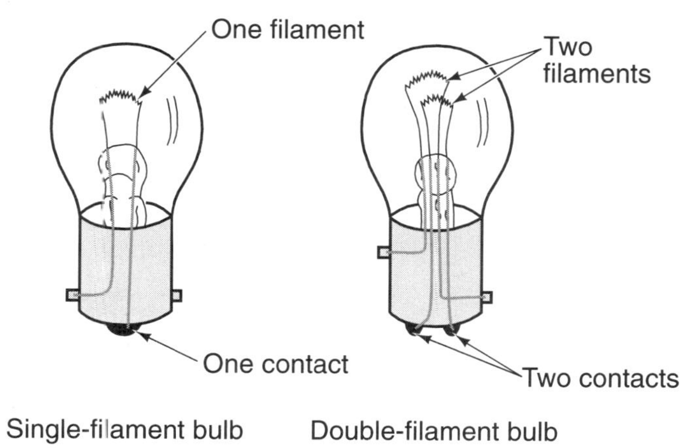

FIGURE 11–33 Single-filament and double-filament bulbs.

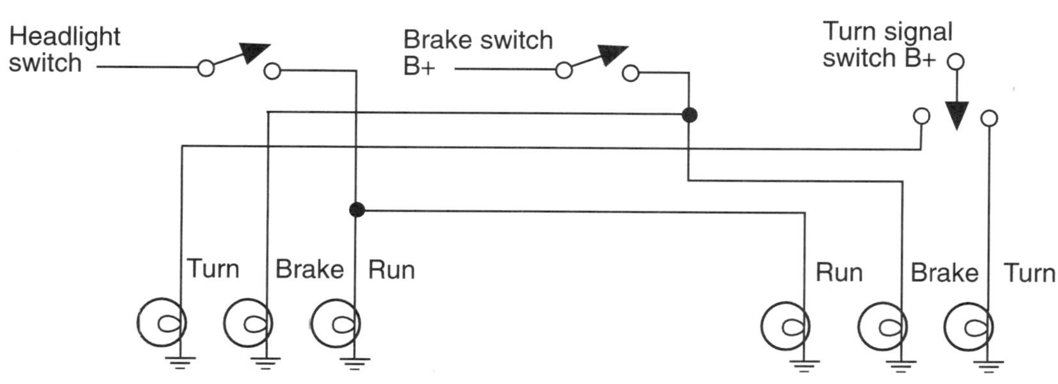

FIGURE 11–34 Typical three bulb taillight circuit.

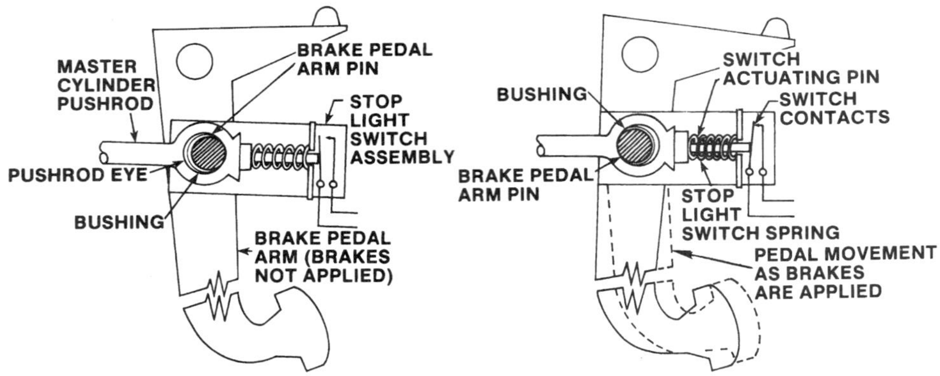

FIGURE 11–35 Typical operation of a brake light switch.

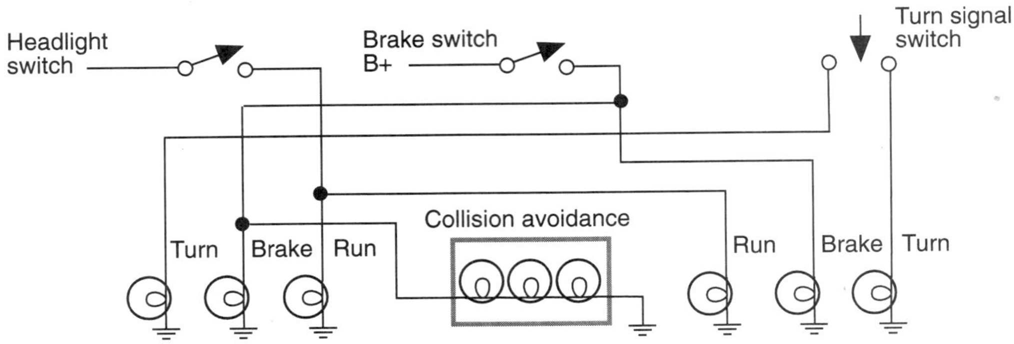

FIGURE 11–36 Brake light circuit with a center high-mounted stoplight (CHMSL).

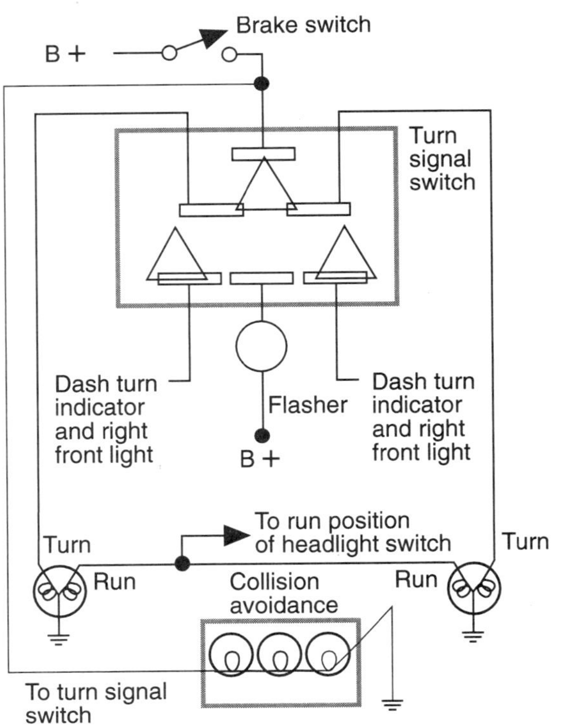

FIGURE 11–37 The CHMSL wired after the brake light switch, but before the turn signal switch.

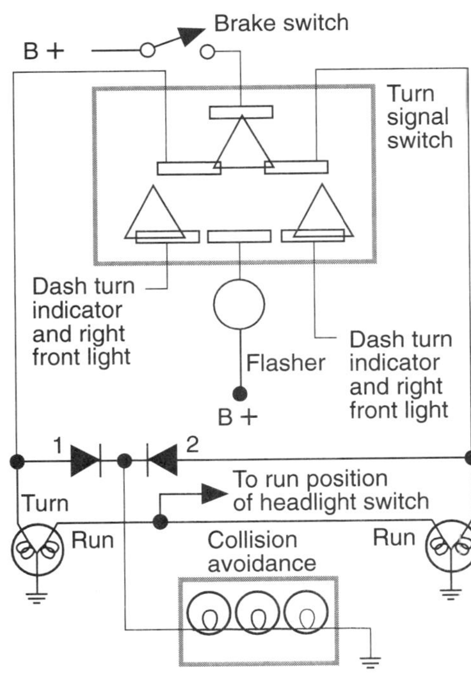

FIGURE 11–38 A CHMSL installed into the two bulb taillight circuit with isolating diodes.

light is isolated out of the brake light circuit. In Figure 11–39, the rectangular bars are stationary contacts that are connected to wires. The triangles are conductive pads that slide over the bars to direct the current flow where indicated.

Figure 11–40 is a simple turn signal switch located in the steering column. Some switches are part of a complex multifunction switch. Refer to the specific vehicle manufacturer for instructions on removal and repair. Vehicles with air bags or that have electronic function controls in the steering wheel need special precautions when servicing the turn signal switch.

Figure 11–41 is an illustration of a turn signal circuit in the neutral or rest position. No current flows past the turn signal flasher to the turn signal switch. When the turn signal switch is moved to the left turn position (Figure 11–42), current flows from the turn signal flasher to the designated left turn circuit. When the turn signal switch is moved to the right turn position (Figure 11–43), the current flows to the designated right turn circuit.

Turn Signal Flasher An integral part of the turn signal and hazard signal circuits is the flasher unit, which is located on or around the fuse panel (Figure 11–44). On most vehicles, the flasher unit for the turn signal circuit is separate from the hazard flasher. Some vehicles, though, may incorporate it into one unit. There are two types of flasher devices—a bimetallic strip flasher and an electronic controlled flasher that is used in most imports. The first flasher type (Figure 11–44) consists of a bimetallic strip and heating element wired in series with the turn signal circuit to control the current flow to the turn signal lamps. When the bimetallic strip bends from the heat, the contact points open the turn signal circuit. The absence of current allows the strip to cool and close the circuit. Electronic flasher units use solid-state circuitry to control the signal circuit by timing the current flow through the system.

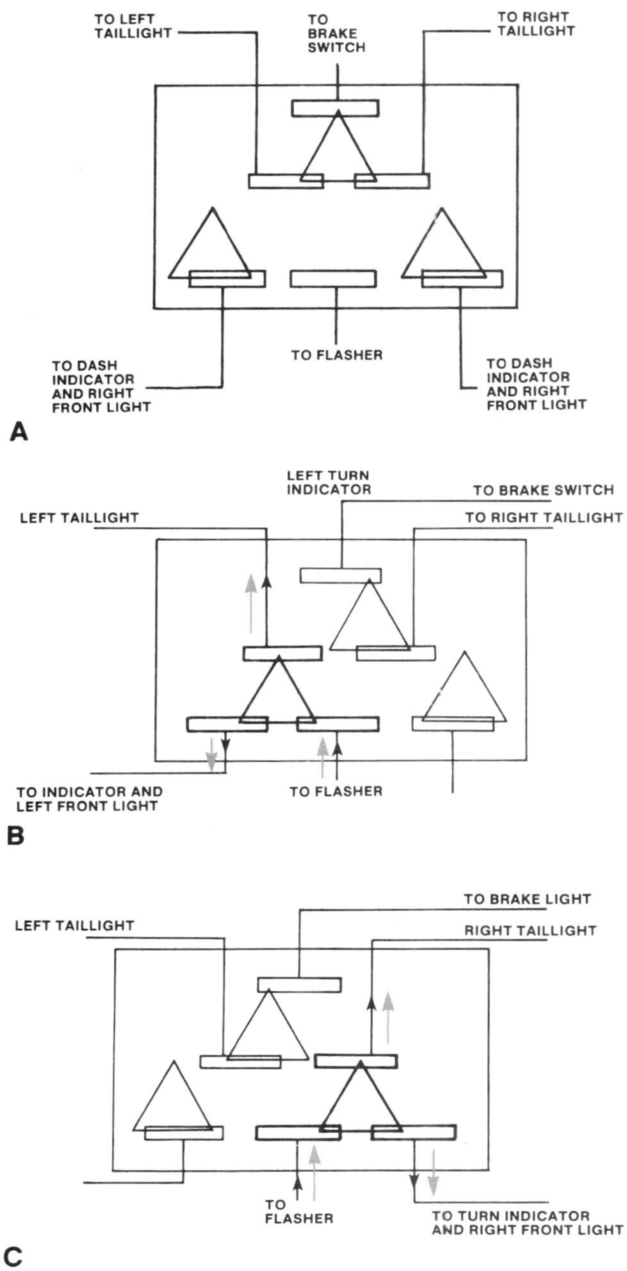

FIGURE 11–39 (A) Turn signal switch in the off position, (B) left turn position, (C) right turn position.

(Courtesy of Ford Motor Company)

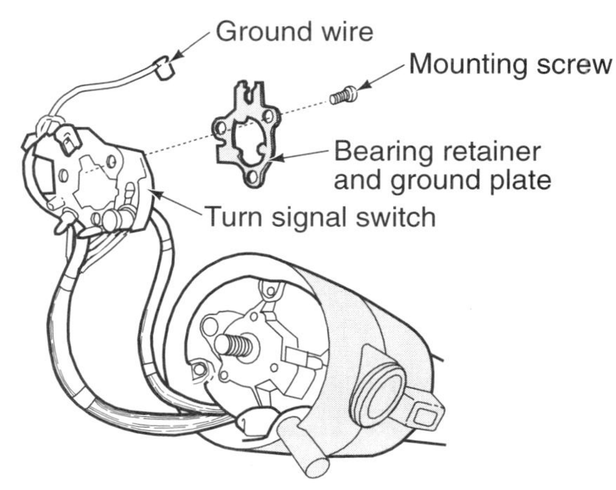

FIGURE 11–40 Typical turn signal switch located in the steering column.

(Courtesy of DaimlerChrysler Corporation)

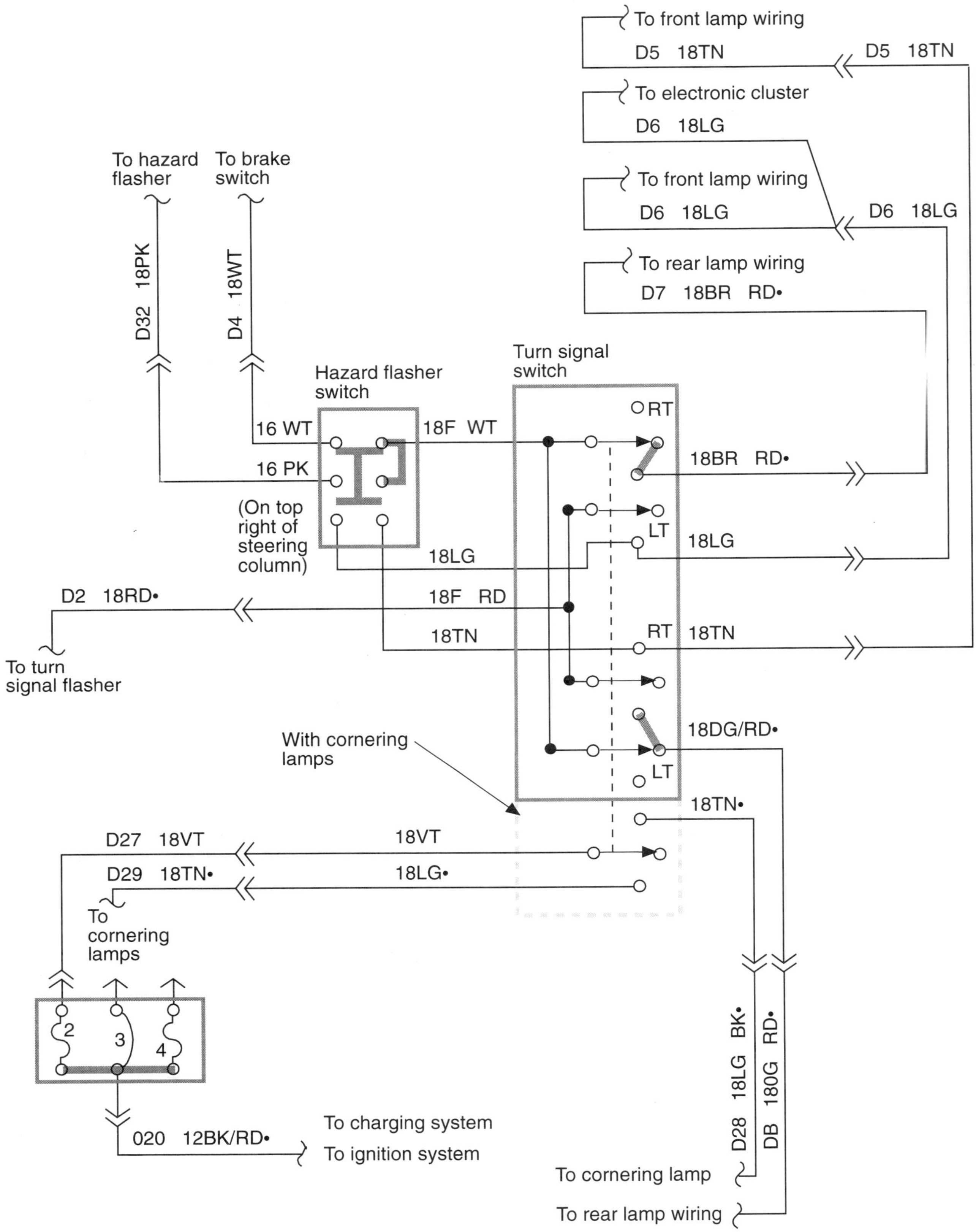

FIGURE 11–41 Turn signal circuit in the neutral or rest position.

(Courtesy of DaimlerChrysler Corporation)

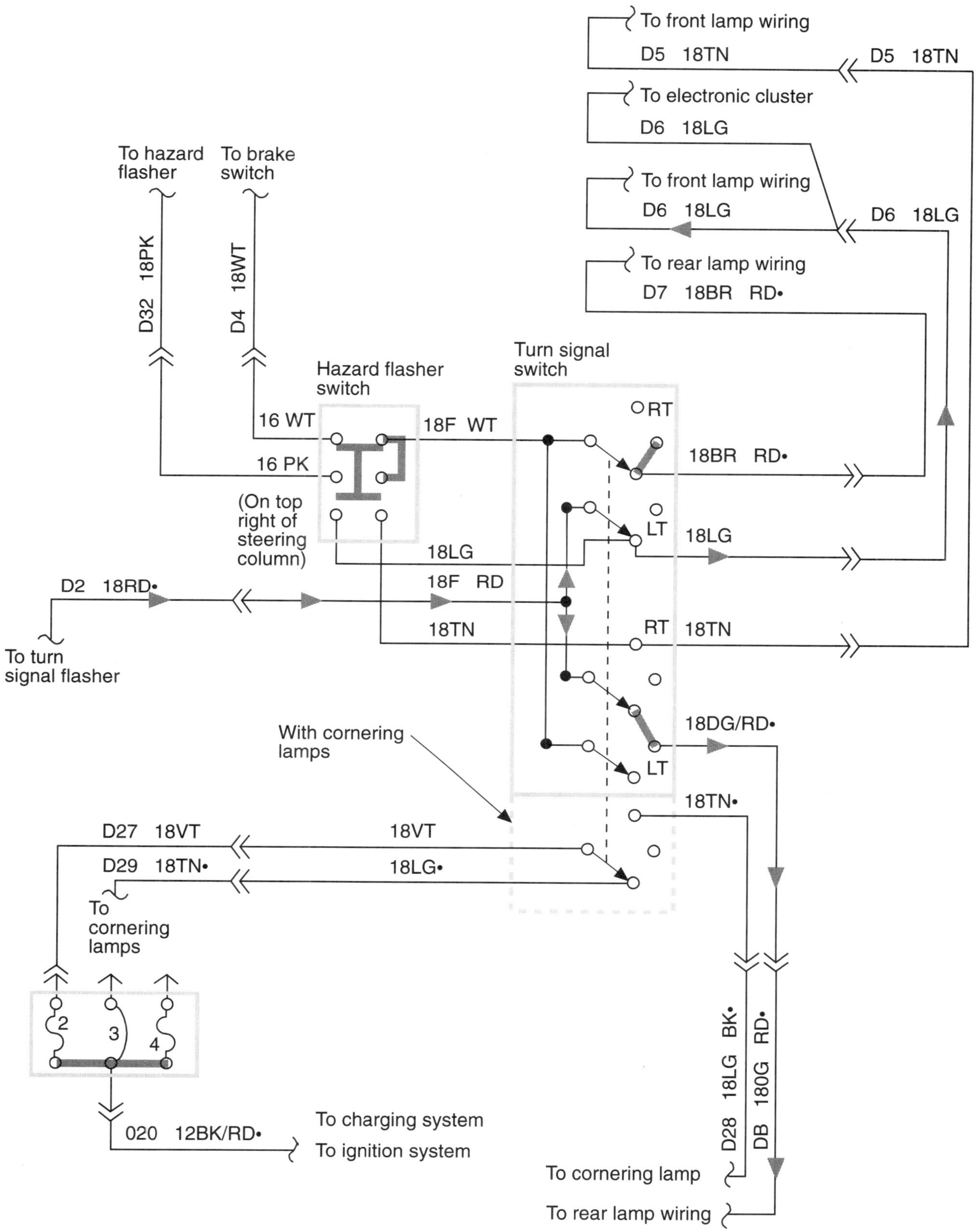

FIGURE 11–42 Turn signal circuit in the left turn position.

(Courtesy of DaimlerChrysler Corporation)

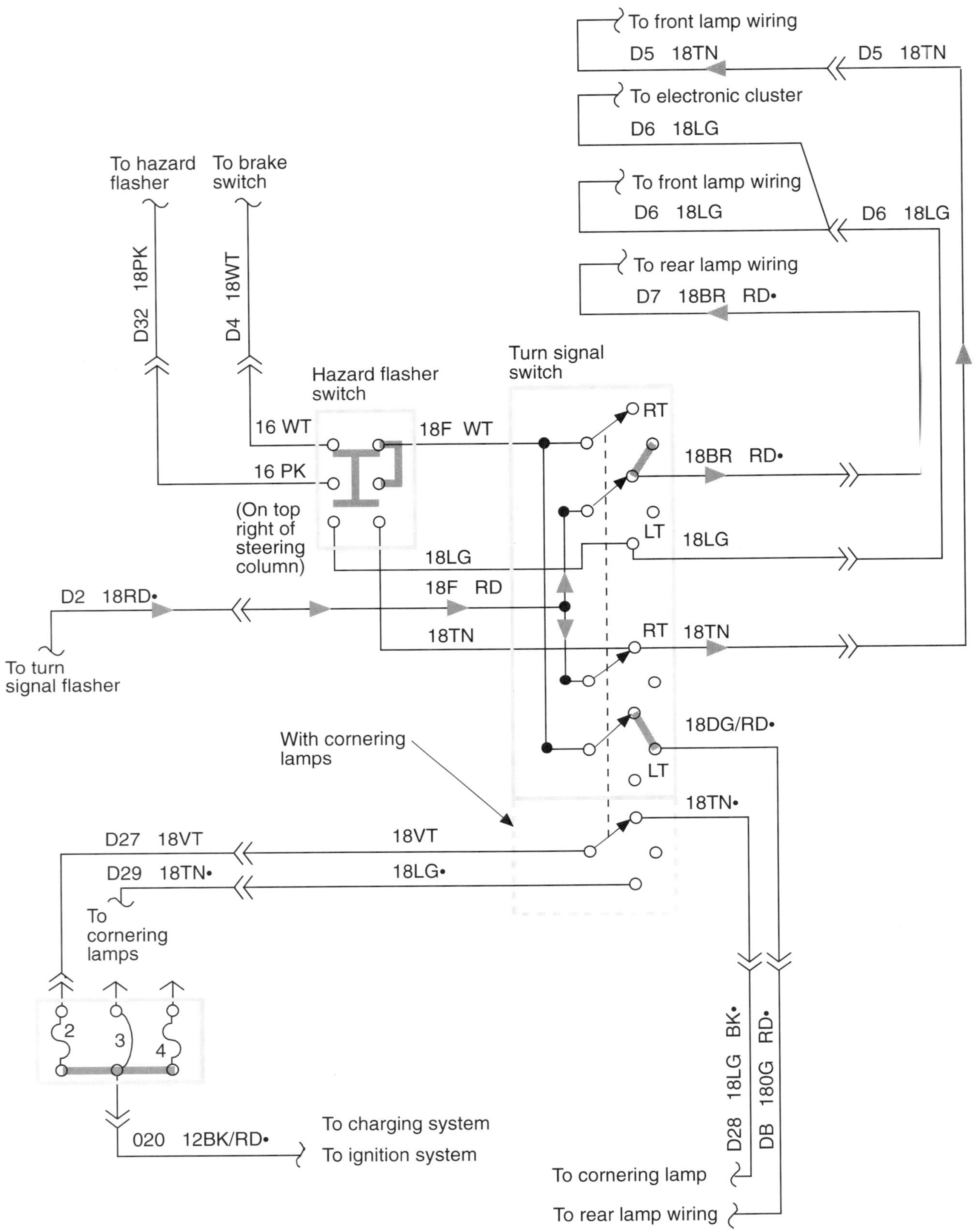

FIGURE 11–43 Turn signal circuit in the right turn position.

(Courtesy of DaimlerChrysler Corporation)

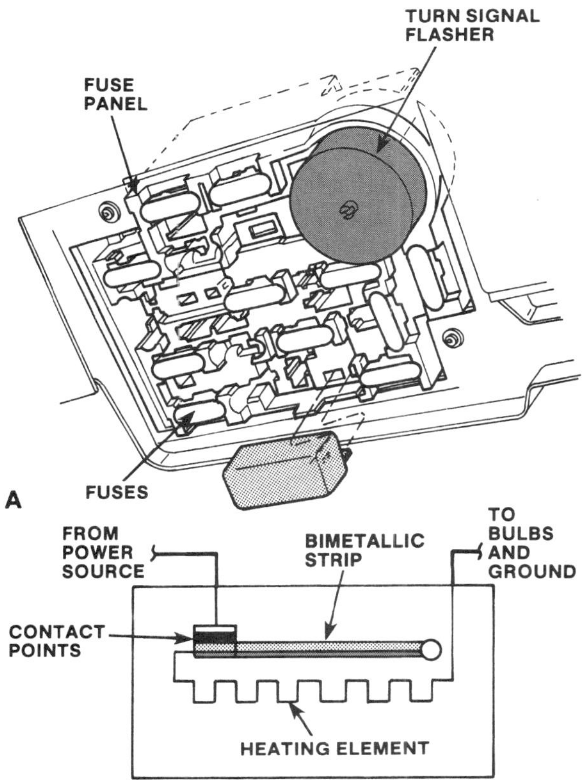

FIGURE 11–44 (A) Typical turn signal flasher location, and (B) flasher bimetallic strip and contacts. (Courtesy of Ford Motor Company)

When flashers fail to flash as fast as originally designed or flash too fast, a bulb may be burned out. Check all the bulbs for proper operation before replacing the flasher unit. When a bulb burns out on a system that uses a bimetallic flasher, the lights normally flash slower than normal. When a bulb burns out on a system with an electronic flasher, the lights flash more rapidly than normal. When trailers are hooked up to a vehicle, the trailer lights are wired in parallel with the vehicle light systems. This could raise the current flow through the flasher unit and cause it to blink faster than normal. Flashers are designed to operate a specific number of bulbs. To correct this problem, heavy duty flasher units are available to correct the blink rate.

Side Marker Lights The side marker lights have been a safety requirement on all vehicles sold in North America since 1969. When a vehicle enters traffic from the side, the marker lamps help the vehicle to be seen. The front side marker lens is amber, and the rear is always red. Many manufacturers wire the side marker lamp circuit in parallel to the parking lamp circuit so they will work anytime the park lights or headlights are on (Figure 11–45).

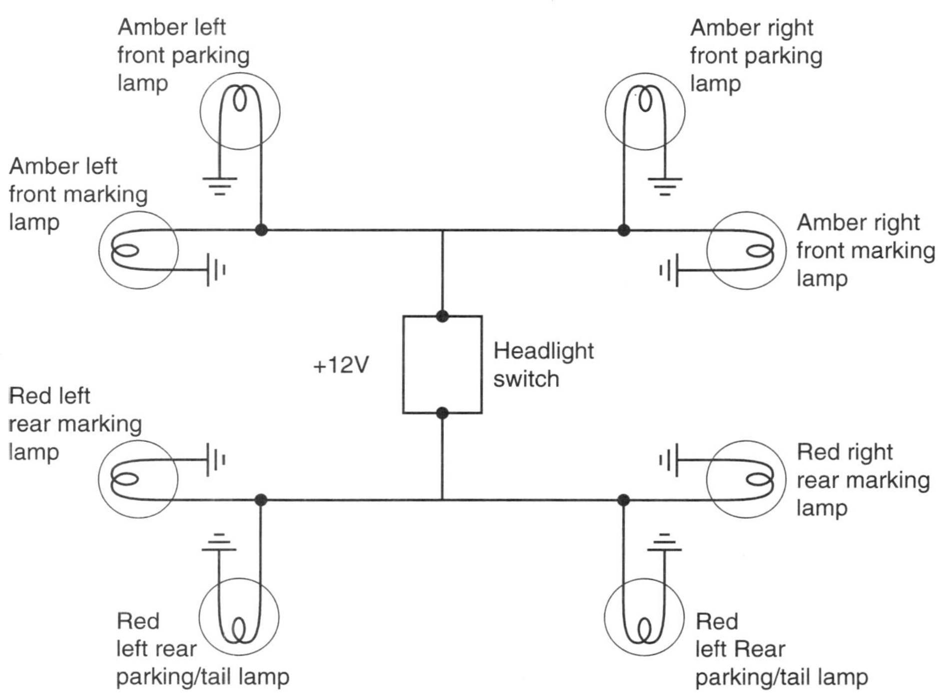

FIGURE 11–45 Typical side marker light circuit.

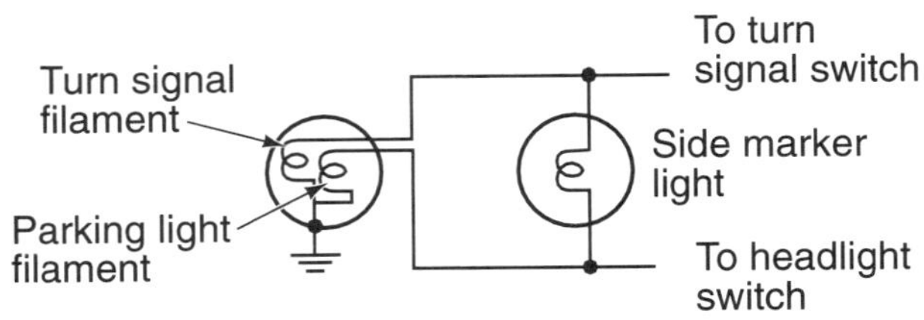

FIGURE 11–46 A side marker light circuit wired between the parking and turn signal circuits.

Some vehicle manufacturers wire the side marker lamps across the parking light and turn signal light circuits (Figure 11–46). In this circuit, the side marker lamp will flash when the turn signals are used. If the parking lights are on, the side marker lamp will use the turn signal lamp as a ground to complete the circuit (Figure 11–47). The turn signal lamp will not light because the voltage drop across the side marker lamp is too high.

If the parking lamps are off, and the turn signals are used, the side marker lamp will flash with the turn signal lamp (Figure 11–48). The side marker lamp uses the parking lamp as a ground to complete the circuit. Because of the high voltage drop across the side marker lamp, the parking lamp will not light.

When the parking lamps are on and the turn signals are activated, the side marker lamp will flash alternately with the turn signal (Figure 11–49). This is due to having equal voltage on both sides of the side marker lamp until the turn signal flasher opens. When this happens the side marker operates (Figure 11–47) until the flasher closes and provides voltage to the turn signal lamp again.

Because the parking lamp circuit and turn signal circuit are connected together through the side marker lamp, feedback of the current flow could occur if a

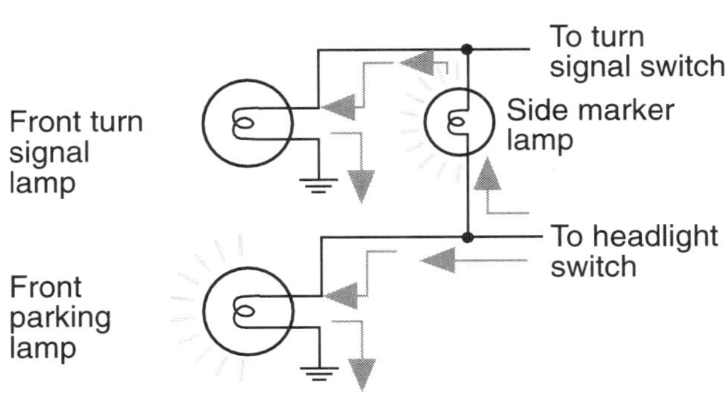

FIGURE 11–47 Current flow through the side marker lamp with the parking light switch *on*.

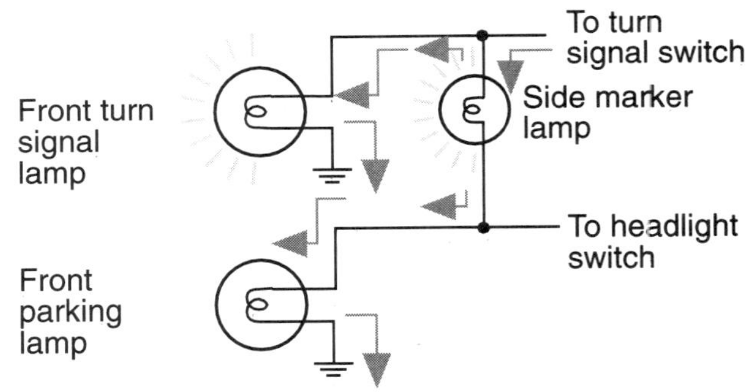

FIGURE 11–48 Current flow through the side marker lamp with the turn signal circuit *on* and the parking light switch *off*.

ground connection goes bad in one of the circuits. This is important to remember when diagnosing a lighting circuit problem.

Fog Light Circuit For additional safety under certain driving conditions, some vehicles have fog lights as optional equipment from the factory or as an after-market add-on. The fog lamp circuit normally has a relay because of the high amperage required by the fog lamps. Figure 11–50 is an example of a fog lamp circuit that can operate only if the headlight switch is in the Park or Headlight position. Some off-road or sport utility vehicles have the fog light circuit wired directly to battery voltage to allow the circuit to operate independent of the headlight circuit. Refer to specific vehicle wiring diagram before attempting any repairs.

Back-Up Light Circuit The back-up light circuit is designed to illuminate the road behind the vehicle. Figure 11–51 illustrates a back-up light circuit for an auto-

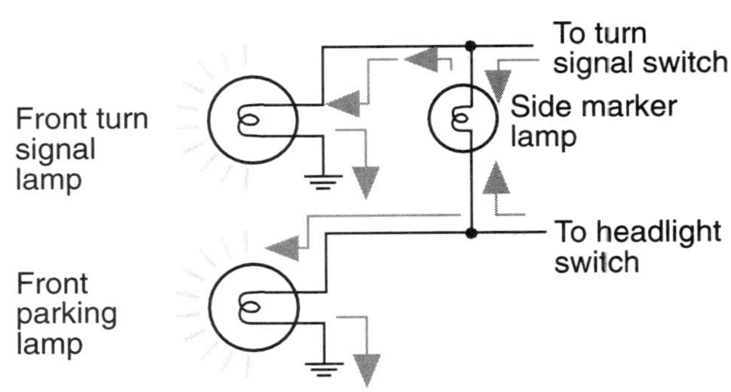

FIGURE 11–49 Current flow through the side marker light *stops* when the turn signal circuit and parking light circuits are both *on*.

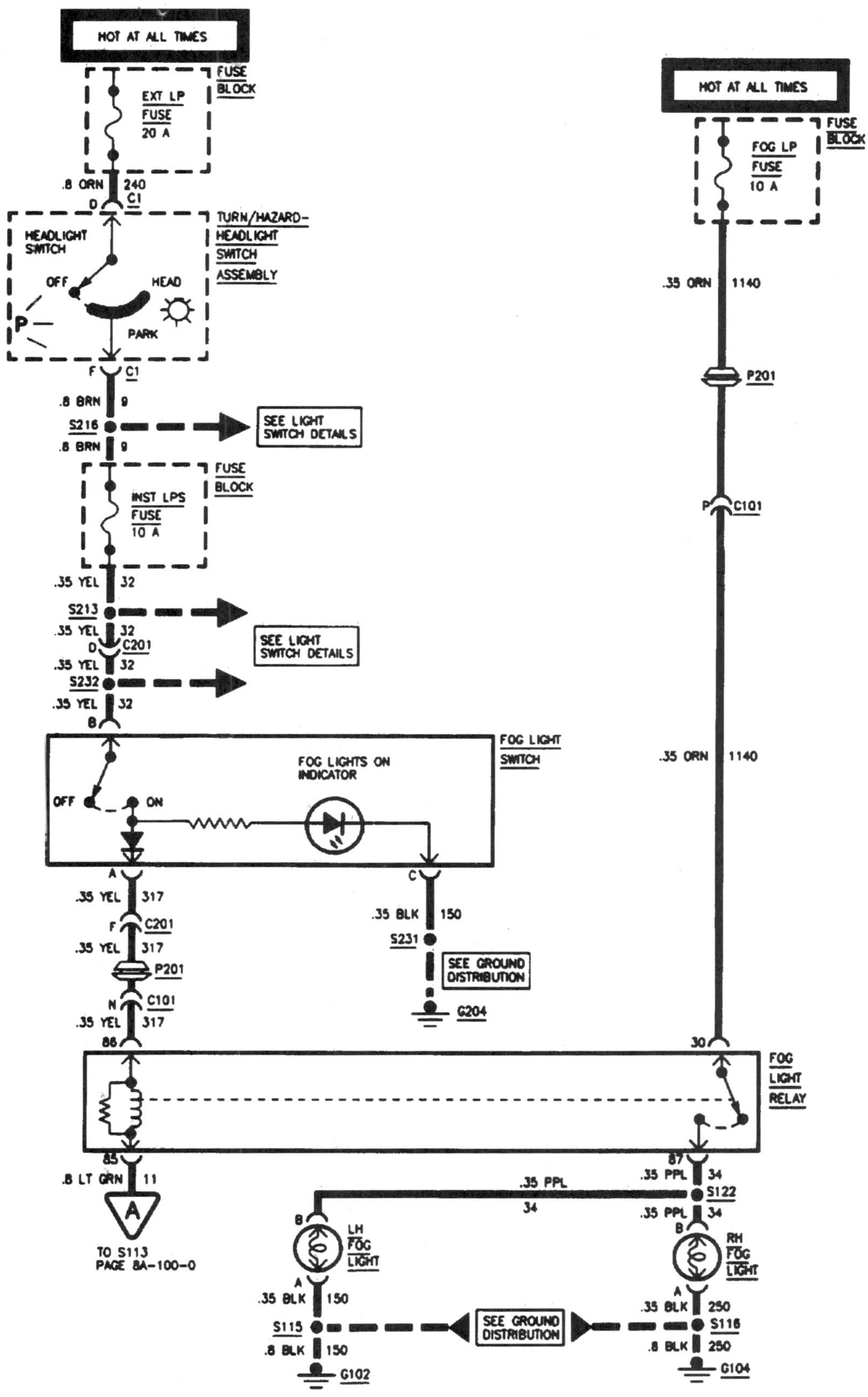

FIGURE 11–50 Typical fog light circuit.

(Courtesy of General Motors Corporation, Service Technology Group)

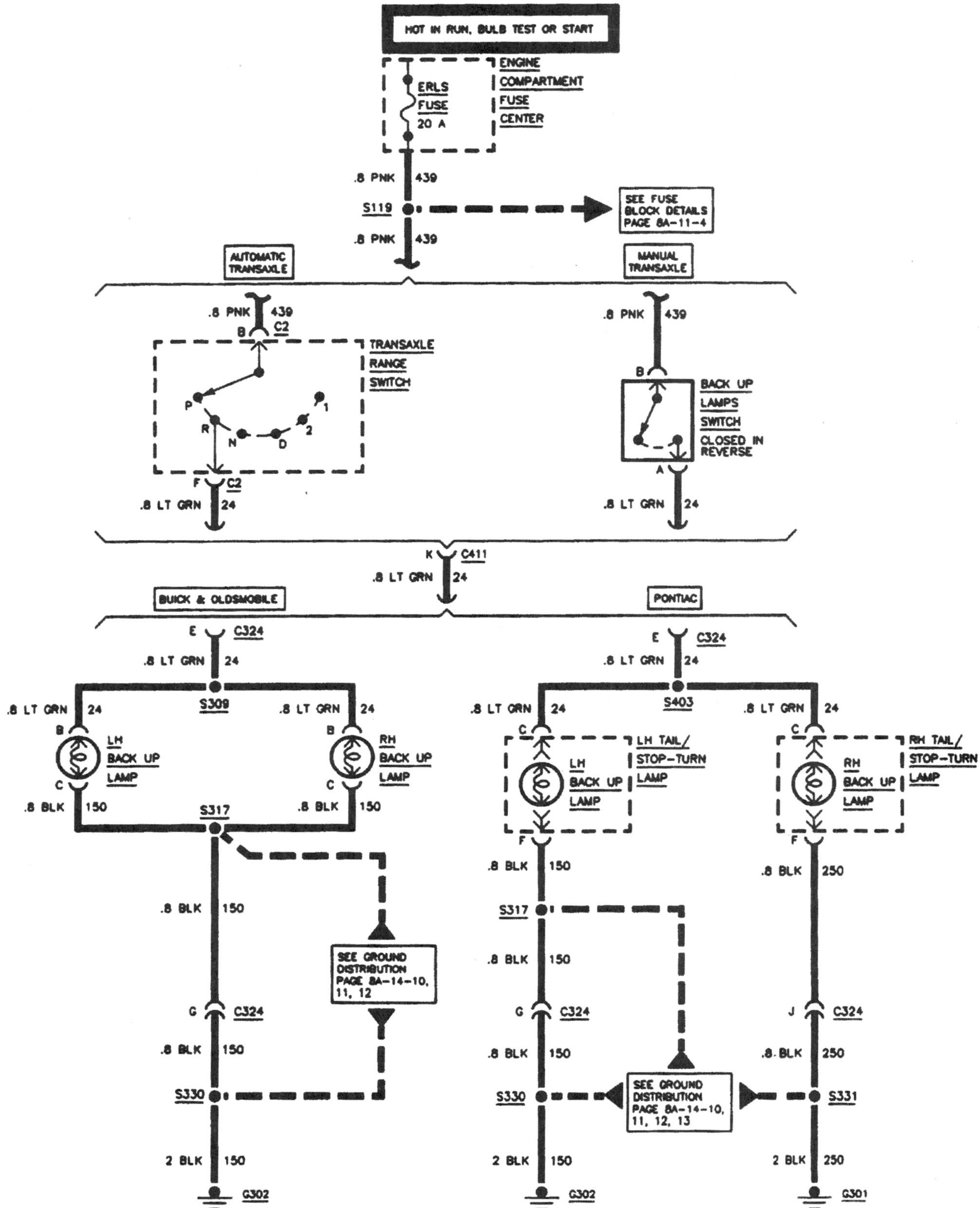

FIGURE 11–51 Back-up light circuit.

(Courtesy of General Motors Corporation, Service Technology Group)

matic transaxle vehicle and a manual transaxle vehicle. Power is supplied to a fuse when the ignition switch is in the Run position only. This prevents the back-up lamps from operating by mistake if the ignition is off. The circuit bulbs will be energized from battery voltage when the transmission is put into reverse with the key on.

The back-up light system is relatively easy to diagnose by performing the following checks.

- Check the system fuse. If there are other accessories that operate off the same fuse, check them for proper operation.
- If power to the switch and the switch check out satisfactory, then check the back-up lamps for burned-out elements, and replace if necessary.
- If lamps check out OK, then check the bulb sockets, connectors, and associated wiring.
- Check for voltage input at the back-up light switch. If voltage is present, bypass the switch with a jumper wire and check the back-up lights for operation. If the system works with the jumper wire, then the switch is faulty and should be replaced.

Interior Lighting Circuits

Interior Light Systems

The interior lighting circuits in today's vehicles vary with the vehicle style and amount of optional equipment. Luxury cars tend to have more interior lighting circuits which operate through another system, such as the body computer. Basic interior lighting includes courtesy lights, reading lights, map or vanity lights, and instrument panel lights.

Courtesy Light Circuit There are several types of courtesy lights on vehicles today. Some courtesy lights are located in the front and rear door trim panels, floor kick panels, under the instrument panel, and in the center of the headliner. Figure 11–52 is a courtesy light circuit with the door jamb switches on the ground side of the circuit. The courtesy light circuit completes the path to ground when either of the doors are opened, or the headlight switch is turned to the interior light position.

The courtesy light circuit can also be designed with the door jamb switches on the battery *positive side* of the circuit (Figure 11–53). These switches are insulated from the ground side of the circuit (chassis) and control the battery voltage to the courtesy lights.

The courtesy light circuit may share its system fuse with other circuits in the vehicle. This may cause a sneak circuit feedback, which is when the current in one circuit seeks an alternate path of flow when the original path is interrupted. This could cause lighting circuits or other accessories to operate when not intended. Figure 11–54 is an example of a circuit that could experience a feedback condition. The dome light, taillight, and brake light share a fuse. The cigar lighter and courtesy light share a fuse. If the dome/tail/brake fuse is blown and the headlight switch is off, whenever the stoplight switch is closed the courtesy light, dome light, and stoplight will all glow dimly because these lights are in series.

Reading, Map, and Courtesy Light Circuits The interior lighting circuit in most vehicles have some branch lighting circuits that are controlled by individual switches. Figure 11–55 is an example of a reading/vanity light circuit. Battery voltage is supplied at connection B from the interior light circuit fuse. Each reading lamp has a control switch to complete the circuit to ground. The vanity mirror lamps have an internal switch that completes the circuit to ground whenever the mirror cover is opened.

Instrument Panel Light Circuit A simple instrument panel light circuit is a perfect example of a series-parallel or compound circuit. In Figure 11–56, the rheostat is connected in a series line to the instrument panel lights. The voltage available at the parallel instrument panel lights depends on the setting of the rheostat control knob. Because the lamps are in parallel, if one lamp burns out, the rest will remain on.

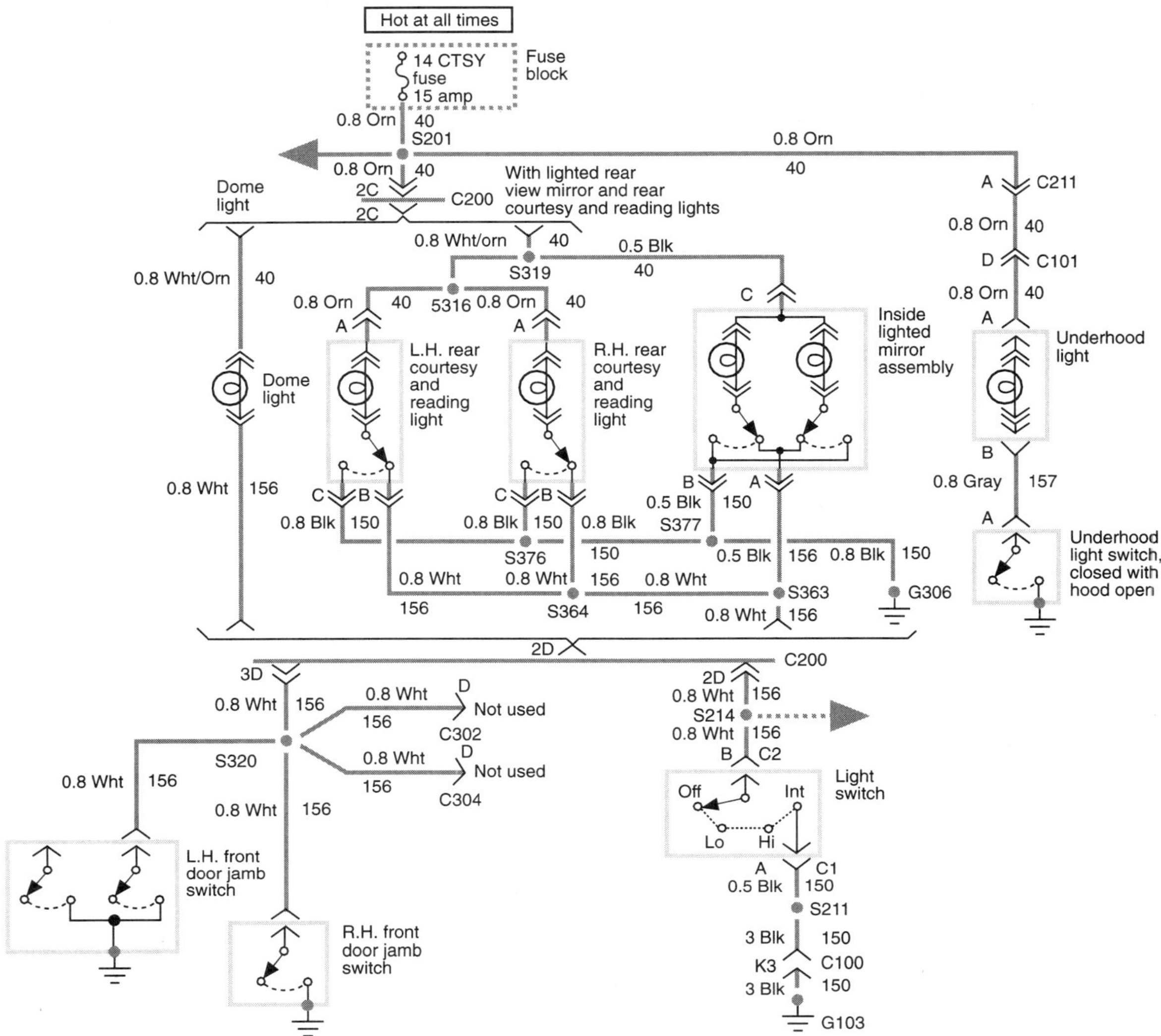

FIGURE 11–52 Typical courtesy light circuit with control switches on the ground side.

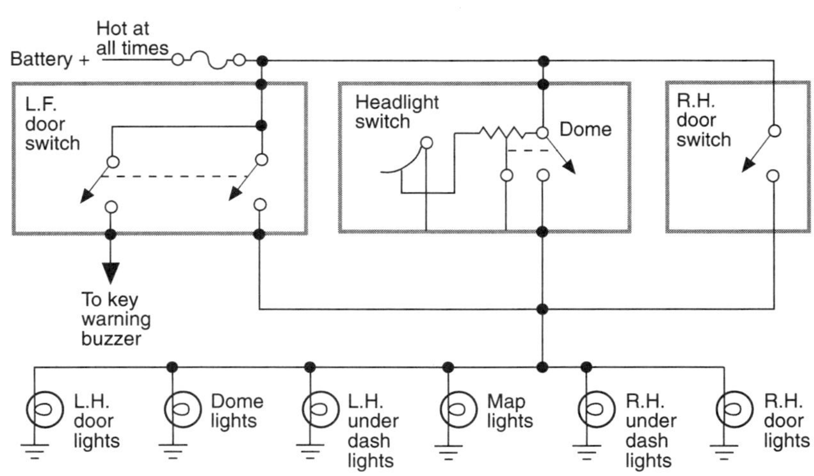

FIGURE 11–53 Typical courtesy light switch with control switches on the positive side.

(Courtesy of Ford Motor Company)

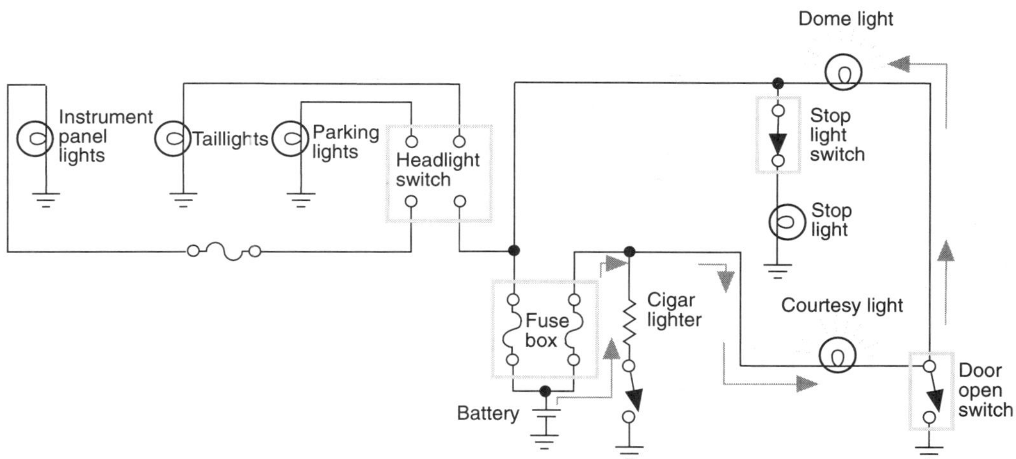

FIGURE 11–54 Circuit feedback if the tail/dome/brake light fuse is blown and the stoplight switch is closed.

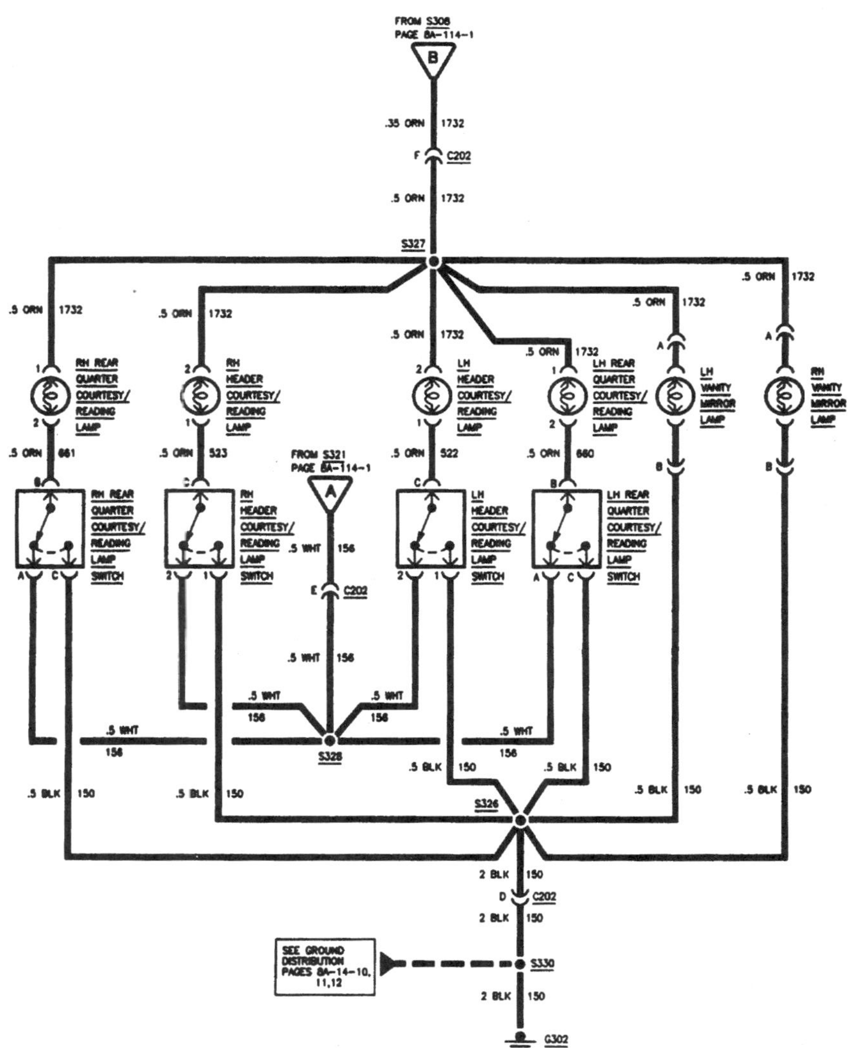

FIGURE 11–55 Typical reading/vanity light circuit.

(Courtesy of General Motors Corporation, Service Technology Group)

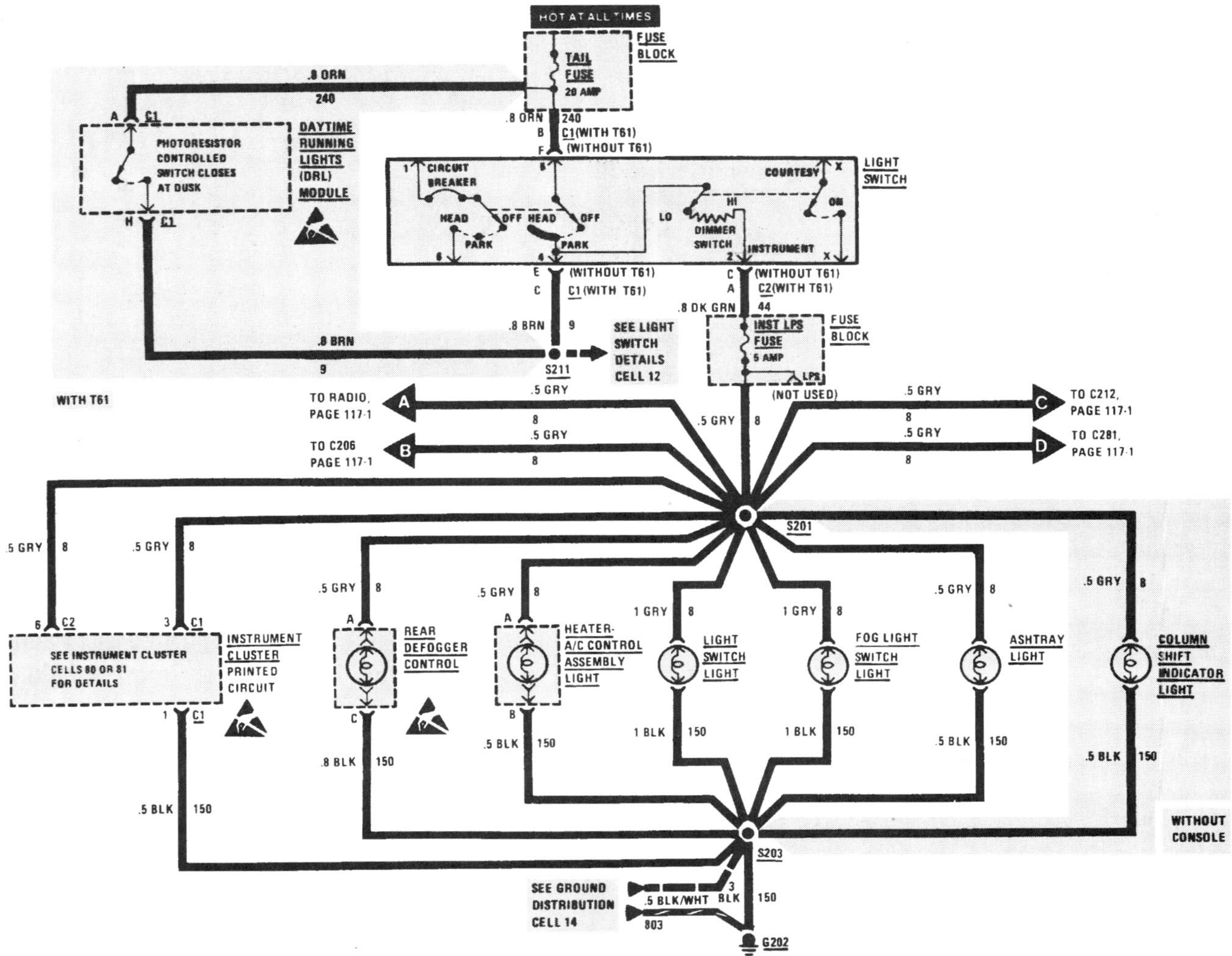

FIGURE 11–56 A series/parallel instrument panel light circuit.

(Courtesy of General Motors Corporation, Service Technology Group)

CASE STUDIES

CASE #1

The following case study is typical of a headlight circuit that technicians often encounter. Knowledge of the circuit and simple test procedures will help you perform the diagnosis quickly and efficiently.

Customer Complaint

A customer brought their vehicle into the shop because the headlights go out when they are switched from low beams to high beams. The customer replaced both headlamps, thinking the high beam bulb elements were burned out, but the problem remained (Figure 11–57).

Known Information

- ❑ Vehicle operating voltage = 14 volts
- ❑ All circuit fuses are OK
- ❑ Both headlamps have been replaced
- ❑ Headlamp switch is OK

Circuit Analysis

Answer the following questions on a *separate* sheet of paper.

1. If the bulbs go out when the beam select switch is switched from low to high beams, what problems could be preventing the high beams from illuminating? ______.
2. If the high beam bulbs operate when a jumper wire is connected from battery power to the high beam circuit connector #30, what would be the *most* likely cause of the problem? ______ ______.
3. What should the voltage be of the L2 14LG wire *low* connection at the headlamp dimmer switch? ______.
4. What is the purpose of the L2 14LG BK wire going to the *high* connection at the headlamp dimmer switch? ______.
5. If there is no voltage at the dimmer switch *high* output, what would be your diagnosis of this circuit problem? ______.
6. How could you verify the circuit problem using a jumper wire? ______ ______.
7. What is the best way to repair this system? ______.
8. What diagnostic steps were helpful when troubleshooting this system or explaining it to the customer? ______.
9. What is your analysis of the above case study? ______ ______.

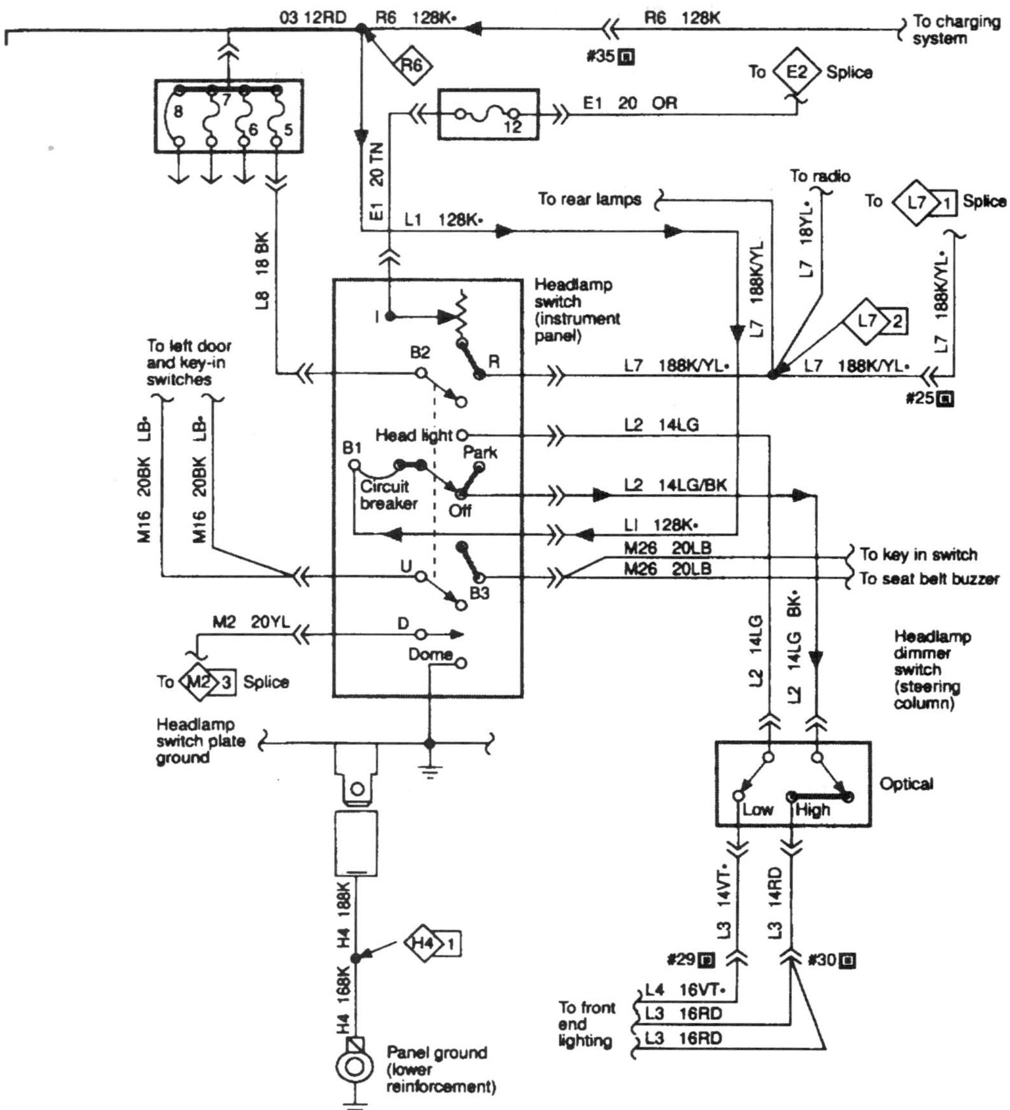

FIGURE 11–57 Case #1–headlights go out when switched from low to high beams.
(Courtesy of DaimlerChrysler Corporation)

CASE #2

The following case study (Figure 11–58) is a turn signal circuit problem. Depending on the wiring design, knowledge of how turn signals operate and where the circuit components are located can help the technician perform the diagnosis of circuit problems quickly and efficiently.

Customer Complaint

The turn signal bulbs do not flash when needed, but stay on when the switch is moved to either the left or right turn positions. The customer replaced a flasher relay located on the fuse panel located under the dash, assuming it was the flasher relay, but the problem remained.

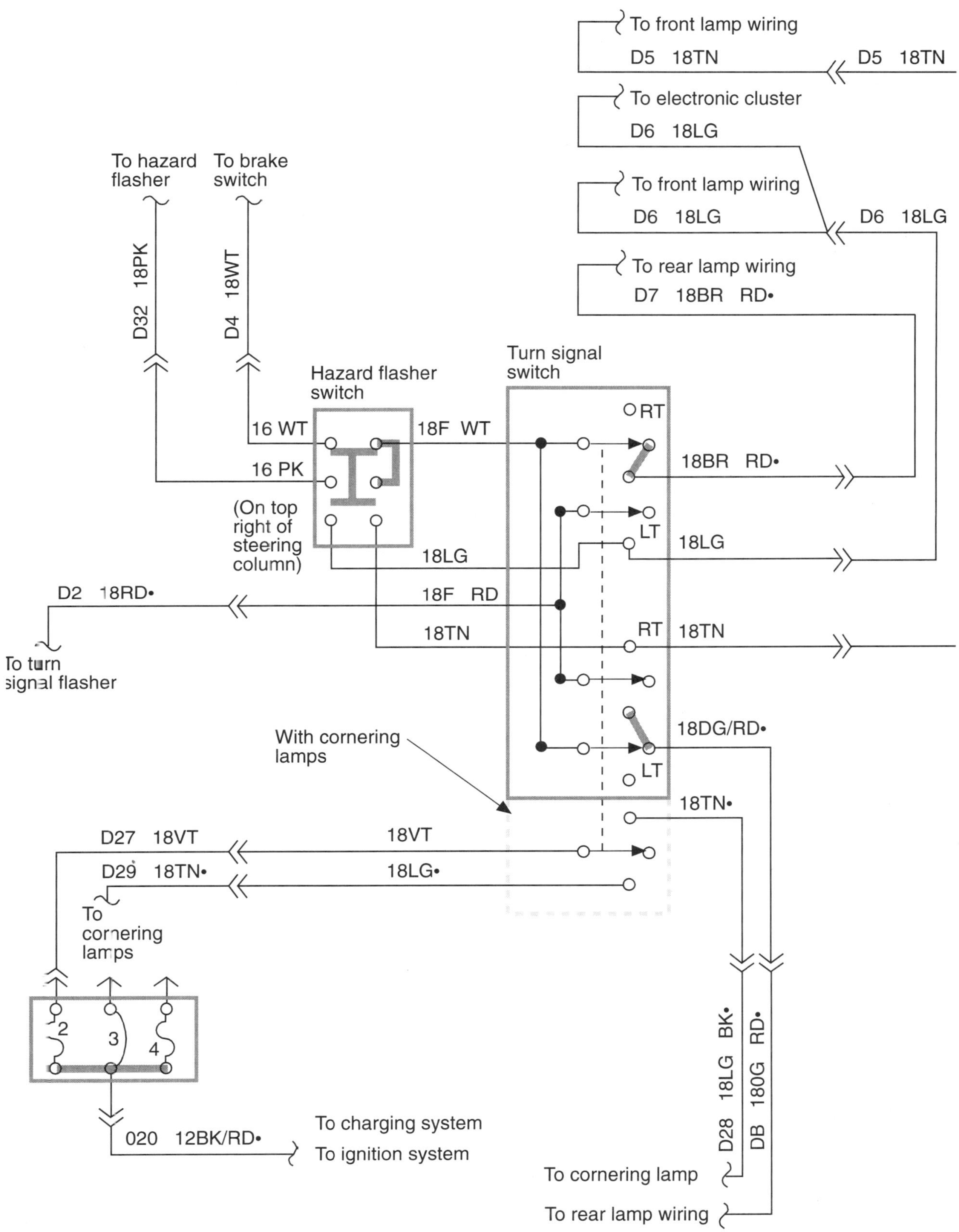

FIGURE 11–58 Case #2–turn signal circuit problem.

(Courtesy of Daimler Chrysler Corporation)

Known Information

- ❑ Vehicle operating voltage = 14 volts
- ❑ Hazard flasher circuit operates OK
- ❑ All system bulbs are OK

Circuit Analysis

Answer the following questions on a *separate* sheet of paper.

1. What is the *most* likely cause of the turn signals to stay *on* ______________________________.
2. How could the technician quickly determine the difference between the turn signal flasher relay and the hazard flasher relay? ______________________________.
3. What should the voltage be from the turn signal flasher to the turn signal switch? ______________.
4. What diagnostic steps would be helpful when troubleshooting a turn signal system? ______________________________.
5. What is your analysis of the above case study? ______________________________.

CASE #3

The following case study (Figure 11–59) is typical of a back-up light circuit problem that technicians often encounter. Knowledge of the circuit and simple test procedures can help you diagnose the circuit safely and efficiently.

Customer Complaint

The back-up lights do not work on a customer's car. The two back-up lamps were both replaced, but the problem remained.

Known Information

- ❑ Vehicle operating voltage = 14 volts
- ❑ Both back-up bulbs have been replaced
- ❑ Circuit fuse is OK

Circuit Analysis

Answer the following questions on a *separate* sheet of paper.

1. What is the *most* likely cause of the back-up lights not working? ______________________________.
2. What should the voltage be at each of the back-up light bulbs? ______________________________.

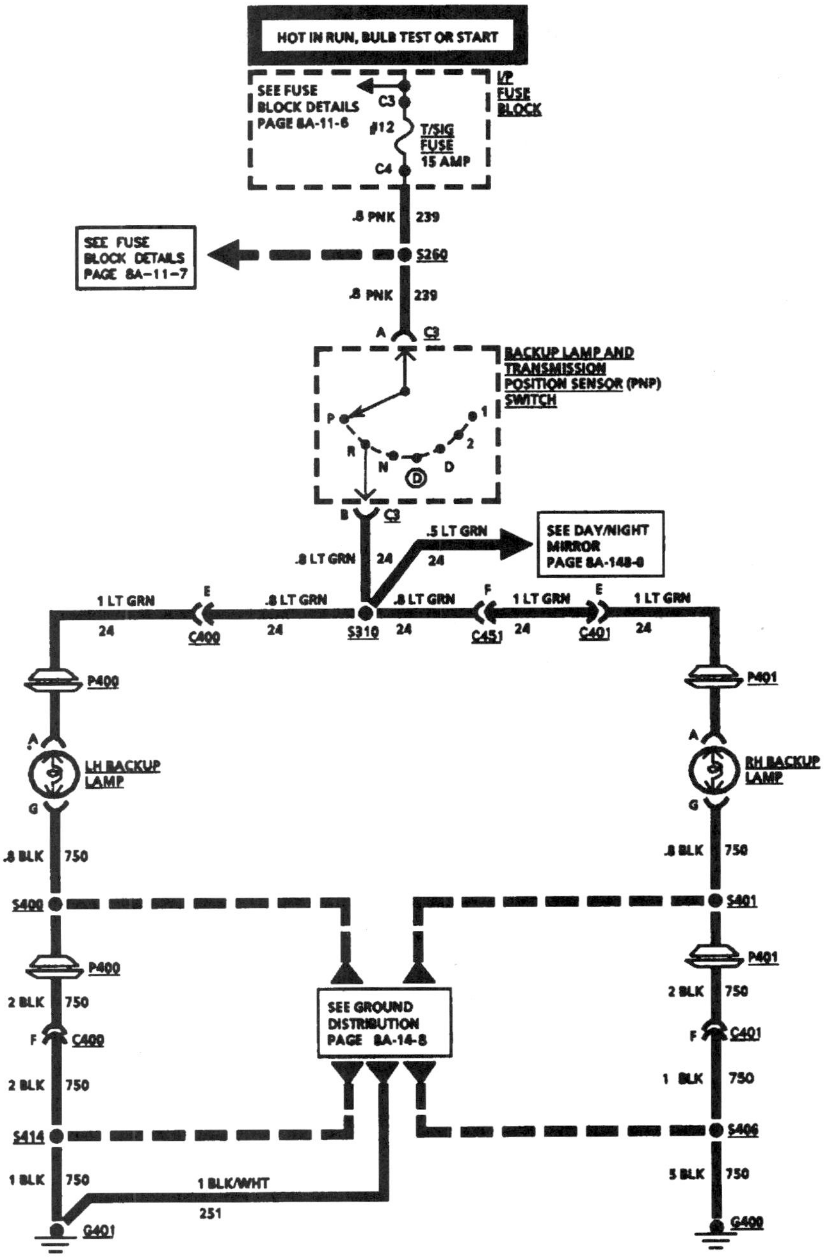

FIGURE 11–59 Case #3–back-up light circuit problem.

(Courtesy of General Motors Corporation, Service Technology Group)

3. How could you verify that the back-up lamp switch is functioning properly using a jumper wire?

   ____.

4. What diagnostic steps would be helpful when troubleshooting this system or explaining the diagnosis to the customer? ______________________________.

5. What is your analysis of the above case study? ______________________________

   ____.

Hands-On Vehicle Tests

The following two hands-on vehicle checks are included in the **NATEF** (National Automotive Technicians Education Foundation) **Task List.** The task reference numbers are the same as the NATEF reference numbers. Complete your answers to the below questions on a *separate* sheet of paper.

Performance Task #1 (NATEF #VI-E-2)

Task Description

Inspect, replace, and aim headlights and bulbs.

Task Objectives

- ❑ Obtain a vehicle that can be used for this task. What model and year of vehicle are you using for this task? ____________________.
- ❑ What type of headlights are on the vehicle you are using? ____________________.
- ❑ What are the pre-alignment checks to be made to the vehicle before adjusting the headlights?
  ________, ________, ________, ________, ________, ________, ________, ________.
- ❑ Describe the procedures used when aligning headlights with mechanical headlight aiming equipment.
  ____
  ____
  ____________________.
- ❑ How is the horizontal adjustment accomplished on the vehicle headlights? ____________________
  ____________________.
- ❑ How is the vertical adjustment accomplished on the vehicle headlights? ____________________
  ____________________.

Task Summary

After performing the above NATEF task, what can you determine will be helpful when aligning headlights with mechanical aiming equipment? ____________________

____________________.

Performance Task #2 (NATEF Task #VI-E-3)

Task Description

Inspect and diagnose incorrect turn signal or hazard light operation. Repair or replace as needed.

Task Objectives

- ❑ Obtain a vehicle that can be used for this task. What model and year of vehicle are you using for this task? ____________________.
- ❑ Locate a wiring diagram for the vehicle that will show the turn/hazard light circuit.

❑ Using the wiring diagram(s), identify areas in the circuit such as connectors, wires, fuses, and components that can cause a possible problem.

Identified Area	Symptom	Cause	Test Procedure

❑ Perform the diagnostic steps and procedures identified in the previous question. Does this help identify the circuit problem(s)?______________________________.

❑ What diagnostic steps were helpful when troubleshooting this system? ____________________.

❑ What is your analysis of using wiring diagrams to help diagnose turn signal circuit problems?

____

____.

Task Summary

After performing the above NATEF task, what can you determine that will be helpful in knowing how to check turn signal and hazard flasher circuits for proper operation?

____

____.

Summary

❑ A dimmer switch is used to change headlights from low beams to high beams.

❑ Typical headlight switches can be mounted in the dash board or the steering column.

❑ A typical headlight switch is a three position style, with Off, Park, and Headlight positions.

❑ A multifunction switch combines the dimmer switch with other function switches in the steering columns.

❑ The flash-to-pass feature of a headlight circuit allows the high beams and/or the low beams to light when the headlight switch is off.

❑ For styling and aerodynamics, some concealed headlights are hidden behind vacuum or electrically controlled doors.

❑ Daytime running lights (DRL) operate the headlights at a reduced voltage during daylight hours for vehicle safety.

❑ A common way to align headlights is with mechanical headlight aiming equipment.

❑ In a two bulb tail/park lamp circuit, a dual-filament bulb is used on each side of the circuit for taillights and brake lights.

- ❑ In a three bulb taillight circuit, an individual bulb is used for each taillight, brake light, and turn signal function.
- ❑ All brake light circuits since 1986 are required to have a CHMSL (center high-mounted stop light).
- ❑ In a two bulb system, the turn signal switch determines which brake light is separated out of the circuit.
- ❑ A typical turn signal or hazard flasher consists of a bimetallic strip and heating element wired in series to the circuit.
- ❑ Side marker lights have been a safety requirement on all vehicles sold in North America since 1969. The front marker lens is always amber and the rear marker lens is always red.
- ❑ In a courtesy light circuit with the door jamb switch on the ground side of the circuit, the circuit is complete when the door is opened.
- ❑ In a courtesy light circuit with the door jamb switch is on the positive side of the circuit, the switches are insulated from the ground side of the circuit.
- ❑ Circuit feedback is when the current in one circuit seeks an alternate path of flow when the original path is interrupted.
- ❑ A typical instrument panel light circuit uses a rheostat to control the intensity of the lamps.

Key Terms

center high-mounted stoplight (CHMSL)
collision avoidance light
daylight running lights (DRL)
flash-to-pass
multifunction switch

Review Questions

Short Answer Essays

1. Describe some of the combined features of a multifunction steering column switch.
2. In a concealed headlight system, when the headlights are turned on, how are the headlight doors opened?
3. Describe the purpose of a CHMSL in a brake light system.
4. What causes a bimetallic turn signal flasher to operate as designed?
5. Describe the most common way instrument panel lights are dimmed.

Fill-in-the-Blanks

1. Front side marker lamp lens is always ______________________ in color.
2. When a bulb burns out in a bimetallic turn signal flasher system, the remaining lights flash ______________________ than normal.

3. A brake light switch is usually attached to the ________________________.

4. Concealed headlight doors are either opened by ________________________ or ________________________.

5. Dimmer headlights are most often the result of ________________________.

ASE Style Review Questions

1. Technician A says the dimmer switch controls the current to the low beams only in a headlight circuit. Technician B says the dimmer switch controls the current to the high beams only in a headlight circuit. Who is correct?
   A. A only
   B. B only
   C. Both A and B
   D. Neither A nor B

2. Technician A says a headlight switch could be mounted in the dash panel or the steering column. Technician B says all headlight switches have a ceramic rheostat that is used to control the dimming of the instrument lights. Who is correct?
   A. A only
   B. B only
   C. Both A and B
   D. Neither A nor B

3. The courtesy and dome light circuit (on the next page) has the dome light on even when the doors are closed. Technician A says the dome light circuit is grounded between the dome light and the door jamb switches. Technician B says a door switch is grounded. Who is correct?
   A. A only
   B. B only
   C. Both A and B
   D. Neither A nor B

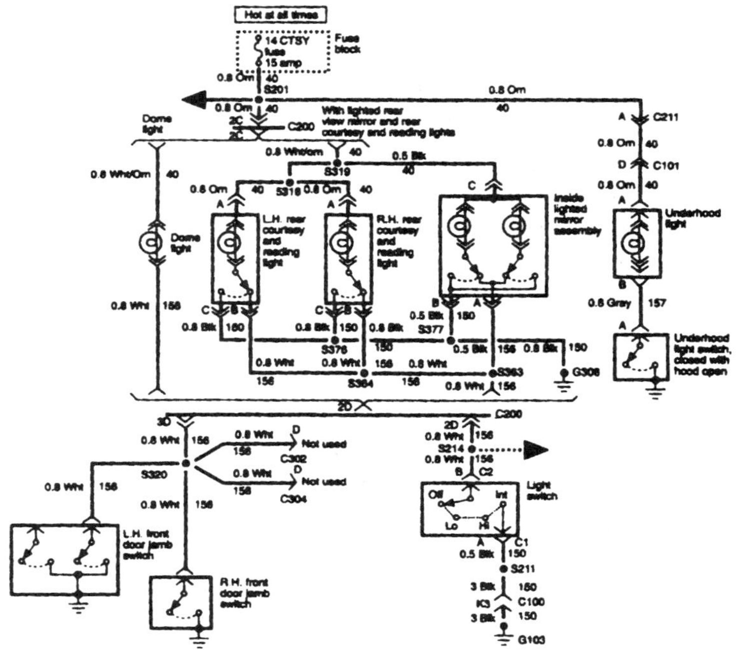

4. Two technicians are discussing a multifunction steering column switch. Technician A says the dimmer switch can be part of a multifunction steering column switch. Technician B says the horn switch could be part of a multifunction steering column switch. Who is correct?

   A. A only
   B. B only
   C. Both A and B
   D. Neither A nor B

5. A flash-to-pass feature is being discussed. Technician A says this system will flash the high beam bulbs at driver command. Technician B says this feature will not work when the headlight switch is in the off position. Who is correct?

   A. A only
   B. B only
   C. Both A and B
   D. Neither A nor B

6. Technician A says concealed headlight doors can be controlled by vacuum motors. Technician B says headlight doors can be opened manually. Who is correct?

   A. A only
   B. B only
   C. Both A and B
   D. Neither A nor B

7. Daylight running lights (DRL) are being discussed. Technician A says the headlights in a DRL system run at maximum designed voltage for brightness. Technician B says a DRL system cannot be over-ridden when the headlight switch is turned on. Who is correct?
   A. A only
   B. B only
   C. Both A and B
   D. Neither A nor B

8. Technician A says that mechanical headlight aiming equipment uses a spirit bubble for adjusting the headlights horizontally. Technician B says the vehicle should be on a flat level surface to properly adjust the headlights. Who is correct?
   A. A only
   B. B only
   C. Both A and B
   D. Neither A nor B

9. Technician A says a typical *three* bulb taillight circuit has individual control for each bulb circuit. Technician B says the mechanical brake light switch attached to the brake pedal arm controls the brake light circuit in a three bulb tail light circuit. Who is correct?
   A. A only
   B. B only
   C. Both A and B
   D. Neither A nor B

10. Technician A says a turn signal switch isolates a brake light from the brake light circuit. Technician B says a turn signal flasher *opens* to let the current flow to the circuit bulbs. Who is correct?
    A. A only
    B. B only
    C. Both A and B
    D. Neither A nor B

11. The back-up light circuit in the figure on the next page does not work when the back-up switch is closed. Technician A says both bulbs could be burned out. Technician B says the fuse could be blown. Who is correct?
    A. A only
    B. B only
    C. Both A and B
    D. Neither A nor B

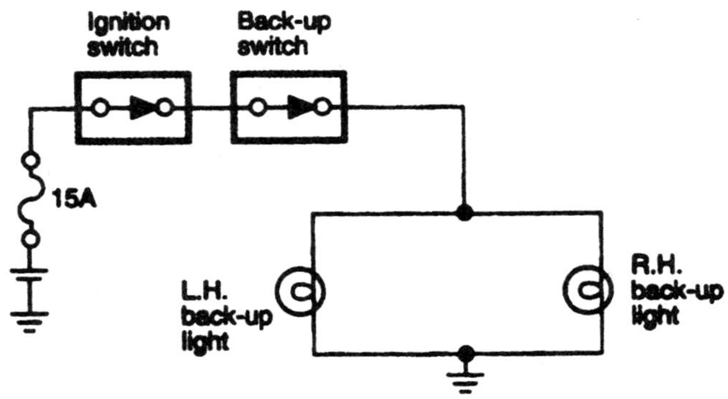

12. Technician A says that a sneak circuit feedback occurs when the current in one circuit seeks an alternate path of flow because the original path is interrupted. Technician B says a sneak circuit feedback can *only* occur when the insulation breaks down between the wires of two separate circuits. Who is correct?
    A. A only
    B. B only
    C. Both A and B
    D. Neither A nor B

13. Technician A says that when one turn signal bulb is burned out, the turn signal flasher may be faulty. Technician B says that the hazard signal circuit and the turn signal circuit normally operate from the same flasher unit. Who is correct?
    A. A only
    B. B only
    C. Both A and B
    D. Neither A nor B

14. A taillight bulb is glowing dimly. A voltmeter across the bulb reads 9.7 volts. Technician A says this is normal in some taillight circuits. Technician B says there is an unwanted voltage drop elsewhere in the circuit. Who is correct?
    A. A only
    B. B only
    C. Both A and B
    D. Neither A nor B

15. Technician A states that most instrument panel lamps are wired in series to allow them to dim evenly. Technician B says that the voltage available to the instrument panel lamps depends on the setting of the rheostat control knob. Who is correct?
    A. A only
    B. B only
    C. Both A and B
    D. Neither A nor B

12 Basics of Electronics and Computers

Introduction

Over the past several years, electronic controls and computer circuits have taken over the control and monitoring of most vehicle functions. Some of these functions are engine operation, climate control systems, anti-lock brakes, electronic suspension systems, electronically shifted transmissions, and charging system regulation. There are textbooks that dedicate their entire content to their study of computer sensors and troubleshooting. This chapter covers the basic concepts of electronics and sensors in the operation of the automotive computer. This will give you the basic background necessary to understand vehicle service manual diagnostic procedures.

Objectives

When you complete this chapter you should be able to:

- ❑ Explain the purpose of electronics and computers in automobiles.
- ❑ Explain the purpose of semiconductors, diodes, and transistors in simple electronic circuits.
- ❑ Describe the difference between analog and digital voltage signals.
- ❑ Describe how the binary code system works in computers.
- ❑ Describe the purpose of amplifiers and converters in computers.
- ❑ Describe the different types of memory in a microprocessor.
- ❑ Explain the purpose and operation of various logic gates in computer circuits.
- ❑ Describe the operation of input and output sensors in an automotive computer control circuit.
- ❑ Identify proper safeguard procedures when working with electronic circuits.

Basics of Electronics

Understanding the basics of electronics begins with knowing the materials used to build the components commonly used in electronic circuits. The material most often used is called a **semiconductor.** As mentioned in Chapter 3, a semiconductor is a stable element with exactly four valence electrons in its structure. The two most used semiconductor materials are silicon (Si) and germanium (Ge), of which silicon is the most commonly used.

Semiconductor materials have a crystal structure called a **covalent bond,** where atoms share the valance electrons of neighboring atoms (Figure 12–1).

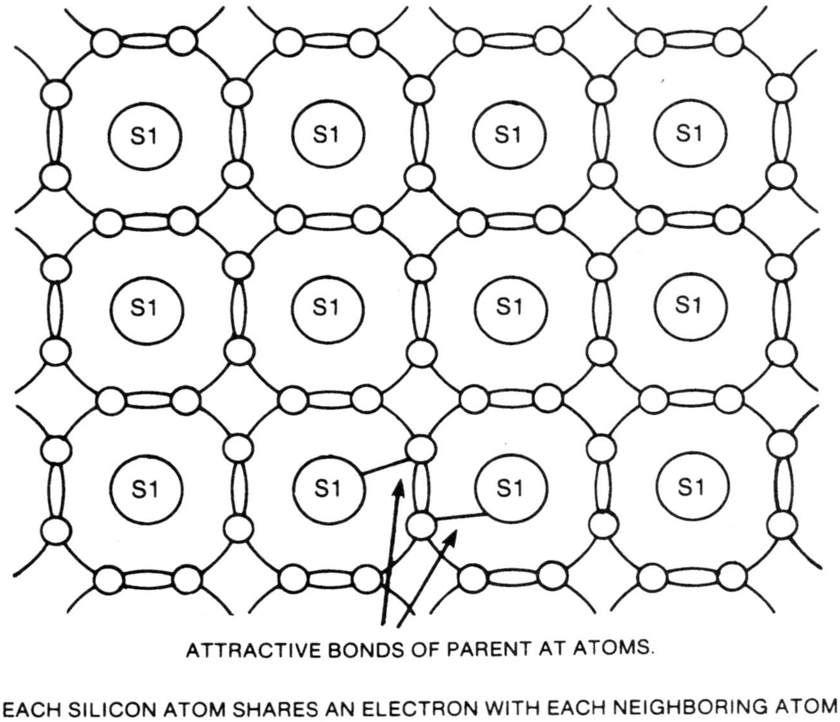

FIGURE 12–1 Silicon crystal covalent bond.

To function as semiconductors, a small amount of trace element impurities are added to the material. There are two types of semiconductor materials produced. The first is an **N-Type** semiconductor, which has loose or excess electrons, and a negative charge (Figure 12–2). N-type semiconductors are produced by adding an impurity with five electrons in the outer ring (called pentavalent atoms such as arsenic, antimony, or phosphorus). Four of these electrons fit into the crystal structure, but the fifth is free. The excess electrons produce a negative charge and enables the semiconductor to carry a current.

The second type of semiconductor material is a **P-Type** semiconductor, which has positive charged material that carry a current (Figure 12–3). P-type semiconductors are produced by adding an impurity with three electrons in the outer ring (called trivalent

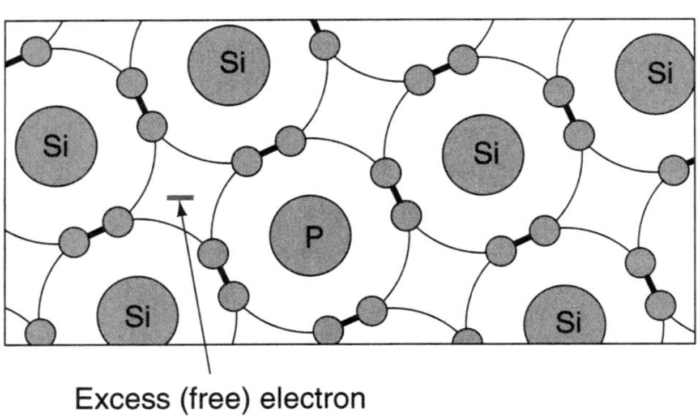

FIGURE 12–2 N-type silicon semiconductor atomic structure.

(Courtesy of General Motors Corporation, Service Technology Group)

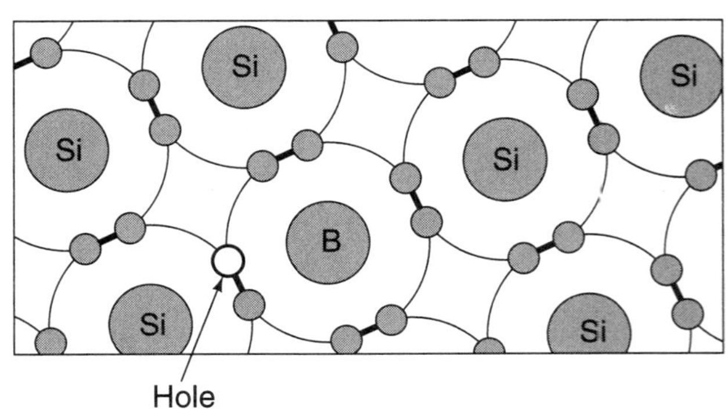

FIGURE 12–3 P-type silicon semiconductor atomic structure.

(Courtesy of General Motors Corporation, Service Technology Group)

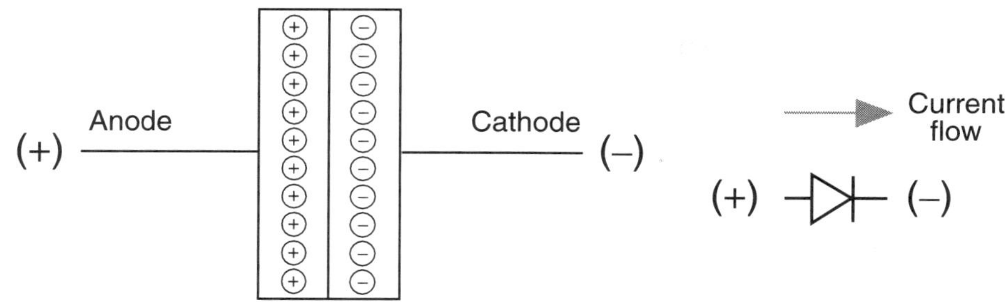

FIGURE 12–4 A diode and its electrical symbol.

atoms such as aluminum, indium, gallium, or boron). When this element is added to silicon or germanium, the three outer electrons fit into the pattern of the crystal, leaving a hole where the fourth would fit. This hole is actually a positive charged empty space, that carries the current in the P-type semiconductor because it attracts electrons. Although the electrons cannot be freed from their atom, they can rearrange their pattern and fill a hole in a nearby atom. Whenever this occurs, the electron leaves a hole, which is filled by another electron, and so on. The electrons move towards the positive side of the structure, and the holes move towards the negative side.

Semiconductors can function as both a conductor and an insulator. Because of this, semiconductors serve as excellent switching and circuit control devices. Two common semiconductor devices used in solid state circuits are **diodes** and **transistors.**

Diodes A **diode** is the simplest semiconductor device made. It is an electrical one-way check valve that allows current to flow in one direction only. It is formed by joining P-type semiconductor material with N-type material (Figure 12–4). The positive side (P) is called the anode, and the negative side (N) is called the cathode. The area where the anode and cathode material join together is called the **PN junction** or **depletion zone.** (Figure 12–5).

A diode allows the current to flow in one direction only. In a DC circuit, when a diode has the positive volt-

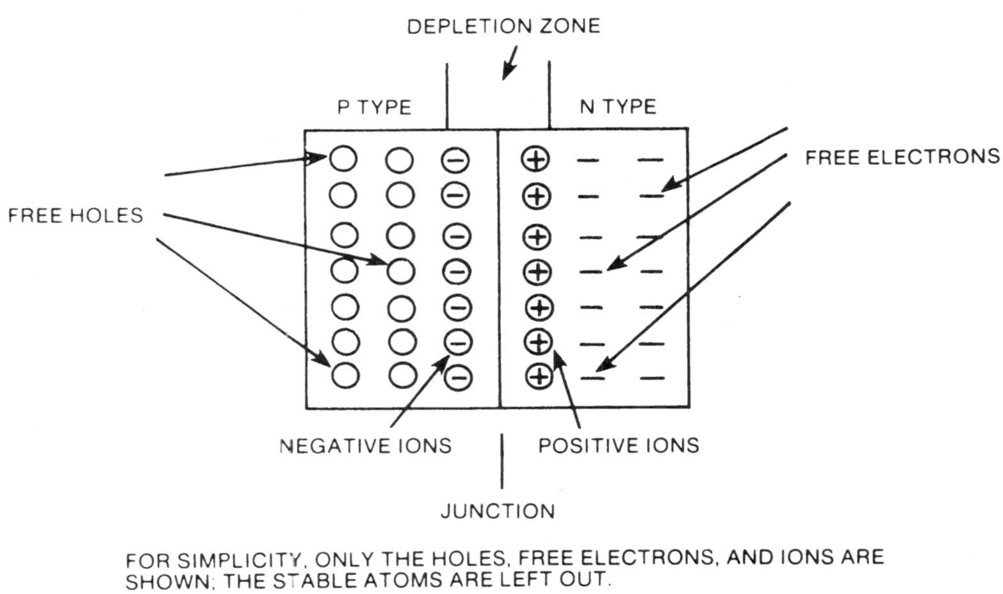

FIGURE 12–5 PN junction of a diode.

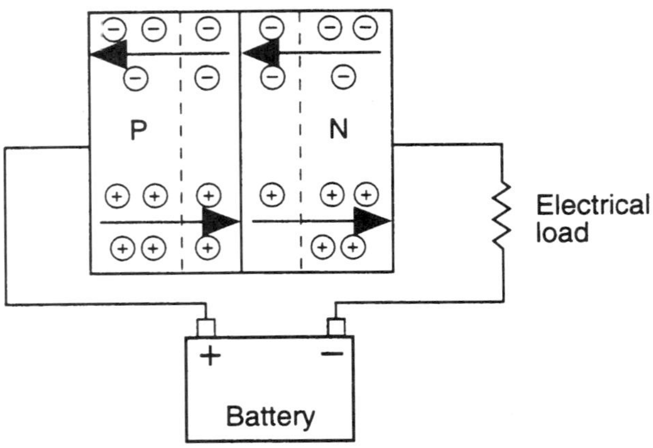

FIGURE 12–6 Diode is forward biased causing current flow.

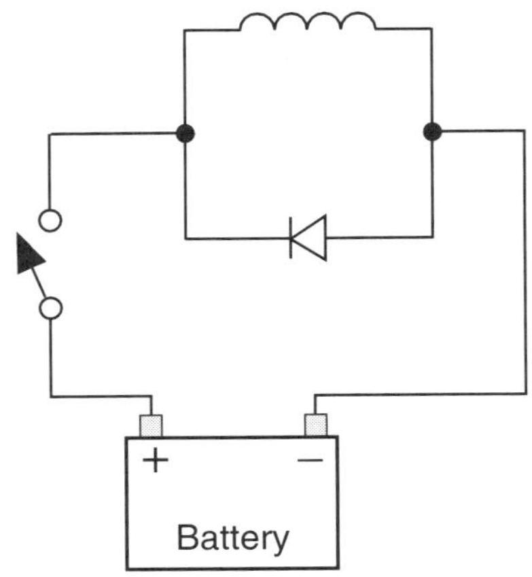

FIGURE 12–8 A clamping diode wired in parallel to a coil prevents voltage spikes when the switch is opened.

age on the P-side or anode, the diode is considered **forward bias** and current will flow through it (Figure 12–6). When a positive voltage is present on the N-side or cathode, the diode is **reverse bias** and current flow is prevented (Figure 12–7).

In an AC circuit, diodes are used to separate the positive and negative voltages. A good example is an AC generator or alternator in a vehicle. The function of diodes in converting AC current to DC current in alternators will be covered in-depth in the charging system chapter of this text.

An application of a diode used as a control device in a circuit is a **clamping diode** (Figure 12–8). When the current flow through a relay coil or solenoid is discontinued, a voltage surge or spike is produced as the collapse of the magnetic field moves across the coil windings. If a reversed-biased diode is wired in parallel to the circuit coil, a by-pass is provided for the electrons in the circuit when the control switch is *opened.* Figure 12–9 is an example of a clamping diode installed in parallel with an air conditioning compressor clutch coil to prevent voltage spikes. Without the clamping diode, when the compressor switch is opened, voltage spikes can reach several hundred volts for a spilt second and damage sensitive electronic control circuits.

Another type of diode that is similar in operation to a regular diode is a **light-emitting diode (LED).** An LED has a small lens built into it and emits light when

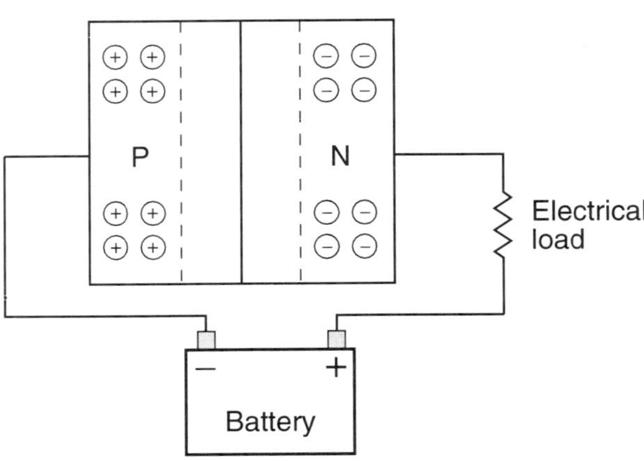

FIGURE 12–7 Diode is reversed biased preventing current flow.

FIGURE 12–9 A clamping diode connected in parallel across the terminals of an air conditioning compressor clutch.

FIGURE 12–10 (A) Light-emitting diode and (B) the symbol for an LED.

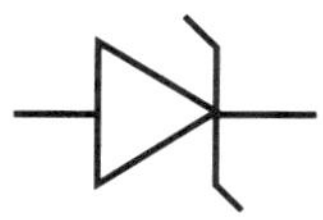

FIGURE 12–11 Typical zener diode symbol.

forward biased with a voltage of 1.5 to 2.5 volts (Figure 12–10). The light from an LED is not heat energy as is the case with other lights, it is electrical energy. LED's are used in instrument cluster digital displays, triggering devices in some ignition systems, and as the lighting source in center high-mounted brake lights.

A **zener diode** is a complex type of diode. They are used to regulate voltage in a circuit by installing them reversed biased (backwards) in a circuit. The depletion zone or PN junction is much narrower in a zener diode, and its barrier voltage becomes very intense when a reverse bias voltage is applied. When a zener diode's specified breakdown voltage level is reached, a breakdown occurs within the PN junction zone and current flows across the diode in reverse direction. Figure 12–11 shows the common symbol for a zener diode, and Figure

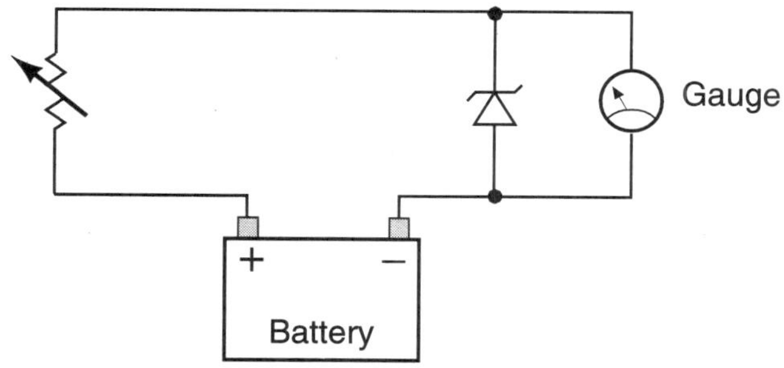

FIGURE 12–12 Zener diode used in a simple gauge circuit to maintain a constant gauge voltage.

12–12 is a simplified gauge circuit with a zener installed to maintain a constant voltage to the gauge. If the gauge voltage must be limited to 9 volts, the 9 volt-specified zener would breakdown and conduct reverse current when the system voltage reached 9 volts. The amount of voltage to the gauge remains 9 volts because the zener diode causes the variable resistor to drop the excess voltage to maintain the voltage limit.

Diode Tests A regular diode can be tested with an ohmmeter or digital multimeter (DMM) equipped with a diode test feature. To test a regular diode for shorted and open conditions with an ohmmeter, connect the meter leads across the diode (Figure 12–13). Observe the ohmmeter reading, then reverse the ohmmeter leads and observe the reading again. There should be a reading close to zero in one direction, and infinite (O/L) in the other direction. A *shorted* defective diode will read close to zero in both directions. An *open* defective diode will read infinite in both directions.

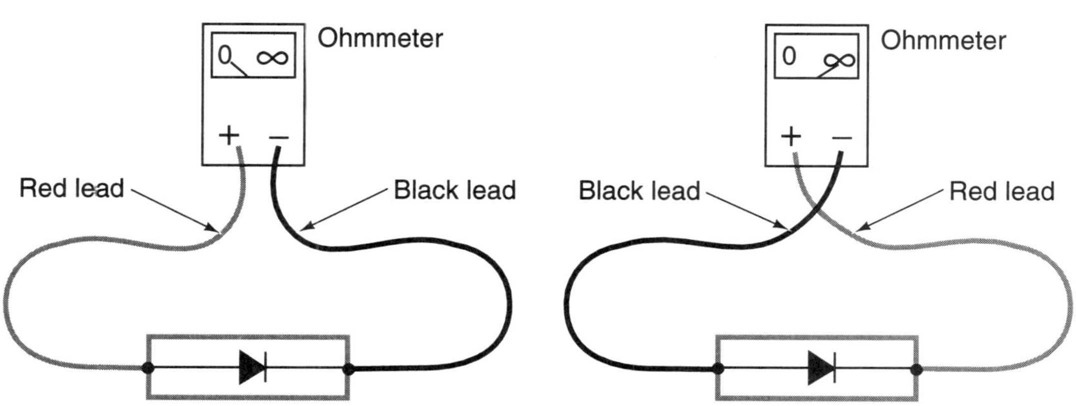

FIGURE 12–13 Testing a diode for an open or short with an ohmmeter.

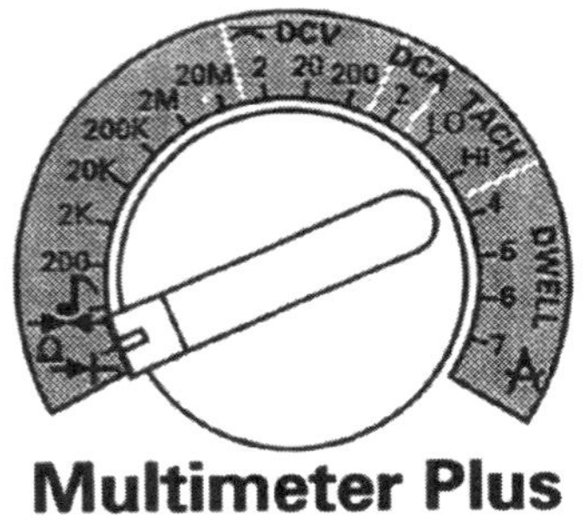

FIGURE 12–14 Diode test function on a DMM.
(Courtesy of Ford Motor Company)

Some high impedance digital multimeters (DMM) can give a false diode test reading when a diode is checked for a forward bias using the ohmmeter function. Many diodes only allow current flow through them when the voltage is at least 0.6 volts, resulting in a diode that shows *open,* when in fact it may not be. Because of this problem, several DMM's are equipped with a diode test function that increases the voltage at the test leads (Figure 12–14).

Transistors The **transistor** is another semiconductor commonly used in solid state circuits throughout the automotive industry. These devices are used in place of mechanical switches and relays to turn circuits on and off. A transistor consists of three layers of semiconductor materials joined together that are designated as **emitter, collector,** and **base.** In effect, a transistor can be thought of as two diodes that share a common center layer (Figure 12–15).

There are commonly two types of transistors used in electronic circuits. The first is an **NPN transistor** that consists of two N-type materials and one P-type material (Figure 12–16). In an NPN transistor, a positive control voltage is applied through a current limiting resistor to the base of the transistor to turn the transistor on. In a *typical* transistor controlled circuit, current flows through the collector and emitter and then through the load device before returning to the negative side of the battery (Figure 12–17).

The second type of transistor is a **PNP transistor** that consists of two P-type materials and one N-type material (Figure 12–18). For a PNP transistor, a negative voltage is applied through a current limiting resistor to the base of the transistor to turn the transistor on. Current then flows from the emitter through the transistor to the collector, then through the load device before returning to the battery (Figure 12–19).

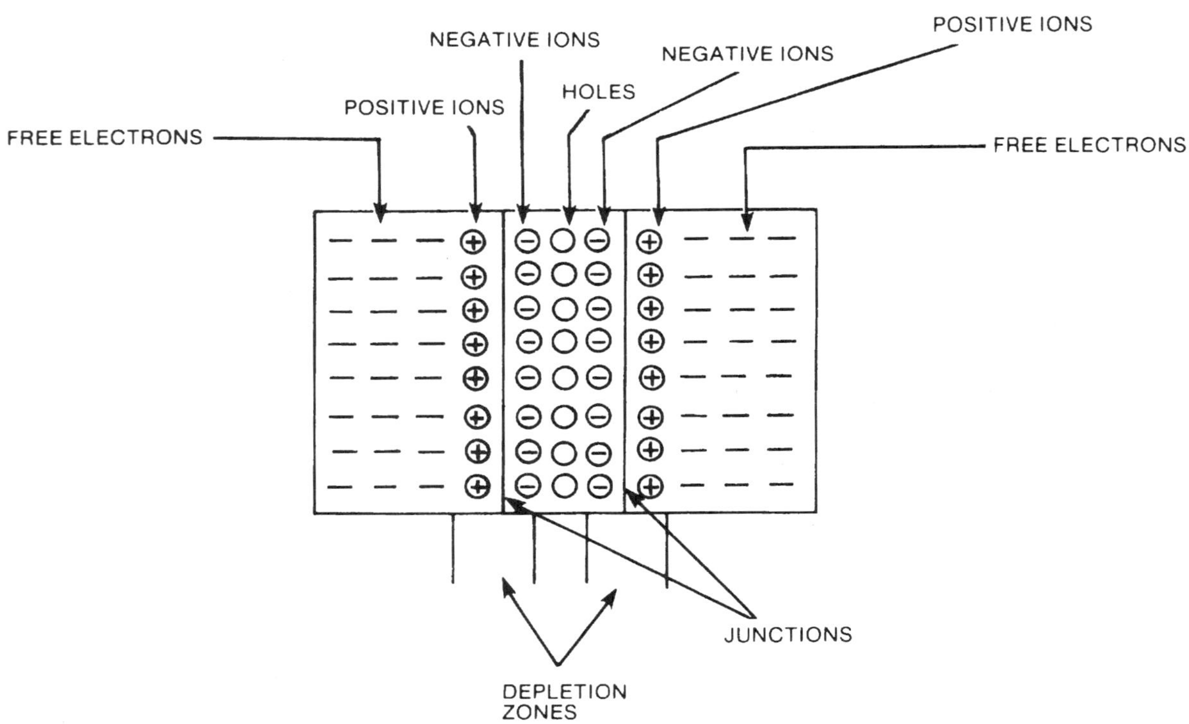

FIGURE 12–15 Unbiased transistor consisting of two diodes that share a common center material.

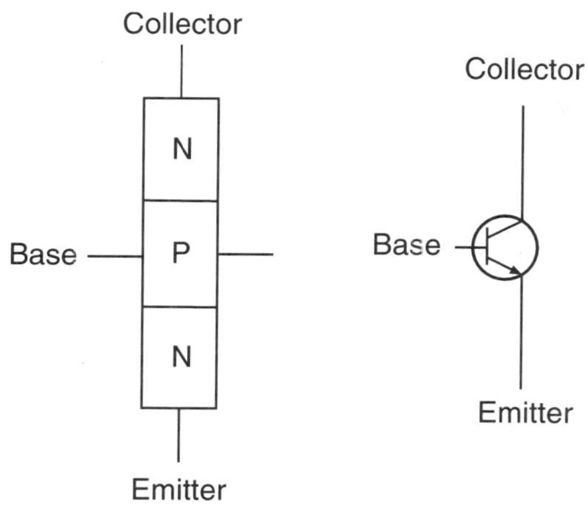

FIGURE 12–16 NPN transistor and its schmatic symbol.

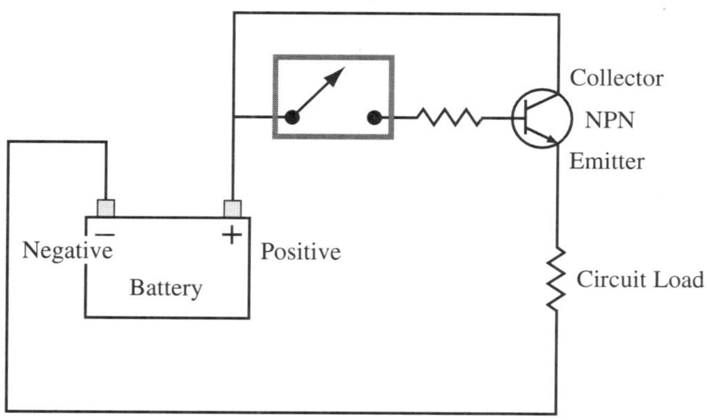

FIGURE 12–17 Simple circuit with NPN transistor.

Automotive Computer Components

Integrated Circuits

A single transistor or diode is limited in its ability to perform complex tasks. However, when many transistors and diodes are combined into a single circuit called an **integrated circuit (IC),** they can perform several simultaneous complex electronic functions. Figure 12–20 is an enlarged illustration of an IC circuit with thousands of transistors, diodes, and resistors. The actual size of an IC can be less than 4 millimeters (Figure 12–21) square. IC chips are constantly shrinking in size, and being designed into vehicles to make logical decisions and issue commands to various sensors and control systems. Some of these systems are the ignition module, fuel injection control, transmission shifting, anti-lock brakes, active sus-

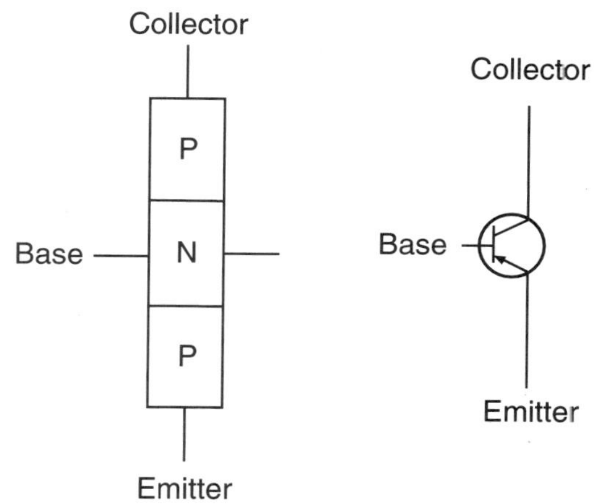

FIGURE 12–18 PNP transistor and its schematic symbol.

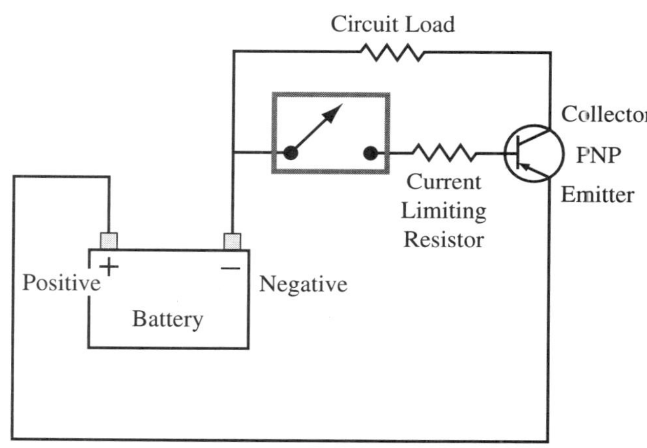

FIGURE 12–19 Simple circuit with a PNP transistor as a load current switch.

pension system, climate controls, and entertainment. With a basic knowledge of the common electronics and IC chips that are used in today's vehicles, the technician will have the diagnostic ability to find faulty electronics.

Basic Voltage Signals

Analog Voltage Signal An **analog voltage signal** is usually produced by an input sensor that has a continuously variable output within a certain voltage range (Figure 12–22). For example, if a rheostat is used to manually pulse a 5 volt bulb on and off, then the corresponding sine wave the rheostat produces would be an analog voltage signal with a 5 volt range. Some of the sensors in automotive computer systems produce analog voltages.

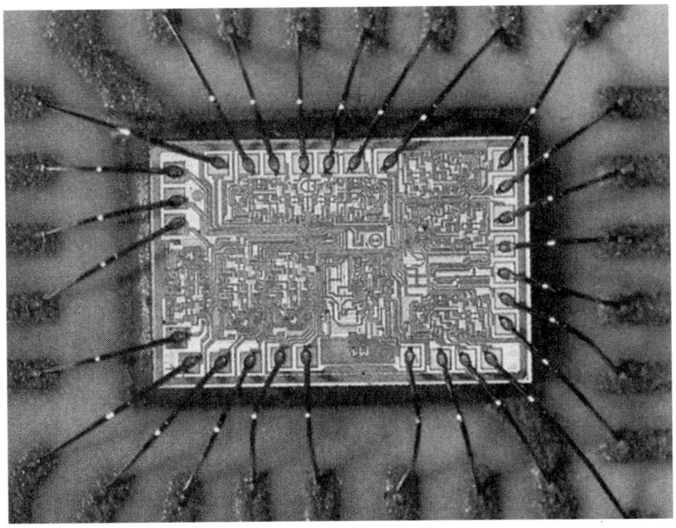

FIGURE 12–20 Greatly enlarged integrated circuit chip from a fuel injection module.

(Courtesy of Robert Bosch Corporation)

Digital Voltage Signal If an on/off switch is connected *before* a 5 volt bulb and the switch is off, then zero volts are available at the bulb. When the switch is turned on, a 5 volt signal is sent to the bulb to light it to full brilliance. Switching this circuit on and off rapidly

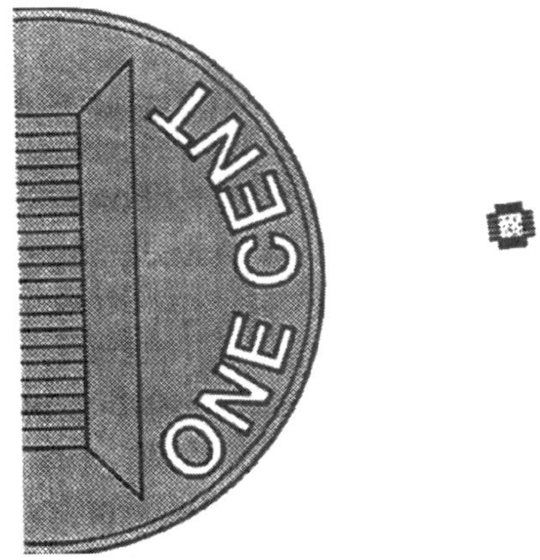

FIGURE 12–21 Size comparison of a typical IC chip.

(Courtesy of Ford Motor Company)

would produce a **digital voltage signal,** also known as a **square wave signal** (Figure 12–23).

The digital voltage signal would be reversed if the on/off switch and a scope are connected *after* a 5 volt bulb (Figure 12–24). When the switch is off (open), 5 volts are available through the bulb to the scope and to the switch. When the switch is turned on (closed), the bulb glows and 0 volts are available to the scope after the bulb.

Digital voltage signals are used to control various relays and components in a vehicle. The computer can

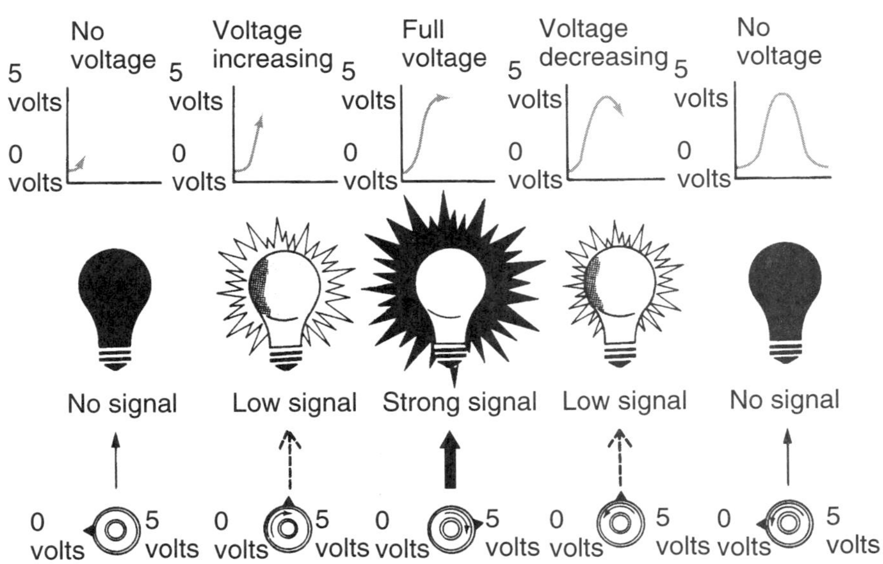

FIGURE 12–22 Typical analog voltage signal.

(Courtesy of Ford Motor Company)

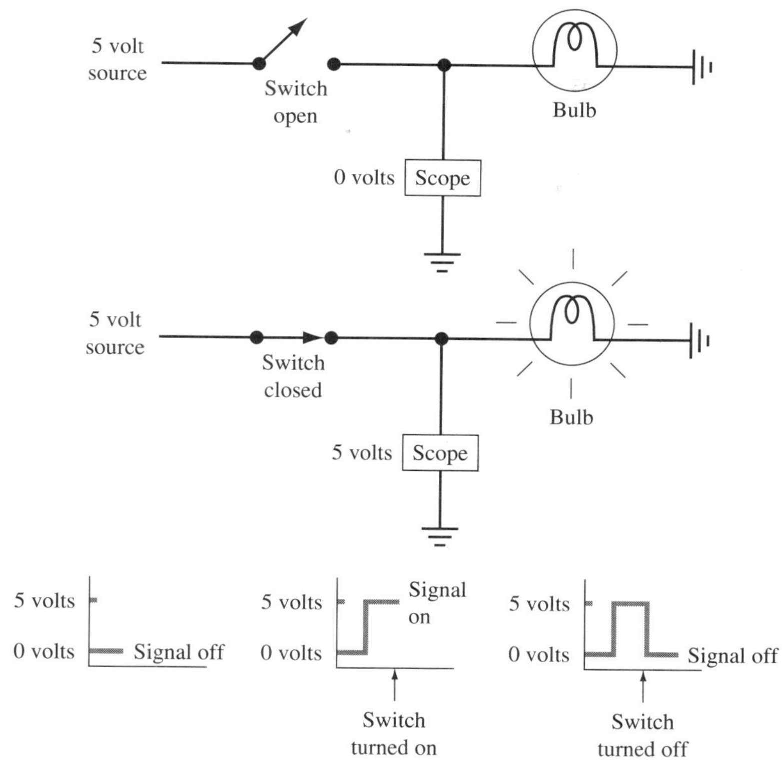

FIGURE 12–23 Typical digital signal voltage.

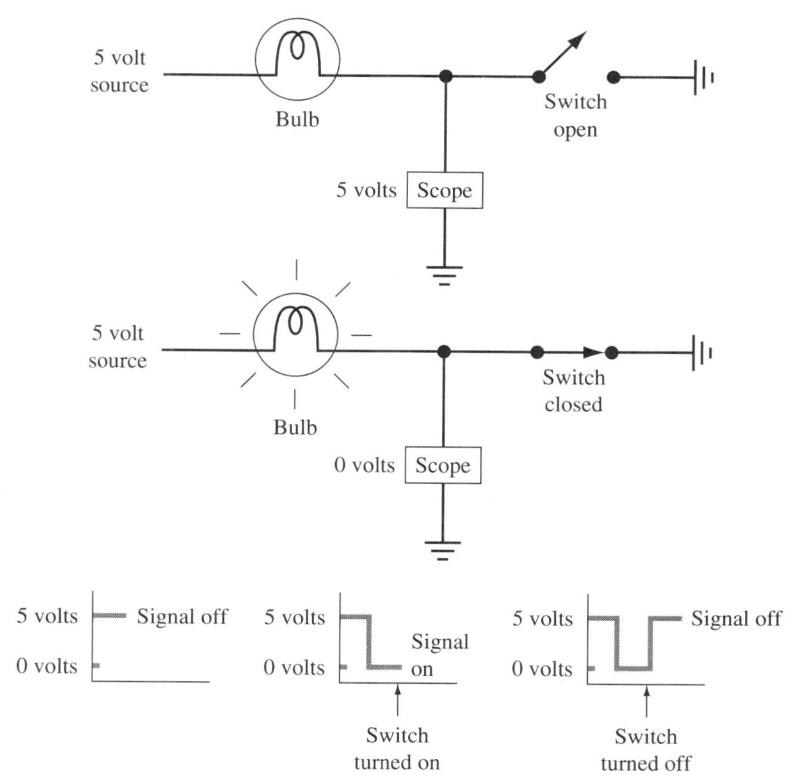

FIGURE 12–24 Reversed digital signal voltage.

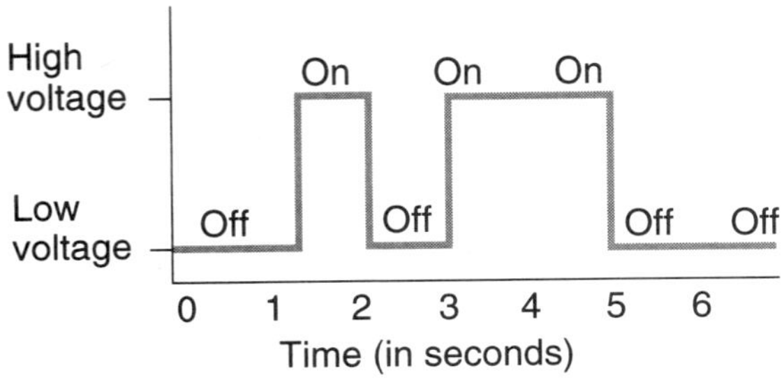

FIGURE 12–25 Variable digital signal voltage.
(Courtesy of Ford Motor Company)

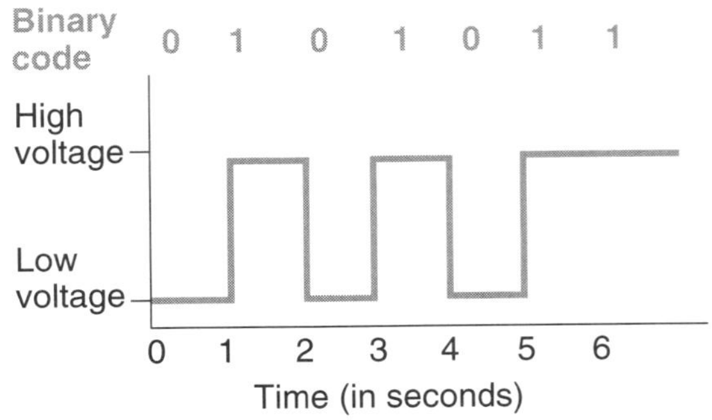

FIGURE 12–26 Assigning a binary code to a digital signal voltage.
(Courtesy of Ford Motor Company)

vary the length of time the digital signal is *high* or *low* for precise component control (Figure 12–25). Depending on the circuit, some digital systems monitor a circuit before a load device, and some after a load device.

Binary Code When a digital signal is on (or off) it is considered HIGH (or LOW). A digital low signal is assigned the numeric value of zero, and a digital high signal is given the numeric value of one. The assignment of numeric values to digital signals is called **binary code** (Figure 12–26). In an automotive computer, all communications are transmitted in binary code form. Each 1 or 0 represents a **bit** of information. The sixteen possible combinations of a group of 4 bits is represented by the numbers 0(0000) to 9, and the letters A to F(1111). Eight bits of information is equal to a **byte.**

Amplification Some input sensors produce a very low output voltage signal of less than 1 volt, along with a low current flow. This type of input signal must be **amplified** through an amplification circuit to increase the strength of the signal so the computer can adequately interpret the information received (Figure 12–27).

Because most of the signals sent to the computer are analog signals, they have to be converted to a digital

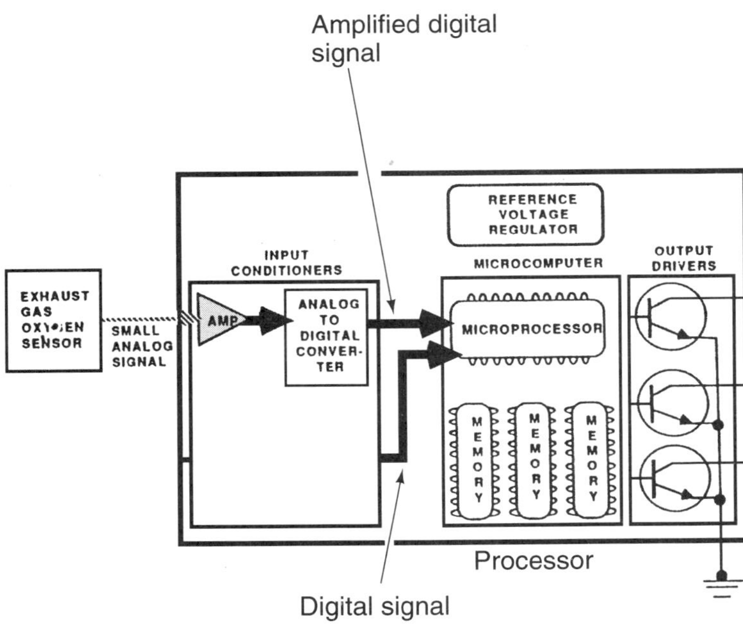

FIGURE 12–27 Use of an amplifier to strengthen the weak analog signal.
(Courtesy of Ford Motor Company)

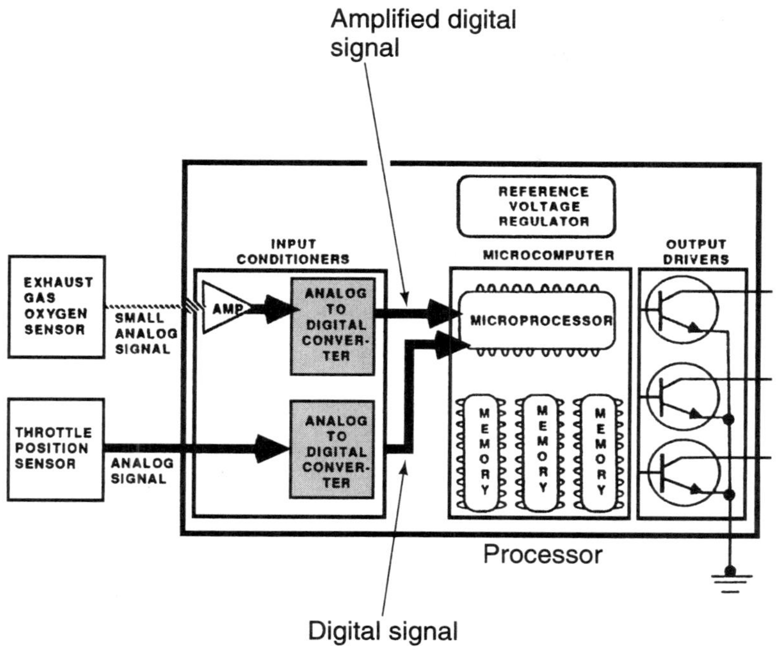

FIGURE 12–28 Analog-to digital (A/D) converter in a computer.

(Courtesy of Ford Motor Company)

signal to be understood by the computer's software program. This is done through an **analog-to-digital (A/D) converter** (Figure 12–28). The A/D converter continually scans the input signals and assigns numeric values to these voltages. Voltages that are above a given value are converted to a 1, and voltages below a given value are assigned a value of 0. In Figure 12–29, the A/D converter changes a series of signals to a binary number made up of 1's and 0's. When eight bits are combined in a specific sequence, they form a byte or word that makes up the basis of the computer's software.

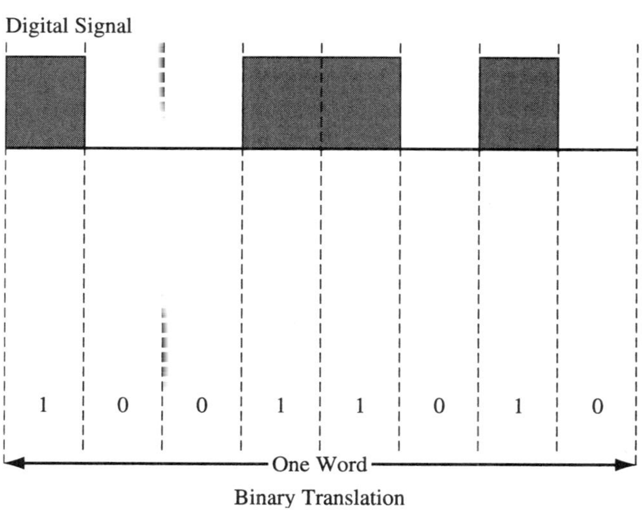

FIGURE 12–29 A/D conversion of a signal and assigning binary code numbers.

(Courtesy of General Motors Corporation, Service Technology Group)

Microprocessors

In an automotive computer, the **microprocessor** is the decision making chip in the computer that contains a large number of transistor and diode switches in an IC that is capable of producing many digital voltage signals per second. The chip containing the silicon IC is mounted in a flat rectangular protective package with metal connecting pins (Figure 12–30).

Computer Memory

When the microprocessor interprets input information, makes calculations, and makes decisions, it requires the use of three specific computer memories (Figure 12–31).

The first memory is called **random access memory (RAM).** It is the note pad for a computer. Information that requires temporary storage is transferred to the RAM memory for use by the microprocessor to read, write, and erase information in relationship to the vari-

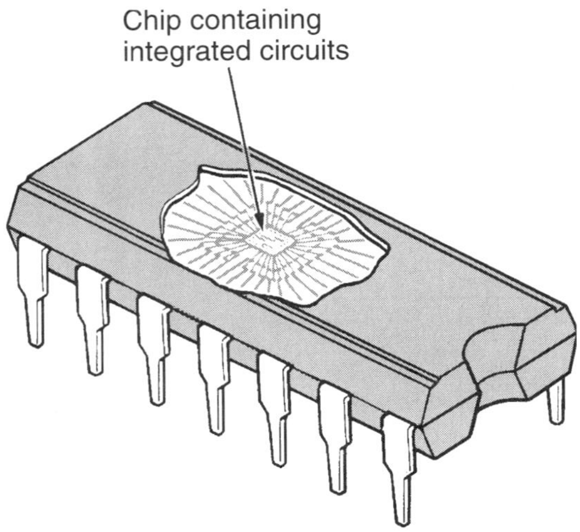

FIGURE 12–30 Typical microprocessor chip with IC mounted in a protective package.

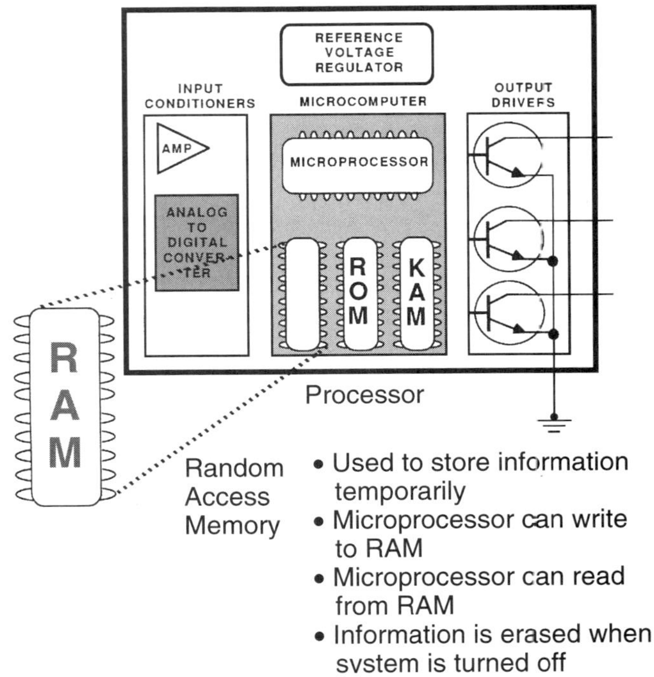

FIGURE 12–32 Random access memory (RAM).

(Courtesy of Ford Motor Company)

ous operating conditions of the vehicle (Figure 12–32). There are two types of RAM memories, volatile and nonvolatile. A volatile RAM maintains information as long as it is connected to a battery power source. The only time the RAM will lose its stored information is if the power source is disconnected. It is also referred to as a **keep alive memory (KAM)** (Figure 12–33). KAM allows variations in the options a vehicle uses, such as electronic transmission shifting patterns, adaptive fuel trim for air/fuel ratios, and idle air control. A nonvolatile RAM will *not* lose its stored memory if the source power

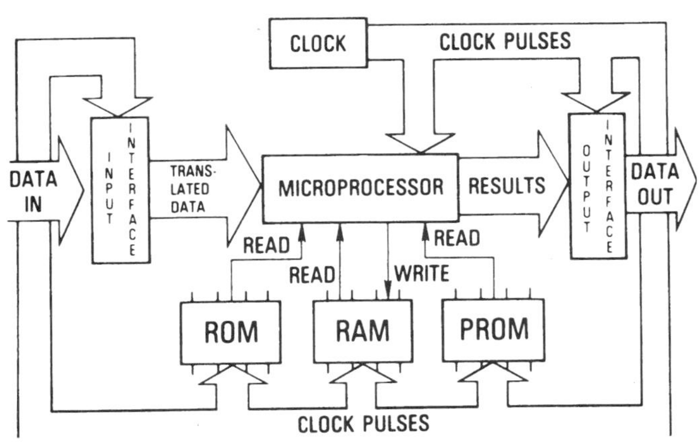

FIGURE 12–31 Various computer memories and their interaction with a microprocessor.

(Courtesy of General Motors Corporation, Service Technology Group)

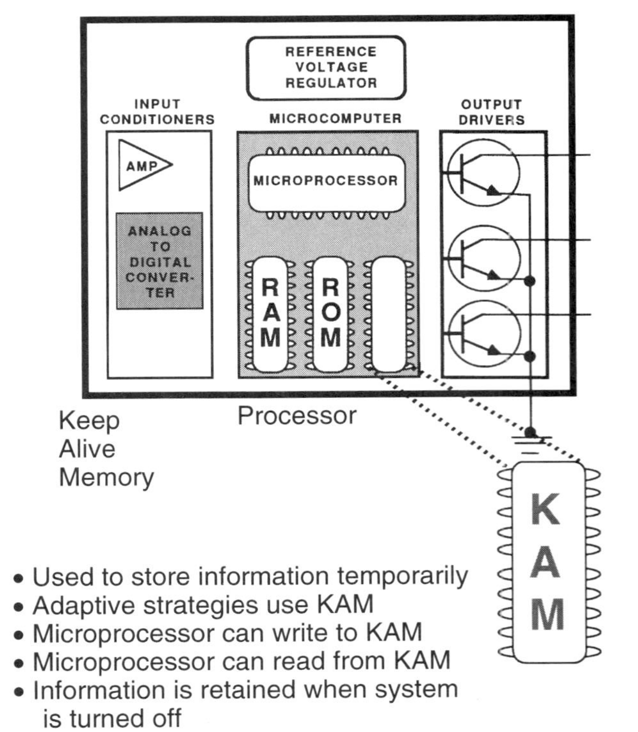

FIGURE 12–33 Keep alive memory (KAM).

(Courtesy of Ford Motor Company)

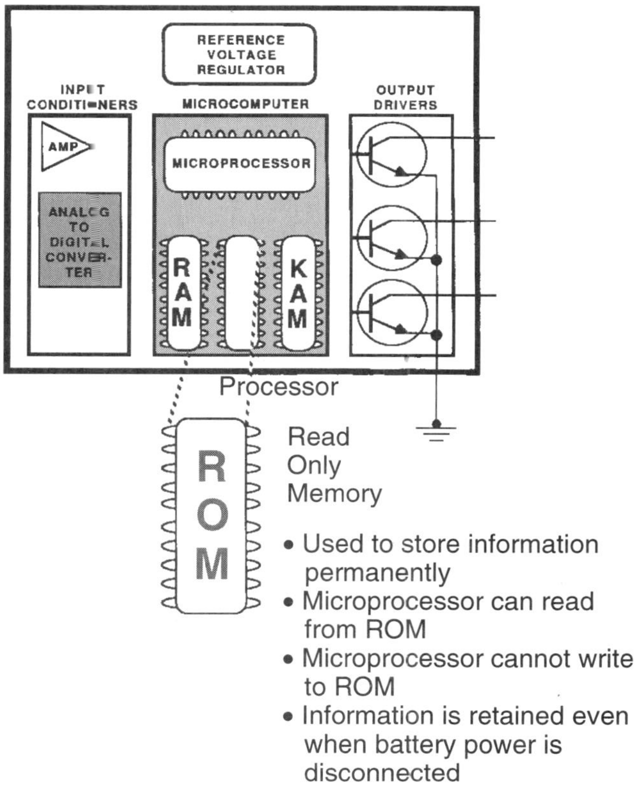

FIGURE 12–34 Read only memory (ROM).
(Courtesy of Ford Motor Company)

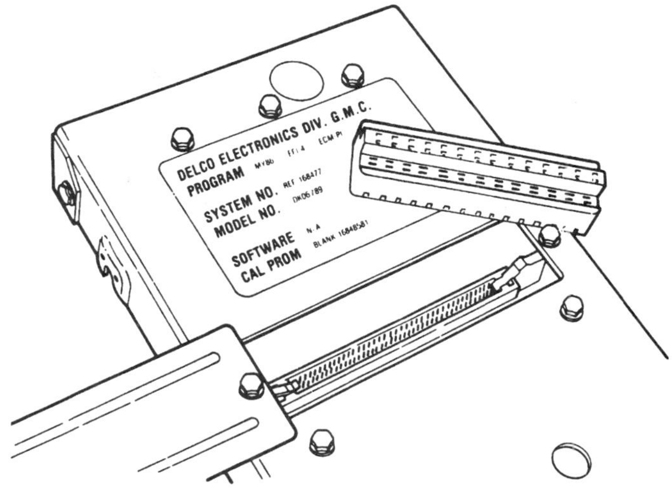

FIGURE 12–35 PROM location in an electronic control module (ECM).
(Courtesy of General Motors Corporation, Service Technology Group)

is lost. Nonvolatile RAM is usually used to store information on vehicles with digital odometers.

The second type of memory is called **read only memory (ROM)** which contains information like a textbook that is permanently stored for use by the microprocessor to determine what to do with the input data (Figure 12–34). The ROM contains specifications and calibration information of various components and systems under a variety of engine operating conditions. The microprocessor compares sensor inputs to the pre-programmed look-up tables and specifications stored in the ROM, and then makes decisions for controlling various engine functions.

Some computers have a removable **programmable read only memory (PROM),** that contains specific calibration programs that pertain to a vehicle model (Figure 12–35). The PROM can be changed easily if revised information is available from the manufacturer. On some of the newer computers the PROM has been changed to an **electronic erasable programmable read only memory (EEPROM).** This type of memory chip can only be reprogrammed with special test equipment available from the vehicle manufacturer.

A basic overview of the systematic processing, and conversion of input information into digital signals, that control various circuits can be seen in Figure 12–36.

Logic Gates

For the computer's microprocessor to make the most informed decisions regarding the operation of various systems, sensor inputs may be sent through various **logic gates,** which incorporate thousands of **field effect transistors (FET)** into the computer's circuitry. A logic gate controls the output voltage signal depending on the different combinations of input signals.

There are five common logic gates used to perform the processing functions within the computer.

1. **NOT Gate** A NOT gate simply reverses binary 1's and 0's and vice versa (Figure 12–37). A high input to a NOT gate results in a low output. A NOT gate is sometime referred to as an inverter.
2. **AND Gate** The AND gate has at least two inputs and one output. The operation of the AND gate is similar to two switches in a series line to a load

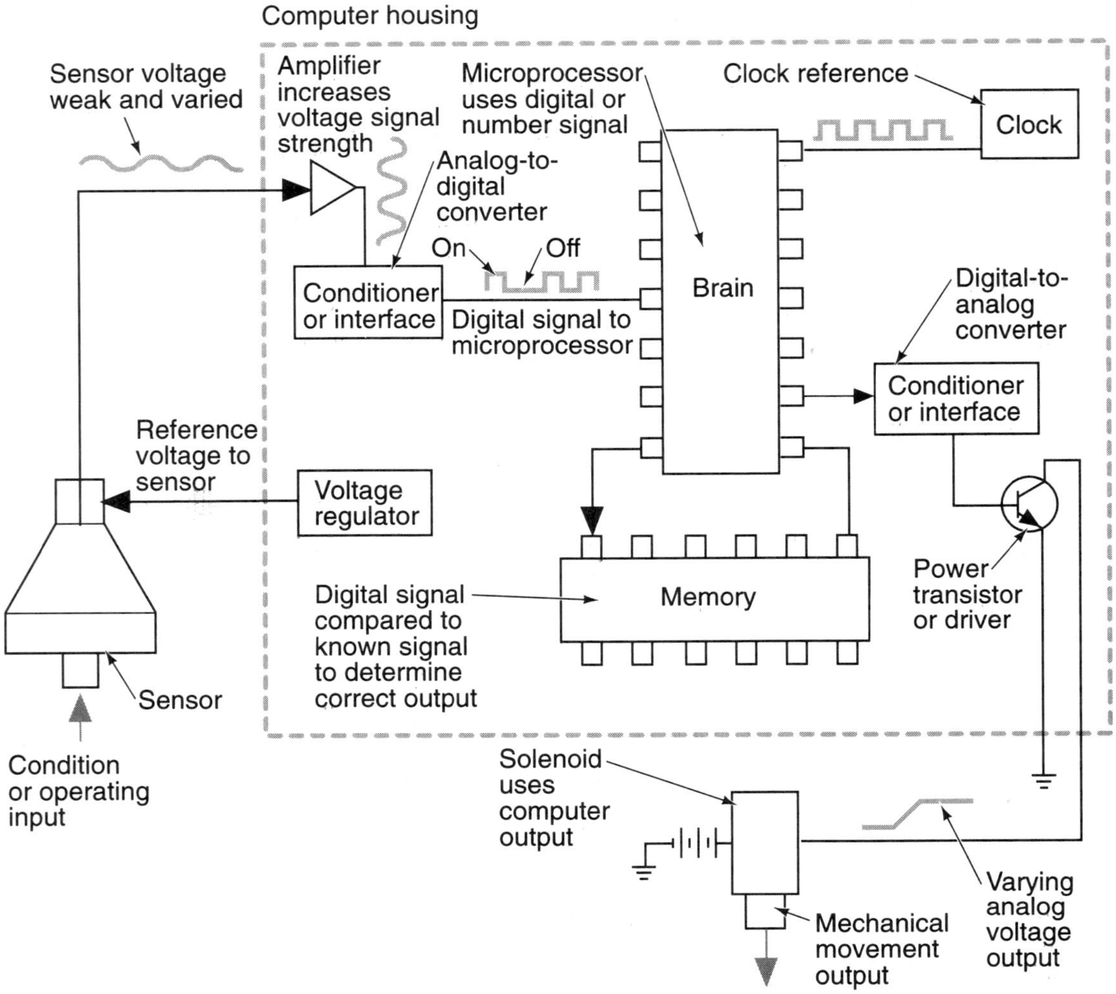

FIGURE 12–36 Computers in automobiles consist of the ability to process analog sensor information and generate digital signals that control various output circuits within specific parameters.

(Courtesy of General Motors Corporation, Service Technology Group)

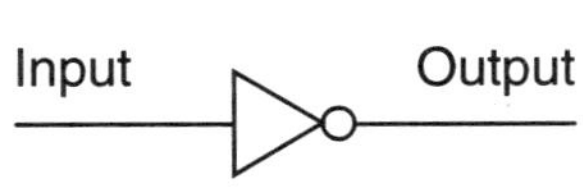

Truth table	
Input	Output
0	1
1	0

FIGURE 12–37 The NOT gate symbol and truth table which shows the NOT gate inverting the input signal. The small circle on the symbol represents an inverting output.

(Figure 12–38). The only way the light will turn on is when both switches are closed. The output of the AND gate will be high only if both inputs are high.

3. **OR Gate** The OR gate operates similarly to two switches that are wired together in parallel (Figure 12–39). If switch A or B is closed, the light will turn on. A high signal to either input will result in a high output.

4. **NAND and NOR Gates** A NAND (or NOR) Gate is the equivalent of an AND (or OR) gate with an inverted output signal (Figure 12–40).

5. **XOR Gate** An Exclusive OR gate is combination of NAND and NOR gates that will produce a

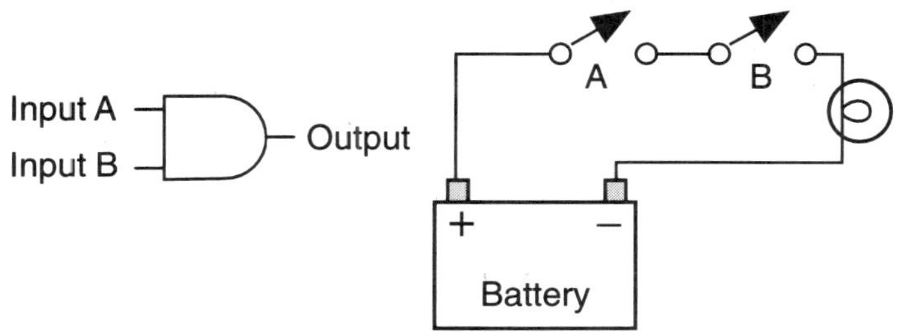

Truth table		
A	B	Output
0	0	0
0	1	0
1	0	0
1	1	1

FIGURE 12–38 The AND gate symbol and truth table.

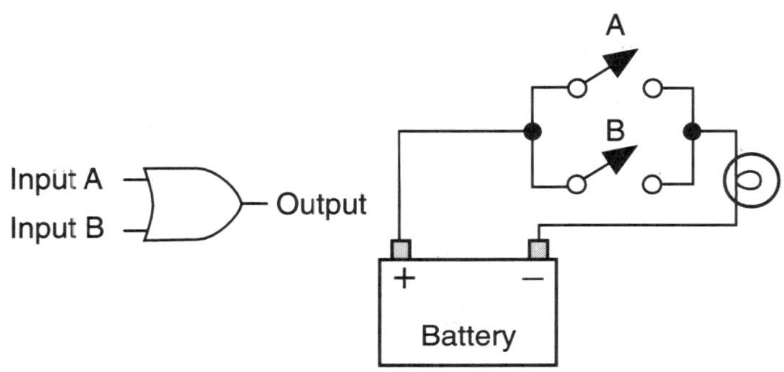

Truth table		
A	B	Output
0	0	0
0	1	1
1	0	1
1	1	1

FIGURE 12–39 The OR gate symbol and truth table.

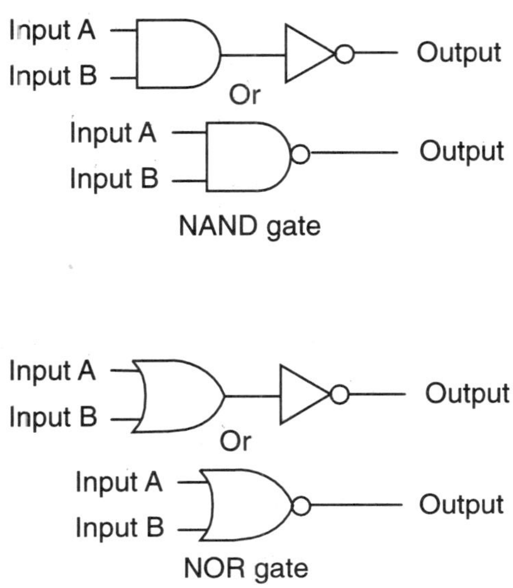

Truth table		
A	B	Output
0	0	1
0	1	1
1	0	1
1	1	0

Truth table		
A	B	Output
0	0	1
0	1	0
1	0	0
1	1	0

FIGURE 12–40 The NAND and NOR gate symbols and truth tables. The small circles represent an inverted output on any logic gate symbol.

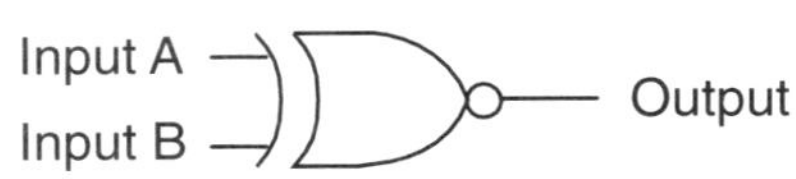

Truth table		
A	B	Output
0	0	0
0	1	1
1	0	1
1	1	0

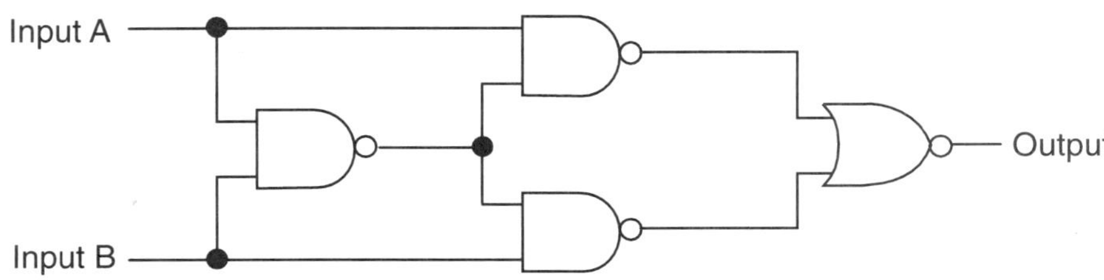

FIGURE 12–41 The XOR gate symbol and truth table.

high output signal only if the inputs are different (Figure 12–41).

The application of the five gate circuits can be combined to form almost any type of logic circuit needed. Figure 12–42 shows an example of a power window circuit combining multiple gate circuits. The input and output characteristics of each gate circuit can help the technician determine the output functions of the overall circuit.

Computer Sensors

Operation of an automotive computer requires several input and output sensors and control actuators (Figure 12–43). **Input sensors** transmit information to the control computer on coolant temperature, air inlet temperature, manifold pressure, barometric pressure, throttle position, vehicle speed, and many others. After gathering, processing, and interpreting the information from the input sensors, the control computer will provide output commands to actuators and controls such as electronic spark timing, air conditioning, emissions devices, and transmission shifting.

Input Sensors

Input sensors are divided into two general categories—sensor *switches* that produce on or off information, and *variable* sensors that provide a wide range of input readings. Input sensor voltage is referred to as the **reference voltage.** This voltage is normally 5 volts or 12 volts depending on the sensor and circuit.

In the first category, there are two common types of input sensor switches used in circuits. The first is a switch-to-power or **pull-up switch** (Figure 12–44). This sensor switch circuit uses an *external* power source to the switch, and a ground circuit *internal* to the control computer. When the signal switch is closed, a *high* reference signal from the external source voltage is sent to the computer. When the switch opens the reference signal is *low.* A pull-up switch can be found on an air conditioning (AC) system to signal the control computer that AC is requested and idle speed should be increased to compensate for the additional engine load.

The second type of input sensor switch is a switch-to-ground or **pull-down switch** (Figure 12–45). This sensor switch circuit uses an *internal* power source from the computer to the switch, and an *external* ground circuit from the switch. When the signal switch is closed a *low* reference signal is sensed in the computer. When the switch opens, the reference signal is *high.* A pull-down switch can be found in a park/neutral transmission switch (Figure 12–46). When the transmission is shifted into gear from park or neutral, the idle speed is increased to compensate for the additional engine load.

In the second category, a variable sensor provides voltage input signals that vary with the control device. An example of these sensors are oxygen sensors, engine

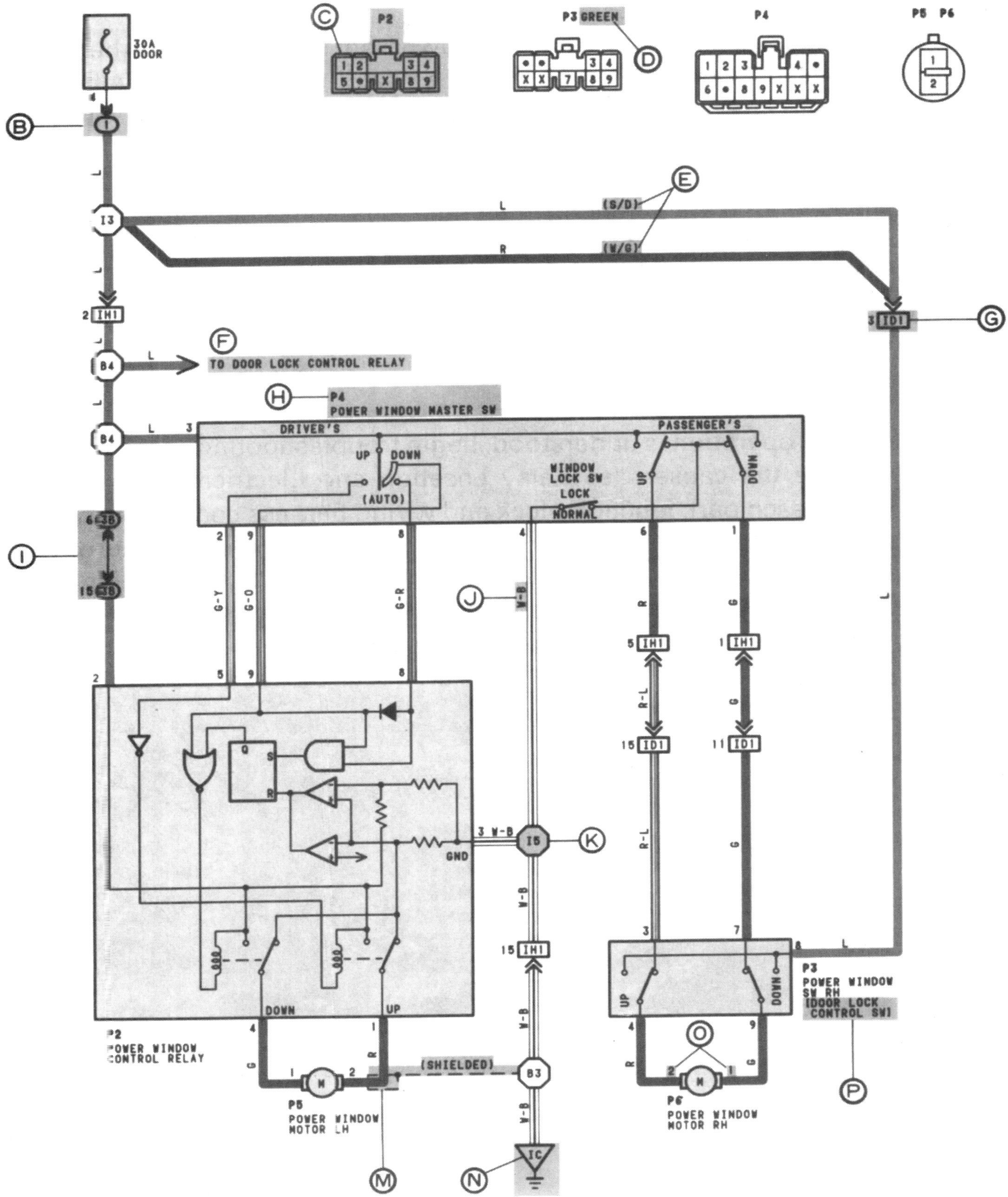

FIGURE 12–42 Power window control circuit combining multiple gate circuits.

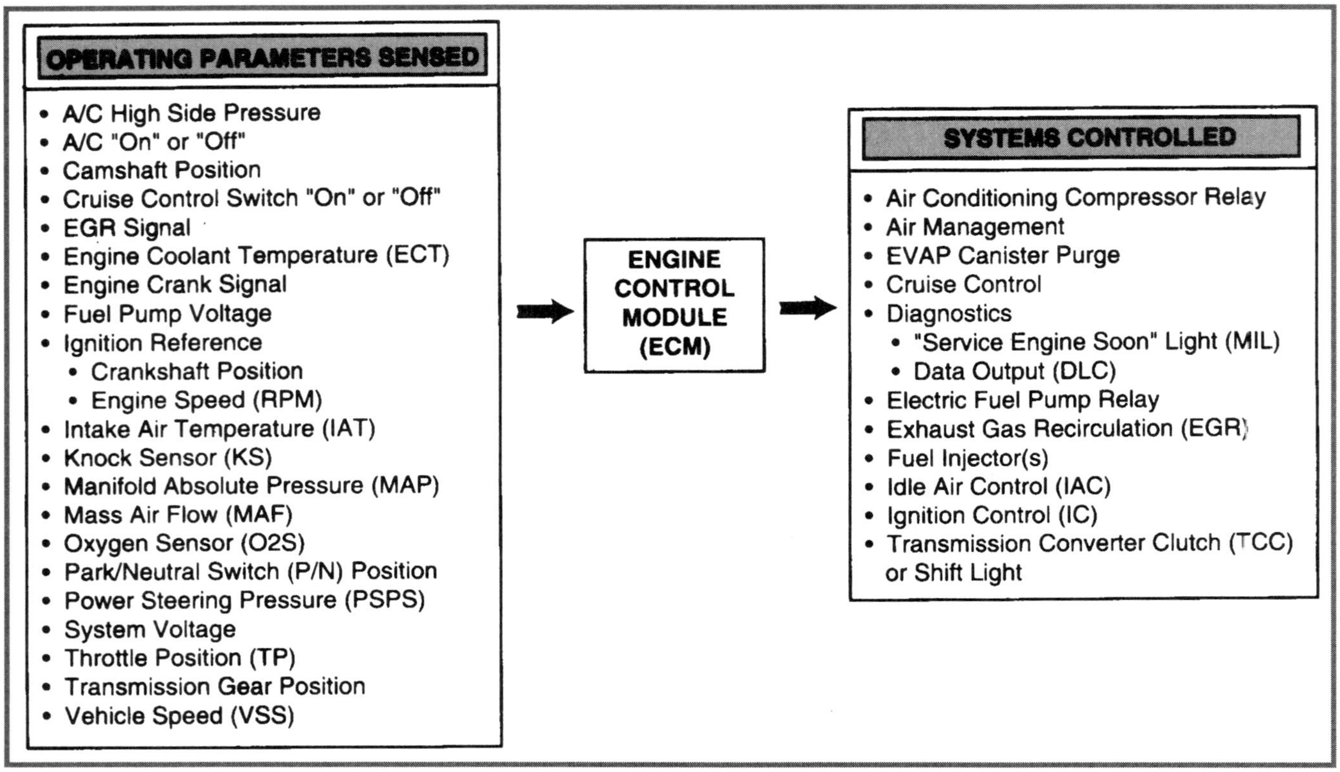

FIGURE 12–43 Typical computer input an output sensors and controls.
(Courtesy of General Motors Corporation, Service Technology Group)

coolant sensor, and throttle position sensor. This type of sensor can be rheostats or thermistors with one or two wires, or potentiometers that contain three wires.

Oxygen Sensors (O_2S) A typical single wire variable sensor is an **oxygen sensor (O_2S)** (Figure 12–47). This sensor uses one wire to carry the 450 millivolt reference voltage from the engine control module (ECM) through the sensor body to ground. The sensor body is constructed of a ceramic **zirconium dioxide** element (Figure 12–48) that reacts with the oxygen content in the exhaust gases. When the exhaust gas with low oxygen

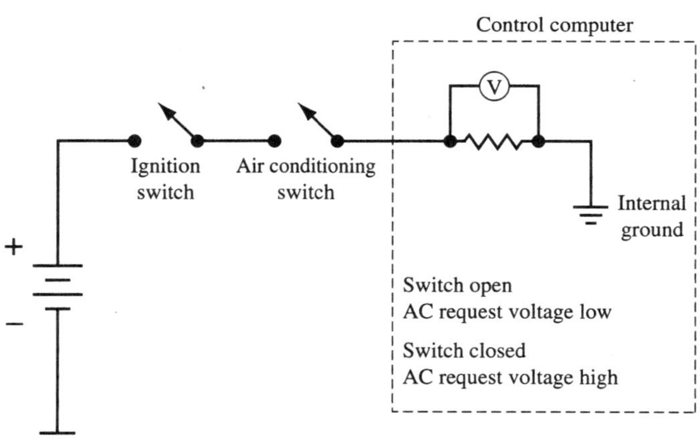

FIGURE 12–44 Typical pull-up switch sensor circuit.

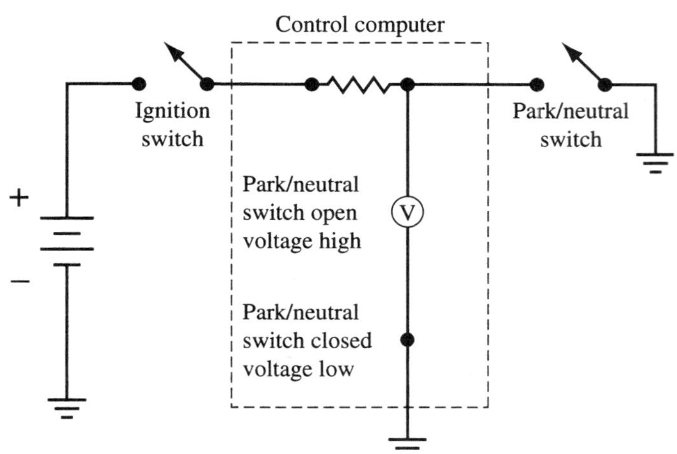

FIGURE 12–45 Typical pull-down switch sensor circuit.

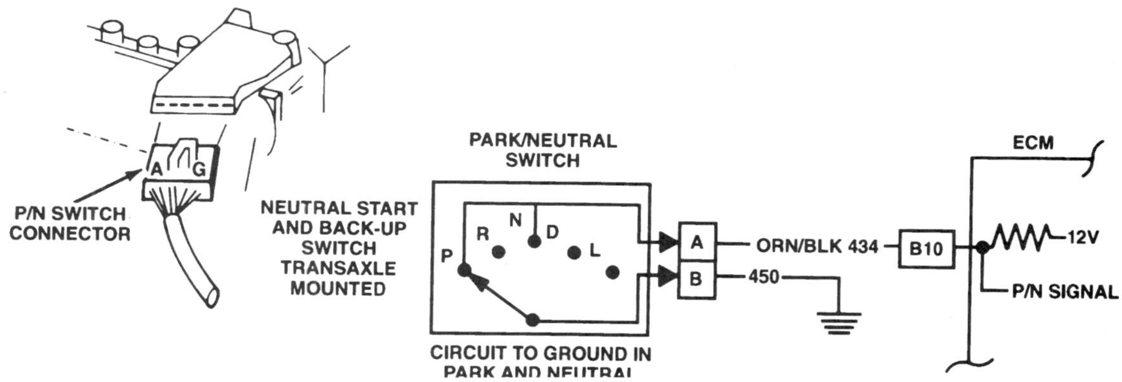

FIGURE 12–46 Typical pull-down park/neutral transmission switch.
(Courtesy of General Motors Corporation, Service Technology Group)

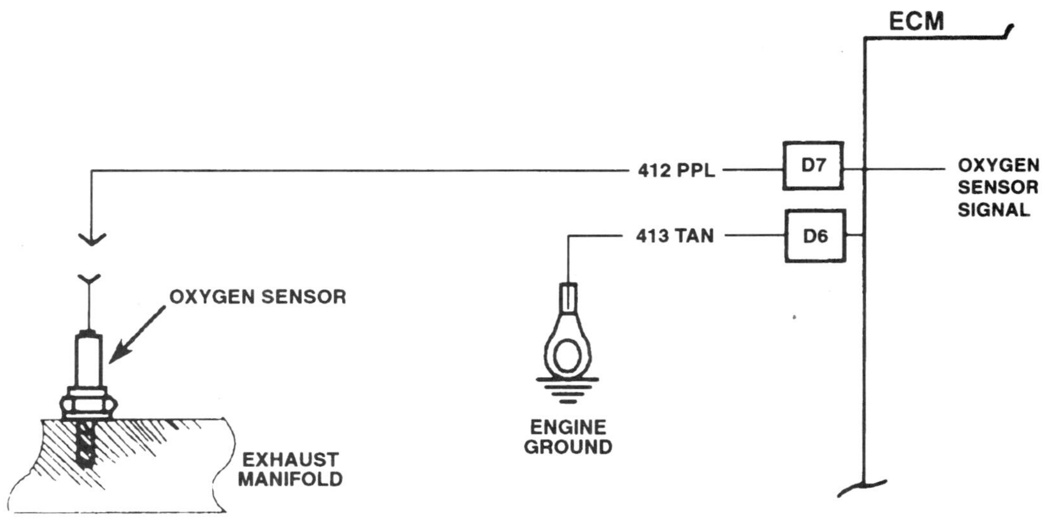

FIGURE 12–47 Single wire oxygen sensor.
(Courtesy of General Motors Corporation, Service Technology Group)

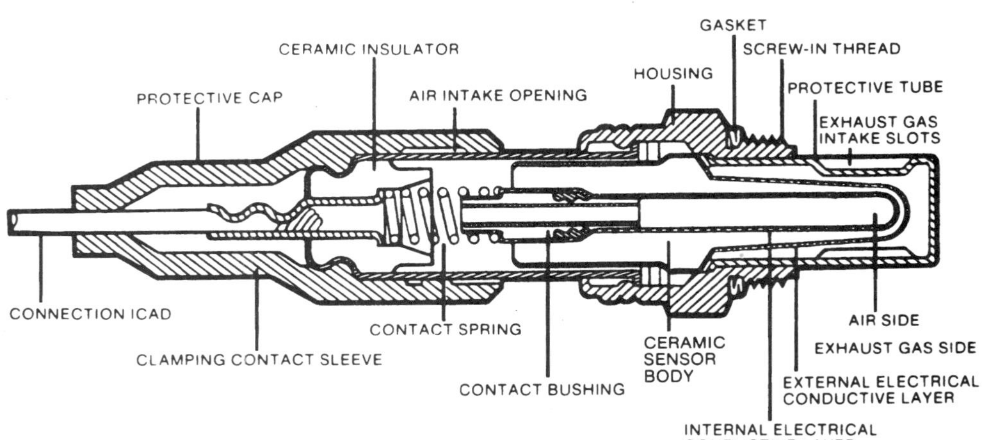

FIGURE 12–48 Ceramic zirconium dioxide oxygen sensor.
(Courtesy of Ford Motor Company)

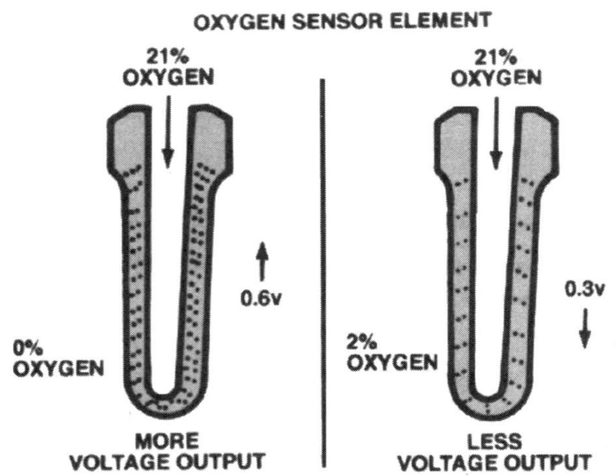

FIGURE 12–49 Reaction of oxygen content in exhaust gas compared to atmosphere.

(Courtesy of General Motors Corporation, Service Technology Group)

content strikes the O_2 sensor, there is a high oxygen content inside the sensor element from the atmosphere. This O_2 difference produces up to 1 volt for a rich mixture, and 0.3 volts for a lean mixture (Figure 12–49).

The atmosphere contains approximately 21% oxygen. The ideal air/fuel mixture is 14.7:1 which is referred to as a **stoichiometric** mixture. This indicates that for every 14.7 pounds of air entering the engine, the fuel system supplies 1 pound of fuel. During normal engine operation the O_2 sensor provides an output voltage to the ECM that continually fluctuates in the closed-loop mode from a *high* (above 450 millivolts) and *low* (below 450 millivolts) (Figure 12–50). When observed on a scope the variable voltage would look like Figure 12–51.

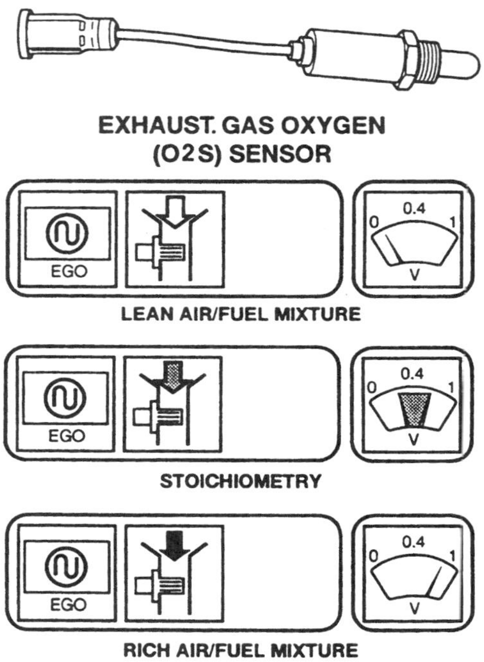

FIGURE 12–50 Output voltage of oxygen sensor in an engine.

(Courtesy of Ford Motor Company)

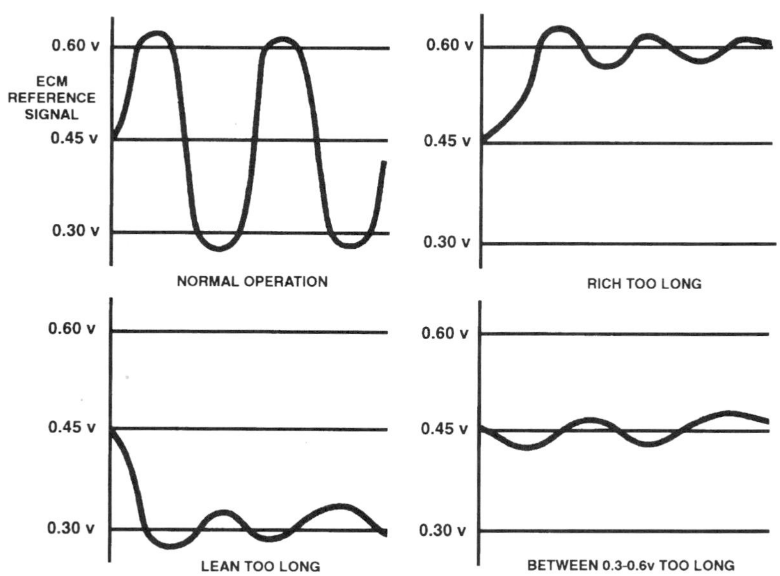

FIGURE 12–51 Oxygen sensor voltage comparison on an oscilloscope.

(Courtesy of General Motors Corporation, Service Technology Group)

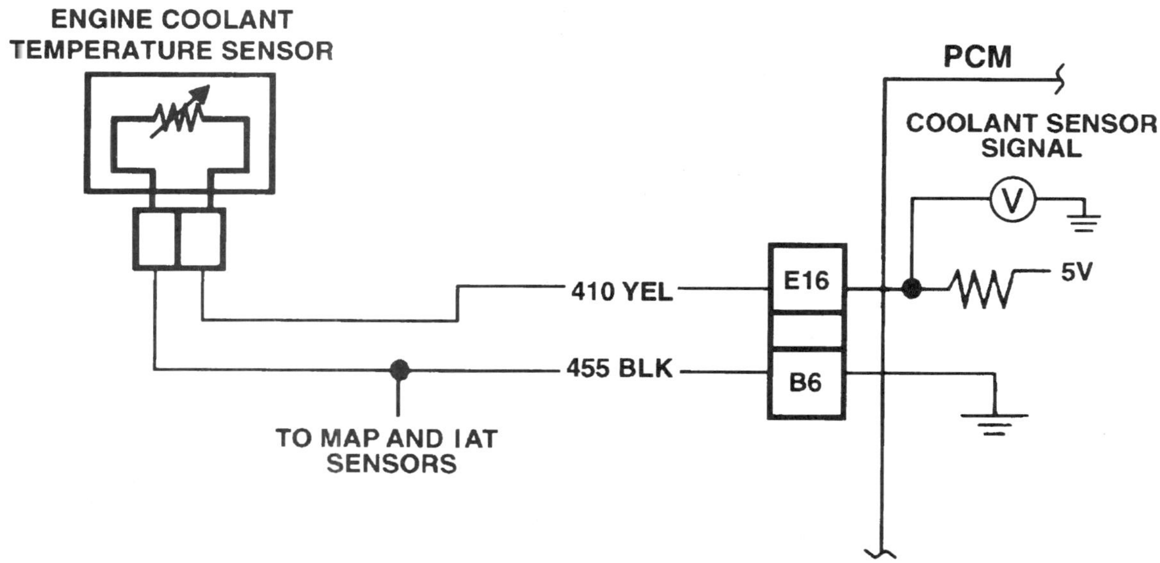

FIGURE 12–52 Two wire engine coolant temperature sensor.
(Courtesy of General Motors Corporation, Service Technology Group)

Engine Coolant Temperature Sensor The **engine coolant temperature (ECT)** sensor is a two wire thermistor sensor (Figure 12–52) and is normally located in a threaded fitting in the intake manifold coolant passage. The ECT provides the ECM with engine coolant temperature information by providing a high resistance, higher voltage signal when cold, and low resistance, lower voltage reading at higher temperature (Figure 12–53) (Figure 12–54).

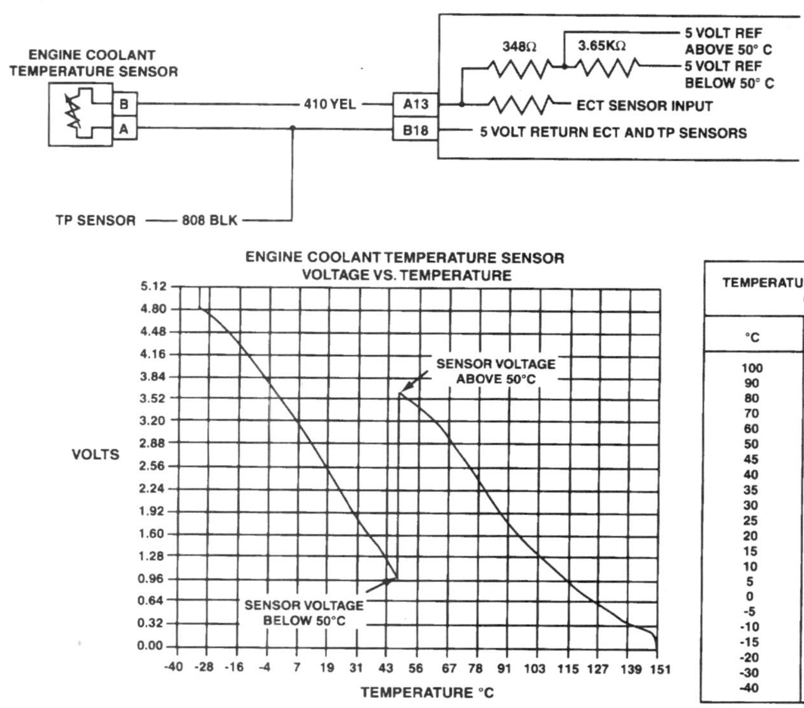

TEMPERATURE VS. RESISTANCE VALUES (APPROXIMATE)		
°C	°F	OHMS
100	212	177
90	194	241
80	176	332
70	158	467
60	140	667
50	122	973
45	113	1188
40	104	1459
35	95	1802
30	86	2238
25	77	2796
20	68	3520
15	59	4450
10	50	5670
5	41	7280
0	32	9420
-5	23	12300
-10	14	16180
-15	5	21450
-20	-4	28680
-30	-22	52700
-40	-40	100700

FIGURE 12–53 Engine coolant temperature (ECT) sensor operating parameters.
(Courtesy of General Motors Corporation, Service Technology Group)

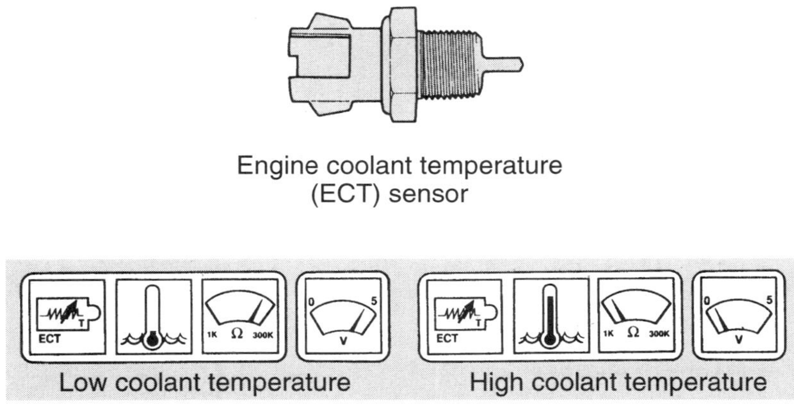

FIGURE 12–54 Engine coolant temperature (ECT) sensor resistance and voltage drop.

(Courtesy of Ford Motor Company)

Engine coolant temperature is a key input for the ECM to calculate several output operating functions. Some of these are air/fuel delivery, idle speed, cooling fan operation, emissions regulation, and ignition control.

Throttle Position Sensor The **throttle position sensor (TP)** consists of a three-wire potentiometer connected to the throttle assembly and connected to the computer (Figure 12–55). The resistance of the circuit changes with throttle position (Figure 12–56). A typical TP sensor at idle position is approximately 1000 Ω and 0.5 volts, and 4000 Ω and 4.5 volts at wide open throttle (WOT) (Figure 12–57).

The voltage of the TP sensor should change gradually from idle to WOT. When the wiper arm on the resistive element has a worn spot (due to extended use in one position), a glitch can occur in smooth engine operation. A digital storage oscilloscope (DSO) can locate this condition (Figure 12–58) by monitoring the exact voltage measurement.

Output Sensors

After gathering information from various input switches and sensors, the control computer will provide electronic output commands through several mechanical motion devices, such as actuators, motors, relays, or solenoids. An example of output devices are relay controlled circuits, idle air control (IAC) valves, and fuel injectors.

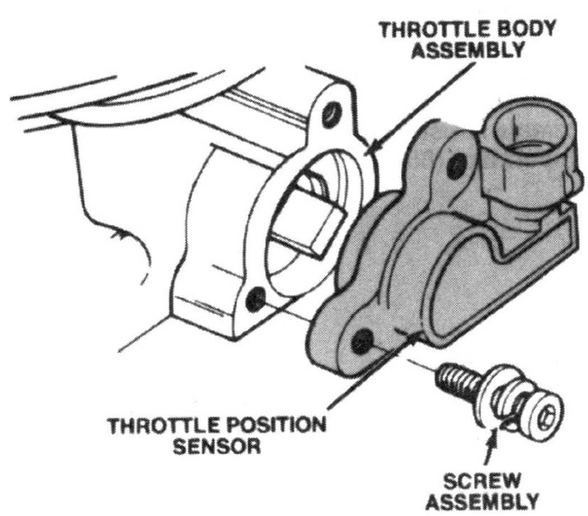

FIGURE 12–55 Typical throttle position (TP) sensor mounted to throttle body assembly.

(Courtesy of General Motors Corporation, Service Technology Group)

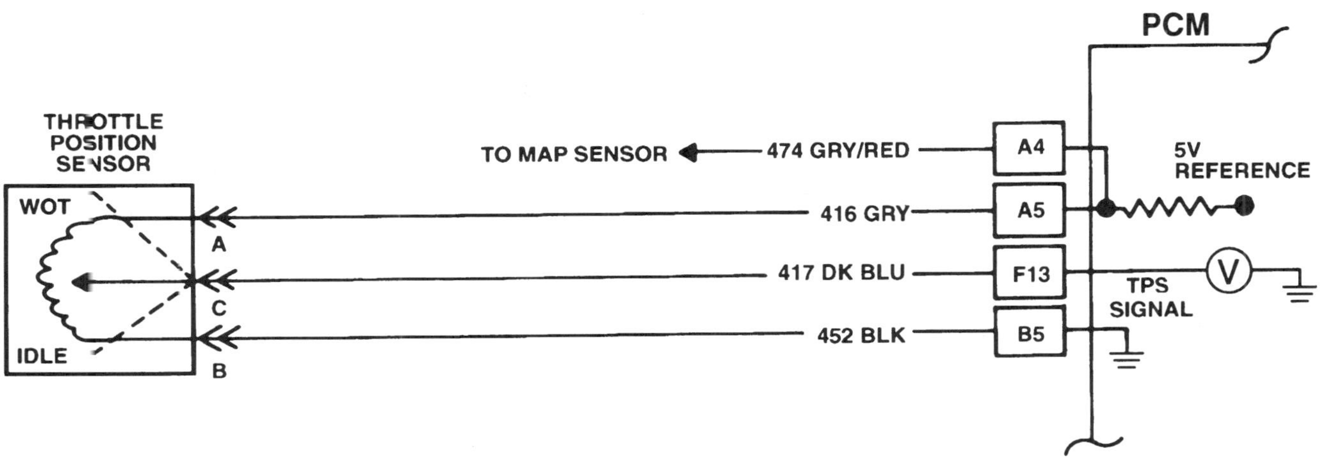

FIGURE 12–56 Typical three wire throttle position (TP) potentiometer sensor.
(Courtesy of General Motors Corporation, Service Technology Group)

Relay Circuits Most relay controlled circuits are electromechanical actuators that mechanically operate components. The control computer usually controls the ground circuit for the relay coil (Figure 12–59). This protects the computer from the high current required to operate some output devices.

The output actuator can be cycled *on* and *off* rapidly to obtain desired results. This is referred to as *pulse-width modulation.* If this is used to control the illumination level of a digital instrument panel light cluster (Figure 12–60), the level of *on* time versus the *off* time would dictate the light intensity.

Fuel Injector Controls A typical fuel injector control circuit controls the fuel that is pressurized in the system manifold. Each injector has an individual coil that is grounded by the control computer to complete the control circuit (Figure 12–61). The quantity of fuel that is injected depends on the amount of time the injector is

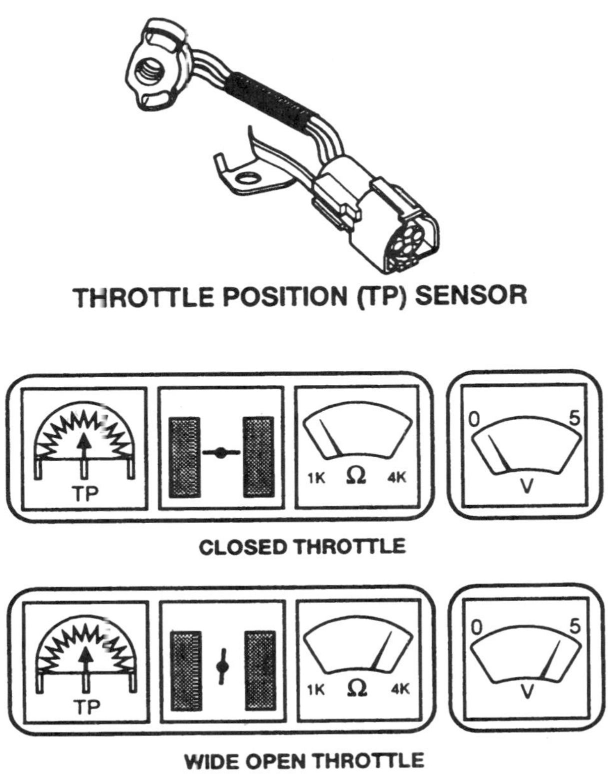

FIGURE 12–57 Engine throttle position (TP) sensor resistance and voltage drop.
(Courtesy of Ford Motor Company)

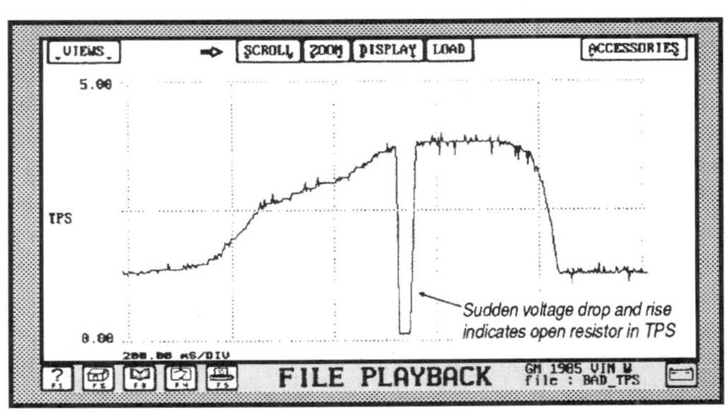

FIGURE 12–58 A throttle position (TP) sensor "glitch" is found using a digital storage oscilloscope (DSO).
(Courtesy of Snap-on Diagnostic Systems)

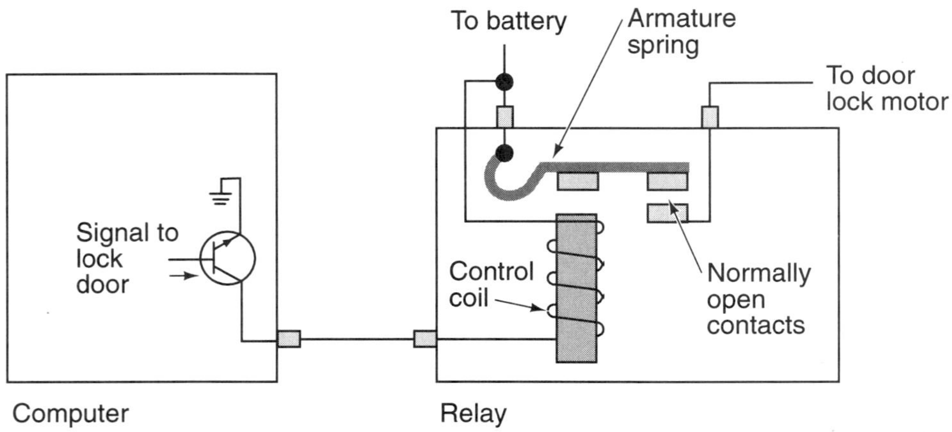

FIGURE 12–59 Computer output driver controls the relay coil ground.

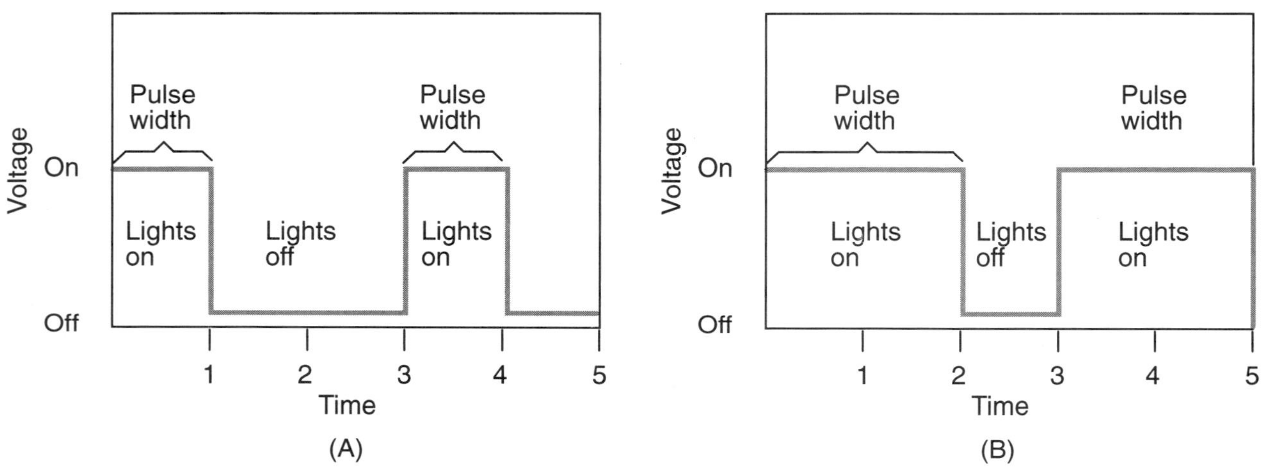

FIGURE 12–60 Pulse width modulating the instrument panel lights to achieve desired illumination level.

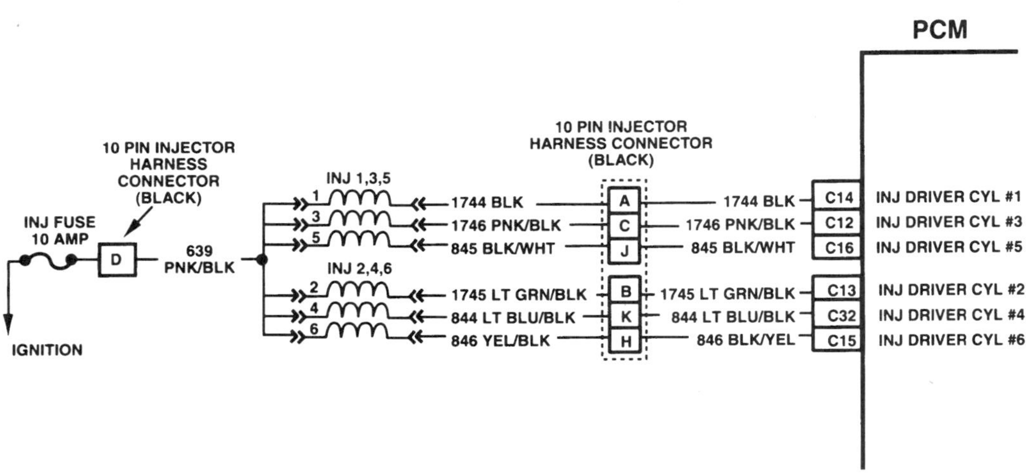

FIGURE 12–61 Fuel injector control circuit.

(Courtesy of General Motors Corporation, Service Technology Group)

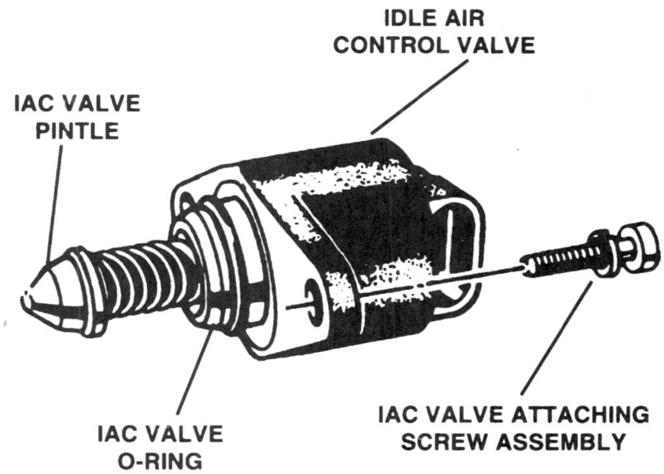

FIGURE 12–62 Idle air control (IAC) valve.
(Courtesy of General Motors Corporation, Service Technology Group)

energized, which is determined by engine temperature and load requirements.

Idle Air Control Valve Many computer-controlled systems will control engine idle speed with an **idle air control (IAC)** valve (Figure 12–62).

The engine control computer will control idle rpm by regulating the idle air passage on the throttle body (Figure 12–63). The IAC motor contains a permanent magnet armature with at least two field coils. It is also referred to as a stepper motor. By applying properly sequenced voltage pulses to selected coils, the armature spindle will move a precise amount to control air flow.

Computer System Diagnostics

Depending on the vehicle, there are numerous computer controlled input and output devices and systems. To perform effective troubleshooting requires the technician to retrieve fault codes stored in the computer memory. Always refer to specific manufacturers procedures when retrieving computer memory codes. Normally this is accomplished in two different ways.

The first method to access trouble codes is a blinking light on the dashboard or computer case that flashes

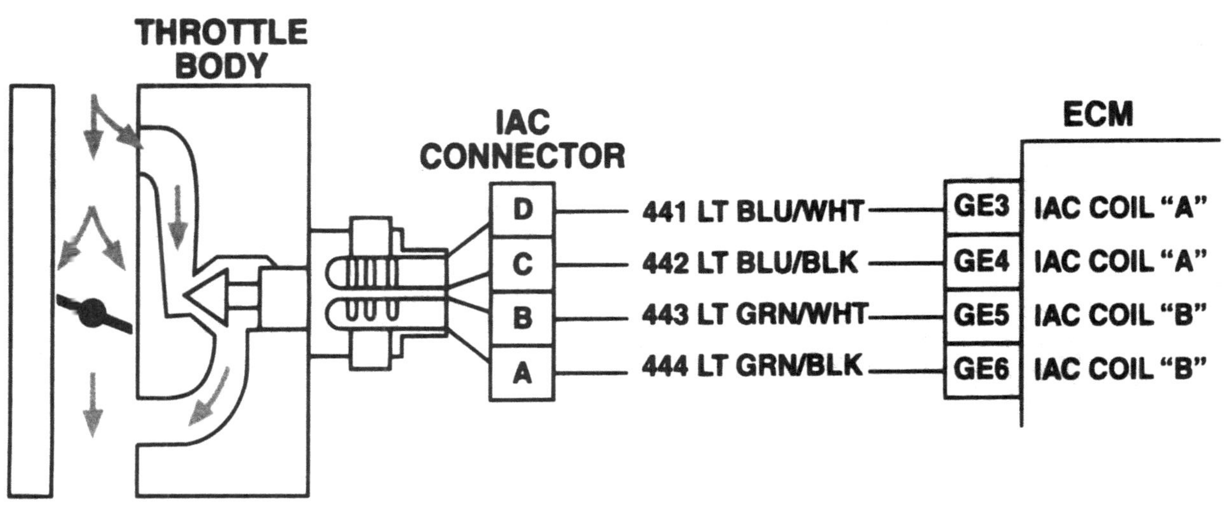

FIGURE 12–63 Typical idle air control (IAC) valve circuit.
(Courtesy of General Motors Corporation, Service Technology Group)

DIAGNOSTIC TROUBLE CODE (DTC) MAIN CODE	DIAGNOSTIC TROUBLE CODE (DTC) SUB-CODE	PROBLEMATIC COMPONENT/ SYSTEM
1	—	ABS pump motor over-run
	2	ABS pump motor circuit problem
	3	High pressure leakage
	4	Pressure switch
	8	Accumulator gas leakage
2	1	Parking brake switch-related problem
3	1	Pulser(s)
	2	
	4	
4	1	Wheel sensor
	2	
	4	
	8	
5	—	Wheel sensor(s)
	4	
	8	
6	—	Fail-safe relay
	1	
	4	
7	1	Solenoid related problem
	2	
	4	

FIGURE 12–64 Typical trouble codes for an antilock brake system.

(Courtesy of American Honda Motor Company, Inc.)

a specific sequence of codes and sub-codes for each code stored (Figure 12–64).

The second method to access trouble codes is through a data-link connector (DLC) located in the wiring harness under the dash, glove box area, or the engine compartment. Between the early 1980's and 1996 the DLC consisted of a 12 pin or less connector depending on the vehicle manufacturer. Beginning in 1996, Federal law required all vehicles meet OBD II (on-board diagnostics, second generation) requirements. (Figure 12–65).

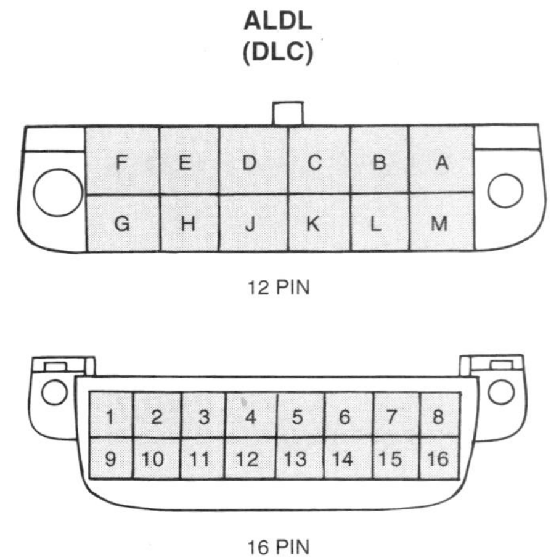

FIGURE 12–65 Older ALDL (assembly line diagnostic link) or newer DLC (data link connector).

A quick way to retrieve a computer code through a DLC is with a special scan tool (Figure 12–66). Fault codes are assigned to each computer circuit and help the technician pinpoint a problem (Figure 12–67).

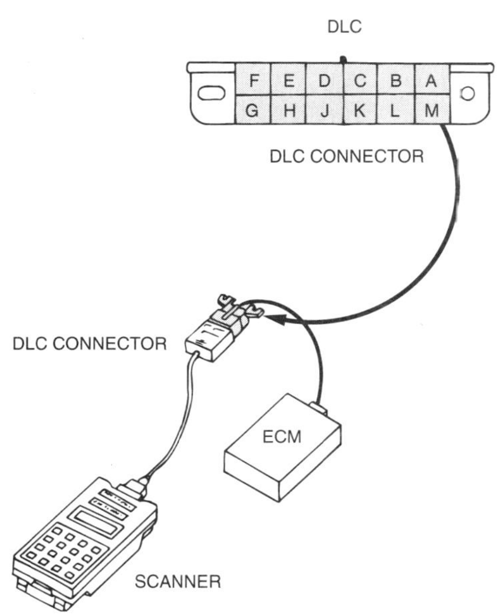

FIGURE 12–66 Hand-held scanner connected to the ALDL or DLC.

FAULT CODES

CODE 12	NO ENGINE SPEED REFERENCE PULSE—This code will appear if the engine is "ON" and the ECM sees no reference pulse from the distributor.
CODE 13	OXYGEN (O2) SENSOR CIRCUIT (OPEN CIRCUIT)—This indicates that the engine is properly warmed up and other conditions have been met but that the oxygen sensor circuit is still open.
CODE 14	COOLANT TEMPERATURE SENSOR CIRCUIT (CTS) (HIGH TEMPERATURE INDICATED)—This code will set if signal voltage indicates a coolant temperature that is above specifications for 90 seconds or more.
CODE 15	COOLANT TEMPERATURE SENSOR CIRCUIT (CTS) (LOW TEMPERATURE INDICATED)—This code will set if signal voltage indicates a coolant temperature that is below specifications for at least 5 minutes after the engine is started.
CODE 21	THROTTLE POSITION SENSOR (TPS) CIRCUIT (SIGNAL VOLTAGE HIGH)—This code will set if the signal from the TPS circuit indicates an open throttle (high voltage) while other signals (engine speed and manifold vacuum) indicate a closed throttle.
CODE 22	THROTTLE POSITION SENSOR (TPS) CIRCUIT (SIGNAL VOLTAGE LOW)—This code will set if the TPS signal voltage is about .2 volt or less (about .4 volt is normal at idle) with the engine running over 20 seconds.
CODE 23	MIXTURE CONTROL (M/C) SOLENOID CIRCUIT (SIGNAL VOLTAGE LOW)—This code will set if the M/C solenoid signal stays low and does not vary. This could indicate an open in the circuit or ground on the ECM side of the mixture control solenoid.
CODE 24A	VEHICLE SPEED SENSOR (VSS) CIRCUIT—This code will set if the vehicle is in gear, the engine is above idle, the throttle is open but the indicated vehicle speed is below 5 mph.
CODE 24B	PARK/NEUTRAL (P/N) CIRCUIT—This code indicates that the P/N circuit signal to the ECM is incorrect compared to the other inputs and does not change as it is moved from park or neutral to a drive range or reverse.
CODE 31	CANISTER PURGE SOLENOID CIRCUIT—This code indicates that the voltage in this circuit does not reflect the command given by the ECM.
CODE 33	MANIFOLD ABSOLUTE PRESSURE (MAP) SENSOR CIRCUIT—This code results when the signal voltage is high, indicating that there is a low vacuum.
CODE 34	PRESSURE SENSOR CIRCUIT (SIGNAL VOLTAGE OUT OF RANGE)—This code will set if the engine is running and the MAP sensor pressure reading is outside a specified range while the engine rpm is less than a specified value for a given amount of time.
CODE 35	IDLE SPEED CONTROL (ISC) SWITH—This code indicates that the ISC switch is shorted. It results when the ECM receives a voltage signal from the ISC switch circuit that indicates over 50 percent throttle for over 2 seconds.
CODE 41	NO DISTRIBUTOR REFERENCE PULSE—This indicates that there is no distributor reference signal when the engine vacuum is present.
CODE 42	ELECTRONIC SPARK TIMING (EST) CIRCUIT—This code will set when the ECM detects an open or grounded bypass circuit or EST circuit.
CODE 43	ELECTRONIC SPARK CONTROL (ESC) CIRCUIT—This code will set when the engine is running and the ESC input signal has been low more than 3.9 seconds.
CODE 44	OXYGEN SENSOR CIRCUIT (LEAN EXHAUST INDICATED)—This code will set whenever the system is in "Closed Loop", TPS voltage is within a specified range (indicating part throttle) and the ECM detects that O2 voltage is lower than specified for longer than a designated time.
CODE 45	OXYGEN SENSOR CIRCUIT (RICH EXHAUST INDICATED)—This code will set whenever the system in in "Closed Loop", TPS voltage is within a specified range (indicating throttle) and the ECM detects that O2 voltage is higher than specified for longer than a designated time.
CODE 51	PROM ERROR (FAULTY OR INCORRECT PROM)—This could indicate a faulty PROM unit, an improperly installed PROM or some PROM pins not making contact.
CODE 53	EGR SYSTEM MALFUNCTION (California Only)—The ECM checks EGR operations by deengergizing the EGR solenoid, which closes the EGR valve. It then looks for an increase in manifold vacuum as indicated by a change in the signal from the MAP sensor. If this change is not within a specified window, the code will set.

FIGURE 12–67 Assigned fault codes help the service technician troubleshoot a computer related system.

(Courtesy of General Motors Corporation, Service Technology Group)

Safety

When repairing or handling computer system components, preventing static electricity is important. As more microprocessors are added to vehicles, the need to inform technicians about how to provide a static free environment is even more important. To effectively work on electronic circuitry, the best precaution is to use a static grounding strap (Figure 12–68). Always refer to specific manufacturer procedures when handling computer related components.

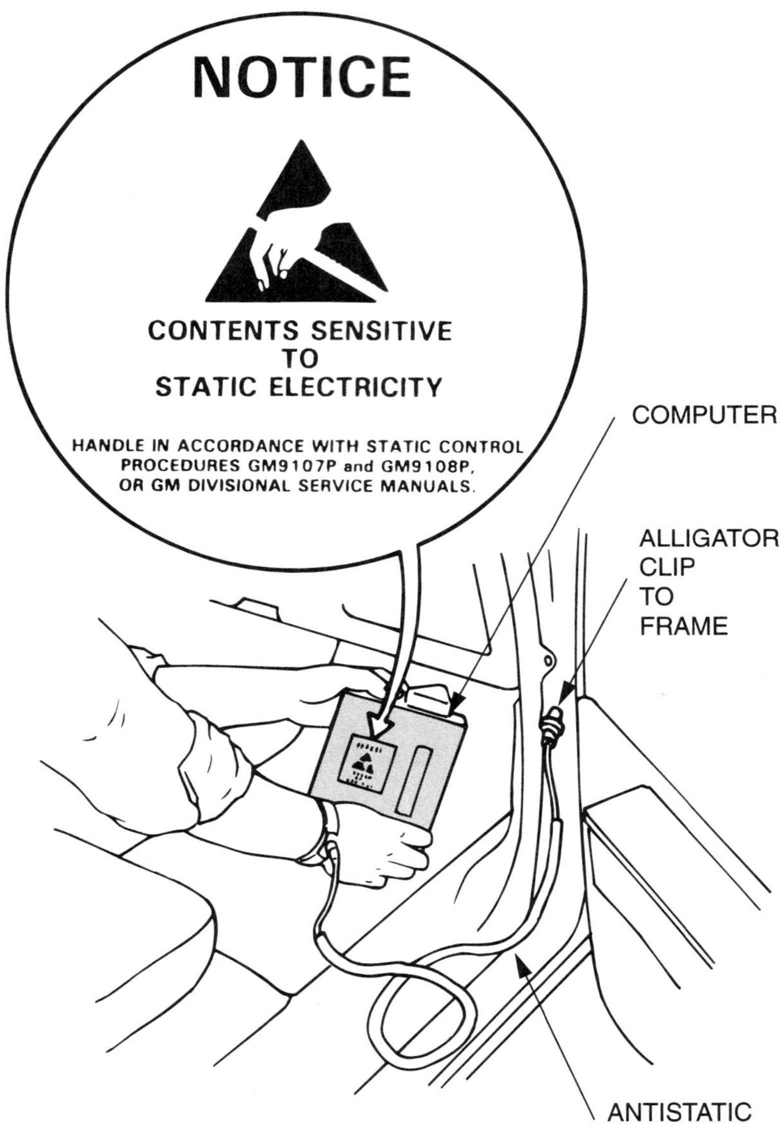

FIGURE 12–68 Typical static grounding strap used to ground the technician and prevent damage to sensitive computer components.

Summary

- ❑ A semiconductor is a stable element with exactly four valance electrons in its structure.
- ❑ The two most common semiconductor materials are silicon and germanium.
- ❑ A covalent bond is atoms that share valance electrons with neighboring atoms.
- ❑ An N-type semiconductor has loose or excessive electrons and a negative charge.
- ❑ A P-type semiconductor has positive charged hole material that carry a current.
- ❑ A diode is an electrical one-way check valve that allows current to flow in one direction only.
- ❑ A PN junction is the area where the anode and cathode material join together.
- ❑ When a diode has the positive voltage on the anode side of the diode, it is forward biased.
- ❑ When a diode has the positive voltage on the cathode side of the diode, it is reversed biased.
- ❑ A clamping diode is used to control voltage surges or spikes in a circuit with a coil.
- ❑ An LED is a light emitting diode that emits an electrical light energy when it is forward biased.
- ❑ A zener diode allows a reversed biased voltage to flow at a designed voltage level.
- ❑ A transistor is an electronic switch that consists of three layers of semiconductor material joined together and designated as the emitter, collector, and base.
- ❑ An NPN transistor consists of two N-type materials and one P-type material.
- ❑ A PNP transistor consists of two P-type materials and one N-type material.
- ❑ An integrated circuit consists of many semiconductors contained on a single circuit.
- ❑ An analog voltage signal is usually produced by an input sensor that is continuously variable within a certain range.
- ❑ A digital voltage signal is a square wave signal that is either high or low.
- ❑ A binary code is the assignment of numeric values to digital signals.
- ❑ Each binary code numeric value equals one bit.
- ❑ Eight bits of information is equal to one byte.
- ❑ An A/D converter converts an analog signal to a digital signal to be used by a computer.
- ❑ A microprocessor is the decision making chip in the computer that contains a large number of miniature transistor and diode switches.
- ❑ A RAM memory is random access memory that is a temporary storage chip for use by the microprocessor to read, write, and erase information.
- ❑ A KAM is a keep alive memory that will be lost if the source battery power is lost.

- ❑ A ROM memory is a read only memory that contains information that is permanently stored for use by the microprocessor.
- ❑ A PROM is a programmable read only memory that contains specific calibration programs that pertain to a model vehicle.
- ❑ A EEPROM is an electronic erasable programmable read only memory that can only be updated by the special manufacturer test equipment.
- ❑ A logic gate controls the output voltage signal depending on the different combinations of input signals.
- ❑ A NOT gate reverses input and output data in a computer circuit.
- ❑ An AND gate has at least two inputs and only one output.
- ❑ An OR gate with a high signal to either of two inputs will result in a high output.
- ❑ A NAND and NOR gate is equivalent to an AND and OR gate with an inverted output signal.
- ❑ An XOR gate is a combination of NAND and NOR gates that will produce a high output signal only if the inputs are different.
- ❑ A computer input sensor transmits information to the control computer on various engine related functions.
- ❑ Computer input sensors can provide either *on/off* information or variable information.
- ❑ A pull-up switch uses a ground circuit internal to the computer and an external power source.
- ❑ A pull-down switch uses a ground circuit external to the computer and an internal power source.
- ❑ An oxygen sensor reacts with the oxygen content in the exhaust gas to produce a voltage for monitoring by the computer.
- ❑ A stoichiometric mixture is the ideal air/fuel mixture of 14.7:1.
- ❑ An engine coolant temperature sensor is a two wire thermistor sensor that provides a voltage signal to the computer depending on the coolant temperature.
- ❑ The throttle position sensor consists of a three-wire potentiometer that monitors the angle of the throttle plate.
- ❑ Output sensors are electromechanical motion devices controlled by the computer after receiving information from the input sensors.
- ❑ The on/off cycling of an output actuator is referred to as pulse-width modulation.

Key Terms

amplified
analog-to-digital (A/D) converter
analog voltage signal
AND gate
base
binary code
bit
byte
clamping diode
collector
covalent bond
depletion zone
digital voltage signal
diode
engine coolant temperature (ECT)
electronic erasable programmable read only memory (EEPROM)
emitter
field effect transistors (FET)
forward bias
idle air control (IAC)
input sensor
integrated circuit (IC)
keep alive memory (KAM)
light-emitting diode (LED)
logic gate
microprocessor
NAND gate
NOR gate
NOT gate
NPN
N-type
OR gate
oxygen sensor (O_2S)
P-type
PN junction
PNP
programmable read only memory (PROM)
pull-down switch
pull-up switch
random access memory (RAM)
reference voltage
reverse bias
read only memory (ROM)
semiconductor
square wave signal
throttle position sensor (TP)
transistor
zener diode
zirconium dioxide

Review Questions

Short Answer Essays

1. Describe how to test a regular diode using a DMM equipped with a diode test feature.
2. Describe the differences between an NPN and a PNP transistor.
3. What is a stoichiometric fuel/air mixture?
4. Describe how the illumination level of digital dash lights be controlled by the rapid cycling on/off of the circuit.
5. Describe how to retrieve a stored computer fault code from a vehicle.

Fill-in-the-Blanks

1. The AND gate is a logic gate that has ______________ input and ______________ output.
2. An engine coolant temperature sensor provides a ______________ voltage signal when cold.
3. A throttle position sensor voltage is ______________ at wide open throttle.
4. A transistor consists of three layers of semiconductor material joined together called ______________, ______________, and ______________.
5. The ______________ is the area where the anode and cathode material join together.

ASE Review Questions

1. Technician A says the diode acts like a one way check valve in an electrical circuit. Technician B says a diode allows current flow in one direction only. Who is correct?
   A. A only
   B. B only
   C. Both A and B
   D. Neither A nor B

2. Technician A says a diode is forward biased if the positive voltage is connected to the cathode side of the diode, and negative voltage is connected to the anode side. Technician B says a diode is reversed biased if the positive voltage is connected to the anode side of the diode, and negative voltage is connected to the cathode side. Who is correct?
   A. A only
   B. B only
   C. Both A and B
   D. Neither A nor B

3. Voltage signals are being discussed. Technician A says a digital voltage signal is either high or low. Technician B says a digital voltage is variable within a certain range. Who is correct?
   A. A only
   B. B only
   C. Both A and B
   D. Neither A nor B

4. Technician A says a zener diode allows current to flow in one direction only. Technician B says a zener diode can be used to regulate circuit voltage. Who is correct?
   A. A only
   B. B only
   C. Both A and B
   D. Neither A nor B

5. Two technicians are discussing the operation of NPN transistors. Technician A says the transistor is *on* when a positive voltage is applied to the base. Technician B says when the transistor is *on*, the current will flow from the collector to the emitter. Who is correct?
   A. A only
   B. B only
   C. Both A and B
   D. Neither A nor B

6. Technician A says a computer pull-up switch uses an external power source to the switch, and a ground circuit that is internal to the control computer. Technician B says in a computer pull-up switch, when the switch is closed, a high reference signal is sent to the computer. Who is correct?
   A. A only
   B. B only
   C. Both A and B
   D. Neither A nor B

7. Technician A says a PNP transistor is *on* when voltage is applied to the base. Technician B says current will flow from the emitter to the collector when the transistor is *on*. Who is correct?
   A. A only
   B. B only
   C. Both A and B
   D. Neither A nor B

8. Technician A says a computer pull-down source uses an external power switch and an external ground circuit. Technician B says in a computer pull-down switch, when the switch opens, the reference signal is low. Who is correct?
   A. A only
   B. B only
   C. Both A and B
   D. Neither A nor B

9. Technician A says RAM memory is a temporary storage area for information used by the microprocessor. Technician B says ROM memory is permanently stored information used by the microprocessor. Who is correct?
   A. A only
   B. B only
   C. Both A and B
   D. Neither A nor B

10. Two technicians are discussing logic gates in computer circuits. Technician A says an AND gate out will be high only if both inputs are high. Technician B says a NOT gate will reverse the input and out signals. Who is correct?
    A. A only
    B. B only
    C. Both A and B
    D. Neither A nor B

13 Accessory Circuits and Basic Instrumentation

Introduction

There are several accessory circuits that are common to most vehicles. Some of them are necessary safety accessories like the horn, windshield wipers and washers; and others are accessory circuits that provide additional safety and comfort, like the blower circuit, rear window defogger, power windows, power seats, power door locks, and power mirrors. Each of these circuits have design features that are common to most vehicles, which helps the technician when diagnosing circuits that do not work.

Objectives

When you complete this chapter you should be able to:

- ❑ Explain the operation and basis diagnosis of various horn circuits.
- ❑ Explain the operation and basic diagnosis of the blower fan control circuit.
- ❑ Explain the operation and basic diagnosis of wiper/washer circuits.
- ❑ Explain the principles of operation and basic diagnosis of power windows, power door locks, and power seat circuits.
- ❑ Explain the operation and basic troubleshooting procedures for window defogger circuits.
- ❑ Explain the operation and basic diagnosis of conventional instrument gauges, warning lights, and sending units.
- ❑ Explain the basic operation of supplemental inflatable restraint (SIR) components.

Accessory Circuits

Horn Circuits

Horn Circuit With No Relay There are two types of horn circuits typically used in newer vehicles. The first type does *not* use a control relay to complete the circuit (Figure 13–1). The current to operate the horn(s) *must* go through the horn switch to complete the circuit from the battery to the horn(s). With this design, the horn switch will fail prematurely if the current to the horn(s) is beyond designed limits.

Horn Circuit With Relay The second horn circuit design—and probably the most common—uses a horn control relay to complete the circuit (Figure 13–2). The relay electrically isolates the horn switch from the horns, and allows a higher amperage to go to the horns assemblies. The horn relay contacts will energize the horns when the horn switch completes the horn relay coil circuit.

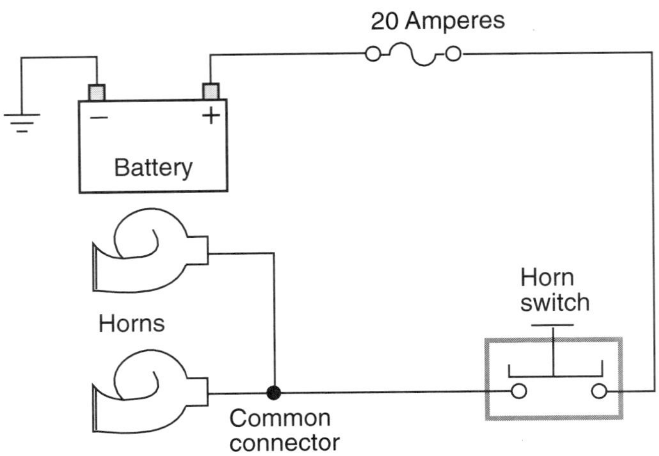

FIGURE 13–1 Simple two horn circuit.

Horn Assemblies Electric horns, operating on an electromagnetic principle, vibrate a steel diaphragm to produce an audible noise. Figure 13–3 is a typical horn with an electromagnet, flexible diaphragm, movable armature, field coil, and a set of normally closed contact points. The internal resistor allows a weak magnetic field to remain after the contact points open, thus reducing the time required to rebuild the magnetic field after the points close again.

To produce an audible noise (Figure 13–4), current flows through the *closed* contact points to the field coil producing a magnetic field, which attracts the armature that is attached to the diaphragm. Movement of the armature will *open* the contact points and return the diaphragm to its original rest position. This event occurs

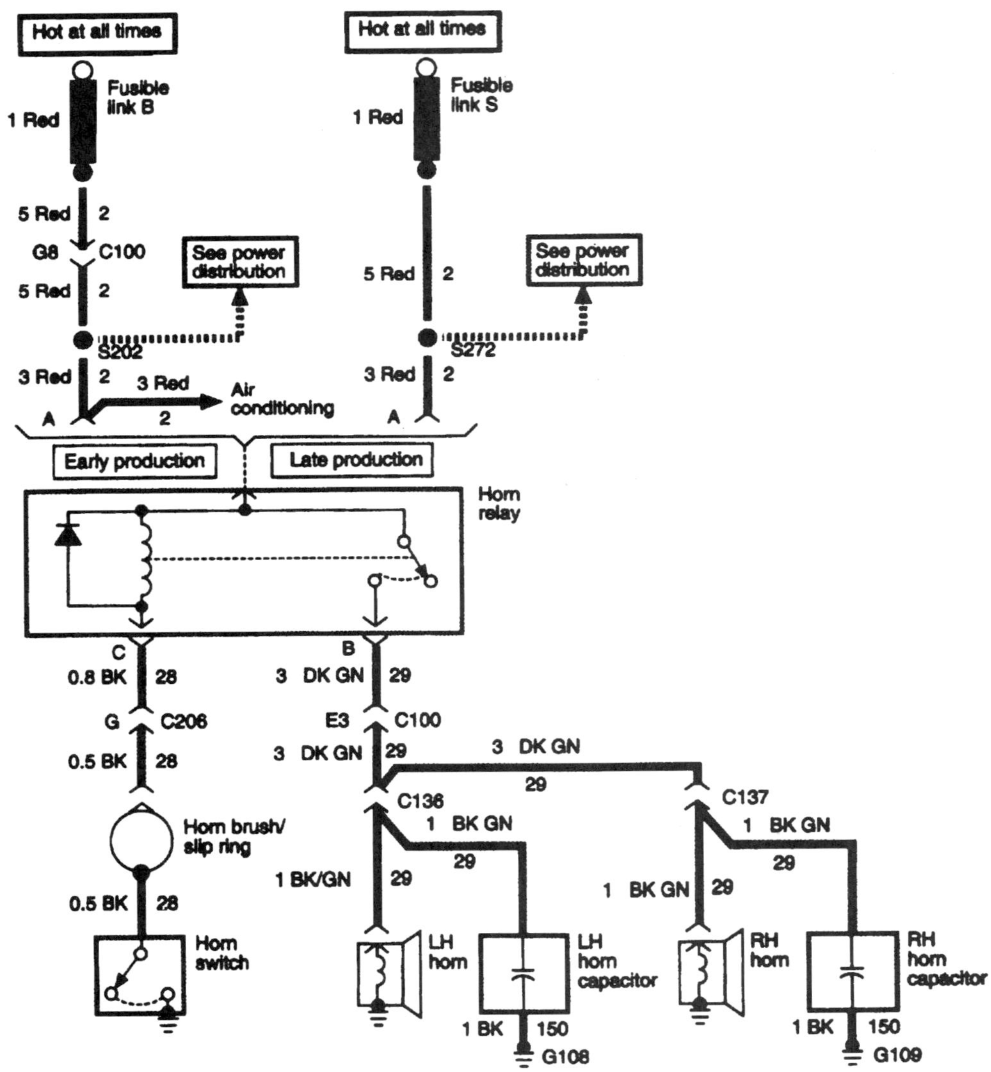

FIGURE 13–2 Typical horn circuit with relay.

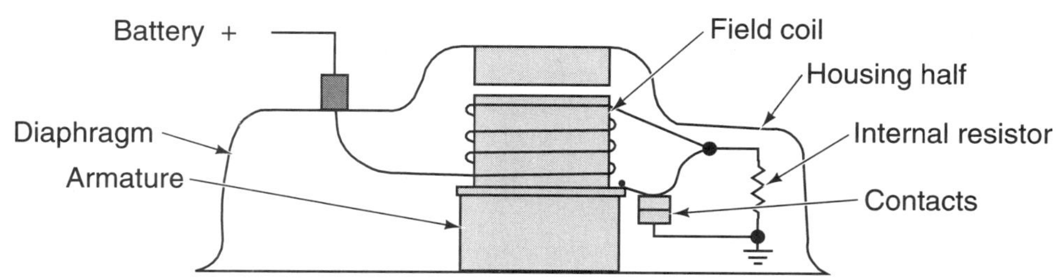

FIGURE 13–3 Typical electromagnetic horn construction.

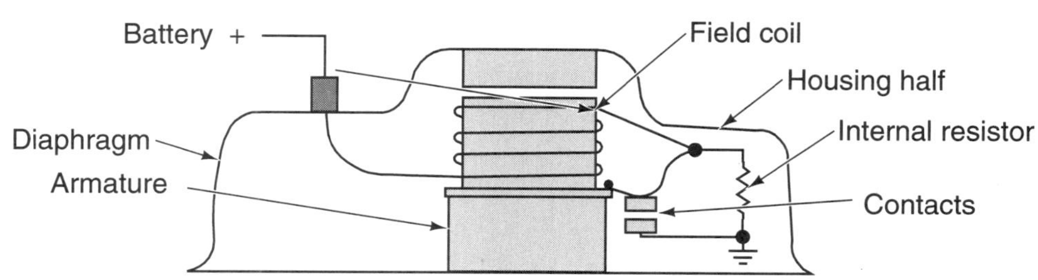

FIGURE 13–4 The magnetic field pulls the armature and diaphragm upward, opening the contacts.

several times per second to cause the diaphragm to vibrate. As the diaphragm vibrates, a column of air in the horn also vibrates to produce the audible noise.

Most vehicles are equipped with two horns, each with a different pitch from the other. The design and shape of the horn will determine the frequency and tone of the sound. The pitch is controlled by the vibrating diaphragm. The faster the vibration, the higher the pitch. A horn's pitch can be adjusted by turning the spring tension screw on the horn case (Figure 13–5).

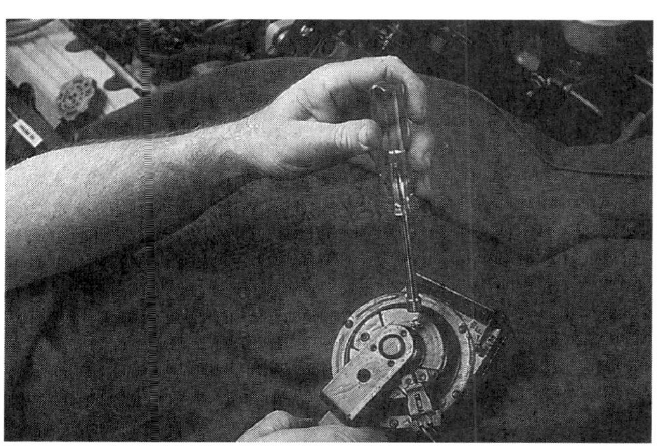

FIGURE 13–5 Changing the horn's pitch with an adjusting screw.

Horn Switch The horn switch in most vehicles is installed in the steering wheel (Figure 13–6), where a *slip ring* and spring-loaded sliding contact provide electrical continuity for the horn switch in all steering wheel positions. The horn switch may also be a push button on a lever in the multi-function switch. Depending on the vehicle, this lever may be used strictly for the horn, or combined with another function like the turn signal or dimmer switch (Figure 13–7).

CAUTION *Remember to follow all manufacturer precautions when working on vehicles with air bags or electronic controls in the steering wheels.*

Blower Control Circuits

The blower control circuit is a good example of a basic series circuit that consists of a blower switch, stepped resistor block, blower motor, and in some circuits a high speed relay in a series line. More expensive vehicles with automatic temperature controls often use solid state electronics to control blower speed.

Blower Circuits Without Relay Control There are two type blower circuits that do *not* use a relay for the blower's high speed setting. Figure 13–8 is an exam-

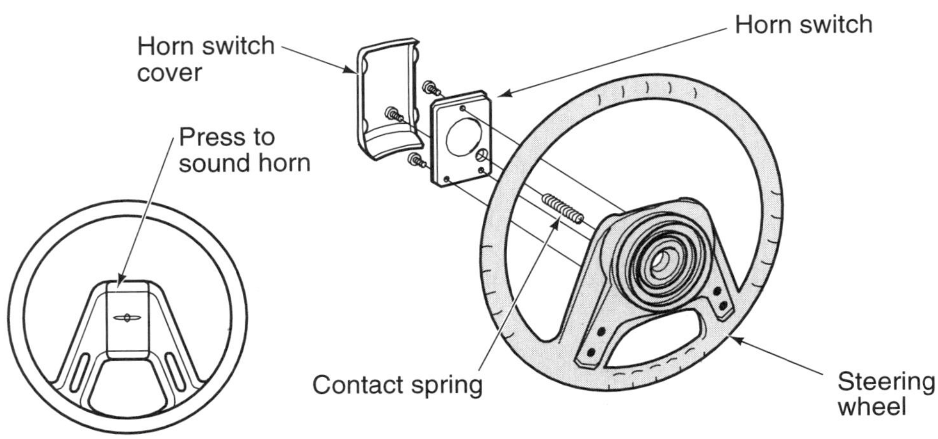

FIGURE 13–6 Single button horn switch installed in the steering wheel.

(Courtesy of Ford Motor Company)

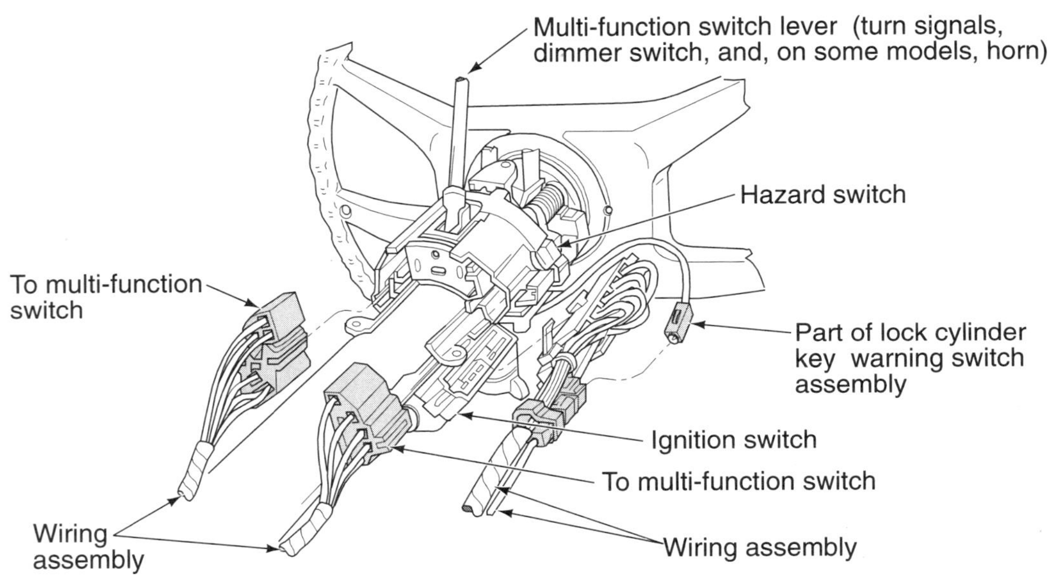

FIGURE 13–7 Horn switch as part of a multi-function switch.

(Courtesy of Ford Motor Company)

ple of a four speed blower motor circuit with the blower switch and the resistor block located before the blower motor, and the ground side of the circuit after the motor. In this circuit, when the air conditioner/heater switch is turned to the AC, Heat, or Def (defroster) settings, battery voltage is available through the C4 16TN wire to the right side of the resistor block when set for low speed blower operation. The battery voltage goes through the three resistors in *series* causing a voltage drop to occur before reaching the blower motor. Remember Ohm's Law states that as resistance goes up, amperage flow goes down. When the blower switch is turned to the M1 speed position, resistor #3 is bypassed through the C5 16LG wire. This lowers the series resistance which allows more current to the blower motor. As the blower switch is moved to the other positions, more resistors will be bypassed. In the High speed position *all* the resistors in the block are bypassed, so all the circuit current goes through the high speed contact in the blower switch to the blower motor.

Figure 13–9 is another blower circuit that does *not* use a relay for the blower's high speed setting. Unlike the previous circuit, this circuit connects the blower motor to the battery voltage side of the circuit, and the resistor

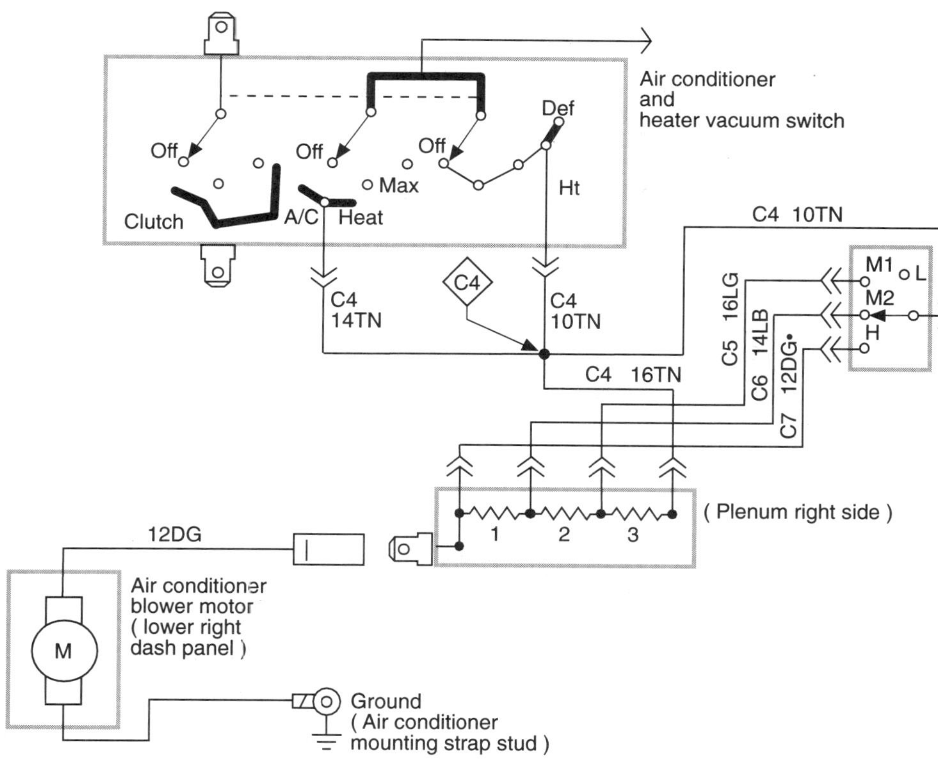

FIGURE 13–8 Typical four speed blower motor circuit with blower motor on the groundside.
(Courtesy of DaimlerChrysler Corporation)

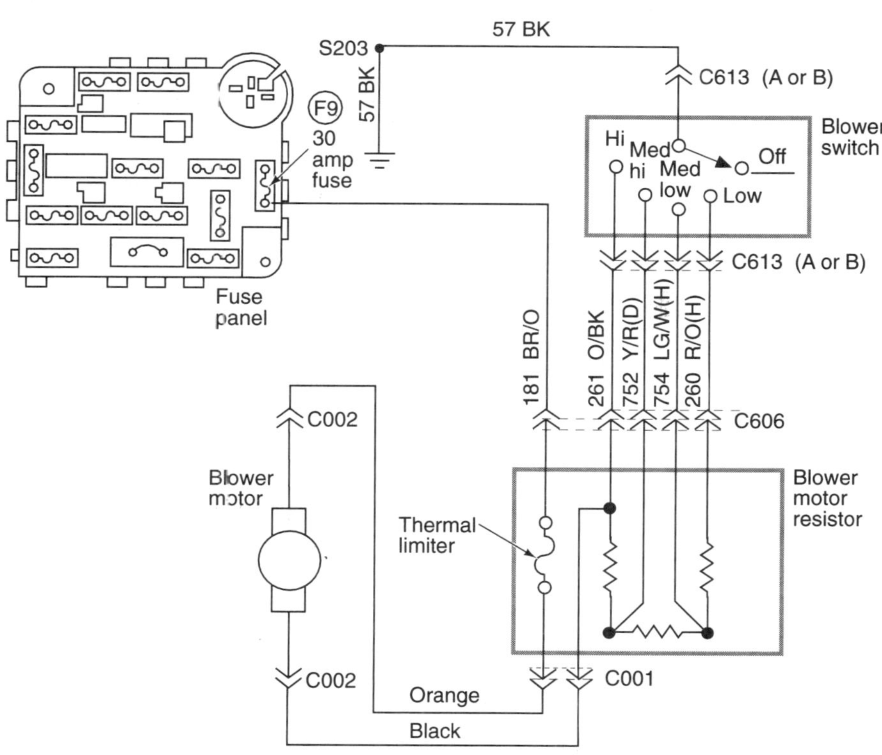

FIGURE 13–9 Typical blower motor circuit with blower motor on the battery voltage side.
(Courtesy of Ford Motor Company)

block controls the circuit ground. The resistor block changes the circuit's resistance the same way as in the previous circuit.

Blower Circuits With Relay Control Because of increased demands on the blower motor in some vehicles, the high speed operation of the motor is controlled through a relay assembly (Figure 13–10). Using the relay helps extend the life of the blower switch by enabling the higher current flow needed for the high speed position to come from a separate circuit. The blower Lo, M1, and M2 switch positions each control the amount of resistance total in the resistor block the same as the previous two circuits.

Resistor Block Testing Two methods are used to determine if a blower resistor block is defective. The first method is to check for voltage drops across each resistor with a voltmeter when the blower switch in the low position. (Figure 13–11). Because the resistors are in series, if one resistor is burned out the circuit will show source voltage up to the open in the circuit.

The second method to test a blower resistor block is to remove the assembly, and after a visual check, check each resistor with an ohmmeter (Figure 13–12). A reading of OL would indicate an open circuit.

Electronic Blower Speed Controls Vehicles with electronic automatic climate control systems (Figure 13–13) typically control blower speed through a body control module (BCM) or computer.

The heater blower speed is controlled by a power module that amplifies computer command signals from the BCM (Figure 13–14A). When a temperature selection is set by the driver through the climate control panel, the BCM monitors temperature inputs from the passenger compartment, ambient air, engine coolant, and air conditioning system before commanding fan blower speed, air delivery mode, and air blend door positioning. Figure 13–14B is a blower motor control schematic consisting of an electronic climate control panel (CCP) to set temperature, a body computer module (BCM) that receives, interprets, and transmits information, an electronic climate control (ECC) power module that amplifies a variable voltage from the BCM for blower speed control, and a blower motor.

The CCP and BCM communicate continually over data circuit line #718 (0.5 dkl/wt). Voltage signals are sent from the BCM to the ECC for amplification over circuit line #760 before being applied to the blower motor through line #65.

Wiper and Washer Circuits

There are several types of front and rear window wiper and washer systems used by vehicle manufacturers today. Most all have main components which consist of a wiper motor with park switch, speed control switch, and washer pump unit. Typical wiper systems can be divided into three design types.

Non-Depressed Park Wiper Circuit The first design is a **non-depressed park** wiper circuit (Figure 13–15). This circuit parks the wiper blades at the end of their normal stroke, usually at the edge of the lower windshield molding. Two sources of battery voltage are available to the speed control switch. Circuit #297 in Figure 13–15 applies battery voltage through an internal circuit breaker to the low or high speed motor brush, and the motor common brush completes the circuit to ground. Figure 13–16 is an example of the most common motor armature brush arrangement. The low speed brush and common brush oppose each other, and the high speed brush is centered between them or is offset. When the speed control switch is turned to the Off position, battery voltage is still available to the motor armature through the integral *Park* switch, located to the right of the wiper motor symbol in Figure 13–15, until a cam opens the park switch contact points. When the contact points open, the wiper blades should be at the end of their normal stroke (edge of lower windshield molding).

CAUTION *Wiper arm levers are very strong when operating. Keep fingers clear when testing the wiper circuit.*

Depressed Park Wiper Circuit The second wiper circuit design is the **depressed park** wiper circuit (Figure 13–17). In vehicles where styling is important, the circuit parks the wiper blades *off* the glass and past the end of their normal stroke—usually below the edge of the lower windshield molding. In this circuit, a second set of contact points are used along with the park switch. When the wiper switch is turned to the Off position, the additional set of park contacts will reverse the rotation of the wiper motor for about 15 degrees after the wiper

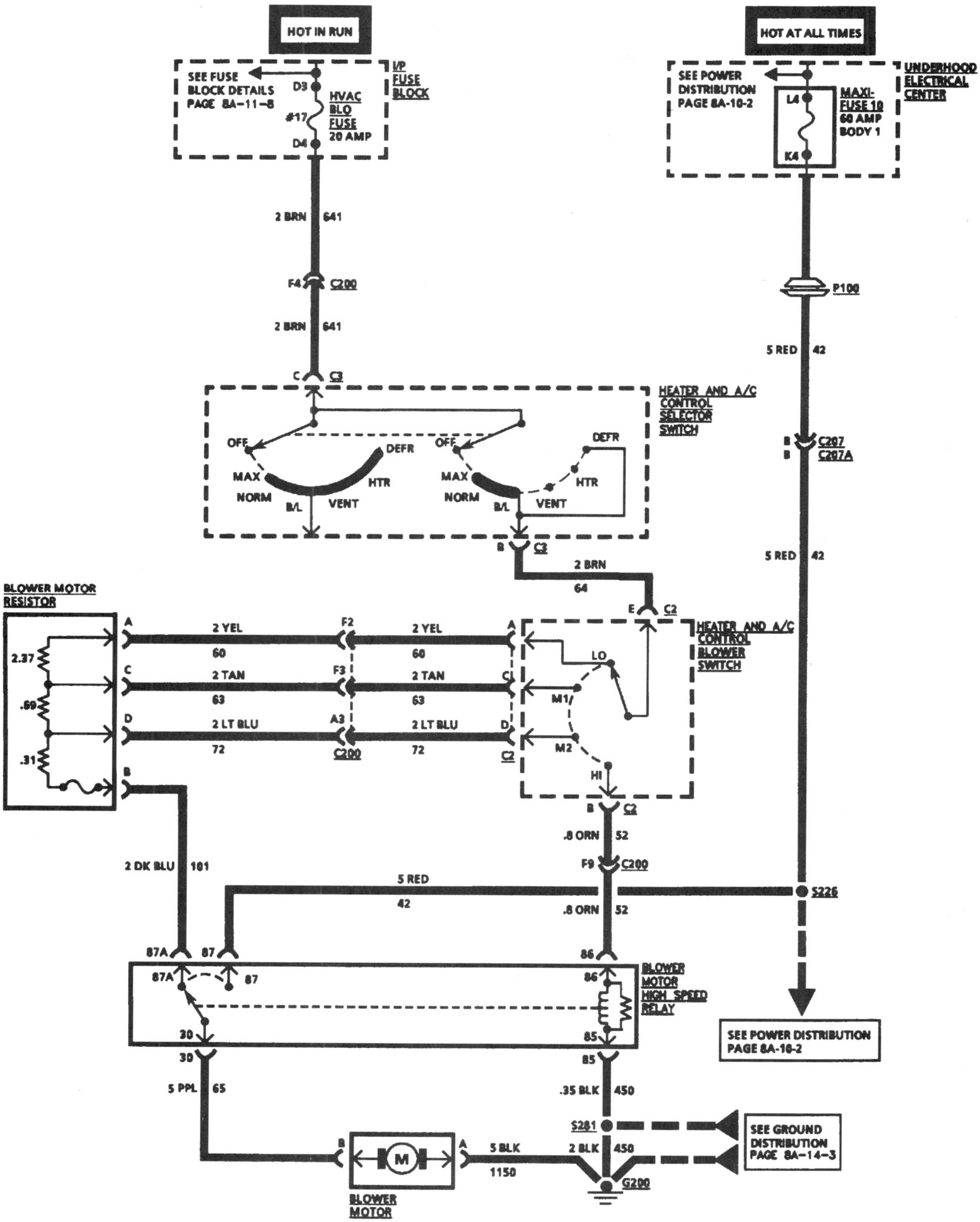

FIGURE 13–10 Blower motor circuit with relay for high speed operation.

(Courtesy of General Motors Corporation Service Technology Group)

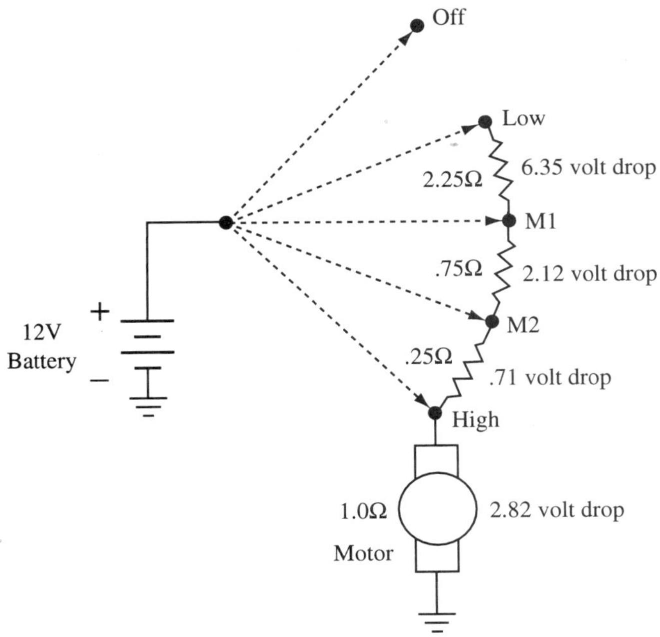

FIGURE 13–11 Checking for voltage drops across a blower resistor block.

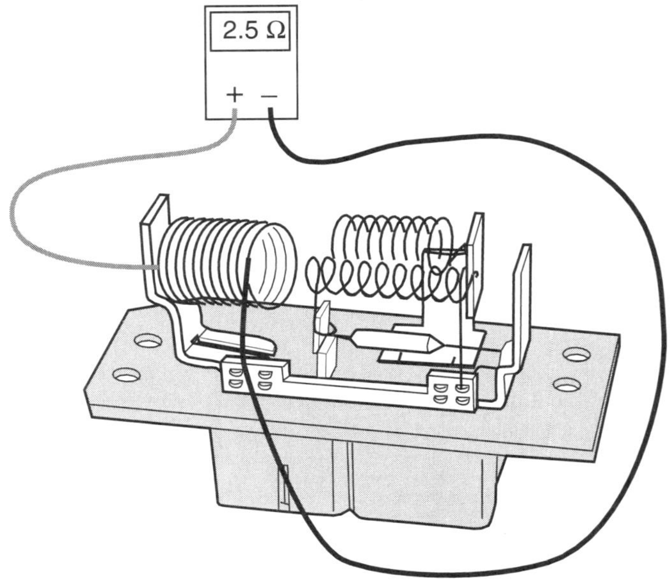

FIGURE 13–12 Checking each blower resistor with an ohmmeter.

(Courtesy of Ford Motor Company)

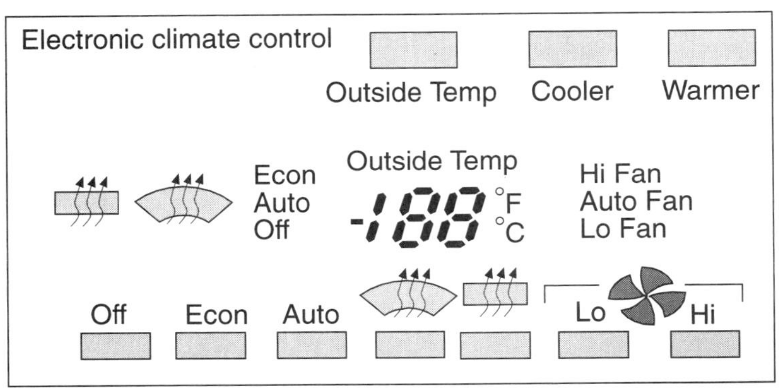

FIGURE 13–13 Electronic climate control unit.

(Courtesy of General Motors Corporation Service Technology Group)

FIGURE 13–14A
Blower circuit power module.

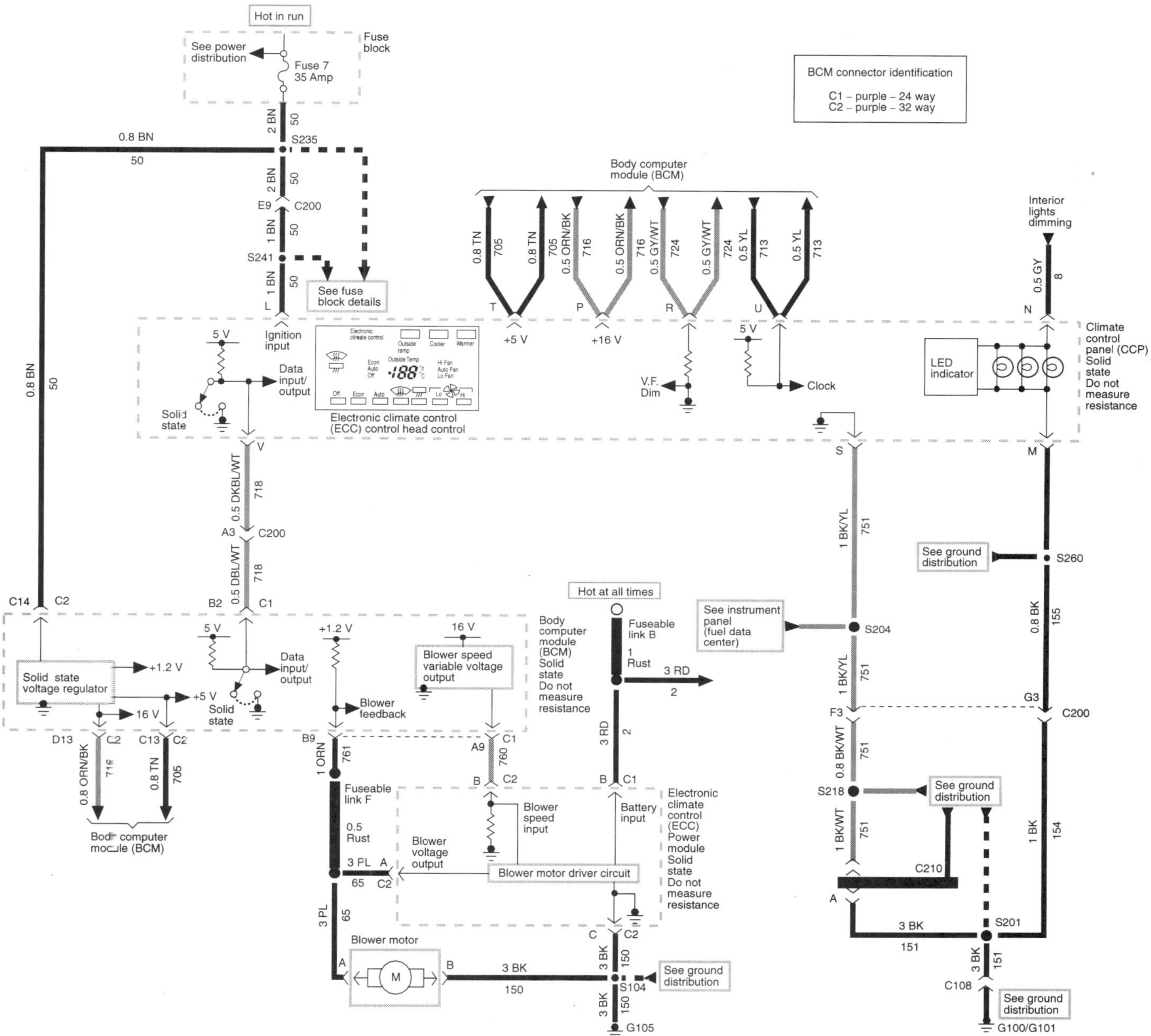

FIGURE 13–14B Electronic-controlled blower schematic.

(Courtesy of General Motors Corporation Service Technology Group)

blades have reached the lower edge of the windshield molding. Current will flow through the internal circuit breaker, through the park switch contact #A, and to wiper switch contact #1. Current then flows through the low speed brush to the motor armature. The ground path is directed to the common brush through wiper switch contact #3, to park switch contact #B, and then to ground. Park switch A opens when the wipers reach the depressed park position.

Faulty park switches with contacts that are *open* all the time causes the wipers to stop whenever the wiper switch is in the off position. In order to get the wipers to the desired park position, the wiper switch must be turned on and off several times. A faulty park switch with contacts stuck or welded closed causes the motor to operate continually, until the wiper motor is disconnected or the ignition switch is turned off.

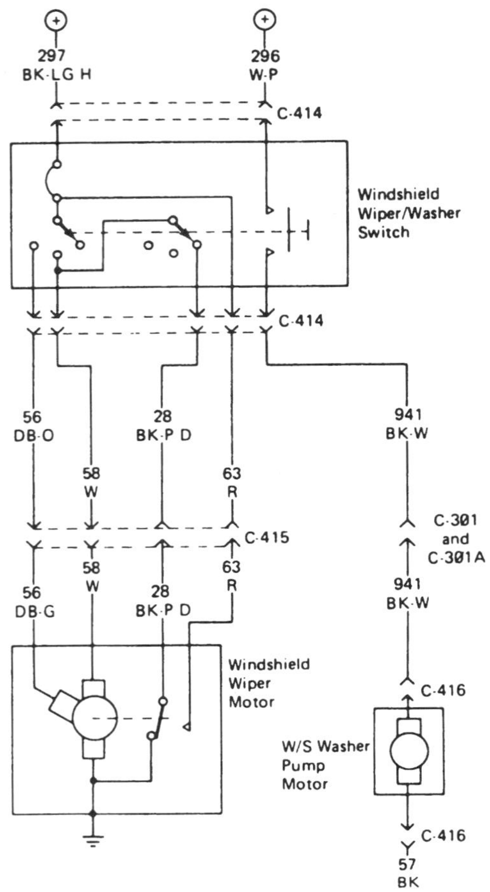

FIGURE 13–15 A typical *non-depressed park* wiper circuit.

(Courtesy of Ford Motor Company)

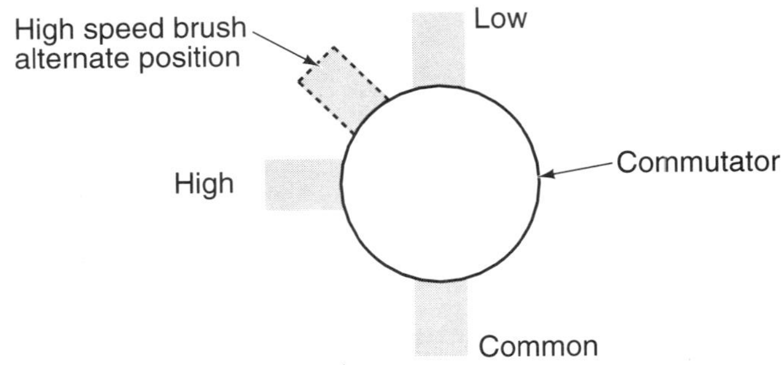

FIGURE 13–16 Typical armature brush arrangement for a two speed wiper motor.

Interval Wiper System The third design of a wiper circuit is an **interval** wiper system. This system is also known as an intermittent wiper system, where the wipers are operated in time intervals depending on the delay setting requested by the driver. Figure 13–18 is an illustration of an interval wiper system activated in the intermittent wiper mode. The internal timer in the wiper module triggers the electronic switch to activate the governor relay contacts which complete the circuit to the low speed brush. The wiper interval is determined by the resistance setting of the interval control knob. The higher the resistance setting, the longer the delay time needed to saturate the capacitor that triggers the electronic switch.

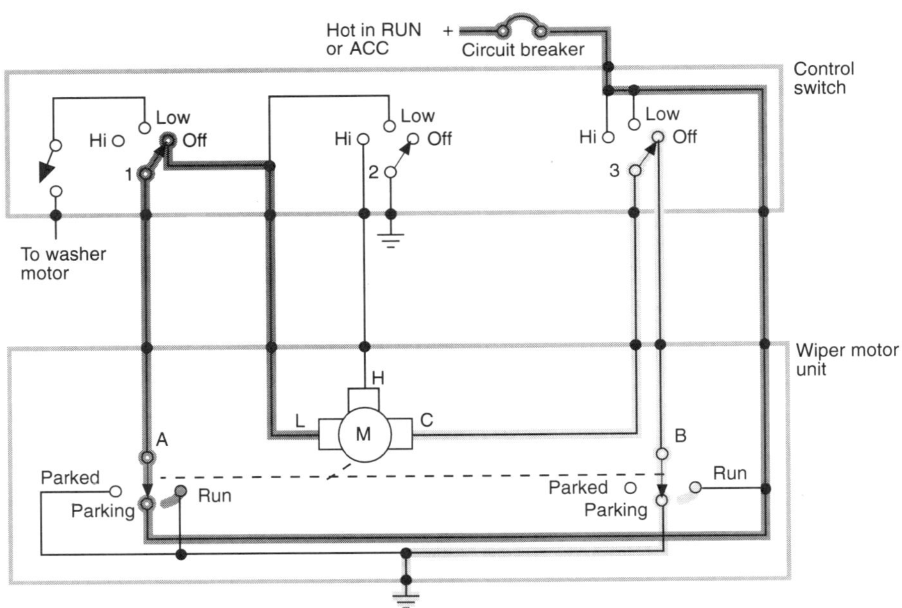

FIGURE 13–17 Typical *depressed park* wiper system with a second set of park contact points.

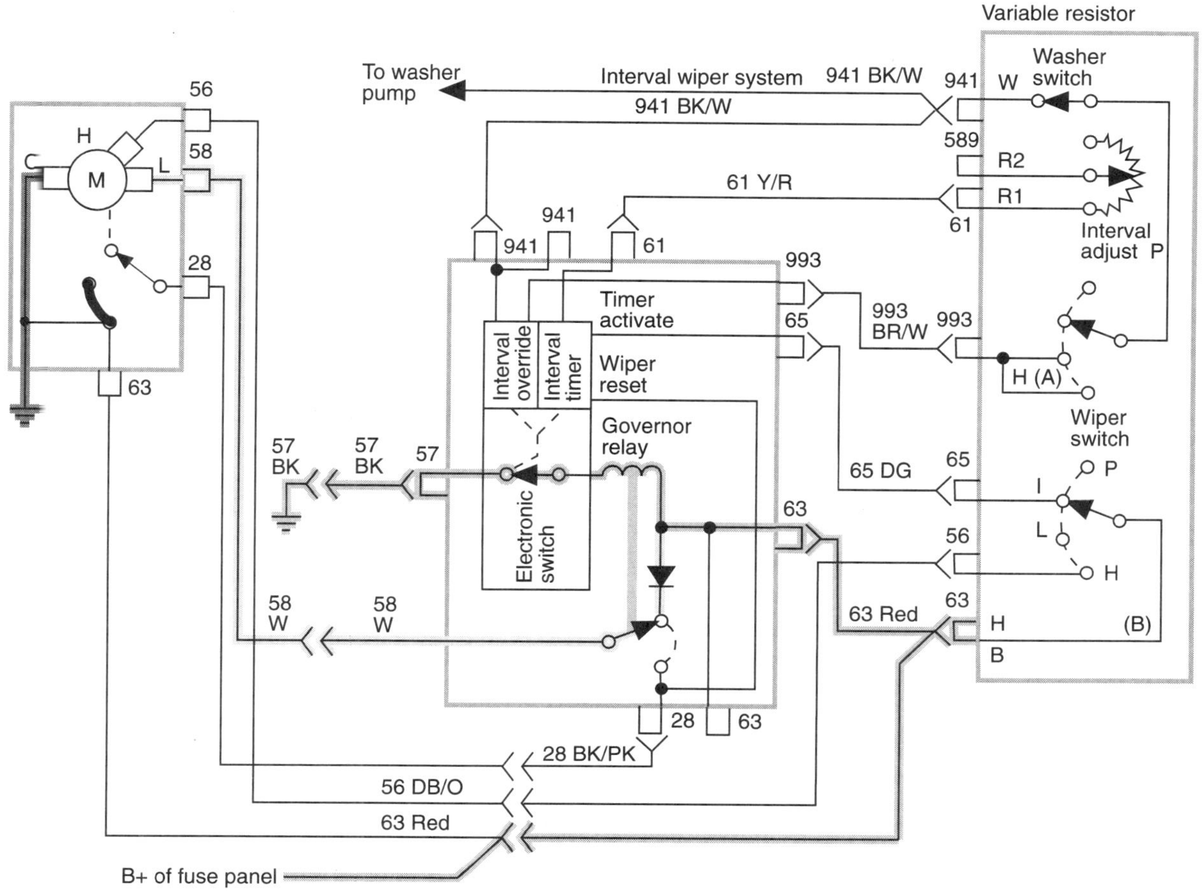

FIGURE 13–18 An interval wiper system activated in the intermittent wiper mode.
(Courtesy of Ford Motor Company)

Washer Pump System The typical washer pump system works in conjunction with the wiper motor by automatically activating the wiper motor low speed setting (Figure 13–19). The washer switch is usually a spring-loaded normally open switch that applies battery voltage to the washer pump when pushed closed. The washer motor can either be located in the fluid reservoir (Figure 13–20), or incorporated with the wiper motor as a pulse-type washer (Figure 13–21).

Window Defogger Circuits

The use of electric window defogger/defroster systems in vehicles has increased over the past several years. The main components in a basic defogger/defroster system are the control switch, relay, indicator light, and the heating grids on the window. Some systems also incorporate a solid state timer to shut off the defogger after a certain time. Figure 13–22 is an example of a simple rear window defogger circuit. When the switch is turned on, the relay applies battery voltage at the L terminal which lights the indicator lamp and energizes the window heating grid. Most vehicles have window heating grid lines that are connected in parallel which allows the heating grids to operate even if one grid is open. The heating grid can draw 20 amps or more, so using a relay is necessary.

Figure 13–23 is an example of a window defogger circuit that uses a solid state timer to control the window heating grid. The timer opens the defogger control switch after the first ten minutes of initial operation, and every five minutes for succeeding operations.

The most common customer complaint about a window defogger system is that only a portion of the heating

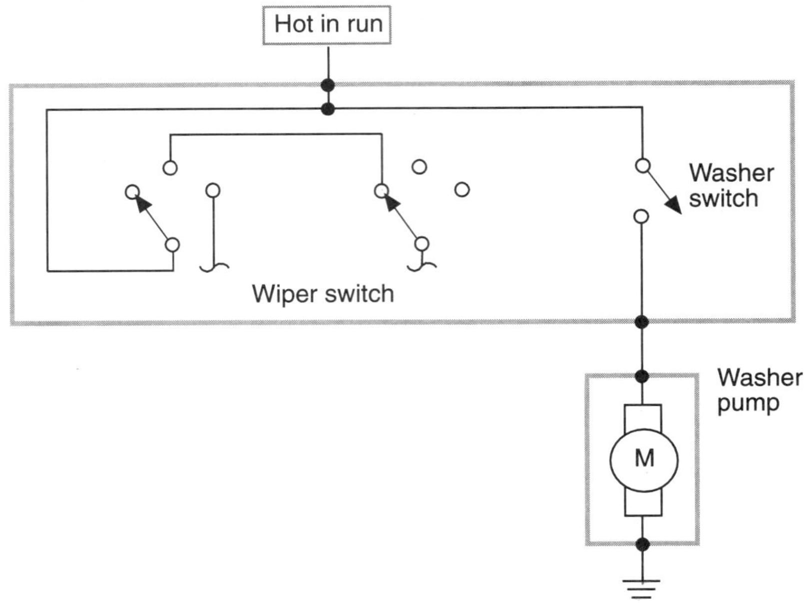

FIGURE 13–19 The typical washer pump works in conjunction with the wiper's low speed setting.

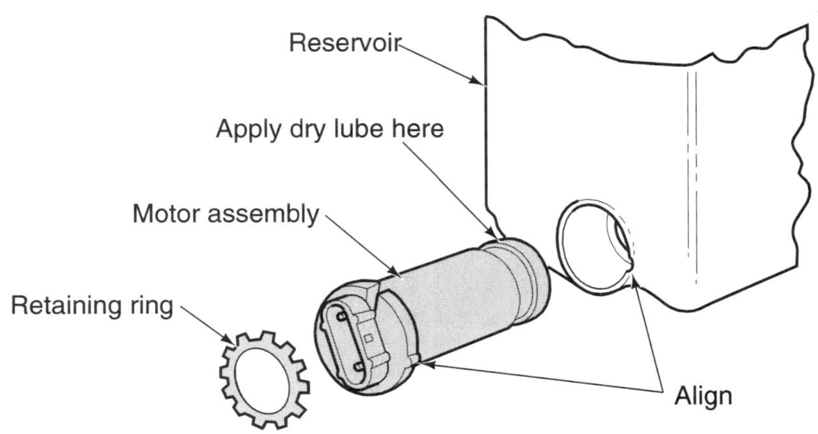

FIGURE 13–20 Washer pump and motor located in the reservoir.
(Courtesy of Ford Motor Company)

FIGURE 13–21 Pulse type washer system incorporated with the wiper motor.

grids are working. This is normally due to broken grids wires in the window. Some breaks in the grid lines are easy to visually detect, but some breaks are too small to see. There are a few methods commonly used by technicians to find a break in a grid line. The first method is using a test light to check the grid wire for a break (Figure 13–24). Because the grid wire is a resistive element that heats up as current is passed through it, the brilliance of the test light varies depending where on the grid line the test light is probed.

If a grid line is broken, the test light will have full brilliance up to the break in the grid and no light after

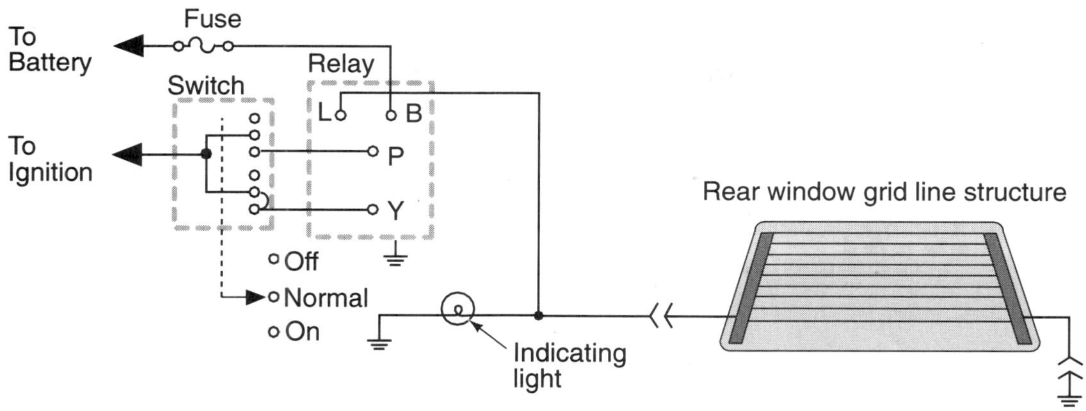

FIGURE 13–22 Typical rear window defogger circuit.

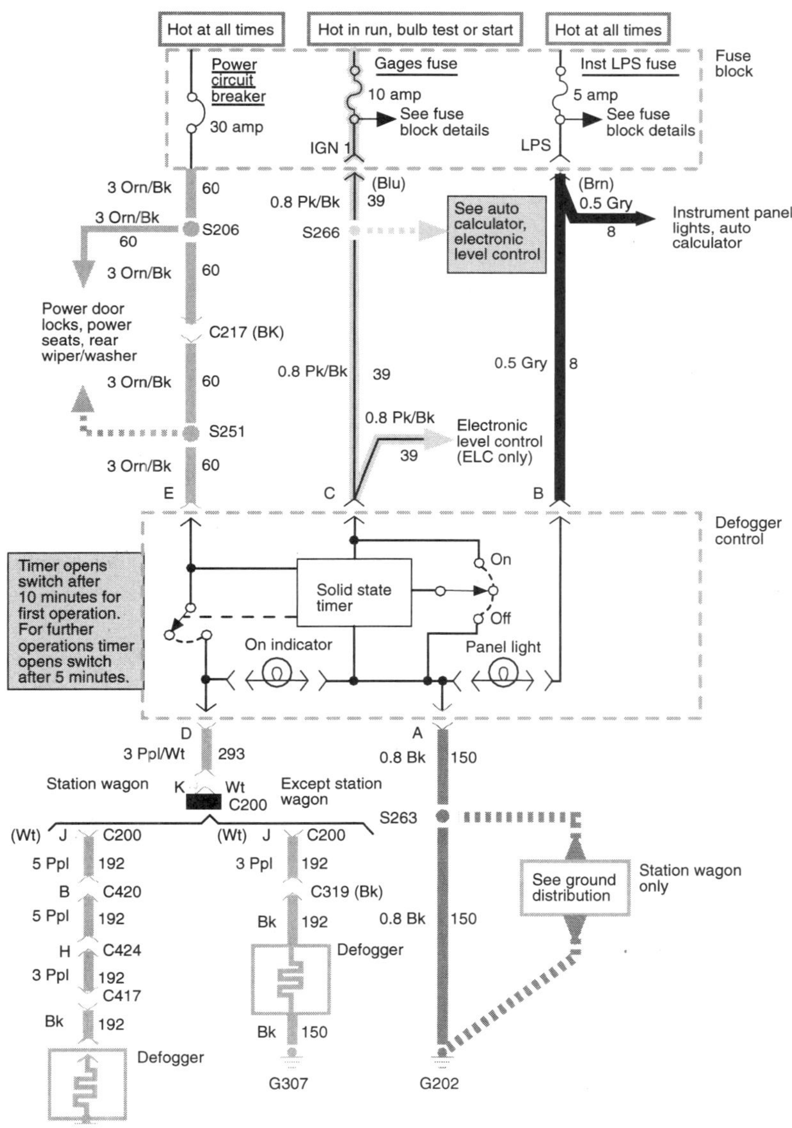

FIGURE 13–23 Solid state timer-controlled rear defogger circuit.

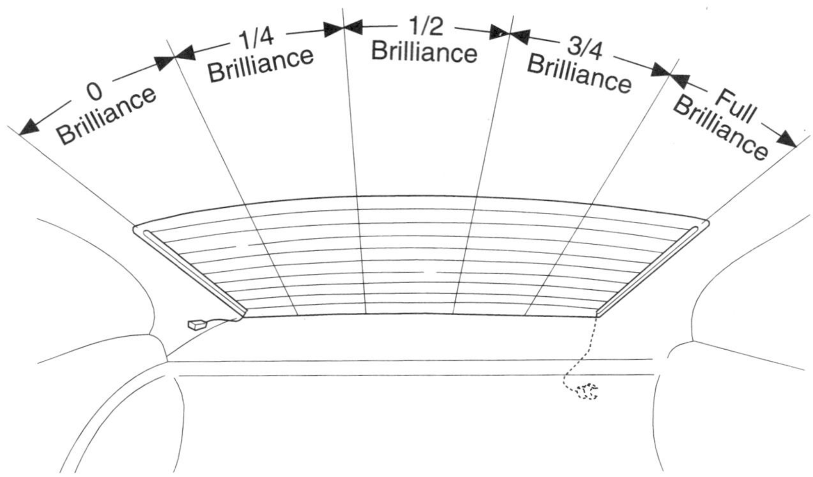

FIGURE 13–24 Checking a test light for brilliance while probing a window grid wire.
(Courtesy of General Motors Corporation Service Technology Group)

the break (Figure 13–25). Remember—the window defogger circuit could be on a timer that shuts off the circuit after a set time (usually around 10 minutes).

CAUTION *Be careful when testing a heating grid wire with a test light. Probing the grid wire too hard with the test light tip might create a break in the grid wire by scratching it.*

Another method to check for breaks in a heating grid wire is to use a digital infrared tester (Figure 13–26) which is either an adapter unit for a DVOM, or a standalone unit with its own digital read-out numbers. To use an infrared tester perform the following simple steps.

1. Turn on the rear window heating grid.
2. Wait approximately two minutes before checking the temperature of grid wires.
3. From the *outside* of the window, point the infrared tester at each grid wire no more than one half inch away. Measure the temperature of wire.
4. Results—if the wire has a break in it, the temperature will be close to room temperature. If the wire is working, the infrared tester will show the heated temperature being produced (normally between 85 degrees and 110 degrees).

Using an infrared tester can be a handy tool to help the technician quickly diagnose broken grid wires from outside the vehicle.

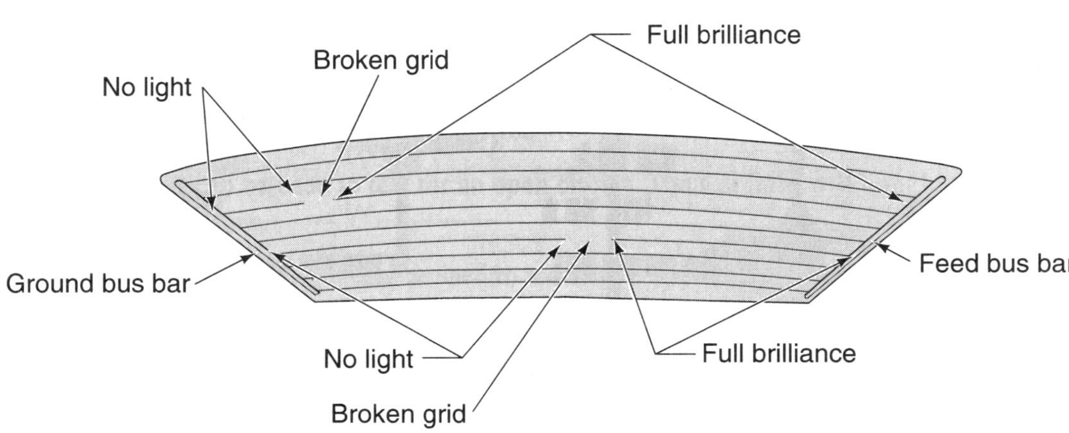

FIGURE 13–25 Finding a grid wire break using a test light.
(Courtesy of General Motors Corporation Service Technology Group)

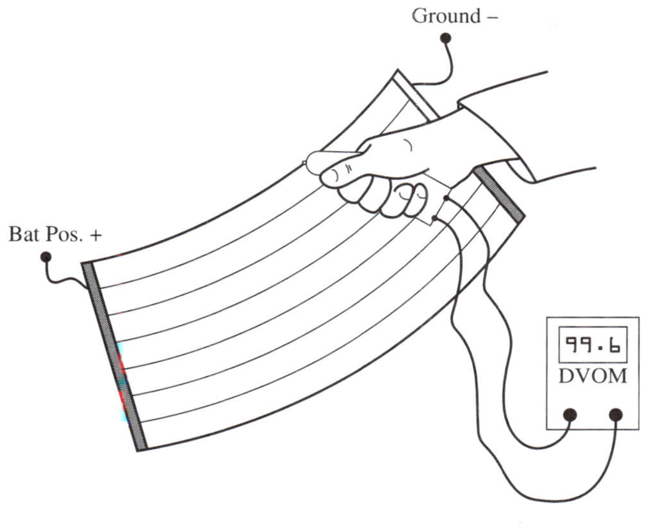

FIGURE 13–26 Using a digital infrared probe to check for broken heating grid wires.

Defogger Grid Wire Repairs In the event a grid wire has a break, a special repair kit can be used to mend the break in the wire. The procedures listed in Photo Sequence #2 is typical for this repair.

Power Window Circuits

There are several types of power window circuits in today's vehicles, depending on the vehicle body style and amount of installed options. The primary components of a power window circuit are a master control switch (normally on the drivers side) with a disable or lock-out switch, individual window control switches, and individual window control motors and sometimes rack-and-pinion gears.

Circuit Controls A simplified power window circuit is shown in Figure 13–27. When the ignition switch is closed, the window motor may be operated by either the window switch or the master control switch. In circuits with PM (permanent magnet) motors, the motor is insulated with the master switch providing the path to ground. There is also a PTC (positive temperature coefficient) circuit breaker installed internally to the motor to protect the electrical circuit from excessive current load. In Figure 13–28, the left rear window motor is being controlled by the master switch in the Up position. When the left rear window switch is used to control the window motor, battery power is supplied from the circuit breaker directly to the LR window switch.

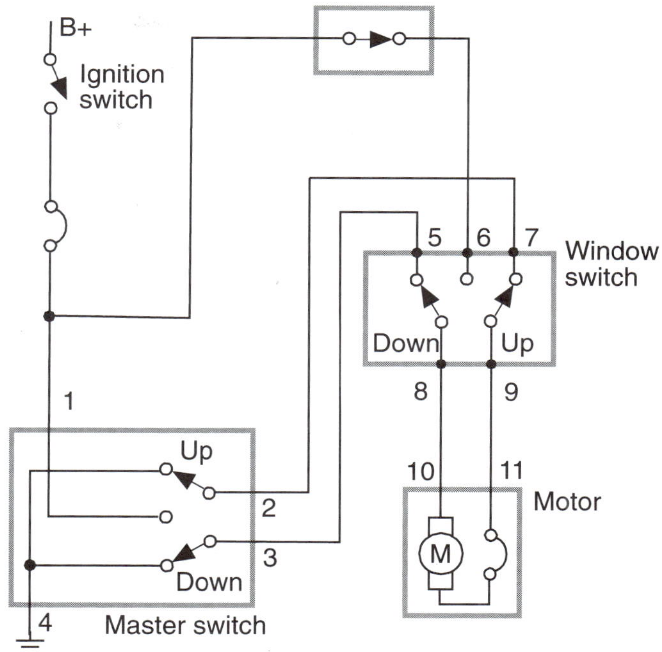

FIGURE 13–27 Simplified power window circuit.

CAUTION *Power window circuits are very strong when operating. Keep fingers clear when testing or working on circuit components.*

Testing a master switch in a power window circuit can be accomplished with a test light (Figure 13–29). Connect the test light between connections #1 and #2 of the master switch. The test light should go on with the master switch in the Off position, but should go out when the master switch is in the Up or Down position. The window switch can be checked in the same manner.

Power Window Motors There are two types of power window motors used to raise and lower a window. The first is a direct drive motor that uses a pinion gear meshed with a sector gear to raise the window regulator (Figure 13–30). When the window lowers, a spiral spring attached to the regulator is wound up. As the window raises, the spiral spring will help the motor raise the window.

The second type of power window motor uses a rack-and-pinion gear set (Figure 13–31). The rack is a flexible strip of gear teeth with one end attached to the window and the other end in a channel. The motor operates the window regulator through the rack assembly.

PHOTO SEQUENCE 2

REPAIRING DEFOGGER GRID WIRES

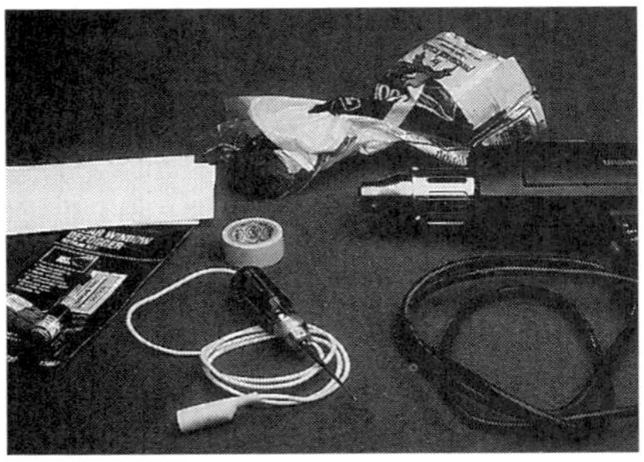

P13–1 Tools required to perform this task include masking tape, repair kit, 500 °F heat gun, test light, steel wool, alcohol, and a clean cloth.

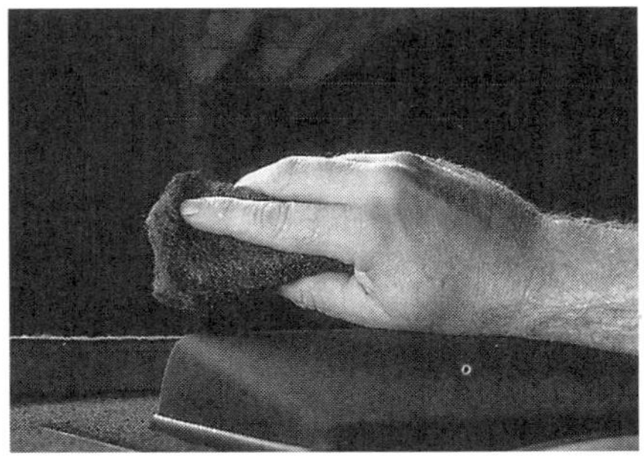

P13–2 Clean the grid line area to be repaired: Buff with steel wool. Wipe clean with a cloth dampened with alcohol. Clean an area about 1/4 inch (6 mm) on each side of the break.

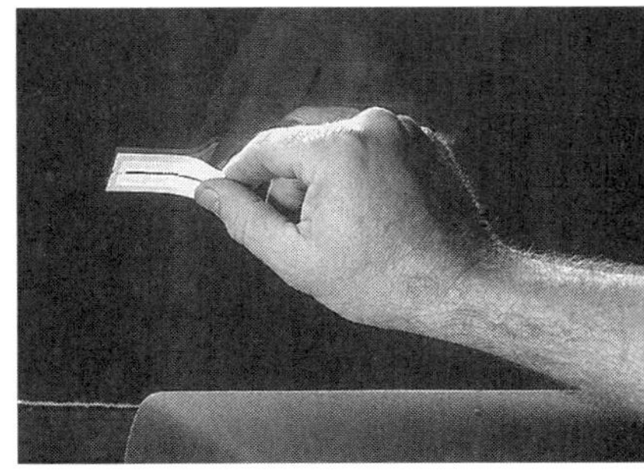

P13–3 Position a piece of tape above and below the grid. The tape is used to control the width of the repair.

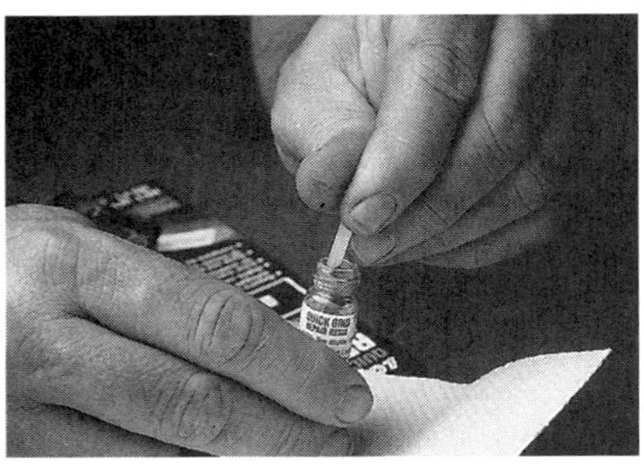

P13–4 Mix the hardener and silver plastic thoroughly. If the hardener has crystallized, immerse the packet in hot water.

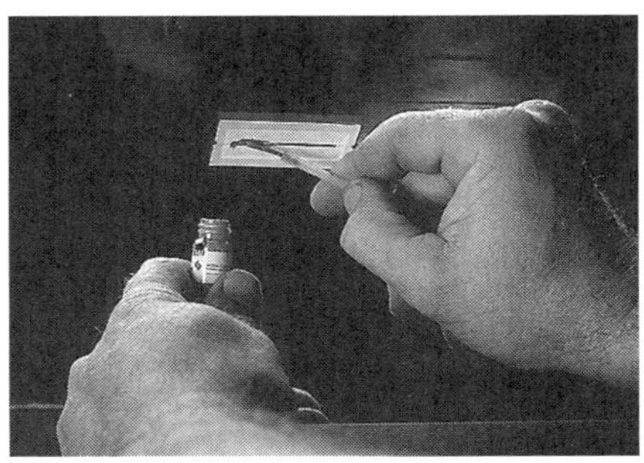

P13–5 Apply the grid repair material to the repair area using a small stick.

P13–6 Remove the tape.

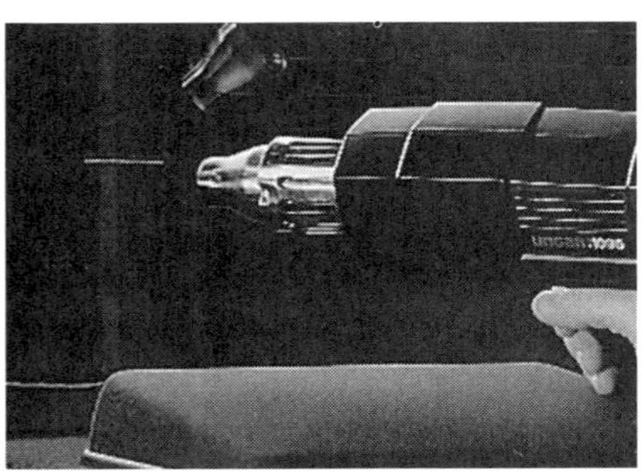

P13–7 Apply heat to the repair area for 2 minutes. Hold the heat gun 1 inch (25 mm) from the repair.

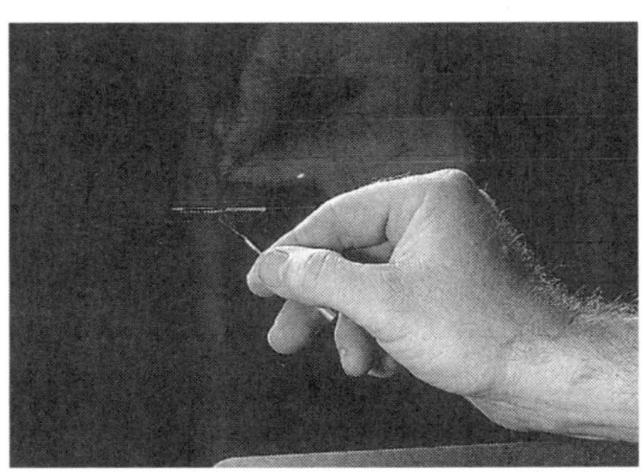

P13–8 Inspect the repair. If it is discolored, apply a coat of tincture of iodine to the repair. Allow to dry for 30 seconds, then wipe off the excess with a cloth.

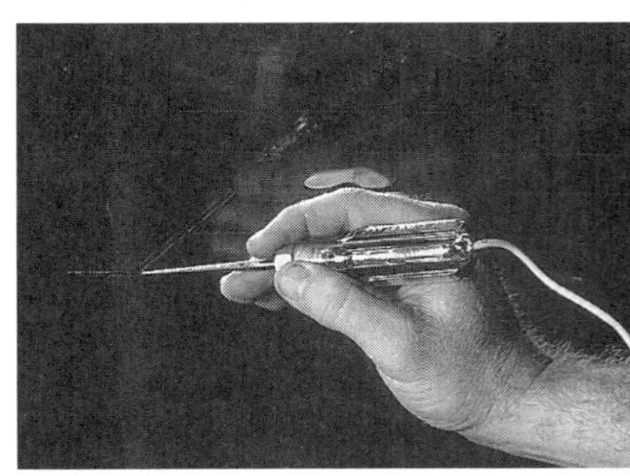

P13–9 Test the repair with a test light. Note: It takes 24 hours for the repair to fully cure.

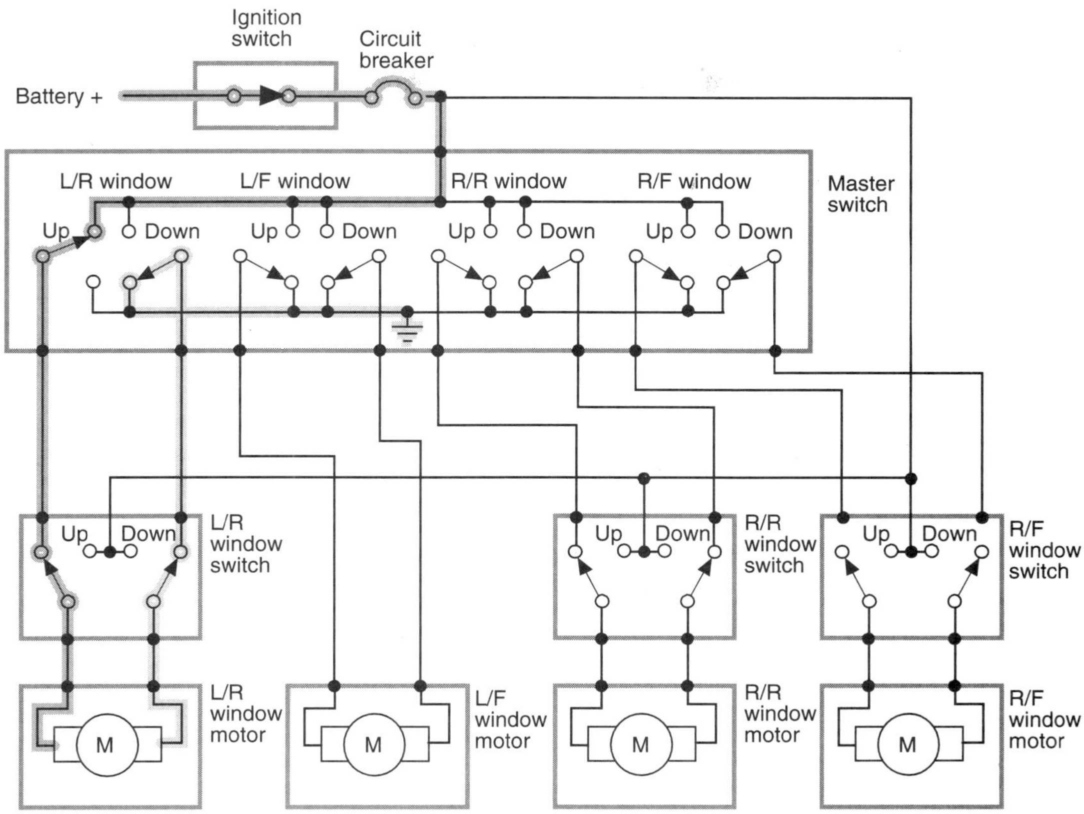

FIGURE 13–28 Master switch controls current to L/R window motor circuit.

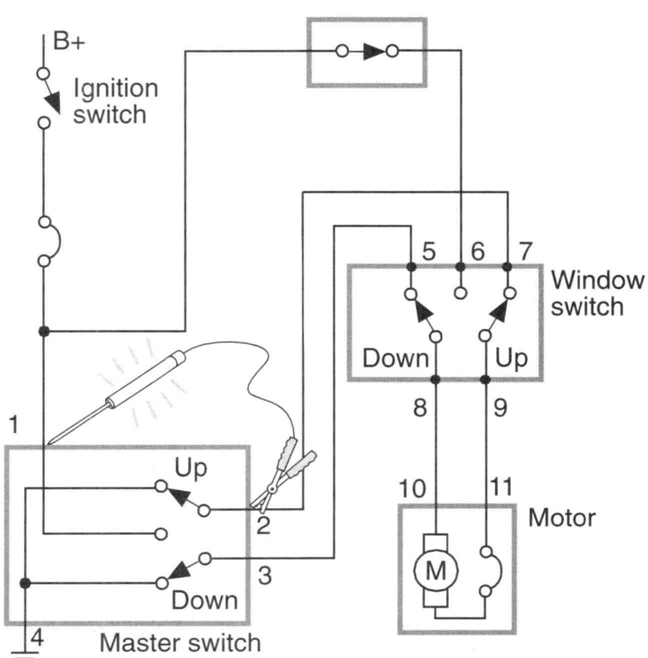

FIGURE 13–29 Using a test light to test the operation of the power window master switch.

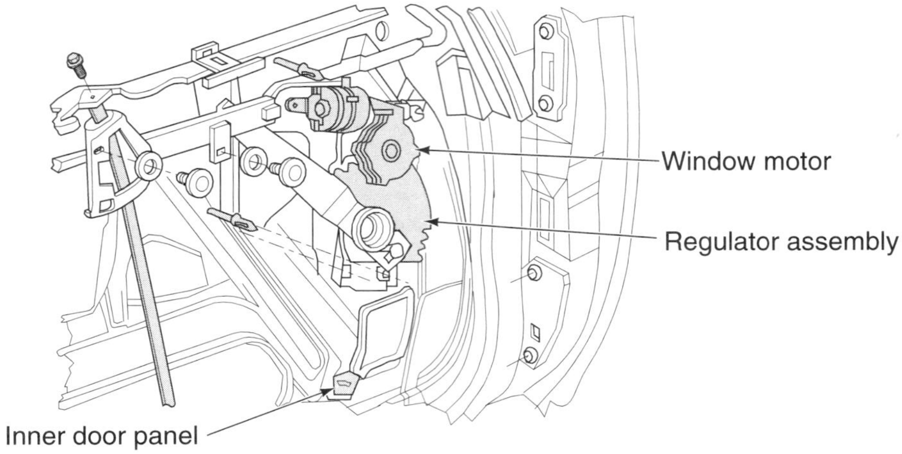

FIGURE 13–30 Direct drive motor and window regulator assembly.
(Courtesy of DaimlerChrysler Corporation)

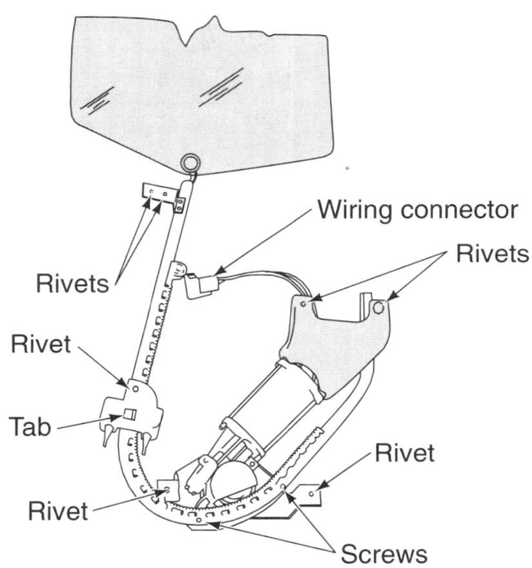

FIGURE 13–31 Rack-and-pinion style power window motor and window regulator assembly.
(Courtesy of DaimlerChrysler Corporation)

Power Seat Circuits

The power seat circuit allows the driver, and in some vehicles, the passenger to adjust the seat for individual comfort (Figure 13–32). Power seats are classified into three basic configurations—*two-way* that moves the seat forward and backwards; *four-way* that move the seat forward, backward, up, and down; and *six-way* that performs the same functions as the four way, but also tilts front and rear.

In Figure 13–33, a reversible permanent magnet motor pack called a **trimotor** controls the function of all six seat positions. Figure 13–34 is an example of a six-way power seat circuit that controls current flow direction through each motor depending on the individual function switches.

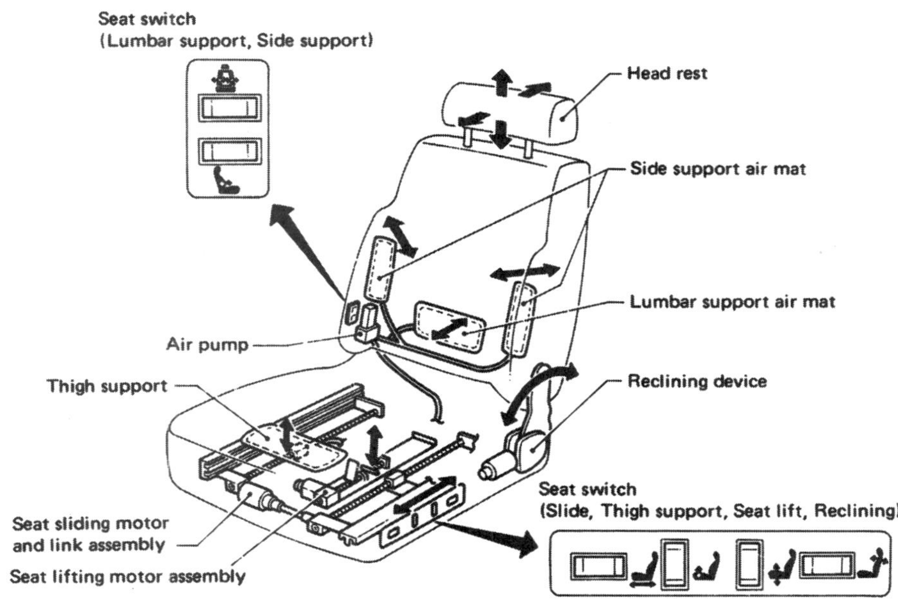

FIGURE 13–32 Typical six-way power seat components.

(Courtesy of Nissan Motor Corporation in U.S.A.)

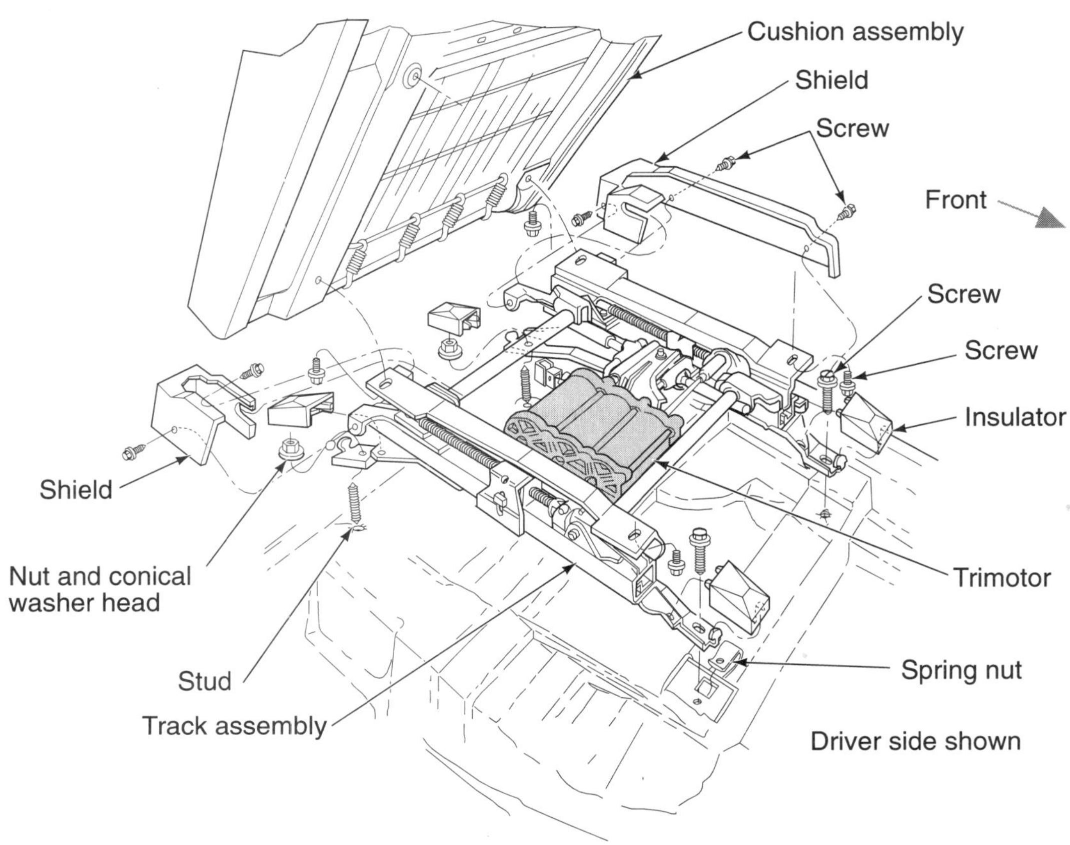

FIGURE 13–33 Location of trimotor pack for six-way adjustable seat.

(Courtesy of Ford Motor Company)

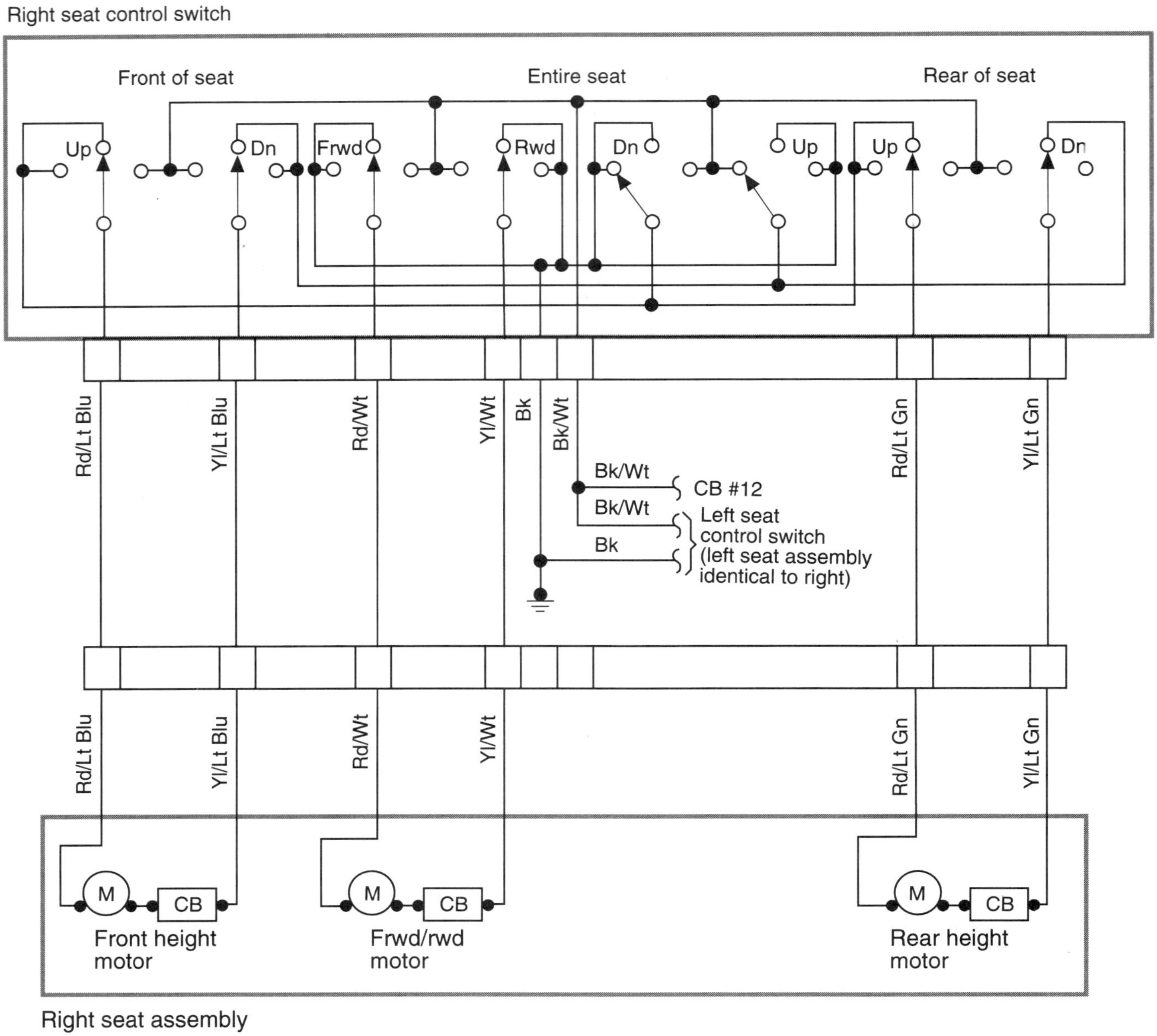

FIGURE 13–34 Typical six-way power seat components.

Power Door Lock Circuits

Modern door locks have small electric motors that operate the door lock mechanisms. In Figure 13–35, a door lock circuit uses a double relay to control door lock motors. When a door lock switch is moved to the lock or unlock position, the relay energizes and sends battery voltage to the door lock motors or actuator solenoids (Figure 13–36).

Some vehicles are also equipped with door locks that activate automatically when the transmission is placed in the drive position, and unlock when the transmission is returned to park.

Power Mirror Circuits

Remote-Controlled Power Mirror A typical power mirror circuit controls the position of the outside mirror(s) through a position switch and individual motors for up/down and right/left functions (Figure 13–37). In Figure 13–38, fused voltage is supplied to the power mirror switch for use by the individual direction switches and the mirror select switch.

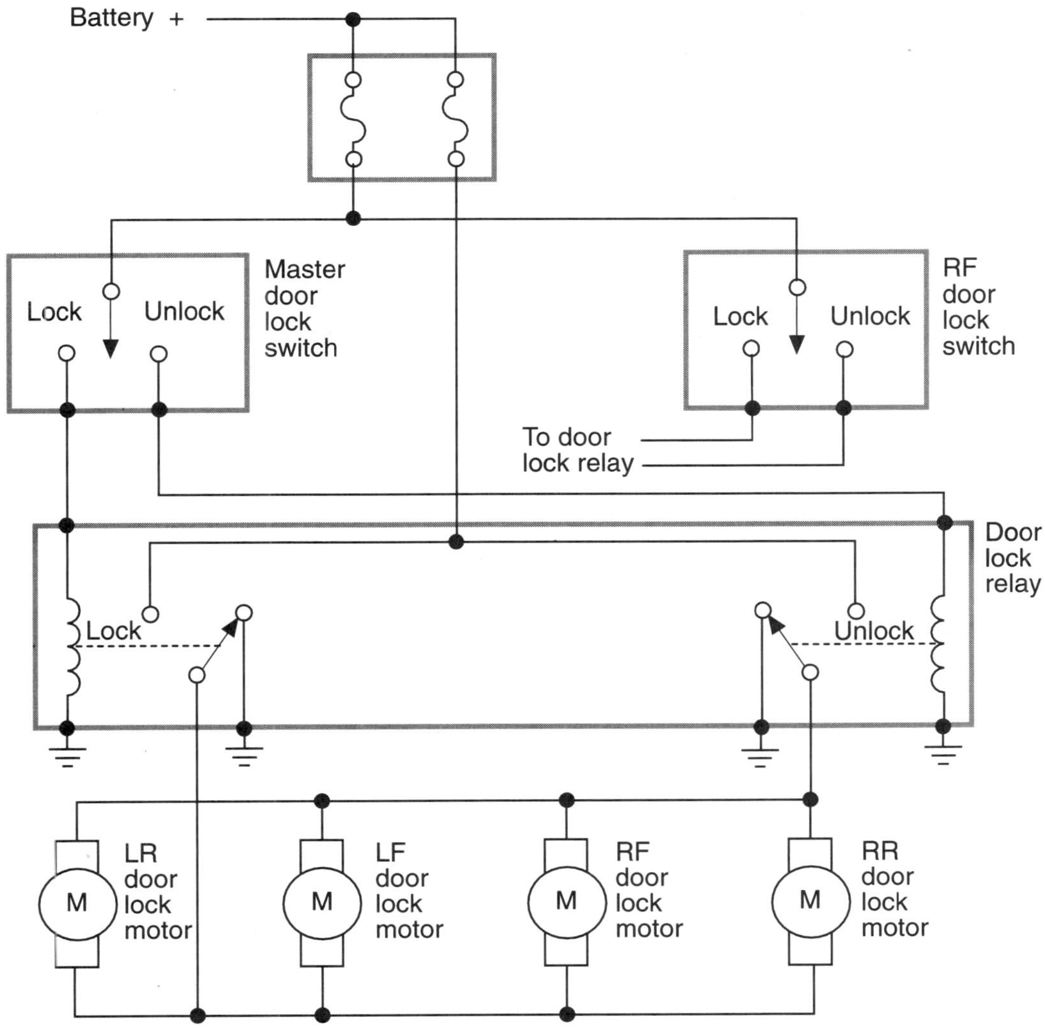

FIGURE 13–35 Typical power door lock circuit using a double relay to control door lock motors.

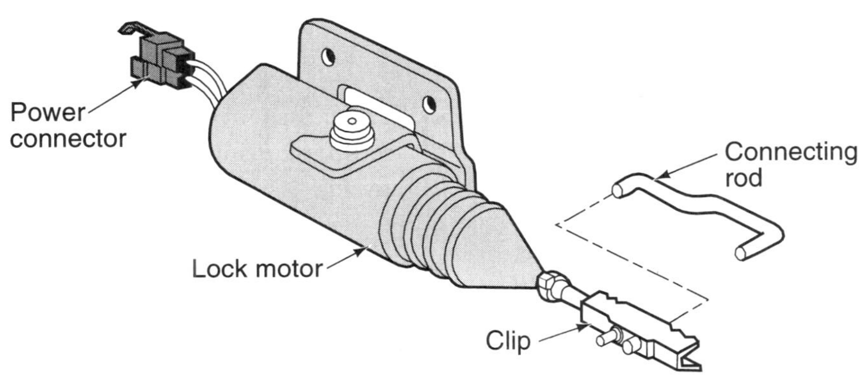

FIGURE 13–36 Typical permanent magnet power door lock motor assembly.

(Courtesy of DaimlerChrysler Corporation)

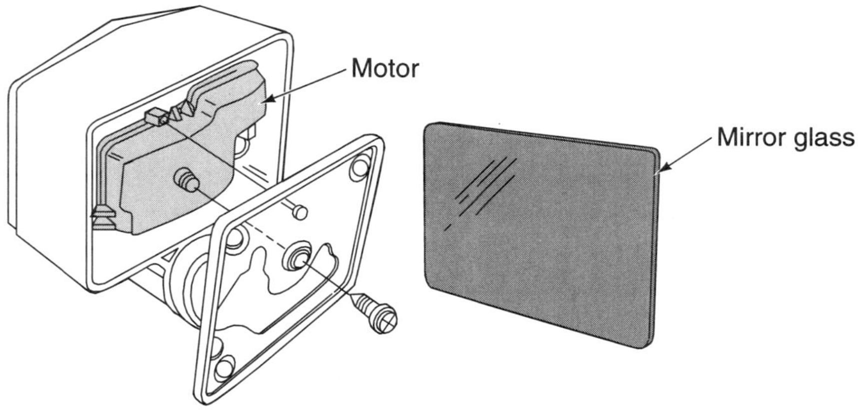

FIGURE 13–37 Remote power mirror components.

(Courtesy of General Motors Corporation Service Technology Group)

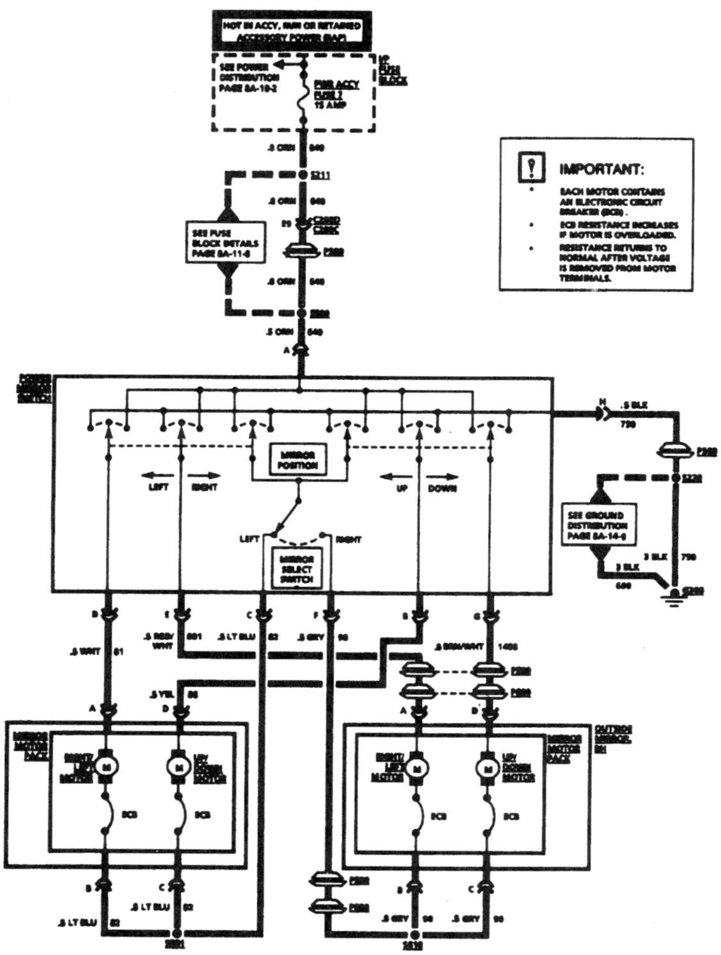

FIGURE 13–38 Power mirror circuit.

(Courtesy of General Motors Corporation Service Technology Group)

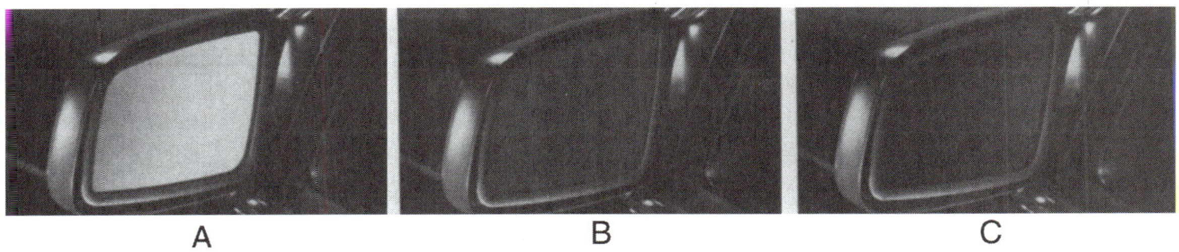

FIGURE 13–39 Electrochromic mirror operation—(A) daytime, (B) mild glare, and (C) high glare.
(Courtesy of Gentex Corporation, U.S.A.)

Electrochromatic Mirrors More advanced power mirror circuits control glare intensity automatically through an electrochromatic process (Figure 13–39). The mirror material is constructed of a micro thin layer of electrochromic material sandwiched between two conductive glass plates. Two internal photo cells measure the light intensity on both sides of the mirror and adjust the electrochromic material to prevent glare (Figure 13–39).

Conventional Instrumentation

Instrument panel gauges and warning lights in today's vehicles are designed to closely monitor the proper function of several vehicle operating systems and provide feedback to the driver. Some of the more complex instrument panel clusters consist of digital circuitry and liquid crystal displays. This chapter looks at a few of the basic analog instrument cluster gauges (Figure 13–40), and some of the input sending units or sensors that control the gauges or warning lights.

Analog Gauges

There are two types of analog gauges used in systems today. The first is a **magnetic gauge** where a magnetic field is used to move the gauge needle. The simplest type of magnetic gauge is called a d'Arsonval gauge, where a coil of wire is wrapped around the base of the pointer needle to form an armature (Figure 13–41). When current flows through the coil of wire, a magnetic

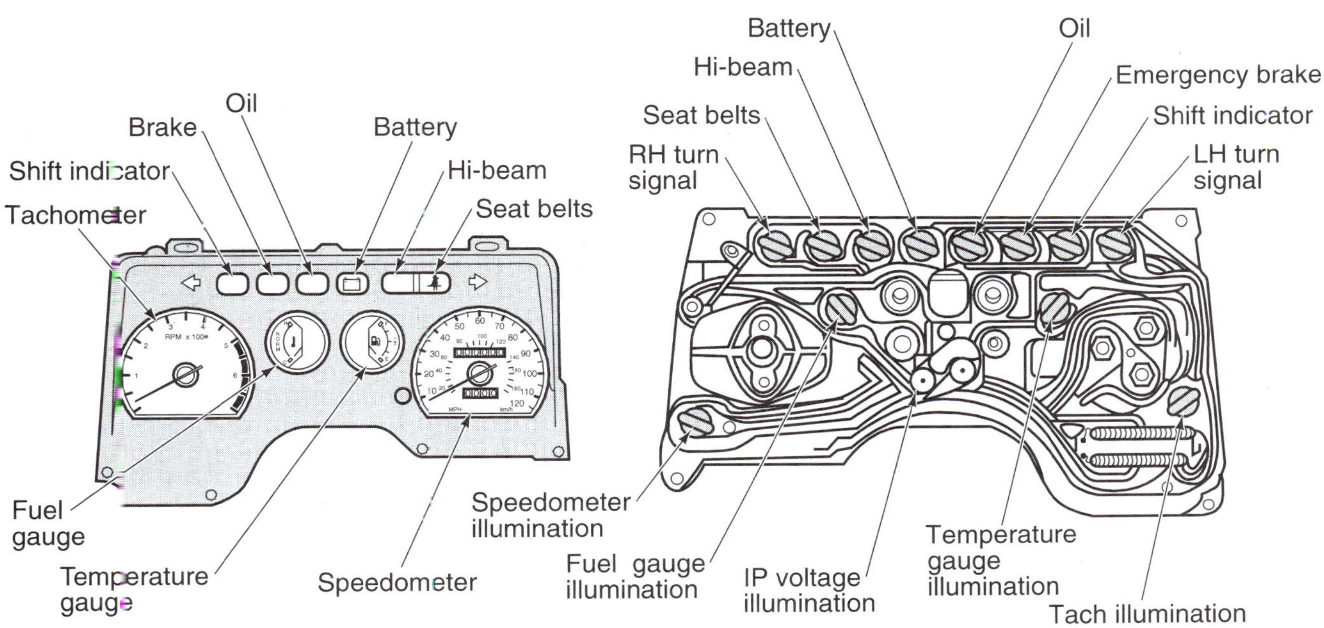

FIGURE 13–40 Instrument panel gauges, warning lights, and printed circuit board.
(Courtesy of Ford Motor Company)

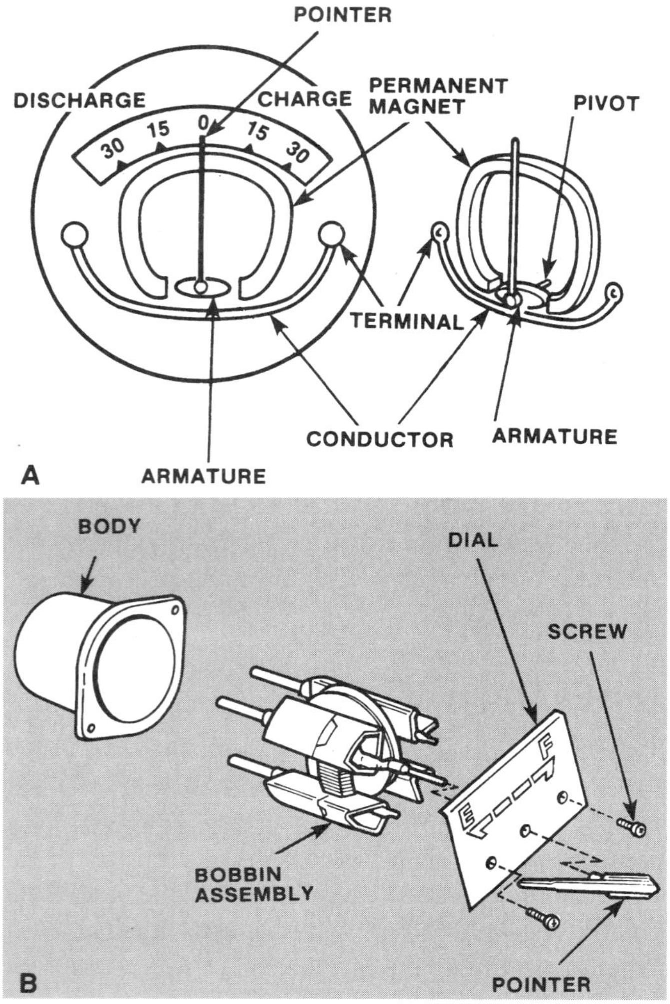

FIGURE 13–41 (A) Simple magnetic d'Arsonval gauge and (B) magnetic bobbin gauge.

(Courtesy of Ford Motor Company)

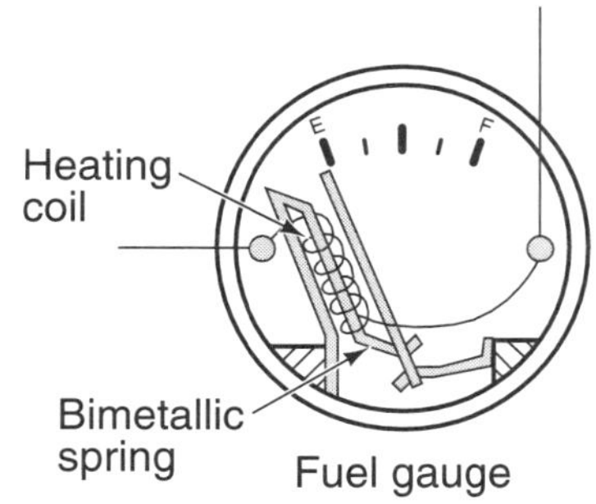

FIGURE 13–42 Simple bimetallic gauge construction.

(Courtesy of Ford Motor Company)

field is produced in the armature that opposes the permanent magnet and causes the pointer needle to swing. Another simple magnetic gauge is a bobbin gauge (Figure 13–41). In this gauge, the bobbin assembly creates a magnetic field that moves the pointer needle attached to the bobbin.

The second type of common analog gauge is a **thermal gauge** (Figure 13–42), also know as a bimetallic gauge. This gauge operates when controlled current flows through a heating coil in the gauge. The generated heat causes the bimetallic strip to bend in proportion to the amount of heat produced.

Instrument Voltage Regulators

Many analog gauges require the use of an **IVR** (**instrument voltage regulator**) that maintains a constant voltage to the gauge under all battery load and charging conditions. In Figure 13–43, a typical IVR consists of a

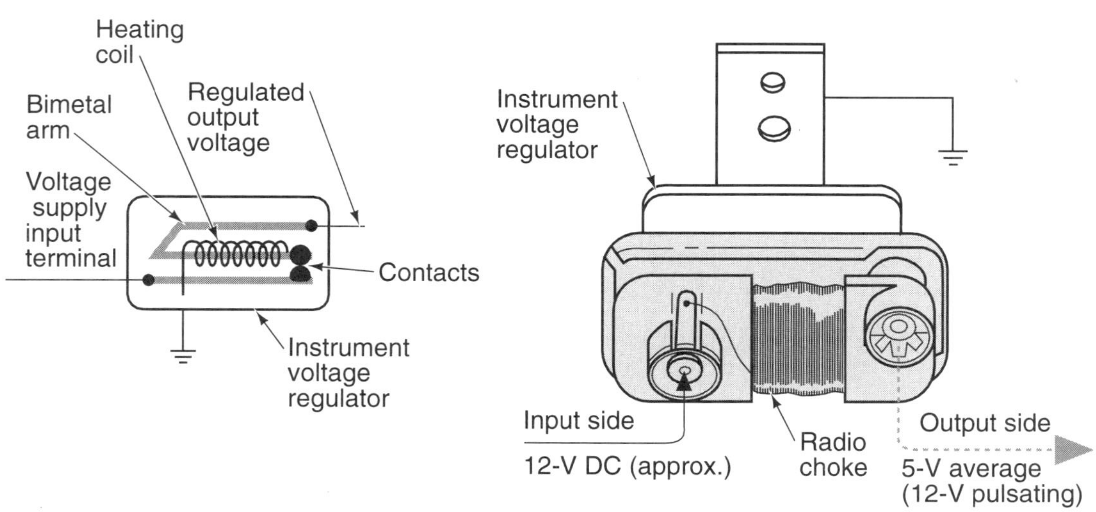

FIGURE 13–43 Typical instrument voltage regulator (IVR).

bimetallic arm, heating coil, and a set of contact points. When current flows through the contact points and heating coil, the bimetal arm bends and momentarily opens the contact points. As the contact points open and close with the heating and cooling of the bimetal arm, a stable output of up to 10 volts is produced, depending on vehicle manufacturer.

Figure 13–44 is an example of a thermal fuel gauge circuit with an instrument voltage regulator (IVR). The fuel level determines the location of the variable resistance sending unit in the fuel tank. The sending unit controls the amount of current flowing through the gauge heating coil. The *higher* the heat produced by the heating coil, the *higher* the pointer needle will read on the gauge scale.

Gauge Sending Units

Depending on the analog gauge circuit, the sending unit is a vital part of the circuit design (Figure 13–45). In the oil pressure switch circuit, the circuit is *open* when the engine is off. When the engine is on and the oil pressure is above a specified level (normally above 4.5-7.5 psi), the switch closes and current flows through the circuit creating a magnetic field for the pointer needle. The oil pressure sending unit used for a gauge (Figure 13–46) consists of a piezo-resistor that is moved against a contact arm when oil pressure is exerted on the flexible diaphragm.

The temperature gauge sending unit in Figure 13–47 is typical for a coolant temperature circuit. Current is sent from the temperature gauge to the sending unit terminal, through the thermistor (variable resistor), and then grounded to the engine. The resistance value of the thermistor depends on the temperature of the engine. As engine temperature rises, the thermistor resistance *decreases,* and the current flow through the gauge circuit *increases,* which raises the pointer needle towards the hot range.

The fuel gauge sending unit is typically a variable resistor that is manually controlled by an arm and float which depends on the fuel level (Figure 13–48). Resistance of the sending unit *increases* as the fuel level *decreases,* and causes the fuel pointer needle to move toward Empty.

When a warning lamp is used in place of a gauge, the sending unit is different than those used for gauges. The sending unit for a warning lamp circuit is either open, or closed to complete the circuit to ground (Figure 13–49). When the engine is off or there is no oil pressure, the contact points in Figure 13–49 are closed to allow a complete circuit for the warning lamp to light. When oil pressure is above the minimum specified pressure (usually 3-5 psi), the diaphragm will move the contacts apart and the warning lamp will go out.

Figure 13–50 is an example of a sending unit and simple circuit for a temperature warning light. The contact points in the temperature sending unit are *normally open* during normal engine temperatures. When a predetermined temperature level has been exceeded, the contacts close and complete the warning light circuit to ground.

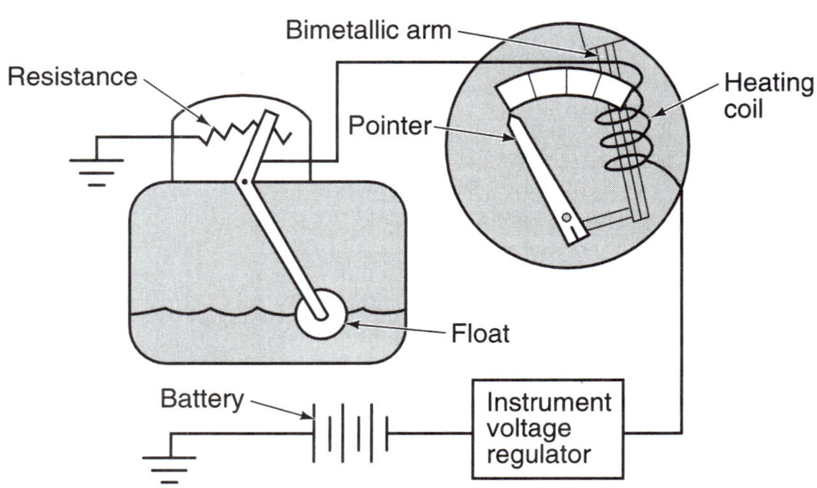

FIGURE 13–44 Thermal fuel gauge circuit with instrument voltage regulator (IVR).

(Courtesy of DaimlerChrysler Corporation)

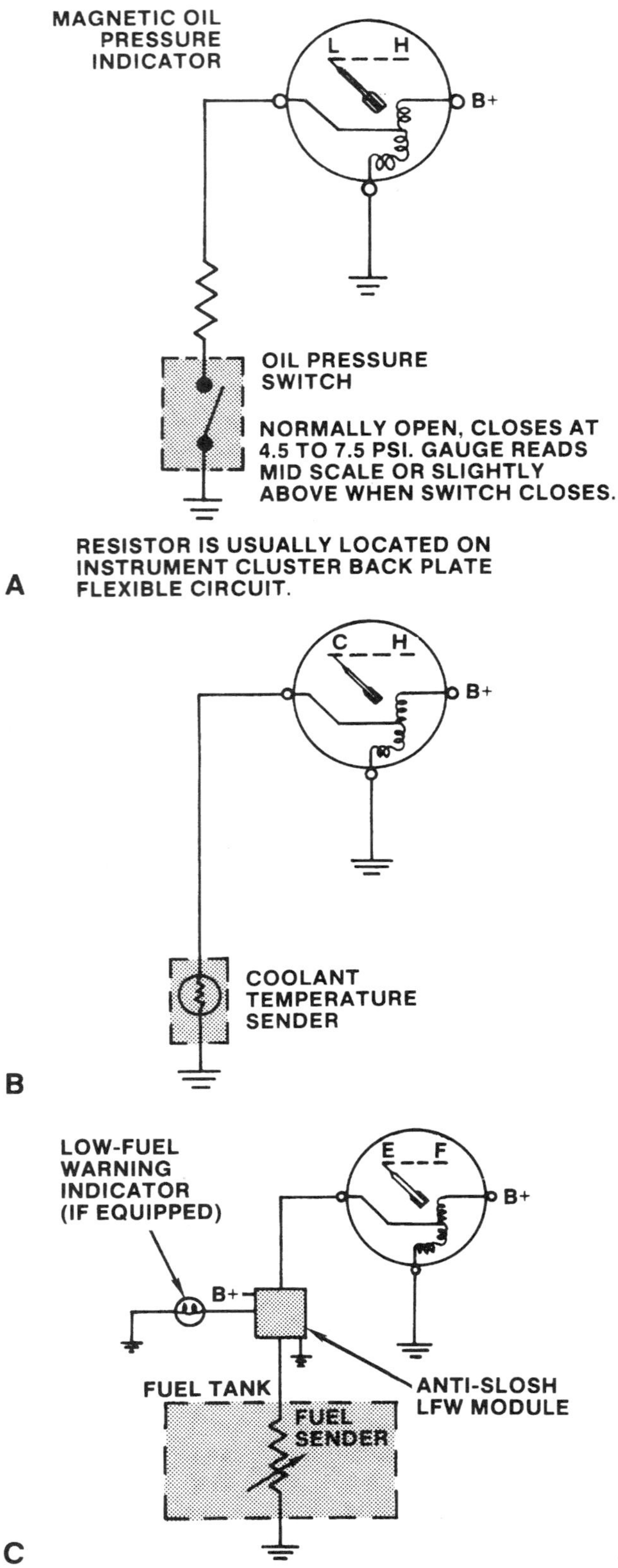

FIGURE 13–45 (A) Typical oil pressure circuit, (B) coolant temperature circuit, and (C) fuel gauge circuit.

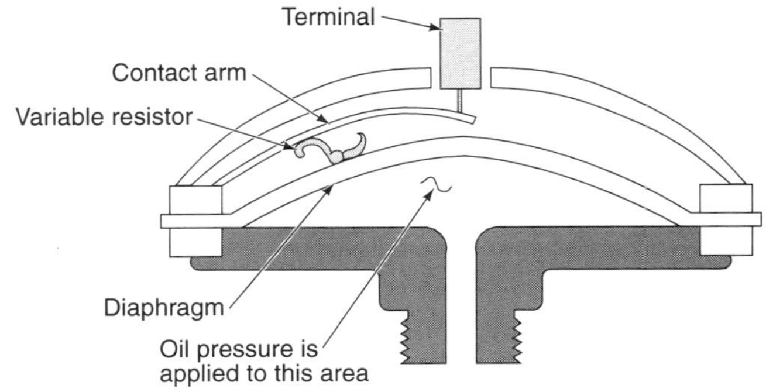

FIGURE 13–46 Typical piezoresistive oil pressure sensor used in an oil pressure gauge circuit.

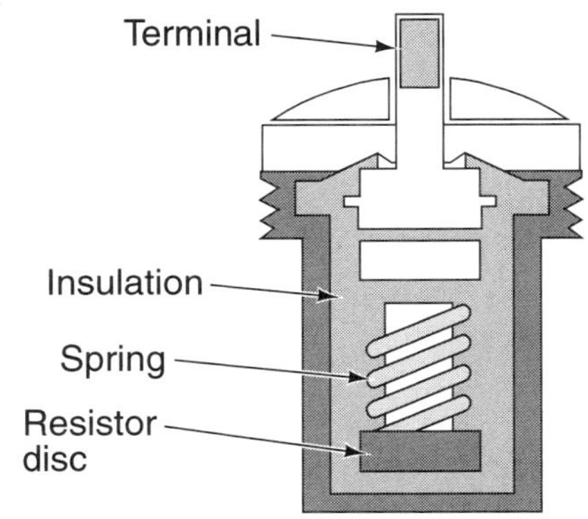

FIGURE 13–47 A typical temperature thermistor resistive sending unit.

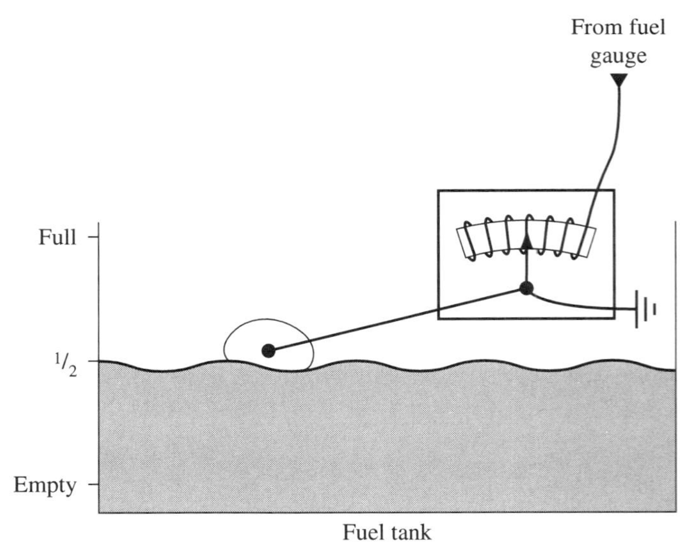

FIGURE 13–48 Typical variable resistance fuel gauge sending unit.

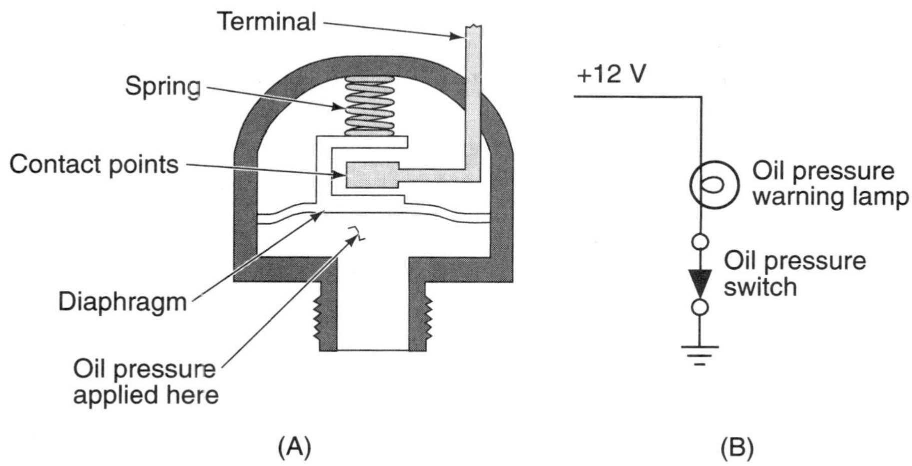

FIGURE 13–49 (A) Oil pressure light sending unit and (B) simple oil pressure warning light circuit.

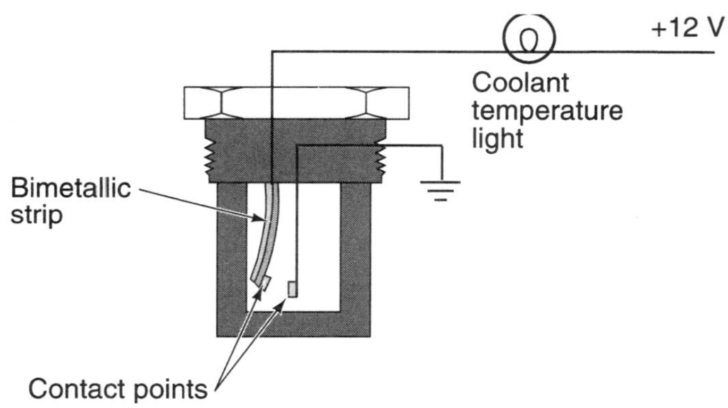

FIGURE 13–50 Typical temperature warning light circuit and sending unit.

Some instrument panels use a combination of gauges and warning lights (Figure 13–51). In the instrument cluster there are individual sending units for each gauge or fluid level indicator, even though they share the same common power source line.

Air Bag Systems

Most newer vehicles are factory equipped with at least one air bag or **supplemental inflatable restraint (SIR)** system (Figure 13–52) that inflates and deflates in milliseconds on vehicle frontal collision impact. A typical single air bag system consists of a steering wheel air bag module, clock spring, diagnostic module, and various sensor arms (Figure 13–53).

Air Bag Module The air bag module contains the inflatable porous nylon air bag assembly and inflator ignitor which are packaged in a single unit (Figure 13–54). When a vehicle frontal crash is indicated, the ignitor spark ignites a canister of zerconic potassium perchlorate (ZPP) that ignites the inflator propellant charge containing sodium azide and copper oxide (Figure 13–55).

As the inflator propellant burns, it rapidly produces nitrogen gas that inflates the air bag in milliseconds.

Clock Spring The clock spring provides electrical continuity to the air bag module through all steering wheel positions (Figure 13–56). Electrical wires connect to one end of the ribbon spring and the other end to the air bag module. The ribbon spring winds and unwinds as the steering wheel is turned, but only for a certain distance. If the steering column is disconnected for service,

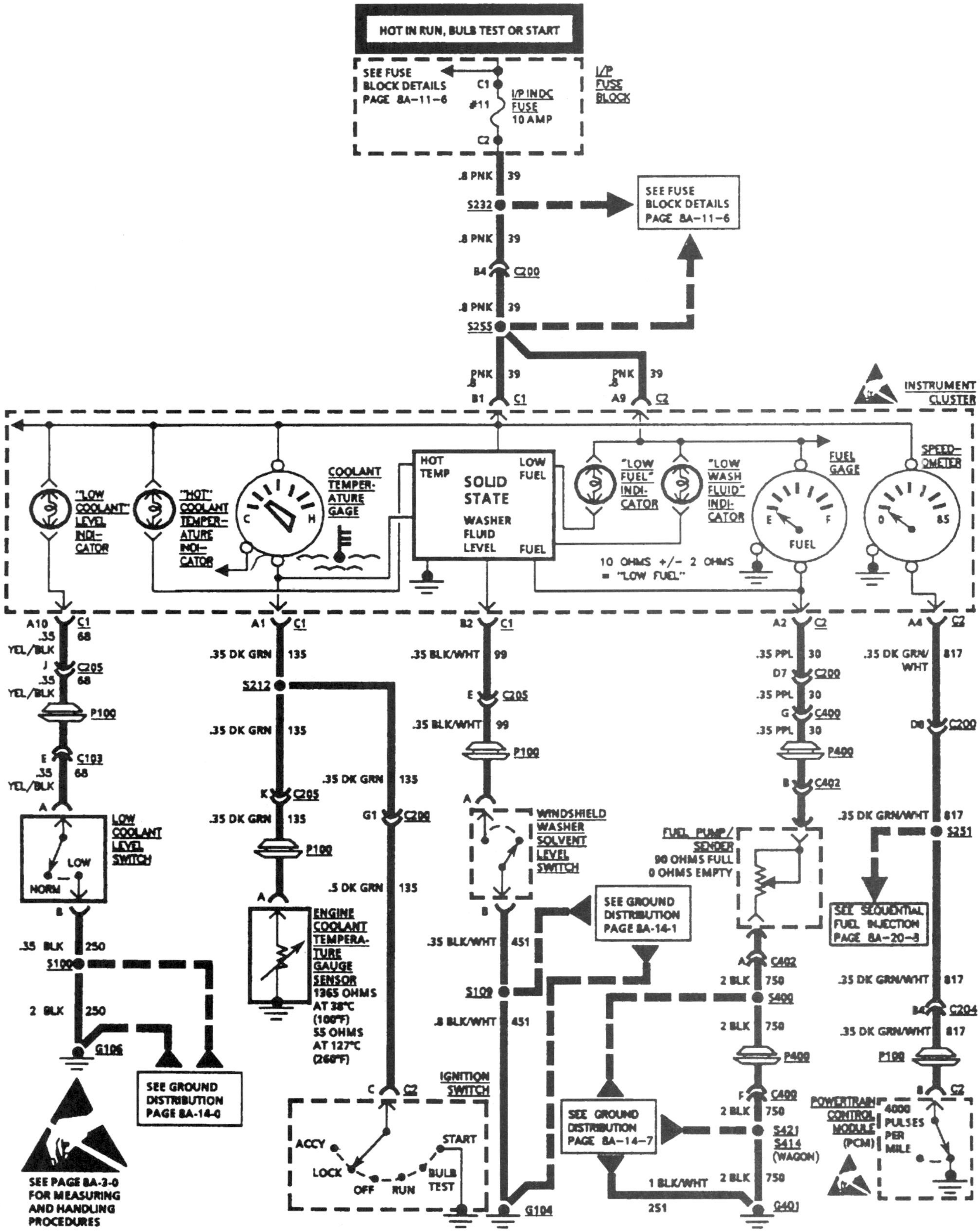

FIGURE 13–51 Instrument panel cluster with gauge and warning lights.

(Courtesy of General Motors Corporation Service Technology Group)

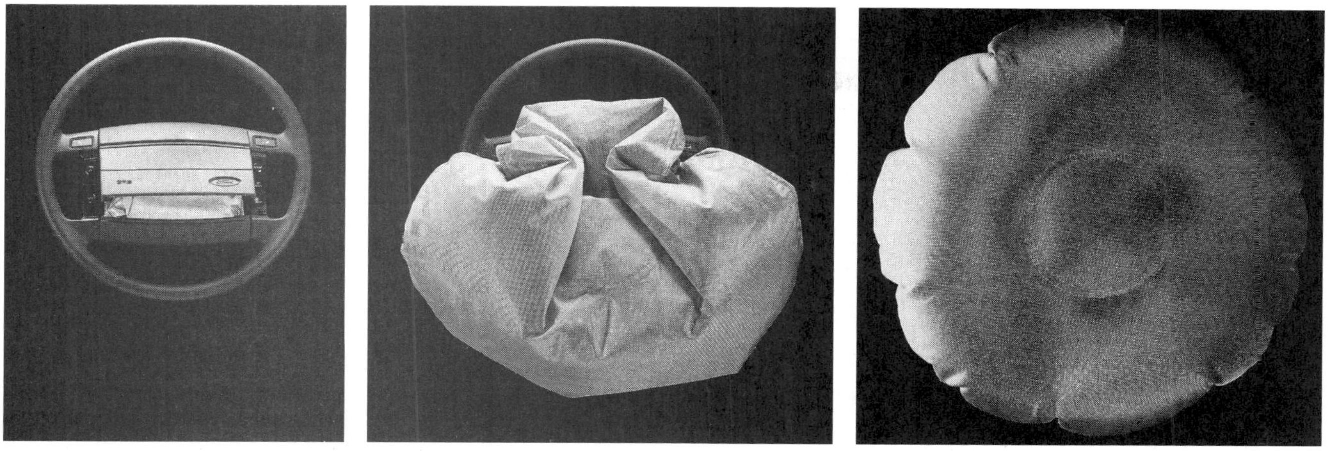

(a)

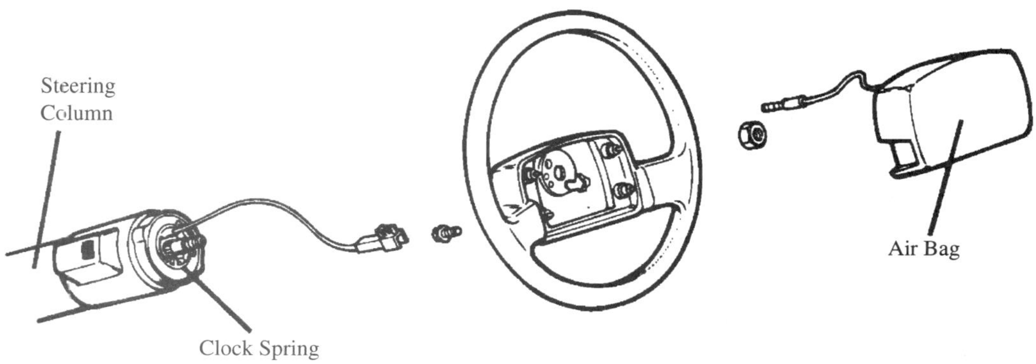

(b)

FIGURE 13–52 (A) Inflation of air bag occurs in a few milliseconds, and (B) basic components of air bag.
(Courtesy of Moog Automotive Inc., St. Louis, MO.)

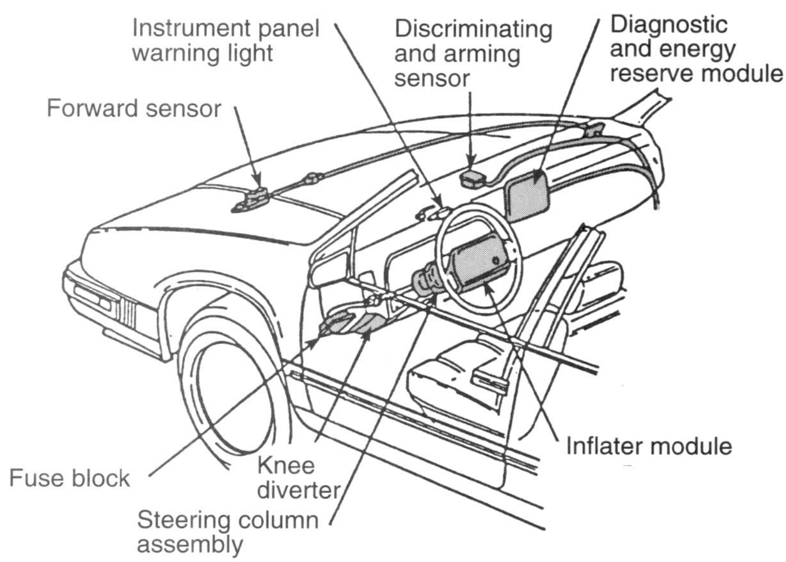

FIGURE 13–53 Location of typical air bag components.
(Courtesy of DaimlerChrysler Corporation)

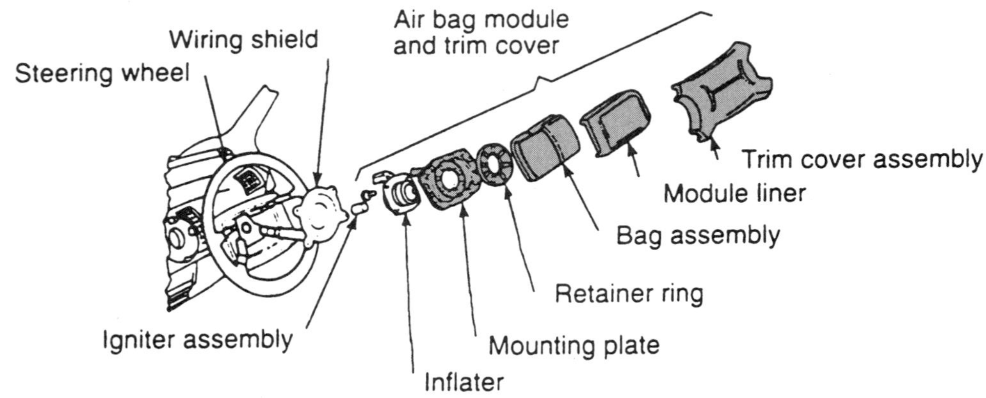

FIGURE 13–54 Detailed components of air bag module.

(Courtesy of DaimlerChrysler Corporation)

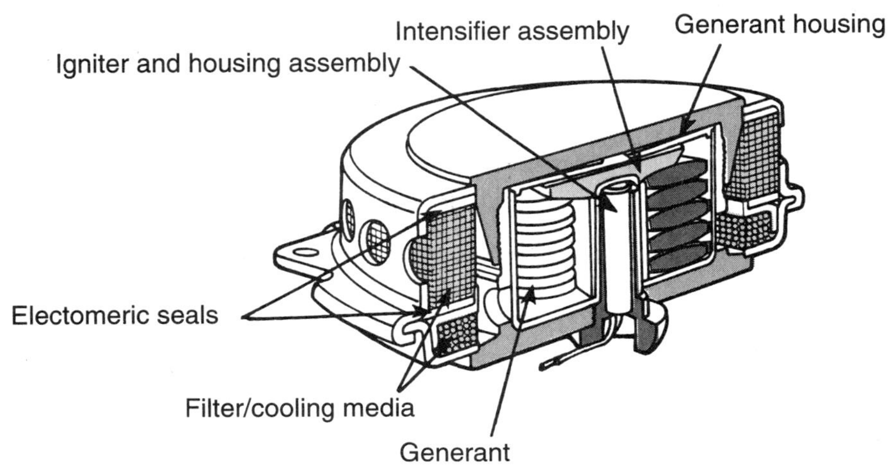

FIGURE 13–55 Air bag igniter assembly.

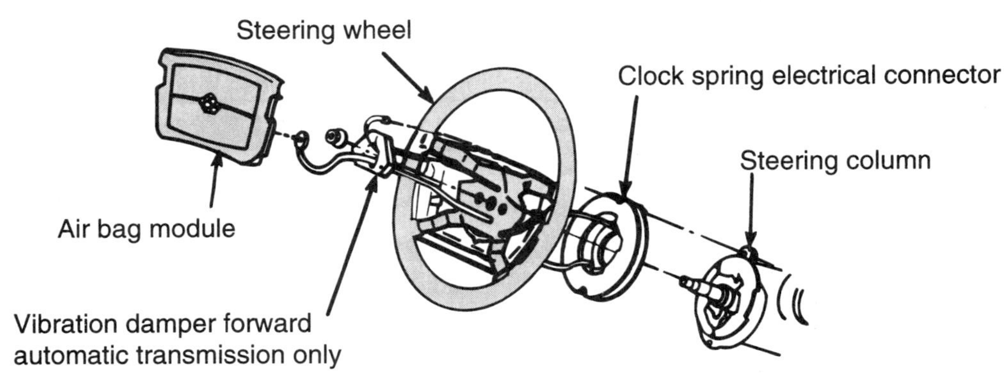

FIGURE 13–56 The clock spring provides a rotating electrical connection to air bag in all steering wheel positions.

(Courtesy of DaimlerChrysler Corporation)

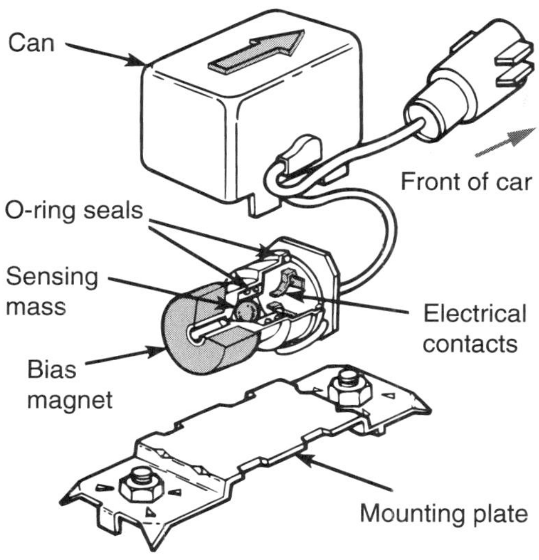

FIGURE 13–57 Mass-type impact sensor.
(Courtesy of DaimlerChrysler Corporation)

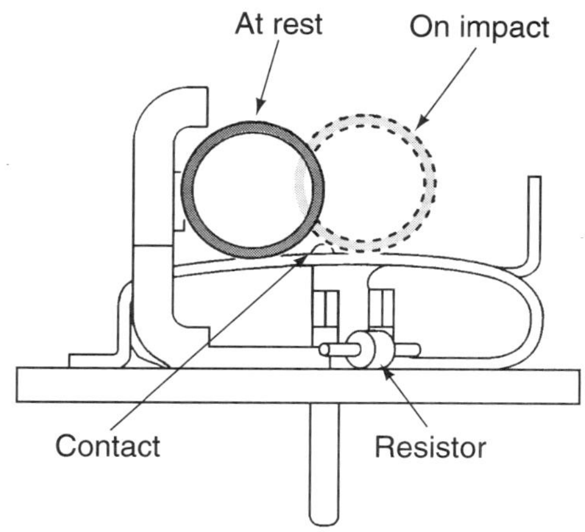

FIGURE 13–58 Roller-type impact sensor.
(Courtesy of DaimlerChrysler Corporation)

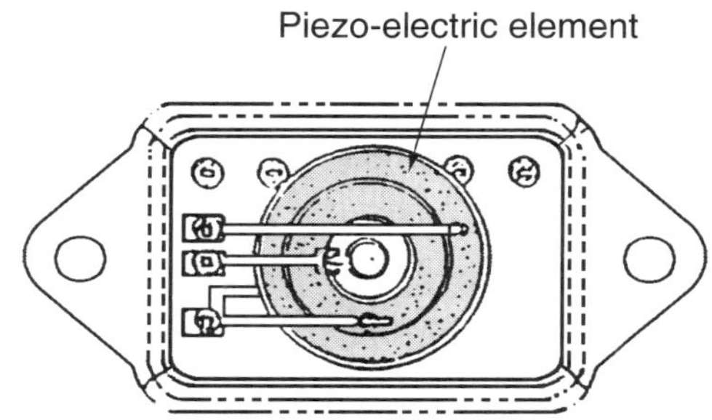

FIGURE 13–59 Accelerometer-type impact sensor with piezoelectric element.
(Courtesy of DaimlerChrysler Corporation)

the column must be locked in place to prevent accidental overwinding of the clock spring.

Impact Sensors. Depending on the SIR system, there may be one to five impact sensors. Most SIR systems with more than one sensor require at least 2 impact sensors to close to deploy the air bag module. There are three common impact sensors used in SIR systems—mass-type, roller-type, and accelerometers.

A mass-type impact sensor (Figure 13–57) uses a gold-plated normally open switch and a gold-plated sensing mass ball that moves in a highly polished cylinder. In a frontal collision exceeding 10 to 15 mph or more, the sensing ball moves forward enough to close the switch contacts.

A roller-type impact sensor uses a spring-loaded roller mass mounted on a ramp (Figure 13–58). In a frontal collision exceeding 10 to 15 mph or more, the roller moves up the ramp and strikes the spring contact to complete the circuit.

The third type of impact sensor is an **accelerometer** (Figure 13–59) which contains a piezoelectric element that distorts in a frontal collision. The voltage generated by the distorted piezoelectric element is processed by the air bag computer which deploys the air bag.

Hybrid Inflator Module All passenger vehicles produced after 1995 are required to have a passenger side air bag in addition to the driver's side (Figure 13–60). Because of the distance between the passenger and dash panel compared to the driver/steering wheel, a larger air bag is needed. Some manufacturers use a container of pressurized argon gas (Figure 13–61) that fills the air bag when the initiator explodes and pierces the propellant container (Figure 13–62).

CAUTION *Because of the explosive force of passenger side air bags,* NEVER *place a rearward facing child safety seat in the front seat. Serious injury or death may result if the air bag is deployed.*

Basic Air Bag Electrical System Figure 13–63 is a simple air bag wiring diagram that requires at least two sensor switches to close before the air bag deploys. Using more than one impact switch helps prevent the inadvertent deployment of an air bag in a minor accident.

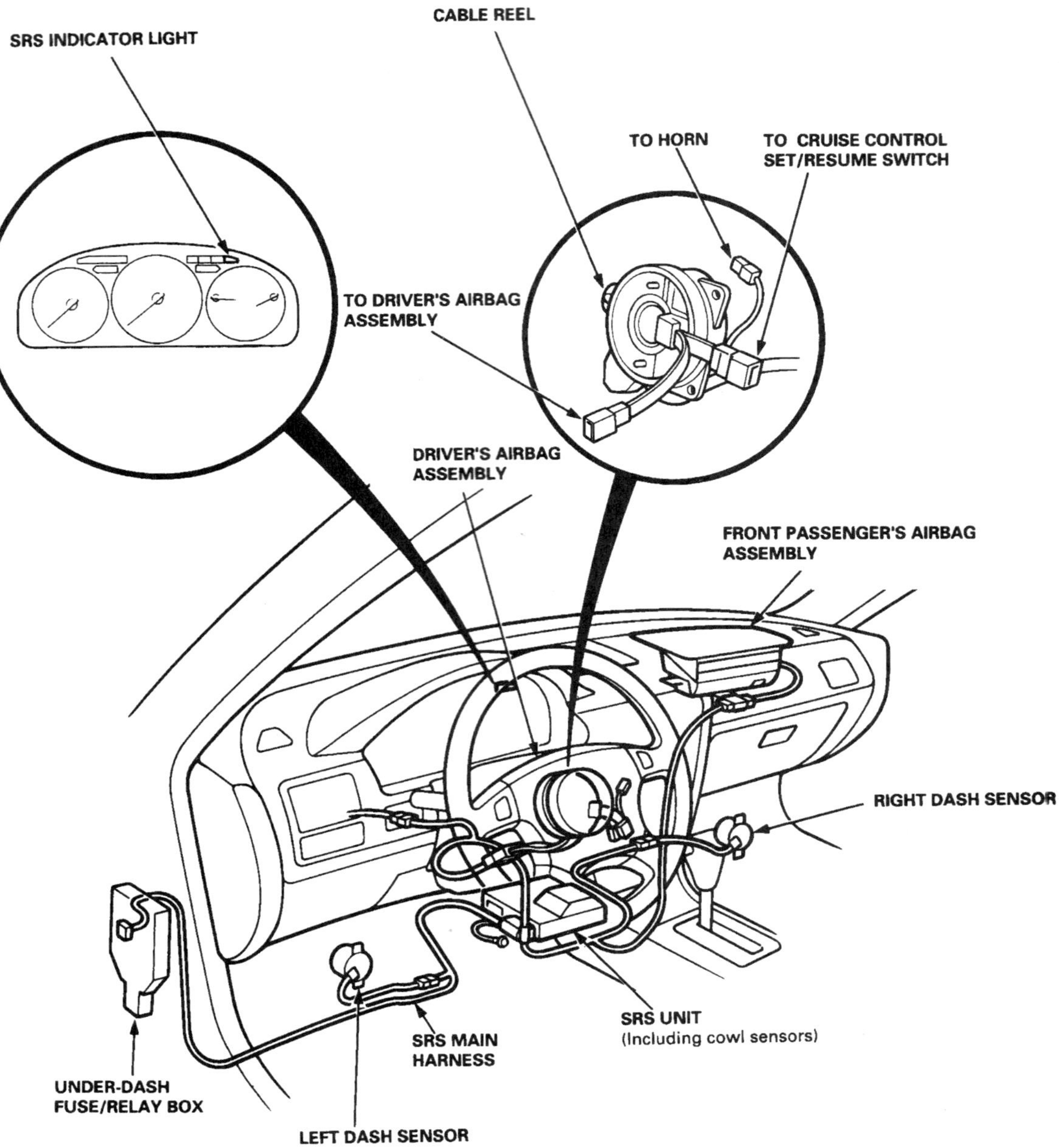

FIGURE 13–60 Location of driver and passenger air bag components.
(Courtesy of American Honda Motor Company, Inc.)

Figure 13–64 is a typical example of a more complex air bag wiring diagram with fault code detecting capabilities built into the system. Depending on the complexity of the vehicle air bag system and the number of components involved, always refer to specific vehicle manufacturer service information before diagnosing or repairing system components.

Air Bag Safety When handling or servicing an undeployed air bag module (Figure 13–65), observe the following safety precautions.

1. Avoid static electricity. Use appropriate static grounding strap when handling air bag components.
2. Always carry the air bag module facing away from you.
3. Store air bag module on clean bench facing upward and in a secure area.
4. Never store an air bag module in a high heat area.
5. Read all manufacturer service information before servicing an air bag system.

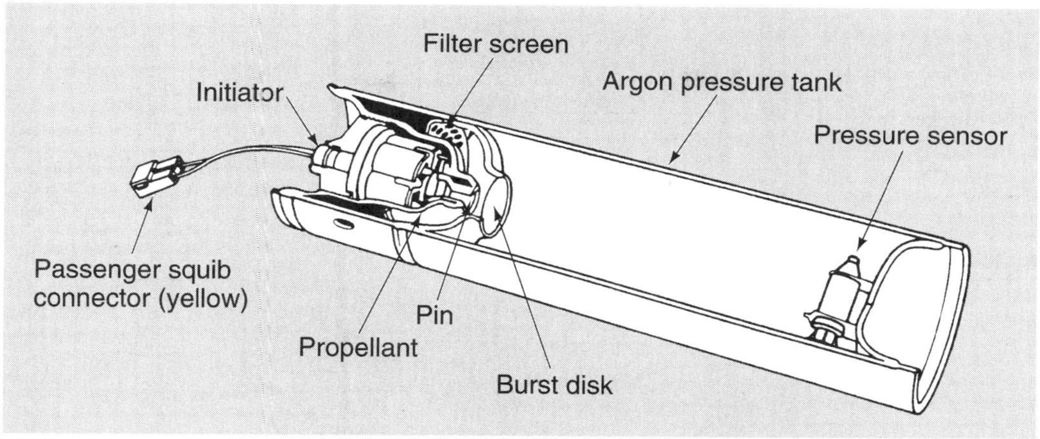

FIGURE 13–61 Hybrid inflator module with argon gas pressure chamber.
(Courtesy of DaimlerChrysler Corporation)

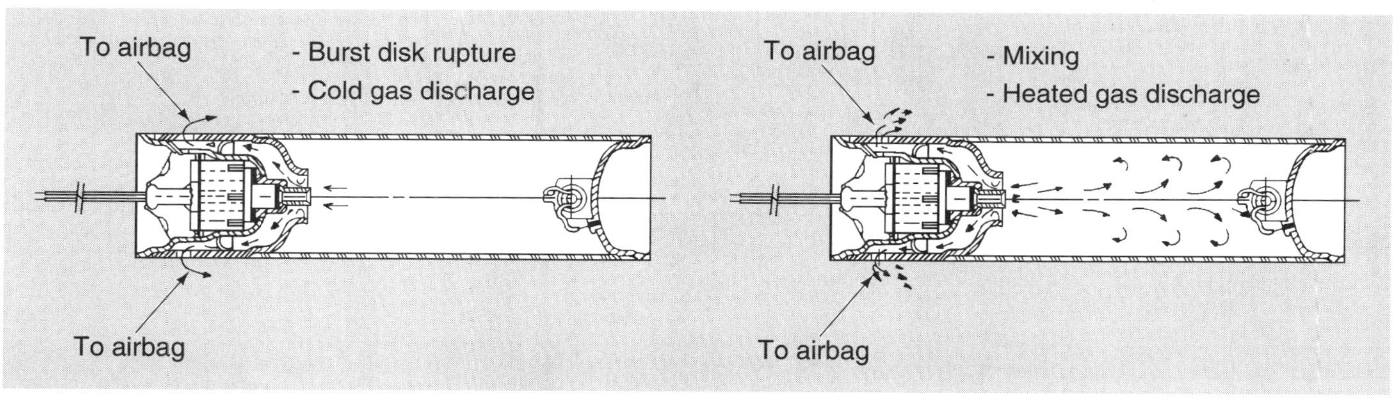

FIGURE 13–62 After initiator explodes and pierces the propellant container, argon gas will rapidly fill the air bag.

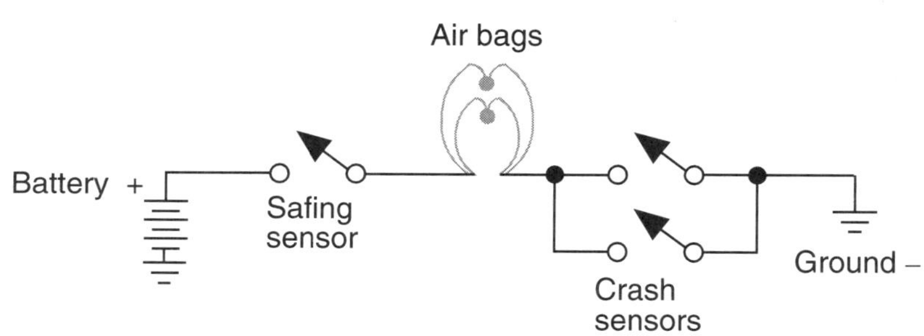

FIGURE 13–63 Simple air bag sensor circuit diagram.

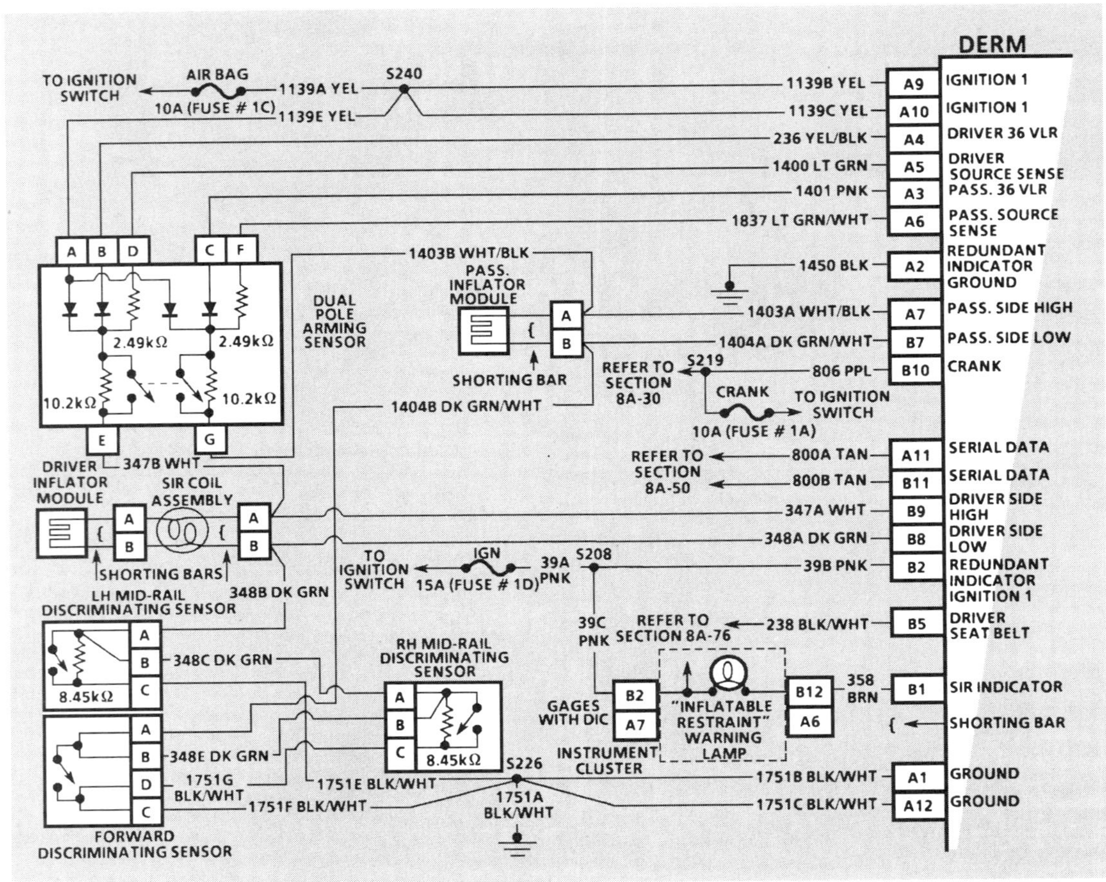

FIGURE 13–64 Complex air bag system diagram.

(Courtesy of General Motors Corporation, Service Technology Group)

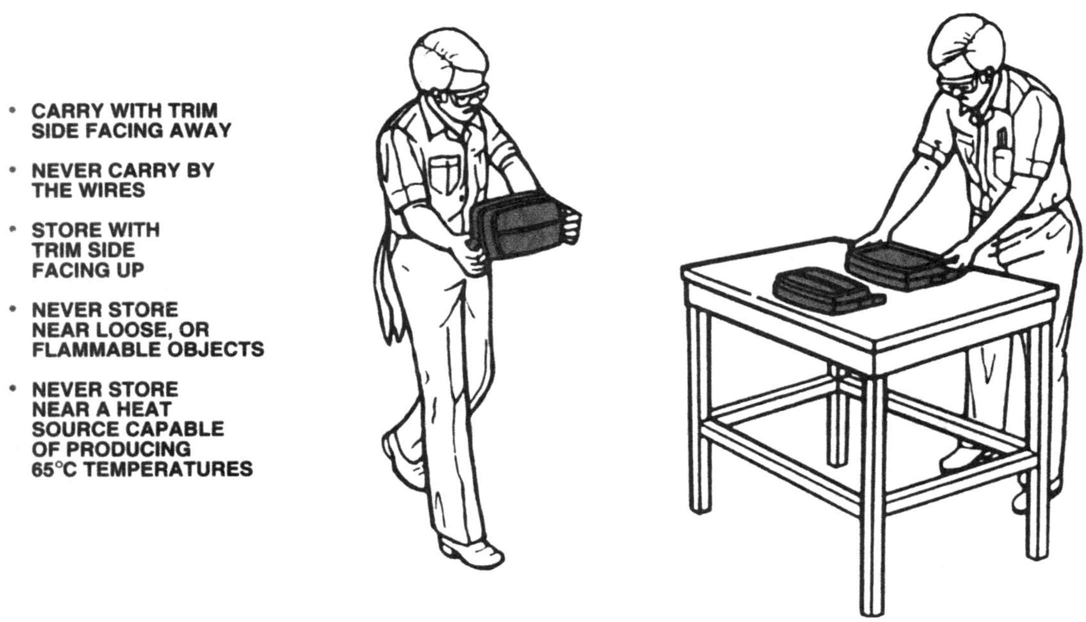

FIGURE 13–65 Proper air bag safety precautions.

(Courtesy of General Motors Corporation, Service Technology Group)

CASE STUDIES

CASE #1

The following case study is typical of a horn circuit problem that technicians often need to diagnose. Knowledge of basic horn circuits and simple test procedures will help you perform the diagnosis quickly and efficiently.

Customer Complaint

A customer brought her vehicle into the shop with a horn system problem. The customer replaced the horn relay, but the problem remained. Figure 13–66 is a diagram of the horn circuit for this vehicle.

Known Information

- ❑ Vehicle operating voltage = 14 volts
- ❑ All circuit fuses are OK
- ❑ Horn relay has been replaced

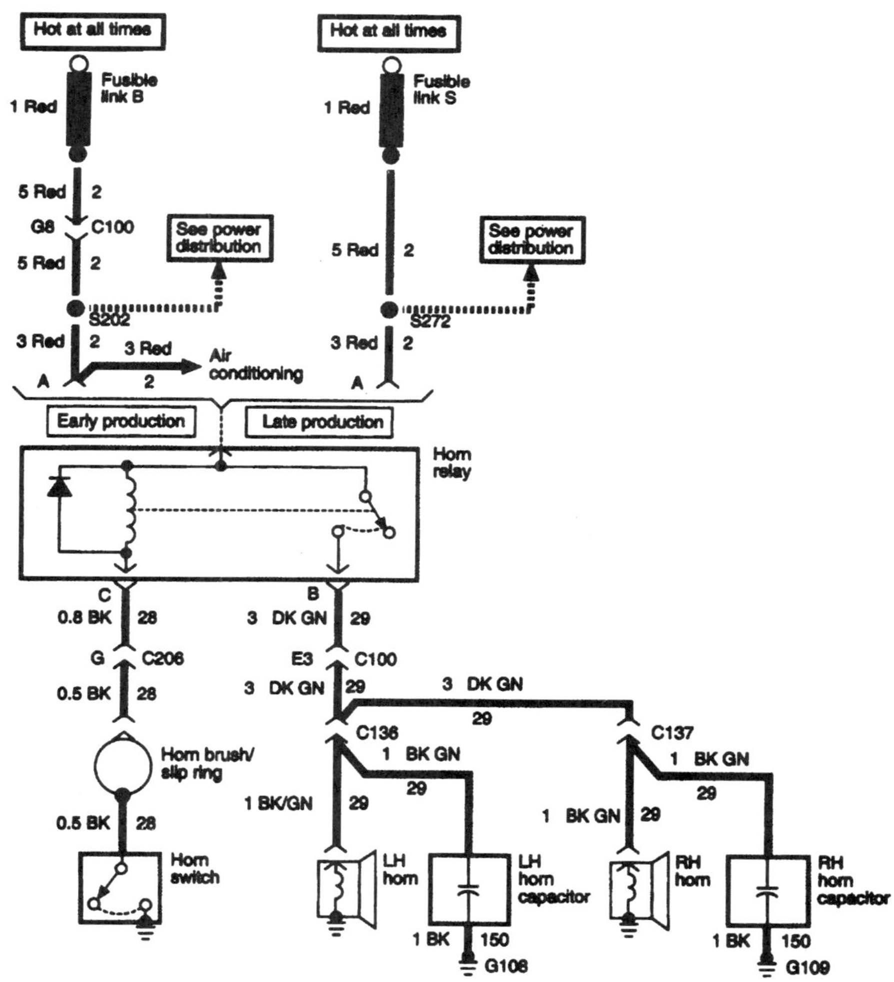

FIGURE 13–66 Diagram of the horn circuit for Case #1.

Circuit Analysis

Answer the following questions on a *separate* sheet of paper.

1. With the known information above, what is the *most* likely cause of the inoperative horn circuit?
   ______________________________.
2. What diagnostic steps and tools would you use to find the suspected problem in this horn circuit?
   ______________________________.
3. If you disconnected the C206 connector between the horn relay and the horn brush slip ring, what *should* happen in the circuit if the relay side of the C206 connector was grounded? ______________________________.
4. If the circuit was working properly, what should the voltage reading be at the C100 connector from the horn relay to the horns? ______________________________.
5. How could you verify the circuit problem using a jumper wire?
   ______________________________.
6. What is the best way to repair this system?
   ______________________________.
7. What diagnostic steps were helpful when troubleshooting this system or explaining it to the customer? ______________________________.
8. What is your analysis of the above case study? ______________________________.

CASE #2

This case study is typical of a window defogger circuit problem that technicians often need to diagnose. Knowledge of defogger circuit components and wiring diagrams will help you perform the diagnosis quickly and efficiently.

Customer Complaint

The rear window defogger does not work when the switch is turned on, but the indicator light illuminates. The relay was replaced by the customer, but the problem remained. Figure 13–67 is a diagram of the window defogger circuit.

Known Information

- ❑ Vehicle operating voltage = 14 volts
- ❑ Defogger relay was replaced
- ❑ System fuse is OK

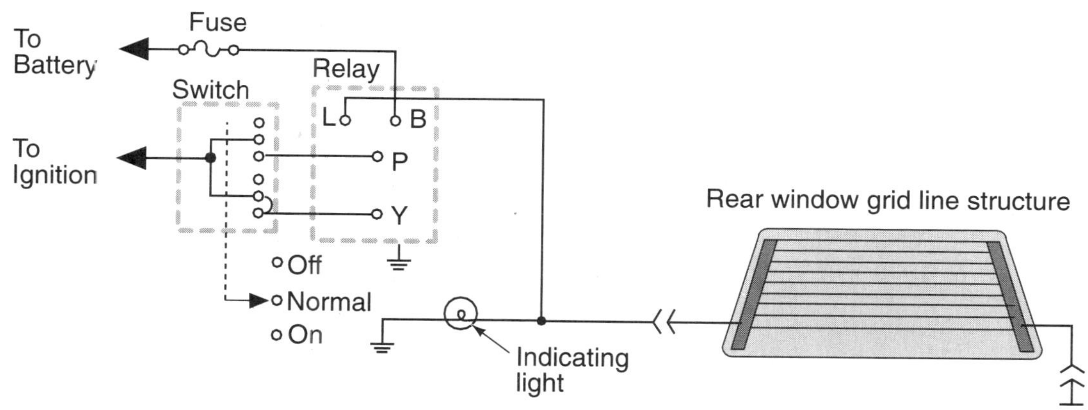

FIGURE 13–67 Diagram of the window defogger circuit for Case #2.

Circuit Analysis

Answer the following questions on a *separate* sheet of paper.

1. With the known information above, what is the *most* likely cause of the inoperative defogger circuit? __________.
2. If the indicator light is working, what does that tell you about the voltage of the wire leading to the connector at the rear window grid assembly? __________.
3. What diagnostic steps and tools would you use to find the suspected problem in this window defogger circuit? __________.
4. If the voltage at the positive bus bar side of the rear window grid is operating voltage, what would be the *next most* likely cause of the circuit problem? __________.
5. How could you verify the circuit problem using a jumper wire? __________.
6. What diagnostic steps were helpful when troubleshooting this system or explaining it to the customer? __________.
7. What is your analysis of the above case study? __________.

CASE #3

The following case study, shown in Figure 13–68, is typical of an instrument gauge problem. Knowledge of gauge designs, wiring diagrams, and simple test procedures can help you diagnose the gauge circuit safely and efficiently.

Customer Complaint

The customer complains that the instrument cluster fuel gauge reads Empty all the time, regardless of the amount of fuel in the tank. The customer would like the system repaired.

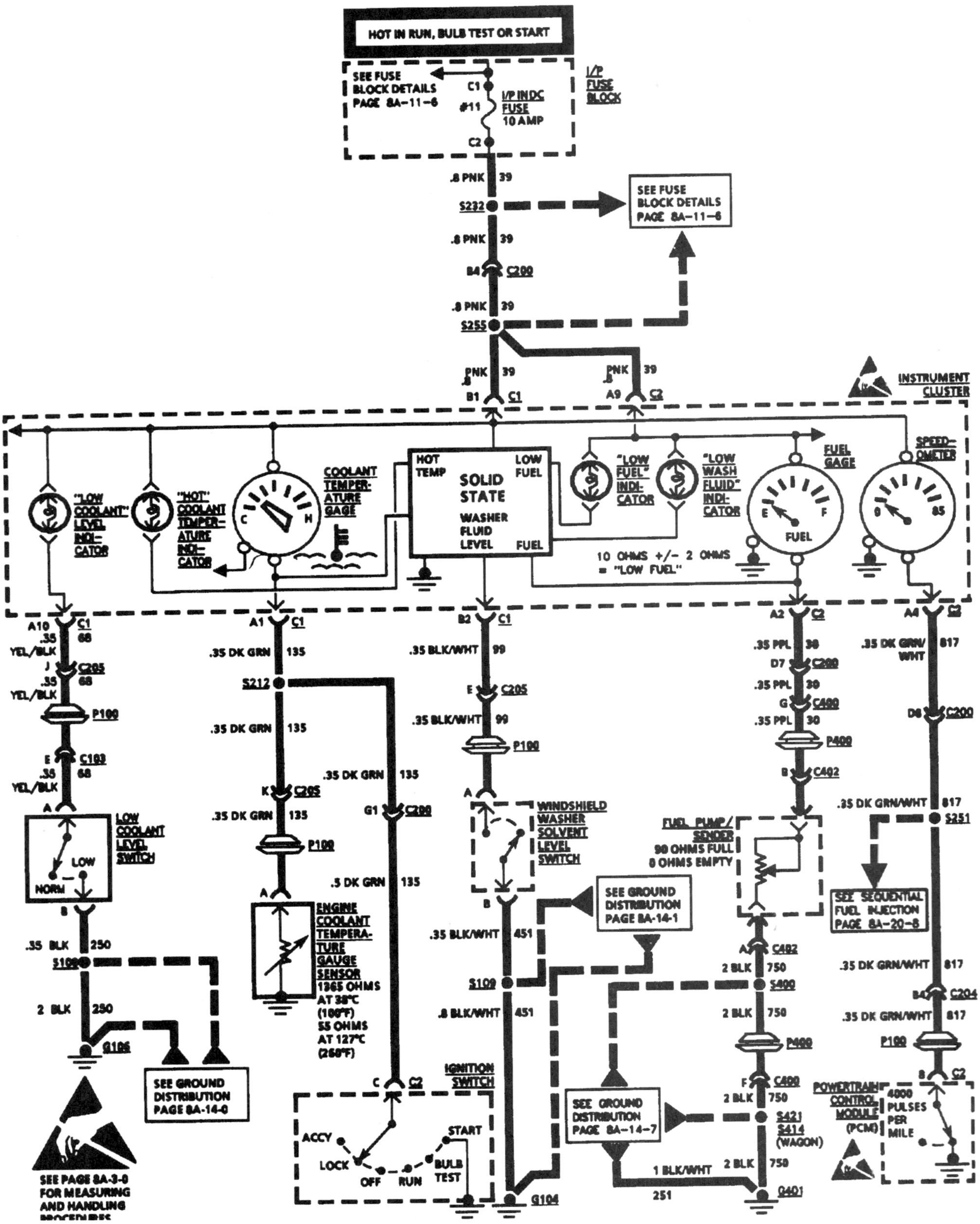

FIGURE 13–68 Diagram of the instrument cluster for Case #3.

(Courtesy of General Motors Corporation, Service Technology Group)

Known Information

- ❑ Vehicle operating voltage = 14 volts
- ❑ Circuit fuse is OK

Circuit Analysis

Answer the following questions on a *separate* sheet of paper.

1. With the known information above, what is the *most* likely cause of the inoperative fuel gauge circuit? ______.
2. What diagnostic steps and tools would you use to find the suspected problem in this fuel gauge circuit? ______.
3. Looking at the wiring diagram, what should the voltage supplied to the fuel gauge be? ______.
4. If you disconnected the C402 connector at the fuel gauge sending unit and grounded the gauge side of the connector with a jumper wire, what should the gauge needle show? ______.
5. In order for the fuel gauge needle to read Full, what should the total fuel gauge circuit resistance be? ______.
6. How is the fuel gauge circuit resistance varied to change the gauge reading needle? ______.
7. If the technician suspected a bad fuel gauge sending unit, what could the technician use to simulate the action of the sending unit to check for proper fuel gauge operation? ______.
8. What diagnostic steps were helpful when troubleshooting this system or explaining it to the customer? ______.
9. What is your analysis of the above case study? ______.

Hands-On Vehicle Tests

The following eleven hands-on vehicle checks are included in the **NATEF** (National Automotive Technicians Education Foundation) **Task List.** The task reference numbers are the same as the NATEF reference numbers. Some tasks are grouped together due to their close association with each other. Complete your answers to the following questions on a *separate* sheet of paper.

Performance Task #1 (NATEF #VI-F-1)

Task Description
Diagnose intermittent, high, low, or no gauge readings.

Performance Task #2 (NATEF #VI-F-2)

Task Description
Test gauge circuit voltage regulators (limiters); replace as needed.

Performance Task #3 (NATEF #VI-F-3)

Task Description
Inspect and test gauges and gauge sending units; replace as needed.

Performance Task #4 (NATEF #VI-F-4)

Task Description
Inspect and test connectors, wires, and printed circuit boards of gauge circuits; repair or replace as needed.

Performance Task #5 (NATEF #VI-F-5)

Task Description
Diagnose incorrect operation of warning devices and other driver information systems.

Task Objectives

- ❑ Obtain a vehicle that can be used for these tasks. What model and year of vehicle are you using for these tasks? ______________________________.
- ❑ Locate a wiring diagram and associated component locator references that will show the instrument cluster gauges, warning lights, associated sending units, wiring, and connectors. List them in the chart below.

Gauge or Warning Light	Sending Unit Location

- ❑ Does the circuit contain an instrument voltage regulator (IVR)? ____________________. How do you check an IVR for proper operation? ______________________________.
- ❑ If a gauge or warning light is not reading or operating properly, what are some simple diagnostic procedures used to determine what circuit component is faulty?

Component	Suspected Fault	Test Procedure

❑ What tools and equipment would be useful to inspect and test circuit wires and connectors? ____.

❑ What diagnostic steps are the most helpful when diagnosing a gauge or warning light circuit problem? ____.

❑ What is your analysis of using wiring diagrams to help diagnose instrument cluster gauge and warning light circuits? ____.

Task Summary

After performing the above NATEF tasks, what can you determine will be helpful in knowing how to check instrument cluster gauges and warning lights for proper operation? ____.

Performance Task #6 (NATEF #VI-G-1)

Task Description

Diagnose incorrect horn operation; repair as needed.

Task Objectives

❑ Obtain a vehicle that can be used for this task. What model and year of vehicle are you using for this task? ____.

❑ Locate a wiring diagram and associated component locators that show the horn circuit, fuses, and various components in the circuit. List them in the chart below.

Circuit Component	Location

❑ If the horn relay is suspected faulty, what test procedures can you use to verify the relay is faulty? ____.

❑ If the horn button switch or slip ring is suspected faulty, what test procedures can you use to verify that the horn button switch or slip ring is faulty? ______________________________

____.

❑ What should the voltage be that is supplied to the horns in the circuit?______________.

❑ To test the horn(s) for proper operation, how could you use a jumper wire to check their normal function? ____.

❑ What diagnostic steps are *most* useful when troubleshooting this system?______________

____.

❑ What is your analysis of using wiring diagrams to help diagnose a horn circuit?______________

____.

Task Summary

After performing the above NATEF task, what can you determine will be helpful in knowing how to check a horn circuit for proper operation? ______________________________

____

____

____.

Performance Task #7 (NATEF #VI-G-2)

Task Description

Diagnose incorrect wiper operation; diagnose wiper speed control and park problems; repair as needed.

Performance Task #8 (NATEF #VI-G-3)

Task Description

Diagnose incorrect windshield washer operation; repair as needed.

Task Objectives

❑ Obtain a vehicle that can be used for this task. What model and year of vehicle are you using for this task? ____.

❑ Locate a wiring diagram that shows the wiper/washer circuit, fuses, and associated components.

❑ Using the wiring diagram(s), identify areas in the circuit such as connectors, wires, fuses, and components, that can cause a possible problem.

Identified Area	Symptom	Cause	Test Procedure

- ❑ *Caution:* Remember safety—keep fingers clear of levers.
- ❑ Perform the diagnostic steps and procedures identified in the previous question. Does this help isolate the possible circuit problem(s)?____________________.
- ❑ Can the windshield washer operate independent of the wiper motor *low* speed selection? __________.
- ❑ What diagnostic steps are helpful when troubleshooting a wiper/washer system? ____________________

  ____

  ____.
- ❑ What is your analysis of using wiring diagrams to help diagnose wiper/washer circuit problems?

  ____

  ____.

Task Summary

After performing the above two NATEF tasks, what can you determine will be helpful in knowing how to check wiper and washer circuits for proper operation? ____________________

____.

Performance Task #9 (NATEF #VI-H-1)

Task Description

Diagnose incorrect operation of motor-driven accessory circuits; repair as needed.

Task Objectives

- ❑ Obtain a vehicle with an electric window circuit that an be used for this task. What model and year of vehicle are you using for this task?____________________.
- ❑ Locate a wiring diagram for the vehicle that will show the electric door window circuit.
- ❑ Using the wiring diagram(s), identify areas in the circuit such as connectors, wires, fuses, and components, that can cause a possible problem.

Identified Area	Symptom	Cause	Test Procedure

- ❑ Perform the diagnostic steps and procedures identified in the previous question. Does this help identify the circuit problem(s)?____________________.
- ❑ What is the circuit voltage supplied to the window motor assembly? ____________________.
- ❑ What is your analysis of using wiring diagrams to help diagnose a motor driven accessory circuit problem?____________________

  ____.

Task Summary

After performing the above NATEF task, what can you determine will be helpful in knowing how to check a motor driven accessory circuit for proper operation? ______________________________

____.

Performance Task #10 (NATEF #VI-H-2)

Task Description

Diagnose incorrect heated glass operation; repair as needed.

Task Objectives

- ❑ Obtain a vehicle that can be used for this task. What model and year of vehicle are you using for this task? ______________________________.
- ❑ Find a wiring diagram that shows the rear window heated glass circuit for this vehicle.
- ❑ Does this heated glass circuit follow the characteristics of a parallel circuit? ______________.
- ❑ Identify the following components and list their location on the vehicle.
  1. Rear window grid ______________________________.
  2. Control switch ______________________________.
  3. Relay (if applicable) ______________________________.
  4. Circuit fuse and fuse #______________________________.
  5. Circuit ground(s) ______________________________.
  6. Circuit indicator light ______________________________.
  7. Circuit timer assembly (if applicable) ______________________________.
  8. Other circuit components ______________________________.

Circuit Checks

- ❑ Check the operation of the indicator light. If operating, proceed to the rear window heating grid. If indicator is not operating, check to see if the rear window grid feels warm. If not, is the circuit fuse OK? ______________________________.
- ❑ Check for circuit source voltage at rear window heater grid.

  DVOM Voltage reading______________________________.
- ❑ What is the voltage from the positive end of the heating grid to the negative end? ______________.
- ❑ If you leave the DVOM hooked to the negative side and slowly moved the positive lead along a grid line, what happens to the voltage readings? ______________________________.

  Why? ______________________________.
- ❑ Why are rear window grid lines wired in parallel? ______________________________.
- ❑ Remove the circuit fuse and install a DVOM in series in its place. What is the amperage of the circuit? ______________________________.

❑ What happens to the total circuit amps as the rear heating grid starts to heat up? ________________.

Task Summary

After performing the above NATEF task, what can you determine that will be helpful in diagnosing a rear window heating grid circuit? ____

____

____.

Performance Task #11 (NATEF #VI-H-3)

Task Description

Diagnose incorrect electric door and hatch/trunk lock operation; repair as needed.

Task Objectives

❑ Obtain a vehicle with an electric door lock or hatch/trunk circuit that an be used for this task. What model and year of vehicle are you using for this task? ____

____.

❑ Locate a wiring diagram for the vehicle that will show the door lock or trunk/hatch circuit.

❑ Using the wiring diagram(s), identify areas in the circuit such as connectors, wires, fuses, and components, that can cause a possible problem.

Identified Area	Symptom	Cause	Test Procedure

❑ Perform the diagnostic steps and procedures identified in the previous question. Does this help identify the circuit problem(s)? ____.

❑ What is the circuit voltage supplied to the door lock solenoid assembly? ________________.

❑ What is the circuit voltage supplied to the trunk/hatch solenoid assembly? ________________.

❑ What is your analysis of using wiring diagrams to help diagnose door lock or trunk/hatch circuit problems? ____

____.

Task Summary

After performing the above NATEF task, what can you determine will be helpful in knowing how to check door locks or trunk/hatch circuits for proper operation? ____

____

____.

Summary

- ❑ Most horn circuits use a relay to separate the high amperage horns from the control switch.
- ❑ A typical horn assembly uses electromagnetism to vibrate a flexible diaphragm and contact points.
- ❑ Horn circuits with two horns normally use a different pitch for each horn.
- ❑ Most horn switches installed in a steering wheel use a slip ring and spring-loaded sliding contact to complete the connection to ground, when needed.
- ❑ A typical blower control circuit uses a *series* resistor block to control the blower motor speed.
- ❑ Many blower circuits use a relay to control the high speed blower motor speed.
- ❑ A non-depressed park wiper circuit parks the wiper blades at the end of their normal stroke by the edge of the lower windshield molding.
- ❑ A depressed park wiper circuit parks the wiper blades past the end of their normal stroke below the edge of the lower windshield molding.
- ❑ An interval, or intermittent, wiper system operates in time intervals depending on the delay setting requested by the driver.
- ❑ Most wiper washer pumps operate in conjunction with the *low* speed wiper motor setting.
- ❑ Typical rear window defogger grid wires are in parallel to each other.
- ❑ Rear window defogger circuits with solid state timers open the control switch after the first ten minutes of operation, and every five minutes on succeeding operations.
- ❑ Direct drive power window motors use a pinion gear meshed with a sector gear to raise the window regulator. Remember safety - Keep fingers clear.
- ❑ Rack-and-pinion window motors use a rack with a flexible strip of gear teeth operating a window regulator.
- ❑ There are three basic configurations of power seat circuits—two-way, four-way, and six-way.
- ❑ A typical power mirror circuit consist of two motors for each mirror—one for up/down and the other for left/right function.
- ❑ There are two types of analog gauge designs used today—the magnetic gauge and the thermal gauge.
- ❑ An instrument voltage regulator (IVR) maintains a constant voltage to the gauge circuit under all battery and charging system conditions.
- ❑ A temperature gauge sending unit is typically a thermistor resistance device that measures coolant temperature.
- ❑ A fuel gauge sending unit is typically a variable resistor that is moved manually by the fuel level in the fuel tank.
- ❑ Warning lamp sending units are normally *open* or *closed* depending on the circuit design.

Key Terms

accelerometer
depressed park
hybrid inflator module interval wipers
instrument voltage regulator (IVR)
interval wipers
magnetic gauge
non-depressed park
supplemental inflatable restraint (SIR)
thermal gauge
trimotor

Review Questions

Short Answer Essays

1. Describe how a blower motor circuit uses a resistor block assembly to control the *low, medium,* and *high* speed operation.
2. Describe how a technician would repair breaks in the grid wire of a rear window defogger.
3. What is the purpose of an IVR in an analog gauge circuit?
4. Describe how a driver's side air bag deploys on a frontal collision impact.
5. What are some simple safety rules when handling air bag systems?

Fill-in-the-Blanks

1. In a horn circuit with a relay, the ________________________ competes the circuit to ground from the relay coil.
2. In most vehicles with rear window defogger circuits, the grid lines are hooked in ________________________ to each other.
3. A reversible permanent magnet motor pack called a ________________________ controls the function of a ________________________-way power seat circuit.
4. When the engine coolant temperature increases, the temperature gauge sending unit thermistor resistance ____________________, and current flow to the gauge circuit ____________________.
5. In an air bag module the ignitor spark ignites a canister of ________________________ to ignite the inflator propellant.

ASE Style Review Questions

1. Technician A says that *all* horn circuits have a relay to prevent excess voltage from flowing through the horn button switch. Technician B says the horn brush/slip ring completes the current path to ground in a typical horn circuit. Who is correct?
   A. A only
   B. B only
   C. Both A and B
   D. Neither A nor B

2. Technician A says the pitch of a horn is caused by a vibrating diaphragm. Technician B says that in a dual horn system, the pitch is normally the same for both horns. Who is correct?
   A. A only
   B. B only
   C. Both A and B
   D. Neither A nor B

3. Technician A says the resistor block in a blower motor circuit is wired in series between the control switch and the blower motor. Technician B says the blower fan speed is controlled by the resistor block and blower switch. Who is correct?
   A. A only
   B. B only
   C. Both A and B
   D. Neither A nor B

4. Two technicians are discussing a blower circuit with a relay. Technician A says the relay controls *high* speed fan operation. Technician B says the blower's *low* speed is controlled by the relay. Who is correct?
   A. A only
   B. B only
   C. Both A and B
   D. Neither A nor B

5. Technician A says a *non-depressed park* wiper circuit will park the wiper blades at the end of their normal stroke. Technician B says a *depressed park* wiper circuit will park the wiper blades below the edge of the lower windshield molding. Who is correct?
   A. A only
   B. B only
   C. Both A and B
   D. Neither A nor B

6. Two technicians are discussing the photo on the next page. Technician A says the technician is adjusting the voltage input for the horn. Technician B says the technician is adjusting the horn pitch. Who is correct?
   A. A only
   B. B only
   C. Both A and B
   D. Neither A nor B

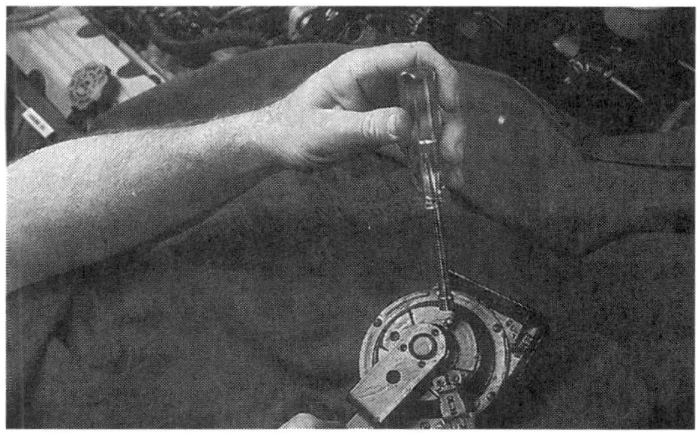

7. Technician A says an interval wiper system will operate in time intervals depending on the delay setting requested. Technician B says an intermittent wiper system uses a spring-loaded pressure switch to time the activation of the wiper motor. Who is correct?
   A. A only
   B. B only
   C. Both A and B
   D. Neither A nor B

8. The rear window defogger circuit shown below is being discussed. Technician A says the indicator light will illuminate even if the relay is defective. Technician B says the defogger switch controls current through the relay coil. Who is correct?
   A. A only
   B. B only
   C. Both A and B
   D. Neither A nor B

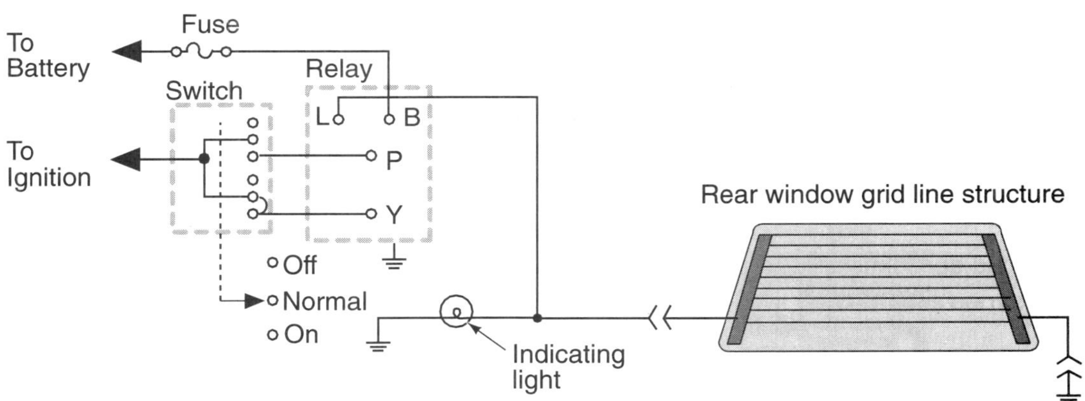

9. Technician A says the master control switch for power windows has total control of all windows in the system. Technician B says the rotation of the power window motor depends on which direction the control switch is moved. Who is correct?
   A. A only
   B. B only
   C. Both A and B
   D. Neither A nor B

10. The testing of a power window circuit below is being discussed. Technician A says the test light across the switch should light when the window switch is *closed* in either direction. Technician B says the test light should light when the master switch is *closed* in either direction. Who is correct?

    A. A only

    B. B only

    C. Both A and B

    D. Neither A nor B

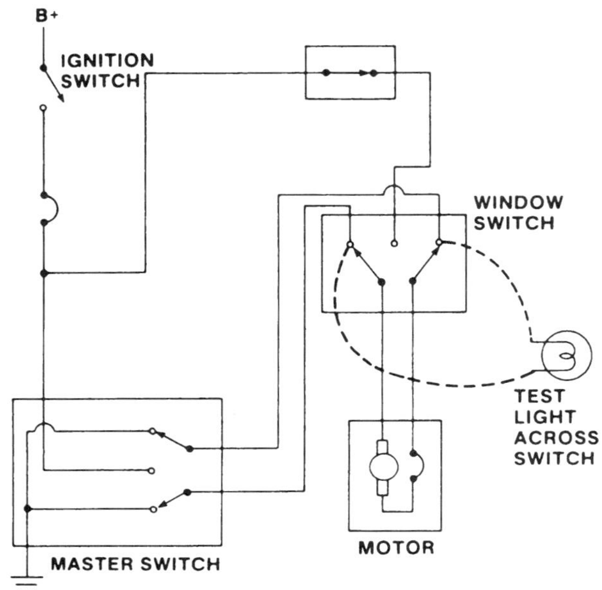

11. Technician A says that *all* power window motors are direct drive with a pinion and sector gear. Technician B says a flexible rack with gear teeth is used to attach the window regulator in *all* power window systems. Who is correct?

    A. A only

    B. B only

    C. Both A and B

    D. Neither A nor B

12. Two technicians are discussing the fuel gauge system on the next page. Technician A says that electromagnetism operates the gauge needle. Technician B says the resistance of the sending unit depending on the fuel level. Who is correct?

    A. A only

    B. B only

    C. Both A and B

    D. Neither A nor B

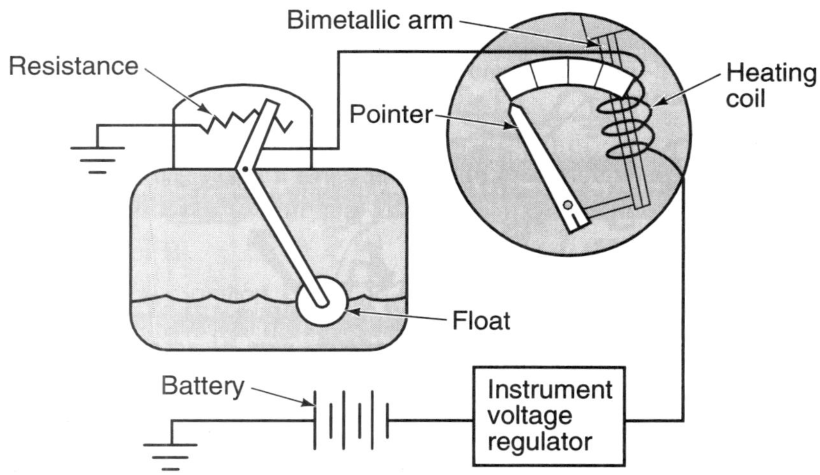

13. Technician A says an IVR can be used on bimetallic gauges. Technician B says the IVR will *open* if the alternator overcharges the battery. Who is correct?
    A. A only
    B. B only
    C. Both A and B
    D. Neither A nor B

14. Technician A says a sending unit for an oil pressure gauge can also be used for an oil pressure warning light system. Technician B says the sending unit for a temperature warning light is *normally open* during normal engine temperatures. Who is correct?
    A. A only
    B. B only
    C. Both A and B
    D. Neither A nor B

15. Technician A says the air bag module is electrically connected in the steering column through a clock spring assembly. Technician B says all air bag systems use a roller-type impact sensor to arm the air bag module. Who is correct?
    A. A only
    B. B only
    C Both A and B
    D. Neither A nor B

14 Batteries and Testing

Introduction

The battery is often referred to as the heart of a vehicle's electrical system. It is defined as a electrochemical device that changes chemical energy to electrical energy by the chemical reaction of two dissimilar plates immersed in a electrolyte solution. The design and testing of the battery has advanced a long way since Alassandro Volta invented the voltaic pile in the late 1700's. Advancements in technology, material improvements, construction design, and increased demands from the vehicle electrical system have all contributed to the batteries that are in use today.

Objectives

When you complete this chapter you should be able to:

- ❑ Describe the main components of a conventional battery.
- ❑ Describe the construction differences and features between conventional, maintenance free, deep cycle, and gel-cell batteries.
- ❑ Explain the chemical reaction that takes place during the discharge cycle and the charging cycle in the battery.
- ❑ Explain the different methods used to rate batteries.
- ❑ Explain the safety cautions required when working around or with batteries.
- ❑ Demonstrate how to test a battery's state of charge using a hydrometer and a voltmeter.
- ❑ Explain the procedures and test equipment used to load test a battery for proper capacity.
- ❑ Demonstrate how to properly use a battery charger and the appropriate load settings.
- ❑ Explain the cautions and procedures used to jump start a vehicle safely.
- ❑ Demonstrate proper battery service procedures.

Battery Types and Construction

There are several types of batteries in use today ranging from devices such as flashlights, watches, and calculators, to automobiles and trucks. The simplest form of battery is a single **voltaic cell** that uses two unlike metals and an alkali, salt, or acid solution to produce a voltage (Figure 14–1). In this example, a nickel and a penny are two unlike metals separated by a paper towel moistened

FIGURE 14–1 A simple voltaic cell.

with a person's saliva. The saliva contains sufficient alkali or acids that act as a suitable electrolyte solution to produce a small voltage. An **electrolyte** is defined as a chemical compound of alkali, salt, or acid solution that is capable of conducting electrical energy.

The amount of voltage produced is determined by the materials used in the battery cell. Figure 14–2 is a table of metals used to produce various types of batteries. The chemical reactions that take place between the metals depends on their ability to receive or transfer free electrons.

The table of metals (Figure 14–2) is divided into the two common types of available cells. The first is a **primary cell,** which is defined as a cell that *cannot be recharged* to its original voltage condition. The chemical reaction that takes place while voltage is produced, erodes away one or both of the cell plates. In Figure 14–3, a carbon-zinc primary cell design is used in flashlight batteries that produces approximately 1.5 volts each. A carbon rod is used as a positive electrode with a zinc container acting as the negative electrode. The electrolyte mixture is a manganese dioxide paste that permits the battery to be used in any position without spillage.

The second common type of battery cell is a **secondary cell,** which is defined as a cell that *can be recharged* once its original stored energy has been

CELL	NEGATIVE PLATE	POSITIVE PLATE	ELECTROLYTE	VOLTS PER CELL
		Primary Cells		
Carbon-zinc (Leclanche)	Zinc	Carbon, manganese dioxide	Ammonium chloride	1.5
Alkaline	Zinc	Manganese dioxide	Potassium hydroxide	1.5
Mercury	Zinc	Mercuric oxide	Potassium hydroxide	1.35
Silver-zinc	Zinc	Silver Oxide	Potassium hydroxide	1.6
Zinc-air	Zinc	Oxygen	Potassium hydroxide	1.4
Edison-Lalande	Zinc	Copper oxide	Sodium hydroxide	0.8
		Secondary Cells		
Lead-acid	Lead	Lead dioxide	Dilute sulfuric acid	2.2
Nickel-iron (Edison)	Iron	Nickel oxide	Potassium hydroxide	1.4
Nickel-cadmium	Cadmium	Nickel hydroxide	Potassium hydroxide	1.2
Silver-zinc	Zinc	Silver oxide	Potassium hydroxide	1.5
Silver-cadium	Cadmium	Silver oxide	Potassium hydroxide	1.1

FIGURE 14–2 Table of cell material and electrolyte.

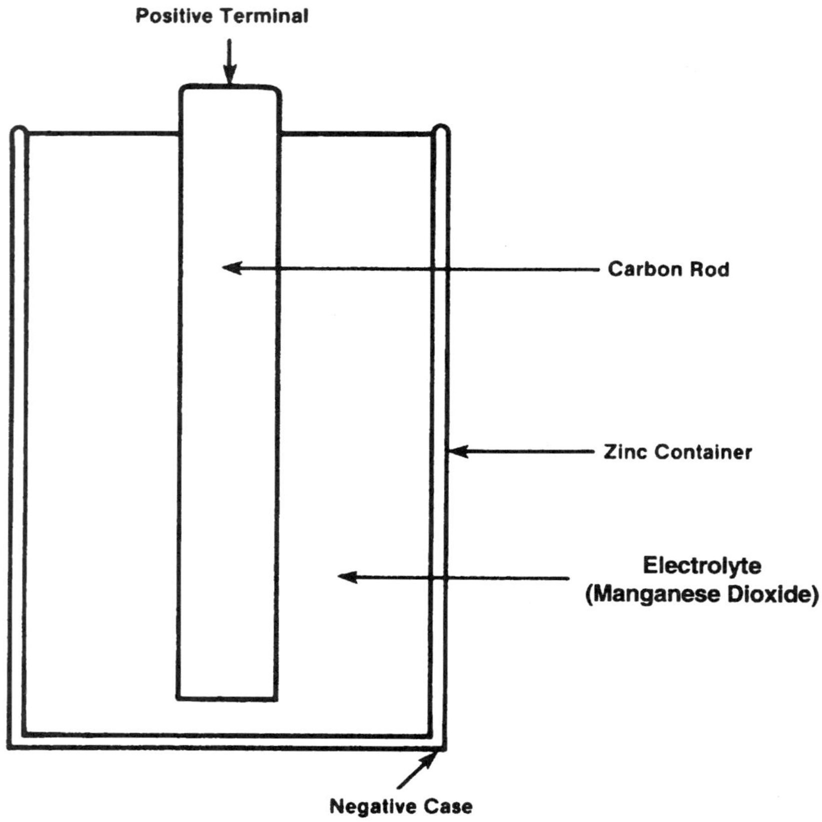

FIGURE 14–3 Simple secondary cell lead acid battery.

depleted. The lead-acid cell of an automotive battery is a perfect example of a secondary cell battery (Figure 14–4). The negative plate is made of pure lead (Pb), and the positive plate is made of lead dioxide (PbO_2). The electrolyte contains a diluted mixture of **sulfuric acid,** a conductive and reactive mixture of H_2SO_4 consisting of 64 percent water and 36 percent sulfuric acid, by weight.

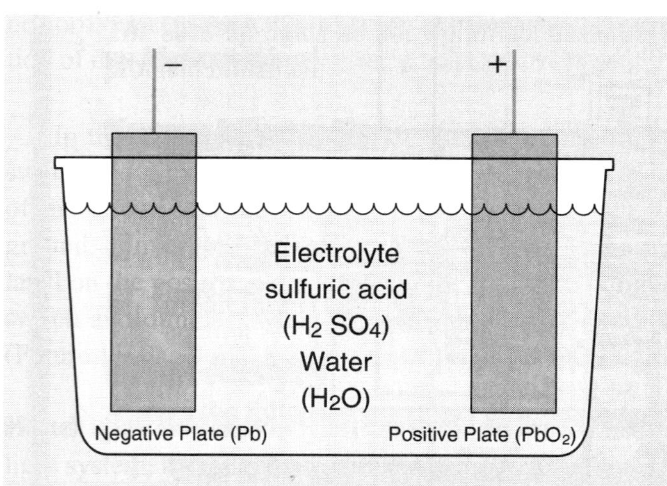

FIGURE 14–4 Simple secondary cell lead acid battery.

Conventional Battery Construction

The basic battery in a vehicle is a **conventional battery** which is a lead acid battery that uses a lead antimony mixture in the cell plates. The battery is constructed of several important parts. The first is the grid plate made of a **lead antimony** that contains up to a ten percent mixture of lead and lead alloys to strengthen the plate grid frame (Figure 14–5). The grid plate has vertical and horizontal bars that hold the *paste* that forms the positive and negative plates.

The active paste material in the positive plate consists of a reddish-brown **lead peroxide** paste—PbO_2. The negative plate paste material consists of a grayish **sponge lead** paste—Pb. In each battery cell there are up to 13 plates arranged alternately to each other (Figure 14–6). Each plate is divided by a **separator** that is a micro-porous material used to prevent contact between

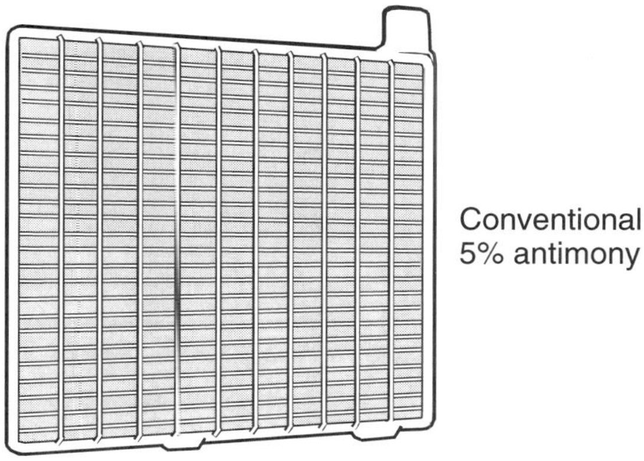

FIGURE 14–5 Typical conventional battery grid plate.

(Courtesy of Ford Motor Company)

each positive and negative plate. When a group of positive plates and negative plates are assembled together with separators, they form an **element.**

Note: Regardless of the number of positive and negative plates in an element, the element produces 2.1 volts. The more plates there are in the element, the longer the battery can supply its 2.1 volts per element.

When six elements are wired in series, positive to negative, positive to negative, etc., the resulting battery produces 12.6 volts for the electrical system (Figure 14–7).

Depending on the manufacturer, the battery case and cover is made of hard rubber, polypropylene, or other plastic materials (Figure 14–7). The case must perform the following functions.

1. Separate and stabilize the elements.
2. Withstand temperature extremes without cracking.
3. Resist vibration from the vehicle under normal operating conditions.
4. Resist the absorption of the electrolyte acid mixture.
5. Provide raised supports in the bottom of the case for sediment and contaminants to fall.
6. Provide vents in the cover for release of any hydrogen gas generated by the charging or discharging process. Non-maintenance free batteries also provide access to the cells for testing or filling the electrolyte (Figure 14–8).
7. Provide a stable platform and seal for the external positive and negative terminals (Figure 14–9).

The battery electrolyte consists of a diluted mixture of thirty six percent sulfuric acid and sixty-four percent water, by weight. When fully charged, the electrolyte in the battery has a specific gravity of 1.265 at 80° F. **Specific gravity** is the weight of a given volume of liquid in

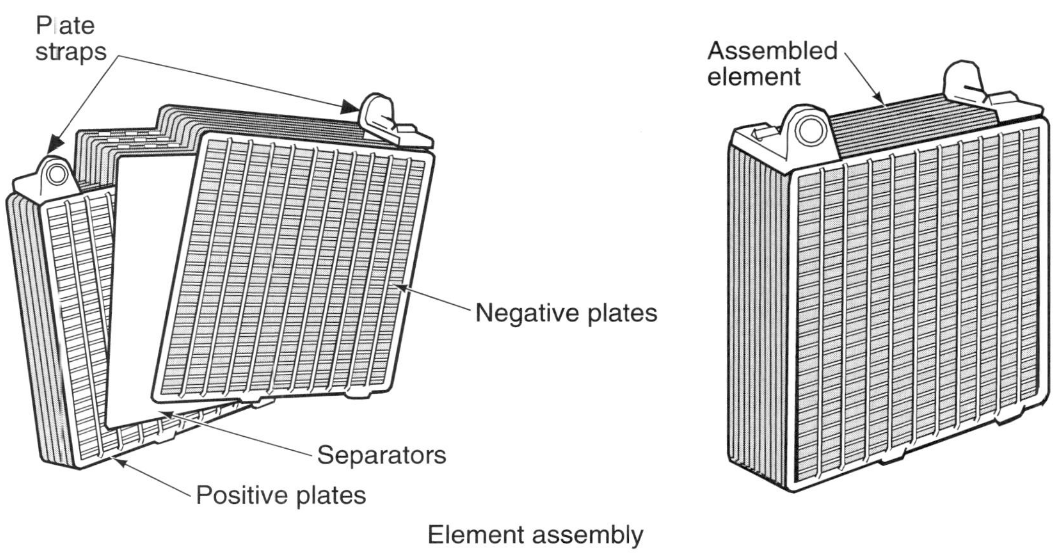

FIGURE 14–6 A battery cell, or element, with positive and negative plates divided with separators.

(Courtesy of Ford Motor Company)

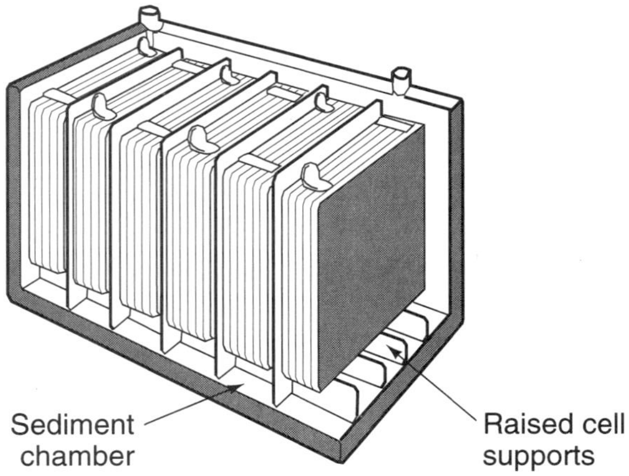

FIGURE 14–7 A 12.66 volt battery consists of six 2.11 volt cells connected in series.

(Courtesy of Ford Motor Company)

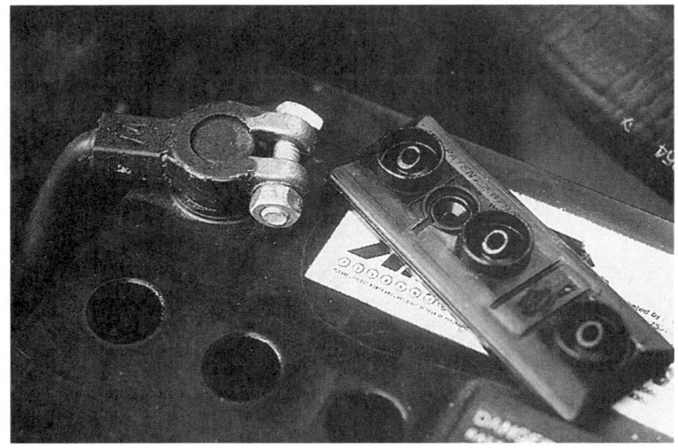

FIGURE 14–8 Conventional battery vent covers to allow escape of gasses.

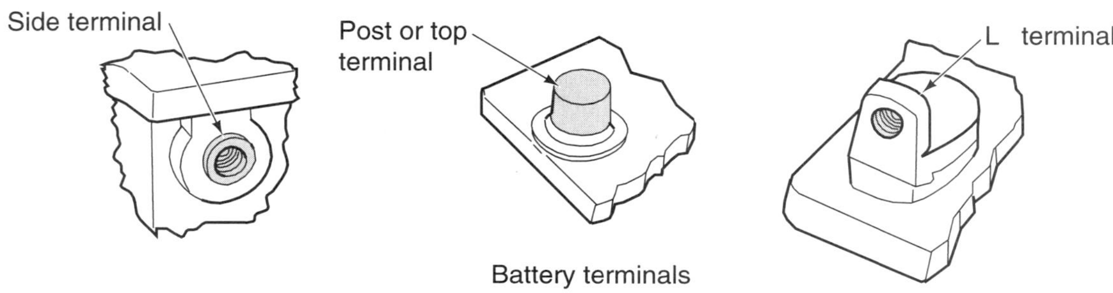

FIGURE 14–9 The most common types of battery terminals.

(Courtesy of Ford Motor Company)

comparison to the weight of the same volume of water with a specific gravity of 1.000. Testing the specific gravity of the electrolyte will be discussed in-depth later in this chapter.

CAUTION *The electrolyte is a very corrosive acid and it can cause severe injury to the eyes or skin (Figure 14–10). Always wear safety glasses and protective gloves and clothing when working around the battery electrolyte. A solution of baking soda and water can be used to neutralize the acid if it comes in contact with skin. If the acid is splashed in the eyes, flush them immediately with a large quantity of cool water, preferably from a emergency eye wash station (Figure 14–11), and then seek medical attention.*

The chemical production of electricity in the battery occurs when an electrical circuit is completed from positive to negative (Figure 14–12) and the discharge process begins. The O_2 from the positive plate enters the electrolyte, and at the same time, the SO_4 from the electrolyte combines with the Pb on the positive plate to

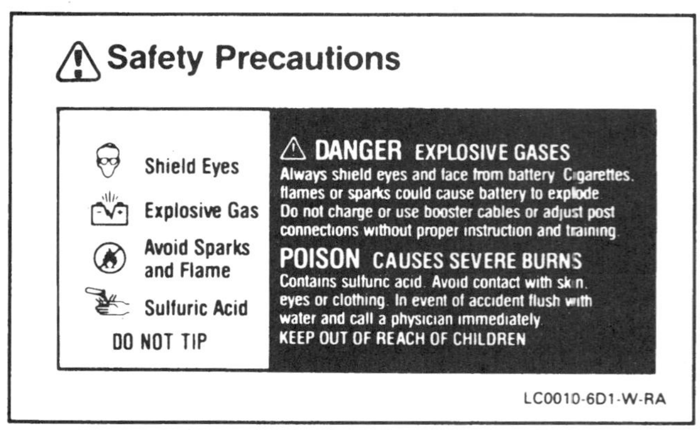

FIGURE 14–10 Safety caution sticker located on battery cover.

(Courtesy of General Motors Corporation, Service Technology Group)

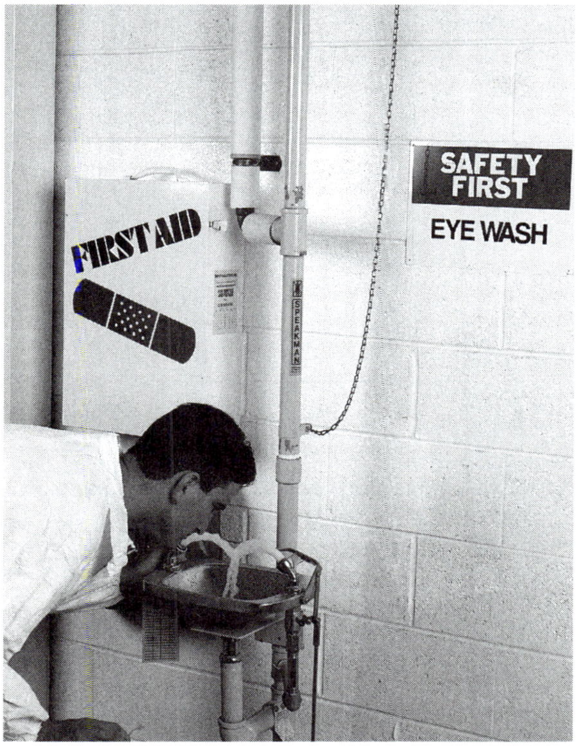

FIGURE 14–11 Typical emergency eyewash station.

form lead sulfate ($PbSO_4$). And, the SO_4 from the electrolyte combines with the Pb on the negative plate to form lead sulfate ($PbSO_4$).

In a discharged battery the electrolyte mixture is only slightly heavier than the specific gravity of water. The following table indicates the specific gravity of a battery at 80° F.

Battery Charge	Specific Gravity
Fully charged	1.265
75% charged	1.225
50% charged	1.190
25% charged	1.155
Discharged	1.120, or less
specific gravity at 80° F	

As the battery is recharged by the vehicle charging system, or externally through a battery charger, the chemical action that took place during the discharge cycle now reverses itself to restore the battery to its original condition of PbO_2 in the positive plate, Pb in the negative plate, and an H_2SO_4 electrolyte mixture of 1.265 (Figure 14–13).

Maintenance Free Battery Construction

A **maintenance free** battery is a lead acid battery that uses a *lead calcium* mixture in the cell plates. Changing the grid plate material and design (Figure 14–14) reduces the amount of gassing of the electrolyte which is prevalent with conventional batteries. **Gassing** or **electrolysis** in a battery occurs when the battery water is converted into explosive hydrogen and oxygen gas.

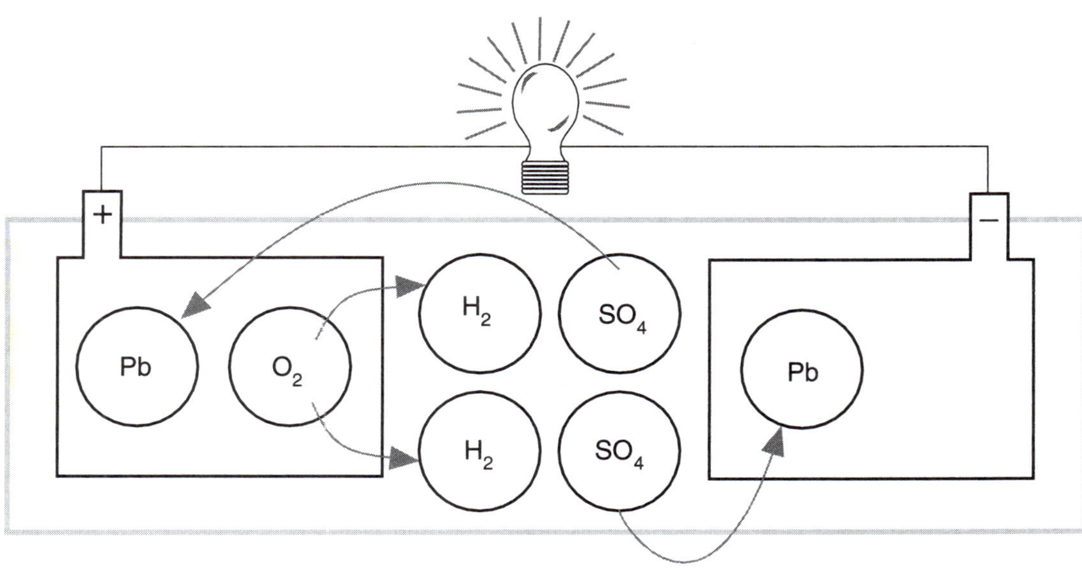

FIGURE 14–12 Chemical action that occurs in the battery cell during the discharge cycle.

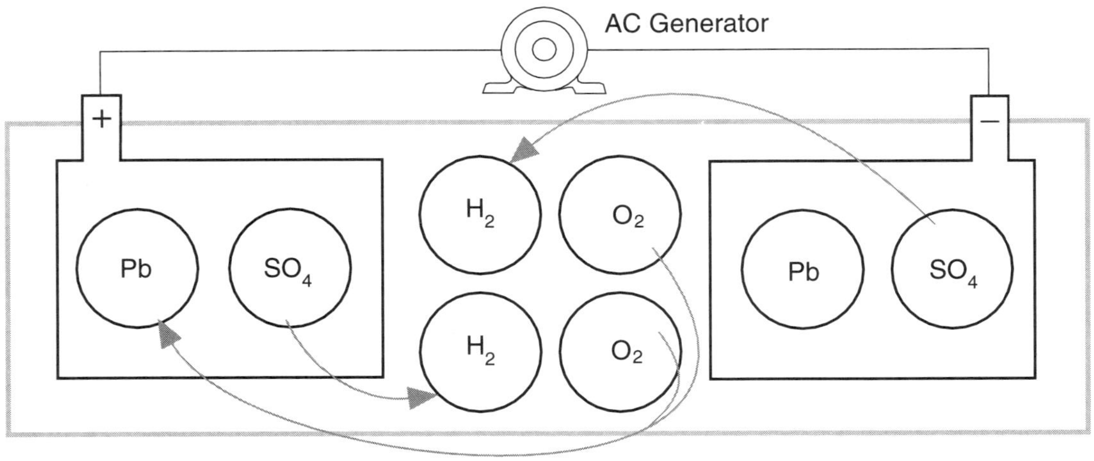

FIGURE 14–13 Chemical action that occurs in the battery cell during the charge cycle.

The **lead calcium** grid contains calcium or strontium to reduce gassing and self discharge.

Because heat and gassing are the main reasons for the vaporization of the electrolyte, using lead calcium grids almost eliminates electrolyte loss or the need to periodically add distilled water to the battery. The battery cover is sealed to prevent electrolyte or vapor loss, and has only a small vent to relieve pressure during excess charging conditions (Figure 14–15).

Compared to conventional batteries, some of the advantages of maintenance free batteries are:

- Larger electrolyte reserve due to a thin polypropylene or plastic case, and a reserve area above the plate cells.
- Longer shelf life of up to two years depending, on the manufacturer.
- Higher cold cranking ampere rating.
- Little or no maintenance to battery terminal connections.

Compared to conventional batteries, there are some important *disadvantages* of maintenance free batteries.

- Shorter overall life expectancy.
- Quick discharge from unwanted parasitic loads.
- Hard to recharge if allowed to completely discharge.

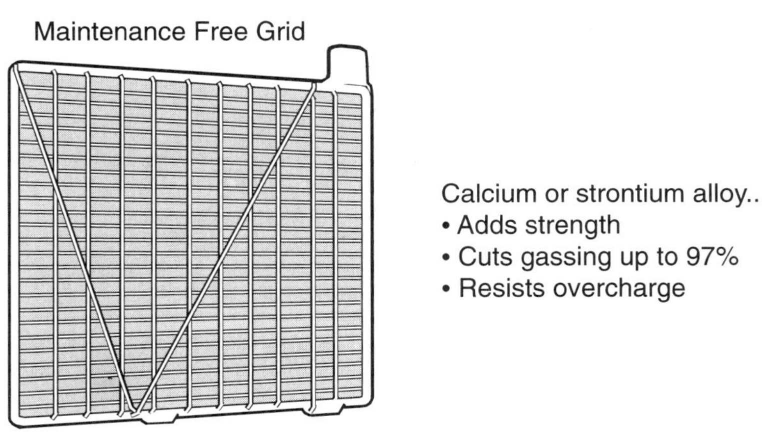

FIGURE 14–14 Typical maintenance free battery grid plate.
(Courtesy of Ford Motor Company)

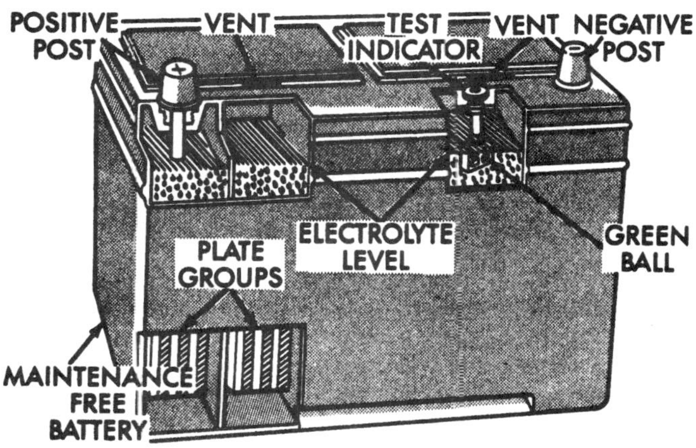

FIGURE 14–15
Typical construction of a maintenance free battery.

- Lack of reserve capacity—especially in cold environments.
- When the battery is exposed to high temperatures, a condition called **grid growth** develops which is a condition that shorts out the plates when little metallic fingers grow through the separators to the plates.

Deep Cycle Battery Construction

A **deep cycle** battery is a lead acid battery that uses approximately 2.75 percent antimony on the positive plate and a calcium alloy on the negative plate. It is also known as a **hybrid** battery. A hybrid battery combines the best grid properties of conventional batteries and maintenance free batteries, allowing it to withstand several deep cycles and still retain the original reserve capacity. The grid construction of a deep cycle battery provides more current at a faster rate because the vertical and horizontal grid bars are arranged in a radial design extending from a lug near the top center of the grid (Figure 14–16). Locating the lug near the center of the grid reduces resistance and provides a shorter path for current to follow to reach the lug.

Recombination Battery Construction

A **recombination battery** is a completely sealed lead acid battery which uses an electrolyte that is in a gel form. This battery is also referred to as a **gel-cell** battery (Figure 14–17) and is one of the most recent advancements in battery technology. In a gel-cell battery, little or no gassing takes place to release hydrogen gas. When oxygen is internally released at the positive plates, it passes through the separators and is forced to recombine with the negative plates. This process virtually eliminates gassing in the battery. Because of these design differences, manufacturer's instructions should be followed anytime testing or charging is performed on the battery. Excessive current drains or charging rates might overheat the cells and cause damage.

Some of the advantages of a gel-cell battery are listed below.

- Has no liquid electrolyte mixture to spill out, even if the case is cracked or mounted upside down. This feature makes the battery very suitable for off-road and 4-wheel drive vehicles.
- Can be deep cycled several times without damage to the reserve capacity.
- Completely sealed case makes battery totally corrosion and maintenance free.
- Outlasts conventional or maintenance free batteries by up to four times.

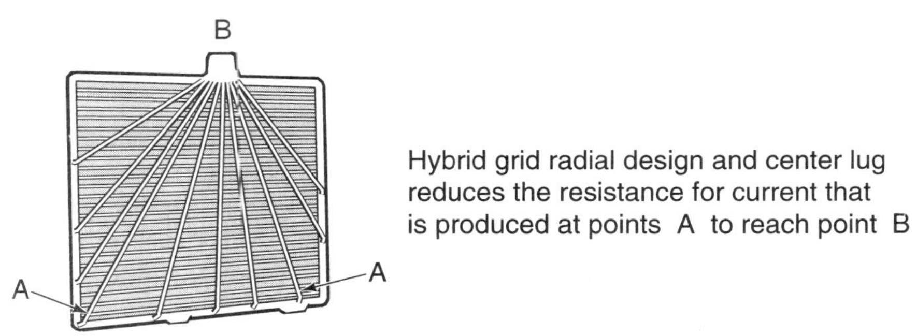

FIGURE 14–16 A typical deep cycle battery grid construction.

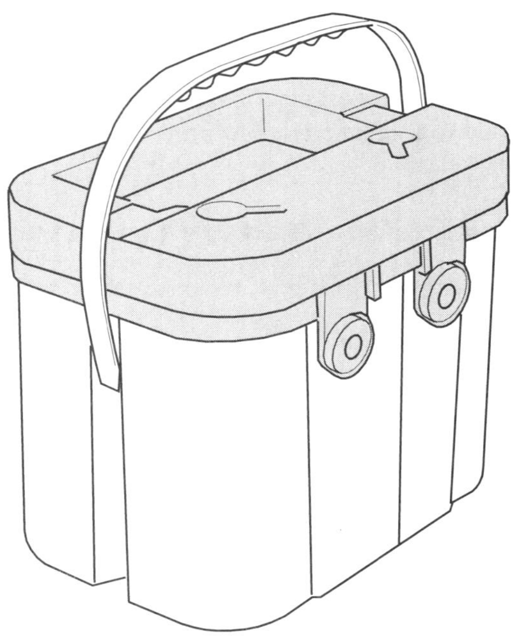

FIGURE 14–17 A typical sealed recombination, or gel-cell battery.
(Courtesy of OPTIMA Batteries)

- Available with ratings of over 800 cold cranking amperes.

A disadvantage of a gel-cell battery is the initial expense—normally two to three times the cost of a conventional or maintenance free battery.

Battery Load Ratings

There are five capacity ratings established by the Battery Council International (BCI) that determine the current capacity of a battery. The current capacity is an indication of the battery's ability to provide high amperage cranking power to the starter, and provide reserve power to the electrical system.

Reserve Capacity Rating

The first battery rating is known as **reserve capacity (RC).** It is defined as the amount of time in minutes a fully charged battery can be discharged at 25 amperes while maintaining a minimum battery voltage of 10.5 volts at 80° F in the event of a charging system failure. For example, a reserve capacity rating of 120 minutes indicates that in the event of a charging system failure, the battery could be discharged at a rate of 25 amps for two hours (60 minutes × 2 = 120 minutes) at 80° F.

Amp-Hour Rating

The second battery rating is the **amp-hour rating (AH).** This rating is the amount of current that a fully charged battery can produce over a twenty hour period at 80° F before the terminal voltage reaches a minimum of 10.5 volts. For example, if a battery can be discharged at a rate of 5 amps for twenty hours before the terminal voltage reaches 10.5 volts then the amp-hour rating would be 100 A/H. This is computed as follows:

$$5 \text{ amps} \times 20 \text{ hours} = 100 \text{ A/H}$$

Most automotive manufacturers use original equipment batteries with an A/H rating of 50 to 120 A/H. A higher A/H rating does not necessarily indicate increased starter cranking capacity, but does improve the ability to provide more circuit current over the 20 hour period.

Cold Cranking Ampere Rating

The third and most common battery rating is the **cold cranking ampere (CCA)** rating. It is defined as the ability of the battery to provide an amperage load for thirty seconds at 0°F, without the terminal voltage dropping below 7.2 volts. The CCA rating on an original equipment battery can range between 300 CCA to slightly over 1000 CCA, depending on the vehicle.

If a battery has an A/H rating, its CCA is determined by multiplying the A/H by 5.25. For example, a 100 A/H battery has a CCA rating of 525 CCA.

$$100 \text{ A/H} \times 5.25 = 525 \text{ CCA}$$

To convert the battery's CCA rating to A/H, simply divide the CCA by 5.25 to get A/H.

$$525 \text{ CCA} \div 5.25 = 100 \text{ A/H}$$

Cranking Amps Rating

The fourth battery rating is the **cranking amps (CA)** rating, which is the battery's ability to deliver a cranking current at 32° F. A battery with a CA rating of 800 may confuse a technician who may assume it is a CCA rating number. To convert CA at 32 °F to CCA at 0 °F, divide CA by 1.25.

Example: An 800 CA rated battery is the same as a 640 CCA rated battery.

$$800 \text{ CA} \div 1.25 = 640 \text{ CCA}$$

Watt-Hour Rating

The final battery rating is the **watt-hour (WH)** rating, which is determined by how many watt-hours of mechanical energy a battery produces at 0° F The watt-hour rating is calculated by multiplying a battery's amp-hour rating by the battery's open circuit voltage. A 120 A/H rated battery of 12.6 volts has a watt-hour rating of 1512 W/H.

$$120 \text{ A/H} \times 12.6 \text{ volts} = 1512 \text{ W/H}$$

Battery Size and Capacity

When a battery needs to be replaced, there are several factors that should be considered to find the best replacement match possible for the vehicle.

- Engine size and type.
- Vehicle electrical options.
- Climate conditions—extreme heat or cold will effect a battery's performance. The requirement for electrical energy to crank the engine increases as the temperature decreases (Figure 14–18).
- Vehicle starting requirements such as frequent start/stops.
- Size and weight limitations—if the battery is too high, it could possibly short out the terminal when the hood is closed. The BCI group numbers are used to indicate the proper size and battery characteristics of the battery (Figure 14–19).

Temperature	% of Cranking Power
80°F (26.7°C)	100
32°F (0°C)	65
0°F (−17.8°C)	40

FIGURE 14–18 A battery's cranking power is affected by temperature.

Grp. Size	Vlt.	Cold cranking power—amps for 30 secs. at 0°F*	No. of mo. warranted	Size of battery container in inches (incl. terminals)		
				Lgth.	Wd.	Ht.
17HF	6	400	24	7¼	6¾	9
21	12	450	60	8	6¾	8½
22F	12	430	60	9	6⅞	8⅛
	12	380	55	9	6⅞	8⅛
	12	330	40	9	6⅞	8⅛
22NF	12	330	24	9½	5½	8⅞
24	12	525	60	10¼	6⅞	8⅝
	12	450	55	10¼	6⅞	8⅝
	12	410	48	10¼	6⅞	8⅝
	12	380	40	10¼	6⅞	8⅝
	12	325	36	10¼	6⅞	8⅝
	12	290	30	10¼	6⅞	8⅝
24F	12	525	60	10¼	6⅞	8⅝
	12	450	55	10¼	6⅞	8⅝
	12	410	48	10¼	6⅞	8⅝
	12	380	40	10¼	6⅞	8⅝
	12	325	36	10¼	6⅞	8⅝
	12	290	30	10¼	6⅞	8⅝
27	12	560	60	12	6⅞	8⅝
27F	12	560	60	12	6⅞	9
41	12	525	60	11 9/16	6 13/16	6 15/16
42	12	450	60	9⅝	6⅞	6¾
	12	340	40	9⅝	6⅞	6¾
45	12	420	60	9½	5½	8⅞
46	12	460	60	10¼	6⅞	8⅝
48	12	440	60	12	6⅞	7½
49	12	600	60	14½	6⅞	7½
56	12	450	60	10	6	8⅜
	12	380	48	10	6	8⅜
58	12	425	60	9¼	7¼	6⅞
71	12	450	60	8	7¼	8½
	12	395	55	8	7¼	8½
	12	330	36	8	7¼	8½
72	12	490	60	9	7¼	8¼
	12	380	48	9	7¼	8¼
74	12	585	60	10¼	7¼	8¾
	12	525	60	10¼	7¼	8¾
	12	505	60	10¼	7¼	8¾
	12	450	55	10¼	7¼	8¾
	12	410	48	10¼	7¼	8¾
	12	380	40	10¼	7¼	8¾
	12	325	36	10¼	7¼	8¾

*Meets or exceeds Battery Council International rating standards.

FIGURE 14–19 BCI battery group numbers that identify battery features.

Battery Testing

To properly check a discharged or weak battery for possible replacement, there are six specific checks that need to be made to ensure the battery is the source of the problem and not a result of another problem. Knowledge of how batteries operate, quality test equipment, and specific testing procedures will help isolate a battery problem.

Battery Terminal Voltage Drop Test

The first testing procedure to perform on a battery is a battery terminal voltage drop test which tests the positive and negative terminal connections for a voltage drop. Figure 14–20 shows using a voltmeter connected from the battery post to the terminal connector while the starter is being cranked (ignition disabled to prevent starting). Voltage drops exceeding .1 volts would indicate that unwanted resistance exists at the terminal connection.

If a voltage drop exists, the battery terminals need to be removed (Figure 14–21) and cleaned. To prevent damage on a top post battery, use a terminal puller to remove the terminal from the post. Remember to *always* remove the *negative* terminal first to prevent the chance of the positive post contacting a ground through the wrench. A terminal cleaning tool or small wire brush should be used to clean the terminals and posts (Figure 14–22).

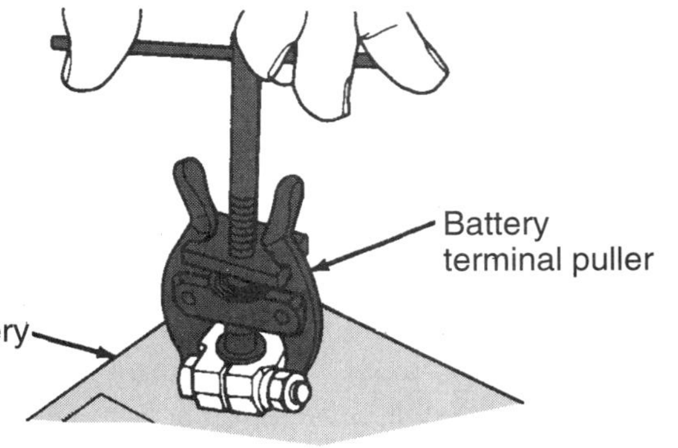

FIGURE 14–21 Removing a battery terminal using a special terminal puller.

(Courtesy of DaimlerChrysler Corporation)

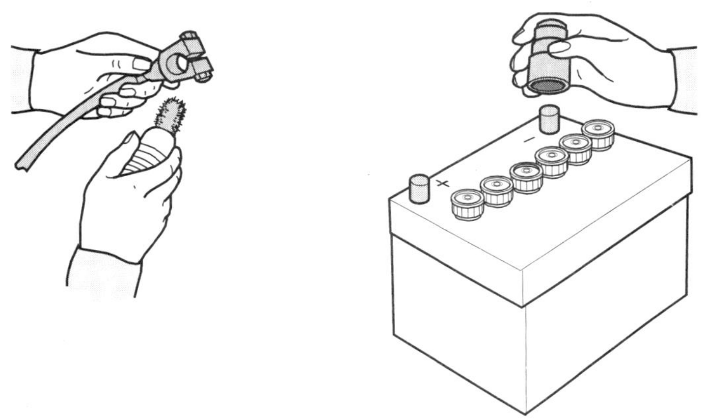

FIGURE 14–22 Using a terminal cleaning tool to clean the battery posts and terminals.

Battery Case Leakage Test

When the top of the battery case becomes wet or dirty, a small unwanted current draw can exist between the positive and negative posts through the dirty top. To find the unwanted current draw, perform a battery case leakage test (Figure 14–23) by connecting the negative lead of a voltmeter to the battery negative post, and touching the positive meter probe to the case top in various locations. A *clean* battery should have *no* voltage reading between the posts and the battery case.

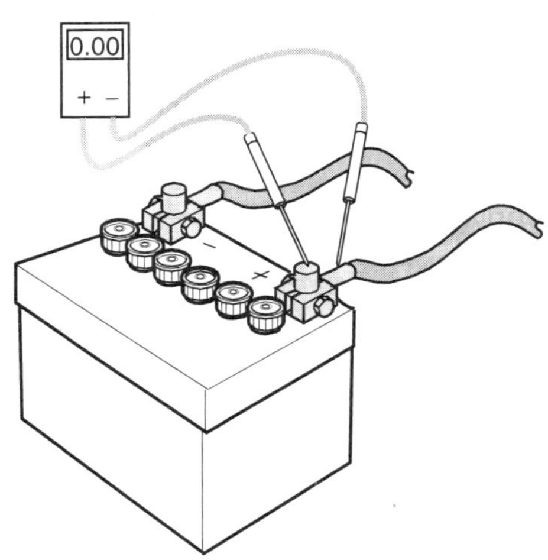

FIGURE 14–20 Performing a battery terminal voltage drop test.

Battery State-Of-Charge Test

The third test to perform on a battery is a state-of-charge test on the electrolyte mixture. This is done by using a **hydrometer** that measures the specific gravity content of the sulfuric acid mixture. When an electrolyte mixture

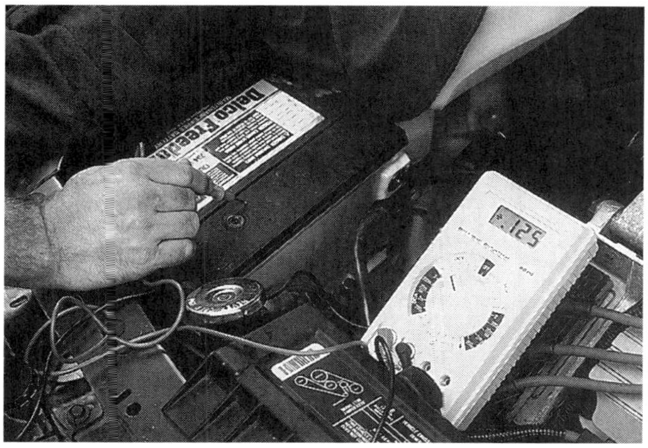

FIGURE 14–23 Checking a battery voltage leakage test on a battery cover using a DMM.

is drawn into the hydrometer (Figure 14–24), the float rises to the specific gravity of the electrolyte solution (Figure 14–25). Because the battery may be tested at various temperatures, using a hydrometer that is temperature corrected will provide a true specific gravity reading (Figure 14–26). For every 10° F above 80° F, add .004 points to the specific gravity reading. For every 10° F below 80° F, subtract .004 points to the specific gravity reading.

A fully charged battery will have a specific gravity of 1.265 to 1.280 depending on the battery. As the battery becomes discharged, the water content of the electrolyte increases which lowers the specific gravity reading. A defective battery can be determined by recording the readings of each cell (Figure 14–27). If the specific gravity reading between the highest and lowest cells varies by 0.050, or more, the battery is considered defective.

Note: Make sure distilled water was not recently added to the battery electrolyte which could give a false specific gravity reading.

On some batteries, a built-in hydrometer is used as a quick reference indicator of a battery's state-of-charge (Figure 14–28). This indicator *only* checks the specific gravity in one cell. When a green dot appears on the indicator, the specific gravity charge is 65% or above. Below 65%, the dot appears dark. When the dot is clear,

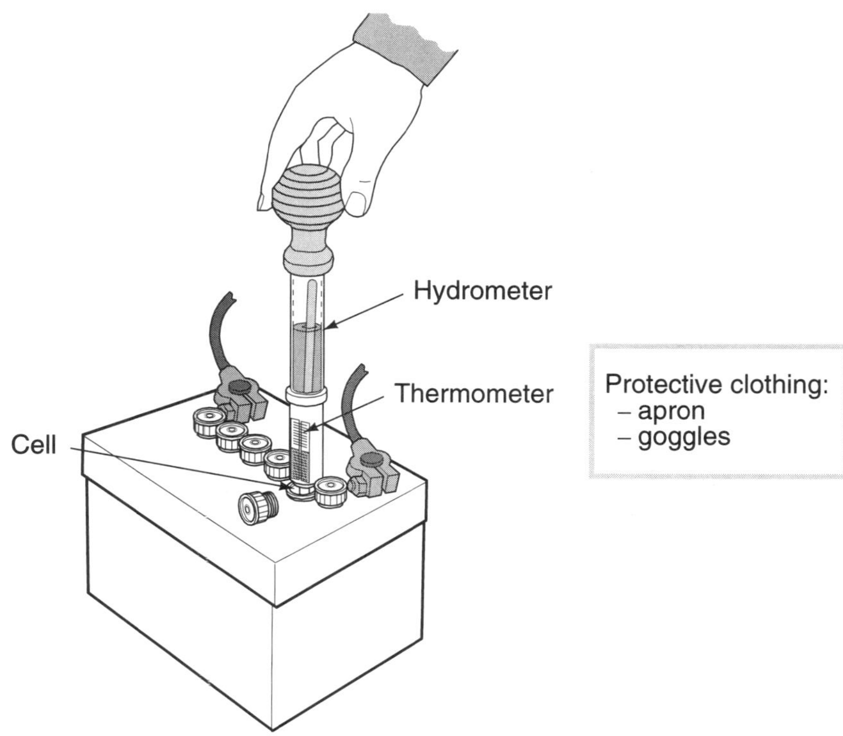

FIGURE 14–24 Drawing electrolyte out of the battery's cell and into the hydrometer.

(Courtesy of DaimlerChrysler Corporation)

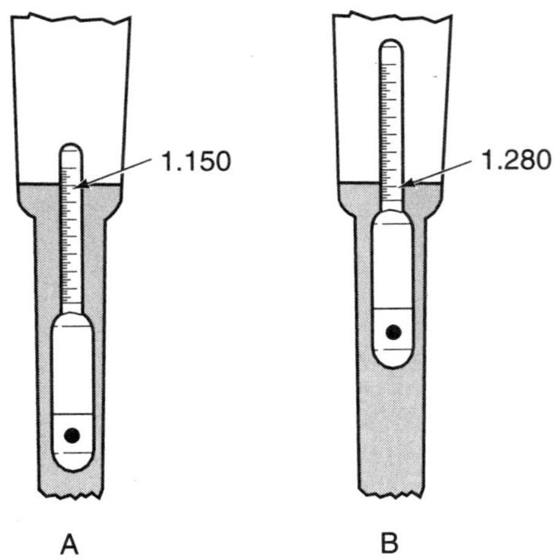

FIGURE 14–25 Where the top of the electrolyte intersects the float is the specific gravity reading for the cell. (A) Float indicates a *low* reading and (B) float indicates a *high* reading.

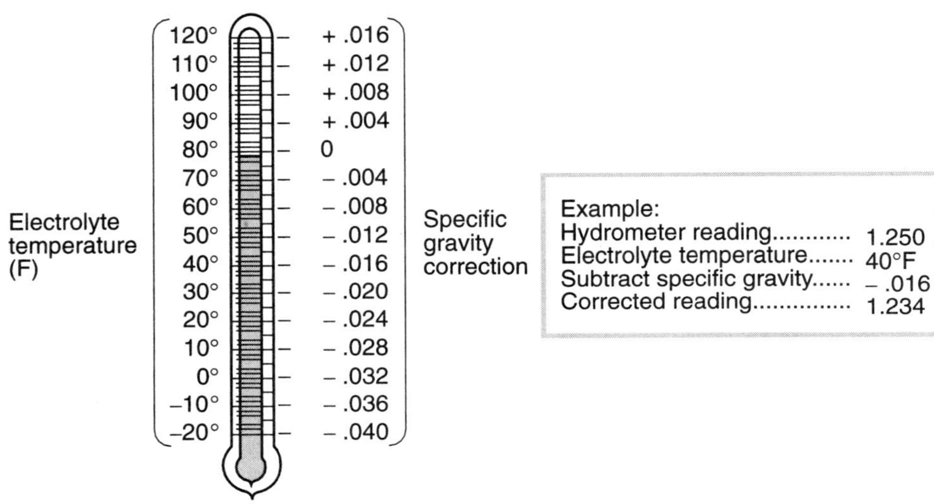

FIGURE 14–26 Temperature correcting the specific gravity reading.

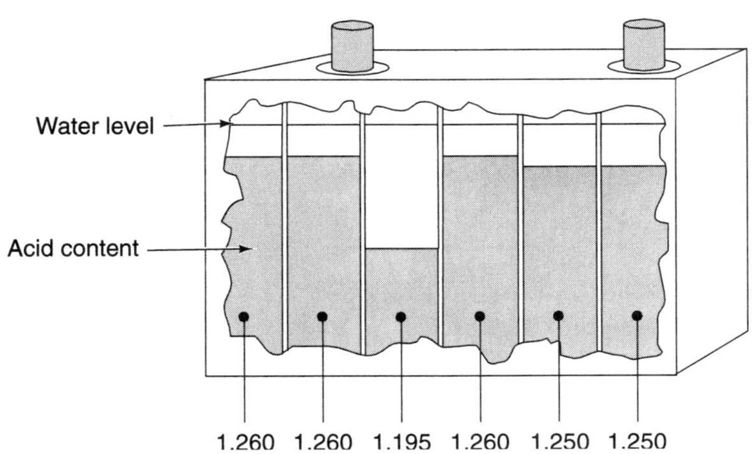

FIGURE 14–27 The specific gravity of a defective cell compared to fully charged cells.

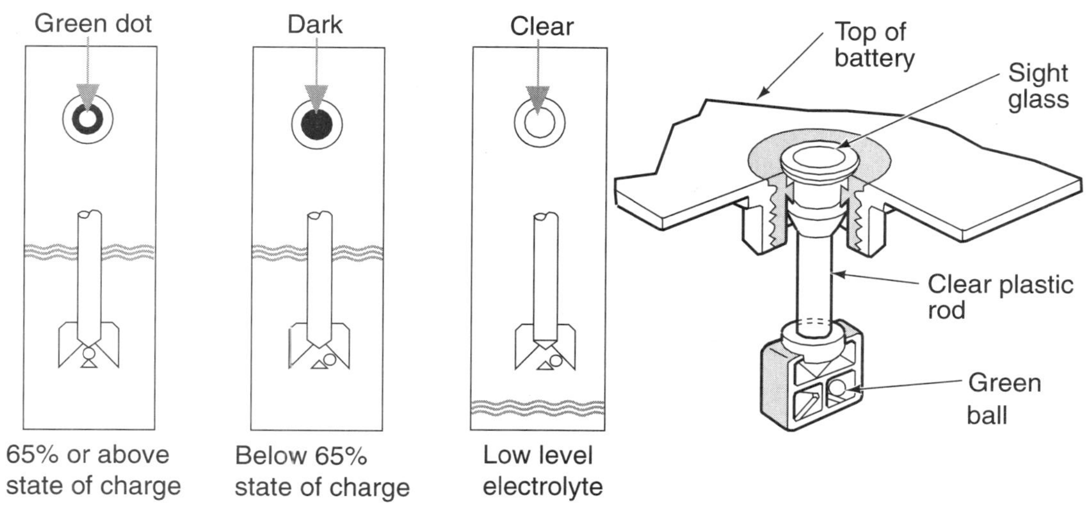

FIGURE 14–28 Built-in hydrometer used to indicate the state-of-charge in a battery.

it indicates the electrolyte level is low and the battery needs to be replaced.

Battery Open Circuit Voltage Test

When a battery has a sealed top, such as a maintenance free battery, a hydrometer cannot be used to check the state-of-charge. In this case, a battery open circuit voltage test can be used to substitute for a hydrometer test. In Figure 14–29, a voltmeter is used to measure the voltage across the battery terminals. For an accurate reading, battery temperature should be between 60° F and 100° F and stabilized for at least 10 minutes with no load applied. If the battery has been recently charged, the surface charge must be removed to obtain an accurate open circuit voltage reading. To remove the surface charge, apply a 20 amp load (high beams and heater blower) for two minutes then shut off the circuits. Let the battery stabilize for five minutes before testing the open circuit voltage.

FIGURE 14–29 Testing the open circuit voltage of a battery.

The relationship between the open circuit voltage and specific gravity is illustrated in Table 14–1.

As you can see by the results in Table 14–1, the voltage difference between a fully charged battery and a discharged battery is only .7 volts. Using a quality digital voltmeter with a DC accuracy of ± 0.1 percent is best to get an accurate reading.

Battery Capacity Load Test

The battery capacity load test is the best test to check a battery's condition. For this test to be accurate and realistic, the battery must be at least seventy-five

TABLE 14–1

Open Circuit Voltage	State Of Charge	Specific Gravity
12.6 volts or greater	100%	1.260
12.4 volts	75%	1.225
12.2 volts	50%	1.190
12.0 volts	25%	1.155
11.9 volts	discharged	1.100

percent charged and have passed the hydrometer state-of-charge test, or the voltmeter open circuit test. The battery needs to supply high amperage to the starting system and still have sufficient voltage to operate the engine computer control systems. Depending on the load test equipment used, certain steps should be followed. If a carbon pile load tester is used (Figure 14–30), the following general steps are used.

- Connect the tester cables following the manufacturer's instructions—normally to the positive and negative posts with the inductive amperage clamp placed around the negative load tester lead (not the negative battery lead).
- Determine the load test specifications for the battery. It is either one of the following.
  1. Battery A/H × 3 = load
  2. Battery CCA ÷ 2 = load
- Apply the calculated load for *15 seconds* while observing the voltmeter. The reading should be 9.6 volts, or above, at 70° F Table 14–2 calculates the minimum load test voltage in relation to the battery temperature. Batteries that fail this test should be charged and retested.

Figure 14–31 shows an example of an automated battery load tester. One brand is called an ARBST (alternator/regulator/battery/starter tester).

This tester performs a load test when the following steps are performed.

TABLE 14–2 Battery Load Test Temperature Chart

Minimum Voltage	at ° F	at ° C
9.6	70+	21+
9.5	60	16
9.4	50	10
9.3	40	4
9.1	30	−1
8.9	20	−7
8.7	10	−12
8.5	0	−18

- Connect the tester's red lead to the positive battery post, and the tester's black lead to the negative post. The tester's gray inductive clamp is *not* used to test batteries, but only used to test starters and alternators.
- Push the battery test button on the ARBST, and input the *actual* CCA of the battery (not one-half of CCA like when using a carbon pile tester).

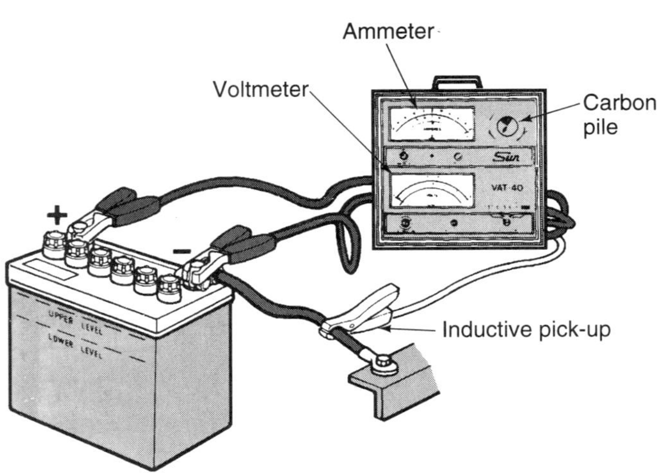

FIGURE 14–30 Connecting a carbon pile load tester to a battery.

(Courtesy of the Saturn Corporation)

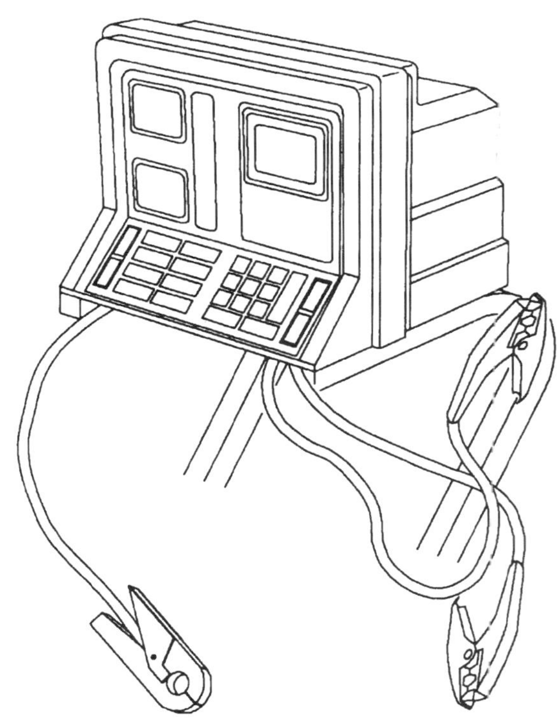

FIGURE 14–31 An automatic carbon pile ARBST load tester.

- Push the enter button on the tester. The load test will count-down through a 50 second test sequence, and then record the results.
- Test results provide the following.

  Good battery—battery is charged and capable of performing as designed.

  Good, low charge—battery is good, but needs to be recharged.

  Bad battery—battery will not hold a charge or perform as designed.

  Charge and retest—tester cannot determine battery condition—charge battery and retest.
- If the wrong CCA is input into the tester, a false reading will be given. Make sure to use the proper battery CCA. Do not assume the CCA by the battery's size!

Battery Three Minute Charge Test

The battery three minute charge test is performed only when a battery fails a capacity load test. As a battery ages, deterioration of the cell plates occurs which is called **sulfation.** The sulfation interferes with the ability of the battery cells to deliver a current and accept a charge (Figure 14–32).

CAUTION *Some battery manufacturers, such as Delco, do not recommend the three minute charge test. Differences in plate design, size and material used in the plates can all affect the results of the test.*

When a battery fails the load test, the three minute charge test is a reasonably accurate method to check for battery sulfation. To conduct the test, perform the following steps.

- Remove the battery ground cable to isolate the vehicles electrical system from the battery.
- Connect the battery charger to the battery, observing polarity.
- Use a battery charger that can provide a continuous 40 amps.
- Maintain a 40 amp charge for three minutes while observing the voltmeter connected to the battery.
- If the voltage reading is less than 15.5 volts, then the battery is not sulfated. If the voltmeter reading is above 15.5 volts, the battery is sulfated. Some maintenance free batteries use 16.5 volts as the divisional line between a good and bad battery. Always refer to the specific battery manufacturer's guidelines before performing this test.

Battery Charging Procedures and Safety

In the event a battery needs to be charged with an external battery charger, several safety precautions and procedures need to be followed.

- Determine the condition of the battery that needs a charge. A battery that is severely sulfated can lead to excessive gassing and heat build-up of the plates

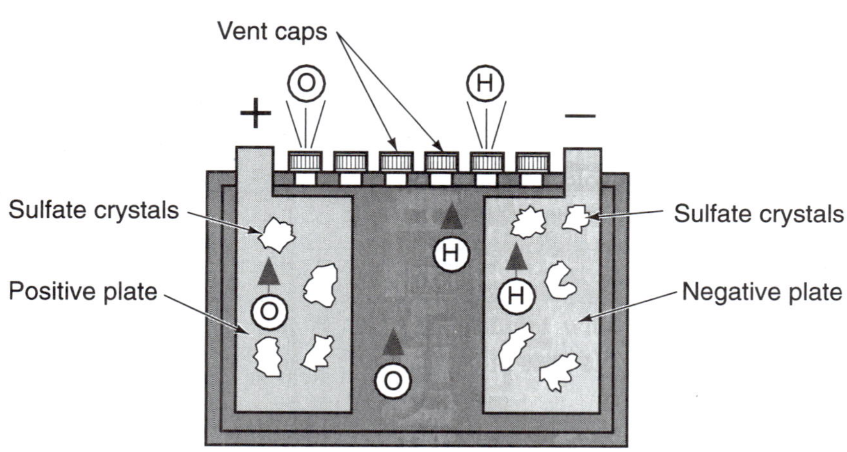

FIGURE 14–32 A sulfated battery condition resulting form sulfate crystals attaching to cell plates.

during fast charging. If a battery is extremely discharged and remains in this state of (dis)charge for an extended period of time, it may be permanently damaged.

- Battery temperatures that exceed 125° F during charging can lead to a thermal runaway, where the internal heat will melt-down the plates and create a short or an explosion.
- Disconnect the negative battery cable to isolate the vehicle's electrical system. If the charging voltage exceed 15 volts during charging, the vehicle computers could be damaged if they were still connected to the battery.
- *Always* ensure the battery charger is *off* when connecting the charger leads to the battery. Hydrogen and oxygen mixtures that may be in the immediate area of the discharged battery could explode if a spark is created.
- Determine the charge rate required to charge the battery. Slow charging a battery will restore the battery grid plates to the fully charged state and minimize the chance of overheating or overcharging the cells. Figure 14–33 determines the rate of slow charge according to the reserve capacity of the battery.
- If adequate charging time is not available, fast charging uses a high current over a short time to boost the battery enough for starting, or to operate the electrical system. Figure 14–34 determines the rate of fast charge. A battery on a fast charge should be monitored at all times to prevent overheating conditions.

Battery Capacity (Reserve Minutes)	Slow Charge
80 minutes or less	10 hrs.@ 5 amperes 5 hrs. @ 10 amperes
Above 80 to 125 minutes	15 hrs.@ 5 amperes 7.5 hrs. @ 10 amperes
Above 125 to 170 minutes	20 hrs.@ 5 amperes 10 hrs. @ 10 amperes
Above 170 to 250 minutes	30 hrs.@ 5 amperes 15 hrs. @ 10 amperes

Courtesy of Battery Council International

FIGURE 14–33 Determining the slow charge rate depending on the reserve capacity.

The difference between slow charging and fast charging a battery affects the amount of chemical transfer between the electrolyte and the positive and negative cell plates (Figure 14–35). The longer the time allowed for battery charging at a slow constant current, the greater the chance for the cell plates to make a complete chemical exchange.

BATTERY HIGH-RATE CHARGE TIME SCHEDULE

SPECIFIC GRAVITY READING	CHARGE RATE AMPERES	BATTERY CAPACITY—AMPERE HOURS			
		45	55	70	85
Above 1.225	5	★	★	★	★
1.200–1.225	35	30 min.	35 min.	45 min.	55 min.
1.175–1.200	35	40 min.	50 min.	60 min.	75 min.
1.150–1.175	35	50 min.	65 min.	80 min.	105 min.
1.125–1.150	35	65 min.	80 min.	100 min.	125 min.

★ Charge at 5-ampere rate until specific gravity reaches 1.250 @ 80°F.

FIGURE 14–34 Determining the fast charge rate depending on the state-of-charge and capacity rate.

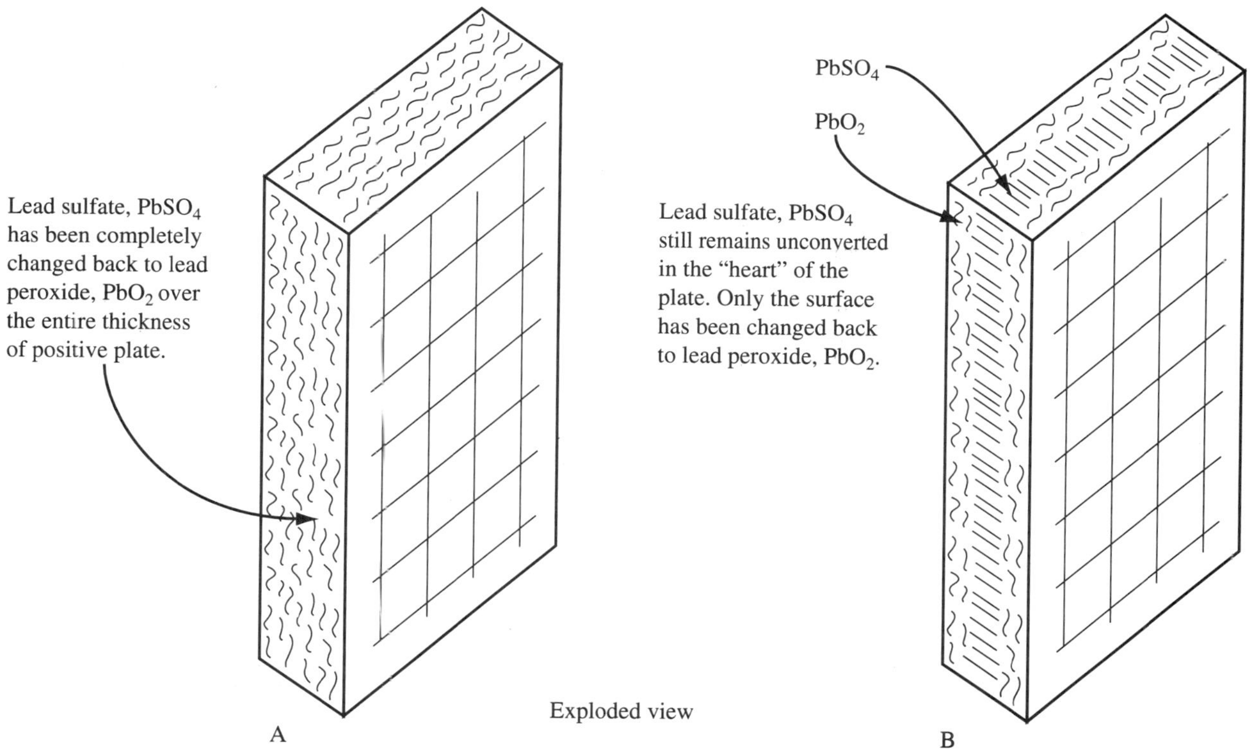

FIGURE 14–35 (A) Effects of a slow charge on a positive plate and (B) effects of a fast charge on a positive plate.

CAUTION *The vehicle's alternator should* not *be used to charge a dead battery that has been jump started. The extra load on the alternator, in addition to the vehicle electrical system, could overload the alternator.*

The percent of charge of a battery can affect the temperature at which the battery will freeze (Table 14–3). Even though a frozen battery can be completely thawed, and then recharged, the stress on the cell plates will significantly shorten the performance and life of the battery. If there is physical damage to the battery case (bloated), charging the battery should not take place.

If a battery needs to be charged or stored on a concrete floor, an insulating mat or clean dry block of wood should be used to insulate the battery. When a battery is stored on a cold floor, there is a difference in temperature inside the cells between the colder bottom section, and the upper section which is at room temperature. A difference in temperature within the same cell causes a **self discharge** to occur, even while the battery is being charged. Self discharge in batteries is a well-known fact

TABLE 14–3 Battery Freezing Temperatures According To The Specific Gravity of the Electrolyte

% Charge	Specific Gravity	Freezing Temp.
100%	1.265	−91° F
50%	1.200	−16° F
25%	1.150	+5° F
	1.100	+19° F
water	1.000	+32° F

in the submarine industry, which use warm sea water cycled in tubes around their batteries to keep all the sections in each cell at the same temperature.

Battery Jump Starting

Various vehicle manufacturers have sufficient differences in recommended jump starting procedures and precautions; jump starting procedures for one manufacturer's vehicle might turn out to be unsafe and costly for another manufacturer's vehicle. If one of the following three conditions exist, the disabled battery should *not* be jump started.

1. Measure the battery terminal voltage on the disabled vehicle. A voltage reading of approximately 10.55 volts could indicate the battery has a *shorted cell* (12.66 volts total − 2.11 volts of one cell = 10.55 volts) and should *not* be jump started. A battery with one shorted cell could put an excessive load on the booster battery as it tries to compensate for the bad battery.
2. A voltage reading of 0.55 volts could indicate the battery has a possible *open circuit* in a cell. This could quickly raise the alternator output on the disabled vehicle to 30 volts or more as the jumper cables are removed. With an *open* battery, the alternator has nothing but itself to use as a reference and the excess voltage spike could damage vehicle computer controls.
3. The built-in hydrometer shows clear or light yellow. In this case, replace the battery.

The following generic safety precautions should be observed for all vehicles.

- Observe specific vehicle service manual(s) for any special procedures.
- Wear safety glasses. Explosive hydrogen gas may be present around the vehicle needing to be jump started.
- Engage the parking brake on both vehicles and put both in neutral—some vehicles have been known to jump out of park, and into reverse!
- Make sure the vehicles *do not* touch. High current should run directly from the battery to the starter through the battery cables. If the vehicles touch, current could flow through the vehicle body through small low-current body ground straps to the starter, and burn up the straps.
- Turn off the ignition switch and *all* accessories on both vehicles before connecting jumper cables. Some manufacturers recommend turning on the heater blower motor on the vehicle with the dead battery to absorb any potentially damaging voltage spikes.
- Do not attempt to use the booster vehicle as a *battery charger*. This could overload the alternator in the booster vehicle. Some vehicle manufacturers are very specific about not using this commonly practiced procedure, as it could severely overheat the alternator and cause premature failure to the windings or bearings—grease melts out of the bearing when it overheats.
- Obtain the ignition keys from the vehicle being jump started. This *ensures* you have control of the whole jump starting process. An overzealous driver trying to start the vehicle before you say it is ready could prove costly to both vehicles.
- Do *not* use more than 16 volts to jump start a vehicle equipped with an engine control computer. The excess voltage may damage the electronic components. Most vehicles built during the last decade fall into this category.
- Do not disconnect jumper cables while the vehicle being started is in the cranking mode. A voltage spike may be generated that could damage electronic components on the booster vehicle.
- To hook-up the jumper cables (Figure 14–36), attach one end of the positive jumper cable to the disabled battery's positive terminal, and the other positive jumper cable to the booster battery's positive terminal. Attach one end of the negative jumper cable to the booster battery's negative terminal, and the other negative jumper cable to a good engine ground on the disabled vehicle at least three feet away from the battery for safety.

The following jump start procedures are recommended on electronic-controlled vehicles only as a last resort if a proper battery charger is not available.

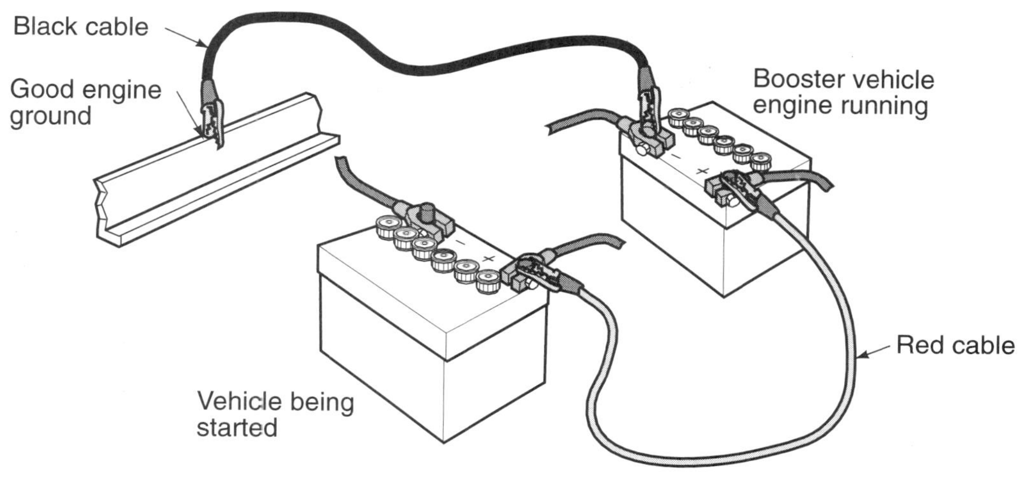

FIGURE 14–36 Connecting jumper cables between the booster vehicle and the vehicle being started.

- Follow any safety precautions that apply from above.
- Shut off booster vehicle and hook up jumper cables (Figure 14–36).
- Attempt to start disabled vehicle. If vehicle will not crank over sufficiently, stop cranking the engine. Start the booster vehicle and raise the idle to approximately 1,200–1,500 rpm for up to five minutes to assist in boosting the disabled battery. Attempt to start the disabled vehicle. If it still cranks slowly, DISCONTINUE JUMP STARTING. Further running of booster vehicle could overload the alternator or cause a voltage spike in the system.

Battery Service Procedures

For premium battery efficiency, a quick visual inspection should be taken to inspect the condition of the battery case for cracking or bloating, the terminal connections and cables for corrosion, and the overall cleanliness of the battery. Corroded battery terminals can reduce the efficiency of a battery by creating an unwanted resistance to current flow for the vehicle's electrical system (Figure 14–37). It could also limit the current used to recharge the battery which could result in lead sulfate crystallizing on the battery plates over an extended period of time.

When battery servicing is necessary and the terminals have to be disconnected, it is important on most vehicles with electronic memory functions in the radio or power train control computers to save this memory. If the electronic memory is lost in some modern vehicles, it may take several drive cycles before the computer memory learns transmission shift points, and engine idle and load conditions. It is not uncommon for these conditions to exist for up to 20 miles or more of driving time.

Electronic Memory Saving

Beginning in the early 1970's, most vehicles were introduced with radios having electronic tuning. A simple way to keep the memory alive in these earlier electronic radios is the use of a 9 volt memory saver adapter plugged into the cigar lighter receptacle (Figure 14–38). To prevent the 12 volt vehicle battery from back feeding into the 9 volt battery when they are both in parallel to the electrical circuit, a diode is placed in series with the adapter plug's positive lead.

Another 9 volt memory saver adapter (Figure 14–39), that can be built with common electronic supply

FIGURE 14–37 A typical corroded battery terminal.

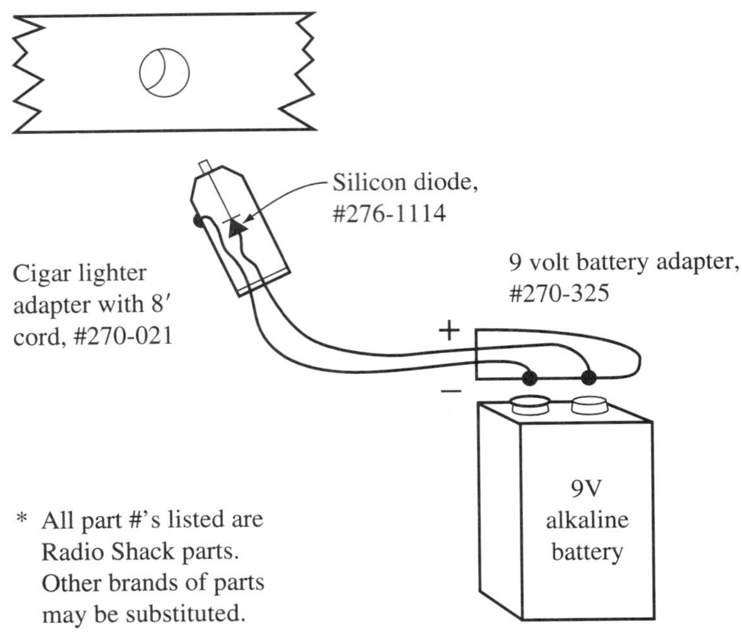

FIGURE 14–38 9 volt battery and cigar lighter adapter for saving vehicle radio memory.

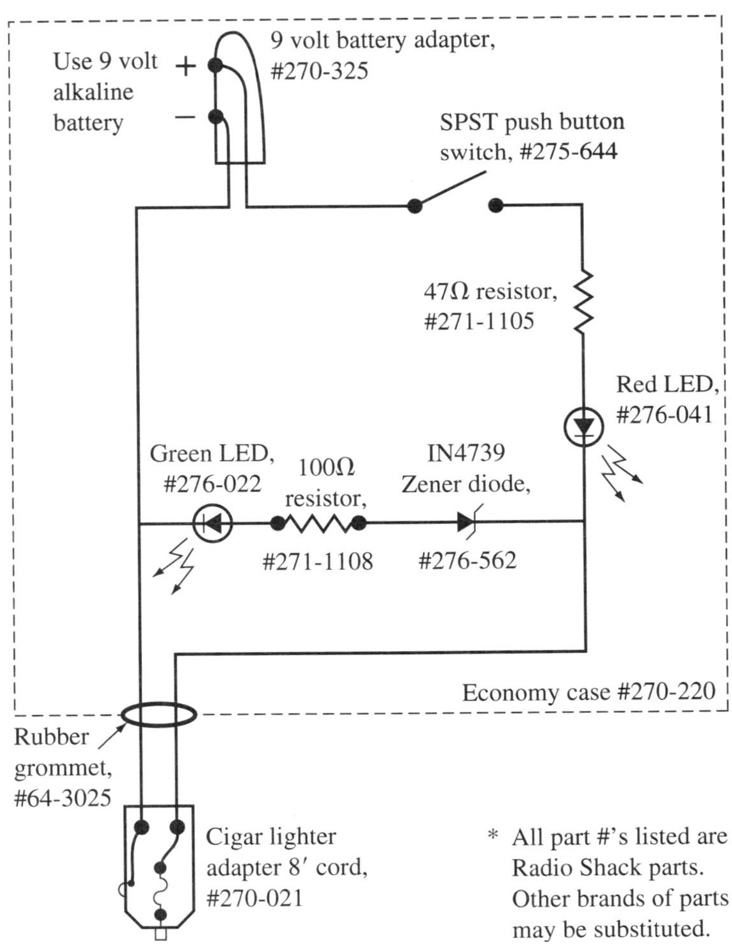

FIGURE 14–39 Memory saver using LED indicator lights, 9 volt battery, and cigar lighter adapter.

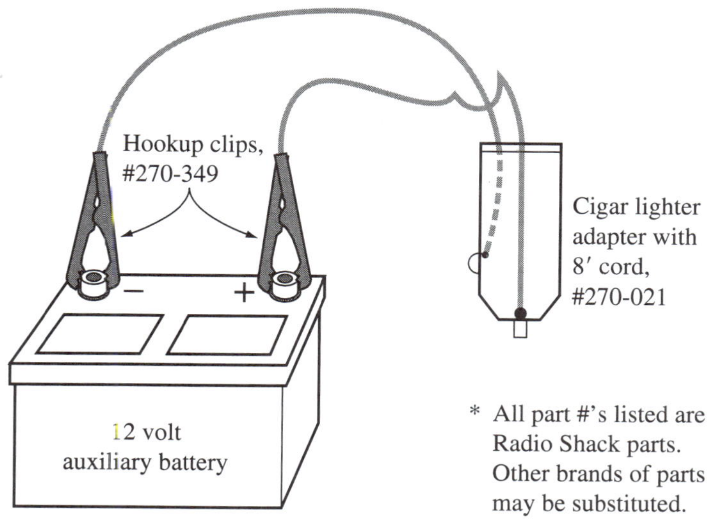

FIGURE 14–40 Saving the vehicle computer memory with a 12 volt auxiliary battery and cigar lighter adapter.

store parts, uses a green LED light to indicate that the vehicle's electrical system is operating off of the vehicle battery. When the battery's negative cable is removed, the memory saver will automatically shift to the 9 volt back-up battery and the red LED light will go on. Either of these memory saver tools are meant to save only the radio memory while the vehicle battery is changed. If a door light or other higher amperage requirement is applied to the circuit, the 9 volt battery will be drained quickly.

On vehicles with computer controls or complex radio programming functions, the 9 volt memory saver should not be used. A possible voltage spike can occur when the negative battery cable is removed or replaced. To prevent this, a secondary 12 volt battery should be put in parallel with the vehicle battery, either directly to the positive and negative battery cables, or through the cigar lighter plug (Figure 14–40). If the cigar lighter plug is used, make sure it is energized when the ignition key is in the Off position.

Battery Removal and Cleaning

The best way to clean a dirty battery is to remove it from the vehicle. Most batteries in modern vehicles are tightly packed into the engine compartment area. Attempting to clean a battery in the vehicle could spread the corrosive acids onto other engine or electrical components.

Begin by removing the battery negative terminal using a battery terminal puller (Figure 14–41), and then remove the battery and carry it using a suitable lifting strap (Figure 14–42).

FIGURE 14–41 To prevent post damage, a terminal puller is used.

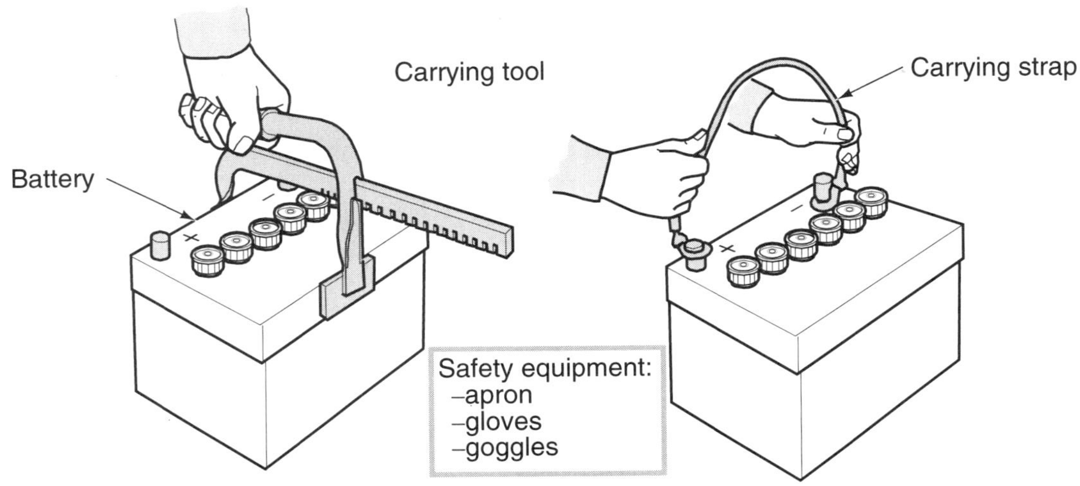

FIGURE 14–42 A battery carrying tool or lifting strap is the preferred method to move a battery.

Apply a solution of baking soda and warm water (Figure 14–43) to the battery case to neutralize any accumulated corrosion on the battery case, and then wipe the battery case dry with paper towels or an old rag. Cotton shop towels from a laundry service should *not* be used to dry a battery because residual acids will eat away the shop towel, and remove it from the counted inventory. The terminals and posts should be cleaned with a cleaning tool or wire brush (Figure 14–44).

When a battery is replaced into the vehicle, make sure the hold-down connectors are secured. A loose battery could short out on the hood or tip over. Install the *positive* connector first, and then the negative connector. Battery terminal clamps should be sprayed with a protective coating to prevent corrosion. Grease or petroleum jelly can also be used to seal the terminals, but will also attract dirt or other contaminants over time. Corrosion protective pads can also be used under the terminals to prevent corrosion (Figure 14–45).

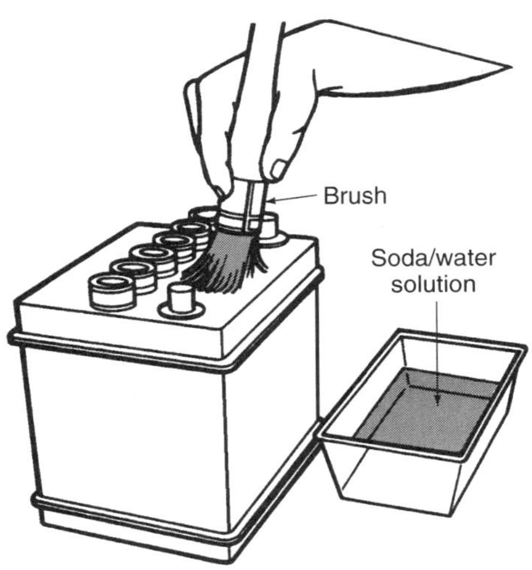

FIGURE 14–43 A solution of baking soda and water will neutralize the battery acid corrosion.

(Courtesy of DaimlerChrysler Corporation)

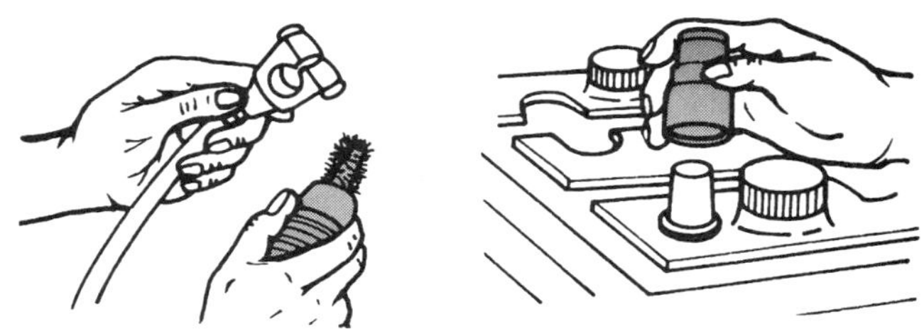

FIGURE 14–44 Cleaning battery posts and terminals with a battery cleaning tool.

FIGURE 14–45 Using corrosion protective pads to prevent terminal corrosion.

CASE STUDIES

CASE # 1

The following case study is typical of a battery related problem that technicians often have to diagnose. Knowledge of how batteries are constructed and of simple test procedures will help in performing the diagnosis quickly and efficiently.

Customer Complaint

A customer brought his vehicle into the shop with a possible battery problem. The vehicle cranks slowly in the morning when cold or under heavy load conditions. The customer changed the alternator, but the problem remained.

Known Information

- ❑ Vehicle operating voltage equals 14 volts
- ❑ Battery is a conventional design
- ❑ Battery is four years old with a CCA rating of 550 CCA
- ❑ Recorded battery hydrometer readings (Figure 14–46)

Circuit Analysis

Answer the following questions on a *separate* sheet of paper.

1. With the known information stated above, what is the *most* likely cause of the battery performance during cold cranking conditions?____________________________________.
2. To properly check a battery for performance under all load conditions, what are the proper steps to take in determining its condition?

   a. ____.

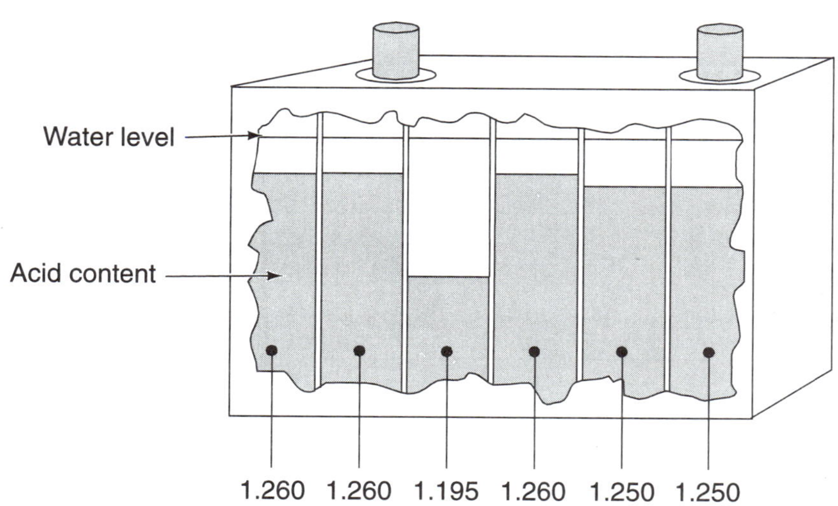

FIGURE 14–46 Hydrometer readings.

b. ____________________.

c. ____________________.

d. ____________________.

e. ____________________.

f. ____________________.

3. What diagnostic steps were helpful when troubleshooting this system or explaining it to the customer? ____________________

____________________.

4. What is your analysis of the above case study? ____________________

____

____________________.

CASE # 2

The following case study concerns jump starting a vehicle. Knowledge of jump start procedures and safety precautions would be helpful.

Customer Complaint

The battery in a vehicle was cranking successfully only an hour previous to the problem. The customer requests a "jump start."

Known Information

- ❑ Battery is only two years old
- ❑ A hydrometer check on the battery showed the specific gravity of all cells were approximately 1.265
- ❑ A DMM was used to measure the open circuit voltage of the battery, which measured 10.55 volts

Circuit Analysis

Answer the following questions on a *separate* sheet of paper.

1. With the known battery information above, what is the most likely cause of the battery problem?

____________________.

2. Can the battery be safely jump started anytime a no-crank or slow crank condition exists? ______.
3. What are some of the precautions that should be followed *before* attempting to jump start any vehicle? ____________________.
4. What diagnostic information is helpful to know before attempting to jump start any vehicle? _____.
5. What is you analysis of the above case study? ____________________

____________________.

Hands-On Vehicle Tests

The following nine hands-on vehicle checks are included in the **NATEF** (National Automotive Technicians Education Foundation) **Task List.** The task reference numbers are the same as the NATEF reference numbers. Some tasks are grouped together due to their close association with each other. Complete your answers to the below questions on a *separate* sheet of paper.

Performance Task # 1 (NATEF # VI-B-4)

Task Description

Inspect, clean, fill, and replace the battery.

Performance Task # 2 (NATEF # VI-B-6)

Task Description

Inspect and clean the battery cables, connectors, clamps, and hold-downs; repair or replace as needed.

Task Objectives

❑ Obtain a vehicle that can be used for these tasks. What model and year of vehicle are you using for these tasks? ____________________.

❑ Describe the general appearance of the battery.____________________

____________________.

❑ Inspect and record the condition of the following battery components:

Component	Condition	Repairs needed
Battery case		
Battery top		
Battery support tray		
Battery hold down		
Battery posts		
Battery cables and clamps		
Surrounding area of battery		

❑ What tools and equipment would be useful in cleaning the battery and associated components?

____________________.

❑ What are the necessary safety precautions that should be observed when working with or near a battery?____________________

____________________.

❑ If the battery needs to be filled, what type of water should be used and how full should it be filled?

____________________.

❑ What procedures are most helpful when inspecting, filling, and cleaning a battery, and associated cables, connectors, and hold downs? ____________________

____________________.

Task Summary

After performing the above NATEF tasks, what can you determine will be helpful in knowing how to check instrument cluster gauges and warning lights for proper operation? ______________________

____

____.

Performance Task # 3 (NATEF # VI-B-1)

Task Description

Perform battery state-of-charge test; determine needed service.

Performance Task # 4 (no NATEF # assigned)

Task Description

Perform battery open circuit voltage test; determine needed repairs.

Performance Task # 5 (NATEF # VI-B-2)

Task Description

Perform battery capacity (load and high-rate discharge) test; determine needed service.

Performance Task # 6 (no NATEF # assigned)

Task Description

Perform battery three minute charge test; determine needed repairs.

Task Objectives

- ❑ Obtain a vehicle that can be used for these tasks. What model and year of vehicle are you using for these tasks? ____.
- ❑ Battery Type ______________________. Battery Rating ______________________.
  Engine Cubic Inches ______________________.
- ❑ Is the battery the sufficient size and rating for the vehicle? ______________________.
- ❑ Perform the following battery state-of-charge test

  Hydrometer Readings (if cells are accessible).

Cell #	1	2	3	4	5	6

- ❑ Test Results.
  1. All cells have a specific gravity of at least 1.265—continue with load test.
  2. All cells have an equally low specific gravity below 1.265—charge and retest battery.
  3. Cell specific gravity readings have a point variation of .050 between the highest and lowest cell—discontinue testing and replace battery.

❑ Perform the following battery open circuit voltage test.

1. Load the battery with 20-25 amps for one minute to remove any surface charge present on cell plates—turn on high beam headlights and heater blower to high.
2. Remove any loads from the battery, and let the battery stabilize for two minutes.
3. Measure the open circuit voltage across the battery terminals with a quality DMM.

❑ Test Results.
Refer to the chart below to convert the open circuit voltage to the state-of-charge or the specific gravity.

Open circuit voltage	State-of-charge	Specific gravity
12.6 volts or greater	100%	1.260
12.4 volts	75%	1.225
12 2 volts	50%	1.190
12.0 volts	25%	1.155
11.9 volts	discharged	1.100

1. If the battery's open circuit voltage indicates a state-of-charge less than 100%, charge and retest battery.
2. If the battery's open circuit voltage indicates a fully charged battery, continue with load test.

❑ Perform the following battery capacity load test.

1. Determine which tester to use to perform the load test. Some testers perform the procedure more automatically than others.
2. Following are the instructions for operating a battery tester with a *manual controlled* carbon pile for load testing the battery.
   - Set tester for the high amperage start position.
   - Determine the load rating of the battery. Calculate the load to apply with the carbon pile, CCA $\div$ 2 $=$ load, or A/H $\times$ 3 $=$ load.
   - Hook up large tester clamps to battery positive and negative terminals. Green inductive pick-up clamp should go around the negative tester lead (not the battery negative lead).
   - Apply the calculated load to the carbon pile for fifteen seconds while watching the voltmeter reading. Turn off load after fifteen seconds.
   - Determine the load test results.
     a. Good battery—reading remained above 9.6 volts at 70° F during the load test. Battery is satisfactory.
     b. Bad battery—battery voltage dropped below 9.6 volts at 70° F during load test. Battery performance failed. Verify condition of cell plates by performing a three-minute charge test. Some manufacturers of maintenance free batteries recommend that this test not be performed due to the differences in construction, materials and size of batteries. Refer to specific manufacturer of battery being tested before proceeding with this test.
3. Following are the instructions for operating a battery tester with an *automatic* load testing feature, such as the ARBST (alternator/regulator/battery/starter tester).

- Hook up the large tester clamps to the battery positive and negative terminals.
- Push the battery test button and enter the CCA rating from the battery.
- Push Enter and then wait for fifty seconds for test results.
- Determine the load test results by reading the digital tester readout.
  a. *Good battery*—battery is charged and capable of performing as designed.
  b. *Good, Low Charge*—battery is good, but needs to be recharged.
  c. *Bad Battery*—battery will not hold a charge or perform as designed.
  d. *Charge and Retest*—tester cannot determine battery condition. Charge battery and retest.

❑ Perform the following battery three-minute charge test.

1. Disconnect the battery ground cable to isolate the vehicle electrical system from higher than normal voltages that might exist during this test.
2. Connect a battery charger to the battery observing polarity and all safety precautions.
3. Connect a DMM across the battery terminals to monitor the battery voltage while the three-minute charge is taking place.
4. Turn on battery charger to 40 amps for three minutes. To get an accurate test, a battery charger that automatically reduces the charging amperage may not be suitable for this test.
5. Observe the voltmeter reading after three minutes.

❑ Three-minute charge test results.

1. Good battery—voltmeter reading *less than* 15.5 volts (16.5 volts for some maintenance free batteries). Battery is *not* sulfated—recharge battery and perform the capacity load test again.
2. Bad Battery—voltmeter reading more than 15.5 volts (16.5 volts for some maintenance free batteries). Battery is sulfated and not able to receive a charge. Battery should be replaced.

Task Summary

After performing the above NATEF tasks, what can you determine that will be helpful in knowing how to check the battery state-of-charge, open circuit voltage, capacity load test, and three minute charge test?____

____

____.

Performance Task # 7 (NATEF # VI-B-5)

Task Description

Perform slow/fast battery charge.

Task Objectives

❑ Obtain a vehicle that can be used for this task. What model and year of vehicle are you using for this task? ____.

❑ What are the safety precautions necessary when slow/fast charging a battery? ______________________

____.

❑ What is the open circuit voltage of the battery needing charged? ______________________.

❑ How do you hook up the battery charger to the battery? ______________________________

____.

❑ What would be the charging rate for this battery with a slow charge? ______________________

____.

❑ What would be the charging rate for this battery with a fast charge? ______________________

____.

❑ How can you determine that the battery is fully charged? ______________________________

____.

Task Summary

After performing the above NATEF task, what can you determine will be helpful in knowing how to slow charge and fast charge a battery? ____

____

____.

Performance Task # 8 (NATEF # VI-B-7)

Task Description

Start a vehicle using jumper cables and a battery or auxiliary power supply.

Task Objectives

❑ Obtain two vehicles that can be used for this task. What model and year of vehicles are you using for this task? ____.

❑ What are the safety precautions necessary when jump starting a battery? ______________________

____.

❑ What is the open circuit voltage of the battery needing jump started? ______________________.

❑ What special precautions are necessary in reference to the open circuit voltage of the battery needing jump started? ____.

❑ What special precautions are necessary when either, or both, the booster vehicle and the dead battery are computer controlled.

❑ List the procedures you used to perform the jump start on this vehicle:

a. ______________________ d. ______________________

b. ______________________ e. ______________________

c. ______________________ f. ______________________

g. ______________________ i. ______________________

h. ______________________ j. ______________________

- ❑ What steps are most helpful when jump starting a vehicle? ____.

Task Summary

After performing the above NATEF task, what can you determine will be helpful in knowing how to jump start a battery? ____.

Performance Task # 9 (NATEF # VI-B-3)

Task Description

Maintain or restore electronic memory functions.

Task Objectives

- ❑ Obtain a vehicle that can be used for this task. What model and year of vehicle are you using for this task? ______________________.
- ❑ What type of memory saving tool are you using to perform this task? ______________________.
- ❑ What precautions are necessary when using a memory saver tool while changing a battery? ______________________.
- ❑ List the steps taken to properly use a memory saving tool to save the vehicle's memory circuits:
  1. ______________________.
  2. ______________________.
  3. ______________________.
  4. ______________________.
  5. ______________________.
- ❑ What is your analysis of using a memory saving tool to save the vehicle's memory circuits? ______________________.

Task Summary

After performing the above NATEF task, what can you determine will be helpful in knowing how to use a memory saving tool? ____.

Summary

- The simplest form of battery is a voltaic cell, with two unlike metals and an alkali, salt, or an acid solution to produce a voltage.
- An electrolyte is a chemical compound of alkali, salt, or acid solution that is capable of conducting electrical energy.
- A primary cell cannot be recharged to its original voltage condition.
- A secondary cell can be recharged to its original stored energy.
- The negative plate in a secondary cell is made of pure lead (Pb).
- The positive plate in a secondary cell is made of lead dioxide (PbO_2).
- A diluted mixture of sulfuric acid contains 64 percent water and 36 percent sulfuric acid, by weight.
- The basic grid plate in a conventional battery is made of a lead antimony mixture.
- The active material in a positive plate is a reddish-brown paste called lead peroxide.
- Battery plates are separated by micro-porous materials called separators.
- Each cell in a lead acid battery produces 2.11 volts, regardless of the number of positive and negative plates used in each cell.
- A fully charged battery will produce 12.6 volts.
- A battery's specific gravity is 1.265 at 80° F when fully charged.
- A maintenance free battery uses calcium in the grid plate material to reduce gassing and self-discharge.
- A deep cycle battery is a lead acid battery that uses approximately 2.75 percent antimony on the positive plate and a calcium alloy on the negative plate.
- A recombination battery is a completely sealed battery that uses the electrolyte mixture in a gel form.
- The reserve capacity (RC) of a battery is the amount of time in minutes a fully charged battery can be discharged at 25 amperes while maintaining a minimum battery voltage of 10.5 volts at 80° F in the event of a charging system failure.
- The amp-hour (A/H) rating of a battery is the amount of current that a fully charged battery can produce over a twenty hour period at 80° F before terminal voltage reaches a minimum of 10.5 volts.
- The cold cranking amperes (CCA) rating is the ability of the battery to provide an amperage load for thirty seconds at 0° F without the terminal voltage dropping below 7.2 volts.
- A battery's size is classified by BCI group numbers.
- A battery case leakage test determines if there is a small current draw between the battery posts through a dirty or contaminated top.
- A battery hydrometer measures the specific gravity content of the sulfuric acid mixture.
- A fully charged battery has a specific gravity of 1.265–1.280.

- ❑ The battery open circuit voltage test can be used to test the battery's state-of-charge when a hydrometer cannot be used.
- ❑ A battery capacity load test will check the ability of the battery to supply sufficient cranking amperage to the starter.
- ❑ The three minute charge test will test some batteries for sulfation on the cell plate surfaces.
- ❑ Self discharge in a battery can occur when the internal temperature between the upper and lower portions of the cell plates are different temperatures.
- ❑ A battery with an open circuit voltage of 10.55 volts should not be jump started because one cell could be shorted.
- ❑ A battery with an open circuit voltage of 0.55 volts could indicate a battery with an open circuit, and should not be jump started.

Key Terms

amp-hour rating (AH)
cold cranking amperes (CCA)
conventional battery
cranking amps (CA)
deep cycle
electrolyte
electrolysis
element
gassing
gel-cell
grid growth
hybrid
hydrometer
lead antimony
lead calcium
lead peroxide
maintenance free
primary cell
reserve capacity (RC)
recombination battery
secondary cell
self discharge
separator
specific gravity
sponge lead
sulfation
sulfuric acid
voltaic cell
watt-hour (WH)

Review Questions

Short Answer Essays

1. Describe the difference between a primary cell battery and a secondary cell battery.
2. What causes the glass bubble to float in a battery hydrometer?
3. Describe how to perform a battery capacity load test using the cold cranking rating method.
4. Describe how would you perform a battery case leakage test.
5. Describe the procedures and safety precautions when charging a battery.

Fill-in-the-Blanks

1. A fully charged battery will have an electrolyte reading of ____________________.
2. When a battery has an open circuit voltage of 12.2 volts, the state-of-charge is ____________________ percent.

3. During a battery three minute charge test, if the battery voltage exceeds ______________________, then the battery is considered sulfated.
4. When battery service is necessary, the ______________________ cable is disconnected first, and then reconnected ______________________.
5. A battery with a CA rating of 600 has a CCA rating of ______________________.

ASE Style Review Questions

As a lead acid battery is being discharged, the positive plate material is converted to:

A. Sulfuric acid
B. Lead sulfate
C. Sponge lead
D. None of the above

2. Technician A says the lead acid battery contains an electrolyte with a mixture of water and sulfuric acid. Technician B says the electrolyte in a lead acid battery is capable of conducting electrical energy. Who is correct?

   A. A only
   B. B only
   C. Both A and B
   D. Neither A nor B

3. Technician A says a primary cell battery can be recharged to its original condition. Technician B says a secondary cell battery cannot be recharged once its original stored energy has been depleted. Who is correct?

   A. A only
   B. B only
   C. Both A and B
   D. Neither A nor B

4. Technician A says that a conventional battery has a grid construction made from a mixture of lead and lead antimony. Technician B says the negative grid in a conventional battery has lead peroxide paste pressed into it to form the negative plate. Who is correct?

   A. A only
   B. B only
   C. Both A and B
   D. Neither A nor B

5. A 500 CCA battery is being load tested with a manual carbon pile load tester. Technician A says the correct load to apply to the load tester is one third of the rated A/H for fifteen seconds. Technician B says the correct load to apply to the tester is 50 percent of the rated CCA for fifteen seconds. Who is correct?
   A. A only
   B. B only
   C. Both A and B
   D. Neither A nor B

6. Technician A says the specific gravity of a fully charged battery is 1.265 at 80° F. Technician B says the specific gravity of a battery is measured with a hydrometer. Who is correct?
   A. A only
   B. B only
   C. Both A and B
   D. Neither A nor B

7. Technician A says a maintenance free battery produces less gassing than a conventional battery. Technician B says a maintenance free battery uses a grid frame made with a lead calcium mixture. Who is correct?
   A. A only
   B. B only
   C. Both A and B
   D. Neither A nor B

8. Technician A says a negative battery cable is always removed first when removing a battery. Technician B says a negative battery cable should always be put on first when replacing a battery. Who is correct?
   A. A only
   B. B only
   C. Both A and B
   D. Neither A nor B

9. Two technician are discussing the temperature correction readings on a hydrometer. Technician A says .004 should be added to the reading for every 10° above 80° F. Technician B says .004 should be subtracted to the reading for every 10° below 80° F. Who is correct?
   A. A only
   B. B only
   C. Both A and B
   D. Neither A nor B

10. Technician A says the battery must be stabilized before performing a battery open circuit voltage test. Technician B says a voltage reading of 12.0 volts is acceptable for a battery open circuit voltage test. Who is correct?
    A. A only
    B. B only
    C. Both A and B
    D. Neither A nor B

11. Two technicians are discussing the results of a three-minute charge test. Technician A says a reading greater than 15.5 volts on a conventional battery indicates a sulfated battery. Technician B says a load of 40 amps should be applied to the battery to get an accurate reading. Who is correct?
    A. A only
    B. B only
    C. Both A and B
    D. Neither A nor B

12. Technician A says a deep cycle battery uses a gel material for the electrolyte mixture. Technician B says a deep cycle battery grid plate has the vertical and horizontal grid bars in a radial design expanding from a center lug for faster amperage flow. Who is correct?
    A. A only
    B. B only
    C. Both A and B
    D. Neither A nor B

13. Technician A says a reserve capacity load rating indicates the amount of time in minutes a battery be discharged at 25 amps while maintaining a minimum battery voltage of 10.5 volts. Technician B says a battery with an RC capacity of 120 minutes will sustain 25 amp load for two hours. Who is correct?
    A. A only
    B. B only
    C. Both A and B
    D. Neither A nor B

14. Technician A says a baking soda and water mixture can be used to neutralize battery surface corrosion. Technician B says baking soda can be used in the battery cells to refresh the sulfuric acid mixture. Who is correct?
    A. A only
    B. B only
    C. Both A and B
    D. Neither A nor B

15. Technician A says a CCA rating on a battery determines the ability to provide an amperage load for thirty seconds at 0° F, with the terminal voltage dropping no lower than 7.2 volts. Technician B says the CCA can be divided be 5.25 to get the approximate A/H rating. Who is correct?
    A. A only
    B. B only
    C. Both A and B
    D. Neither A nor B

16. Technician A says the battery capacity rating is determined by the battery case size. Technician B says the battery case is always made of a hard rubber material. Who is correct?
    A. A only
    B. B only
    C. Both A and B
    D. Neither A nor B

17. Technician A says a voltage terminal drop test of 0.50 volts maximum is satisfactory. Technician B says the battery terminal voltage drop test is performed by removing the battery cable and hooking the voltmeter in series between the post and the terminal. Who is correct?
    A. A only
    B. B only
    C. Both A and B
    D. Neither A nor B

18. Technician A says battery temperature above 125° F helps the battery charge faster. Technician B says the battery negative cable should be disconnected before charging to isolate the vehicle's electrical system from voltage conditions above 15.5 volts. Who is correct?
    A. A only
    B. B only
    C. Both A and B
    D. Neither A nor B

19. Technician A says a battery will self discharge if left on a cold floor. Technician B says a battery will self discharge if the internal temperature of the cell plates vary. Who is correct?
    A. A only
    B. B only
    C. Both A and B
    D. Neither A nor B

20. Technician A says a 9 volt battery and cigar plug adapter can be used to save the radio memory while the vehicle battery is being changed. Technician B says an auxiliary 12 volt battery hooked in parallel to the vehicle battery is preferred for computer controlled vehicles to save memory when changing the vehicle battery. Who is correct?

    A. A only

    B. B only

    C. Both A and B

    D. Neither A nor B

15 Starting System Theory

Introduction

The starter circuit in modern vehicles is a high amperage circuit that combines electrical and mechanical components together to crank the engine for starting. The electrical energy produced by the battery is converted to mechanical energy through the interaction of magnetic fields in the starter motor. The typical small internal combustion engine of today requires 90 to 150 amperes to turn the engine over at 400 rpm's, or more, to be effective.

Objectives

When you complete this chapter you should be able to:

- ❑ Describe the main components of the starter motor system and their purpose.
- ❑ Explain the principle theory of magnetism in a DC starter motor.
- ❑ Explain the differences and operation of various solenoid control switches.
- ❑ Explain the differences between starter drive mechanisms.
- ❑ Describe the main features and operation between different types of starter motors.
- ❑ Explain the differences and operation of various starter safety switches.
- ❑ Perform a systematic diagnostic check on a starter circuit.
- ❑ Perform system and individual tests on the starting system components.
- ❑ Remove and reinstall a starter motor.
- ❑ Disassemble, clean, inspect, provide necessary repairs, and reassemble a starter motor.

Magnetism and Motor Principles

Learning about the operation of a DC starter motor begins with a thorough understanding of magnetism and the interaction of magnetic fields in the starter as the electrical energy from the battery is converted into mechanical energy to crank the engine. If you were able to see the magnetic fields that surround a permanent magnet (Figure 15–1), you would find that *like* poles *repel* each other, and *unlike* poles *attract* each other. This magnetic repulsion and attraction forms the basis of the electromagnetic motor.

When current flows through a conductor in a complete circuit, a weak magnetic field is produced that surrounds the conductor (Figure 15–2). The concentration

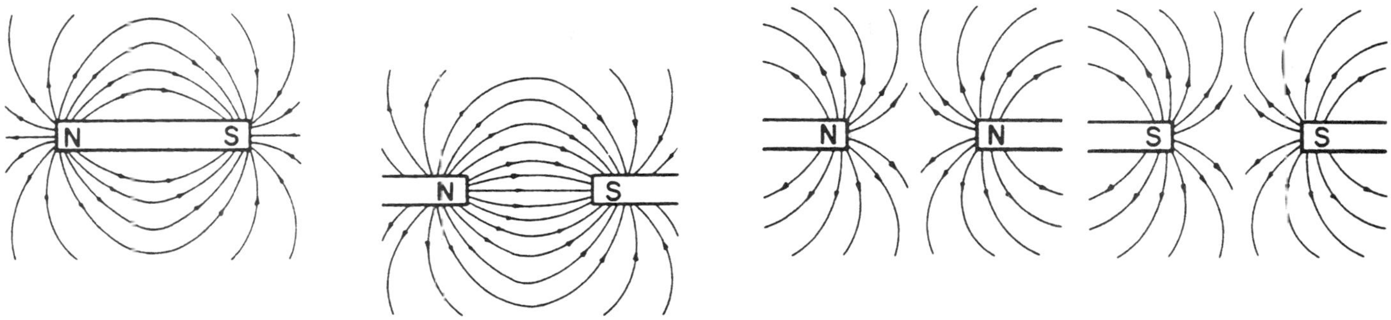

FIGURE 15–1 Principles of magnetism. (A) Magnetic north and south poles, (B) unlike poles attract each other, and (C) like poles repel each other.

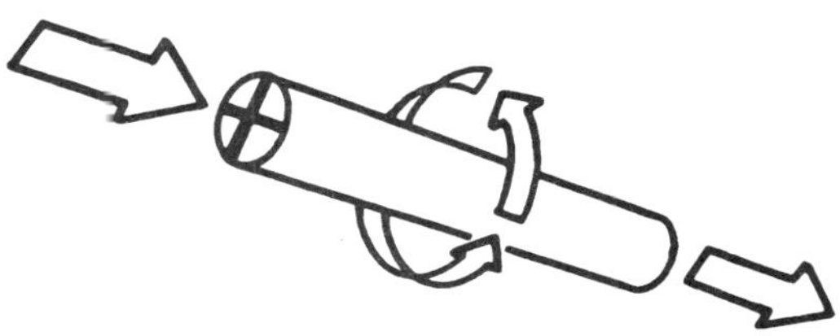

FIGURE 15–2 A weak magnetic field surrounds a conductor that has current flowing through it.

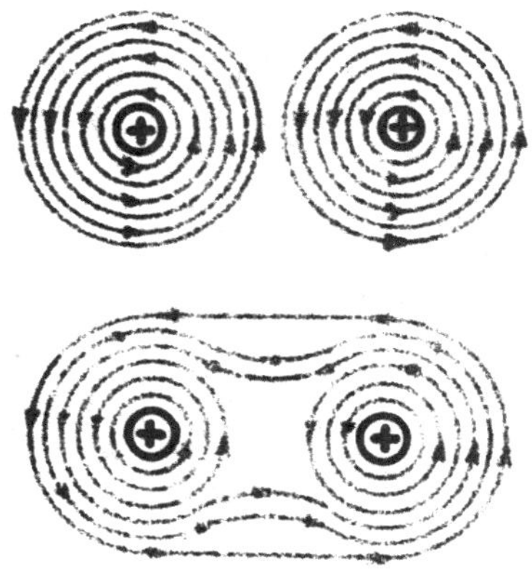

FIGURE 15–4 The magnetic lines of force join together when two current-carrying conductors are placed close together.

of magnetic lines of force is called **flux density.** If a current-carrying conductor is looped in a circle (Figure 15–3), where the conductor is placed side-by-side, then the flux density and magnetic field will both increase (Figure 15–4).

When a current-carrying conductor is placed within the magnetic field of a horse shoe magnet (Figure 15–5),

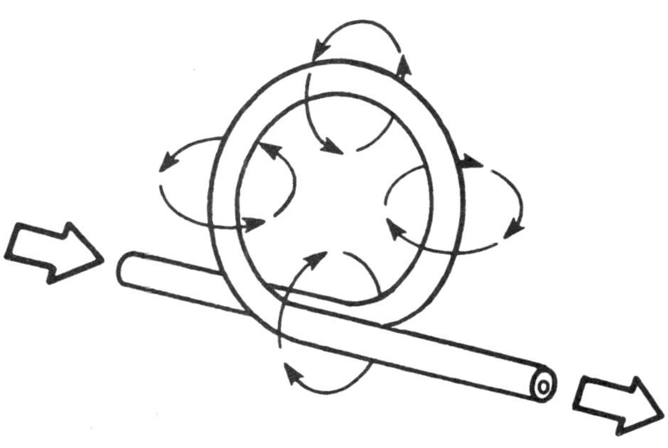

FIGURE 15–3 The flux density and magnetic field increases when a conductor is looped in a circle.

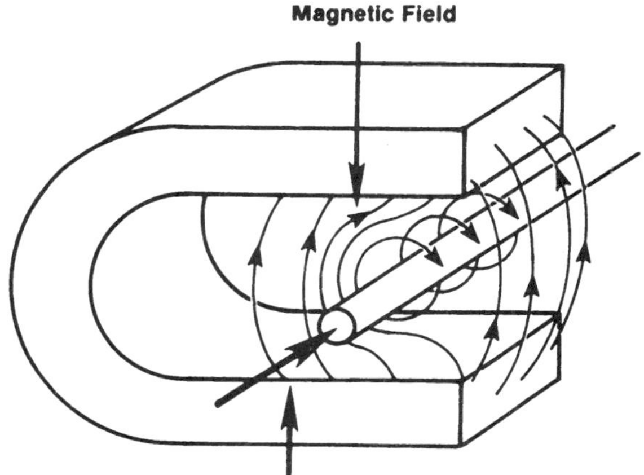

FIGURE 15–5 Placing a magnetic current-carrying conductor in the magnetic field of a horseshoe magnet.

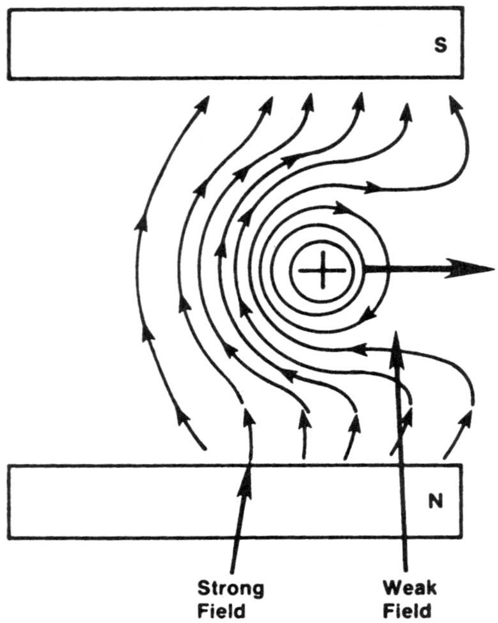

FIGURE 15–6 The conductor movement in a magnetic field will tend to move towards the weaker outside magnetic field.

the conductor moves towards the weaker magnetic field on the outside of the magnet (Figure 15–6).

A simple electromagnetic starter motor consists of several important components that form a motor strong enough to crank an engine (Figure 15–7).

The first important component in a starter motor is the **field coils** which are constructed of heavy copper wire wrapped in a coil around an iron core to form an electromagnet (Figure 15–8). The soft iron core in the field coil is called the **pole shoe,** which helps to concentrate and direct the lines of force in the field coil.

The second important component is the **armature** assembly that rotates within the stationary field coils. It consists of several heavy loops of copper wire wrapped around, and insulated from, a laminated iron core (Figure 15–9). At one end of the armature is the **commutator** which is a series of split ring copper segment bars that form a sliding contact with a set of electrically conductive brushes. The **brushes** are made of a mixture of carbon and copper that conduct high amperage through the armature windings.

When current is applied to the field coils and the armature windings, both will produce magnetic flux lines (Figure 15–10). The current flowing though the conductor will interact with the north and south pole magnetic field coils and cause a rotational force of the armature to occur in the direction of the weaker magnetic field. To keep the armature spinning, the spit ring copper commutator attached to the end of the armature will keep the current flowing in one direction until the current is turned off. The commutator, in fact, is what converts the alternating current produced in the armature into a direct current.

When the battery supply voltage pushes current through the starter motor, electromagnetic induction within the starter windings will produce a **counter electromotive force** or **CEMF.** This induced voltage acts against the supply battery voltage and opposes the flow of current through the starter motor. There is a relation-

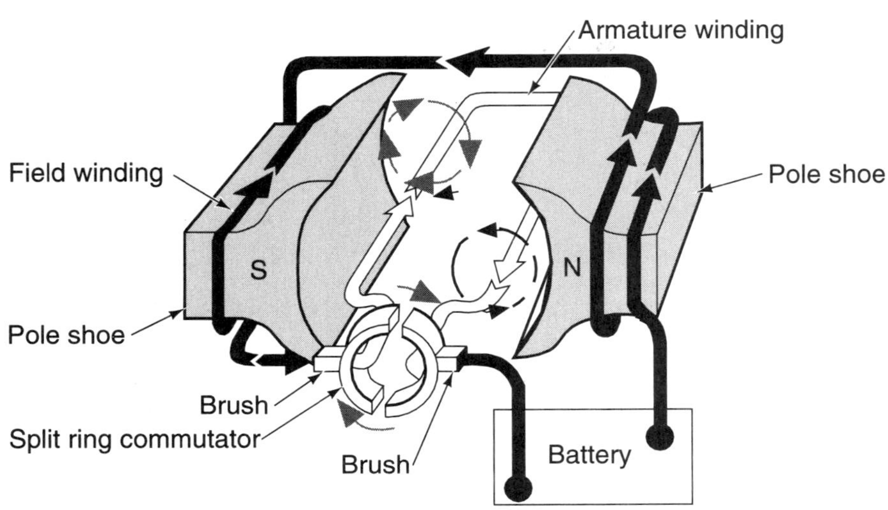

FIGURE 15–7 A simple electromagnetic starter motor.

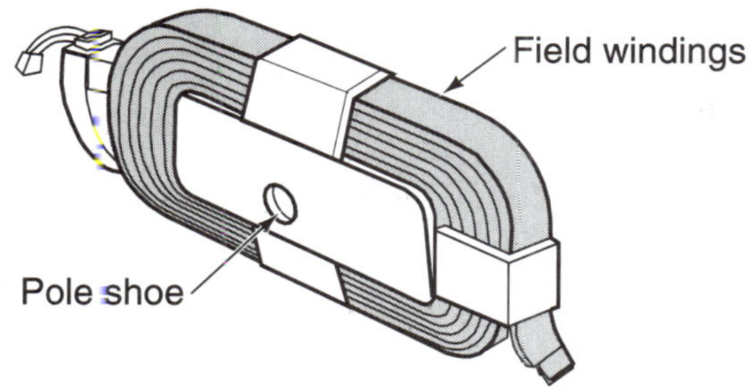

FIGURE 15–8 Heavy copper wire wrapped around a soft iron core called a pole shoe.

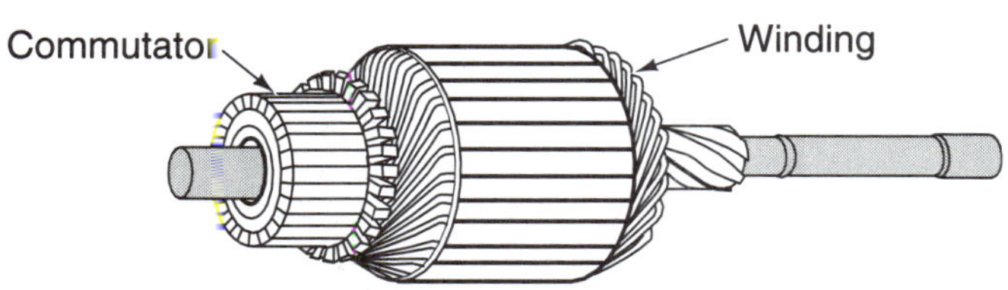

FIGURE 15–9 Starter armature assembly.

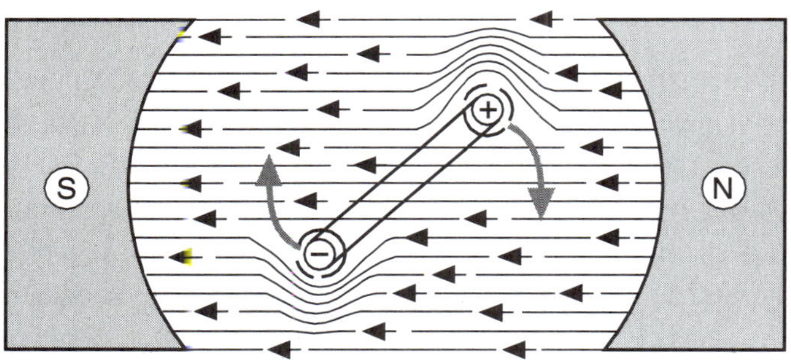

FIGURE 15–10 Rotation of the conductor is in the direction of the weaker magnetic field.

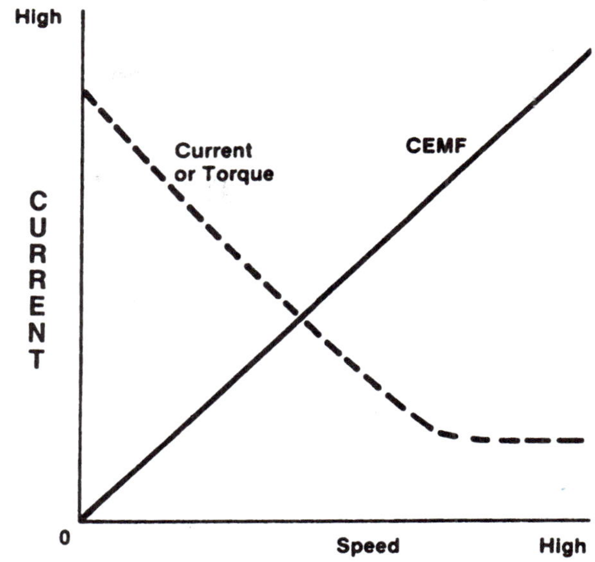

FIGURE 15–11 Relationship between CEMF, starter motor speed, and current draw.

ship between CEMF, motor speed, and current draw (Figure 15–11). As the starter speed and CEMF increases, the current or torque decreases.

The designed operating characteristics of the starter depend on how the field coils are wound inside the starter housing. There are four common basic types of starters, each using a different type of winding.

The first type is a **series wound** starter (Figure 15–12), where the current flows through the field windings, through the positive brushes to the commutator, and then through the armature windings to the commutator and negative brushes before reaching negative ground. A series wound motor develops its maximum torque output during initial starter cranking. As the speed increases, the torque output of the starter motor decreases because of the CEMF produced in the field windings.

The second type of starter is a **parallel wound** starter (Figure 15–13), also known as a **shunt wound** starter, where the current flows in parallel through the field windings, through the positive brushes and commutator, to the armature windings, then to the commuta-

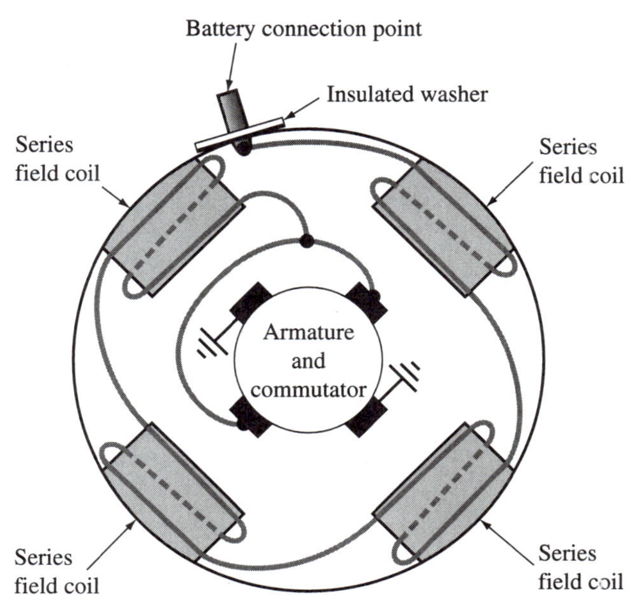

FIGURE 15–12 A series wound starter motor.

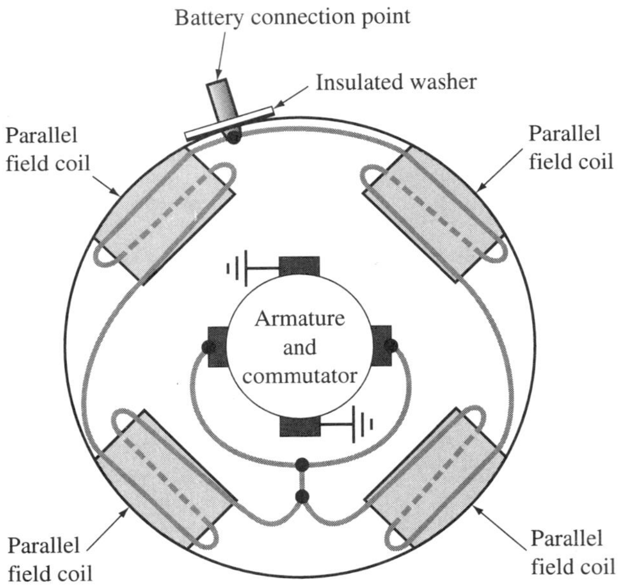

FIGURE 15–13 A parallel wound, or shunt wound, starter motor.

tor and negative brushes to negative ground. A parallel wound motor does not produce a high cranking torque, and spins at a slower cranking speed. The CEMF produced in the field windings does not decrease the field coil strength, even as the starter motor speed increases.

The third type of starter motor is a **compound wound,** or series/parallel wound starter (Figure 15–14) where most of the field coils are connected in series to

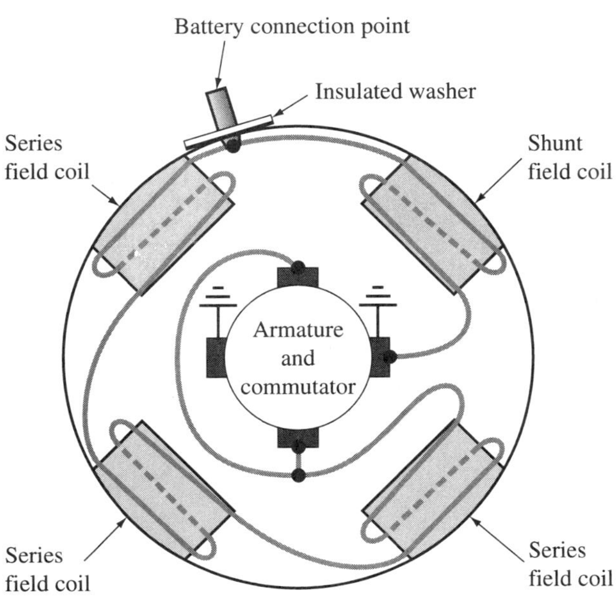

FIGURE 15–14 A compound wound starter motor.

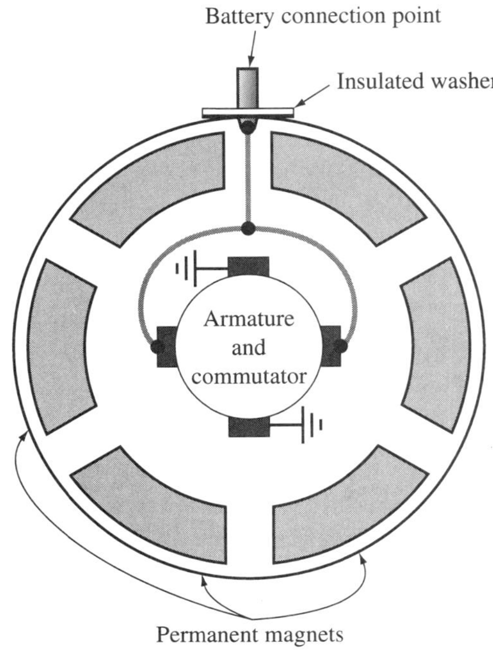

FIGURE 15–15 A permanent magnet starter motor.

the armature, and a single field coil is connected to negative ground. This starter combines the best properties of a series wound starter and a parallel wound starter into a starter with good starting torque and a constant operating speed. The field coil that is connected to negative ground is called the shunt coil, and limits the speed of the starter motor.

The fourth type is a **permanent magnet starter** (Figure 15–15), that uses permanent magnets in place of the electromagnet field coils. The permanent magnets are extra high strength in comparison to normal permanent magnet motors for other applications. Using permanent magnets instead of field coils reduces the weight of the starter, and allows for a significantly smaller motor housing.

Starting Circuit Components

A vehicle starting system is a combination of electrical and mechanical parts that operate together to crank the engine fast enough for starting. The typical starting system consists of the following main components (Figure 15–16)—battery, battery cables, starter motor, ignition switch, harness wiring, starter relay or solenoid, pinion drive, flywheel ring gear, and a neutral safety switch.

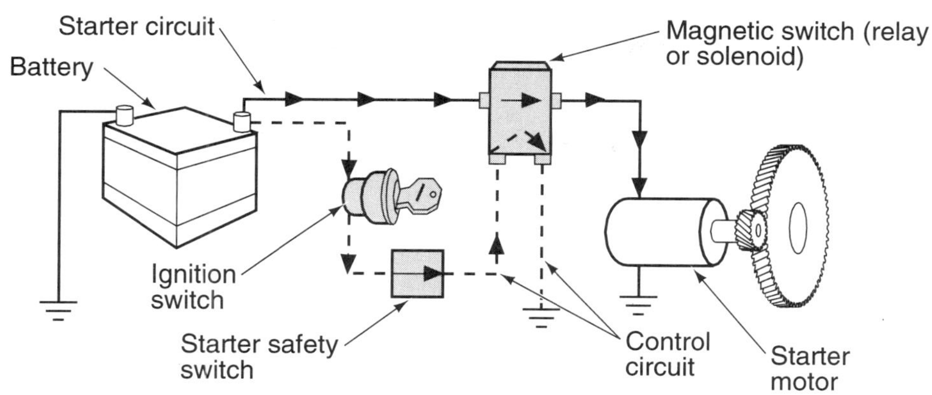

FIGURE 15–16 The major components of a starter motor system. The solid line indicates the high amperage cranking circuit, and the dashed line indicates the starter control circuit.

Starter Pinion Drives

There are two types of starter pinion drives, in use today, that mesh with the flywheel ring gear (Figure 15–17). The engagement of the drive pinion gear into the flywheel ring gear is critically timed to prevent grinding of the teeth of one or both gears. The drive pinion must be engaged into the ring gear *before* the starter motor spins.

Bendix Inertia Drive The *least* common type of starter drive pinion is the **Bendix inertia drive** (Figure 15–18), which depends on the inertia movement of the starter armature spinning to mesh the drive pinion forward into the ring gear teeth. When the engine starts and is rotating faster than the starter, the pinion gear will back down its own shaft and disengage from the flywheel teeth.

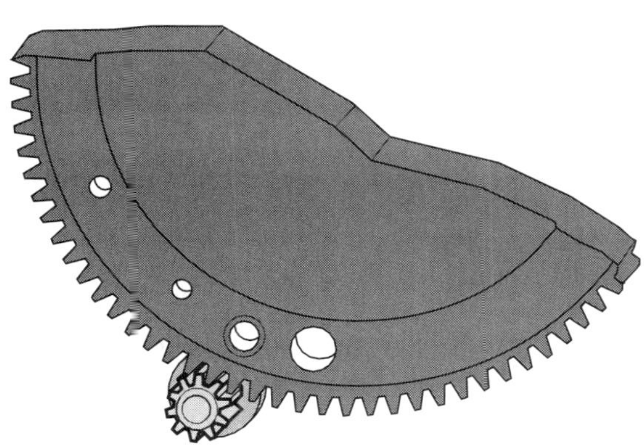

FIGURE 15–17 The starter drive pinion gear is engaged with the flywheel ring gear before the starter motor turns.

Overrunning Clutch Drive The *most* common type of starter drive pinion is the **overrunning clutch drive** (Figure 15–19), which is a roller type clutch that wedges spring-loaded rollers between an inner and outer housing to engage the pinion teeth with the flywheel teeth. The inner clutch housing is splined to the starter armature shaft, and the pinion barrel contains the free spinning pinion teeth. When the starter armature turns, the spring-loaded rollers are wedged tightly into the small ends of the tapered slots to transfer starter motor torque to the ring gear (Figure 15–20). When the engine starts and is rotating faster than the starter, the spring-loaded rollers unlock and allow the pinion gear to rotate freely around its shaft until the starter motor is de-energized.

Starter Relays and Solenoids

A starter motor can require a current draw of up to 250 amperes or more to crank the engine fast enough for starting. To control this high amperage requirement, an

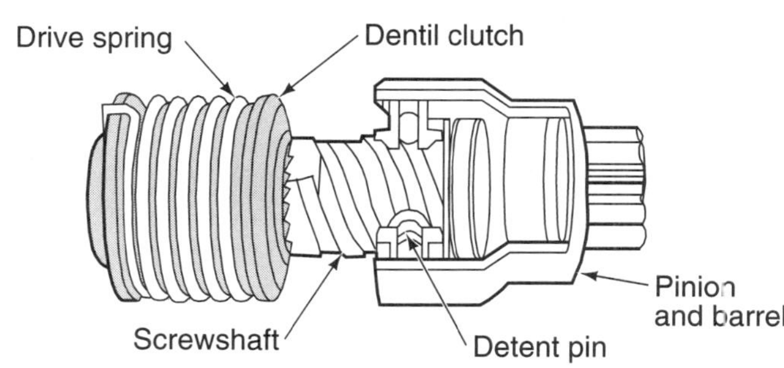

FIGURE 15–18 Bendix inertia drive pinion.
(Courtesy of General Motors Corporation, Service Technology Group)

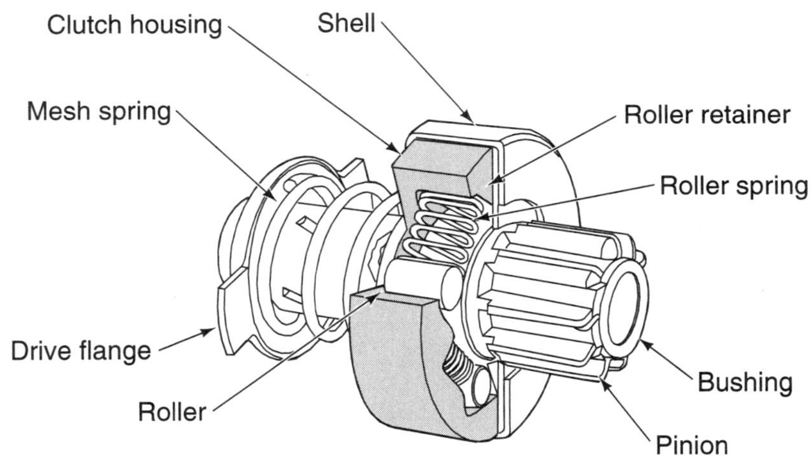

FIGURE 15–19 Overrunning clutch drive pinion.
(Courtesy of Ford Motor Company)

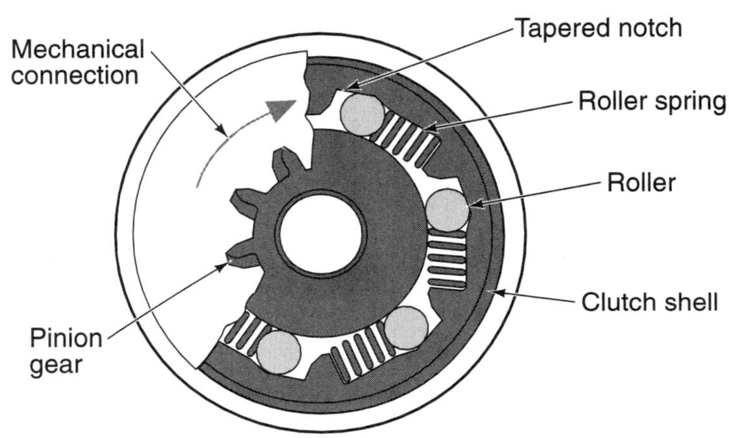

FIGURE 15–20 As the armature turns, the spring-loaded rollers are wedged into the smaller ends of the tapered slots.
(Courtesy of Robert Bosch Corporation)

electromagnetic switch is used. There are two basic types of switches used today to perform this function—the starter solenoid and the remote starter relay.

Starter Solenoid In a **starter solenoid**-actuated starter system, the solenoid is attached directly to the starter motor assembly (Figure 15–21) and controls two important functions sequenced in a specific order. The first function is to move the solenoid plunger, which is connected through a linkage to the pinion gear that meshes with the flywheel ring gear. The second function occurs after the teeth of the pinion are engaged into the flywheel—the solenoid plunger connects the battery B+ terminal to the motor M terminal, which cranks the starter motor (Figure 15–22). This sequence must take place to prevent the overrunning clutch from spinning before contacting the flywheel ring gear, and thus damaging the teeth on one or both.

The schematic in Figure 15–22 consists of two separate solenoid windings—the **pull-in winding** and the **hold-in winding.** Both are energized when battery voltage going through the ignition switch in *start* mode to the solenoid S terminal, electro-magnetically pulls in the solenoid plunger. When the copper disc at the end of the solenoid plunger contacts the B+ and the M terminals, the pull-in windings are de-energized because equal bat-

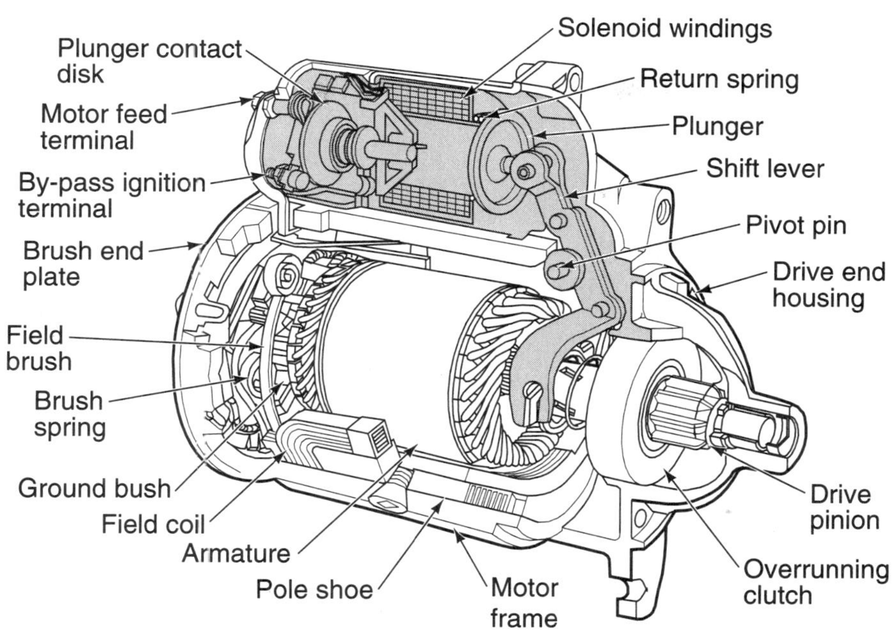

FIGURE 15–21 A solenoid-operated starter motor has the solenoid mounted directly to the motor.

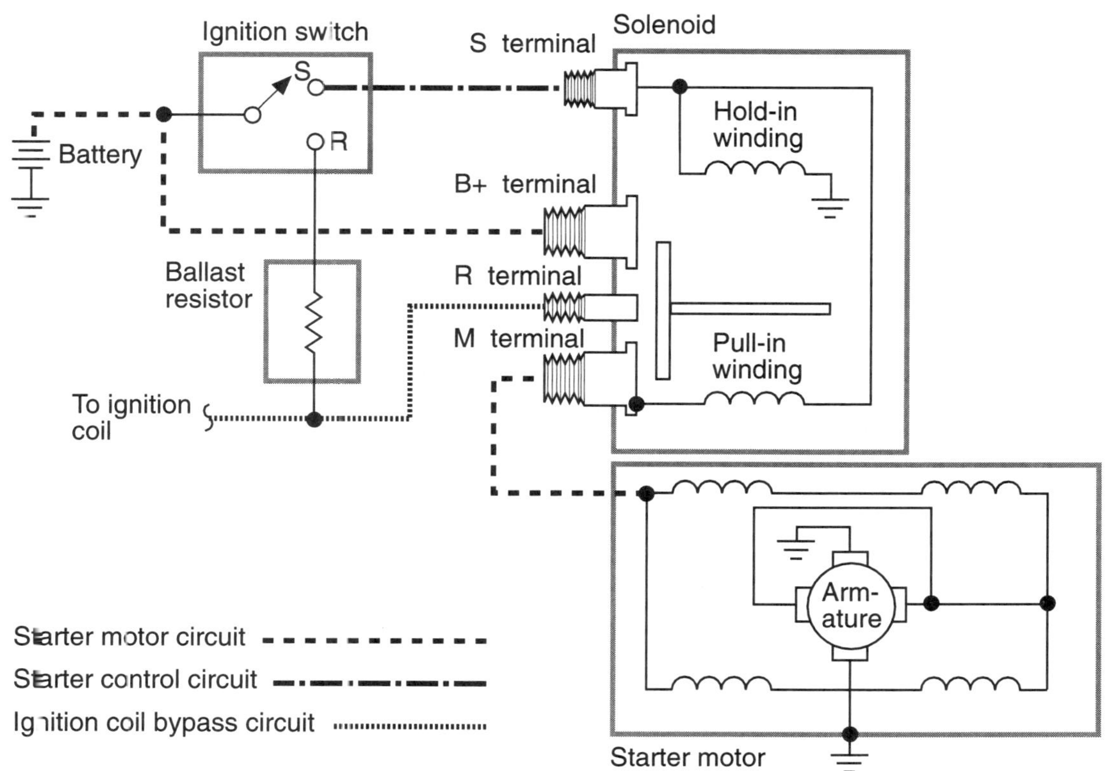

FIGURE 15–22 Schematic of a solenoid-operated starter motor circuit.

tery voltage is present on both sides of the pull-in windings. The hold-in winding continues to hold the copper disc that is connected between the B+ and the M terminal, which allows the starter motor to crank.

The R terminal is an **ignition bypass terminal** that supplies full battery voltage to the ignition coil during engine cranking conditions only. This bypass circuit normally applies to older vehicles that have point-style ignition systems which use a ballast resistor or a resistance wire to control the ignition coil primary voltage to be less than full battery voltage. Modern electronic ignition systems normally do not have a bypass terminal because their ignition systems operate on full battery voltage.

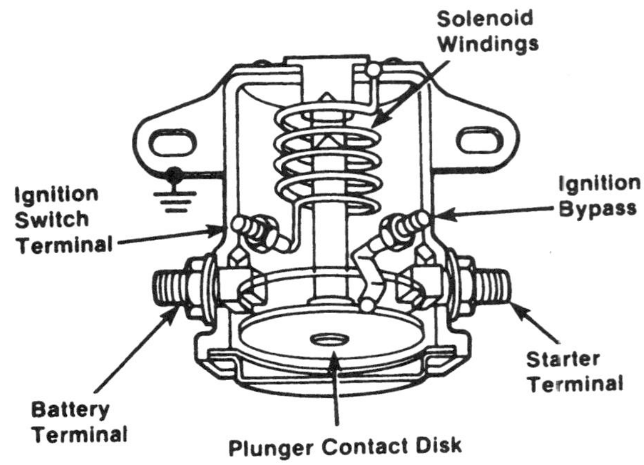

FIGURE 15–23 Remote mounted starter relay.

Starter Relay In a **starter relay**-actuated starter system, the relay is normally mounted separately from the starter, and controls the high current to the starter motor (Figure 15–23). This type of starter relay controls only the high current to the starter motor, and does not control the pinion gear meshing into the flywheel teeth. When the ignition switch is in start mode, battery voltage is supplied to the S terminal (Figure 15–24) of the ignition switch and current then flows through the relay coil. The relay coil pulls the plunger disc downward into the battery and starter terminals (Figure 15–23) to supply battery current to the starter motor. There is an I terminal that serves the same purpose as the ignition bypass R terminal in the starter solenoid.

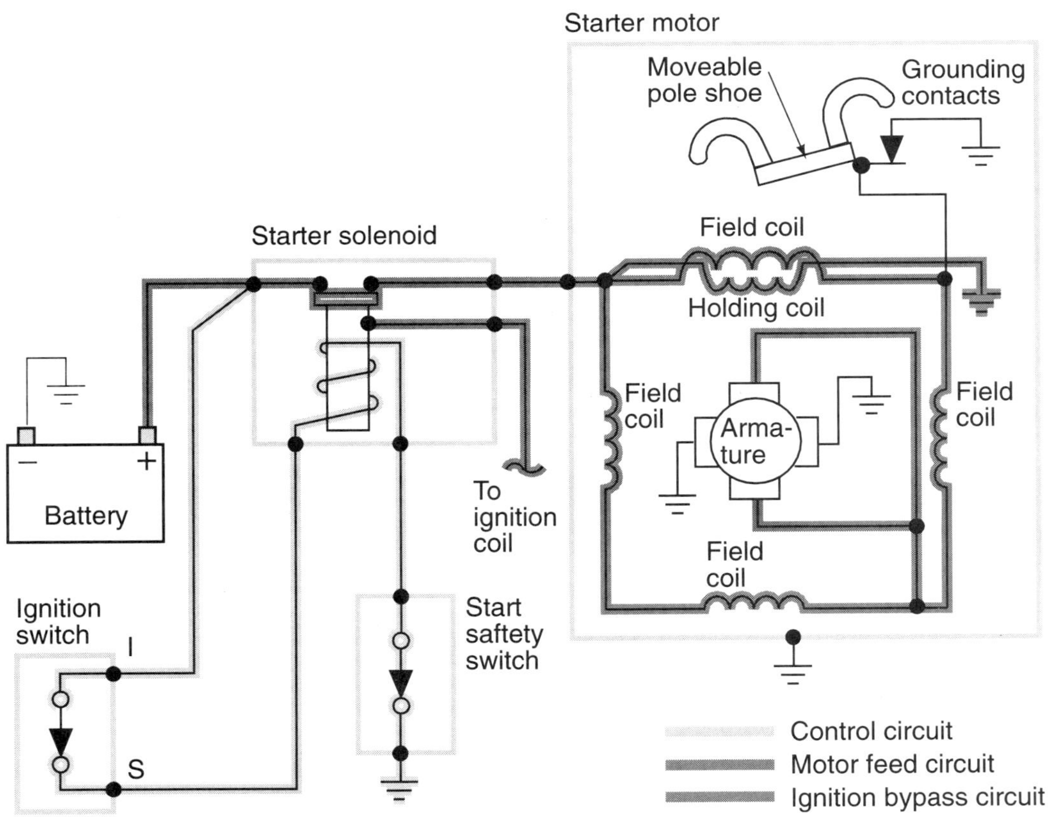

FIGURE 15–24 Current flow in a starter circuit with a remote mounted starter relay.

Starter Motor Designs

There are four starter motor designs used in today's vehicles that use the overrunning clutch pinion drive to crank the engine. Depending on the engine application, some manufacturers use the same basic starter, with only slight variations, for several different models of vehicles. Be sure the starter replacement is the specific one designed for the vehicle being worked on.

Direct Drive Starter The **direct drive starter** (Figure 15–25) design is probably the most commonly used in vehicles today. This starter design uses a solenoid directly attached to the starter motor to control the pinion gear movement and the high amperage required for the starter motor windings. When the solenoid is energized, it will first engage the overrunning clutch pinion in to the flywheel, and then will energize the starter motor windings so they will turn.

Positive Engagement Starter The **positive engagement starter** (Figure 15–26) uses an internal movable pole shoe, which is connected through linkage to the pinion gear, that in turn meshes with the flywheel ring gear. This type of starter uses a separate starter relay (Figure 15–23) to control the high amperage from the battery to the drive coil. When current flows from the relay to the drive coil, electromagnetism pulls a movable pole shoe that is attached through a linkage to the overrunning clutch pinion. This action engages the pinion gear into the flywheel teeth, and then opens a set of normally closed grounding contacts which allows current to flow to the starter motor windings.

Gear Reduction Starter The **gear reduction starter** (Figure 15–27) contains a small gear reduction drive gear between the starter motor and the drive pinion. The design of the starter solenoid, armature assembly, and overrunning clutch is similar to those on the direct drive starter, with the exception of the starter's physical size and a gear reduction assembly driving the overrunning clutch pinion. The reduction ratio is normally between 2:1 and 3.5:1 depending on the application. The higher the gear ratio, the greater the starting torque available for the same size starter housing. Many smaller vehicles use a smaller motor running at a higher operating speed to produce the necessary gear reduction torque.

Permanent Magnet Starters The **permanent magnet starter** (Figure 15–28) is the newest design of starter, and creates its magnetic field from four to six strong permanent magnets instead of electromagnetic field coils. The starter amperage that previously was supplied to the field coils and armature assembly, now supplies only the armature assembly through the

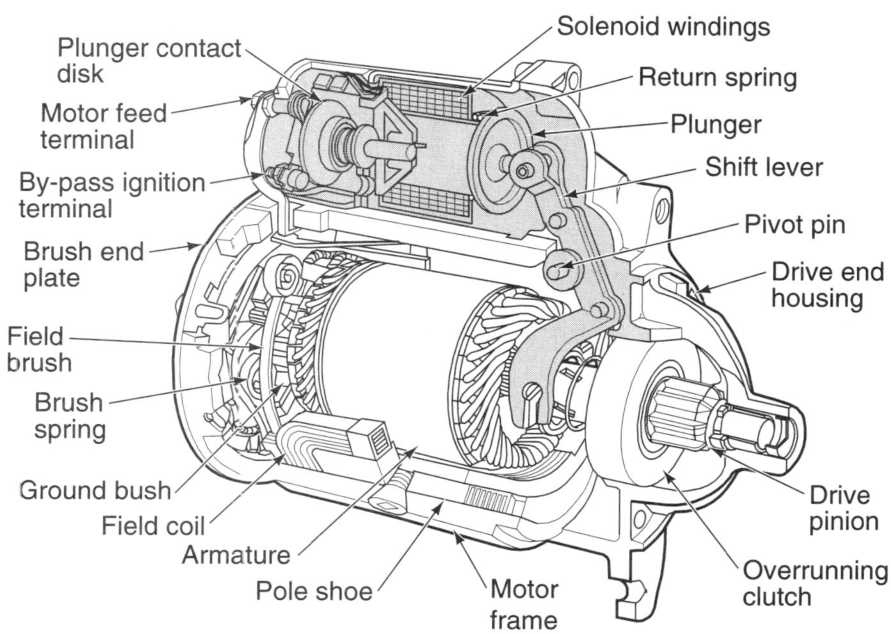

FIGURE 15–25 A direct drive starter motor.

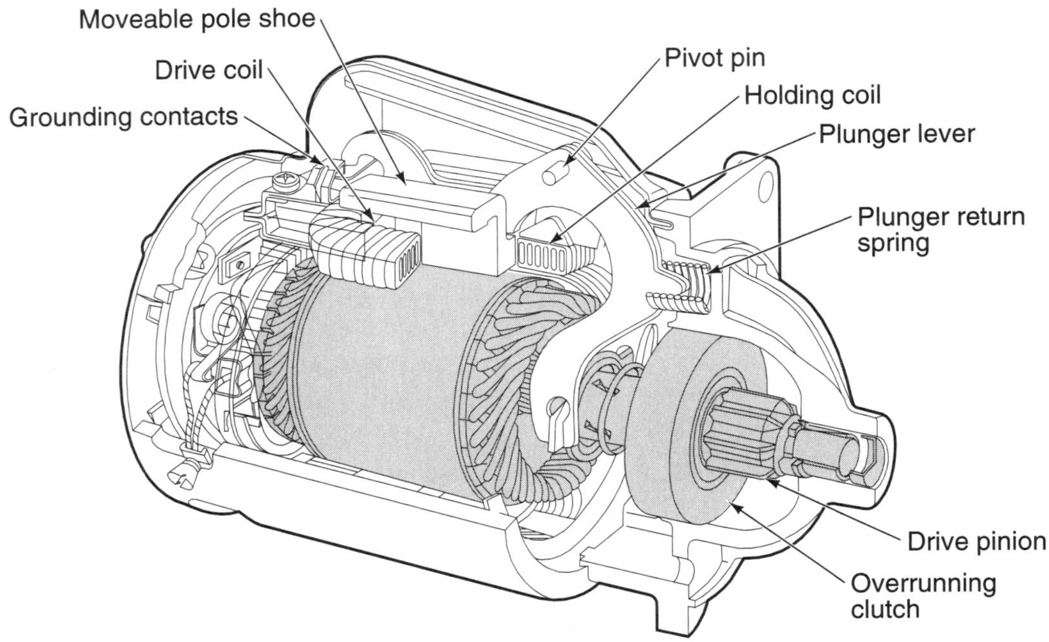

FIGURE 15–26 A positive engagement starter motor.
(Courtesy of Ford Motor Company)

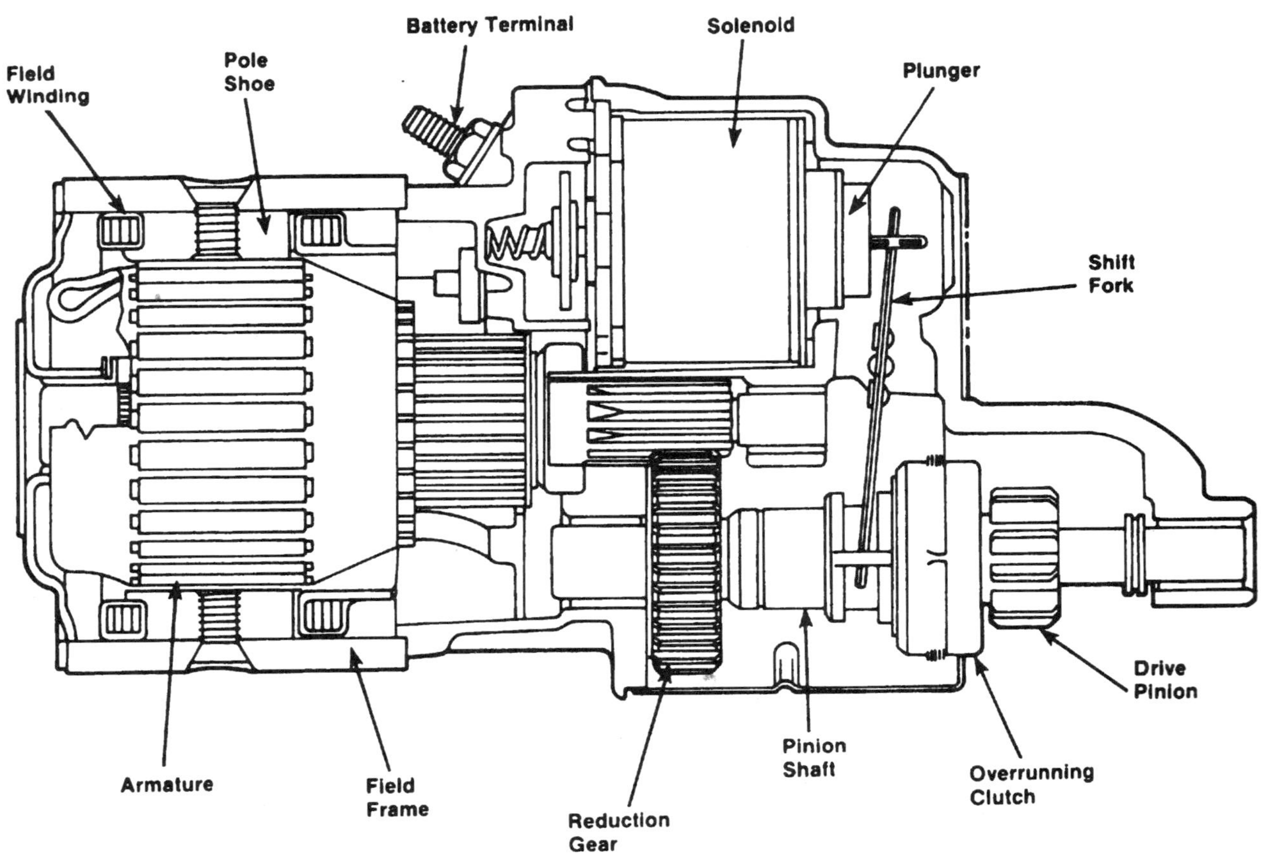

FIGURE 15–27 A gear reduction starter motor.
(Courtesy of Ford Motor Company)

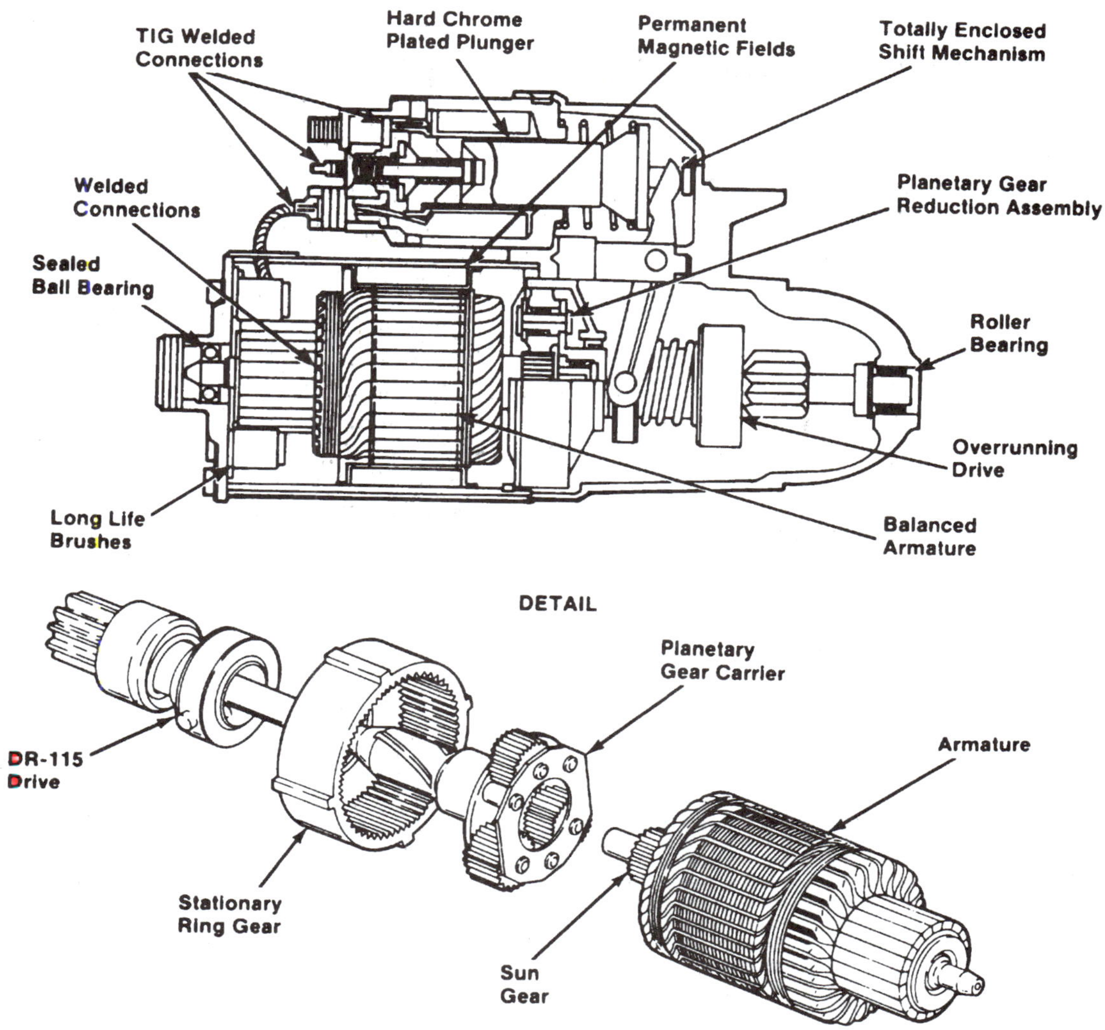

FIGURE 15–28 A permanent magnet starter motor.
(Courtesy of General Motors Corporation Service Technology Group)

brushes and commutator. Combined with a planetary gear reduction set that produces a 4:5:1 reduction, the physical size of the starter is smaller, overall current demand on the battery is less, and starting torque output is increased.

Neutral Safety Switch

The **neutral safety switch (NSS)** is a safety switch that prevents a starter from cranking in an automatic transmission vehicle unless the gear selector is in park or neutral, or a manual transmission has the clutch pushed in. Figure 15–29 is a starter schematic in an automatic transmission vehicle that has a safety switch attached to the gear shift assembly.

The neutral safety switch may be located in several places depending on the vehicle and manufacturer's preferences. The safety switch may be located on the side of the transmission housing (Figure 15–30), and may be combined with the back-up light switch and/or the transmission range (TR) sensor that tells the PCM (powertrain control module) the position of the gear shift lever.

Another location of the neutral safety switch on an automatic transmission vehicle is through a mechanical linkage that blocks the movement of the ignition switch lock cylinder unless the transmission is in park or neutral (Figure 15–31).

In a manual transmission vehicle, a popular location for the neutral safety switch is attaching it to the gear selector or the clutch pedal assembly (Figure 15–32).

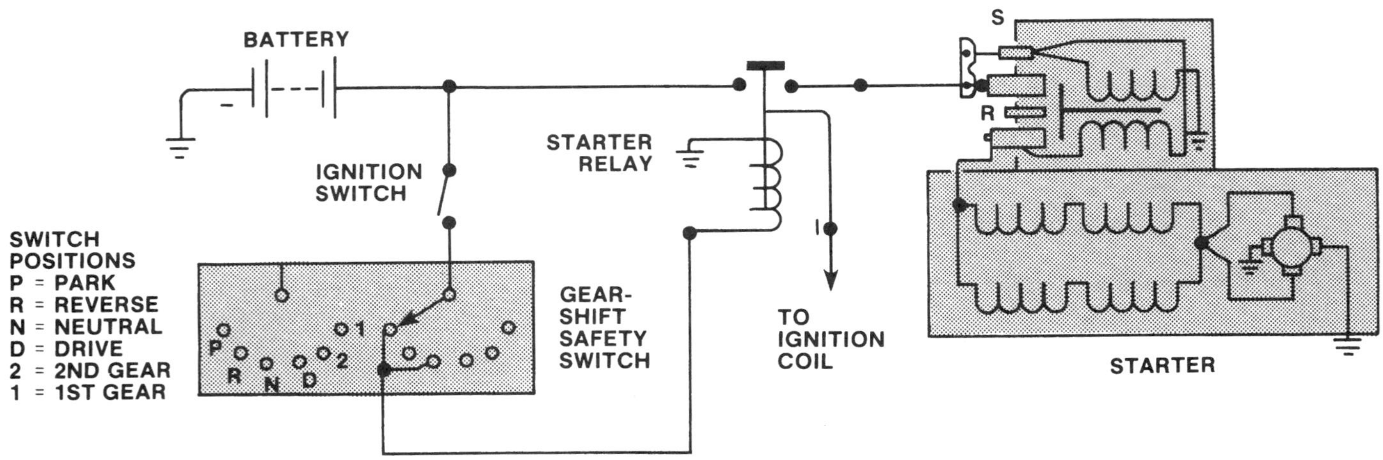

FIGURE 15–29 The neutral safety switch (NSS) must be closed to complete the starter control circuit.

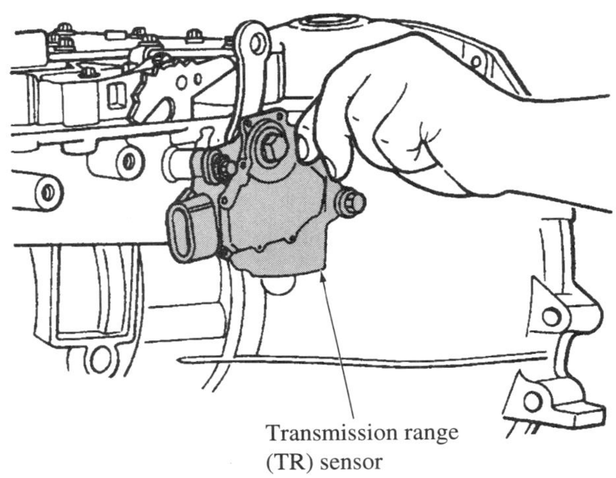

FIGURE 15–30 The neutral safety switch may be combined with the back-up light switch and the transmission range (TR) sensor.

(Courtesy of Ford Motor Company)

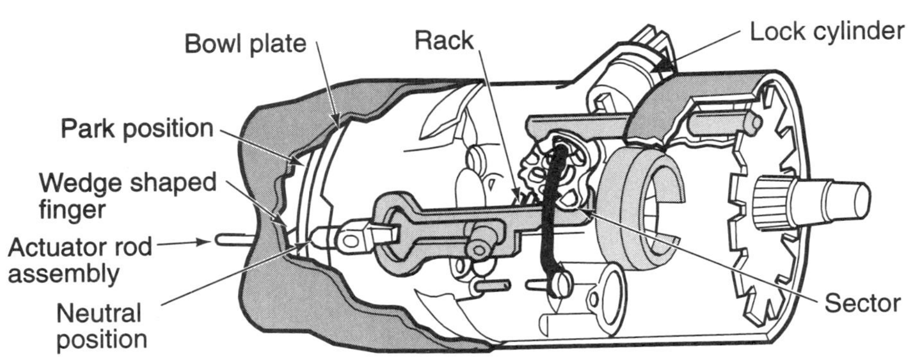

FIGURE 15–31 Mechanical linkage in the steering column prevents the ignition lock cylinder from turning when the transmission is in gear.

(Courtesy of General Motors Corporation, Service Technology Group)

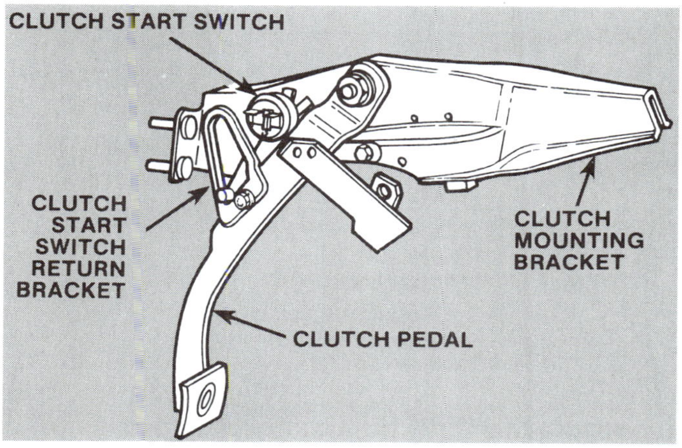

FIGURE 15–32 A starter safety switch located on the clutch pedal linkage for a manual transmission vehicle.

The safety switch contact points are closed only when the clutch pedal is fully depressed.

Basic Starter System Testing

The starting system in a vehicle normally provides years of trouble free service, but can produce aggravating experiences to the driver when it fails to operate as designed. To properly test and diagnose a starting system circuit, systematic troubleshooting procedures should be followed to prevent a mis-diagnosis. This is important if you consider that over three-quarters of the starters returned for warranty replacement test satisfactorily at the rebuilding facility, and were probably returned because of a bad battery, a bad, loose, or corroded cable or connection, a bad switch or relay, or a circuit wiring fault.

Before testing a starting system there are three important preliminary vehicle checks to be made.

1. Condition of the battery—a three-quarter, or greater, charged battery *must* be used to accurately test the starting system. When a partially charged battery is used, the starter may crank the engine slower than normal or fail to engage completely into the flywheel ring gear. Refer to the previous chapter about batteries and testing them to perform any battery tests.
2. Engine mechanical condition—satisfactory condition of the engine internally should allow the starter motor to crank the engine at 200–400 rpm, depending on the vehicle and the engine. The following mechanical problems could effect cranking conditions.
    - Engine crankshaft bearing failed or is too tight, causing the crankshaft to seize.
    - Broken internal engine components, such as connecting rods and loose bearing caps are wedged against the crankshaft and prevent rotation.
    - Hydrostatic lock in one or more cylinders. Liquid antifreeze cannot be compressed, and if allowed to fill the engine cylinder(s) through a bad head gasket, for example, could prevent the piston from rising in the cylinder, creating a hydrostatic lock.
3. Other conditions—the engine should be filled with a proper weight oil as recommended by the vehicle manufacturer for specific weather conditions. If a heavier weight oil is used in combination with cold weather conditions, the cranking RPM required to start the engine could be below the designed minimum that is acceptable.

Engine mechanical conditions that create a no-start condition, can be verified by rotating the engine using the crankshaft pulley nut. The engine should be rotated, without excessive resistance, for at least two rotations in a clockwise direction only, which prevents un-threading the crankshaft pulley nut.

If the above preliminary conditions check out satisfactorily, then the following five starter system tests should be performed.

- Starter circuit isolation test.
- Starter current draw test.
- Starter circuit voltage drop test.
- Starter open circuit test.
- Starter bench testing, if necessary.

Starter Circuit Isolation Test

If the starter does *not* crank the engine, and the preliminary engine mechanical condition checks out satisfactory, then a quick isolation test can be performed on

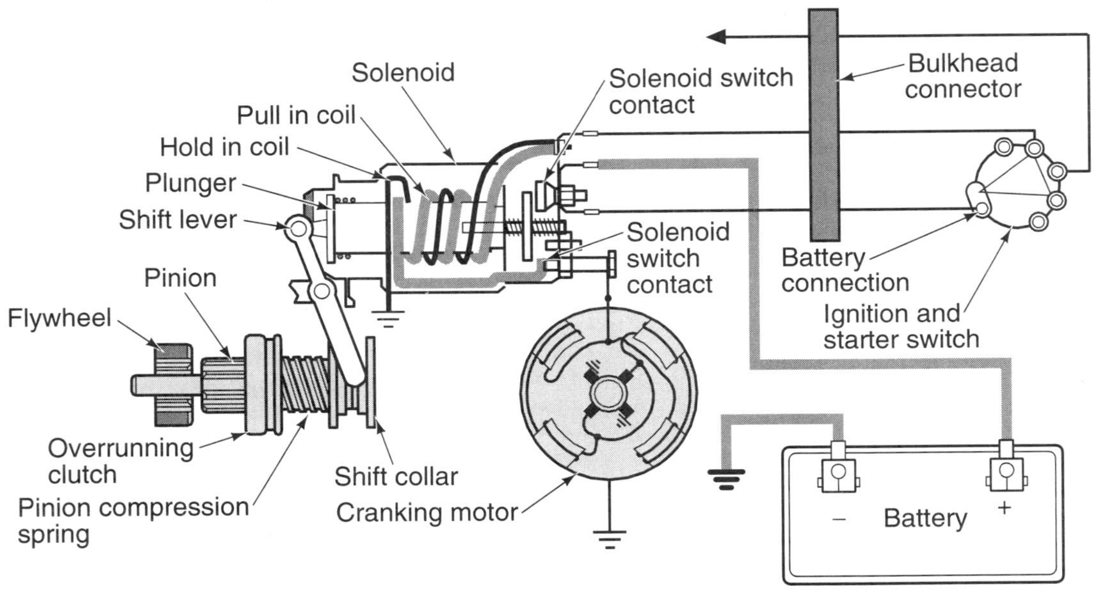

FIGURE 15–33 Potential connections and components that could cause a no-crank condition.

the starter system to determine if the starter motor, solenoid, relay, or control switch circuit is at fault (Figure 14-33).

To perform a quick isolation test on a starting system, make sure the parking brake is set and the transmission is in neutral. Turn on the headlights, and turn the ignition switch to the start position while watching the headlight intensity. Depending on the results of this headlight intensity test, three conditions can be determined.

1. The headlights go out—this result is most likely caused by a poor battery terminal connection due to corrosion or looseness, which should be corrected before proceeding with any other tests.

2. The headlights go dim—this could be caused by a discharged battery, engine mechanical condition, or internal starter motor damage. If the battery and engine mechanical condition check out satisfactory during the preliminary tests, the problem would most likely be internal to the starter—armature shorts, worn bearings, shorted field windings, or loose or dragging conditions. A starter in this condition will draw excessive amperage from the battery during starter cranking.

3. The headlight intensity remains the same—this could be caused by an open in the circuit if the starter makes *no* sound. If, however, there is a heavy clicking noise, either the solenoid or the control circuit could be at fault, which can be verified by connecting a remote starter switch across the starter solenoid (Figure 15–34) to bypass the ignition control and safety circuits. If the starter motor cranks while using the remote starter switch, the control circuit is at fault. If the starter does not rotate, and the lights do not go dim, then the solenoid or relay is at fault.

Starter Current Draw Test

If the starter *does,* at least, crank the engine, a current draw test should be performed with a battery/starter/charging system tester (Figure 15–35) that is capable of recording several hundred amperes.

The following steps are used with a VAT40 tester to perform a current draw test on a starter. Most other starter current draw testers have procedures similar to these to perform this test.

- Connect the large red lead to the battery positive post and the black tester lead to the battery negative battery post. Select the *INT 18 Volt* meter scale. Place the test selector knob in the *starting* position (0-600 amp), then zero the ammeter needle. Connect the green inductive probe around the battery ground cable (Figure 15–36).

- Turn off all vehicle accessories so the battery can provide its full amperage to the starter.

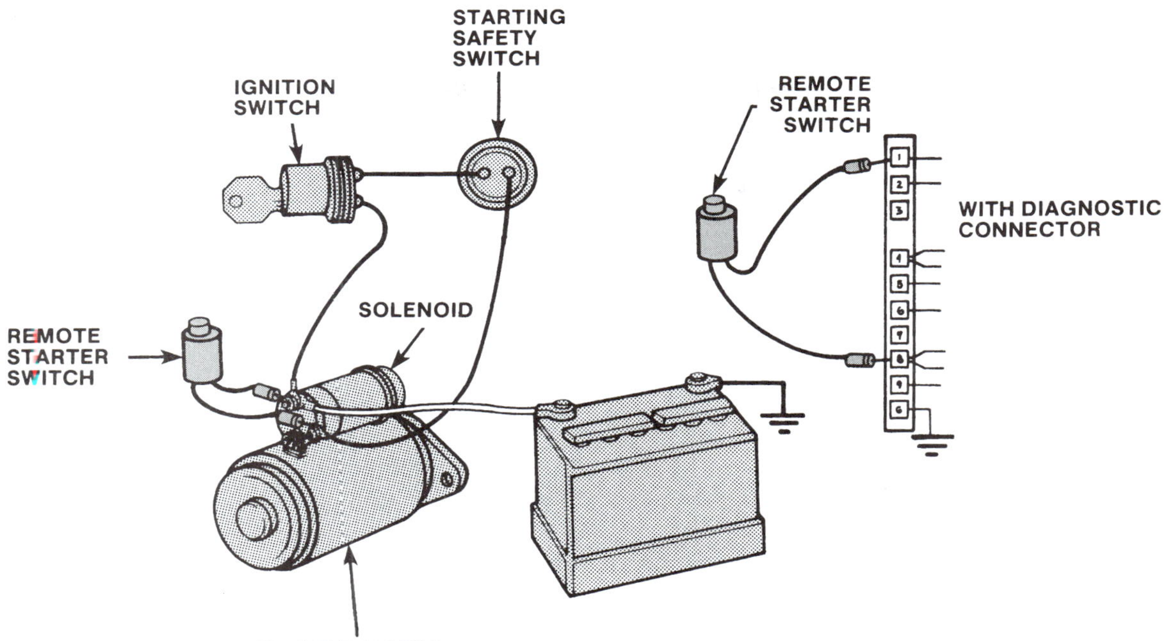

FIGURE 15–34 A remote starter switch can be used to bypass the starter control circuit.

- Disable the ignition system to prevent the engine from starting by grounding the coil secondary wire (Figure 15–37), or by disconnecting the battery power to the electronic high energy ignition (HEI) distributor as shown in Figure 15–38. Refer to specific manufacturer instructions if unsure of ignition disabling procedures. On vehicles with electronic ignitions, if the coil is *not* properly grounded, the coil could seek a ground through the ignition module causing damage to the module.
- While cranking the engine for 10-15 seconds, observe the voltmeter reading and the starter cranking amperage draw. Minimum starter cranking voltage should remain at 10.5 volts or higher. Minimum starter cranking rpm's should be at 200-400 rpm's, depending on the engine. Typical starter amperage draw is dependent on the size of the engine, number of cylinders, special modifications to the engine, and severe weather conditions. If manufacturer's specifications are not available, refer to the amperage chart in Table 15-1 as a guide to determine typical starter amperage draw.

FIGURE 15–35 A starter-current draw tester that is capable of measuring several hundred amps.

TABLE 15–1 Typical Starter Current Draw Test Results

Engine size	Amperage draw range	Normal
4 cylinder	75-150 amps	125 amps
6 cylinder	100-175 amps	150 amps
Small V-8	125-200 amps	150 amps
Large V-8	150-300 amps	200 amps

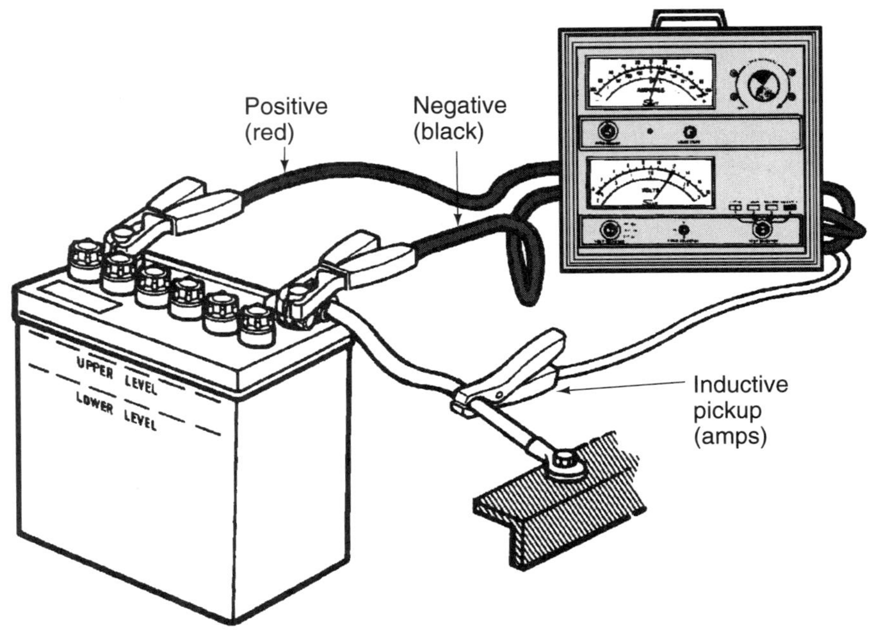

FIGURE 15–36 Connecting starter current draw tester leads to a starting system.

Results Of Starter Current Draw Test If the starter's current draw exceeds the standard range for the engine size, this could indicate one or more of the following possible causes.

- A short in the starter motor windings.
- Spinning resistance in the starter due to worn bushing or dragging armature assembly.
- Broken, binding, or misaligned starter motor components.
- Engine mechanical problems.
- Ignition timing excessively advanced beyond specifications.

If the starter's current draw is below the standard range for the engine size, this could indicate the following possible causes.

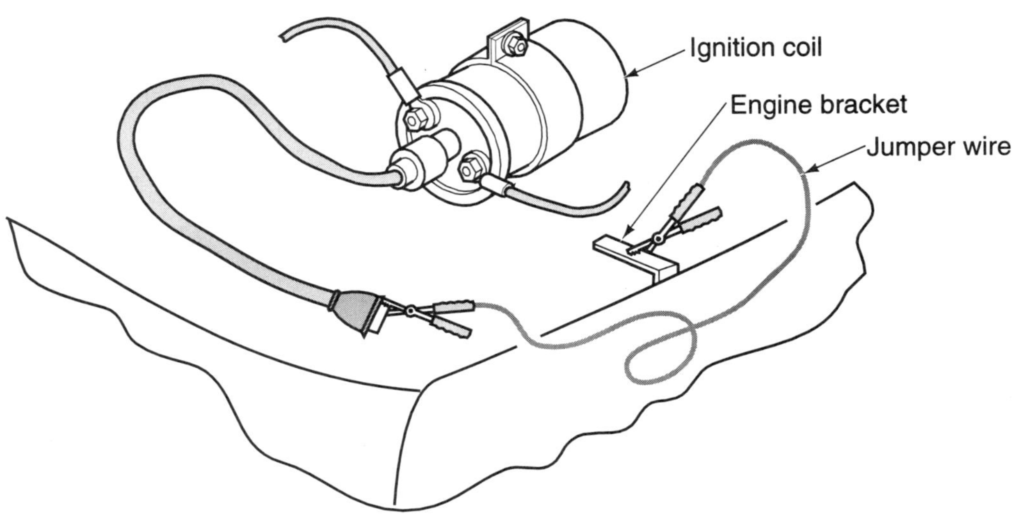

FIGURE 15–37 Grounding the secondary coil wire to prevent the engine from starting during a starter cranking test.

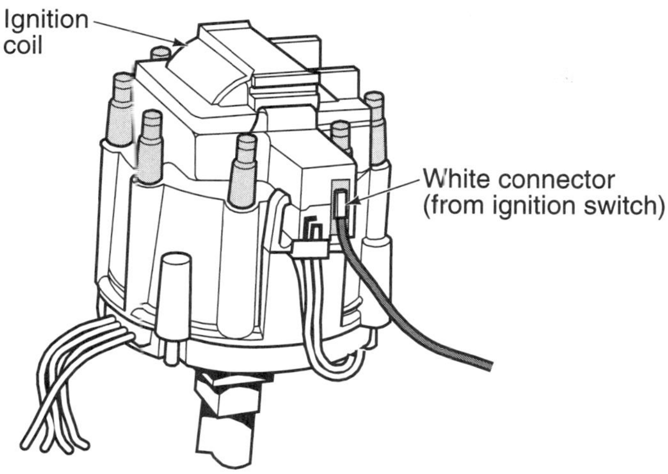

FIGURE 15–38 Disconnecting the battery wire lead from an HEI distributor to prevent the engine from starting during a starting cranking test.

- Engine components excessively worn, such as from a high mileage engine.
- Voltage loss in starter circuit due to bad cables, connections, or control solenoid or relay. A voltage drop test on the insulated side and the ground side of the starter circuit should be performed.

Starter Circuit Voltage Drop Test

In a normal starter motor circuit there are several connections, terminals, cables, switches, and wires that could develop *unwanted* resistance. This resistance could drop the voltage that is available to crank the engine to a level below the minimum that is required to crank the engine fast enough for starting. This test must be performed while the starter is cranking the engine. Remember that voltage is the pressure that pushes the amperage though a completed circuit. A voltage drop indicates there is unwanted resistance reducing the current flow in the circuit. Figure 15–39 is an example of a positive engagement starter circuit and the individual voltage drop test points that should be checked for an unwanted resistance.

Each of the test points in Figure 15–39 is described as follows.

- Test points A to B—tests connection between the battery positive post and terminal conductor.
- Test points B to C—tests the battery positive cable #1.
- Test points C to D—tests the connection from the battery positive cable to the solenoid.
- Test points D to E—tests the contact points inside the solenoid.
- Test points E to F—tests the connection from the solenoid to the battery positive cable #2.
- Test points F to G—tests the battery positive cable #2.
- Tests points G to H—tests the connection from the battery positive cable #2 to the starter motor.
- Tests points H to I—tests the actual voltage drop across the starter motor.
- Test points I to J—tests the ground connection at the engine block.

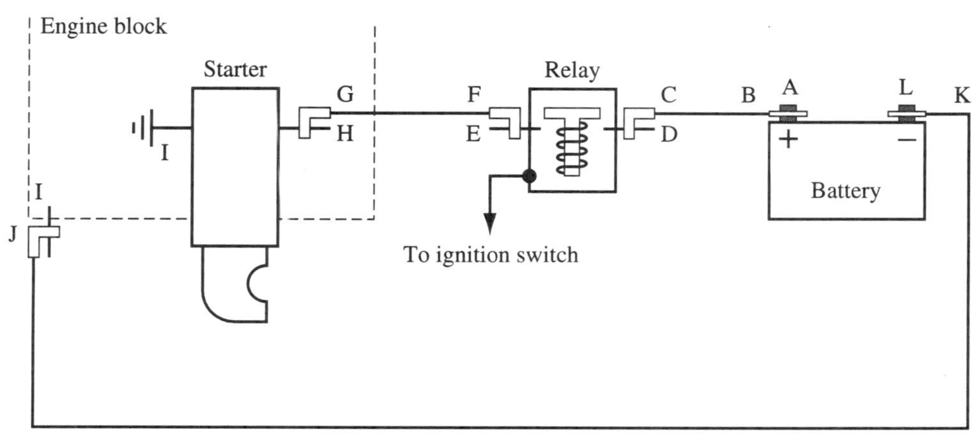

FIGURE 15–39 Starter circuit voltage drop test points.

- Test points J to K—tests the battery negative cable.
- Test points K to L—tests the connection between battery negative post and terminal connector.

If the amounts for allowed voltage drops are not available from the vehicle manufacturer, the following general specifications can be used as a general rule for circuit voltage drops.

Connection Point	Allowed Voltage Drop
Ground connections	0.1 volts
Switch contacts	0.2 to 0.4 volts
System wiring	0.2 volts
Terminal connections	0.0 volts
Computer sensor connections	0.0 to 0.05 volts

The starter circuit is designed to operate on the entire voltage supplied by the battery. A voltage drop in any one of these connections that exceeds the amount allowed, could effect the proper operation of the starter motor.

Starter Open Circuit Test

When a starter system has a *no-crank* problem, a voltage drop test cannot be performed. The most common no-crank problem is caused by an *open* circuit in the starter or the control circuit. The simplest way to diagnose a no-crank problem is with a high wattage test light. A custom made high wattage test light can be made from a #1156 bulb, a socket, and connection leads. Standard commercial test lights are insufficient for this test because they are 3 watts or less, and need only 0.25 amp or less to glow brightly. To perform the test on a starter system with the solenoid mounted on the starter motor (Figure 15–40), connect the test light positive lead to the *M* terminal and the test light negative lead to a good engine ground. The test light should go on brightly when the ignition switch is turned to the start position, *and* if the control circuit is operating properly. If the test light does not come on, or on very dimly, there is an open or high resistance in the circuit somewhere between the M terminal and the battery. The test light should be used to back track though the system from the starter to the battery to find the open circuit.

To test the ground side of the starter motor circuit, connect the test light positive lead to the starter M terminal and the test light negative lead to the starter housing. The test light should come on brightly when the ignition switch is turned to the start position, which indicates a good ground. If the light comes on brightly, then a bad starter motor is suspected.

Starter Bench Testing

In the event a starter has failed the previous tests, or a new starter is going to be installed, a free running (no-load) bench test should be performed to determine the following.

- Starter free running rpm.
- Starter current draw.

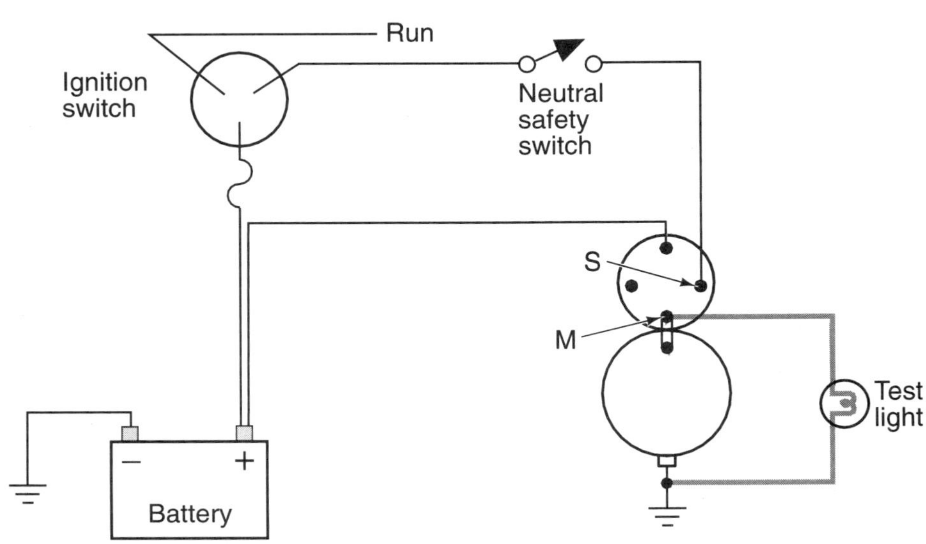

FIGURE 15–40 Using a test light to determine a no-crank starter condition.

- Operation of the overrunning clutch pinion drive.
- Operation of the solenoid or relay, if attached to the starter.

To perform a starter free running, no-load bench test, the starter should be fastened in a vice snugly, but not too tight as to crack the frame or pole shoes (especially on a permanent magnet starter). Connect a starter load tester (VAT 40 for example, Figure 14–41) to *load* the battery until a voltage reading of 10 volts is obtained. This prevents the starter from over speeding the armature and causing damage to the armature windings. If available, a manual hand tachometer can be used to determine starter motor rpm. Energize the starter solenoid with a remote starter switch and observe the current draw. Typical results should be as follows.

- Starter free running rpm equals 6000-12,000 rpm.
- Starter current draw equals 60-100 amperes.

If the starter fails the free running bench test, there are four categories to help determine the cause.

1. If the rpm is slower than specified and the current draw is excessive, then the cause could be worn starter bushings, shorted or grounded armature, shorted field windings, bent armature, or a combination of the above.
2. If the rpm is slower than specified and the current draw is lower than normal, then excessive resistance is present in the starter. This is normally caused by a poor connection between the commutator and the brushes, or poor connections in the starter.
3. If the rpm and current draw are both higher than normal, check for a shorted field winding in the starter motor.
4. If the starter did not rotate and there is no current draw, the cause could be open field windings, open armature coils, or broken brush springs or brushes.

The final starter bench test to perform is a load check on the overrunning clutch pinion. Using a piece of hard wood, such as an old hammer handle, push against the overrunning clutch gear while the starter is engaged. If the clutch is slipping, the hardwood handle will stop the gear from turning.

Because some starters are difficult to replace, some manufacturers recommend that the replacement starter be bench tested to ensure proper operation before installation.

Starter Pinion Gear Clearance

On some General Motors starter motors, there is a specified clearance between the overrunning clutch pinion gear and the flywheel ring gear. To check this clearance, insert a flat blade screwdriver into the access hole on the side of the pinion drive housing and pry the drive pinion into the engaged position (Figure 15–42). Use a piece of

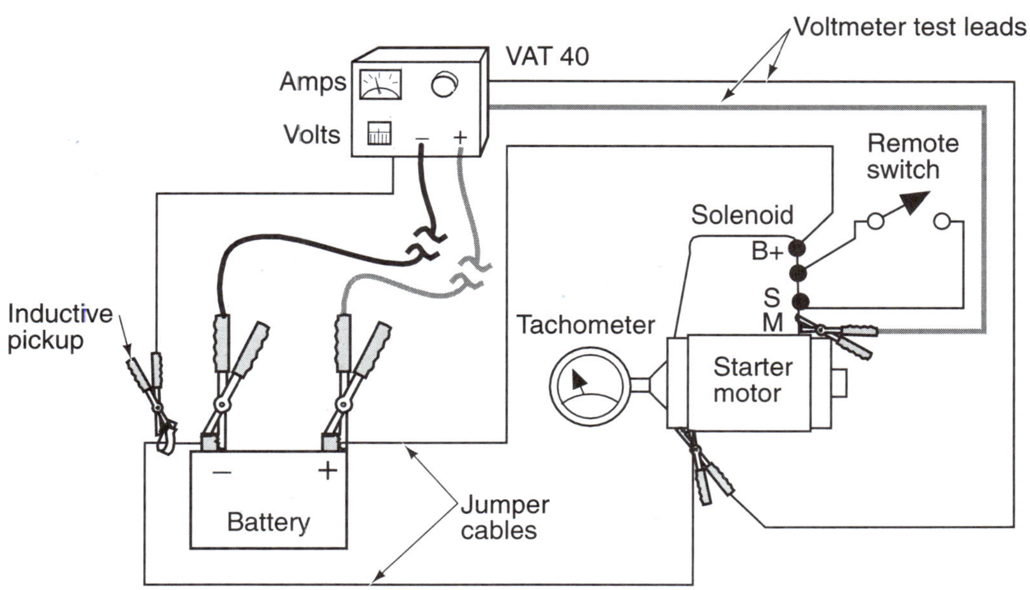

FIGURE 15–41 Bench testing a starter motor with a battery and a starter system tester.

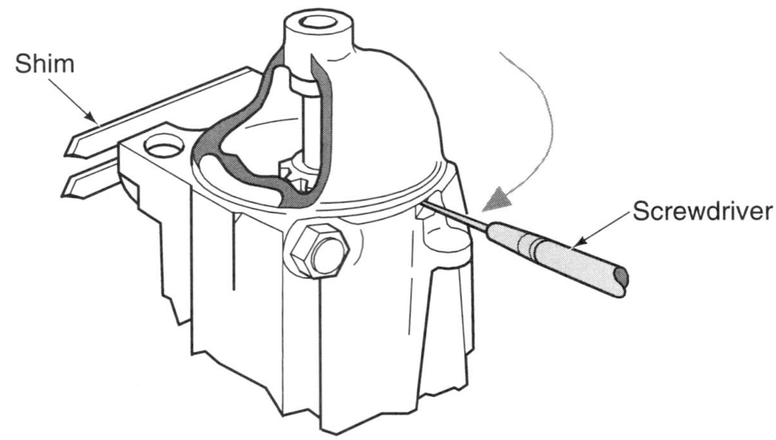

FIGURE 15–42 Prying the pinion into the engaged position to check clearance with ring gear.
(Courtesy of General Motors Corporation, Service Technology Group)

.020 inch diameter wire to check the clearance between the gears (Figure 15–43). If the clearance is excessive, the starter will produce a high pitched whine while the engine is being cranked. If the clearance is too small, the starter will make a high pitched whine after the engine starts and the ignition switch is returned to the run position. This is also the major cause of a broken pinion drive housing.

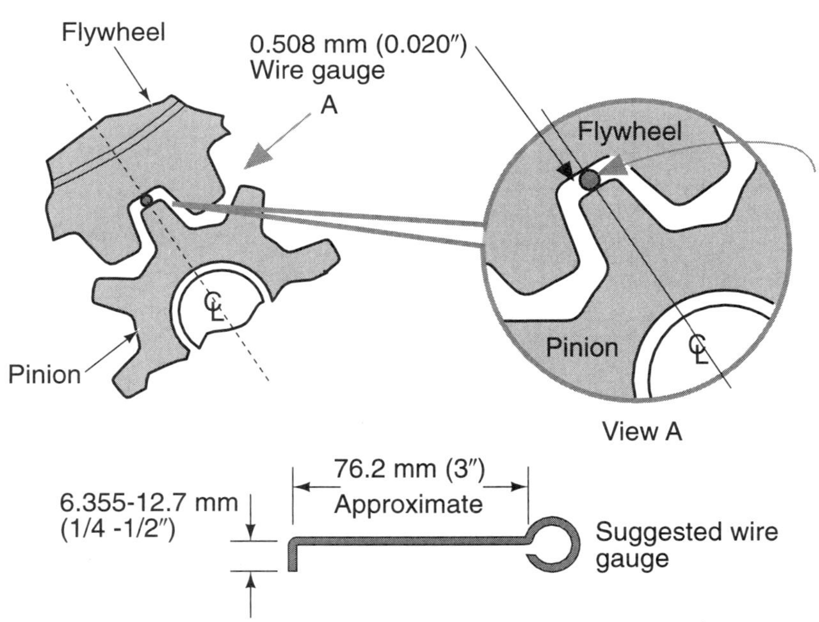

FIGURE 15–43 Checking the clearance between the pinion drive gear and the flywheel ring gear.

Starter Motor Service

Starter motor removal and replacement, and overhaul, varies with manufactures and vehicle brands. Refer to service manuals specific to the vehicle being worked on for detailed starter and component repairs. The following general information can be used as a guide though if specific information is not available. Figure 15–44 is typical of the parts in a direct drive starter. Many of the parts listed can be replaced or rebuilt depending on several factors.

1. What is preferred by the customer.
2. Shop warranty policies.
3. Rebuild equipment available in the shop.
4. Specific knowledge of the shop employees.
5. Cost of a new, or remanufactured, starter compared to an in-house rebuild.
6. Time required to repair or rebuild the starter.
7. Complexity of the starter needing service.

If the decision is made to rebuild the starter motor, then the manufacturer's service manual should be consulted. Photo Sequence #3 illustrates a typical procedure for disassembly a Delco Remy starter motor.

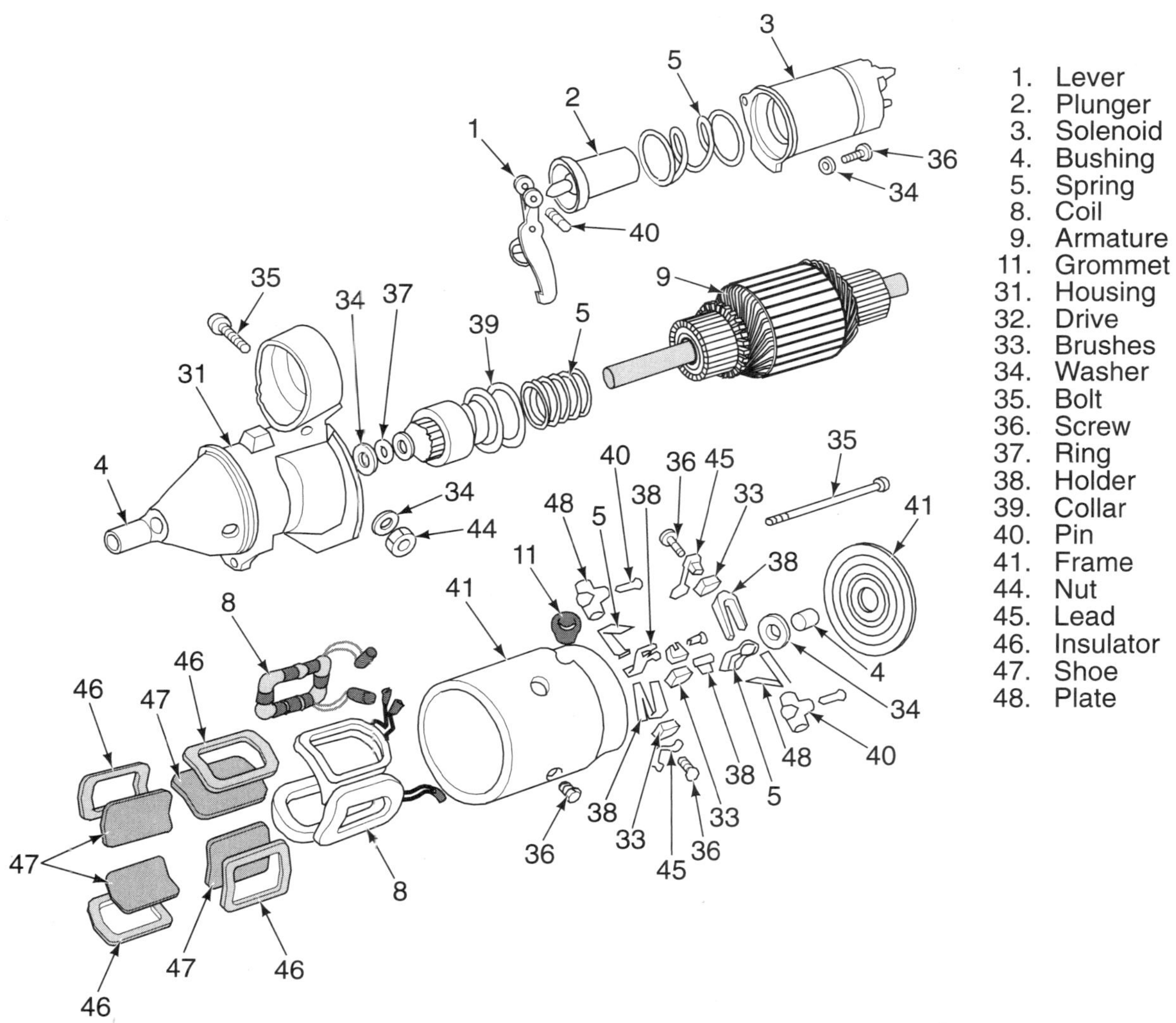

FIGURE 15–44 Typical parts in a direct drive starter.

PHOTO SEQUENCE 3

Typical Procedure for Delco Remy Starter Disassembly

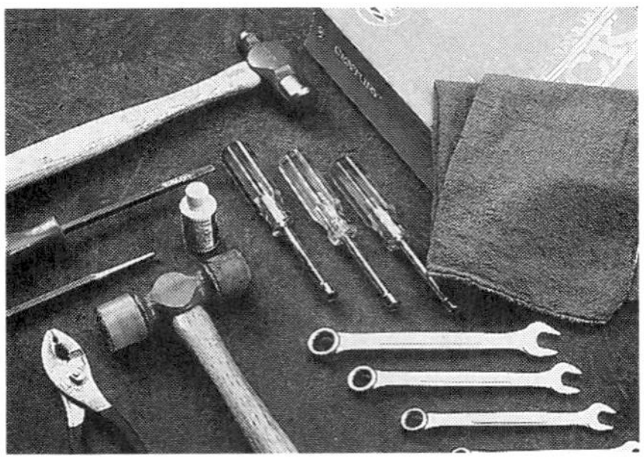

P15–1 Always have a clean and organized work area. Tools required to disassemble the Delco Remy starter: rags, assorted wrenches, snap ring pliers, flat blade screwdriver, ball-peen hammer, plastic head hammer, punch, scribe, safety glasses, and arbor press.

P15–2 Clean the case.

P15–3 Scribe reference marks at each end of the starter end housings and the frame.

P15–4 Disconnect the field coil connection at the solenoid's M terminal.

P15–5 Remove the two screws that attach the solenoid to the starter drive housing.

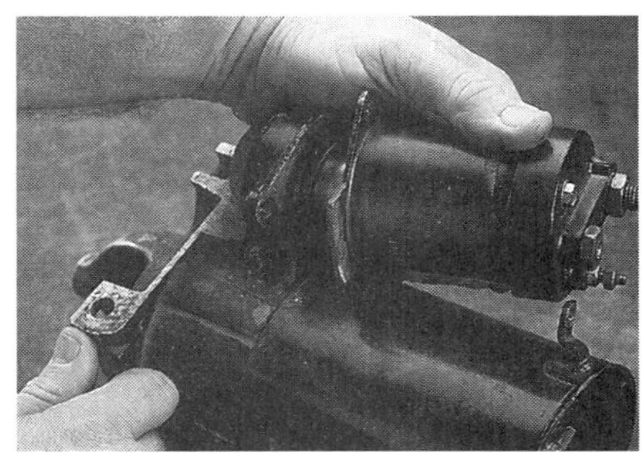

P15–6 Rotate the solenoid until the locking flange of the solenoid is free. Then remove the solenoid.

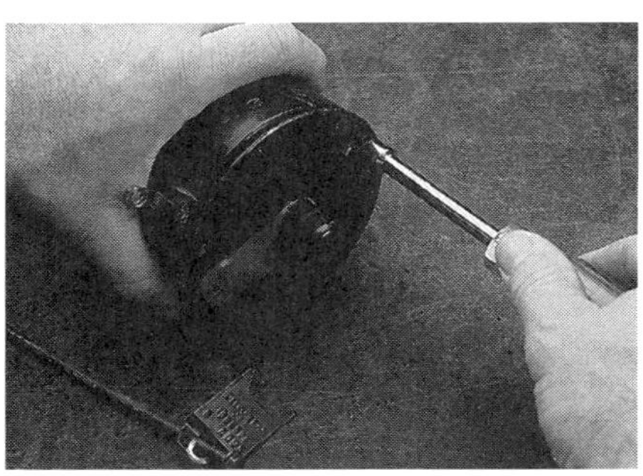

P15–7 Remove the trough bolts from the end frame.

P15–8 Remove the end frame.

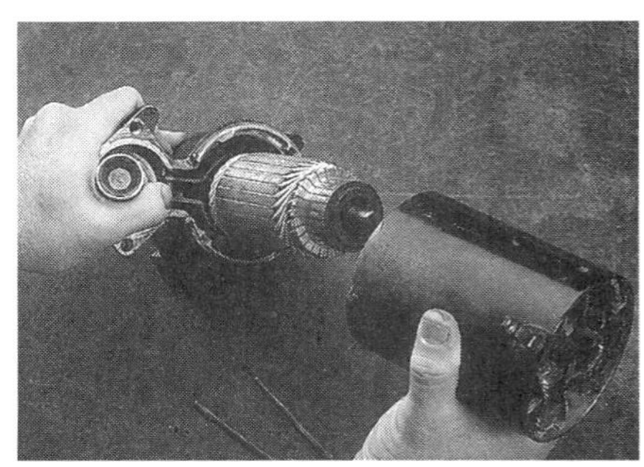

P15–9 Remove the frame.

PHOTO SEQUENCE 3

Typical Procedure for Delco Remy Starter Disassembly

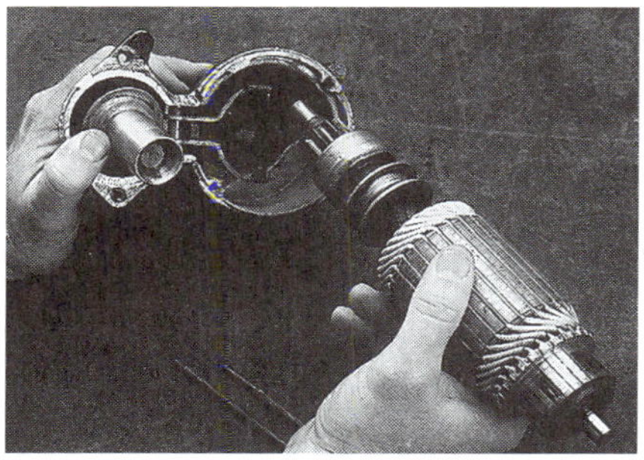

P15–10 Remove the armature from the drive housing. Note: On some units it may be necessary to remove the shift lever from the drive housing before removing the armature.

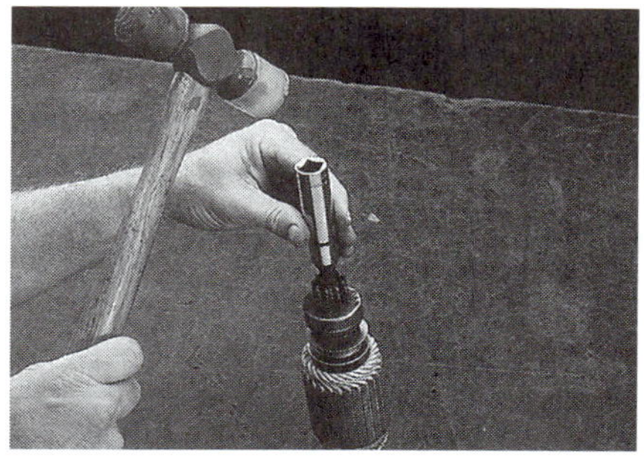

P15–11 Place a 5/8" deep socket over the armature shaft until it contacts the retaining ring of the starter drive.

P15–12 Tap end of socket with a plastic hammer to drive the retainer toward the armature. Move it only far enough to access the snap ring.

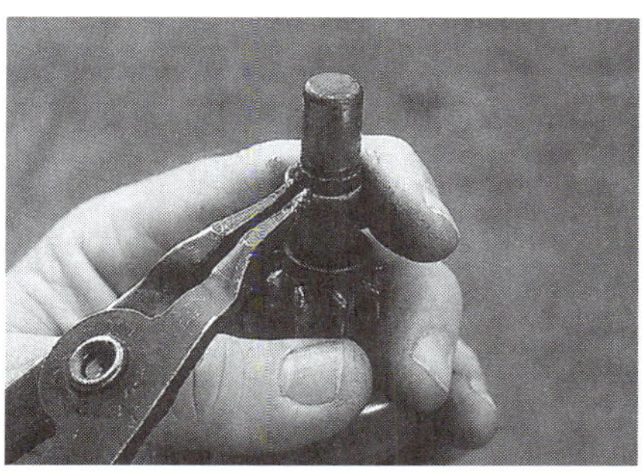

P15–13 Remove the snap ring.

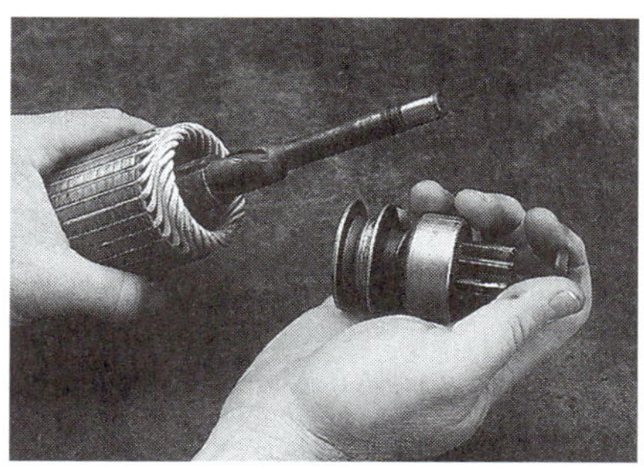

P15–14 Remove the retainer form the shaft and remove the clutch and spring from the shaft. Press out the drive-housing bushing and the end-frame bushing.

FIGURE 15–45 A growler is used to test an armature for shorts.

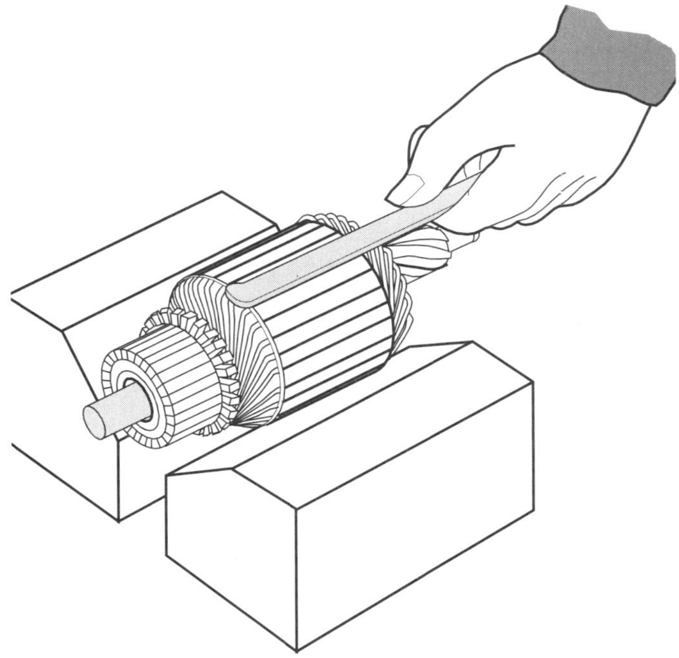

FIGURE 15–46 If a short exists in the armature, the hacksaw blade will vibrate when over the shorted area.

After the starter motor is disassembled, the parts should be cleaned and tested. Begin by testing the armature for shorts using a *growler* (Figure 15–45). The armature is placed in the growler and slowly rotated with a steel blade held parallel to the armature core (Figure 15–46). If the blade vibrates, the armature is shorted and should be replaced.

The armature can be tested for opens using an ohmmeter. Figure 15–47 shows connecting the ohmmeter test leads to any two commutator bar sections. There should be *zero* ohms of resistance in a good armature. If there is resistance, the armature will have to be replaced.

The armature can be tested for grounds using an ohmmeter. Figure 15–48 shows connecting the ohmmeter test leads between each commutator bar and the armature shaft. There should be an infinite reading (no continuity). If there is continuity, the armature will have to be replaced.

The field coil windings can be tested for both opens and grounds using an ohmmeter (Figure 15–49). When the field coil windings are disconnected, they should

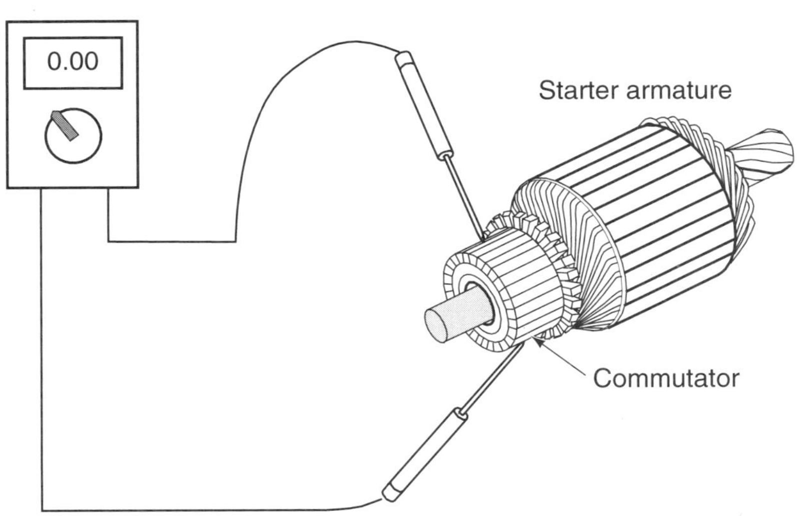

FIGURE 15–47 Testing an armature for opens using an ohmmeter.

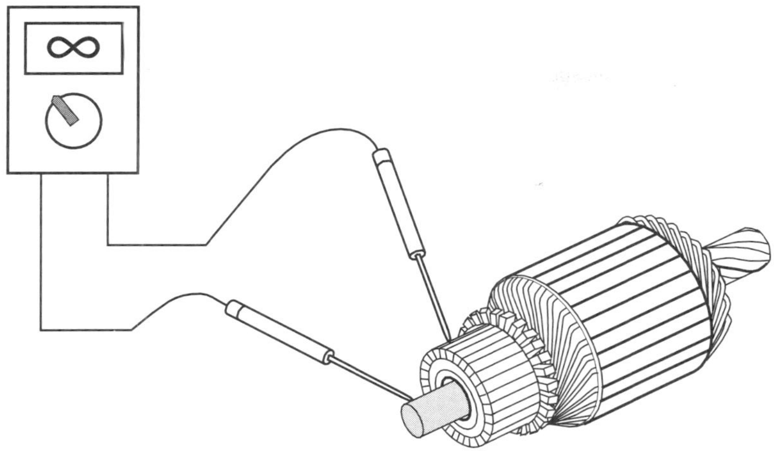

FIGURE 15–48 Testing the starter field coils for a grounded condition.

show an infinite reading from the coil windings to the starter housing. The winding should read zero ohms through each field coil in the circuit. An infinite reading would indicate an open circuit. Depending on how the field windings are wired, if one of these conditions exist, the starter is not serviceable.

On the type of starter in Figure 15–49, the pinion gear-to-drive housing clearance should be checked for clearance (Figure 15–50) with a feeler gauge. Depending on the starter, specifications range from 0.010 to 0.140 inches.

Note: Remember to perform the free running bench test after starter repair or rebuild.

FIGURE 15–49 Checking the drive pinion gear-to-drive housing clearance.

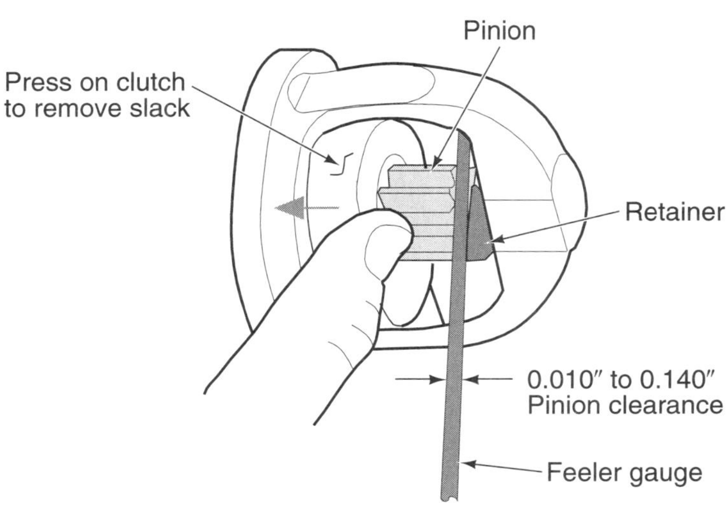

FIGURE 15–50 Pinion gear clearance specifications.

(Courtesy of General Motors Corporation, Service Technology Group)

CASE STUDIES

CASE #1

The following case study is typical of a starter circuit problem that technicians often have to diagnose. Knowledge of how individual starter circuit components work and simple test procedures will help you perform the diagnosis quickly and efficiently.

Customer Complaint

A vehicle was towed into the shop with a no-crank condition. The battery and starter were replaced, but the condition remained.

Known Information

- ❑ New battery that has an OCV of 12.66 volts
- ❑ Starter circuit is a positive engagement design with a remote solenoid relay assembly
- ❑ Starter motor was replaced
- ❑ Starter does not crank the engine when ignition switch is turned to the start position
- ❑ Engine is free when turned with a socket wrench attached to the front pulley nut

Circuit Analysis

Answer the following questions on a *separate* sheet of paper.

1. With the known information above, what is the most likely cause of the starter system problem above? ____

   ____.
2. What tests can you perform on the starter system to determine whether the problem is with the starter motor or the control circuit? ____

   ____.
3. What diagnostic procedures are helpful to know to determine the cause of the no-crank condition?

   ____.
4. What is your analysis of the above case study? ____

   ____

   ____.

CASE #2

The following case study is typical of a slow crank condition that technicians often have to diagnose. Knowledge of the starter circuit being tested and the simple test procedures discussed in this chapter will help you perform the diagnosis quickly and efficiently.

Customer Complaint

A vehicle with a direct drive starter cranks slowly all of the time. The customer replaced the starter motor and solenoid assembly, the battery, and the alternator, with no improvement in starter cranking speed. The customer was thinking of replacing the engine when the vehicle was towed into the shop for one final diagnosis.

Known Information

- ❑ New battery that has an OCV of 12.66 volts
- ❑ Starter motor and solenoid assembly is new and has been bench tested as OK
- ❑ Alternator has been changed (out of desperation!)
- ❑ Battery positive and negative cables and terminals *look* OK, with only a slight amount of corrosion at the terminal ends

Circuit Analysis

Answer the following questions on a *separate* sheet of paper.

1. With the known information above, what is the *most* likely cause of the slow crank problem above?

   ______________________________.

2. What tests can you perform on the starter system to determine whether the problem is with the starter motor, the control circuit, or the cables and connections?______________

   ______________________________.

3. What diagnostic procedures are helpful to know to determine the cause of the *slow* crank condition?

   ______________________________.

4. What is your analysis of the above case study? ______________

   ______________________________.

Hands-On Vehicle Tests

The following six hands-on vehicle checks are included in the **NATEF** (National Automotive Technicians Education Foundation) **Task List.** The task reference numbers are the same as the NATEF reference numbers. Some tasks are grouped together due to their close association with each other. Complete your answers to the below questions on a *separate* sheet of paper.

Performance Task # 1 (NATEF # VI-C-1)

Task Description

Perform starter current draw and circuit voltage drop test; determine needed repairs.

Performance Task # 2 (NATEF # VI-C-2)

Task Description

Inspect and test starter relays and solenoids; replace as needed.

Performance Task # 3 (NATEF # VI-C-5)

Task Description

Inspect, test, and repair or replace switches, connectors, and wires of starter control circuits.

Task Objectives

- ☐ Obtain a vehicle that can be used for these tasks. What model and year of vehicle are you using for these tasks? ____________________.
- ☐ Set the vehicle emergency brake and place the vehicle in neutral. Observe all standard shop safety precautions.
- ☐ Perform a visual inspection of the vehicle starter motor circuit. What type of starter circuit does the vehicle have, and list the main components of the circuit below.

  Type of starter circuit ____________________.

  Main starter circuit components

  1. ______________ 5. ______________
  2. ______________ 6. ______________
  3. ______________ 7. ______________
  4. ______________ 8. ______________

- ☐ What are the preliminary vehicle checks that should be made to any vehicle before performing tests on a starter circuit? ____________________

  ____________________.

- ☐ List the equipment you would use to perform diagnostic tests on a starter system.

Diagnostic equipment	Purpose of test equipment
1.	
2.	
3.	
4.	

- ☐ List the steps required to perform a current draw test on a starter motor using the equipment you have. ____________________

  ____

  ____________________.

- ☐ Perform a current draw test on the vehicle being tested. List the results below.

  Starter circuit voltage ______________.

  Starter amperage draw ______________.

- ☐ What are the current draw specifications for the vehicle being tested? If this information is unavailable, what is the standard current draw limits for the size of the engine in the test vehicle?

  ____________________.

❑ How do you test a starter relay and/or solenoid for proper operation? ______________________________
____.

❑ What can you use to bypass a solenoid or relay that you suspect is faulty? ______________________
____.

❑ Bypass the solenoid or relay. What happened to the starter motor? ____________________________.

❑ How could this test help you when diagnosing circuit related problems? ________________________
____.
____

❑ Perform a voltage drop test on all starter circuit components and record the results below.

Component	Voltage drop found	Industry limits
1.		
2.		
3.		
4.		
5.		
6.		
7.		
8.		
9.		
10.		

❑ How would voltage drops in the circuit affect the performance of the starter motor?______________
____
____.

❑ What procedures are most helpful when inspecting and testing the starter motor circuit? __________
____.

Task Summary

After performing the above NATEF tasks, what can you determine will be helpful in knowing how to check a starter motor system and individual components for proper operation? ____________________
____
____
____.

Performance Task # 4 (NATEF # VI-C-3)

Task Description

Remove and replace/reinstall starter.

Performance Task # 5 (NATEF # VI-C-6)

Task Description

Disassemble, clean, inspect, and test starter components; replace as needed.

Performance Task # 6 (NATEF # VI-C-4)

Task Description

Perform starter bench tests; determine needed repairs.

Task Objectives

- ❑ Obtain a vehicle that can be used for these tasks. What model and year of vehicle are you using for these tasks? ______________________.
- ❑ List the procedures for removing and replacing the starter motor according to the manufacturer of the vehicle being worked on. ______________________ ______________________.
- ❑ What tools and equipment did you use to remove and replace the starter motor? ______________________ ______________________.
- ❑ List the steps that you used to disassemble, clean, and inspect the starter motor. ______________________ ______________________.
- ❑ Perform individual starter motor component tests according to the manufacturer of the specific vehicle being worked on. Record the results below.

Starter component	Test equipment used	Results of test
1.		
2.		
3.		
4.		
5.		
6.		

- ❑ What are the steps and precautions necessary when performing a free running bench test of a starter motor? ______________________.
- ❑ Perform a free running bench test on the vehicle starter motor. Record the results.

  Starter RPM ______________. Starter current draw ______________.
- ❑ Does the starter meet standard industry specification for a free running bench test? ___________.

Task Summary

After performing the above NATEF tasks, what can you determine will be helpful in knowing how to check a starter motor system and individual components for proper operation? ______________________ ______________________.

Summary

- ❑ The magnetic repulsion and attraction of electromagnets form the basis of a DC starter motor.
- ❑ Flux density is the concentration of the magnetic lines of force.
- ❑ Field coils are several turns of heavy copper wire wrapped around an iron core called a pole shoe.
- ❑ The commutator is a series of split ring copper segment bars at one end of the armature that provide the sliding contact to a set of electrically conductive carbon brushes.
- ❑ CEMF is the counter electromotive force that is induced within the starter windings when current flows through the windings.
- ❑ A series wound starter directs current through the starter field winding in a series line.
- ❑ A parallel wound, or shunt wound, starter directs current through the starter field windings in two parallel paths.
- ❑ A compound wound starter directs most current through the starter in a series line, but has a separate field winding in parallel to control starter speed.
- ❑ A permanent magnet starter uses strong permanent magnets in place of electromagnets for the field coils.
- ❑ The typical starter system consists of a battery, battery cables, starter motor, ignition switch, harness wiring, starter relay or solenoid, pinion drive, flywheel ring gear, and a neutral safety switch.
- ❑ The bendix inertia drive is an older style drive pinion that uses inertia speed to engage the pinion teeth into the flywheel ring gear.
- ❑ The overrunning clutch drive is the most common type of starter drive pinion. It uses a roller type clutch that wedges spring-loaded rollers between an inner and outer housing to engage the pinion teeth with the flywheel teeth.
- ❑ A starter relay, or solenoid, is the high-amperage switch that controls the battery amperage going to the starter motor.
- ❑ The solenoid pull-in and hold-in windings both energize to engage the overrunning clutch into the flywheel ring gear.
- ❑ The ignition bypass terminal supplies full battery voltage to the ignition coil only during engine cranking conditions.
- ❑ The direct drive starter is the most popular style of starter that uses a solenoid attached directly to the starter motor.
- ❑ The positive engagement starter uses an separate relay to control an internal movable pole shoe in the starter to engage the pinion gear.
- ❑ A gear reduction starter contains a small gear reduction drive gear between the starter motor and the drive pinion.
- ❑ A permanent magnet starter uses strong permanent magnets instead of electromagnets to form the field coils.

- ❑ The neutral safety switch (NSS) is used to prevent the starter motor from cranking on an automatic transmission when the transmission is in gear, or unless the clutch is pushed in for a manual transmission.
- ❑ A starter circuit isolation test is a quick test that determines if the starter motor, solenoid, or control circuit is at fault during a no-crank test.
- ❑ A starter current draw test tests the starter current draw during cranking conditions.
- ❑ A starter circuit voltage drop test checks for unwanted resistance in the circuit that creates a voltage drop condition which could affect cranking voltage available to the starter motor.
- ❑ A starter open circuit test is performed on a starter with a no-crank complaint.
- ❑ Starter bench testing will check the starter motor rpm, current draw, and overrunning clutch operation during no load conditions.

Key Terms

armature
Bendix inertia drive
brushes
commutator
compound wound
counter electromotive force (CEMF)
direct drive starter
field coil
flux density
gear reduction starter
hold-in winding
neutral safety switch (NSS)
overrunning clutch drive
permanent magnet starter
parallel wound
pole shoes
ignition bypass terminal
pull-in winding
series wound
shunt wound
starter relay
starter solenoid
positive engagement starter

Review Questions

Short Answer Essays

1. When an engine starts, what causes the starter motor overrunning clutch to disengage itself for safety before the ignition key is returned to the run position?
2. Describe how the pull-in and the hold-in windings work in a typical starter solenoid.
3. What is the purpose of the neutral safety switch in a starter system?
4. Describe the procedures used to perform a starter current draw test using a VAT-40 tester?
5. What are the allowed voltage drops in a typical starter motor circuit?

Fill-in-the-Blanks

1. The soft iron core in the field coil in the starter motor is called the ____________________.
2. On a starter solenoid the *R* terminal is an ____________________, that supplies battery voltage to the ____________________ during engine cranking conditions.
3. A gear reduction starter contains a ____________________ between the starter motor and the drive pinion.

4. To perform a starter system test, the battery must be ________________.
5. Normal starter current draw for a small V-8 engine should be ________________ amps.

ASE Style Review Questions

1. Technician A says the ignition coil is part of the starter system. Technician B says the ignition switch is part of the starter system. Who is correct?
   A. A only
   B. B only
   C. Both A and B
   D. Neither A nor B

2. Technician A says the commutator is a series of conducting segments located around one end of the armature. Technician B says the armature is the stationary component that consists of several turns of copper wire wrapped around a pole shoe. Who is correct?
   A. A only
   B. B only
   C. Both A and B
   D. Neither A nor B

3. Technician A says the starter solenoid moves a plunger to engage the overrunning clutch. Technician B says the two windings of the starter solenoid are called the pull-in and hold-in windings. Who is correct?
   A. A only
   B. B only
   C. Both A and B
   D. Neither A nor B

4. Technician A says the CEMF is present only in high amperage starters. Technician B says CEMF is caused by excessive resistance in the starter motor field windings. Who is correct?
   A. A only
   B. B only
   C. Both A and B
   D. Neither A nor B

5. Technician A says the neutral safety switch controls *only* the starter circuit on all vehicles. Technician B says the neutral safety switch prevents the starter from cranking when the automatic transmission is in gear. Who is correct?
   A. A only
   B. B only
   C. Both A and B
   D. Neither A nor B

6. Two technicians are discussing the results of a starter current draw test on a four cylinder engine. Technician A says the minimum voltage should not drop below 9.6 volts. Technician B says the amperage draw of 125 amps is considered satisfactory. Who is correct?
   A. A only
   B. B only
   C. Both A and B
   D. Neither A nor B

7. Technician A says a permanent magnet starter uses strong permanent magnets in place of electromagnetic field coils. Technician B says a permanent magnet starter uses a planetary gear set for maximum gear reduction in a small starter. Who is correct?
   A. A only
   B. B only
   C. Both A and B
   D. Neither A nor B

8. Technician A says the maximum allowed voltage drop for a battery cable is 0.5 volts. Technician B says a voltage drop at a connection should not exceed 0.3 volts per connection. Who is correct?
   A. A only
   B. B only
   C. Both A and B
   D. Neither A nor B

9. Technician A says if starter pinion gear clearance is excessive, the starter will produce a high pitched whine while cranking the engine. Technician B says if starter pinion gear clearance is excessive, the starter drive housing will break. Who is correct?
   A. A only
   B. B only
   C. Both A and B
   D. Neither A nor B

10. An engine cranks slowly. Technician A says a possible cause could be a low battery charge. Technician B says a possible cause could be incorrect ignition timing. Who is correct?
    A. A only
    B. B only
    C. Both A and B
    D. Neither A nor B

11. Technician A says the purpose of the starter relay is to complete the circuit from the battery to the starter motor. Technician B says the purpose of the starter solenoid is to complete the circuit from the battery to the starter motor. Who is correct?
    A. A only
    B. B only
    C. Both A and B
    D. Neither A nor B

12. A starter system has a no-crank condition. The headlights remain on with the same intensity when the ignition switch is turned to the Start position. Technician A says this could be caused by a dead battery. Technician B says this could be caused by a faulty solenoid switch. Who is correct?
    A. A only
    B. B only
    C. Both A and B
    D. Neither A nor B

13. A vehicle is being tested for a slow-crank condition. A starter current draw test indicates 450 amps. Technician A says this could be caused by a battery with a CCA rating too large for the system. Technician B says this could be caused by a pinion drive that slips when engaged with the flywheel ring gear. Who is correct?
    A. A only
    B. B only
    C. Both A and B
    D. Neither A nor B

14. Technician A says the *R* terminal on the starter solenoid is used for the reactor windings in the solenoid. Technician B says the pull-in windings become de-energized when the overrunning clutch engages into the flywheel ring gear. Who is correct?
    A. A only
    B. B only
    C. Both A and B
    D. Neither A nor B

15. A slow-crank condition exists in a starter system. Technician A says corroded conditions at the battery could be the cause. Technician B says excessive battery CCA capacity could be the cause. Who is correct?
    A. A only
    B. B only
    C. Both A and B
    D. Neither A nor B

16 Charging System Theory

Introduction

The purpose of the automotive charging system is to convert the mechanical energy of the engine into electrical energy to recharge the battery when necessary, and to provide all the current needed to operate the vehicle electrical accessories while the engine is running. Accessory electrical demands on the vehicle have increased almost five-fold to 100 amps or more over the past three decades. To keep up with this demand, charging systems have made significant design improvements since the DC (direct current) generator was discontinued in the mid 1960's and AC (alternating current) generators became the predominant charging device.

Objectives

When you complete this chapter you should be able to:

- ❑ Explain the purpose of the charging system in a vehicle.
- ❑ Explain the principles of magnetism in the operation of a charging system.
- ❑ Explain the differences between DC generators and AC generators in a charging system.
- ❑ Identify and explain the function of the major components of the charging system.
- ❑ Explain the function of regulation devices and circuits in charging systems.
- ❑ Perform charging system output tests to diagnose an undercharge, overcharge, or no-charge condition, and determine needed repairs.
- ❑ Perform tests on individual charging system components and determine needed repairs.
- ❑ Perform charging system voltage drop tests and determine needed repairs.

Charging System Principles

All vehicle charging systems use the principles of electromagnetic induction to generate the current necessary to charge the battery and operate the vehicle accessory circuits. The electromagnetic principle states that a voltage will be produced if motion between a conductor and a magnetic field occurs (Figure 16–1). The amount of voltage produced in the conductor is dependent on three things.

1. The strength of the magnetic field.
2. The speed at which the conductor is cutting through the magnetic field.
3. The number of conductors passing through the magnetic field.

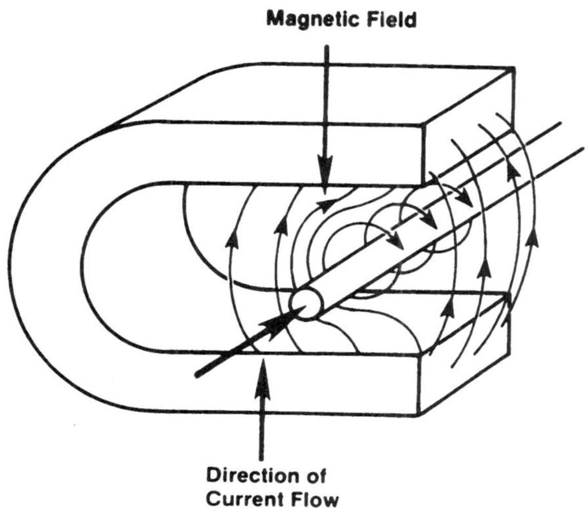

FIGURE 16–1 Conductor interacting within a magnetic field.

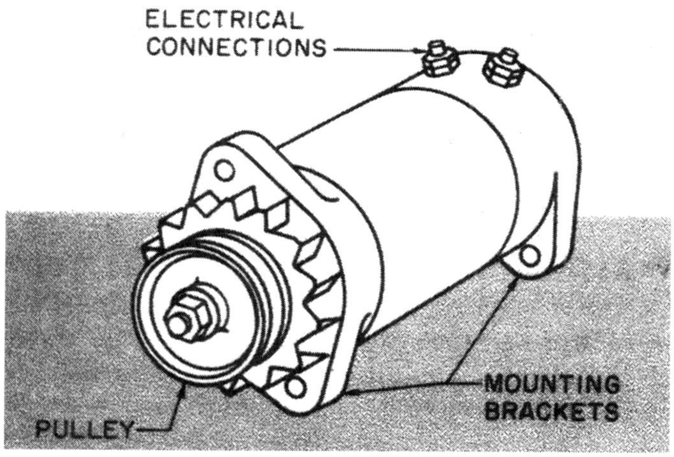

FIGURE 16–2 A simple DC motor.

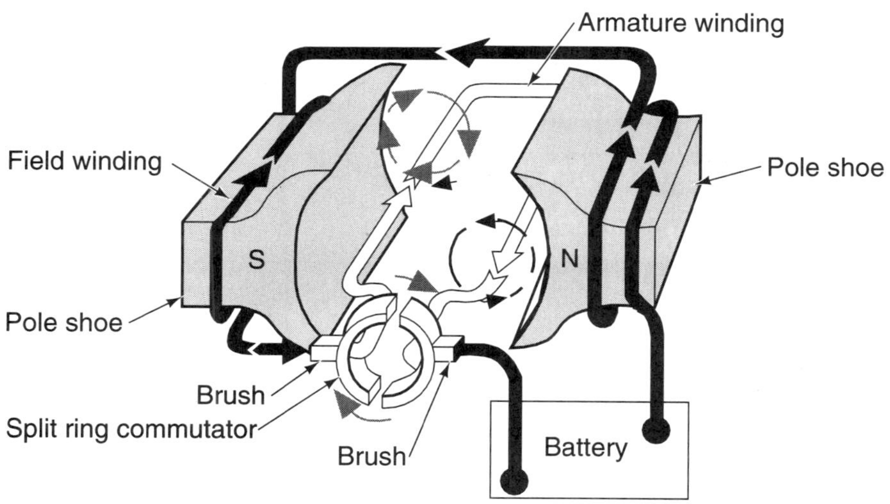

FIGURE 16–3 Simple DC motor with two stationary field coils and rotating armature output.

Direct Current (DC) Generators

Although not used in modern automobiles, knowledge of DC generator principles will help you understand the benefits of an AC generator. The DC generator (Figure 16–2) is similar to the DC starter motor discussed in Chapter 15. Figure 16–3 is a simple DC motor that contains two stationary field coils that create the north and south magnetic poles. Output voltage is generated in the wire loop(s) of the armature as it rotates through the magnetic field. The more wire loops there are in the armature, the more voltage will be produced at the same rpm and strength of the magnetic field (Figure 16–4).

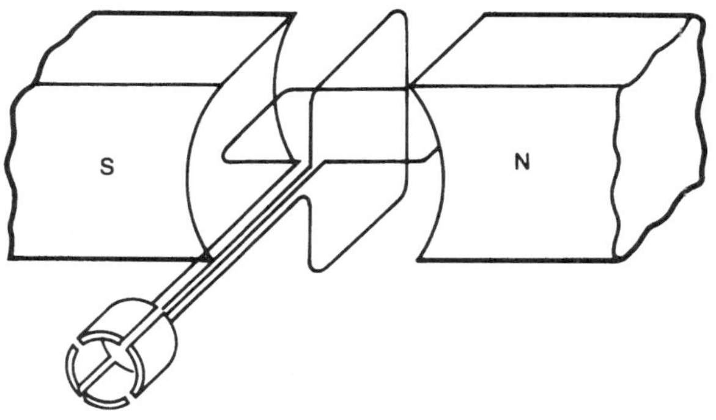

FIGURE 16–4 Strengthening the voltage output by increasing the number of conductors.

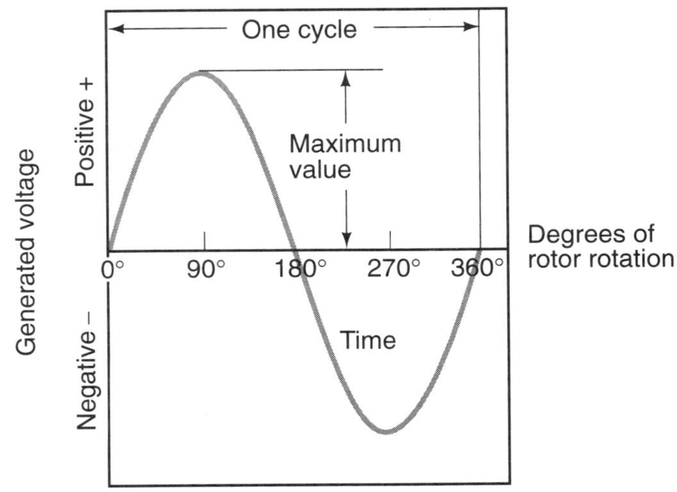

FIGURE 16–5 Sine wave produced in one revolution of a conductor in a magnetic field.

In theory the output voltage of a conductor passing through a magnetic field from two directions will produce an alternating current (AC) **sine wave,** where there is positive and negative voltage generated during one revolution (Figure 16–5). In a DC generator the AC sinewave is changed to a pulsing DC sinewave by the armature commutator bars (Figure 16–6).

Alternating Current (AC) Generators

The AC generator is widely used in all modern vehicles because of its ability to produce sufficient amounts of current at lower engine RPM's. The speed at which a conductor is cutting through the magnetic field is important when electrical accessory load demands are high, and the engine rpm is low. Figure 16–7 shows the components of a typical AC generator which include the following.

- Front and rear housings.
- Front and rear bearings.
- Cooling fan.
- Rotor assembly.
- Stator assembly.
- Rectifier assembly.
- Brush holder and brushes.

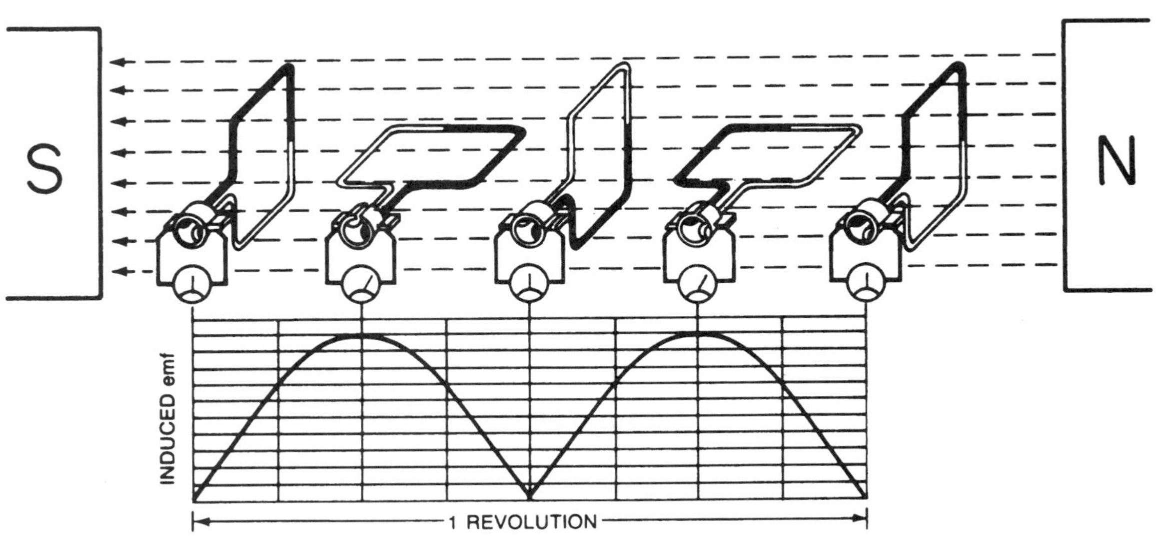

FIGURE 16–6 AC sine wave is converted to a pulsing DC sine wave by the armature commutator bars.

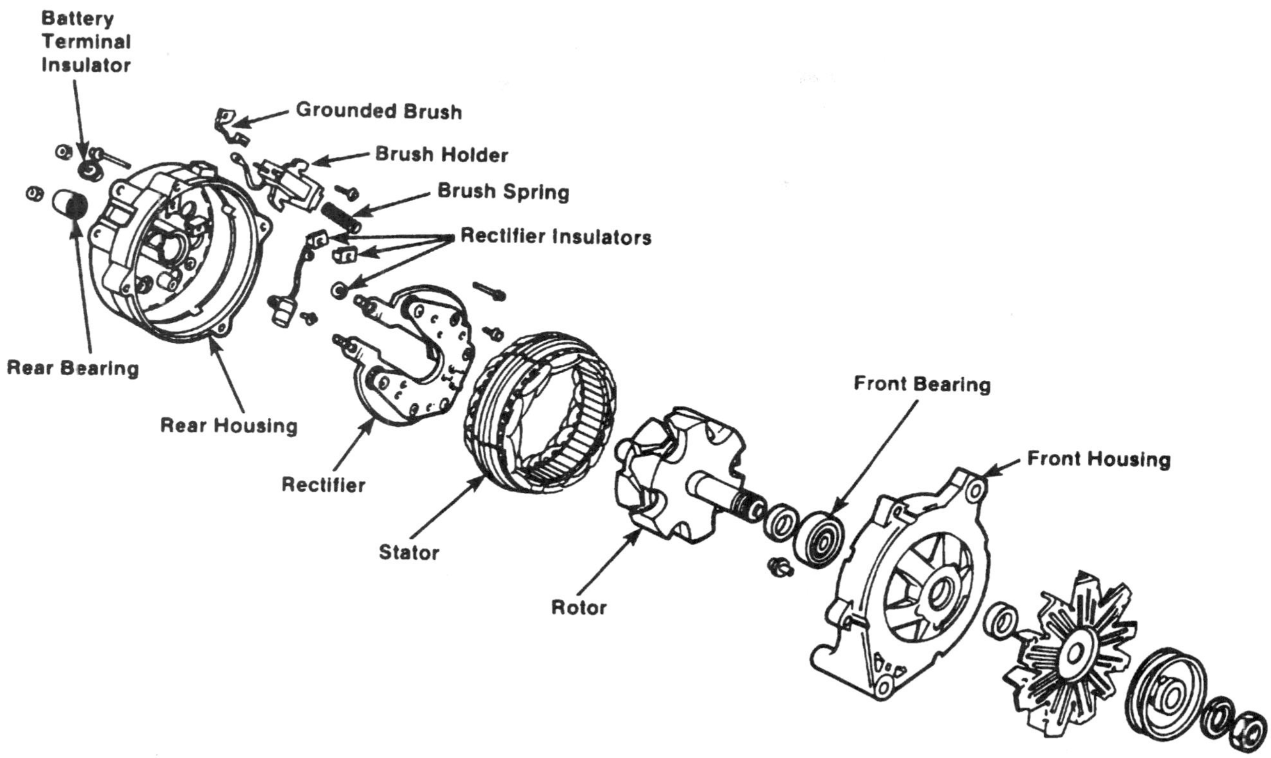

FIGURE 16–7 Exploded view of an alternator assembly.
(Courtesy of Ford Motor Company)

Charging System Components and Operation

Rotor Assembly

The **rotor** assembly is considered the heart of the AC generator. It is a rotating electromagnetic field with several north and south poles. The rotor is constructed of many turns of insulated copper wire wrapped around an iron core called a field coil. The field coil is surrounded on each end by soft metal pole pieces with fingers that interlace, but do not touch each other (Figure 16–8). When current passes through the coil, the pole pieces become magnetized and take on a north and south pole polarity (Figure 16–9). The poles will alternate their

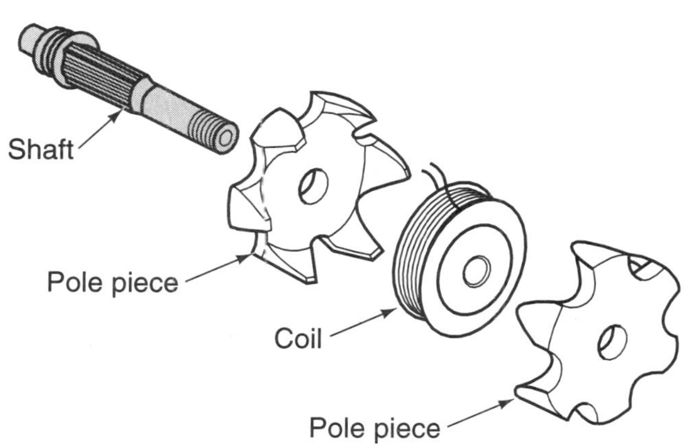

FIGURE 16–8 Typical components of an alternator rotor.

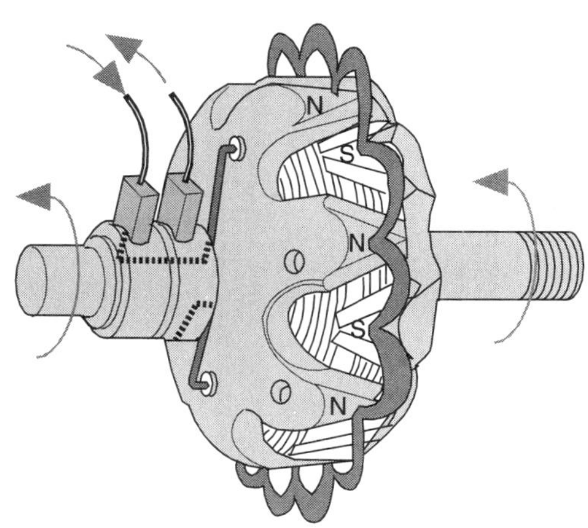

FIGURE 16–9 Multiple north and south poles of a rotor.

FIGURE 16–10 Magnetic flux lines move in opposite directions between the rotor poles.

(Courtesy of Ford Motor Company)

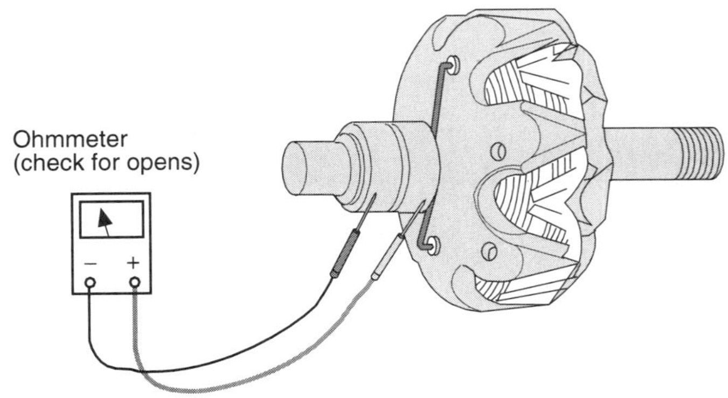

FIGURE 16–12 Testing a rotor field coil for an open or shorted condition.

north and south pole arrangement around the rotor coil, and cause the magnetic flux lines to move in opposite directions between adjacent poles (Figure 16–10).

The rotor **slip rings** provide a current path connection from the insulated carbon brushes to the rotor coil assembly (Figure 16–11). The slip rings are insulated from the rotor shaft and from each other. When the circuit is energized, regulated current flows through the first brush and slip ring to the field coil. Current then passes through the field coil and the second slip ring and brush to the ground in the alternator or regulator.

There are three basic checks to perform on a rotor that has been removed from the AC generator.

1. The first check is a complete visual inspection of the rotor field windings for overheating or discoloration. The slip rings should be smooth and flat with no grooves. A rotor indicating one of these conditions is unusable.
2. The second check is to measure the resistance between the slip rings (Figure 16–12) for an open condition. Always refer to the vehicle manufacturer service manual for correct specifications. Typical resistance values are between 2.4 Ω and 6.0 Ω. If the resistance value is below the specified value, the rotor coil is shorted. If there is an infinite resistance (O/L) the rotor is open.
3. The third check is to test the slip rings and field coil for a short to the rotor shaft (Figure 16–13).Using an ohmmeter from each slip ring and the rotor shaft will check for this condition. Ohmmeter reading should be infinite (O/L) resistance. If the reading is low, the field coil or one of the slip rings is shorted and the rotor is unusable.

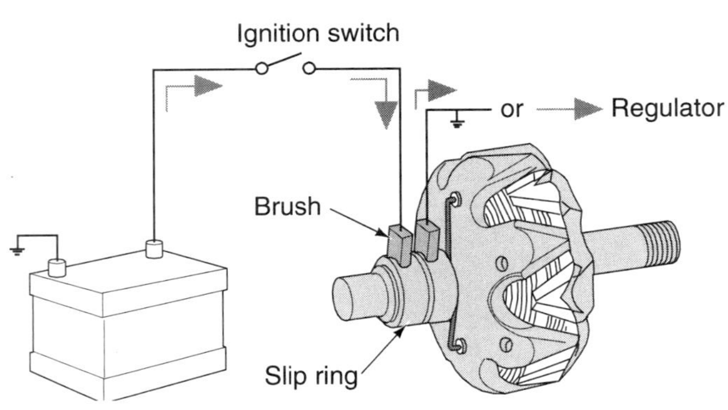

FIGURE 16–11 The slip rings and brushes provide an insulated current path to the rotor coil.

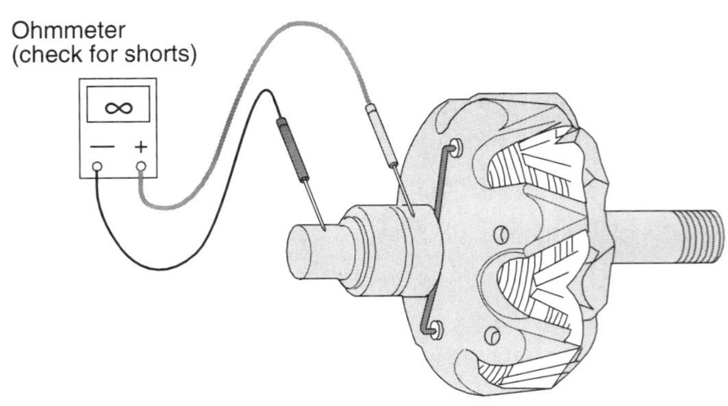

FIGURE 16–13 Testing the slip rings and field coil for a short to ground with the rotor shaft.

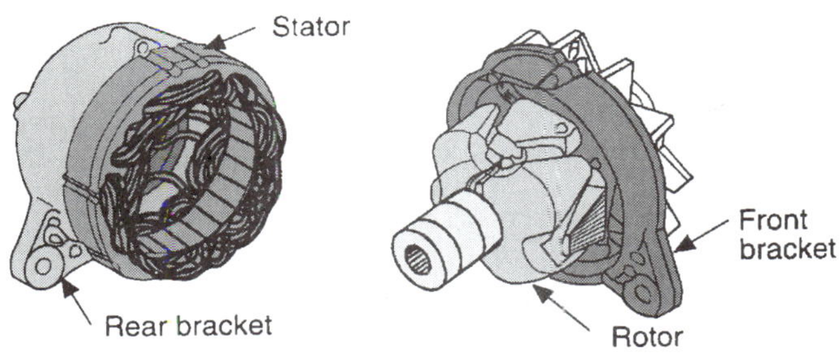

FIGURE 16–14 Location of stator in alternator assembly.

(Courtesy of daimlerchrysler Corporation)

Stator Assembly

The **stator** assembly in an AC generator is the stationary coil windings in which electricity is induced. It is located between the front and rear halves of the alternator shell (Figure 16–14). The stator consists of three sets of insulated windings (Figure 16–15), with coils equal to the number of north and south poles on the rotor. The windings are wrapped in insulated slots around a laminated circular soft iron frame, and overlapped to produce the required phase angles. (Figure 16–16).

Depending on the electrical requirements of the vehicle, there are two basic configurations of stator winding connections used on AC generators in modern vehicles. The first is a **wye wound** stator winding (Figure 16–17), which consists of a common junction and three output connections that branch out in a *Y* pattern. A wye wound alternator is usually found in applications that do not require high amperage output.

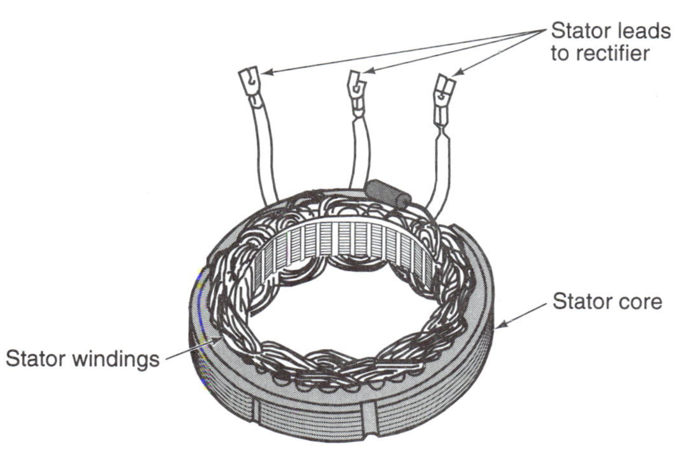

FIGURE 16–15 Components of a typical stator assembly.

(Courtesy of DaimlerChrysler Corporation)

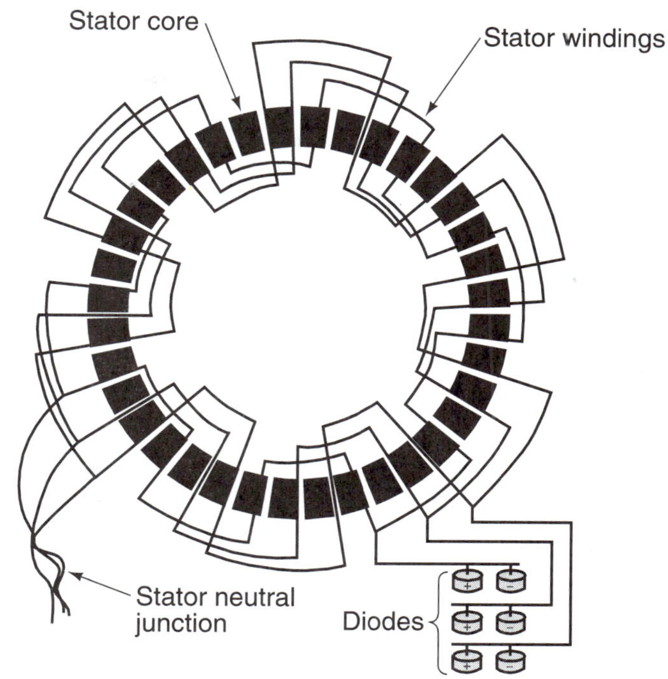

FIGURE 16–16 Overlapping stator windings produce the required phase angles.

The second stator winding configuration is a **delta wound** winding (Figure 16–18), which consists of three windings connected together in a triangle shape with three output leads. The delta connection is the most common style in use today because of its high amperage output for its small size. The reason for the increased amperage output will be explained in the diode section of this chapter.

There are three basic checks to perform on a stator winding which has been removed from the AC generator.

1. The first check is a complete visual inspection of the stator phase windings for overheating or discoloration. Phase windings that have blackened varnish insulation indicate overheating has occurred and the stator should be replaced.

2. The second check is to check for an open condition in any combination of the stator windings (Figure 16–19). Connect an ohmmeter to any two stator output leads—three combinations in all. The results should show a resistance of less than 1.0 Ω. If the ohmmeter reading is infinity (O/L) between any two leads, the stator has an open and must be replaced.

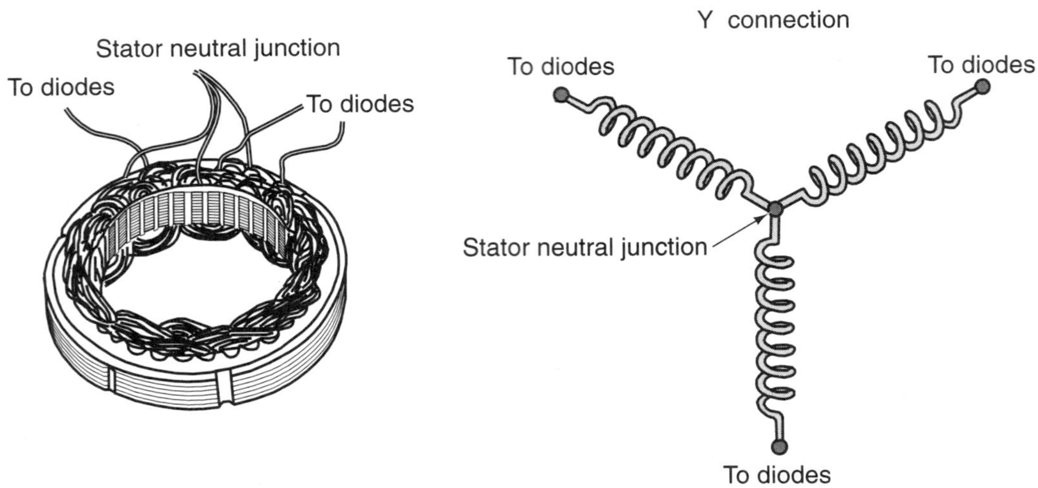

FIGURE 16–17 Typical wye wound stator windings.

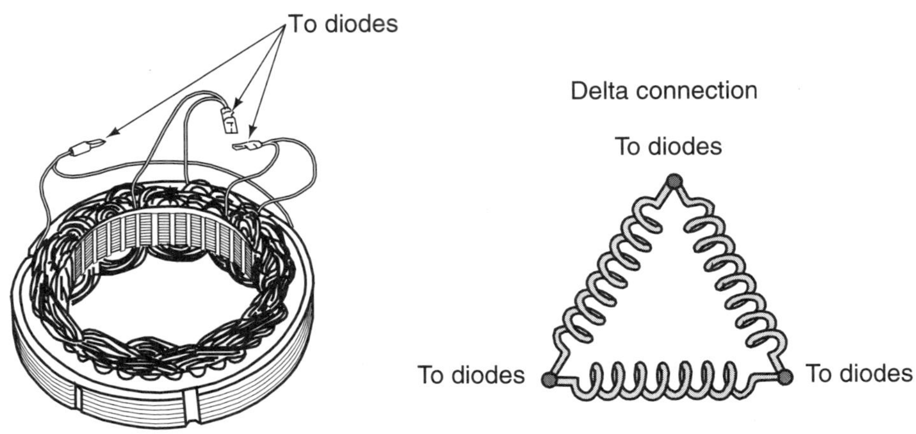

FIGURE 16–18 Typical delta wound stator windings.

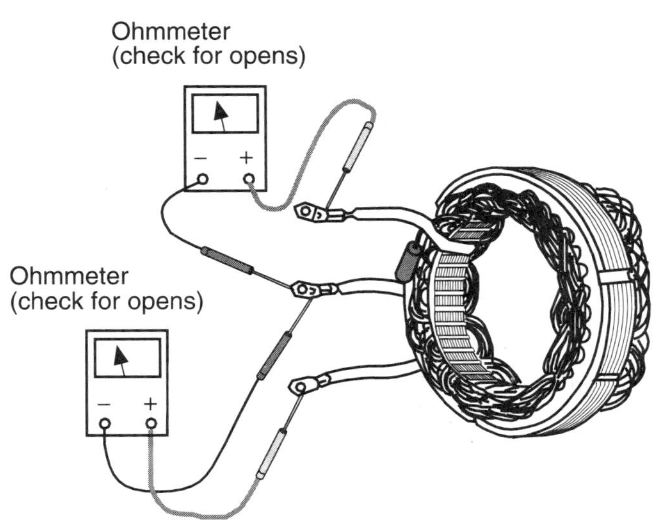

FIGURE 16–19 Testing the stator assembly for opens.

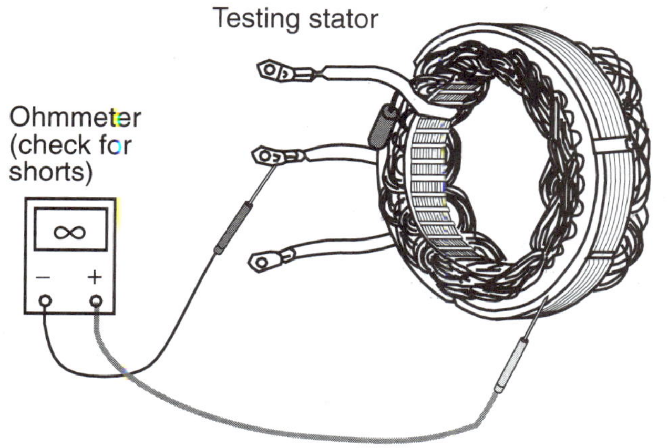

FIGURE 16–20 Testing the stator assembly for shorts to ground.

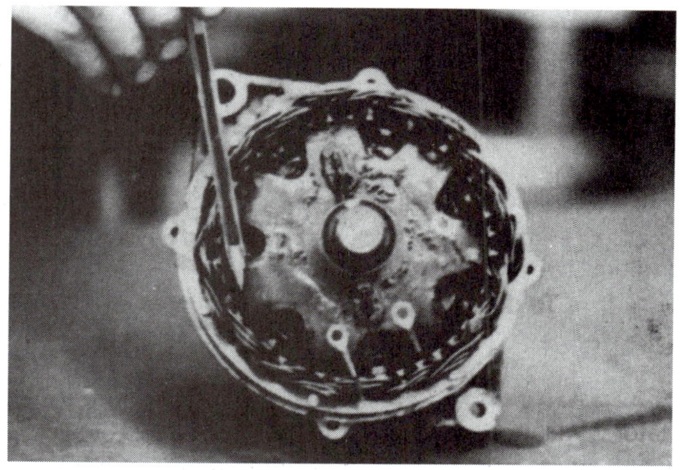

FIGURE 16–21 A small air gap between the rotor and stator maximizes the magnetic force.

3. The third check is to test the stator windings for a short to ground (Figure 16–20) by connecting the ohmmeter to each stator output lead and the support frame. The ohmmeter should read infinity (O/L) on all three stator leads. For any reading less than infinity, the stator is shorted to ground and is unusable.

The rotor assembly is fitted inside the stator with a clearance of approximately 0.015 inch of air gap, which maximizes the magnetic force of the rotor field to all of the stator windings at the same time (Figure 16–21).

In a simple alternator (Figure 16–22), a single wire is induced with an alternating current by the rotating north and south poles. This can be seen on an oscilloscope as an alternating current sine wave (Figure 16–23) that begins at (A) zero volts, when the phase winding is parallel to the magnetic field. When the magnetic field rotates 90 degrees, the magnetic field is at a right angle to the phase winding and produces (B) its *maximum positive* voltage. When the magnetic field turns another 90 degrees, being parallel with the phase winding, the voltage returns to (C) zero volts.

A third 90 degree rotation results in the magnetic field being reversed at the top phase winding, and produces its (D) *maximum negative* voltage. The last 90 degree rotation will complete the 360 degree revolution and returns the voltage to (E) zero.

In a wye wound and a delta wound stator winding, each of the three groups of windings occupies one third (120 degrees) of the circle. As the rotor, with its multiple north and south poles, rotates inside the stator windings, each loop of the stator produces a voltage with a different phase angle each 120 degrees apart form the others (Figure 16–24). The resulting overlap of the three phases

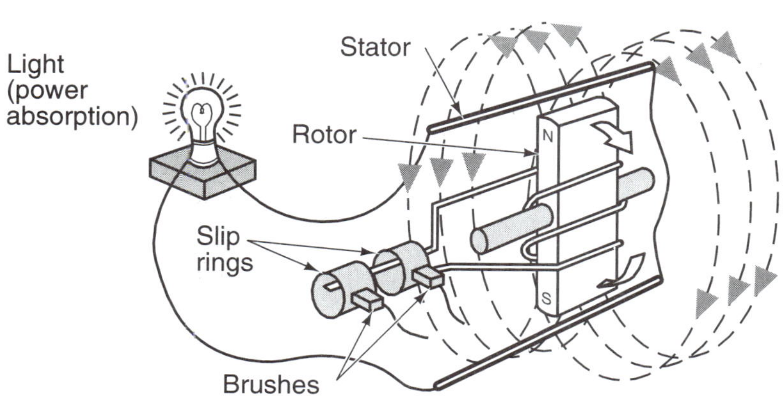

FIGURE 16–22 Simplified alternator showing electromagnetic induction.

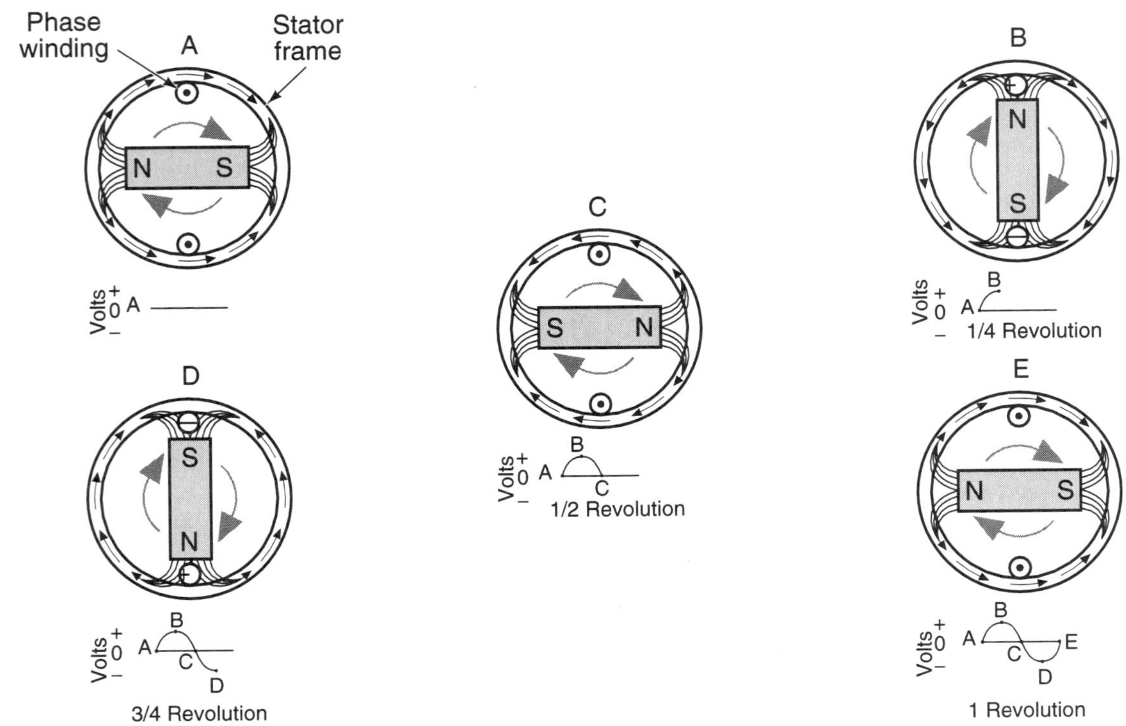

FIGURE 16–23 Producing a simple alternating current sine wave.

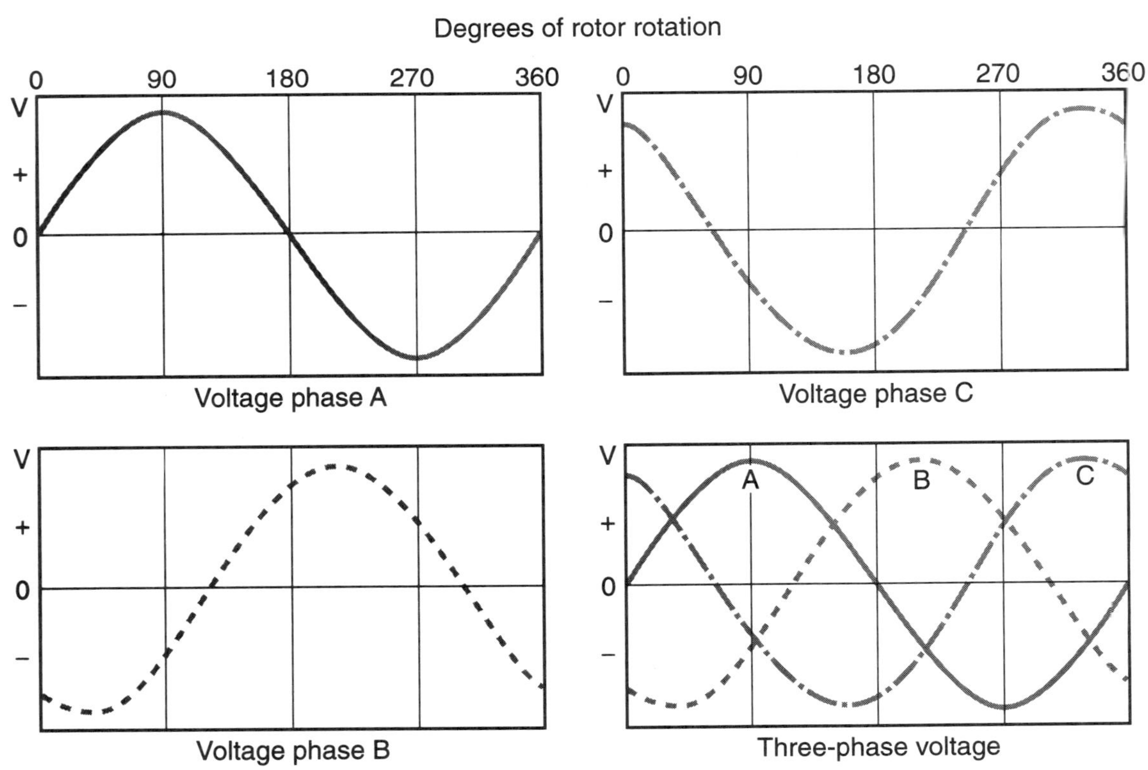

FIGURE 16–24 Overlapping of three stator windings creates a three phase voltage output.

of each stator winding group produces a **three phase voltage** output.

Alternating Current Rectification

A vehicle's electrical circuits and battery operate with direct current (DC). The AC current produced by the alternator needs to be converted to a DC voltage. This process is called **rectification.** In a DC generator, the split ring commutator is used to rectify the AC voltage to DC voltage. In an AC generator, the AC voltage is rectified by diodes that act as one-way check valves, allowing current to flow in one direction only. In a simple circuit, if AC voltage passes through a positively biased diode, the diode will block the negative portion of the sine wave. The result, as seen on the oscilloscope pattern in Figure 16–25, would be a pulsing DC current called **half-wave rectification** (Figure 16–25).

To also rectify the negative half of a simple AC sine wave requires four diodes (Figure 16–26). The direction of current flow through the phase windings determines the path of current flow through the diode circuit. The final result is current that flows through the load circuit and battery in one direction. The process of converting a complete AC waveform to a positive DC waveform is called **full-wave rectification.**

To rectify the three phase AC output of an alternator to a DC output, a **diode rectifier bridge** is attached to the stator output leads to provide the constant DC voltage necessary for the vehicle's electrical system and battery. The diode rectifier bridge (Figure 16–27) consists of a series of six diodes—one pair for each stator winding. Three of the diodes are positive biased and are mounted in a **heat sink** (Figure 16–28), and the three remaining diodes are negative biased and attached directly to the alternator rear housing, which also acts

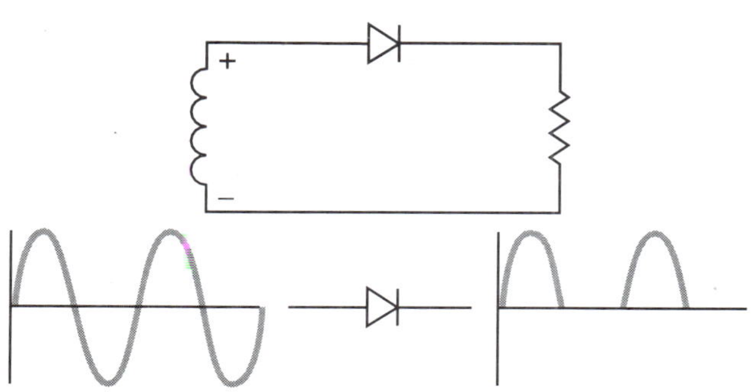

FIGURE 16–25 Half-wave rectification of an AC current to create a pulsating DC current.

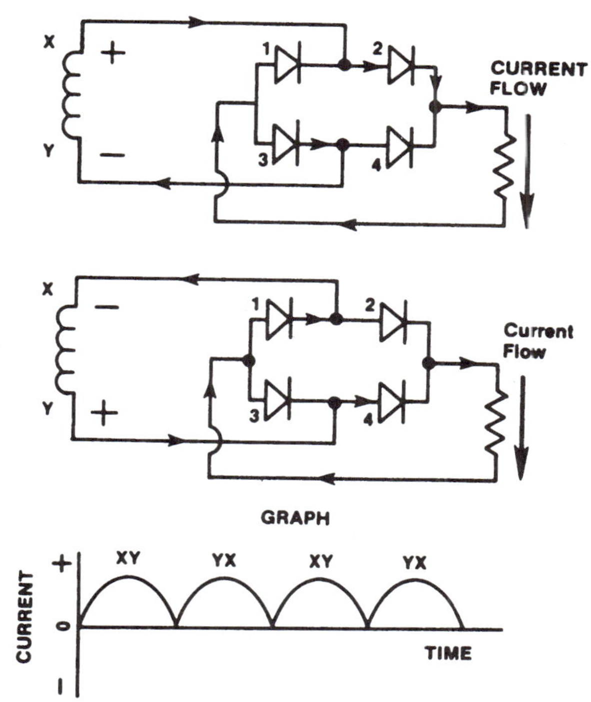

FIGURE 16–26 Full-wave rectification of a simple AC current requires four diodes.

like a heat sink (Figure 16–29). The heat sink absorbs the heat produced by the diodes and dissipates it to the generator frame.

Testing a rectifier bridge consists of checking the six individual diodes for continuity in one direction only (Figure 16–30). Using an ohmmeter suitable for diode testing, each diode will show infinite (OL) in one direction and low resistance in the opposite direction (Figure 16–30).

FIGURE 16–27 Typical diode rectifer bridge installed in the rear housing of an alternator.

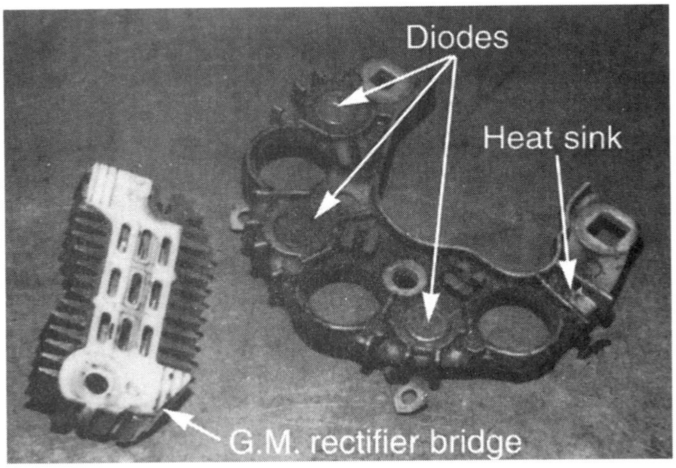

FIGURE 16–28 The positively biased diodes are mounted into a heat sink for heat dissipation.

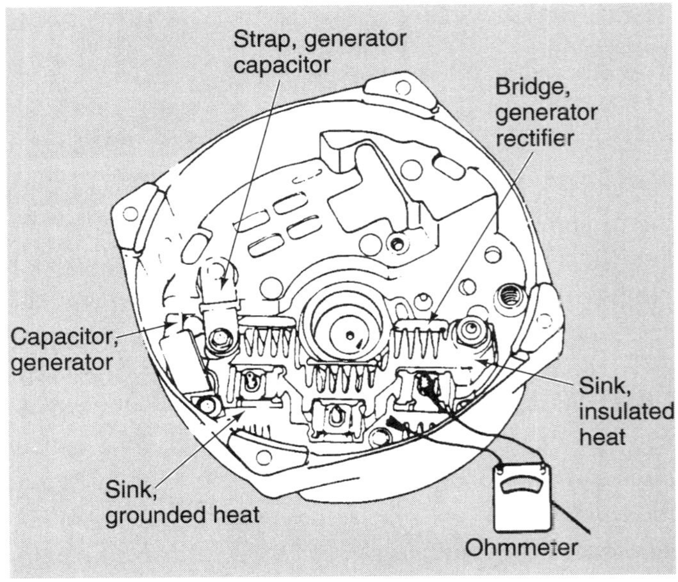

FIGURE 16–30 Testing rectifier bridge diodes with an ohmmeter.

(Courtesy of General Motors Corporation, Service Technology Group)

Full-wave rectification of a wye wound stator requires a six-diode rectifier bridge, which redirects the current from the stator windings so that all the current flows in one direction through the battery (Figure 16–31). When the rotor turns, two stator windings will be aligned in series to supply voltage, and the third winding is always connected to diodes that are reversed biased. There are six possible paths for current to follow depending on the position of the rotor. In Figure 16–32, current is supplied to the positive battery terminal from stator terminal *A* through diodes two and three, which are forward biased, and then back to stator terminal *B* to complete the circuit. Stator winding *C* does not produce

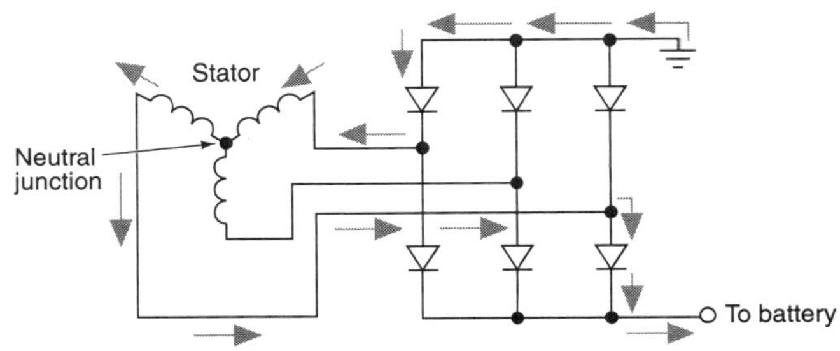

FIGURE 16–31 Current flow through a wye would stator and diode assembly.

FIGURE 16–29 The negatively biased diodes are pressed into the alternator housing for heat dissipation.

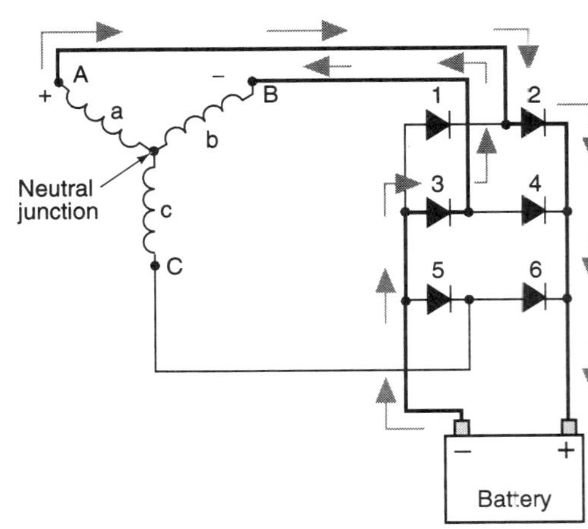

FIGURE 16–32 Current flow through circuit when stator terminals A and B are lined up.

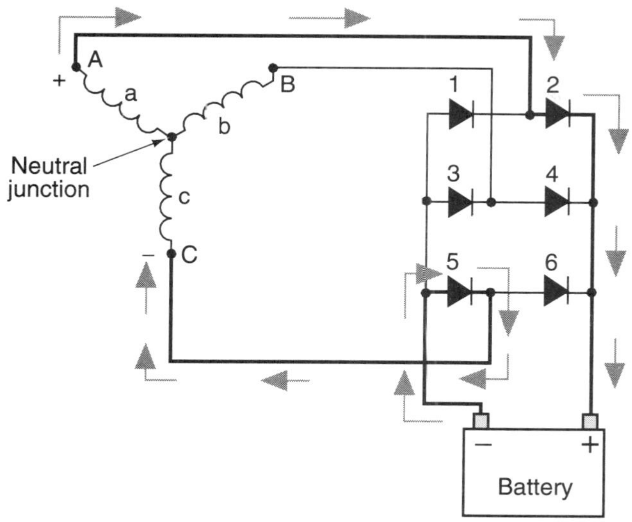

FIGURE 16–33 Current glow through circuit when stator terminals A and C are lined up.

current because it is connected to diodes that are reversed biased.

This process continues through the other five possible current paths before it repeats itself (Figure 16–33 to Figure 16–37). Notice that the current flow through the battery is always in the *same* direction.

Full-wave rectification in a delta wound stator is accomplished with the same diode rectifier bridge as in a wye wound stator, except all the stator outputs are in use at the same time (Figure 16–38). The three outputs of the delta stator are in parallel, so current flows from each winding all the time—this increases the alternator

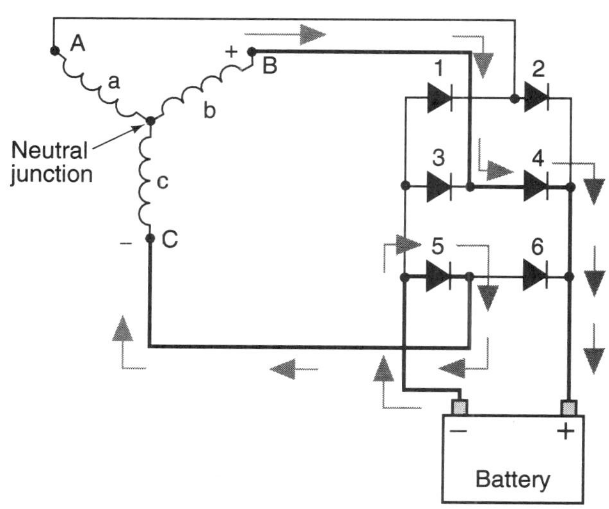

FIGURE 16–34 Current flow through circuit when stator terminals B and C are lined up.

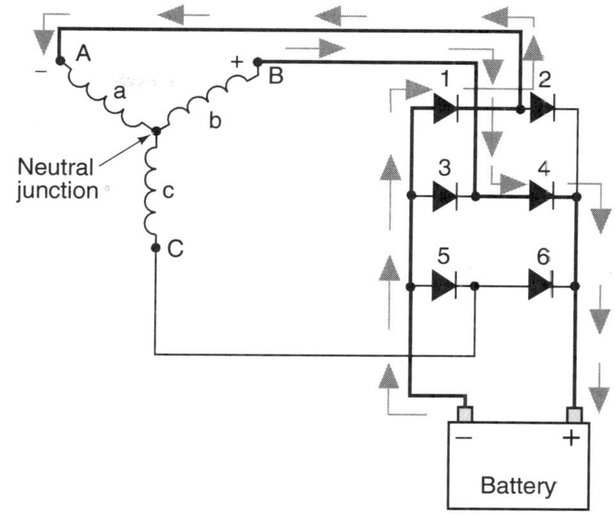

FIGURE 16–35 Current flow through circuit when stator terminals B and A are lined up.

output current potential compared to the wye wound alternator.

Some manufacturers of AC generators use a **diode trio,** which rectifies the current from the stator in order to create the needed magnetic field excitation in the rotor field coil (Figure 16–39). In order to excite the field coil, the voltage on the anode side of the diode trio (Figure 16–40) must be at least 0.6 volts more positive than the cathode side. When the ignition switch is turned on, the current load of the warning lamp acts as a small magnetizing current because it pre-excites the field coil

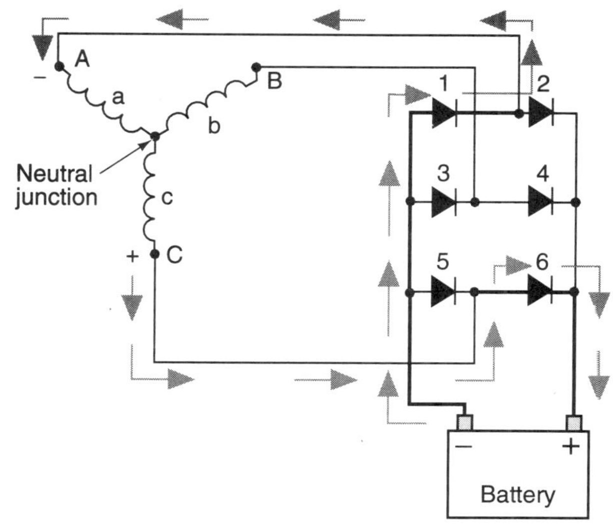

FIGURE 16–36 Current flow through circuit when stator terminals C and A are lined up.

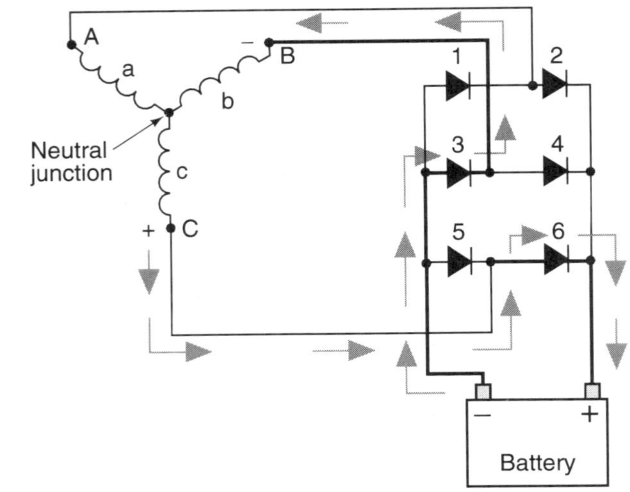

FIGURE 16–37 Current flow through circuit when stator terminals C and B are lined up.

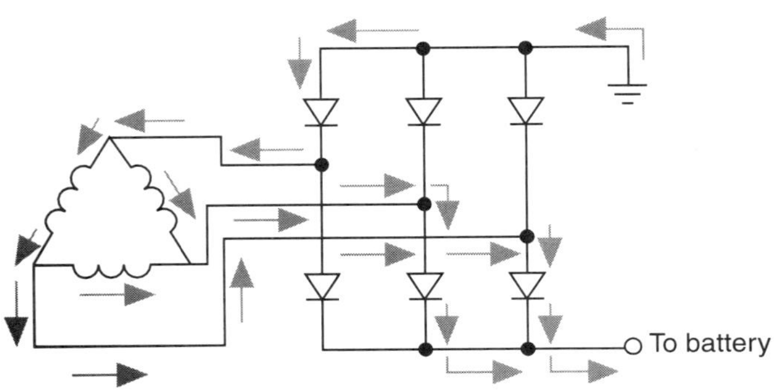

FIGURE 16–38 Current flow through a delta wound stator and diode assembly.

until the rotor turns (Figure 16–41). Each stator output is connected to a diode in the diode trio, which rectifies the current entering the voltage regulator, which in turn controls the strength of the magnetic field in the rotor.

Testing a diode trio for opens and shorts is similar to testing a diode rectifier bridge. Disconnect all wires to the diode trio and connect one lead of an analog ohm-

FIGURE 16–39 Diode trio assembly.

meter to the single large connector, and the other lead to each of the small diode connectors (Figure 16–42). Record the ohmmeter reading while connected to each small lead before repeating with the ohmmeter leads reversed. The ohmmeter should read less than 300 Ω in one direction and above 300 Ω in the other direction. A diode trio that is not within specifications should be replaced. A digital multimeter (DMM) with a *diode check* function may also be used for this test.

Charging System Regulation

A vehicle's electrical system and battery must be protected from unstable or excessive charging system voltage. Computer control circuits are especially sensitive to voltage fluctuations that may develop as the alternator adjusts for varying electrical load demands while at the same time keeping the battery fully charged. If the charging current is lower than the vehicle's electrical demands, the battery will gradually run down over time. If the charging current is higher than needed, the battery or electrical system components could be damaged.

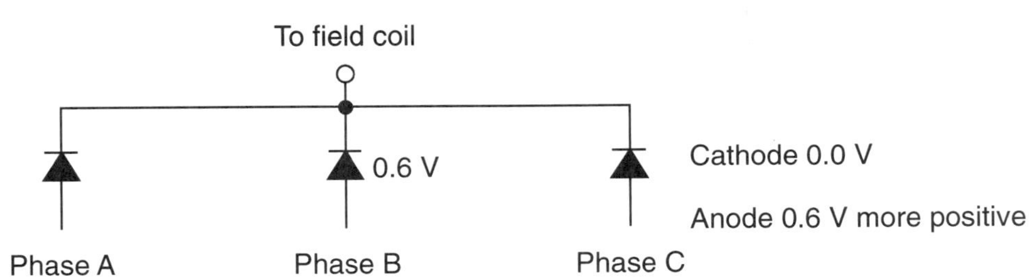

FIGURE 16–40 A typical diode trio connects the stator phase windings to the field coil to create field excitation.

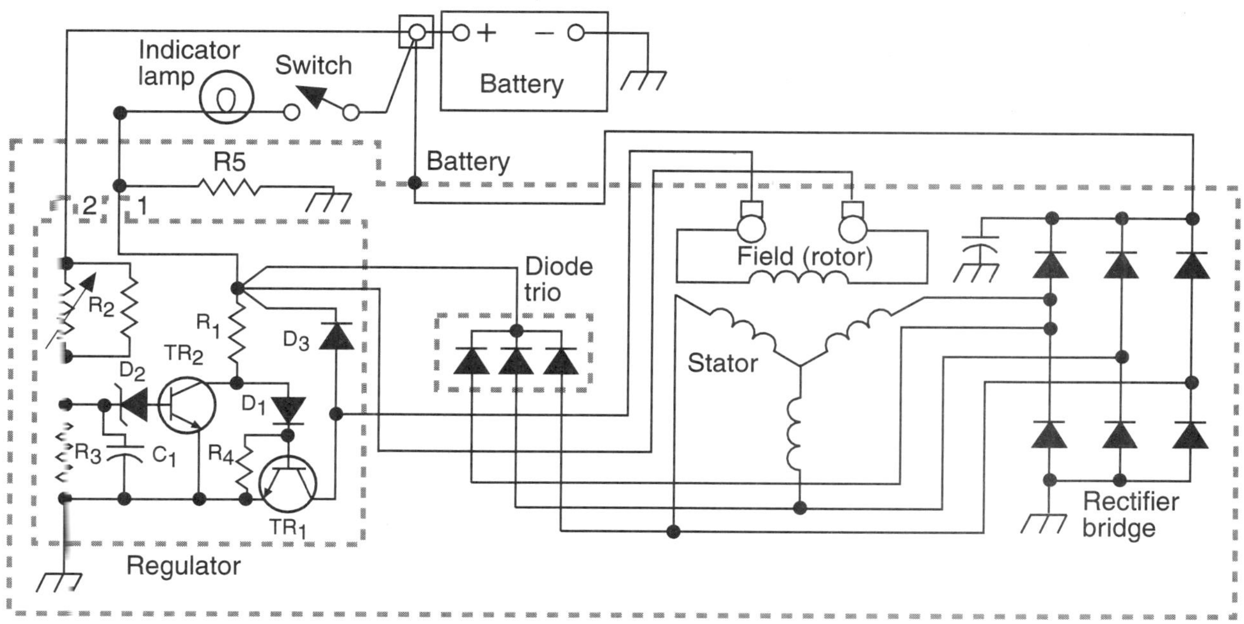

FIGURE 16–41 Schematic of a charging system with a diode trio.

The **voltage regulator** controls and stabilizes the charging system voltage by (A) controlling the resistance in series to the rotor field coil or (B) toggling the field circuit very rapidly between open and closed (Figure 16–43). The amount of regulated **field current** going to the rotor coil varies depending on system voltage needs. Voltage regulators use a **sensing voltage** input from the battery, interpret the voltage, and then regulate the alternator output voltage between 13.5 and 14.5 volts.

In addition to the sensing voltage input, another input that affects regulation is temperature. Regulators are temperature compensated because ambient temperature influences the rate of charge that a battery can accept. Figure 16–44 shows the relationship between temperature and charge rate. Batteries are more reluctant to accept a charge at lower ambient temperatures, so the regulator compensates for this by increasing the system voltage until it is at a high enough level that the battery will accept a charge quickly.

There are three basic types of field circuits used in charging systems. Knowledge of each type will help you in diagnosing basic charging system problems.

1. The first type of field circuit is called the **A-circuit** which has the regulator on the ground side of the

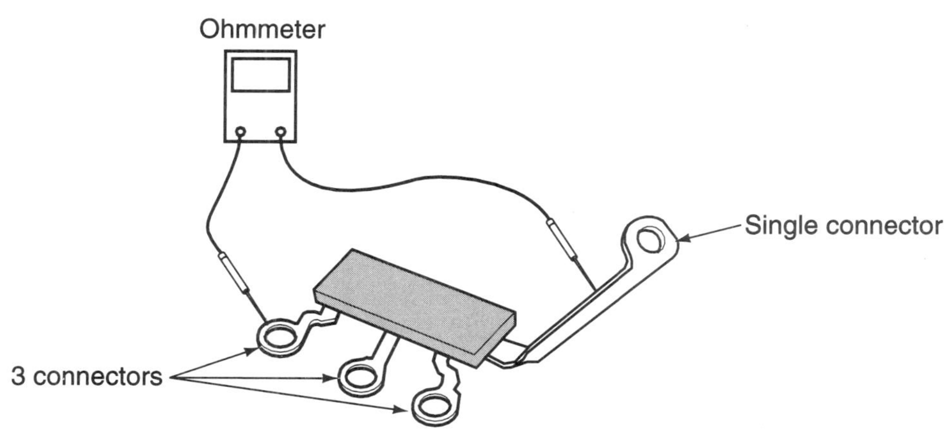

FIGURE 16–42 Testing a diode trio for opens and shorts with a suitable ohmmeter.

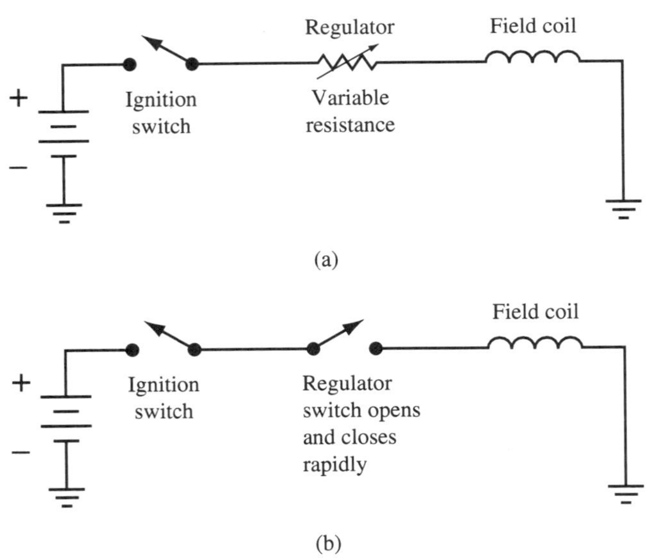

FIGURE 16–43 The regulator controls the rotor field by (A) controlling the resistance in series to the rotor coil, or (B) field circuit it toggled on and off very rapidly.

field coil (Figure 16–45). It is also referred to as an *externally grounded field circuit.* The B+ used for the field coil is picked up from within the AC generator assembly. The regulator is located in series on the ground side and controls the field current by varying the current flow to ground.

2. The second type of field circuit is called the **B-circuit** which has the regulator controlling the power input side of the field circuit (Figure 16–46). It is also referred to as an internally grounded circuit. The field coil completes the circuit by being grounded within the AC generator assembly.

3. The third type of field circuit is called an **isolated field circuit** because it picks up both B+ and ground externally. It can be easily identified by the *two* field wires attached to the outside of the alternator case. Isolated field circuit alternators can have the regulator located on A-circuit (ground) side or the B-circuit (positive) side (Figure 16–47).

	Volts	
Temperature	Minimum	Maximum
20° F	14.3	15.3
80° F	13.8	14.4
140° F	13.3	14.0
Over 140° F	Less than 13.3	–

FIGURE 16–44 Relationship between ambient temperature and charging rate.

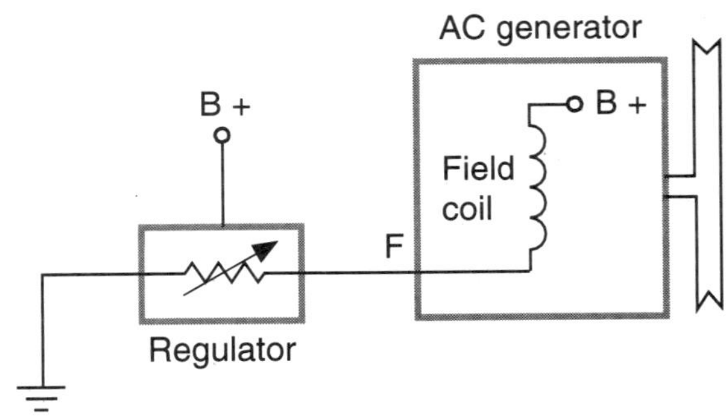

FIGURE 16–45 A simplified A-circuit style field circuit.

Regardless of the type of field circuit the alternator has, it is designed to control the amount of current flowing through the field windings. There is a relationship between the field current, rotor speed, and the regulated voltage (Figure 16–48). As the rotor speed increases, the field current must be reduced to maintain a regulated voltage.

Electromechanical Regulators On most older vehicle charging systems an external electromechanical regulator with two vibrating electromagneticly operated contact point relays are used to control the charging circuit (Figure 16–49). One relay coil is the field relay with two contact points and the other is the voltage regulator with three contact points (Figure 16–50).

Looking at Figure 16–50, when the ignition switch is closed current flows through the resistor illuminating the indicator lamp, then through the lower contacts shown in the regulator section, out the F terminal, through the

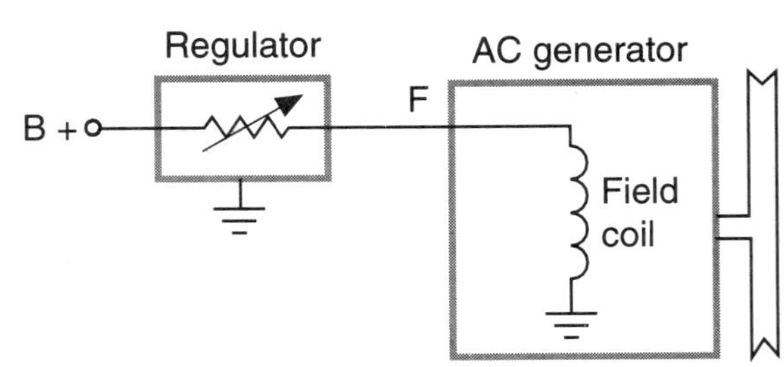

FIGURE 16–46 A simplified B-circuit style field circuit.

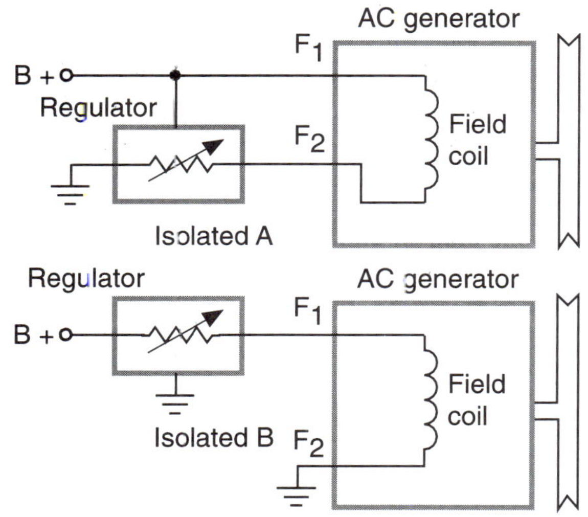

FIGURE 16–47 In an isolated field circuit for an AC generator, the regulator can be installed on either side of the rotor field coil.

alternator field coil, and then to ground. When the engine has started and the alternator rotor is turning, system voltage may be below 13.5 volts—however, some voltage is produced at the S terminal which energizes the field relay. When the field relay is energized, the field relay points close magnetically, and the indicator bulb goes out because battery voltage is now present on both sides of the light circuit.

At the same time, the voltage regulator relay—also known as the **voltage limiter** relay—automatically determines what the field circuit voltage will be for the required amount of charging. When battery voltage is below 13.5 volts, the voltage limiter relay coil does not produce enough electromagnetism to pull the voltage

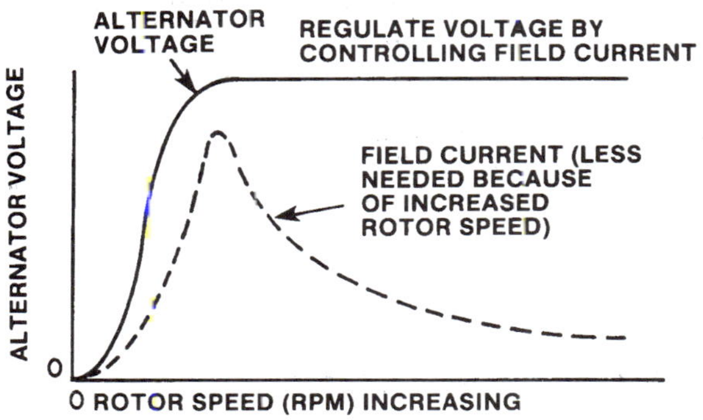

FIGURE 16–48 The relationship between the field current, rotor speed, and regulated voltage changes depending on the vehicle electrical loads.

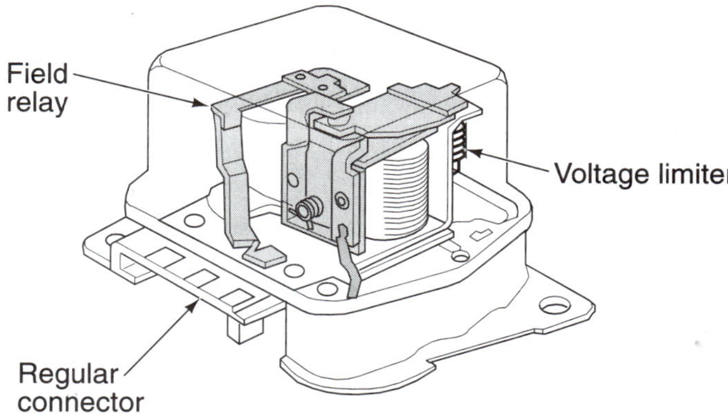

FIGURE 16–49 An external electromechanical contact point regulator.

(Courtesy of Ford Motor Company)

limiter relay contact points from closed to open. Maximum charging conditions are allowed when the voltage limiter relay points are closed. As the battery charges and its voltage rises, the increased voltage strengthens the electromagnetic field of the voltage limiter relay coil and causes the voltage limiter contact points to open, which grounds the field circuit. The contact points constantly vibrate open and closed several times per second to maintain the system voltage output between 13.5 and 14.5 volts.

After the engine shuts off, there is no current flow from the STA terminal so the field relay coil is de-energized. This opens the points and prevents the battery from discharging.

Electronic Regulators Modern vehicle charging systems use an electronic regulator to provide precise control of the field current. All of the control circuitry and components are located on a single silicon chip sealed in a plastic module normally located inside the alternator (Figure 16–51). There are no moving parts in the regulator which enables it to cycle several thousand time per second, if necessary, for accurate control of the field circuit voltage (Figure 16–52).

The on and off cycling of the field circuit is an example of **pulse width modulation.** The cycling modulation controls the alternator output by varying the amount of time the field coil is energized based on the vehicle's electrical demand. For example, if a 100 amp alternator on a vehicle had an electrical system demand of 50 amps, the regulator would energize the field coil fifty percent of the cycle time on, and off fifty percent of

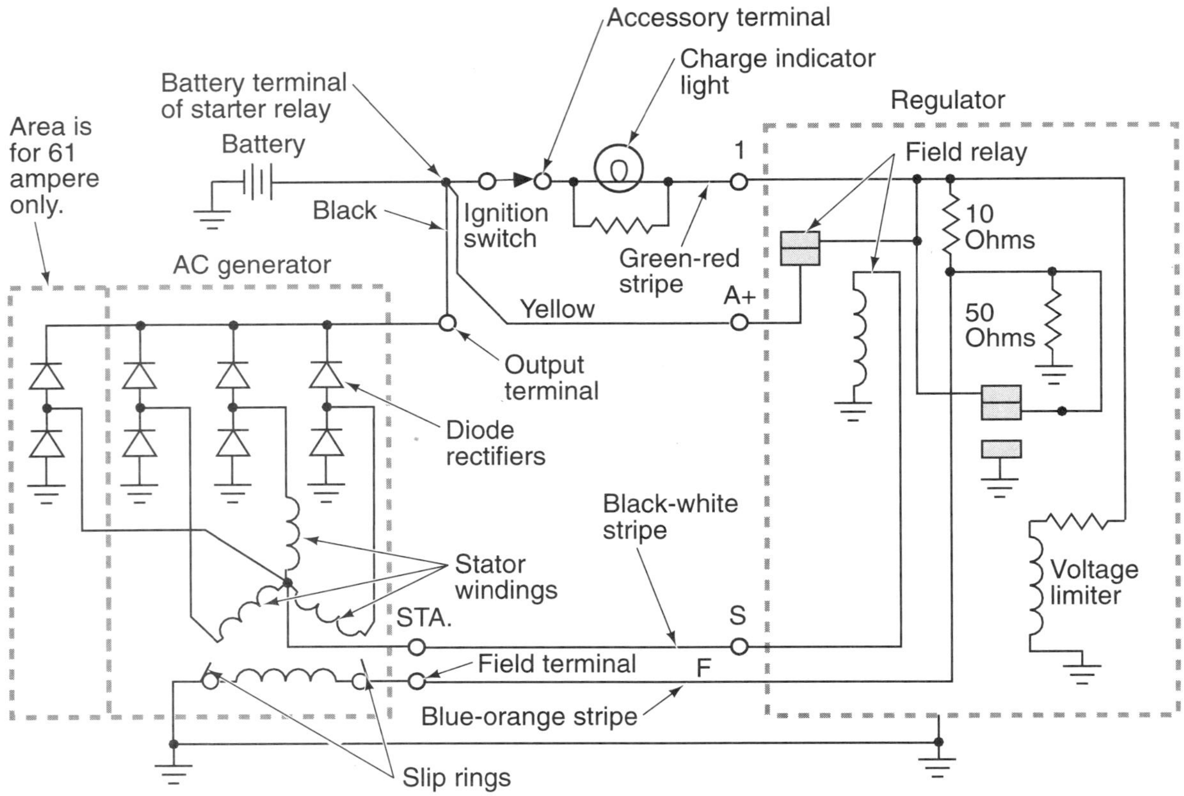

FIGURE 16–50 Complete charging system circuit with electromechanical regulator. (Courtesy of Ford Motor Company)

the cycle time (Figure 16–53). If the amperage demands on the vehicle's electrical system increased to 75 amps, the regulator would energize the field coil seventy five percent of the cycle time. The period of time for each cycle does not change, only the amount of *on* time in each cycle changes. The percent of on time relative to the total cycle time is known as **duty cycle.**

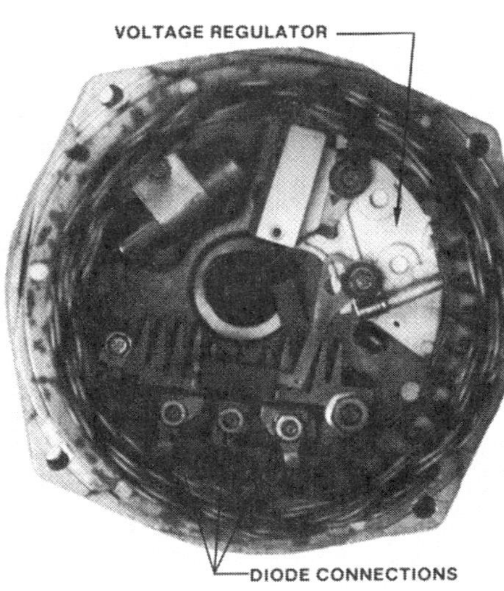

FIGURE 16–51 Internally mounted voltage regulator.

In an electronic regulator, one of the main components is a zener diode (Figure 16–54), which is similar to the diodes in other electronic circuits except it has the ability to conduct current in the reverse direction without damage. The zener diode in an electronic regulator is designed to conduct in reverse once a predetermined voltage of 13.5-14.5 volts has been achieved. In Figure 16–55, the zener diode has B+ on the anode side—side with the arrow—and no current flowing through the diode. When the upper design voltage limit (approximately 14.5 volts) is reached, the zener diode "breaks down" which then allows a voltage signal to occur at the base of transistor #1. This turns on transistor #1 and switches off transistor #2 which is in control of the field current and alternator output. Transistor #2 will modulate the field circuit on and off as many time per second as necessary to maintain the designed voltage of the system. The thermistor shown in the upper left of Figure 16–55 changes circuit resistance according to tempera-

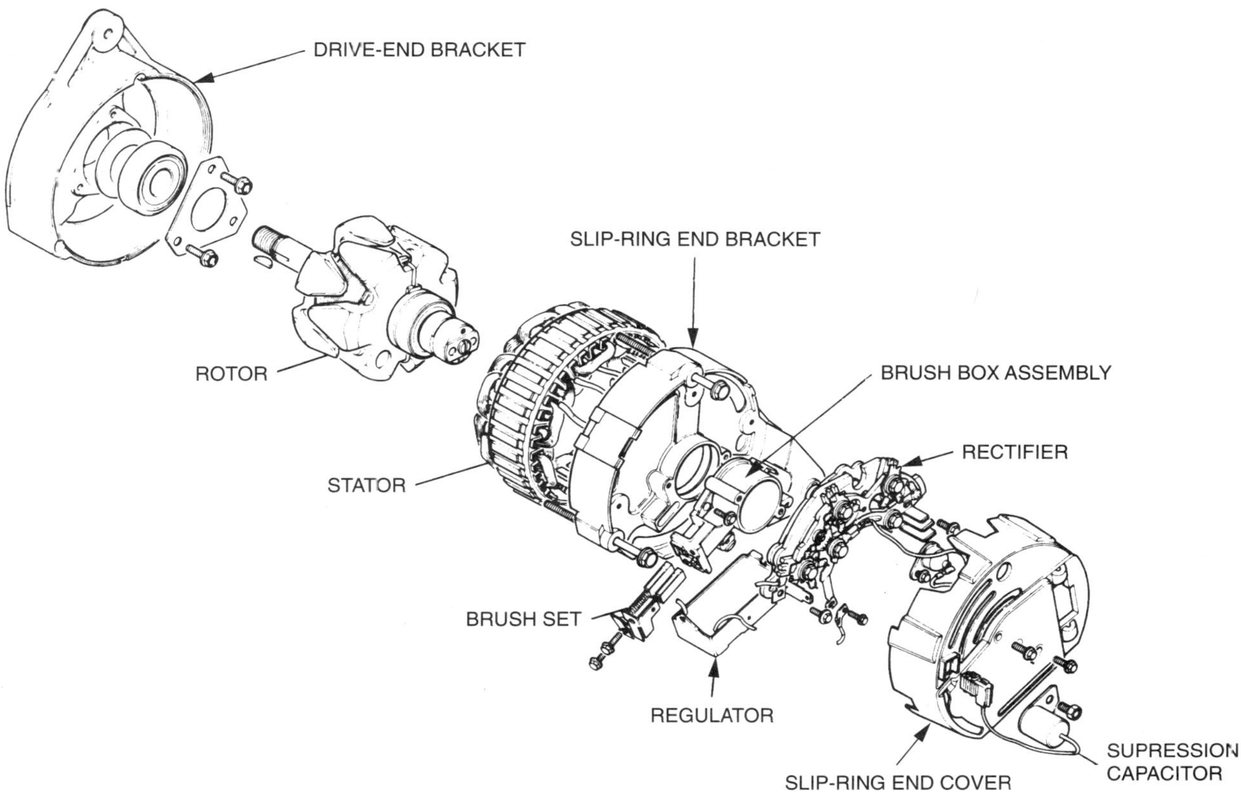

FIGURE 16–52 Component locations of an alternator with an internally mounted voltage regulator.
(Courtesy of Lucas Aftermarket Operations)

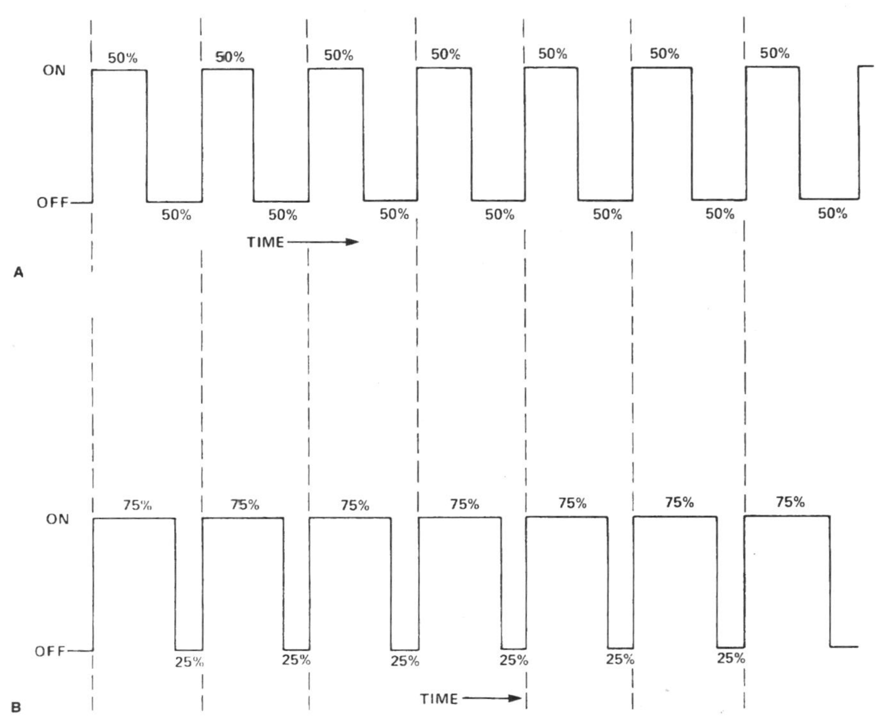

FIGURE 16–53 Examples of pulse width modulated field current.

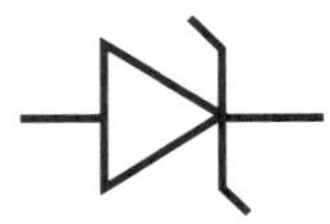

FIGURE 16–54 Zener diode symbol.

ture changes, thus providing for the temperature-related voltage change necessary to keep the battery charged in cold weather conditions.

An application of a zener diode in the electronic regulator of a charging system is shown in Figure 16–56. In this particular system there are only two input connections, which are located in the upper left of Figure 16–56. Input #1 receives B+ that comes through the ignition switch, and through the indicator lamp and resistor that are in parallel. Input #2 is for connecting the sensing circuit in the regulator to the battery.

When the ignition switch is turned on, B+ is available at input #1, and at the field circuit through the common battery connection at R1. Transistor TR1 is on and conducts the field current from the field coil to ground, producing a weak magnetic field in the field coil. When the alternator begins to produce a current output, the diode trio conducts and B+ is available for the field and for terminal #1 at the common battery connection at R1. This turns off the indicator lamp because there is B+ on both sides of the lamp.

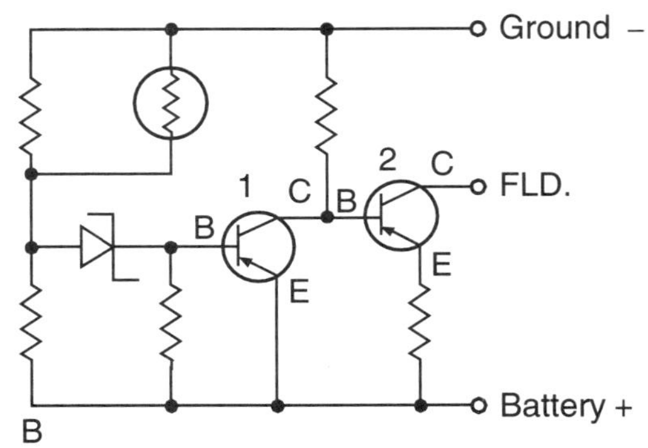

FIGURE 16–55 Zener diode and a simple transistor regulator circuit.

As the alternator output reaches its designed charging voltage, sufficient voltage at the sensing circuit that is connected to input #2 is applied to the zener diode so it will "break down" and turn on transistor TR2 which turns off transistor TR1. This process continuously turns the switching transistor TR1 on and off as the voltage rises above and below the designed voltage.

Computer Controlled Regulators On many newer vehicles the regulator function has been incorporated into the (PCM) powertrain control module (Figure 16–57). Operation of the regulator is similar to the internal electronic voltage regulator, except a logic module controls the duty cycling of the switching transistor. This

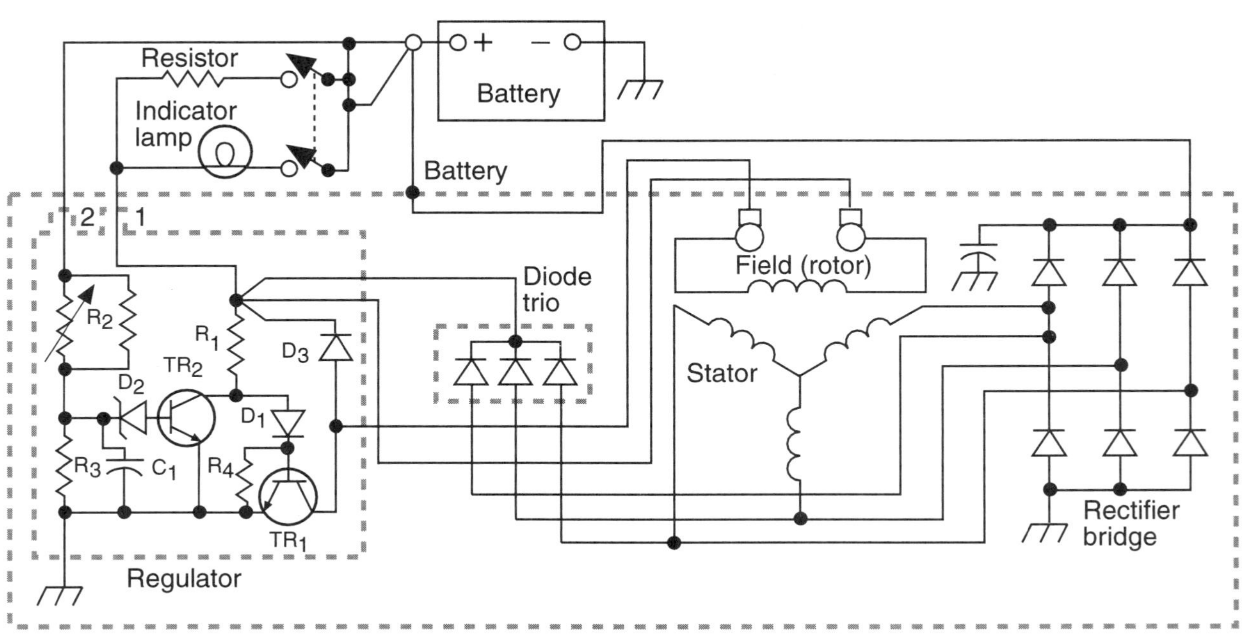

FIGURE 16–56 Charging system schematic with an internal regulator and an indicator lamp.

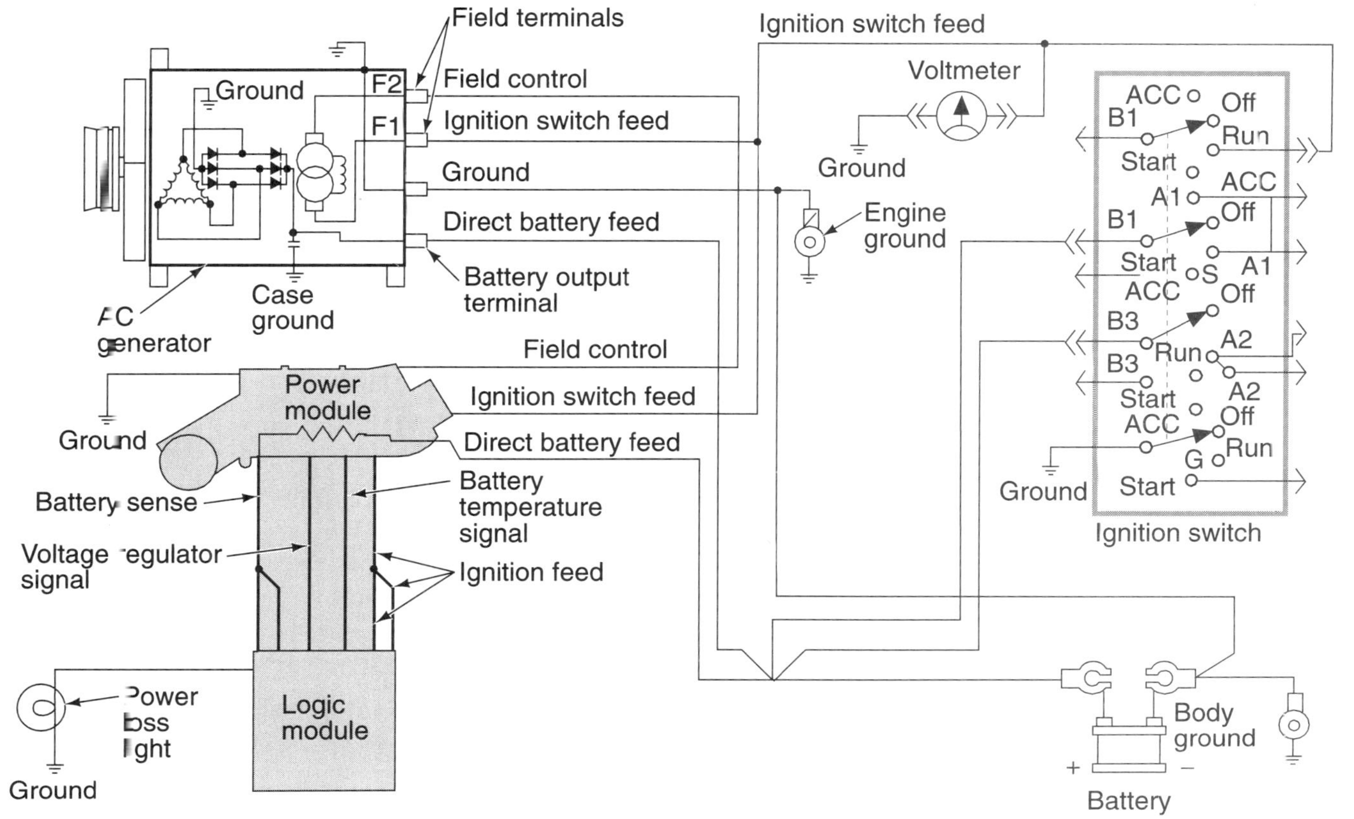

FIGURE 16–57 Computer-controlled voltage regulator circuit.
(Courtesy of DaimlerChrysler Corporation)

type of system does not control rotor field current by modulating the transistor circuit. Instead, the computer pulses the field current on and off at a fixed frequency of about 400 cycles per second. By pulse width modulating the duty cycle, an average field current is produced to provide designed alternator output. At high engine speeds with minimum system load requirements, the field circuit on time may be as little as ten percent (Figure 16–58). At low engine speeds with maximum system load requirements, the field circuit on time may be seventy-five percent or more.

A computer controlled regulator has the ability to precisely manage the charging rate according to vehicle electrical requirements, ambient temperature, and vari-

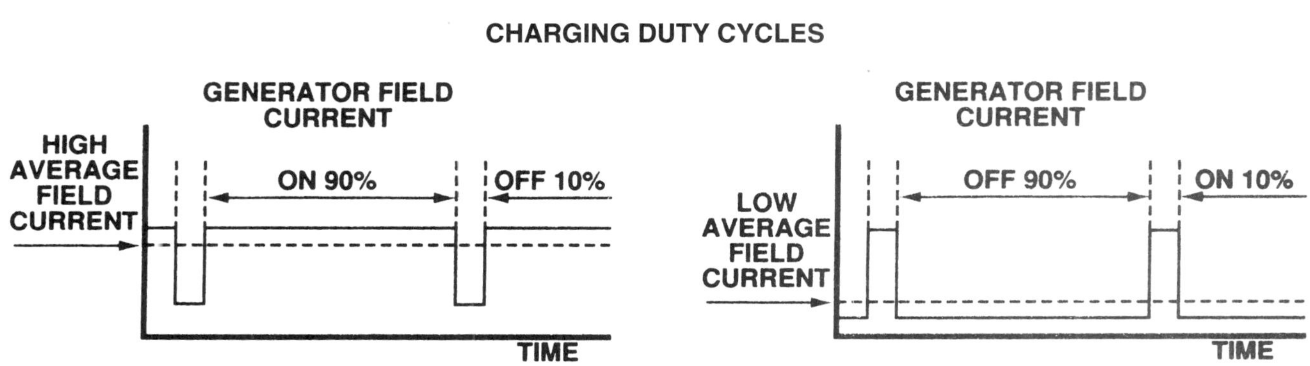

FIGURE 16–58 On/off duty cycling of the alternator.
(Courtesy of General Motors Corporation, Service Technology Group)

ous sensor inputs (Figure 16–59). As a result, the following list are some of the benefits of computer regulation.

- Small lightweight storage battery.
- Reduced magnetic drag on alternator, resulting in increased engine horsepower.
- Increased gas mileage.
- Reduced rough idle speed problems caused by parasitic voltage loss.
- Allows the computer's diagnostic capabilities to be used in troubleshooting charging system problems.

Charging System Indicators There are three common methods used in charging systems to inform the driver of the charging system's condition or of a potential problem.

1. The first method is an indicator lamp that operates, in most circuits, on the basis of opposing voltages (Figure 16–60). If the alternator output is less than battery voltage, there is an electrical potential difference in the lamp circuit and the lamp will light. If the output voltage is equal to the battery voltage, then the equal voltage potential will result in the lamp not lighting (Figure 16–61). Figure 16–62 is an example of an electronic regulator that uses an indicator lamp to warn of a charging system problem. If there is no stator output through the diode trio, then the indicator lamp circuit is completed to ground through the rotor field and transistor TR1.

2. The second method is an ammeter wired in series between the alternator and battery (Figure 16–63). The ammeter is normally mounted in the dash and gives the driver more information compared to an indicator lamp (Figure 16–64). If the charging system is not generating sufficient current to keep up with vehicle electrical load demands, or the charging system is inoperative, the needle points to dis-

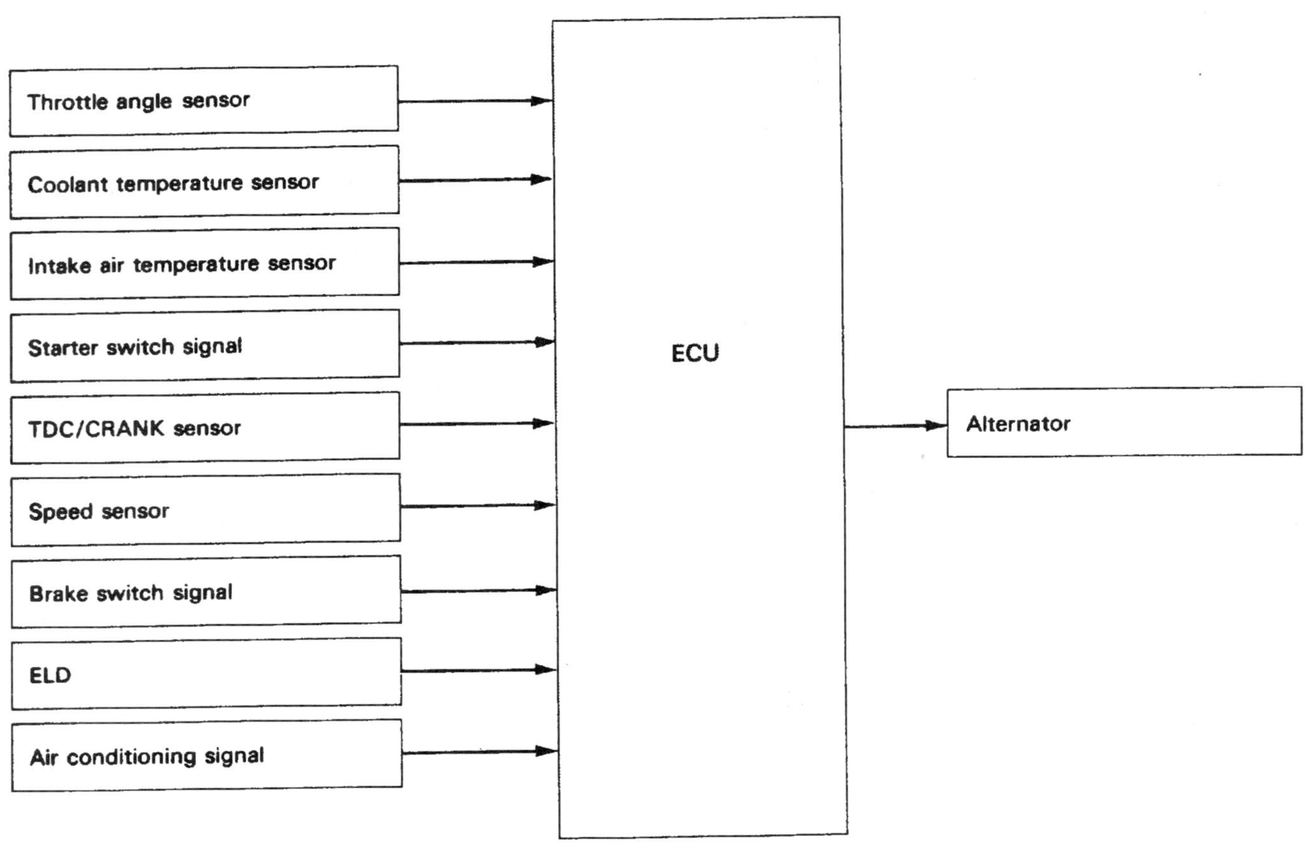

FIGURE 16–59 Alternator regulation through the on-board computer and various input sensors.

(Courtesy of American Honda Motor Company, Inc.)

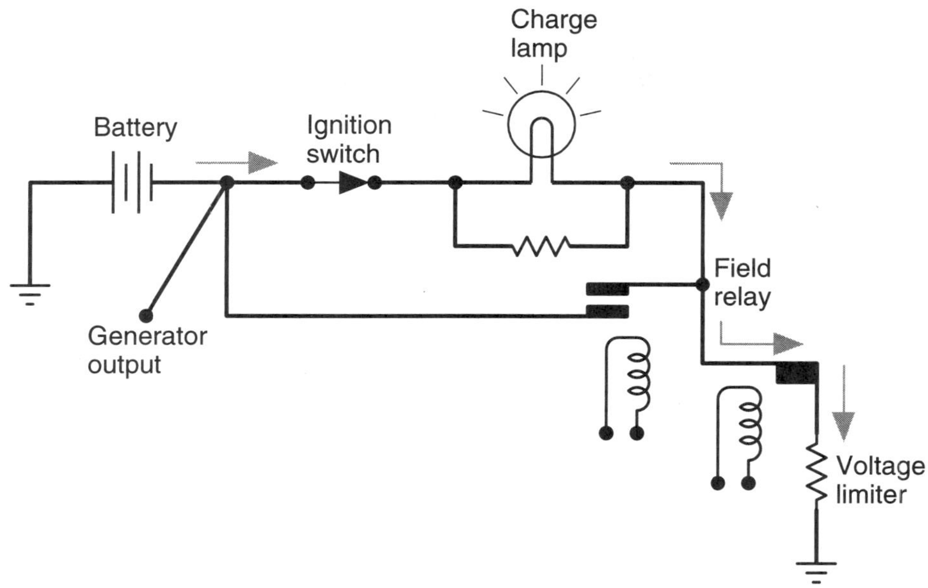

FIGURE 16–60 Electromechanical regulator indicator lamp is on when there is no alternator output.

charge. After the engine initially starts or the electrical load demand is high, the gauge needle will indicate a high rate of charge until designed system voltage is maintained.

3. The third method is a voltmeter connected in the battery electrical feed circuit to monitor system operation (Figure 16–65). With the ignition switch on and engine off, the voltmeter should show the voltage available to the electrical circuit from the battery—normally around 12.66 volts. When the engine is started, it is normal for the voltmeter to indicate a reading between 13.2 volts and 15.2 volts, depending on the vehicle. If the voltage level is below 13.2 volts, the charging system is not providing sufficient current for the electrical load demands. If the voltmeter indicates a voltage reading

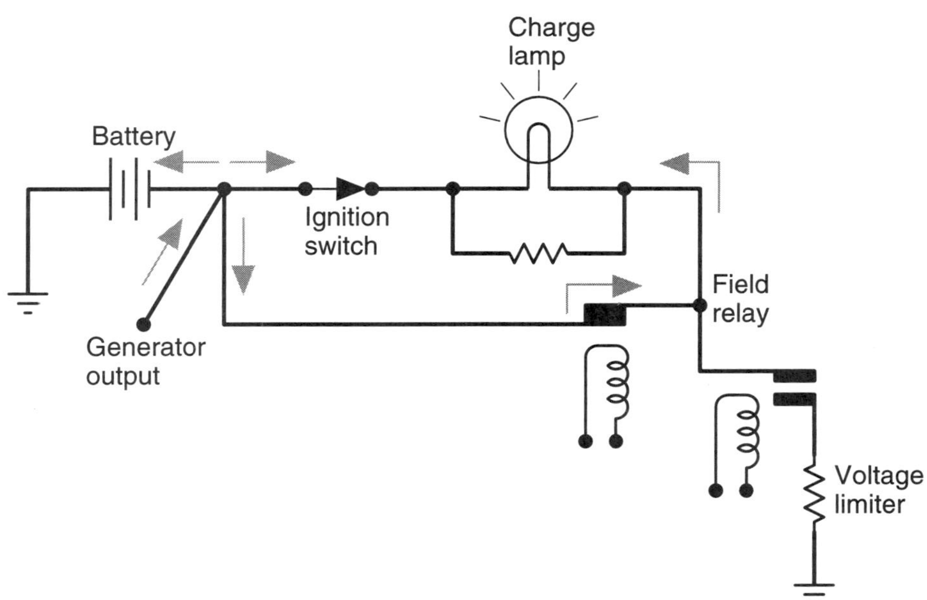

FIGURE 16–61 Electromechanical regulator indicator lamp is off when there is alternator output.

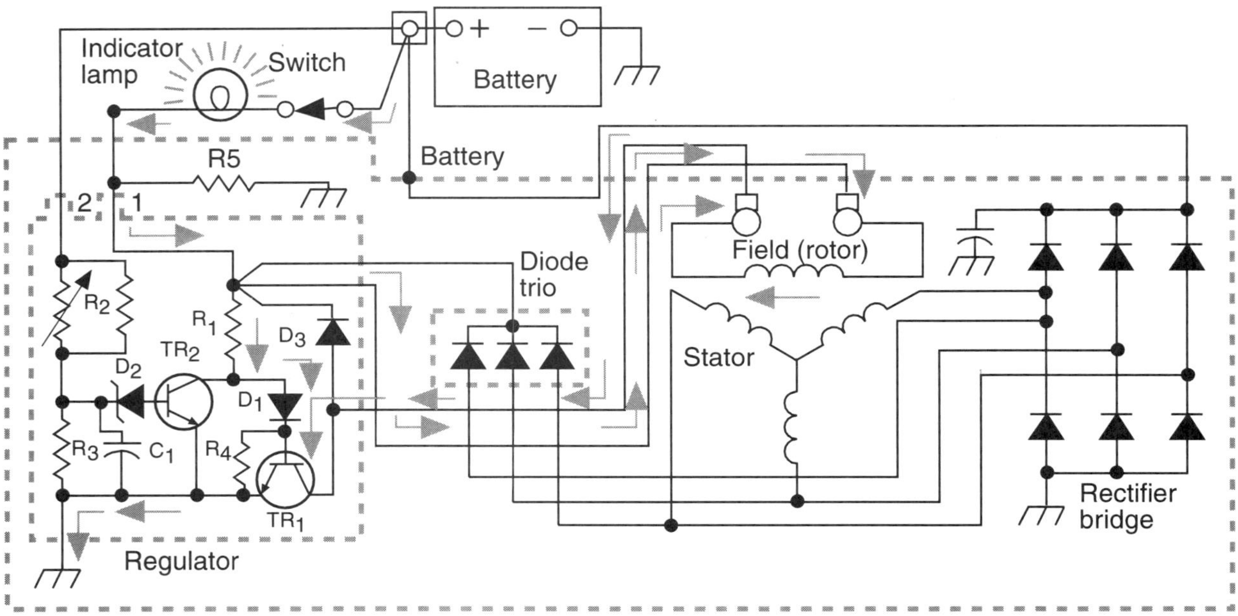

FIGURE 16–62 Example of an electronic regulator circuit with an indicator lamp on when there is no alternator output.

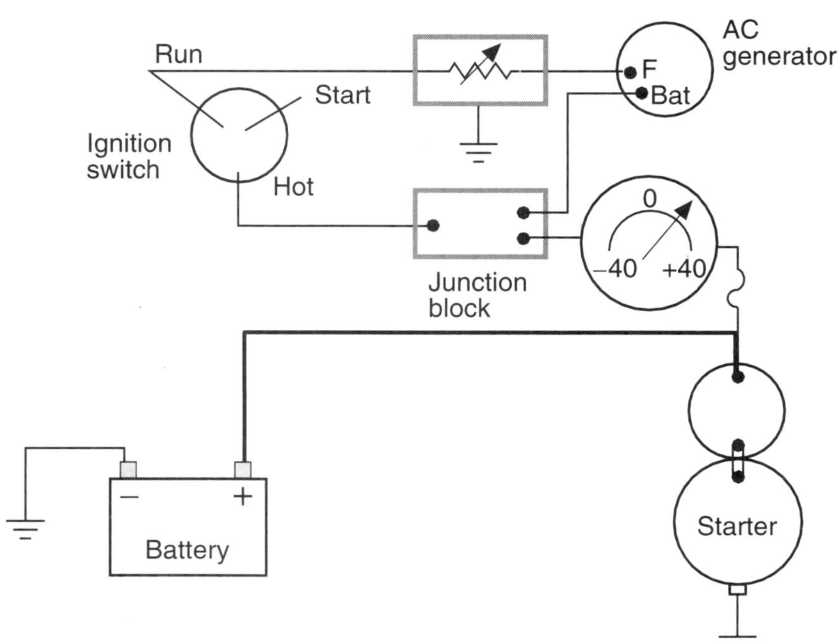

FIGURE 16–63 Ammeter connected in series to indicate charging system operation.

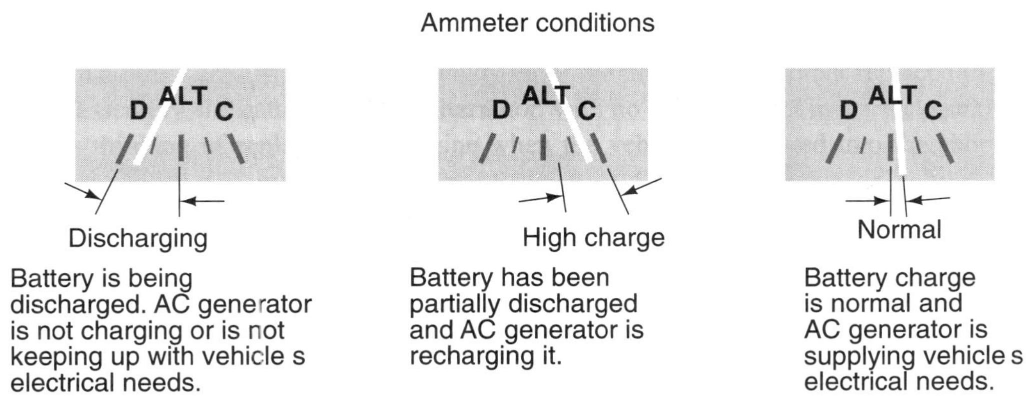

FIGURE 16–64 Dash mounted ammeter needle movement indicates charging system conditions. (Courtesy of Ford Motor Company)

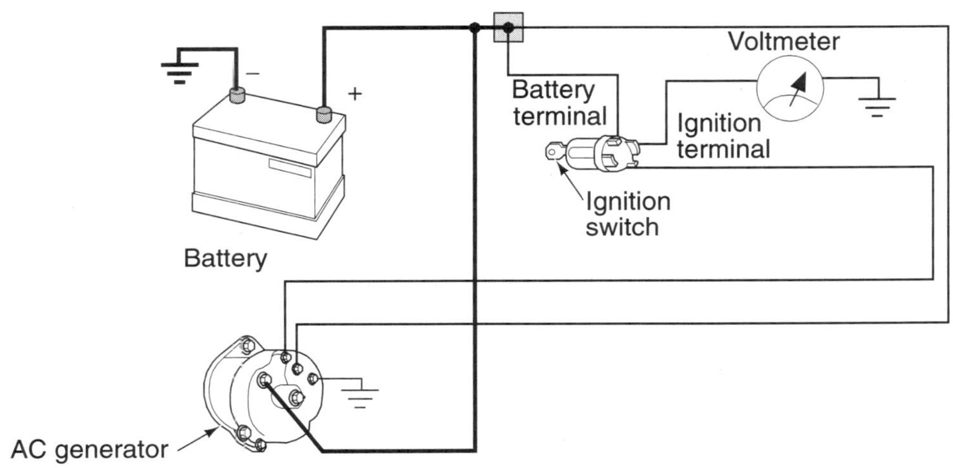

FIGURE 16–65 Voltmeter connected in parallel to the charging circuit to monitor operation.

above 15.2–15.5 volts, depending on manufacturer, there is an overcharge condition occurring that will damage the battery and sensitive electrical circuit components.

Basic Charging System Testing

Reliable testing, diagnosis, and repair of a charging system problem is determined by several factors—type of charging system, knowledge of the circuit being checked, service manual information available, detailed preliminary checks of the system, type of charging system test equipment available, actual charging system tests, and availability of repair parts.

CAUTION *Due to the complexity of modern charging systems, always refer to specific charging system diagnostic procedures to verify possible problems. Do not assume that general diagnostic tests that work for one charging system will work for a different system.*

Preliminary Charging System Checks

The following preliminary check list should be followed before attempting to diagnose charging system problems.

- Battery condition. The charging system is designed to maintain the charge in a battery. It is not designed to charge a dead battery. The state of charge and condition of the battery must be determined and corrected if necessary before performing charging system tests.

- Condition of alternator drive belt(s). A drive belt(s) that is loose, worn, glazed, damaged, or oil soaked (Figure 16–66) may cause the alternator belt to squeal when under a heavy charging condition. If the drive belt condition checks out satisfactory, the belt tension should be checked with a tension gauge and adjusted to manufacturers specifications (Figure 16–67).
- Condition of electrical cables and terminal connections. Loose, corroded, or disconnected electrical connections to the alternator, regulator, or battery could damage the charging system. The battery acts as a buffer to stabilize voltage spike or surge conditions in the electrical system. Loose or corroded connections could give the charging system regulator a false reading on the battery state of charge and cause an overcharge or undercharge condition.

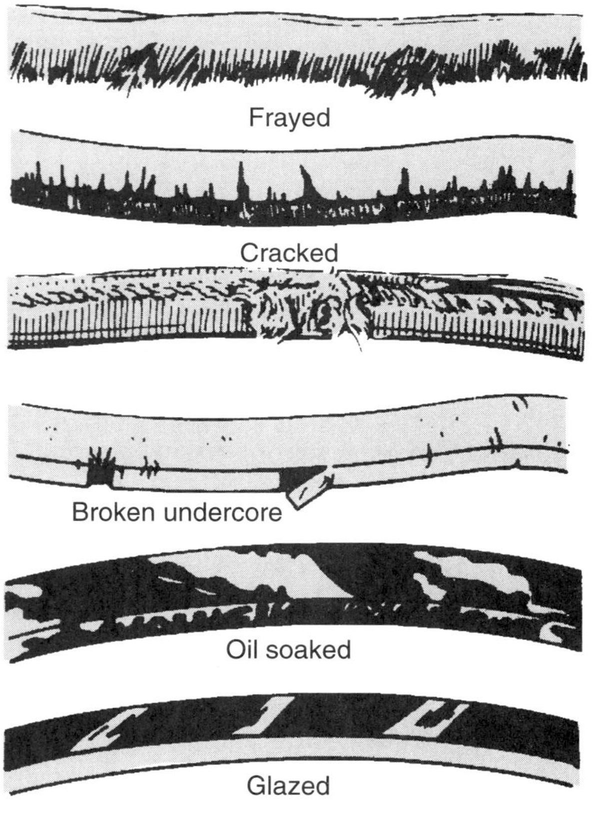

FIGURE 16–66 Drive belt conditions that need replacing.

FIGURE 16–67 Checking belt tension with a belt tension gauge.

On-Car Charging System Tests

Vehicle manufacturers are very specific about the procedures used for performing charging system tests. Refer to service manual information to identify circuit component locations, wiring color codes, component specifications, and *actual* test procedures.

There are four basic charging system checks used to determine the alternator and regulator condition—charging system voltage output test, charging system current output test, regulator by-pass test, and diode pattern test.

1. The first charging system check is a voltage output test. To perform this test, connect a voltmeter in parallel to the battery cable connections. Note the open circuit voltage (OCV) reading which should be around 12.66 volts. Start the engine and raise the rpm to around 1,500. With no electrical load, the charging system voltage should be approximately 2 volts higher than the OCV reading of 13.5 to 15.0 volts, depending on specific manufacturer. A reading of less than 13.0 volts indicates a charging system problem that will drain the battery over time. A reading of 16 volts, or more, indicates severe overcharge is occurring. If the unloaded charging system voltage is within specifications, test the voltage output under a load condition. Operate the engine rpm at 2,000–2,500 rpm, and turn on the headlamps, heater fan, rear window defogger, and other high current accessories. The charging system output under a heavy electrical load condition should be at least 13.2 volts or more.

2. The second charging system check is a current output test. Using a professional starting/charging/battery carbon pile tester is the preferred method for performing this test. There are several good testers. Always refer to the tester's operator's manual when performing any tests. Figure 16–68 is an example of the tester connections for performing most charging system checks. The charging system must be loaded with the carbon pile to obtain the alternator current output at 12 volts. Typical testing procedures are as follows.

   - Hook up charging system test as shown in Figure 16–68, or for specific tester you are using. Some testers specify the inductive pickup lead be attached around the alternator output lead to obtain an accurate reading.
   - Start the engine and operate at 1,500–2,000 rpm.
   - Load the system with the carbon pile until the highest ammeter reading is obtained without dropping the voltmeter reading below 12.6 volts. The maximum current produced under these conditions should not be lower than ten percent below alternator specified output plus the ignition current draw. For example, if the ignition system draw is 8 amps with the key on and engine off, and the alternator was rated at 100 amps, then 10% × 100 amps + 8 amps = 18 amps. Minimum alternator output should be no less than 82 amps, 100 amps − 18 amps. Output readings lower than specifications indicate further tests are needed.

3. The third charging system check is a regulator bypass test, also referred to as a full-field test. This test is performed by fully fielding the rotor field windings in the alternator using battery supply voltage with the regulator by-passed.

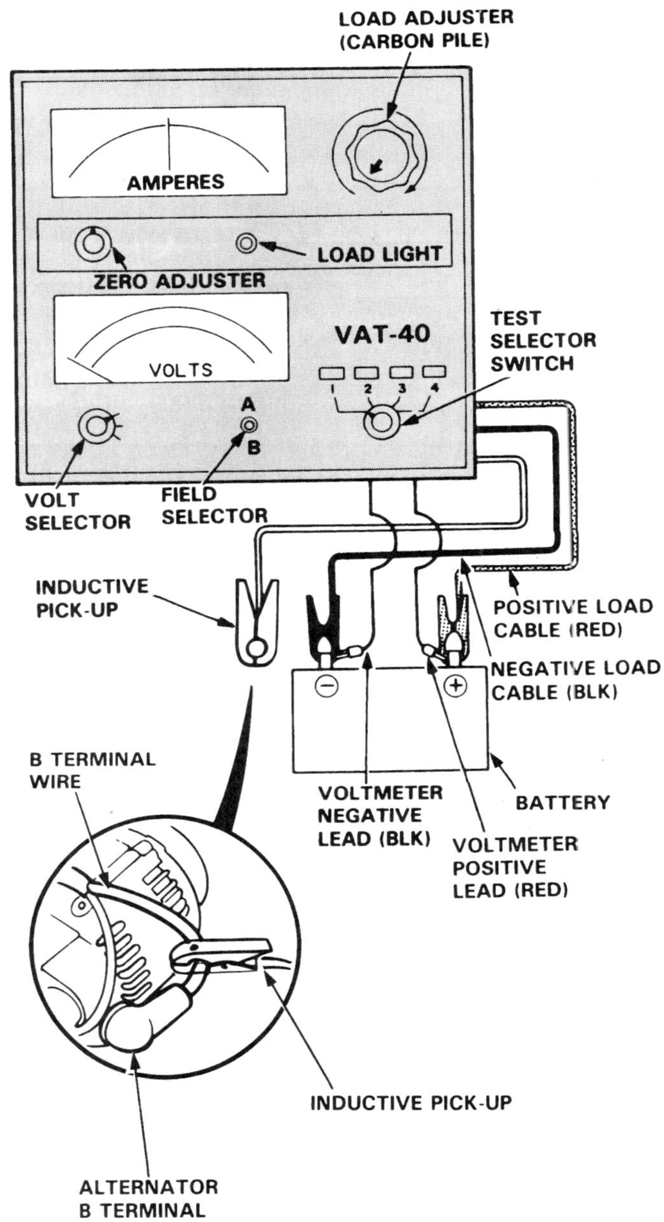

FIGURE 16–68 Typical tester connections for checking the charging system.

(Courtesy of American Honda Motor Company, Inc.)

CAUTION *Every vehicle manufacturer has specific procedures that should be followed when performing a full-field test. Not all alternators can be fully fielded. Check the manufacturer's procedures before attempting to perform this test.*

The full field test produces maximum alternator output at the available engine rpm. Do not allow the voltage output to exceed 16 volts in order to protect electrical components and circuits. Figure 16–69 is an example of a full field test on a General Motors 10SI alternator with an internal regulator. Inserting a small screwdriver about one half inch into the D-hole in the slip ring housing and grounding it to the alternator housing—for 10 seconds maximum—will fully field the system. If the current output is within specifications, the regulator is at fault.

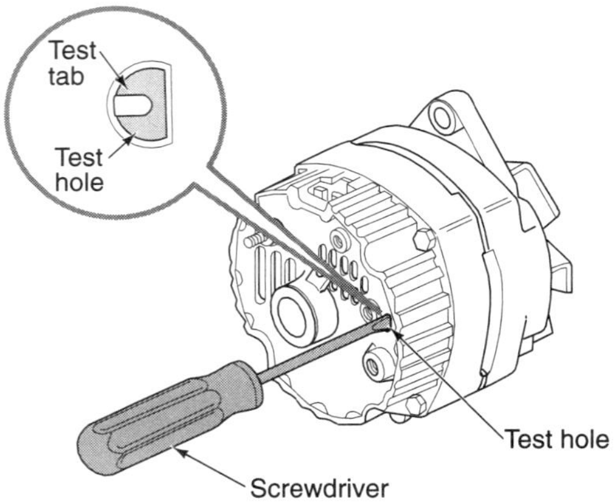

FIGURE 16–69 Fully field the GM1OSI alternator by grounding the test tab through the D-hole.

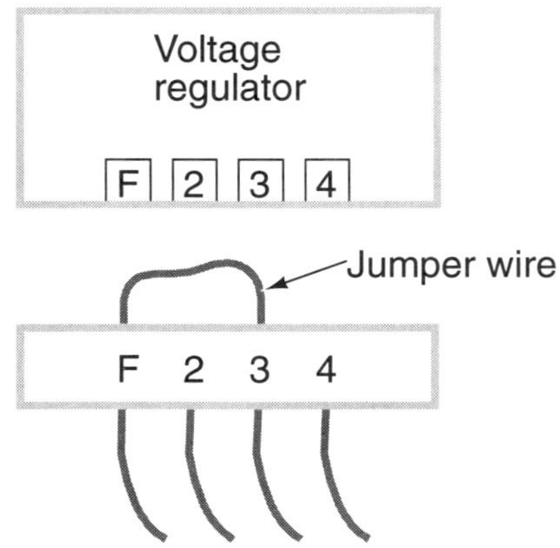

FIGURE 16–70 By-pass the voltage regulator to fully-field the charging circuit.

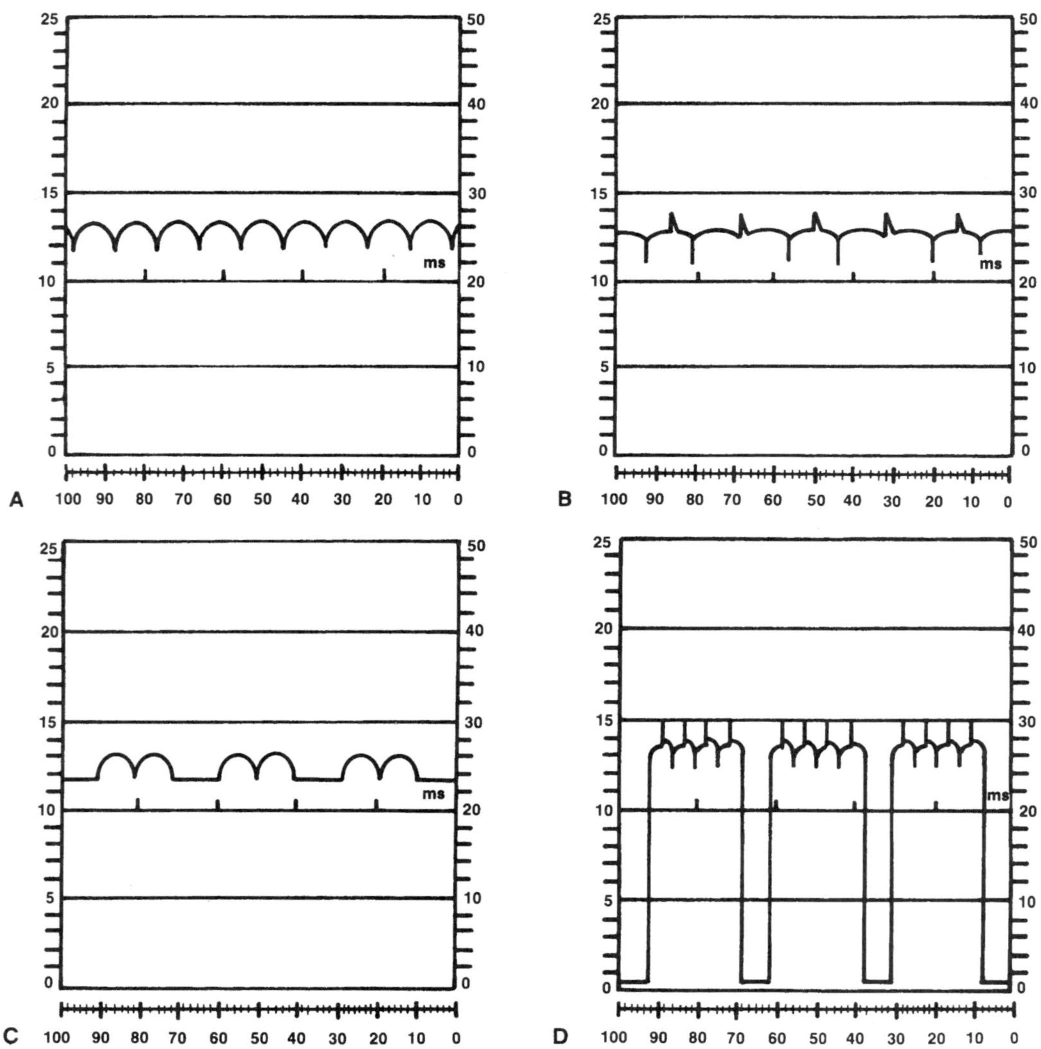

FIGURE 16–71 Typical alternator oscilloscope patterns. (A) Good alternator under a full load, (B) good alternator under no load, (C) alternator with a shorted diode and/or stator winding under a full load, and (D) alternator with an open diode trio.

On a typical GM charging system with an external regulator, the full-field test is performed by disconnecting the regulator from the system and using a jumper wire between the F (field) connector lead and the connector #3 lead (Figure 16–70). While the engine is running, if the current output is within specifications the regulator is at fault.

The fourth basic charging system check is a diode pattern test. The diodes are responsible for transferring the alternating current (AC) from the stator into the direct current (DC) used by the electrical system.

An alternator may have an open diode, yet be close to its manufacturer's specifications for alternator current output. Failing to find a bad system diode may result in a repeat failure of new replacement components. One way to test for an open diode requires an oscilloscope to visually see the diode patterns under a *load* condition. Figure 16–71 shows four examples of diode test patterns that illustrate comparisons between good diode patterns and ones that have problems.

Another way to quickly check an alternator for an open diode is to use the special diode test feature on a DMM (Figure 16–72). With the alternator electrically disconnected, one meter test probe is touched to the alternator housing and the other probe to the alternator output terminal. A meter reading of approximately 0.8 volts is satisfactory and a meter reading of approximately 0.4 volts indicates a shorted diode. The measurements are repeated with the meter probes reversed. The meter will exhibit a continuous tone when two diodes shorted. To determine which diodes are faulty, the alternator must be dismantled and each diode checked individually.

After diagnosing a faulty alternator, the preferred method of repair in many shops is simply to replace it with a new or rebuilt unit. If your shop prefers to rebuild the unit, follow specific service manual procedures when removing, disassembling, repairing, reassembling, and testing the alternator.

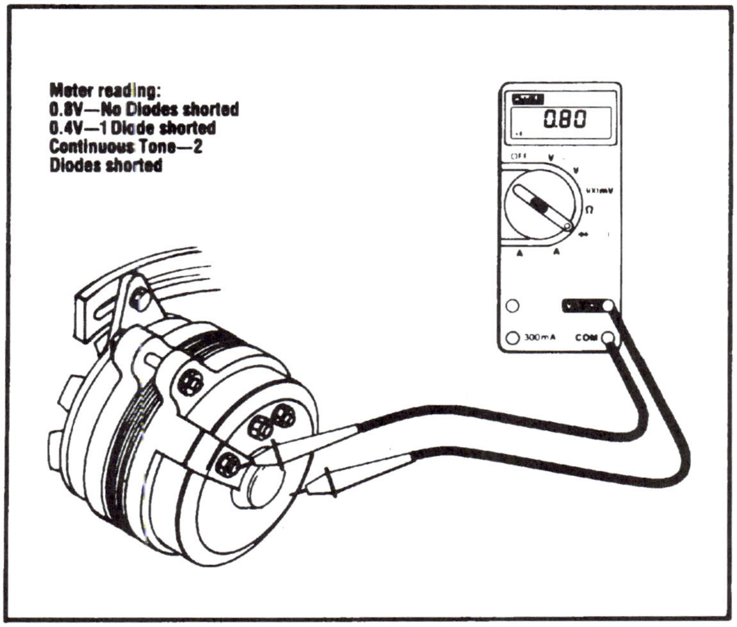

FIGURE 16–72 Alternator diode test using the diode feature of a DMM.

(Courtesy of The Fluke Corporation)

CASE STUDIES

CASE #1

The following case study is typical of a charging circuit problem that technicians often need to diagnose. Knowledge of various charging systems, testing equipment, and simple test procedures will help you perform the diagnosis quickly and efficiently.

Customer Complaint

A vehicle was driven into the shop with a complaint that the battery runs dead if the vehicle is driven with many accessories on. The battery was replaced by the customer, but the condition returned after several hours of driving.

Known Information

- ❑ New battery has an OCV of 12.2 volts
- ❑ Charging system indicator lamp is off while the engine is running
- ❑ Upon examination, the alternator belt appears loose and glazed

Circuit Analysis

Answer the following questions on a *separate sheet* of paper.

1. With the above information, what could be the *most* likely cause of the charging system problem above? ______________________________.
2. What preliminary checks should you perform before proceeding with any charging system tests?

   ______________________________.
3. After the above checks and repairs are done, what diagnostic system tests would you perform to verify the condition of the charging system? ______________________________

   ____

   ______________________________.
4. What is your analysis of the above case study? ______________________________

   ____

   ______________________________.

Hands-On Vehicle Tests

The following six hands-on vehicle checks are included in the **NATEF** (National Automotive Technicians Education Foundation) **Task List.** The task reference numbers are the same as the NATEF reference numbers. Some tasks are grouped together due to their close association with each other. Complete your answers to the questions below on a *separate* sheet of paper.

Performance Task #1 (NATEF #VI-D-1)

Task Description

Diagnose charging system problems that cause an undercharge, no-charge, or overcharge condition.

Performance Task #2 (NATEF #VI-D-3)

Task Description

Inspect and test voltage regulator; replace as needed.

Performance Task #3 (NATEF #VI-D-6)

Task Description

Perform charging circuit voltage output test; determine needed repairs.

Task Objectives

- ❑ Obtain a vehicle that can be used for these tasks. What model and year of vehicle are you using for these tasks? ______________________________.
- ❑ Set the vehicle emergency brake and place the vehicle in neutral. Observe all standard shop safety practices and precautions.
- ❑ Perform a preliminary charging system check on the vehicle and determine repairs needed.

Preliminary check made	Repairs needed
1.	
2.	
3.	
4.	

- ❑ What diagnostic equipment are you going to use to test the condition of the charging system?

  ______________________________.

- ❑ List the steps used to perform a voltage output test on a charging system. ______________________________

  ____

  ______________________________.

- ❑ Perform a voltage output test on the charging system. Record the results below.

  Battery OCV test ______________________________.

  Charging system output voltage at 1,500 rpm ______________________________.

- ❑ List the steps used to perform a current output test on a charging system. ______________________________

  ____

  ______________________________.

- ❑ Perform a current output test on a charging system. Record the results below.

  System specifications ______________________________.

  Charging current output ______________________________.

- ❑ List the steps used to fully field a charging system and determine if the regulator is performing as designed. ____
____.

- ❑ What are the precautions to observe when performing a full-field alternator test? ______________
____.

- ❑ Perform a full-field regulator by-pass test and record the results below.

  Charging system current output before full-field test. ______________________.

  Charging system current output during a full-field test. ______________________.

- ❑ What procedures are most helpful when inspecting and testing the charging system circuit? ______
____.

Task Summary

After performing the above NATEF tasks, what can you determine will be helpful in knowing how to check a starter motor system and individual components for proper operation? ____________________
____
____.

Performance Task #4 (NATEF #VI-D-4)

Task Description

Remove, inspect, and replace/reinstall alternator.

Performance Task #5 (NATEF #VI-D-5)

Task Description

Disassemble, clean, inspect, and test alternator components; replace as needed.

Performance Task #6 (NATEF #VI-D-2)

Task Description

Inspect and adjust alternator drive belts; replace as needed.

Task Description

- ❑ Obtain a vehicle that can be used for these tasks. What model and year of vehicle are you using for these tasks? ____.

- ❑ List the procedures for removing and replacing the alternator according to the manufacturer of the vehicle being worked on. ____
____
____.

- ❑ What tools and equipment did you use to remove and replace the alternator? ____________________
____.

- ❑ List the steps that you used to disassemble, clean, and inspect the alternator. ______________________

  ____.

- ❑ Perform individual alternator component tests according to the specific vehicle being worked on. Record the results below.

Alternator component	Test equipment used	Results of test
1.		
2.		
3.		
4.		
5.		
6.		

- ❑ What procedures and special tools did you use to check the alternator belt tension after replacing the alternator? ____.

Task Summary

After performing the above NATEF tasks, what can you determine will be helpful in knowing how to check a starter motor system and individual components for proper operation? ______________________

____

____.

Summary

- ❑ The amount of voltage produced in a conductor is dependent on three things—the strength of the magnetic field, the speed at which the conductor is cutting through the magnetic field, and the number of conductors passing through the magnetic field.
- ❑ The DC generator has two stationary field coils that create the individual north and south magnetic poles.
- ❑ Output voltage in a DC generator is produced in the wire-looped armature as it rotates through the magnetic field.
- ❑ In a DC generator, the AC voltage is changed to a DC voltage by the armature commutator bars.
- ❑ The rotor assembly in an AC generator—alternator—contains several rotating north and south poles.
- ❑ The rotor slip rings provide a current path connection from the insulated carbon brushes to the rotor field coil assembly.
- ❑ The stator assembly in the AC generator is the stationary coil windings in which electricity is produced.
- ❑ The wye wound stator consists of a common junction and three output connections.
- ❑ The delta wound stator consists of three windings connected together in a triangular shape with three output leads.

- ❑ Half-wave rectification of an AC sine wave converts the positive portion of the sine wave through a positively biased diode, while blocking the negative portion.
- ❑ Full-wave rectification of an AC sine wave is the process of converting the complete AC waveform to a positive DC waveform.
- ❑ A diode rectifier bridge consists of six diodes—three positive biased diodes mounted in a heat sink and three negative biased diodes mounted in the alternator rear housing.
- ❑ A diode trio rectifies current from the stator to create the magnetic field excitation in the rotor field coil.
- ❑ A voltage regulator controls and stabilizes the charging system voltage by controlling the resistance in series with the rotor field coil.
- ❑ The A-circuit regulator has field circuit control on the ground side of the field coil.
- ❑ The B-circuit regulator has field circuit control on the power input side of the field circuit.
- ❑ An isolated regulator field circuit picks up both B+ and ground externally.
- ❑ An electromechanical regulator uses two vibrating electromagnetically operated point contact relays to control the charging circuit.
- ❑ An electronic regulator controls the field circuit by modulating the on and off cycle of the field circuit, depending on the electrical circuit loads.
- ❑ When the field circuit in an electronic regulator modulates, it is called pulse width modulation.
- ❑ The zener diode in an electronic regulator is designed to conduct current in a reverse direction at a designed-in voltage level.
- ❑ Computer controlled regulators incorporate a powertrain control module (PCM) with a logic module to modulate the field circuit.
- ❑ A charging system indicator lamp operates on the principle of opposing voltage to control lamp function.
- ❑ An ammeter is sometimes wired in series to monitor the function of a charging system.
- ❑ A voltmeter is sometimes wired in parallel to the battery feed circuit to monitor the function of a charging system.
- ❑ A voltage output test can be used to compare the battery OCV with the charging system voltage.
- ❑ A current output test uses a carbon pile tester to determine the alternator current output while the electrical system is under a load condition.
- ❑ A regulator by-pass test is also referred to as a full-field test, and determines whether the alternator or regulator is faulty.

Key Terms

A-circuit
B-circuit
delta wound
diode trio
diode rectifier bridge
duty cycle
field current
full-wave rectification
half-wave rectification
heat sink
isolated field circuit
pulse width modulation
rectification
rotor
sensing voltage
sine wave
slip rings
stator
three-phase voltage
voltage limiter
voltage regulator
wye wound

Review Questions

Short Answer Essays

1. List the parts of a typical AC generator.
2. Describe the difference between a wye wound stator and a delta wound stator.
3. Describe how the AC current produced by the alternator is converted to DC current.
4. What is the purpose of the voltage regulator in a typical charging system?
5. Describe the purpose of a zener diode in a charging system electronic regulator.

Fill-in-the-Blanks

1. The rotor ________________________ provide a current path connection from the insulated carbon brushes to the rotor coil.
2. A wye wound stator winding has a common junction and ________________________ output connections. The delta wound stator winding has the windings connected together in a ________________________ shape with ________________________ output connections.
3. The process of converting a complete AC waveform to a positive DC waveform is referred to as ________________________.
4. The on and off cycling of the field circuit is called ________________________.
5. Normal charging system output under a heavy load condition should be at least ________________________ or more.

ASE Style Review Questions

1. Two technicians are discussing the magnetic principles of charging systems. Technician A says the output is rotating in a DC generator. Technician B says the output is stationary in a AC generator. Who is correct?
   A. A only
   B. B only
   C. Both A and B
   D. Neither A nor B

2. Technician A says the battery has no affect on the charging system operation. Technician B says the charging system is used only to restore the electrical power to the battery that was used up during engine starting. Who is correct?
   A. A only
   B. B only
   C. Both A and B
   D. Neither A nor B

3. Technician A says the slip rings on an alternator rotor convert the AC voltage to a DC voltage. Technician B says the field coil on an alternator rotor must be insulated from the rotor shaft. Who is correct?
   A. A only
   B. B only
   C. Both A and B
   D. Neither A nor B

4. Technician A says a wye wound stator winding consists of three output connections and a common neutral junction. Technician B says a wye wound stator winding produces a higher amperage output than a delta wound stator winding. Who is correct?
   A. A only
   B. B only
   C. Both A and B
   D. Neither A nor B

5. Before adjusting the alternator drive belt tension, Technician A checks for proper pulley alignment. Technician B refers to the vehicle's service manual to determine the proper belt tension. Who is correct?
   A. A only
   B. B only
   C. Both A and B
   D. Neither A nor B

6. Technician A says the charging system regulator uses a sensing voltage input from the battery to interpret the alternator output voltage. Technician B says an A-circuit charging system regulator controls the field circuit on the power side of the circuit. Who is correct?
   A. A only
   B. B only
   C. Both A and B
   D. Neither A nor B

7. Technician A says a delta wound winding consists of three windings connected together in a triangular shape with three output leads. Technician B says a delta wound winding provides a high amperage output for it's size. Who is correct?
   A. A only
   B. B only
   C. Both A and B
   D. Neither A nor B

8. The charging system voltage of a vehicle with an external regulator is 16.5 volts at 1,500 rpm. Technician A says the alternator's field circuit may have excessive resistance. Technician B says the regulator sensing circuit may have excessive resistance. Who is correct?
   A. A only
   B. B only
   C. Both A and B
   D. Neither A nor B

9. Technician A uses only a current output test to check alternator output. Technician B uses only a voltage output test to check alternator output. Who is correct?
   A. A only
   B. B only
   C. Both A and B
   D. Neither A nor B

10. Technician A says the positive biased diodes in a rectifier bridge are attached directly to the alternator rear housing. Technician B says the negative biased diodes in a rectifier bridge are mounted in a heat sink. Who is correct?
   A. A only
   B. B only
   C. Both A and B
   D. Neither A nor B

11. Technician A says a diode trio rectifies current from the stator to create the magnetic field excitation in the field coil. Technician B says the diode trio attaches to the rotor slip rings. Who is correct?
    A. A only
    B. B only
    C. Both A and B
    D. Neither A nor B

12. Technician A says an isolated field circuit can be easily identified by the two field wires attached to the outside of the alternator case. Technician B says an isolated field circuit has two B+ field wires and grounds internally. Who is correct?
    A. A only
    B. B only
    C. Both A and B
    D. Neither A nor B

13. Technician A says the pulse width modulation in an electronic voltage regulator is the on and off modulation of the field circuit. Technician B says the modulation in a electronic voltage regulator varies the amount of time the field coil is energized depending on the vehicle's electrical demand. Who is correct?
    A. A only
    B. B only
    C. Both A and B
    D. Neither A nor B

14. Technician A says a slipping belt can cause a squealing noise to happen when the alternator is under a heavy load. Technician B says a slipping belt can become glazed over a period of time. Who is correct?
    A. A only
    B. B only
    C. Both A and B
    D. Neither A nor B

15. Technician A says when an alternator is full-fielded, the alternator will put out maximum designed amperage. Technician B says an alternator that charges 16.5 volts or less indicates a good alternator. Who is correct?
    A. A only
    B. B only
    C. Both A and B
    D. Neither A nor B

17 Ignition System Theory

Introduction

Over the past several decades automotive ignition system designs have made a steady growth in complexity from the pre 1970's point-style distributor to the computer-controlled ignitions of today. Demand for more fuel efficient and cleaner burning engines have been primary reasons for ignition system design changes. The need for maintenance free ignition systems that perform to exact engine combustion efficiency standards is important to prevent tail pipe emissions from exceeding federal vehicle pollution standards. There is a tremendous variation of ignition systems in use today, even within a single vehicle manufacturer. This chapter provides a look at basic ignition system theory, components, diagnosis, and typical service needs. Diagnosis and service of a specific ignition system should be referred to appropriate manufacturer's service manual information.

Objectives

When you complete this chapter you should be able to:

- ❑ Describe the basic function of an ignition system.
- ❑ Name the components and their function in the primary side of an ignition system.
- ❑ Name the components and their function in the secondary side of an ignition system.
- ❑ Describe the operation of and basic tests on an ignition coil.
- ❑ Describe the function of spark plugs and plug wires in an ignition system.
- ❑ Identify combustion chamber conditions by observing spark plug wear patterns.
- ❑ Describe the purpose and operation of advancing units in ignition systems.
- ❑ Describe the operation and characteristics of distributorless ignition systems.

Ignition System Principles

The ignition system in an internal combustion engine is designed to supply three things.

1. Step-up voltage high enough to jump the spark plug gap.
2. Control ignition timing based on engine speed and load.
3. Distribute spark to each cylinder based on firing order.

Combustion of the fuel/air mixture in the cylinder will occur in milliseconds depending on several design and operating factors. Because of this, the combustion process must begin **before-top-dead-center (BTDC)** near the end of the compression stroke. This ensures that the fuel/air mixture burns completely and delivers its maximum compression push when the piston is at ten to twenty-three degrees **after-top-dead-center (ATDC)** (Figure 17–1).

Calculating when the ignition spark should occur for a controlled burning of the fuel/air mixture depends on several design and operating factors. Some of these are engine load, intake air temperature, barometric pressure, humidity, cylinder compression, and the fuel/air mixture turbulence in the cylinder.

Much has been learned over the past two decades concerning the tendency of an engine to knock, or ping, under abnormal combustion conditions. Knowledge of these conditions is important because it may not be ignition related problems affecting the engine operation. There are two typical combustion conditions that may cause a knock or ping.

The first is engine **detonation.** It can cause a knock to occur when unburned fuel/air mixture self-ignites and collides with the normal combustion flame front (Figure 17–2). Detonation that is allowed to continue can cause severe piston, cylinder head, or spark plug damage over time (Figure 17–3).

The second is a **pre-ignition** combustion condition caused by an early ignition of the fuel/air mixture by a source other than the spark plug and ignition system (Figure 17–4). The most common causes of pre-ignition are hot carbon deposits or overheating of sharp edge surfaces in the combustion chamber. In a pre-ignition condition the fuel/air mixture is ignited by a hot carbon

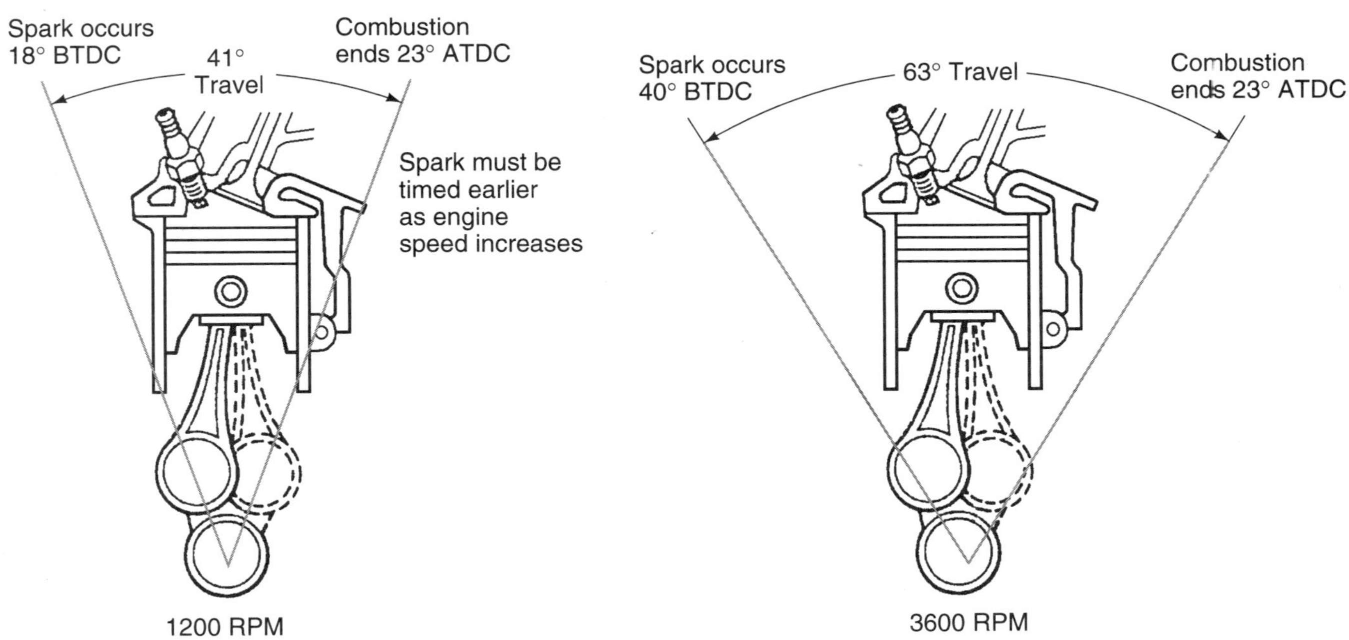

FIGURE 17–1 As piston speed increases, ignition spark must be advanced to obtain maximum combustion pressure by twenty-three degrees ATDC.

(Courtesy of Ford Motor Company)

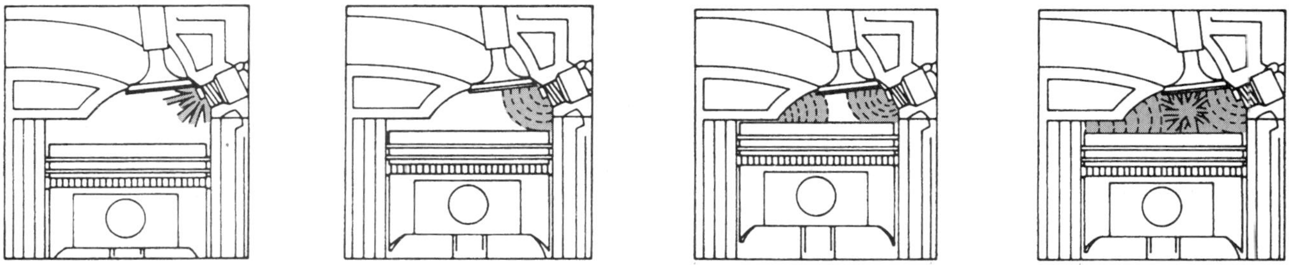

FIGURE 17–2 Detonation stages in a combustion chamber.

deposit that creates a flame front that collides with the normal spark ignition flame front. A pinging or light tapping noise will result from an engine with this condition.

A good way to remember these conditions is that detonation occurs *after* the start of normal spark ignition conditions, and pre-ignition occurs *before* the start of normal spark ignition conditions.

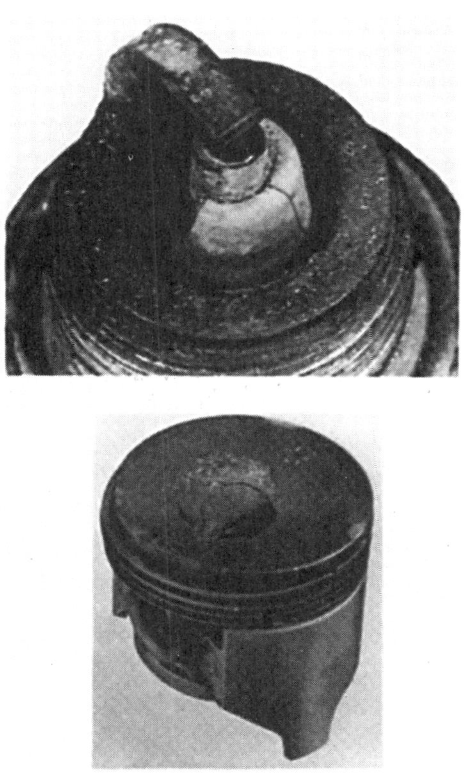

FIGURE 17–3 The effects of detonation on a piston and spark plug.

(Courtesy of Champion Spark Plug)

Ignition System Components and Operation

All ignition systems consist of two interconnected electrical circuits and components designed to work together to produce the necessary high voltage to jump the spark plug gap. The two systems are a low voltage **primary circuit,** and a high voltage **secondary circuit** (Figure 17–5).

Depending on the exact type of ignition system, the following typical components are part of the low voltage primary circuit in a distributor-type ignition system.

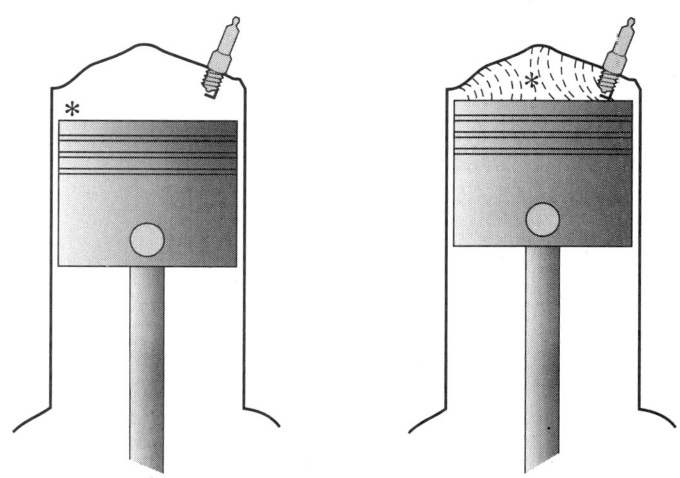

FIGURE 17–4 Pre-ignition stages in a combustion chamber.

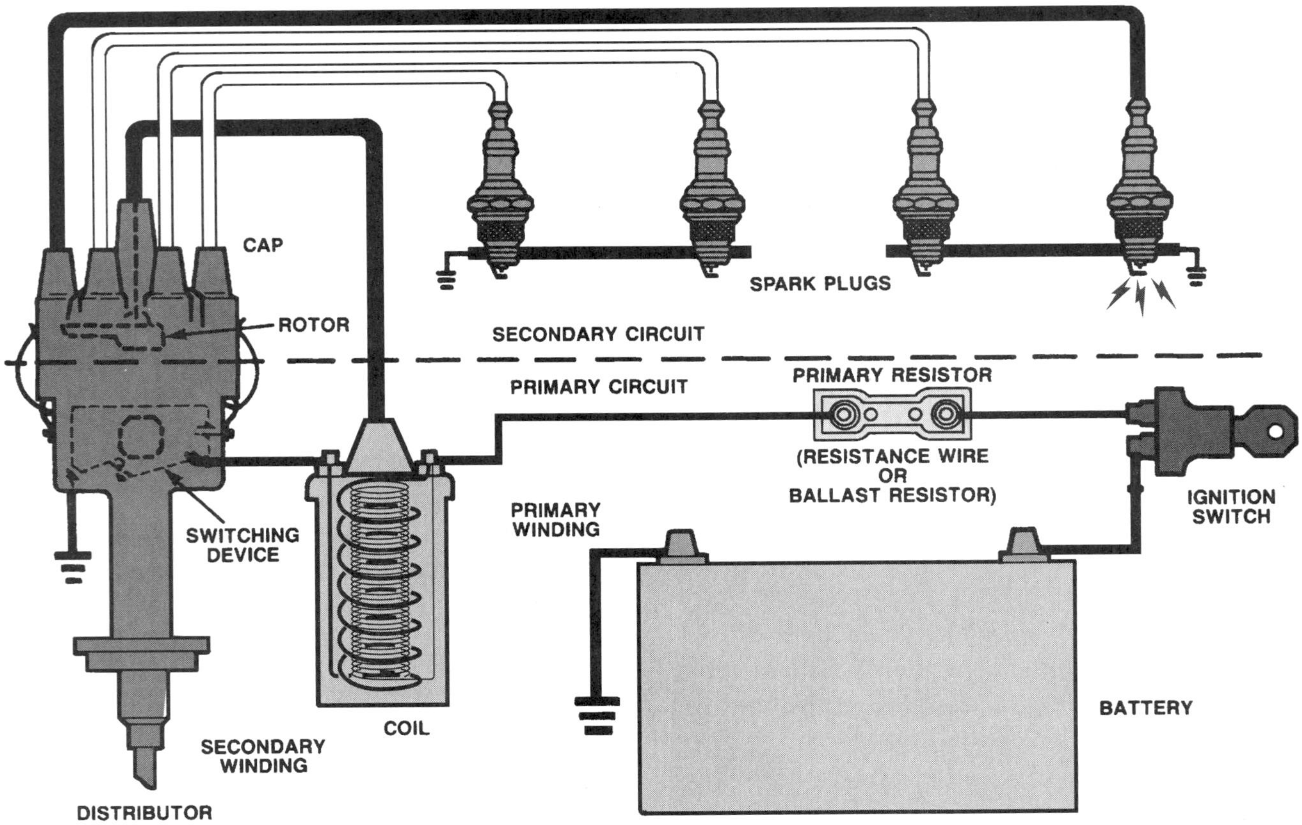

FIGURE 17–5 Typical primary and secondary ignition circuits.
(Courtesy of Champion Spark Plug)

- Battery.
- Ignition switch.
- Ballast resistor or resistance wire.
- Ignition coil primary windings.
- Ignition by-pass circuit during starting mode.
- Points or electronic triggering device.
- Electronic control module or switching device.
- Ground connection.

The typical secondary circuit components in a distributor-type ignition system include the following.

- Ignition coil secondary windings.
- Coil wire and spark plug wires.
- Distributor cap and rotor.
- Spark plugs.
- Ground connection.

Ignition Coil

The heart of ANY ignition system is the ignition coil which can be referred to as a **pulse transformer** that generates a high voltage when a magnetic field moves across a wire. The design function of the ignition coil is to transform 12 volts from the battery to a high voltage potential capable of jumping across the spark plug gap. This voltage may reach 50,000 volts, or more, in some modern systems.

Construction of a typical ignition coil (Figure 17–6) begins with a couple hundred turns of twenty to twenty-four gauge copper wire connected between the coil positive and negative posts to form the primary coil windings. The secondary coil windings consist of up to 20,000 turns, or more, of very fine copper wire connected normally between the coil positive or negative post and the coil center tower. Both sets of copper windings are coated with an insulating enamel for protection. The soft laminated iron core has a low **inductive reluctance** that freely expands or strengthens the magnetic field around it. The core and windings are immersed in a

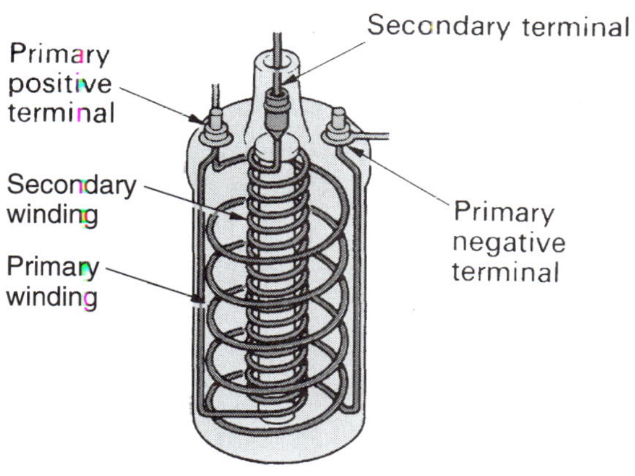

FIGURE 17–6 Cutaway of a typical ignition coil.

metal case that is filled with oil to dissipate heat generated from the windings. The top of the coil is hermetically sealed to prevent the oil from leaking out, or moisture and condensation from entering inside.

Another common ignition coil design used mostly in electronic ignition systems is an air-cooled epoxy resin-sealed **E-coil** (Figure 17–7). This type of coil consists of a soft iron laminated core shaped like the letter "E". The primary and secondary windings wrap around the middle core of the E-coil and are then sealed with epoxy resin.

The operation of an ignition coil involves three important principles of magnetism in the primary and secondary windings—electromagnetism, **mutual induction,** and **self induction.**

The first principle is called electromagnetism. We have discussed this principle before in previous chapters. Anytime a current flows through a wire, magnetic lines of force create an electromagnet. The strength of the electromagnet depends on the number of wire loops and the amount of current flowing through these loops (Figure 17–8).

The second principle is called **mutual induction** which occurs as the magnetic lines build up around the primary windings. The magnetic lines cut across the secondary windings and mutually induce a voltage (Figure 17–9).

The third principle is called **self induction** which occurs when the primary windings induce a voltage onto itself. This self induction creates a counter voltage in the primary winding called counter electromotive force

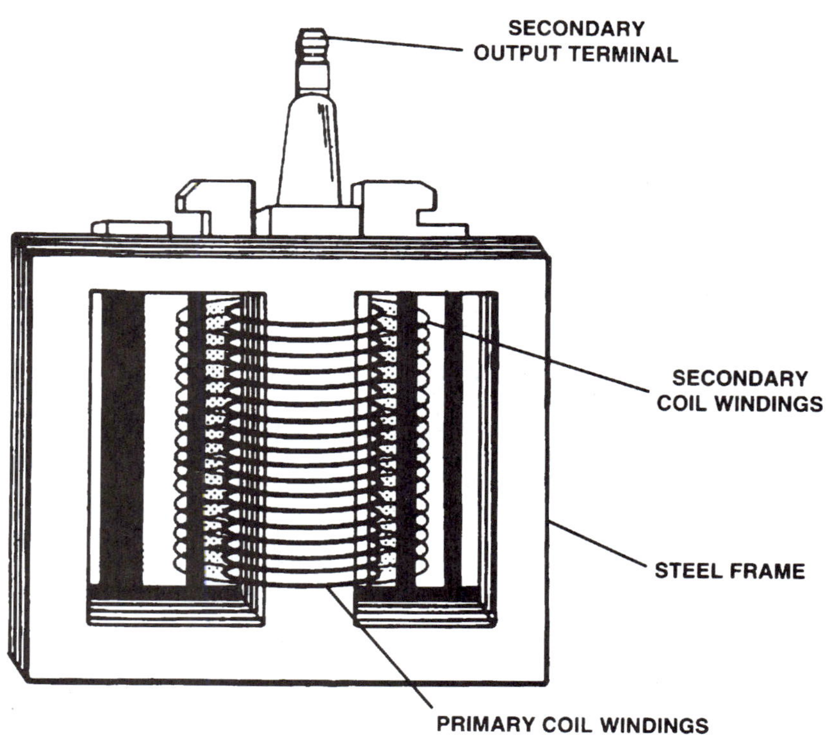

FIGURE 17–7 Typical air-cooled epoxy resin-sealed E-coil.
(Courtesy of General Motors Corporation, Service Technology Group)

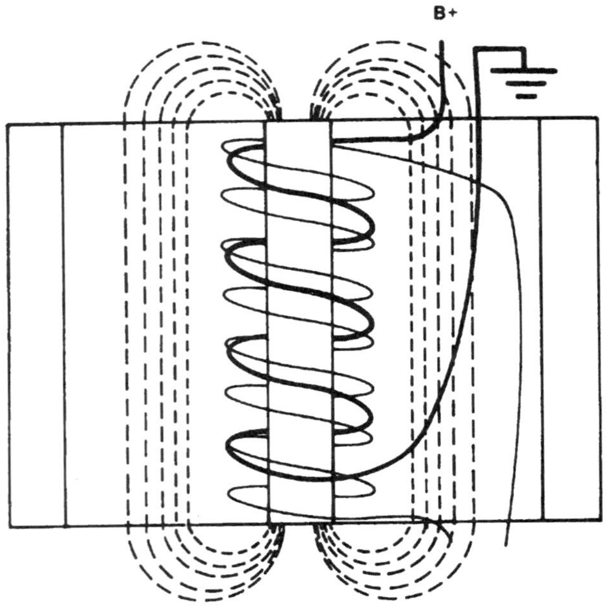

FIGURE 17–8 Magnetic field develops as current flows in the primary windings.

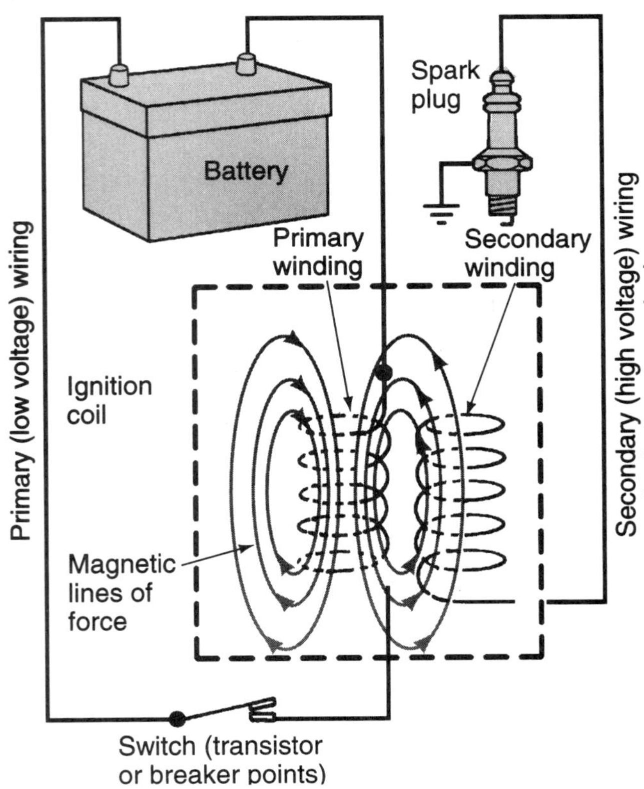

FIGURE 17–9 Magnetic lines of force cutting across the secondary windings.

(CEMF) that tries to resist the electromotive force (EMF) from the battery. The effects of CEMF on the primary windings allow time for the coil to become fully magnetized with voltage from the battery. The build up of the magnetic field in the coil is called the **saturation** time. The longer the time that current flows through the primary windings, the stronger the magnetic field will become.

When the primary circuit is suddenly opened by the primary control device, such as a transistor switch or breaker points, the magnetic field instantly collapses. This sudden collapse produces a very high voltage in the secondary windings that rises high enough to jump the spark plug gap. On some ignition systems, to protect the primary switching circuit and speed up the collapse of the magnetic field, a **capacitor**—also known as a condenser—is wired in parallel to the circuit to absorb the surges of electrons induced by the collapsing magnetic field (Figure 17–10 and Figure 17–11).

The steps of ignition coil operation can be summarized as follows.

1. Battery current flows through the primary windings when the primary control device closes.
2. Current flowing through primary windings creates an electromagnetic field that mutually induces a voltage onto the secondary windings.
3. Current flowing through the primary windings induces a CEMF upon itself called self induction.
4. The coil's magnetic field becomes fully saturated as the battery EMF overcomes the CEMF in the primary coil windings.

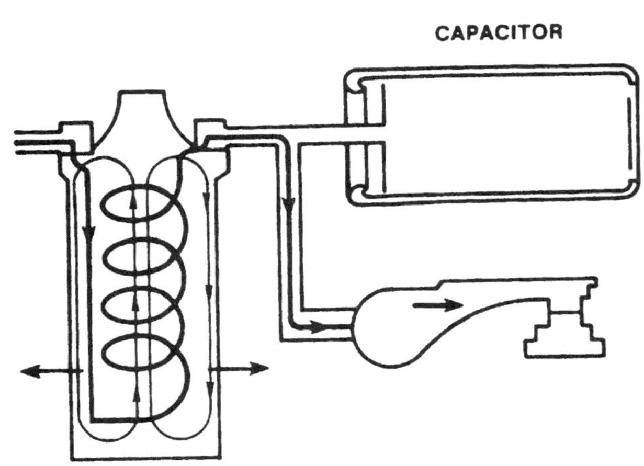

FIGURE 17–10 Current flow through primary circuit with points closed.

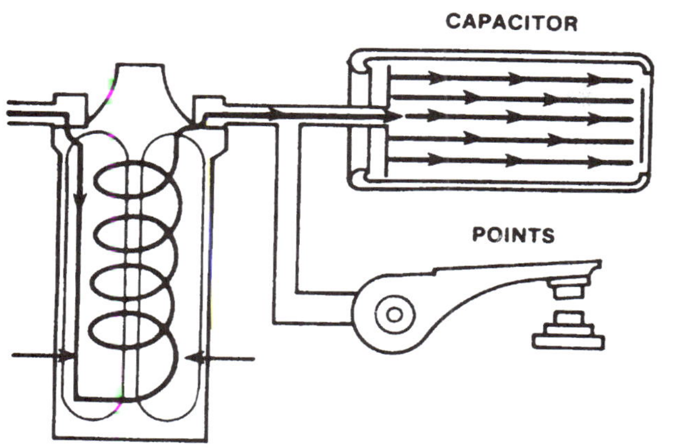

FIGURE 17–11 Capacitor absorbing surges of electrons as points open.

5. The primary control device opens the primary circuit, stopping the current flow.
6. The magnetic field rapidly collapses and the magnetic lines of force cut across the secondary windings.
7. The system capacitor absorbs the surges of electrons induced in the primary windings by the collapsing magnetic field.
8. The moving magnetic lines of force induce a voltage into the secondary windings that is high enough to jump the spark plug gap.
9. The high voltage—several thousand—jumps the spark plug gap and completes the circuit to ground.
10. Sequence repeats itself several thousand times per second.

Ignition Coil Testing There are three basic tests of the coil windings that can be performed with an ohmmeter. The first test checks the resistance of the primary coil windings. With the key off and the wiring disconnected from the coil, connect the ohmmeter leads across the positive and negative terminals of the coil (Figure 17–12). Primary resistance usually ranges from 0.5 to 2.0 Ω maximum. Always refer to the vehicle's service manual for detailed specifications.

The second ignition coil test checks the resistance of the secondary windings. Again, with the key off and the wiring disconnected from the coil, connect the ohmmeter leads across the positive terminal and the coil's high voltage center tower connection (Figure 17–12). Secondary resistance usually ranges from 5,000 to 20,000 ohms maximum. Always refer to the vehicle's service manual for actual specifications.

The third ignition coil test checks for grounding of the primary and secondary windings to the coil case. There are three individual ground tests to perform while the circuit wiring is disconnected from the coil. The reading in each test should indicate infinite (O/L). A resistance reading would indicate a shorted primary or secondary winding to the coil case.

1. Connect one ohmmeter lead to the coil case, and the other to the primary positive terminal.
2. Connect one ohmmeter lead to the coil case, and the other to the primary negative terminal.
3. Connect one ohmmeter lead to the coil case, and the other to the secondary center post.

The ignition coil tests give a general view of the condition of the primary and secondary coil windings. These tests do not guarantee that the coil will work as designed. Some defects can only be detected while the coil is in operation and using an oscilloscope that checks for defective insulation on the windings, and secondary tracking or arcing shorts to the case (Figure 17–13).

Ignition Primary System Resistance

Depending on the requirements of the ignition system, there are three methods to implement the primary system resistance that is found in most ignition systems

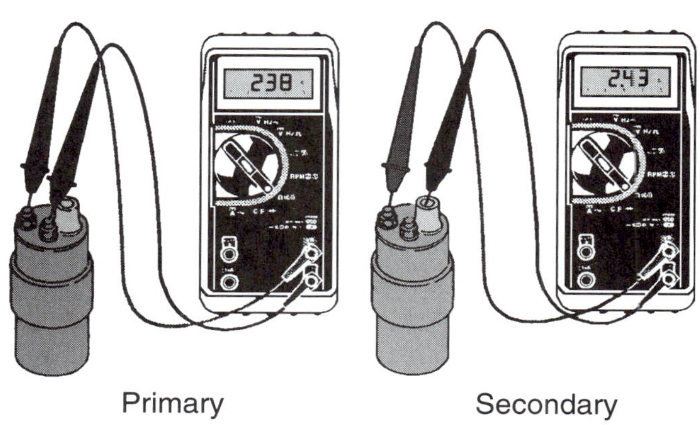

FIGURE 17–12 Checking the resistance of ignition coil primary and secondary winding using an ohmmeter.

(Courtesy of The Fluke Corporation)

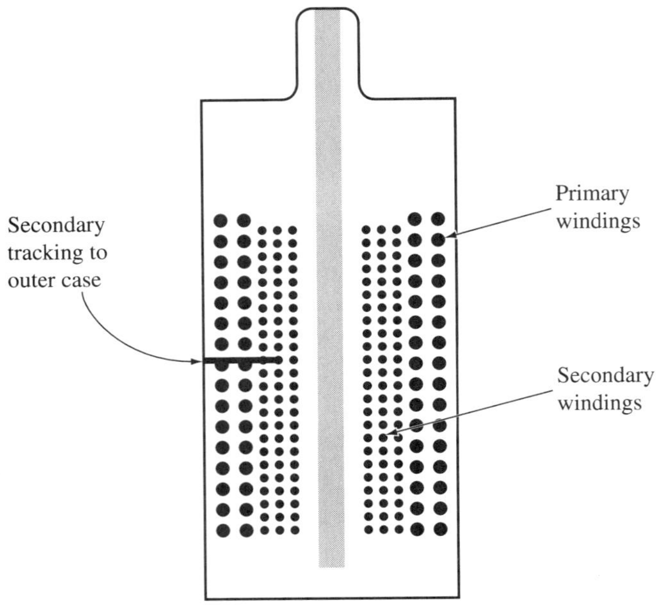

FIGURE 17–13 Secondary tracking or arcing to the coil case.

today. The purpose of having a primary resistance is to control the current to the primary control switch in the system. Many older ignition systems had breaker points that could only handle 2 or 3 amps maximum. Excessive circuit current would cause the points to arc as they opened and closed.

The first method is to use an ignition coil that has internal resistance added to its primary winding circuit (Figure 17–14). Use of this coil is limited to the system it is designed for. If the coil is accidentally installed in another system, it could create ignition system troubles.

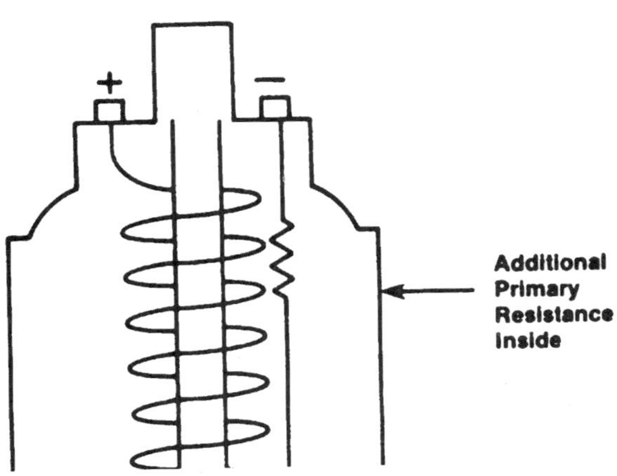

FIGURE 17–14 Primary resistance mounted internally in the coil.

The second method is to use a primary resistance that is external to the ignition coil (Figure 17–15). This can be accomplished by either a resistance wire or a **ballast resistor** that consists of a resistive element encased in a ceramic block assembly. Either one will control the current flow through the primary system and reduce coil primary voltage to 10 volts or less.

On ignition systems that operate on a reduced voltage level, it is necessary to by-pass the resistance during engine starting conditions. Figure 17–16 is an example of a typical by-pass circuit being controlled by the ignition switch. Full battery voltage is applied to the ignition coil only when the ignition switch is turned to the start mode. After the switch returns to the run position, the primary circuit returns to being ballast resistance-controlled. The bypass circuit may also come from the starter solenoid as explained in Chapter 15.

Primary System Control Switching Devices

There have been major advancements in ignition primary circuit switching devices over the past two decades. Even though there are tremendous variations between vehicle manufacturers, the central purpose has never changed. And that is, to break the current flow in the coil primary windings at a precise time so the saturated field may collapse. In this section, we will study the simplest form of switching device that was used for decades, up to the complex computer-controlled systems in use on newer vehicles.

Breaker Point Ignition For over 60 years, the **breaker point ignition** was the main method of controlling the primary system circuit. It consists of a mechanical contact point set and capacitor (condenser) mounted on a breaker plate inside the distributor (Figure 17–17). The contact point set consists of a fixed contact point that is grounded to the breaker plate base, and a movable contact point connected to the negative terminal of the coil's primary winding. The movable contact points are connected to a rubbing block that rubs against the rotating distributor cam causing the points to open and close a set distance (Figure 17–18).

When the contact points are closed, current flows into the primary circuit windings and coil saturation occurs. The amount of time that the contact points are

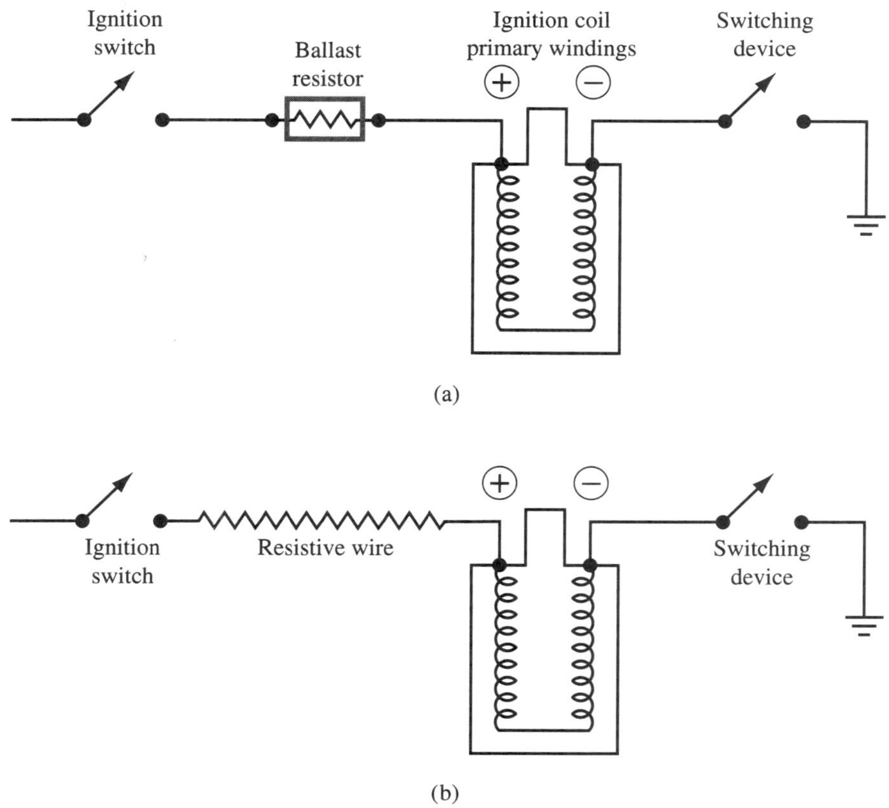

FIGURE 17–15 Two methods of current control in an ignition primary circuit. (A) Using a ballast resistor and (B) using resistive wiring.

closed is called the **dwell,** or cam angle as it also referred to (Figure 17–19). Ignition dwell is adjusted to the manufacturer's specifications by adjusting the point gap. Moving the points away from the distributor cam will increase the dwell, while moving the points closer will decrease the dwell.

Although still common in older vehicles still on the road, the use of breaker point ignition has been abandoned in an effort to decrease auto emissions and increase fuel efficiency.

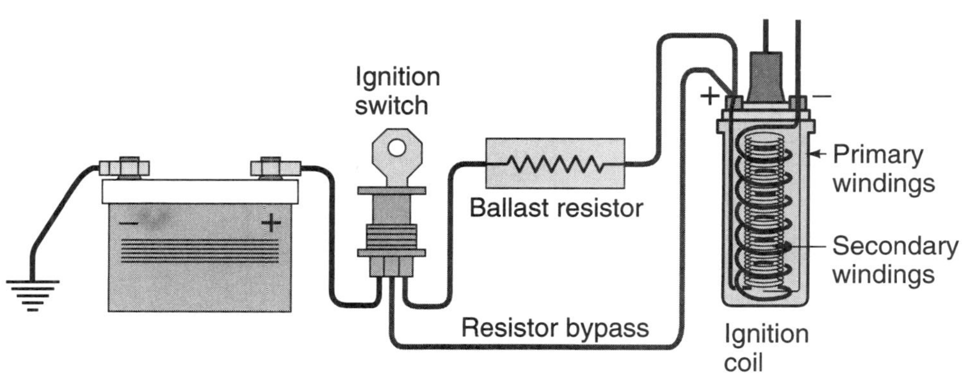

FIGURE 17–16 Typical ignition resistor by-pass circuit.

(Courtesy of Snap-on Diagnostics)

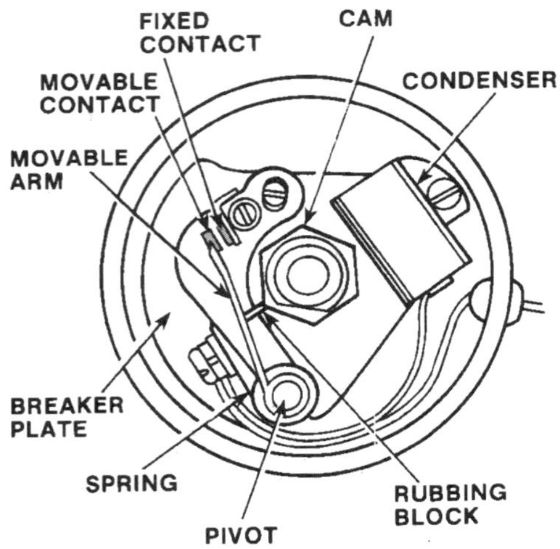

FIGURE 17–17 Typical older ignition system using breaker points as the primary switching device.

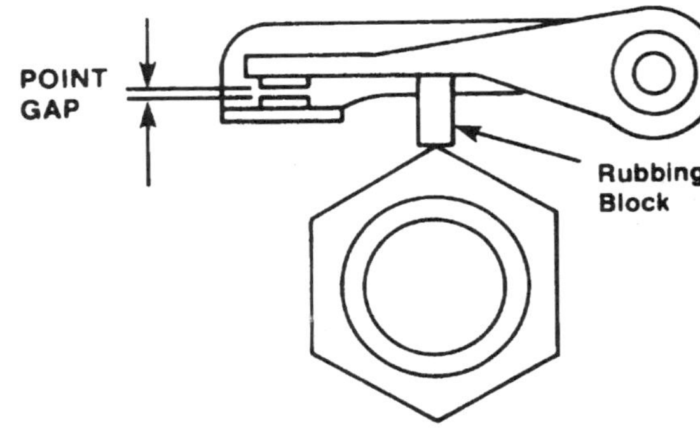

FIGURE 17–18 Typical contact point set and distributor cam.

Solid State Ignitions There are several designs of electronically controlled primary ignition circuits in use today. Breaker points have been replaced with electronic triggering devices that are connected to a **solid state ignition** control module that precisely controls the on and off time of the primary circuit (Figure 17–20).

There are four common electronic triggering systems used by major vehicle manufacturers—magnetic pulse generator system, metal detection sensor system, hall effect sensor system, and a photoelectric sensor system.

Magnetic Pulse Generator The **magnetic pulse generator** consists of a timing disc and a pick-up coil assembly. The timing disc may also be called a reductor, trigger wheel, pulse ring, armature assembly, or timing

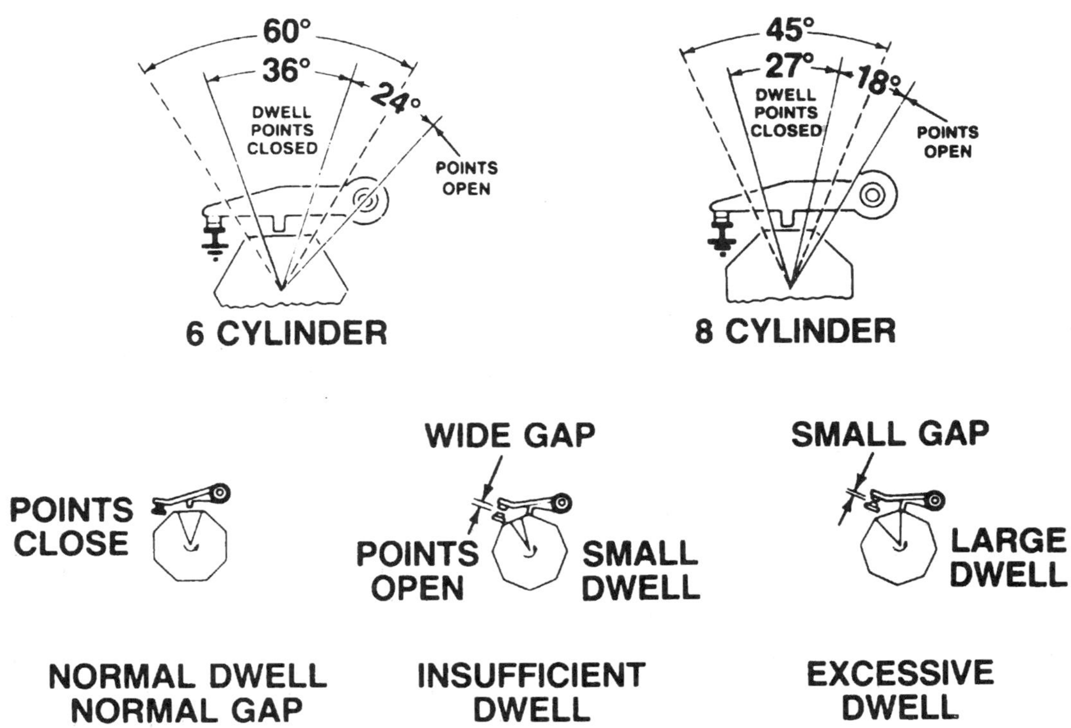

FIGURE 17–19 Relationship between contact point gap and dwell.

(Courtesy of Ford Motor Company)

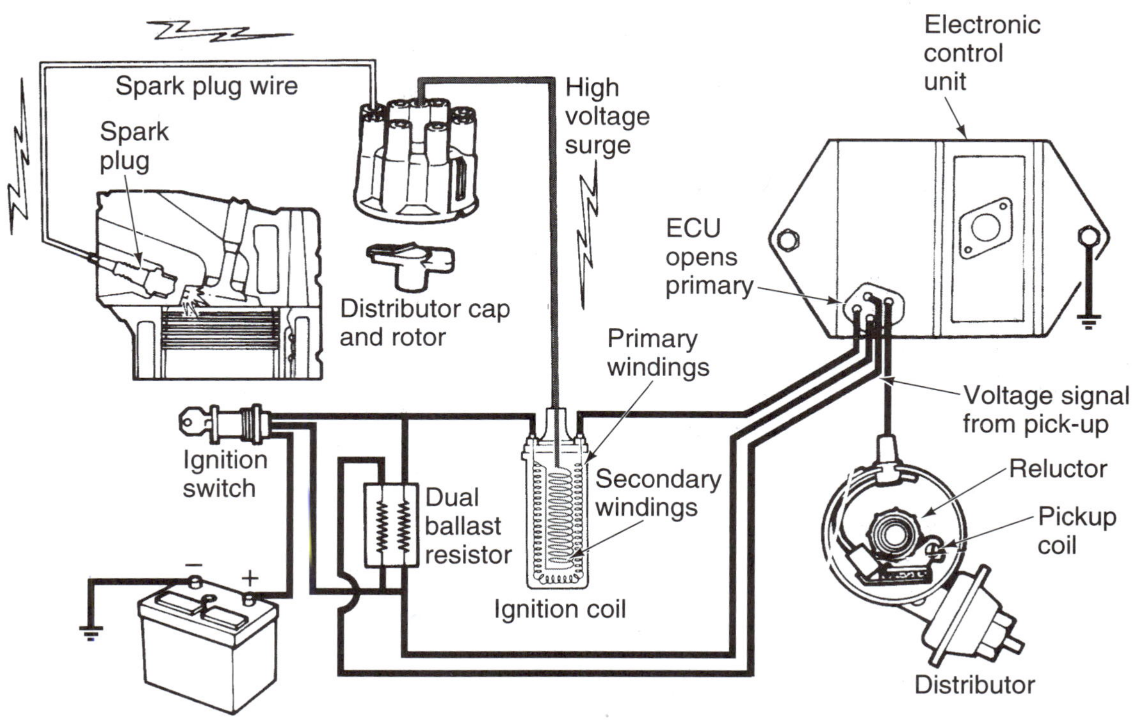

FIGURE 17–20 Typical solid state ignition system.
(Courtesy of DaimlerChrysler Corporation)

core depending on the specific ignition system manufacturer. The timing disc may be mounted on the distributor shaft in the distributor (Figure 17–21), on the engine flywheel drive plate (Figure 17–22), or on the crankshaft vibration damper (Figure 17–23).

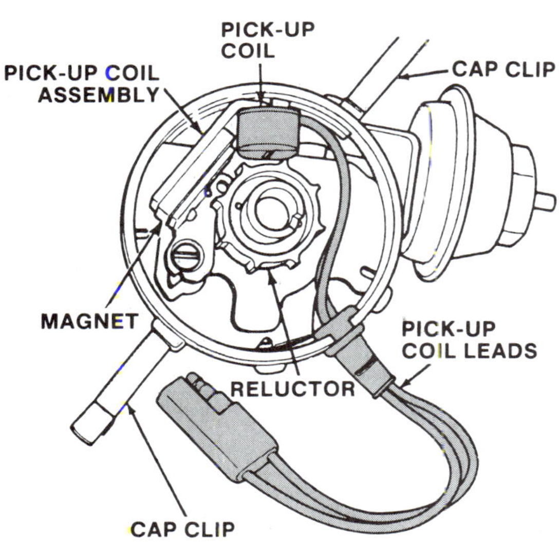

FIGURE 17–21 Magnetic pulse generator assembly located inside a distributor.

The magnetic pulse generator or **PM generator** operates on the basis of electromagnetism. The magnetic field for the PM generator is provided by the **pick-up coil** assembly, and the movement of the timing disc, or reductor, through the magnetic field. In Figure 17–24, the high points of the left reductor are out of alignment with the pick-up coil so a weak magnetic field and voltage exists in the pick-up coil. When the reductor tooth is directly in line with the pick-up coil (B in Figure 17–24), the magnetic field is not expanding or contracting so voltage in the pick-up coil at this precise moment drops to zero. The ignition module switching device reacts to the zero voltage and turns off the ignition primary circuit current. When the reductor high point passes the pick-up coil, the magnetic field expands again and produces another voltage signal. The increasing and decreasing of the voltage in the pick-up coil, in relation to zero, is what triggers the ignition module to turn the primary circuit on and off. Figure 17–25 is an example of a typical electronic control module mounted external to the distributor.

The pick-up coil of a PM generator can be checked for an open or shorted condition (Figure 17–26). With the ohmmeter #2 connected as shown between the two pick-up coil leads, the pick-up coil resistance can be

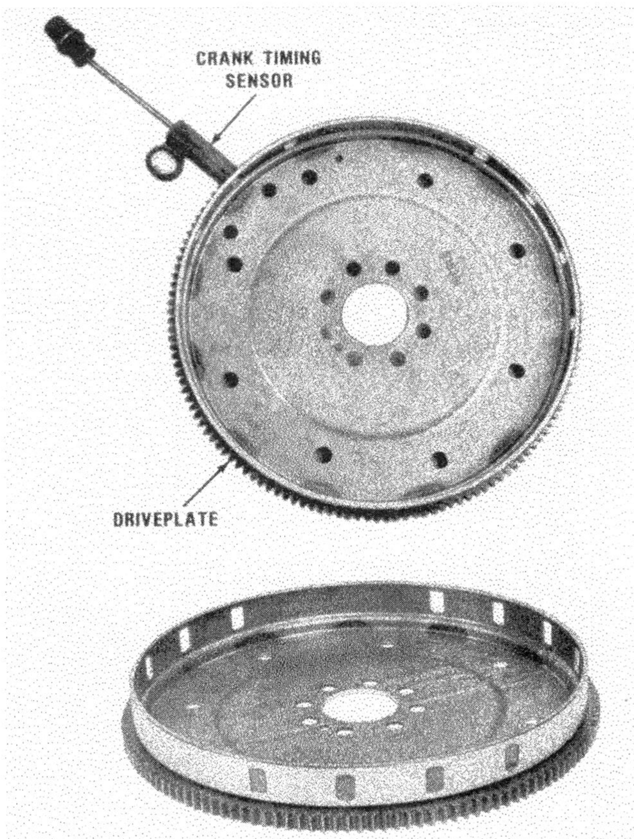

FIGURE 17–22 Magnetic pulse generator positioned to sense flywheel rotation through drive plate cutouts.

(Courtesy of DaimlerChrysler Corporation)

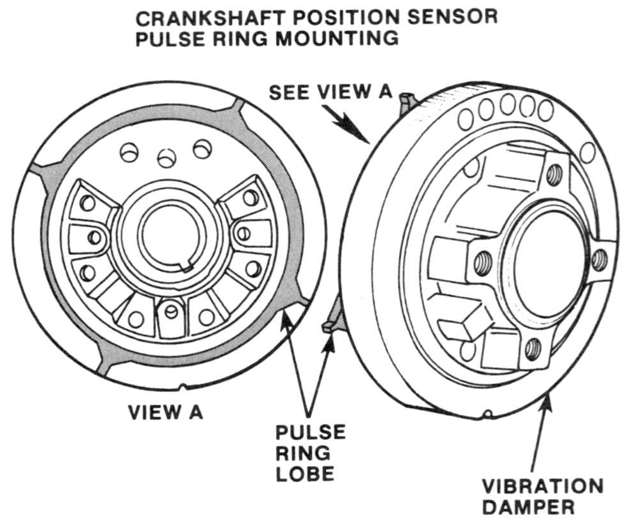

FIGURE 17–23 Pulse ring located behind the crankshaft vibration damper.

checked. Most pick-up coils have a resistance between 150 and 900 Ω but always refer to the manufacturer's specifications for allowed resistance values. If the pick-up coil is open, the reading would be infinite (O/L).

With ohmmeter #1 connected as shown between the distributor housing and one of the pick-up coil leads, the pick-up coil can be checked for a short-to-ground condition. The ohmmeter reading should be infinite (O/L).

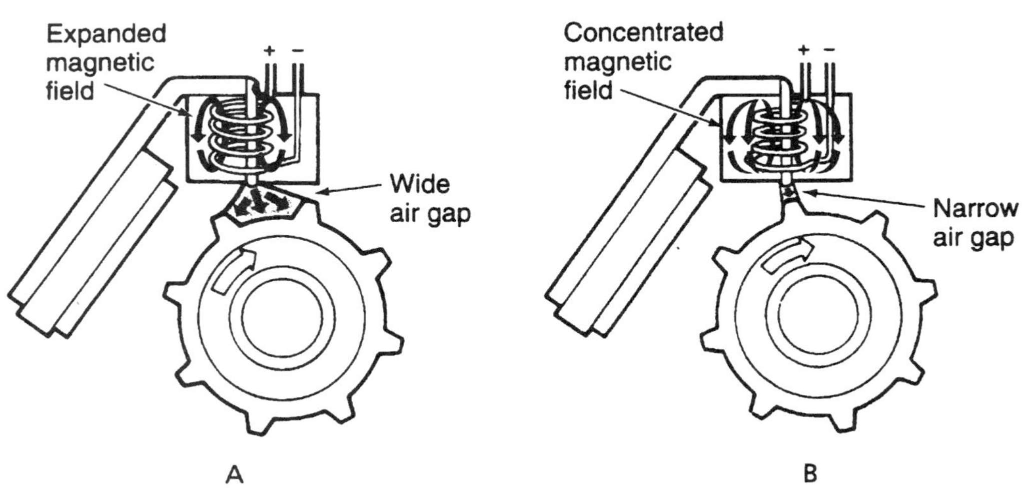

FIGURE 17–24 Action of a magnetic field in a PM generator.

(Courtesy of DaimlerChrysler Corporation)

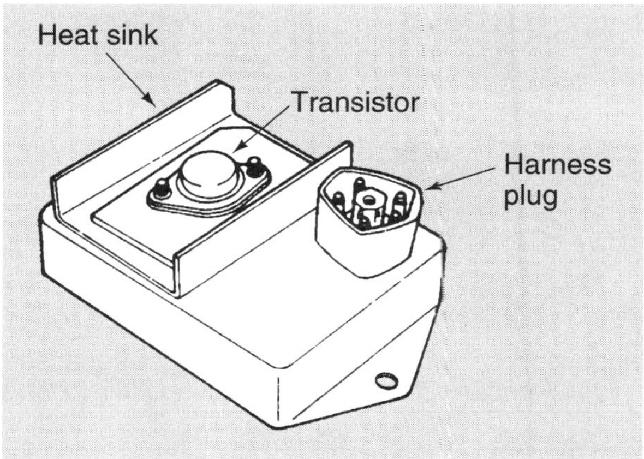

FIGURE 17–25 A typical electronic control module.
(Courtesy of DaimlerChrysler Corporation)

The pick-up coil operation can also be tested with an oscilloscope (Figure 17–27). With the scope leads hooked to the pick-up coil leads, a pulsing AC waveform should appear on the screen while the distributor is turning. The trace is not a true sine wave, but it should have both a positive and negative pulse. This test can also be performed with a voltmeter set on the AC scale.

Metal Detection Sensors On a few early electronic ignition systems, **a metal detection sensor** is used to control the ignition primary circuit (Figure 17–28). The metal detection sensor works similarly to a PM generator, but with the exception that the pick-up coil of a metal detection sensor does not have a permanent magnet. Instead, the pick-up coil is a weak electromagnet with a low-level current supplied to the pick-up coil by the electronic control unit. The operation of the rest of the system is the same as the PM generator described earlier.

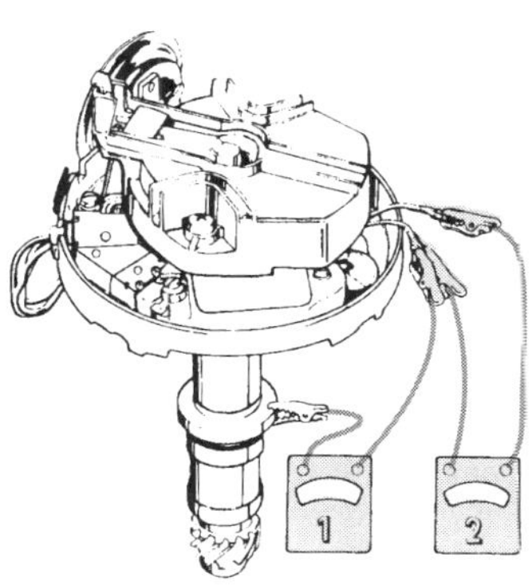

FIGURE 17–26 Ohmmeter #1 tests for shorts to ground, and ohmmeter #2 tests the resistance of the pick-up coil.
(Courtesy of General Motors Corporation, Service Technology Group)

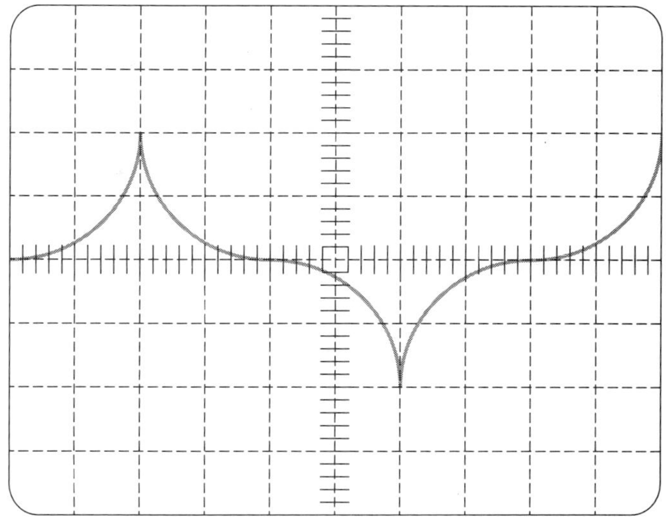

FIGURE 17–27 An AC pulse waveform of a PM generator pick-up unit.

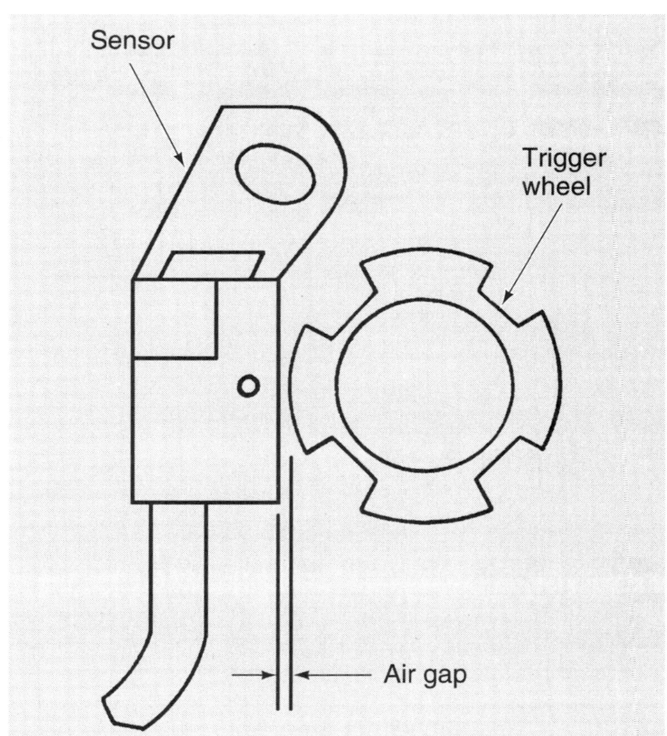

FIGURE 17–28 Typical metal detector sensor.

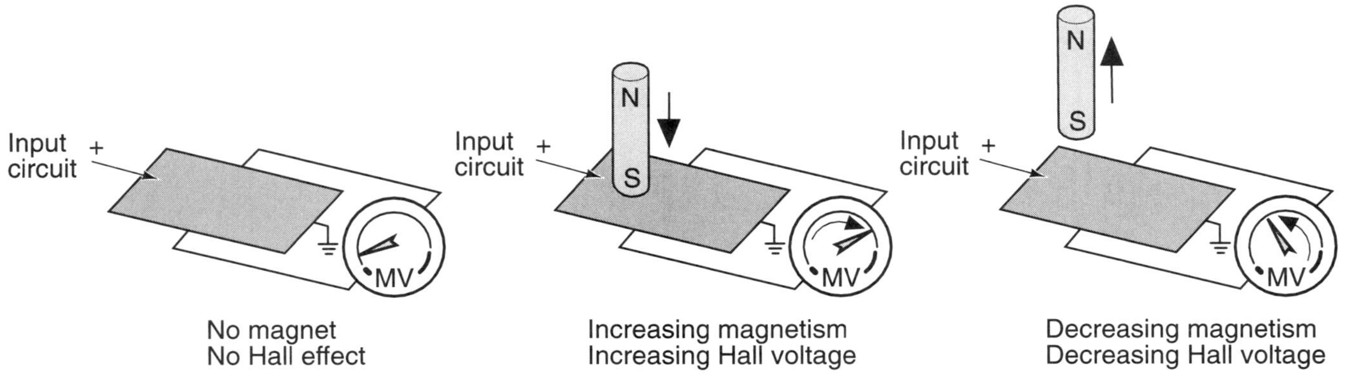

FIGURE 17–29 Hall-effect principles of voltage induction.

Hall Effect Sensor Since early in 1982, the **Hall effect sensor** has worked well as a precise triggering mechanism for the primary ignition system. Unlike the PM generator or the metal detection sensor, the Hall effect sensor produces an accurate square wave voltage signal throughout the entire rpm range of the engine. The Hall effect switching method is based on the principle that if a current is allowed to flow through a thin conducting material, and when that material is exposed to a magnetic field, another voltage is produced (Figure 17–29).

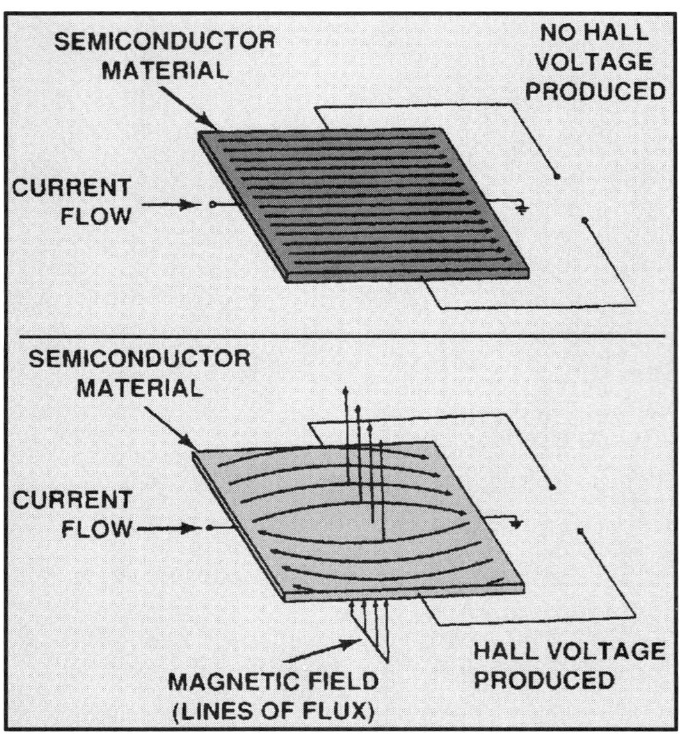

FIGURE 17–30 Hall-effect semiconductor material producing a small voltage.

(Courtesy of Ford Motor Company)

The heart of the Hall effect system is a thin semiconductor gallium arsenate crystal with one positive and one negative terminal that provides source current for the Hall transformation, and two additional terminals that form the signal output from the Hall layer. Directly across from the semiconductor element is a small permanent magnet positioned so its magnetic flux lines bisect the Hall crystal at right angles to the current flow (Figure 17–30).

When current is passed through the Hall layer, a voltage is produced perpendicular to the direction of current flow and magnetic flux (Figure 17–31). The signal voltage that is produced is the direct result of the magnetic field's effect on the electron flow. As the magnetic lines of force collide with the electrons in the supply current, a weak voltage potential is produced through the crystal from end to end.

Anytime the Hall switch is being exposed to a magnetic field, it is on and producing a very weak voltage signal. To shield the Hall switch from the magnetic field, a shutter wheel (Figure 17–32) consisting of a series of alternating windows and vanes passes between the Hall layer and permanent magnet to create a magnetic shunt that blocks the magnetic field (Figure 17–33). As a result, the magnetic field is concentrated in the shutter blade, and the electrons in the supply current are no longer disrupted and return to a normal state. This results in a low voltage potential in the signal circuit portion of the Hall switch.

Before the Hall signal can be of any use, it must be modified (Figure 17–34). The weak voltage signal is first sent to an amplifier which strengthens and inverts it

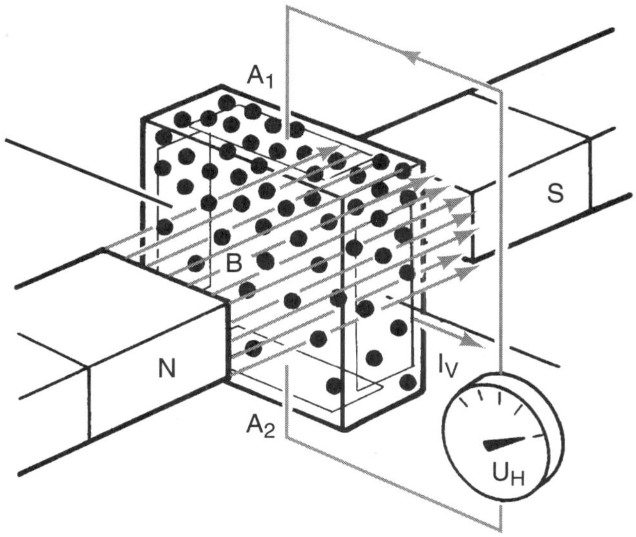

U_H = Hall voltage
B = Magnetic field (flux density)
Ic = Constant supply current
A_1, A_2 = Hall layer

FIGURE 17–31 Action of a Hall-effect switch.

so the signal reads *high* when it is actually coming in *low* and vice versa. Once it has been inverted, the signal goes through a pulse shaping device called a **schmitt trigger** where it is cleaned up to be a square wave digital signal. The square wave signal is then fed to the base of a

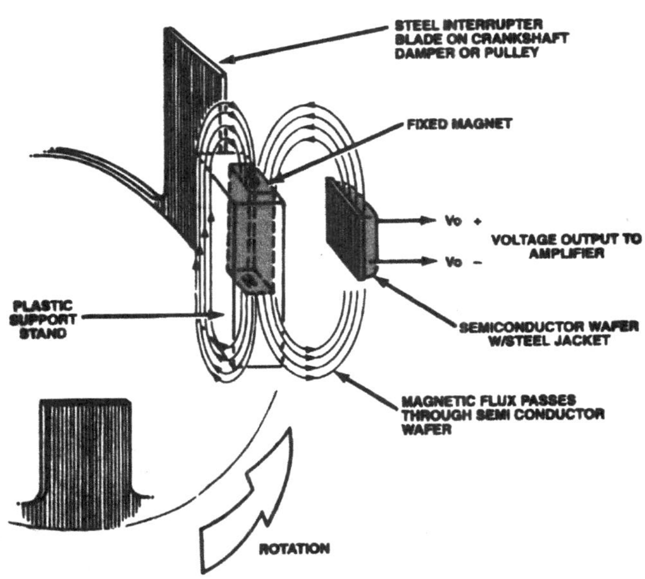

FIGURE 17–32 A Hall-effect switch and fixed magnet.

(Courtesy of General Motors Corporation, Service Technology Group)

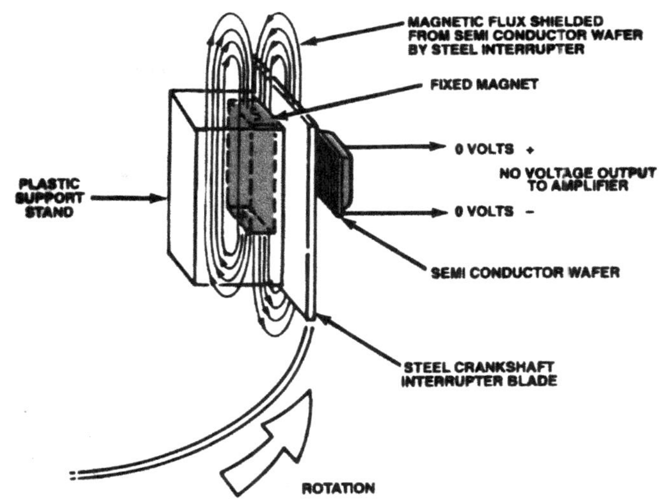

FIGURE 17–33 The shutter blade creates a magnetic shunt that blocks the magnetic flux lines.

(Courtesy of General Motors Corporation, Service Technology Group)

switching transistor to turn it on and off in response to the signals generated by the Hall switch assembly. The computer's voltage sensing circuit senses the switching transistor being turned on and off, and then determines the frequency of the signals and calculates engine speed.

Most Hall-effect sensors can be tested by connecting a 12 volt battery across the positive and negative supply voltage terminals of the Hall switch, and a voltmeter across the negative and signal voltage terminals. With the voltmeter hooked up, insert a feeler blade between the Hall switch and the permanent magnet to shield the switch from the magnetic field (Figure 17–35). The voltmeter should read within 0.5 volts of battery source voltage. When the feeler gauge is removed, the voltage should read less than 0.5 volts.

The Hall-effect switch activity can also be viewed on a running engine with an oscilloscope. Set the scope on an AC voltage scale and connect the scope positive lead to the Hall signal lead, and the scope negative lead to ground or the ground terminal at the Hall switch connector. With the engine running, the scope pattern should be a square wave pattern ranging from approximately zero volts to 12 volts (Figure 17–36). If the pattern is distorted or the range is out of specifications, the Hall switch should be replaced.

Photoelectric Sensor The last type of sensor used to control the ignition primary circuit is the **photoelectric**

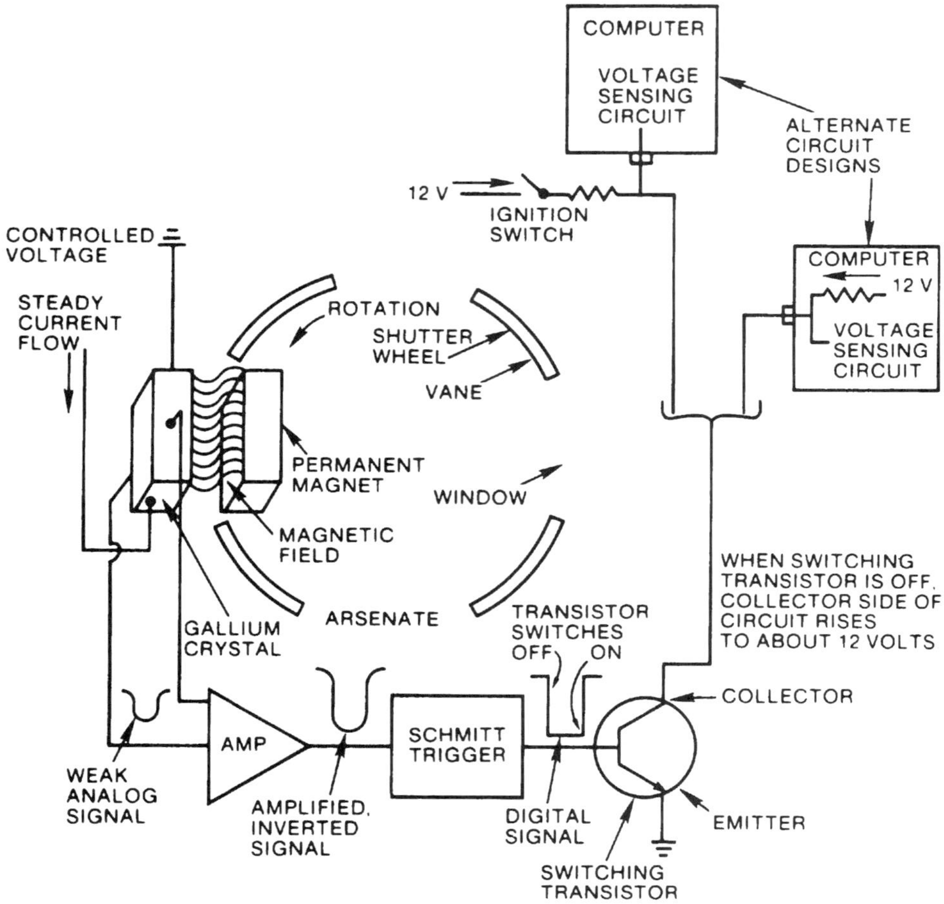

FIGURE 17–34 Typical Hall-effect switch circuit.

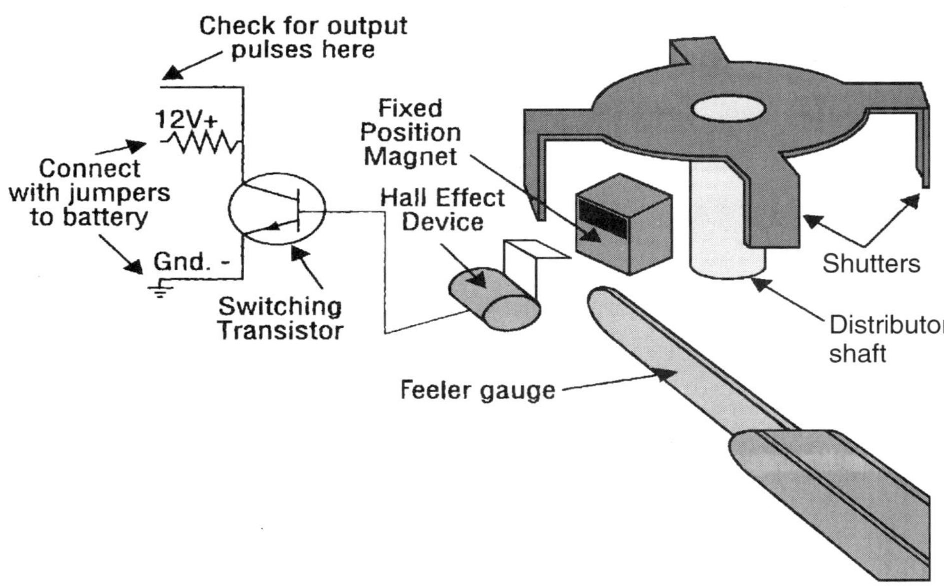

FIGURE 17–35 Testing a Hall-effect switch with a feeler gauge, voltmeter, and a 12 volt battery.

(Courtesy of The Fluke Corporation)

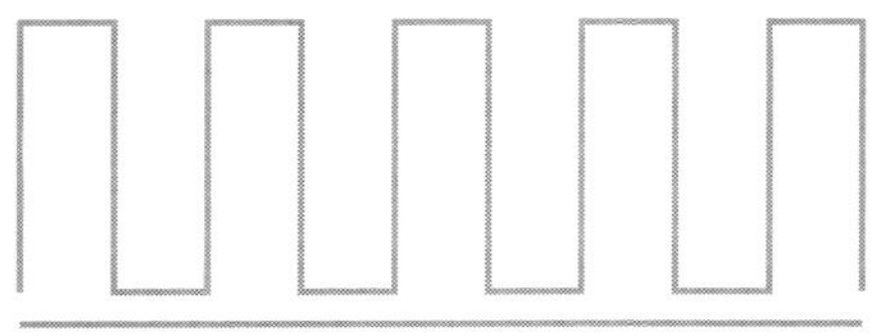

FIGURE 17–36 Square wave trace produced by a normal Hall-effect switch.

sensor. It consists of a light emitting diode (LED), light sensitive photo cells, and a slotted disc called a light beam interrupter (Figure 17–37). The slotted disc is attached to the distributor shaft and rotates between the LED and the photo cells. Light from the LED shines through the slotted discs to the photo cells which translate the intermittent flashes of light into a voltage signal pulse similar to a Hall-effect sensor. When the voltage signal occurs, the control unit turns the primary ignition system on until the voltage signal ceases from the photo cell.

The photoelectric sensor system in Figure 17–38 uses two photo diode cells: one for the 360 degree camshaft signal and the other a sixty degree signal corresponding to the number of cylinders and crankshaft degrees. The slotted disc would be for a six cylinder engine with a cylinder firing every 120 degrees of crankshaft rotation.

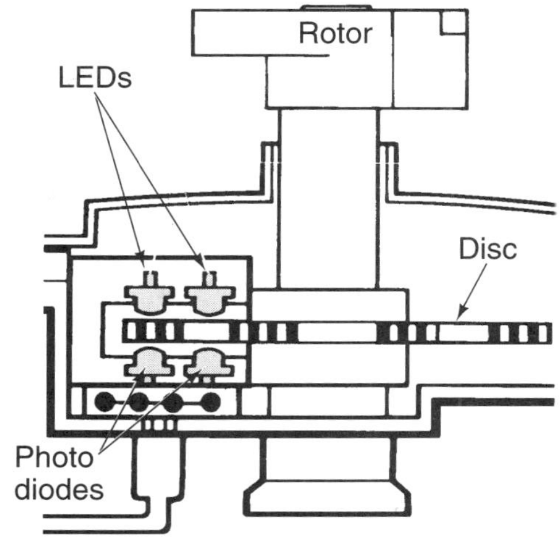

FIGURE 17–37 A distributor with a photoelectric sensor.

(Courtesy of DaimlerChrysler Corporation)

Timing Advance Mechanisms

Early electronic ignition distributors changed the ignition timing mechanically the same way that breaker point distributors did. The need to change ignition timing is determined by the burn time of the fuel, the engine speed changes, and load fluctuations. The ending result is the peak combustion power must occur just after the piston passes top-dead-center (TDC). There are two mechanical timing advance systems that accomplish this, centrifugal advance and vacuum advance.

Centrifugal Advance The **centrifugal advance** mechanism consists of a set of weights and springs connected to the distributor shaft and armature plate assembly (Figure 17–39). During idle speeds, the springs keep

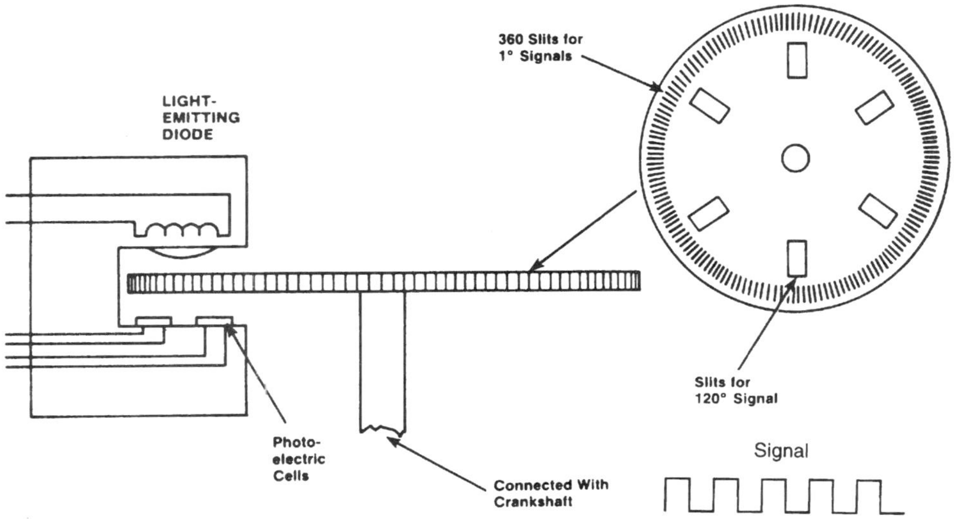

FIGURE 17–38 A distributor with a dual-photoelectric sensor.

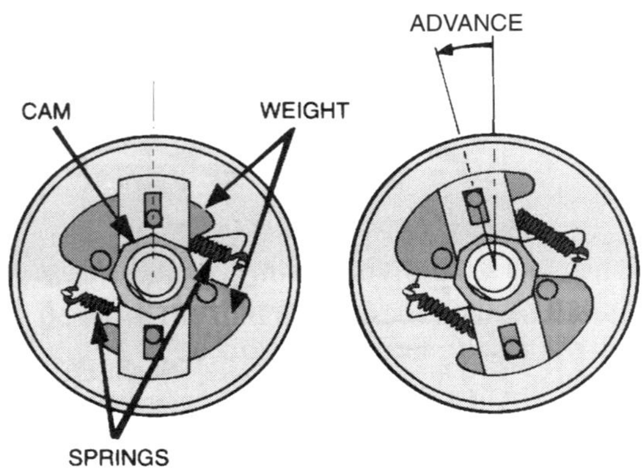

FIGURE 17–39 Typical distributor centrifugal advance operation.

(Courtesy of Allied Signal Automotive Aftermarket/Autolite)

the weights in their resting position and the armature and distributor shaft rotate as one assembly. When engine speed increases, centrifugal force causes the weights to move outward which allows the armature assembly to move ahead of (advance) the distributor shaft rotation. The ignition's triggering device is mounted to the armature plate assembly. The amount of centrifugal advance depends on engine speed, and usually follows a designed power curve (Figure 17–40).

Vacuum Advance When the engine is at part-throttle operation, the **vacuum advance** mechanism advances the ignition timing according to the vacuum that exists in the intake manifold. The burn rate of the fuel/air mixture at a part throttle and load is slower than at full throttle and heavy load. To ensure that peak combustion pressure occurs just after the piston passes TDC, the vacuum advance mechanism will advance the plate in the distributor that holds the primary switching device in relation to the distributor shaft (Figure 17–41). The heart of this system is the spring-loaded diaphragm in the vacuum advance unit. It is connected to a movable plate in the distributor containing the primary circuit's triggering switch components. When vacuum is applied to one side of the diaphragm in the advance unit, atmospheric pressure pushes on the other side of the diaphragm. This action causes the movable plate to turn and advance the ignition timing.

Control of the vacuum advance assembly is normally accomplished through a vacuum connection at the throttle plate. Depending on the ignition system design, there are two different sources for the vacuum advance. The first is **manifold vacuum,** which is the vacuum that is below the throttle plate (Figure 17–42). With manifold vacuum control, the vacuum advance unit is activated anytime there is high manifold vacuum.

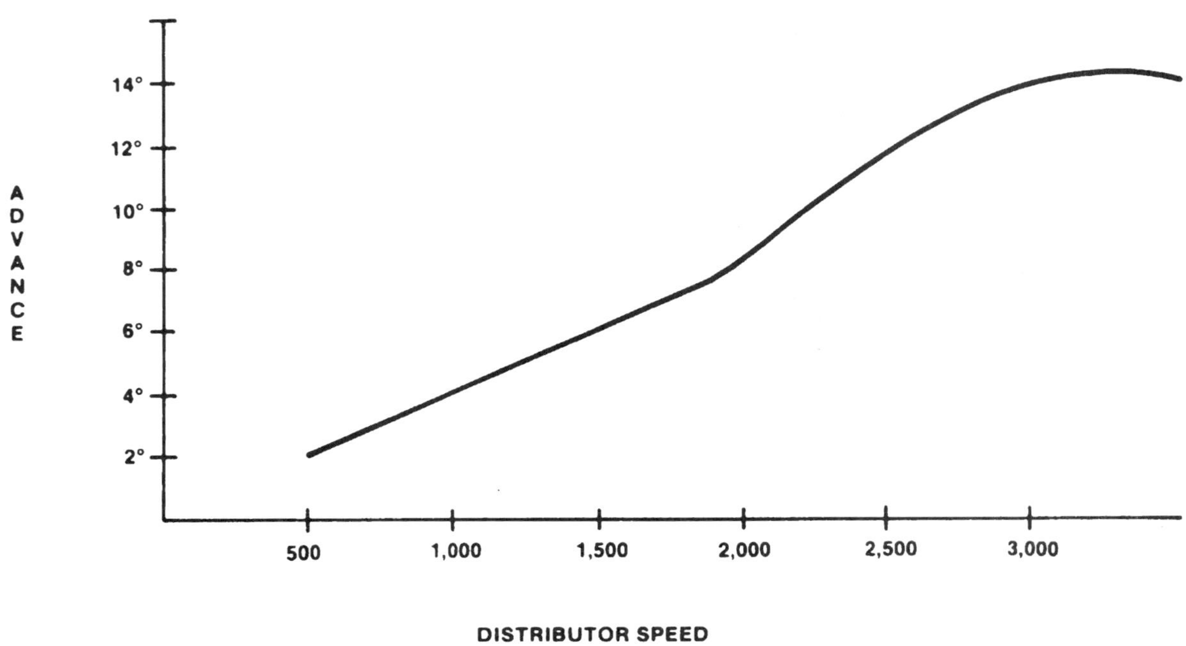

FIGURE 17–40 Engine speed affects centrifugal advance operation.

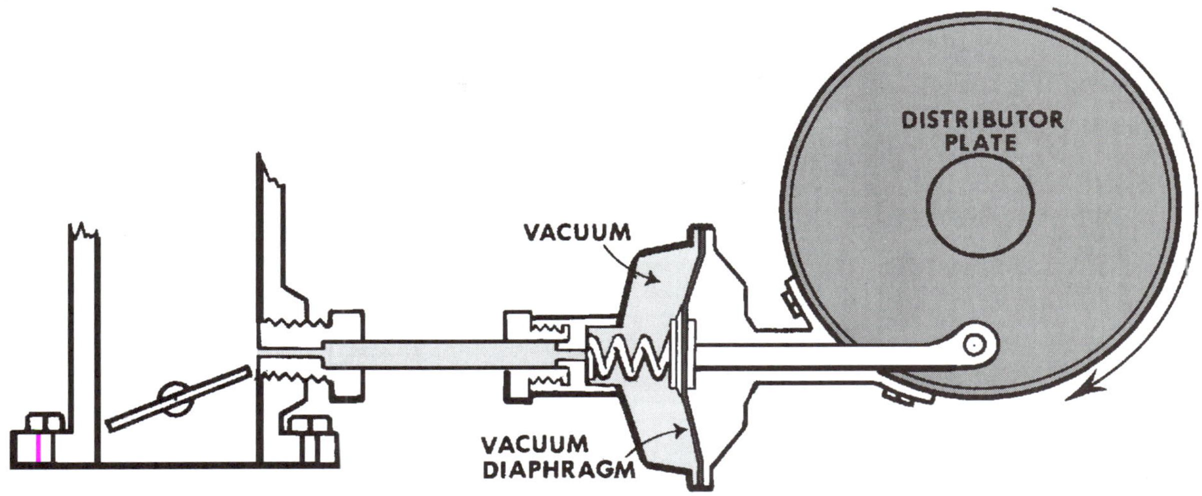

FIGURE 17–41 Typical action of a vacuum advance mechanism.
(Courtesy of The Dana Corporation)

The second source for vacuum advance is **ported vacuum,** which is the vacuum that is above the throttle plate (Figure 17–43). With ported vacuum control, the vacuum advance unit is activated only when the throttle plate opens and exposes the port to vacuum. This vacuum control gives the correct timing in relation to the quantity of air/fuel needed in the engine.

In some vehicles, for emission control purposes, the ignition systems use an additional vacuum diaphragm that retards the ignition timing (Figure 17–44). This type of system usually applies manifold vacuum to the retard diaphragm, and ported vacuum to the advance diaphragm.

Electronic-Controlled Timing Advance In an effort to increase engine fuel efficiency and improve tailpipe emissions, the centrifugal and vacuum advance mechanisms have been replaced in most modern automobiles by computer-controlled systems, especially those produced in North America. One of the main reasons for the change is the inability of the mechanical or vacuum advance systems to be able to recognize the individual timing needs of each cylinder.

With the elimination of the mechanical advance systems, the distributor's sole purpose is to generate the primary circuit switching signal and distribute the secondary voltage to the spark plugs. The ignition sys-

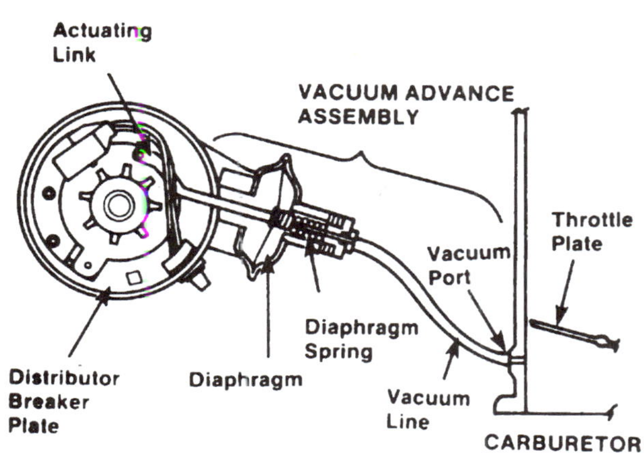

FIGURE 17–42 Vacuum advance assembly controlled by manifold pressure.

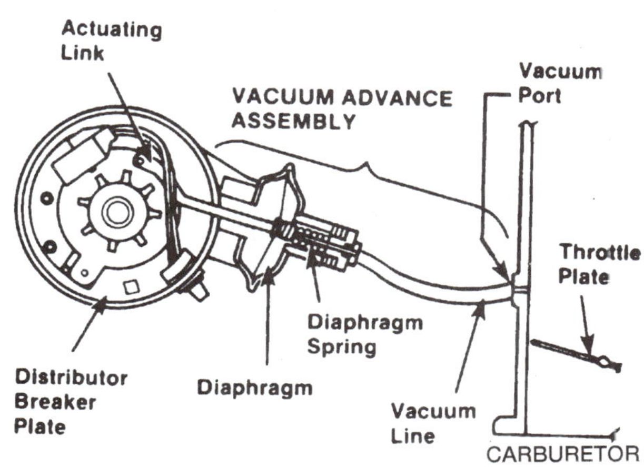

FIGURE 17–43 Vacuum advance assembly controlled by ported vacuum.

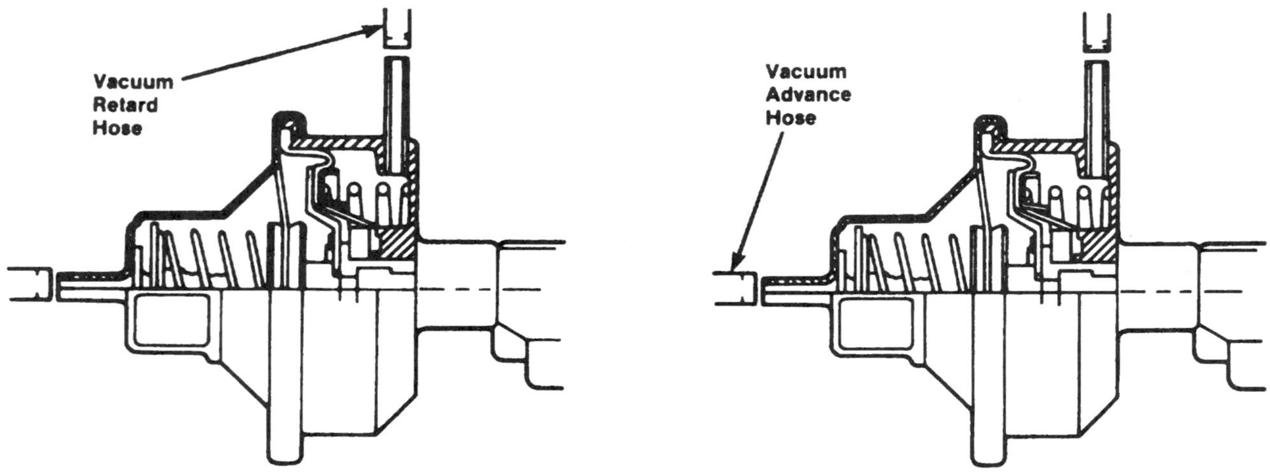

FIGURE 17–44 Typical vacuum advance/retard unit.
(Courtesy of Ford Motor Company)

tem in Figure 17–45 is an example of a distributor that has no centrifugal or vacuum advance units, but only the primary switching function. Timing advance is controlled totally by a microprocessor or computer. Some of these systems have even removed the primary switching mechanism from the distributor by using a crankshaft position sensor. The function of the distributor in this case is only to distribute the secondary voltage to the spark plugs.

Electronic Ignition Systems

In an effort to replace the ignition distributor with systems that have no moving parts, **electronic ignition (EI)** systems have emerged to electronically perform all the functions once performed by the distributor (Figure 17–46). In an EI system, the basic purpose is to provide the same ignition functions that a distributor has done in the past, but with no moving parts. The earlier EI sys-

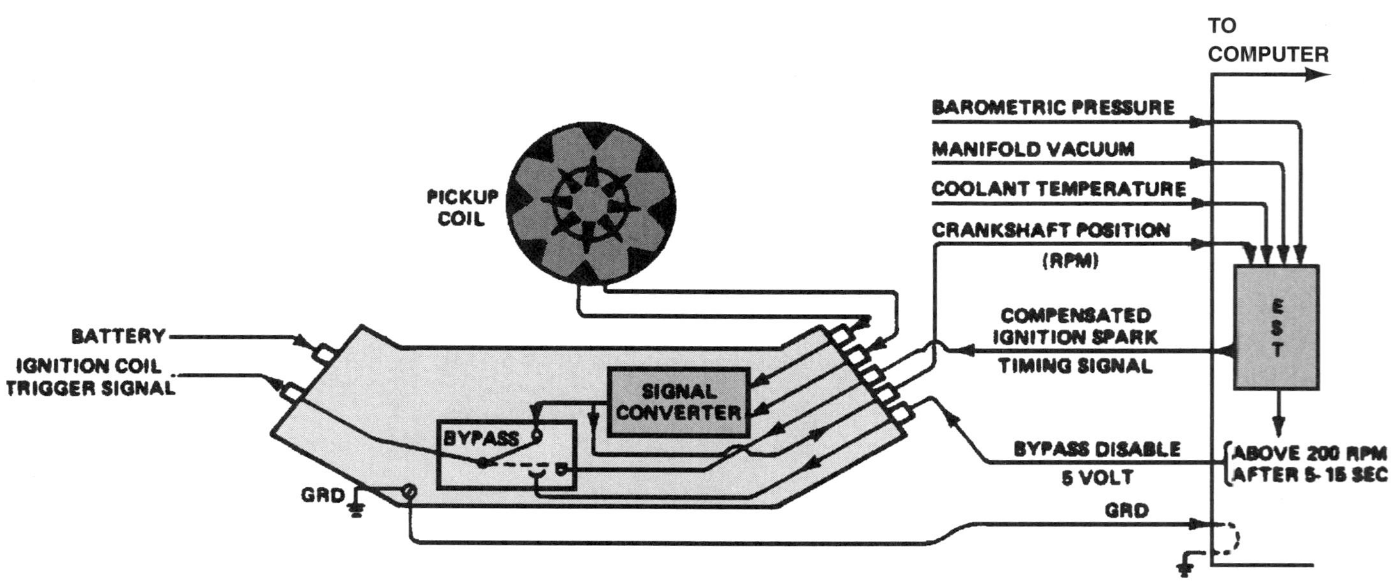

FIGURE 17–45 Computer-controlled ignition system.
(Courtesy of General Motors Corporation, Service Technology Group)

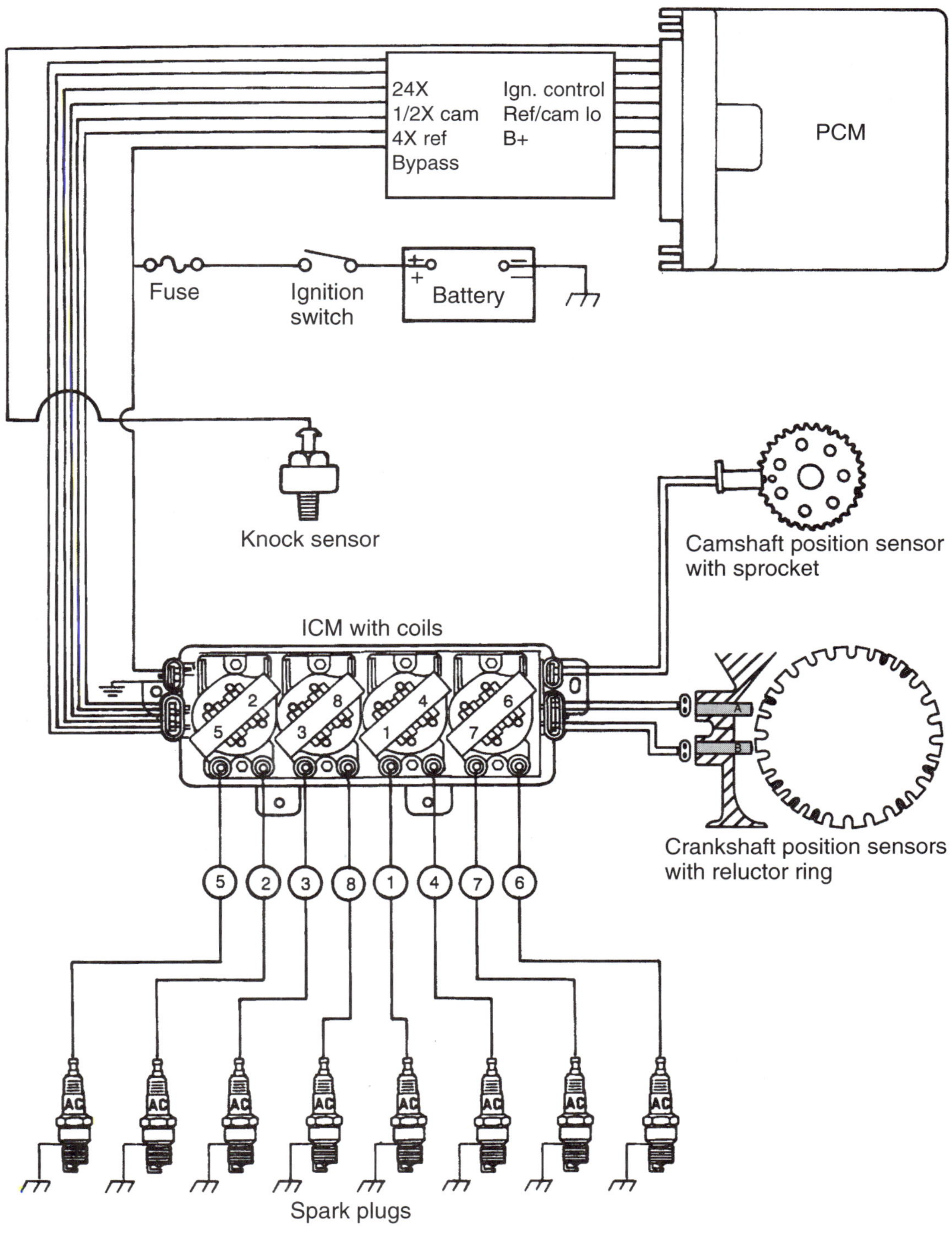

FIGURE 17–46 A typical electronic ignition (EI) system.
(Courtesy of General Motors Corporation, Service Technology Group)

tems were called distributorless ignition systems (DIS), but with the standardization of the SAE J1930 terminology, the term electronic ignition (EI) replaces all previous terms for distributorless ignition systems.

In an EI system the function of an ignition coil is the same as in the previously discussed ignition systems. The coil transforms battery voltage to a high voltage potential that jumps the spark plug gap. On most EI sys-

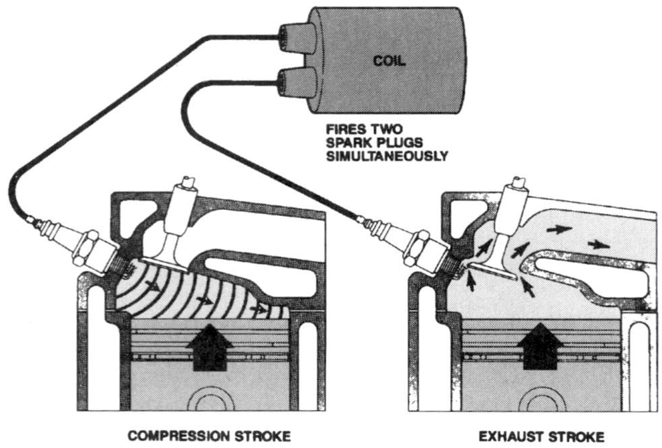

FIGURE 17–47 A typical waste spark ignition system.
(Courtesy of Champion Spark Plug)

tems, there is one coil for every two spark plugs. This system is called the **waste spark** method of spark distribution. Each coil secondary winding is connected to a pair of spark plugs in those cylinders that have pistons that rise and fall together (Figure 17–47). When one cylinder is on the end of its compression stroke, the other is on the exhaust stroke. Both cylinders get spark simultaneously, but only one spark generates power while the other is wasted out the exhaust.

When the induced voltage that cuts across the primary and secondary windings collapses, one spark plug fires in the normal direction from the center electrode to the outer electrode. The other spark plug fires in the reverse direction from the outer electrode to the center electrode to complete the series circuit. Each plug always fires in the same direction on both the compression and exhaust stroke (Figure 17–48).

Even though there is a reversed polarity in one of the spark plugs, the coil is able to produce a voltage as high as 100,000 volts to fire the spark plugs. There is very little resistance across the plug gap on the exhaust stroke, so the plug requires very little voltage fire; thereby providing the plug on the end of the compression stroke with enough available voltage.

In a typical EI system (Figure 17–49), the ignition module uses crank/cam sensor data to control the timing of the primary circuit in the coils. The computer collects and processes this information to determine the ideal amount of spark advance for the operating conditions. The ignition module synchronizes the coils firing sequence in relation to the crankshaft position and firing order of the engine. The ignition module is the component that takes the place of the distributor to distribute the high voltage spark to the cylinders.

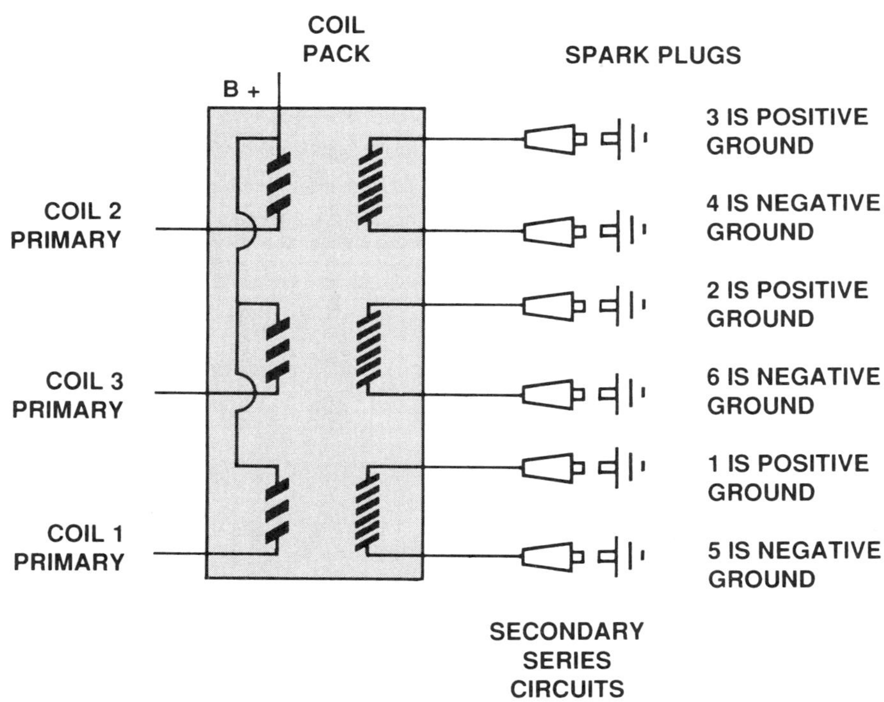

FIGURE 17–48 Spark plug firing for a 6-cylinder EI system.
(Courtesy of Ford Motor Company)

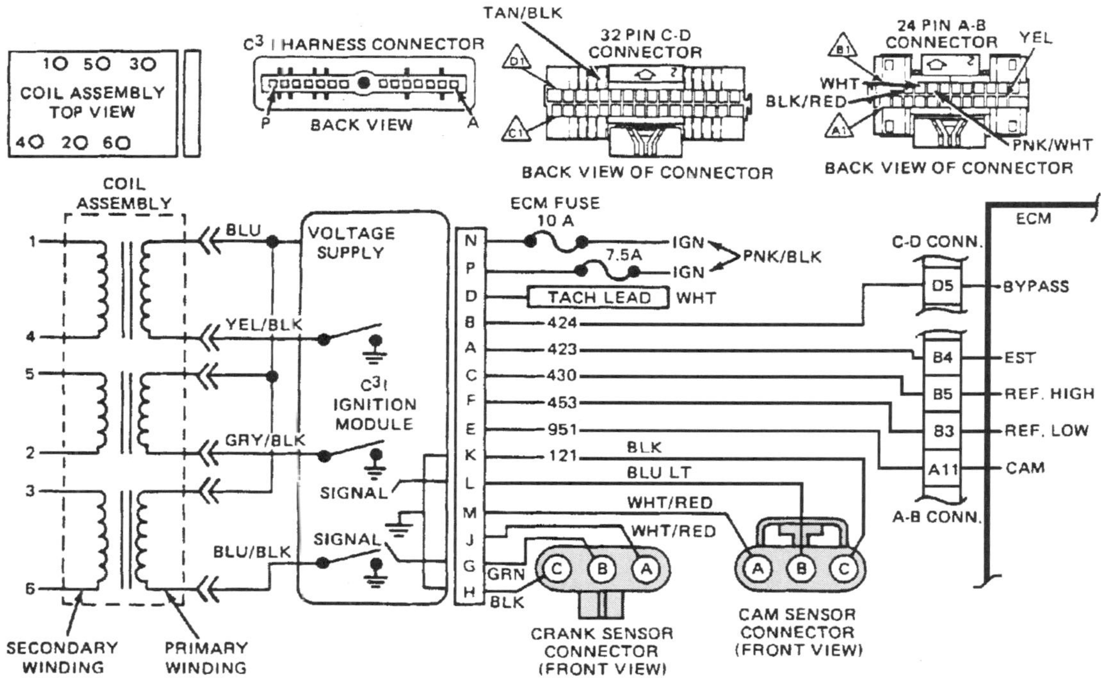

FIGURE 17–49 Schematic of a typical electronic ignition (EI).
(Courtesy of General Motors Corporation, Service Technology Group)

Figure 17–50 shows an example of a waste spark system called the **coil-over-plug** system where the coils are mounted directly over the spark plugs. This eliminates the need for plug wires between the coils and plugs. This type of system functions the same as the other waste spark systems.

On other EI systems, the coil packs are mounted remotely from the spark plugs, which requires using high tension secondary wires from the coil pack to the spark plug wires (Figure 17–51).

Spark Plugs and Plug Wires

Spark Plugs Every type of ignition system uses spark plugs to ignite the fuel/air mixture in the cylinder. There are three main parts to a spark plug—the steel shell, the ceramic insulator, and a pair of electrodes (Figure 17–52). The steel shell holds the ceramic core and electrodes in a gas-tight assembly, and contains the threads and sealing seat.

The center core on most spark plugs contains a resistor between the terminal and the center electrode. The purpose of this resistor is to control **radio frequency interference (RFI)** by suppressing AC voltage spikes while the spark plug is firing. Always use a resistor spark plug when one is specified for the engine. The additional radio frequency interference produced by a non-resistor plug could interfere with the operation of the radio or computer sensor.

Automotive spark plugs are available in either 14 or 18 millimeter diameters. All 18 mm plugs feature a tapered seat that seals with a mating seat in the cylinder head. The 14 mm plug can have either a tapered seat or a flat metal gasket for sealing (Figure 17–53).

Spark plug reach is an important design characteristic that could cause severe engine damage if it is wrong. The spark plug must be properly placed in the combustion chamber to adequately expose the electrodes to the fuel/air mixture (Figure 17–54). If the spark plug reach is too short, the electrodes may not fire adequately in the fuel/air mixture, and carbon deposits may develop in the formed pocket. If the spark plug reach is too long, the spark plug may interfere with the operation of the piston and/or valves, causing severe engine damage. Exposed plug threads may also get so hot that pre-ignition of the fuel/air mixture may occur.

The spark plug heat range is another important design feature that is important for the proper operation and long life of the spark plug (Figure 17–55). The heat range indicates the ability of the spark plug to conduct

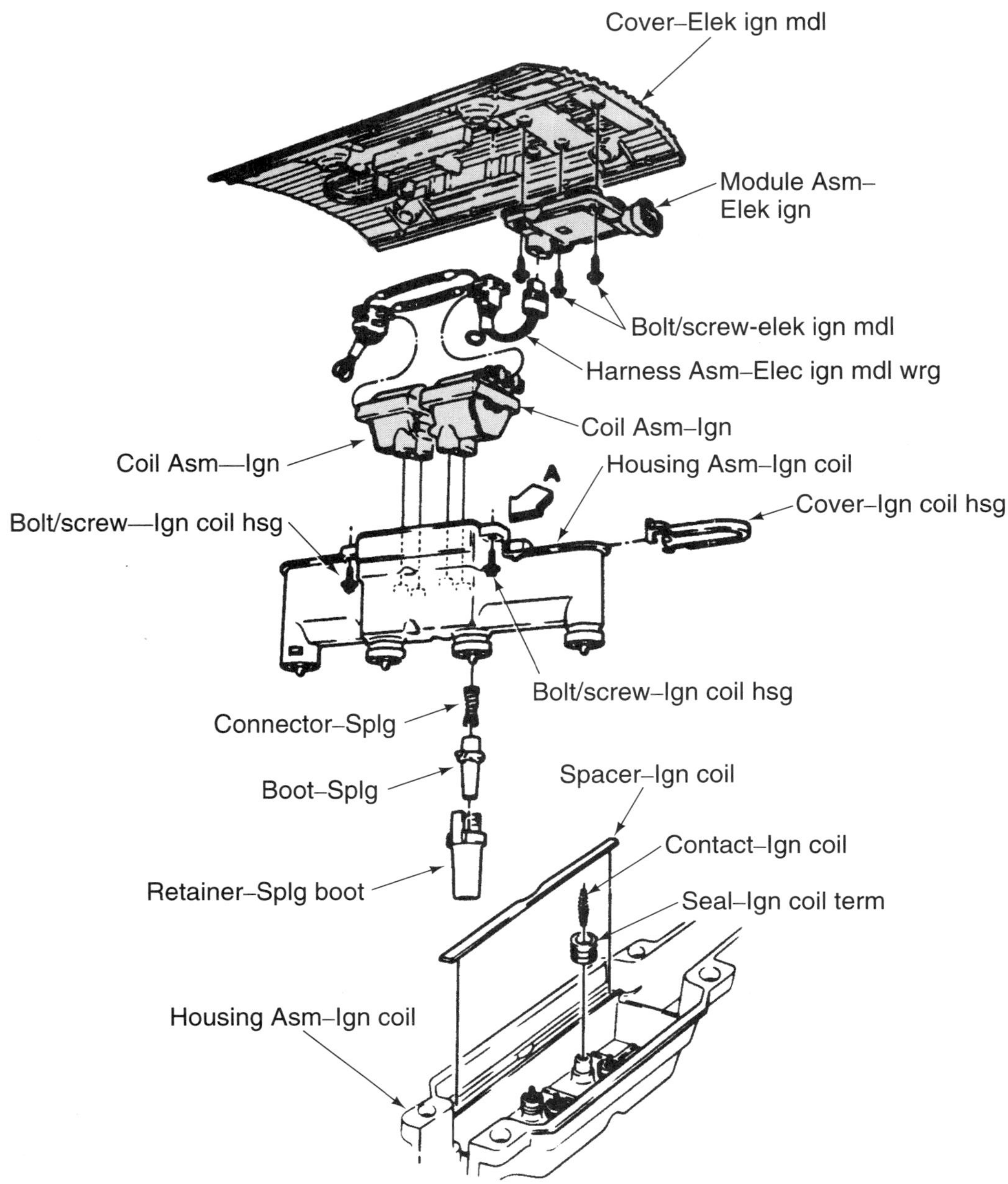

FIGURE 17–50 GM's Quad-4 waste spark system with ignition coils connected directly to the spark plugs. (Courtesy of General Motors Corporation, Service Technology Group)

heat away from the center electrode tip. A cooler plug will transfer away the heat rapidly. A hotter plug retains the heat longer. Installing a plug with the correct heat range is important. The spark plug must remain hot enough to burn away deposits and prevent fouling, but not hot enough to melt the plug or cause a pre-ignition condition. Continuous heavy driving conditions require a cooler plug than if the vehicle does a lot of stop-and-go driving.

The gap of the spark plug between the center electrode and the grounding electrode is important for proper operation of the plug. Figure 17–56 shows an example of a correct spark plug gap being set by a gap gauge. When the plug is new, the electrode surfaces are squared off and perpendicular to one another. Using a spark plug with an incorrect gap (Figure 17–57) will cause the outer electrode to be bent at an angle and shorten the plug's service life.

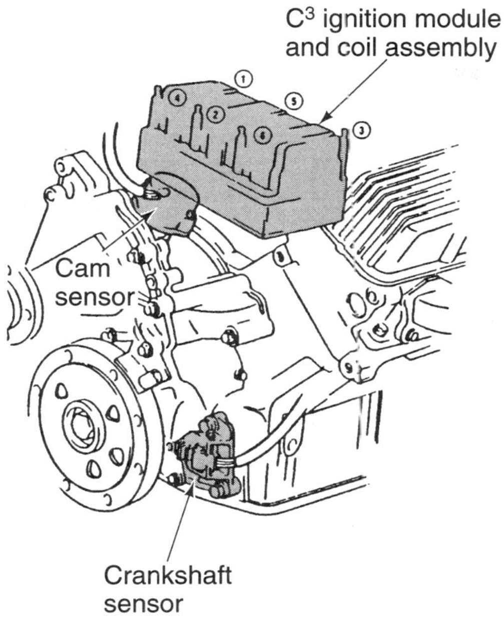

FIGURE 17–51 Coil pack and sensor locations in an electronic ignition (EI) system.

(Courtesy of General Motors Corporation, Service Technology Group)

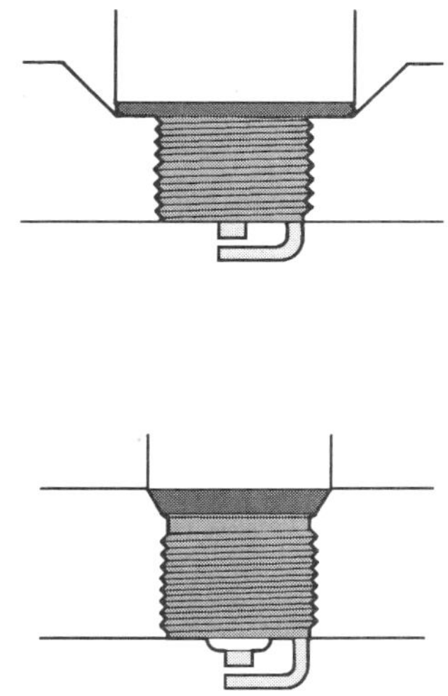

FIGURE 17–53 Spark plug sealing seat designs.

(Courtesy of General Motors Corporation, Service Technology Group)

Spark plugs can provide a lot of information about the internal condition of the engine combustion chamber (Figure 17–58). A normally worn spark plug has minimal deposits on it, and is light tan or gray in color. The electrode gap on a used plug should have no more than a 0.001 inch change for every 10,000 miles of engine operation compared to original gap setting.

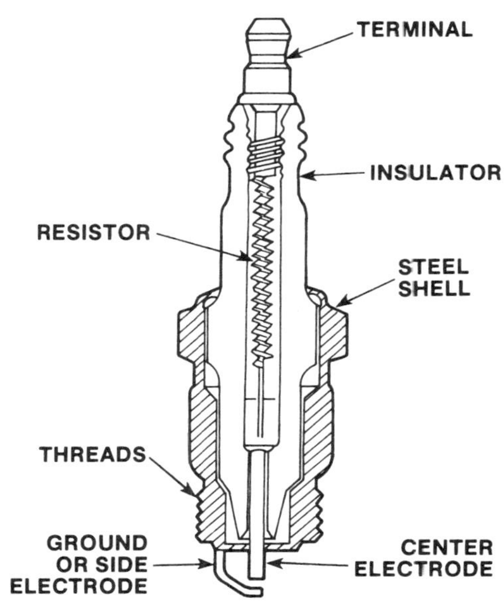

FIGURE 17–52 Components of a typical spark plug.

Plug Wires The ignition cables, or plug wires, deliver the high voltage from the ignition coil to the individual spark plugs. Because this voltage can reach up to 100,000 volts on some of the modern electronic ignition (EI) systems, the insulation surrounding the conductor is just as important as the conductor. Figure 17–59 shows an example of a flexible nylon core wire that is impregnated with carbon. The use of carbon in the core suppresses RFI because it offers resistance to the spark and prevents an AC current from being generated as the spark plug fires. An ohmmeter placed at the two ends of this wire can read several thousand ohms per foot,

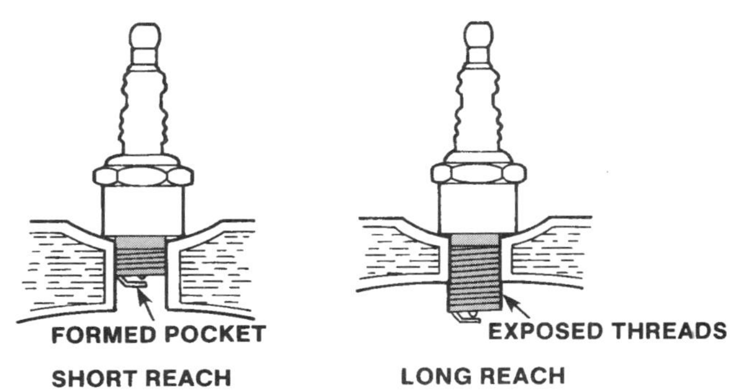

FIGURE 17–54 Incorrect spark plug reach conditions in combustion chamber.

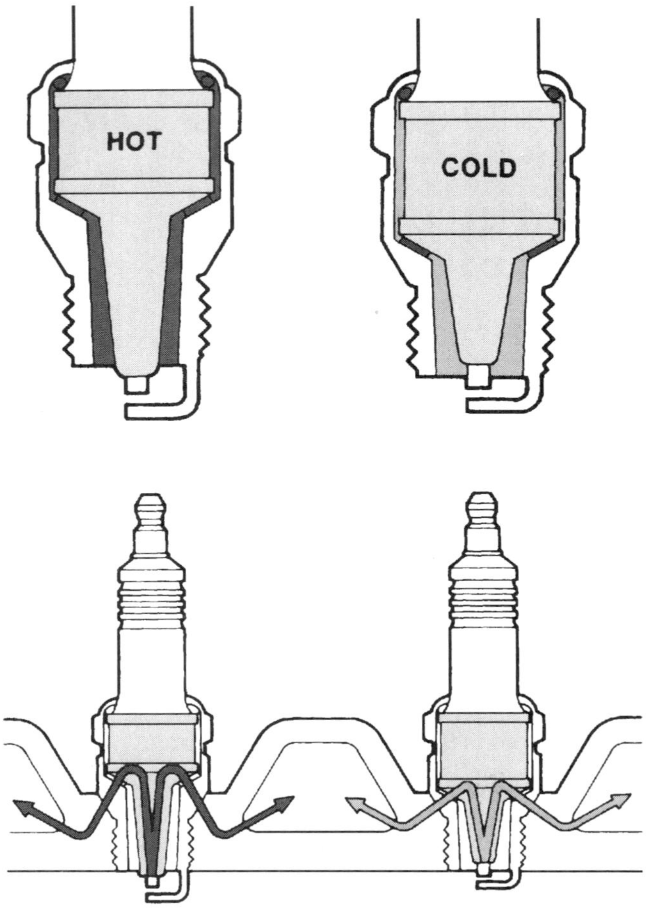

FIGURE 17–55 Heat range transfer of spark plugs.

(Courtesy of General Motors Corporation, Service Technology Group)

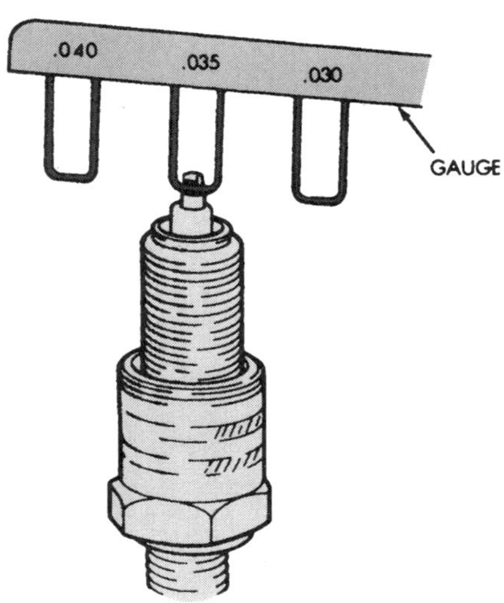

FIGURE 17–56 Setting the gap on a new spark plug.

(Courtesy of DaimlerChrysler Corporation)

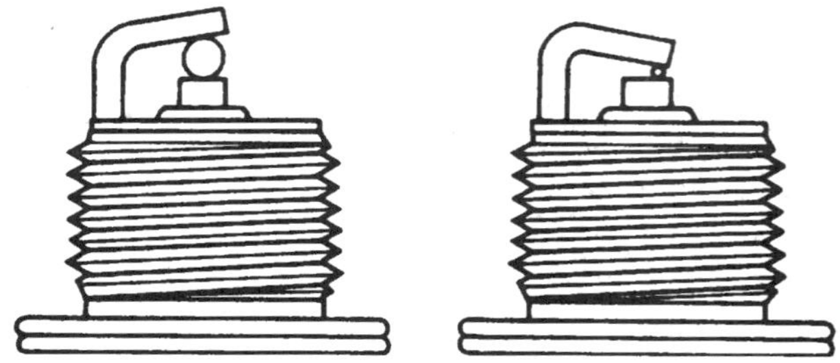

FIGURE 17–57 Incorrect spark plug gaps.

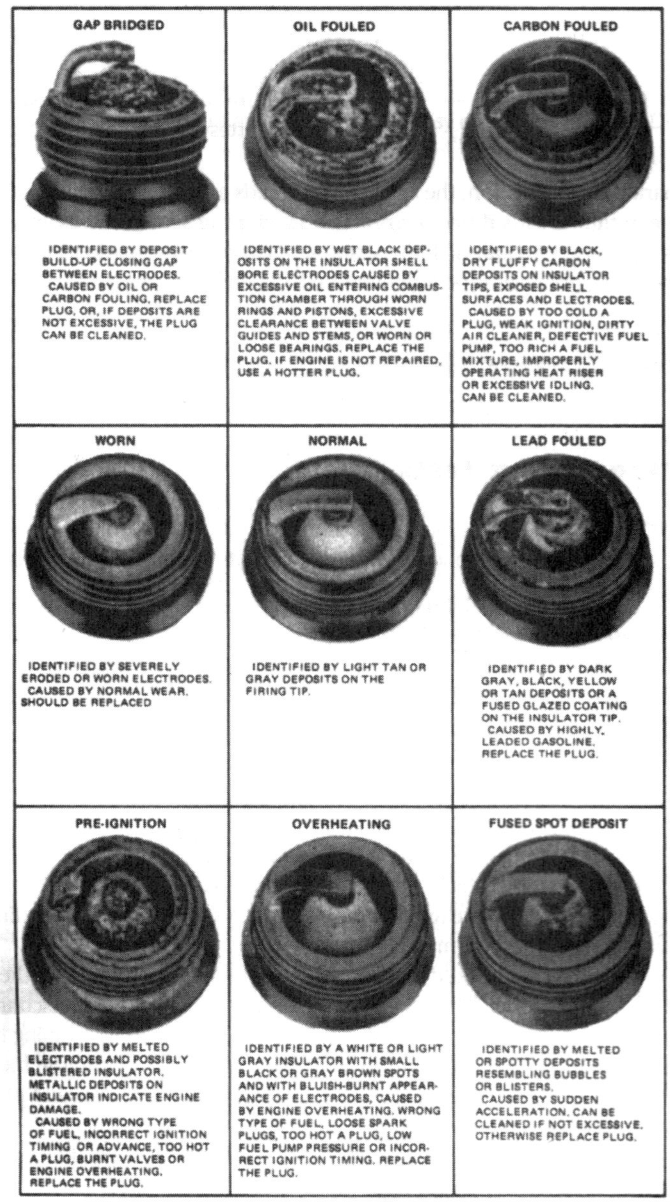

FIGURE 17–58 "Reading" a spark plug to determine internal engine condition.

(Courtesy of Ford Motor Company)

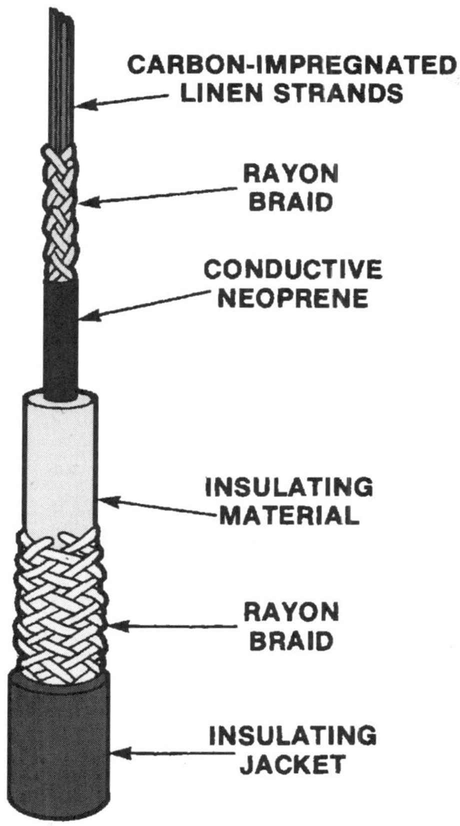

FIGURE 17–59 Typical carbon-impregnated secondary cable.

(Courtesy of General Motors Corporation, Service Technology Group)

depending on the manufacturer. Always refer to the vehicle service manual for detailed specifications.

Another type of spark plug wire that is used to control RFI is **magnetic suppression** wire (Figure 17–60). The principle behind this type of plug wire is the wire that is looped around the conductor in the center of the plug wire. This wound wire shields the magnetic field that is produced as current flows through the plug wire, thus preventing RFI from leaving the surface of the plug wire. A magnetic suppression wire may look the same as a typical plug wire, but the resistance specification is different. An ohmmeter placed at the two ends of the plug wire, will typically read zero ohms on a good wire, so refer to the vehicle service manual for actual specifications of plug wire.

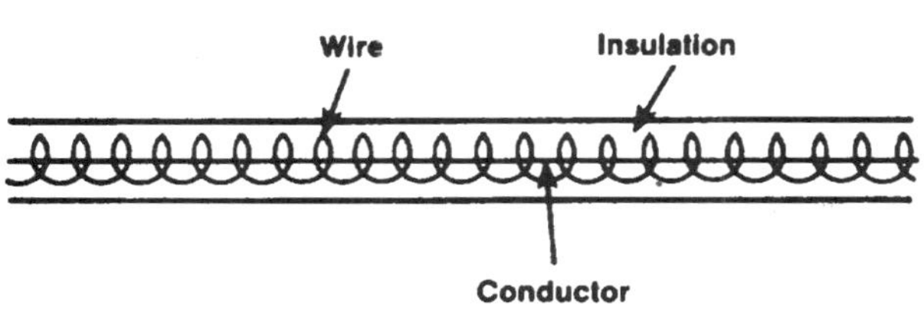

FIGURE 17–60 Typical magnetic suppression secondary cable.

Secondary Voltage Distribution Most ignition systems that have a distributor for distribution of the secondary voltage, perform this important function by using a distributor cap and rotor assembly. The distributor cap fits tightly to the top of the distributor housing with hooks, screws, clips, or screws (Figure 17–61).

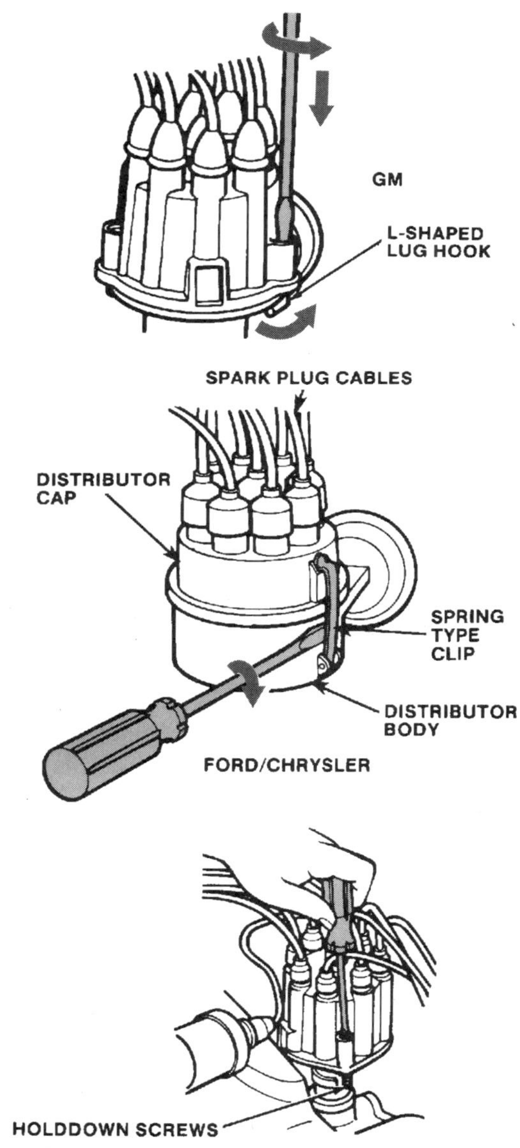

FIGURE 17–61 Typical distributor cap hold downs.

(Courtesy of DaimlerChrysler Corporation)

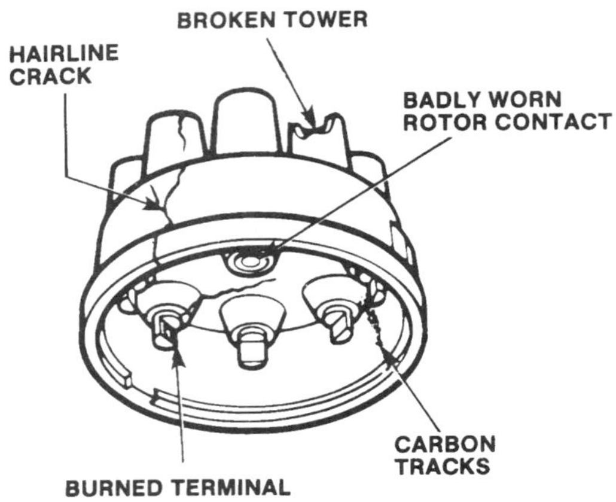

FIGURE 17–62 Typical distributor cap damage.

The distributor cap is typically made of a plastic or resin material that can be easily inspected for damage (Figure 17–62). Electrical damage from the secondary voltage can be corroded or burned metal terminals, and carbon tracking inside the cap. Carbon tracking is the formation of a carbonized dust line between the high voltage terminals across the cap surface.

The **rotor** assembly is mounted to the top of the distributor shaft inside the distributor. It is inspected and serviced the same as the distributor (Figure 17–63). As the distributor turns, the rotor directs the high voltage received from the ignition coil to the individual cylinders depending on the engine **firing order** (Figure 17–64). Spark plug wires from consecutively firing cylinders should cross each cylinder rather than run parallel to one another to prevent a cross firing condition.

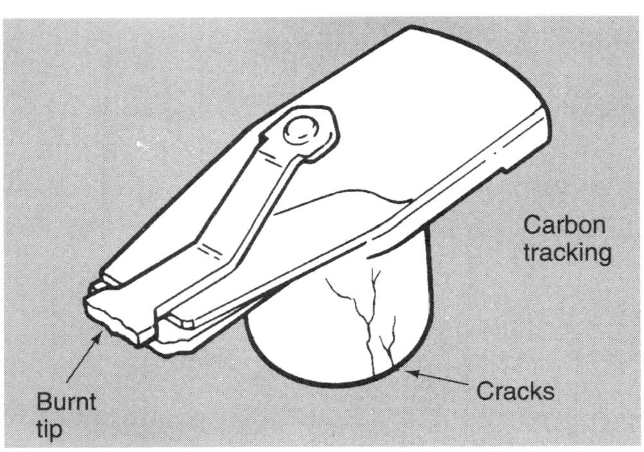

FIGURE 17–63 Typical rotor damage.

(Courtesy of Ford Motor Company)

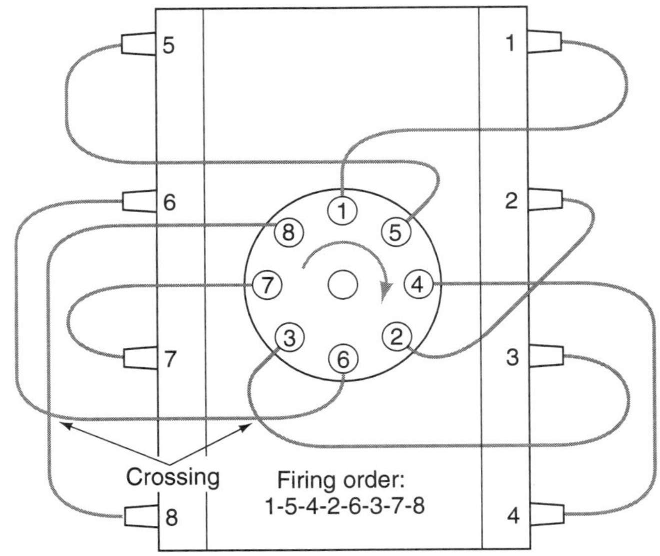

FIGURE 17–64 Typical firing order for a V-8 engine.

Basic Ignition System Testing

Setting Ignition Timing

For an ignition system to operate as designed, it must be timed to the engine. Vehicles with distributors normally operate from a base timing that is set while the engine is at idle. Refer to the vehicle's service manual for exact procedures for setting the ignition timing. Typical timing marks are shown in Figure 17–65. To read the timing marks requires using a special timing light (Figure 17–66) that emits a strobe light to illuminate the timing marks. Adjustment of the ignition timing is normally done by turning the distributor housing in the appropriate direction (Figure 17–67).

Oscilloscope Testing

An excellent way to check the operation of the ignition system is with an oscilloscope. The job of the oscilloscope is to convert the electrical signals of the ignition system into a visual image showing voltage changes over a period of time (Figure 17–68).

Oscilloscopes normally have four leads that connect to the ignition system (Figure 17–69). The #1 spark plug is normally the reference cylinder.

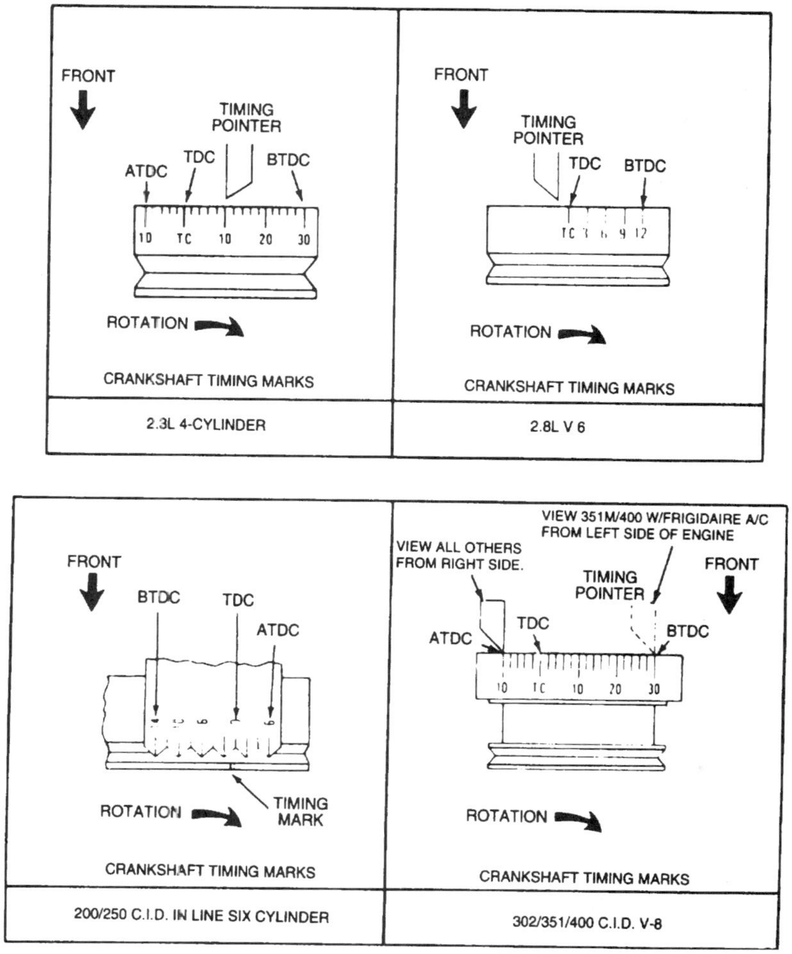

FIGURE 17–65 Typical ignition system timing marks.

(Courtesy of Ford Motor Company)

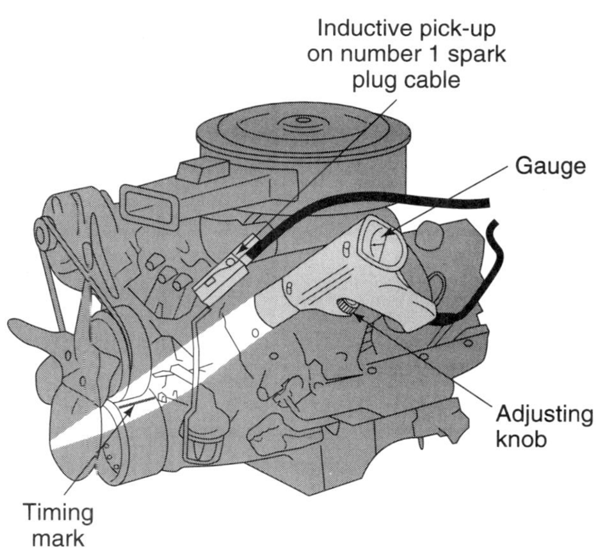

FIGURE 17–66 Using a timing light to observe ignition timing marks.

(Courtesy of Snap-on Diagnostics)

Depending on the oscilloscope function selected by the technician, the scope can display the voltage versus time pattern for either the secondary or primary circuit (Figure 17–70). The secondary pattern displays the following types of information.

- Spark plug firing voltage.
- Spark duration.
- Coil and condenser oscillations.
- Primary switching operation of transistor or breaker points.
- Dwell and cylinder timing accuracy.
- Secondary circuit accuracy.

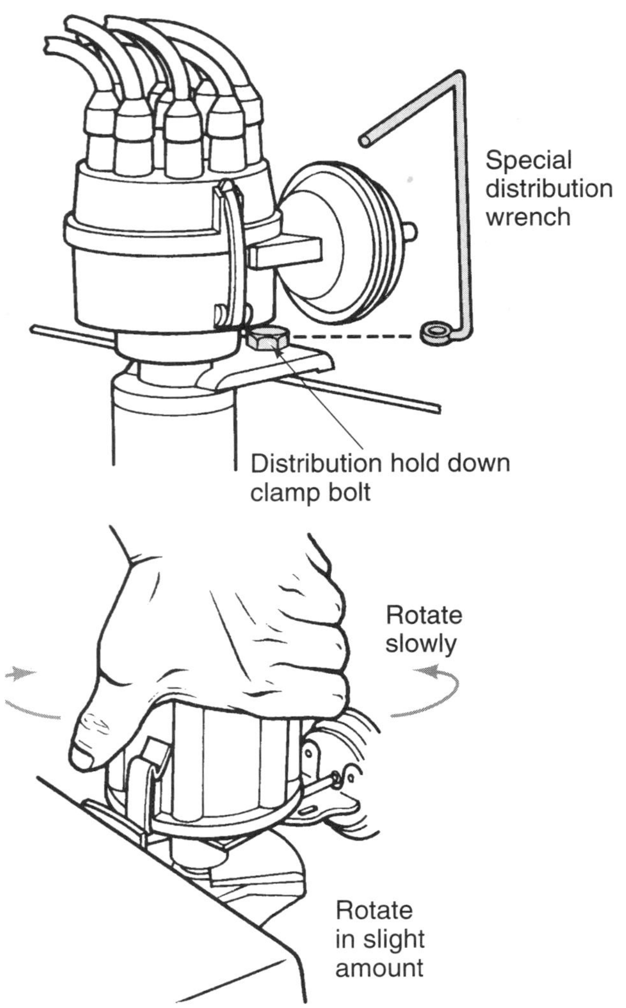

FIGURE 17–67 Setting the ignition timing by rotating the distributor housing.

(Courtesy of DaimlerChrysler Corporation)

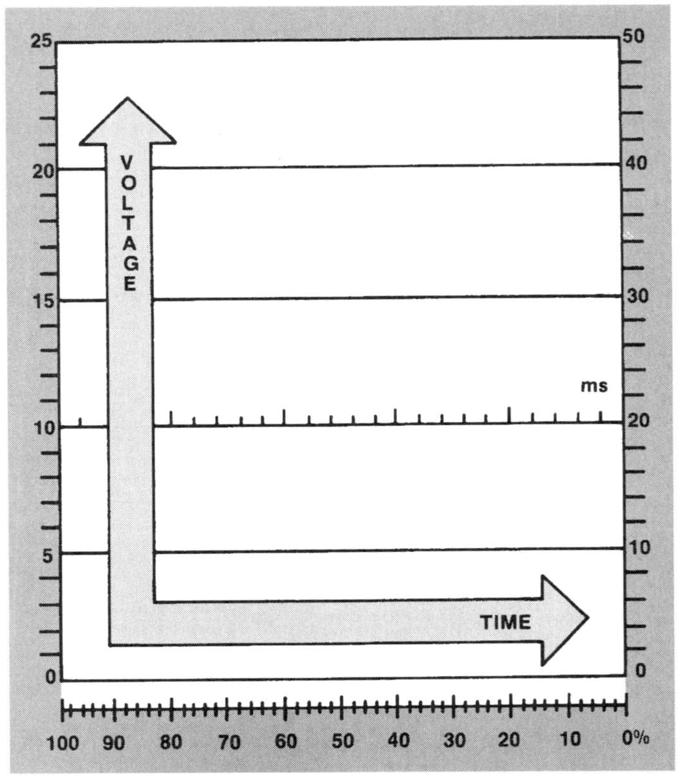

FIGURE 17–68 Scope display showing voltage and time scales.

The oscilloscope has several ways to display the voltage patterns of the primary and secondary circuits. When the display pattern is selected, the scope displays the patterns of all the cylinders in a row, from left to right (Figure 17–71).

Another scope pattern that can be used is a raster pattern that stacks the voltage patterns of each cylinder above each other for easy comparison (Figure 17–72). In a raster pattern, the patterns for each cylinder are displayed across the width of the graph, beginning with the spark line and ending with the firing line. This allows for close inspection of the voltage and time trends of the pattern.

A third scope pattern is the superimposed pattern that displays all the patterns on top of each other the full width of the screen (Figure 17–73). A superimposed pattern allows the technician to compare one cylinder's pattern to the others. A worn or bent distributor shaft would be easily seen with this pattern because one or more of the cylinder pattern lines would be out of line with the rest.

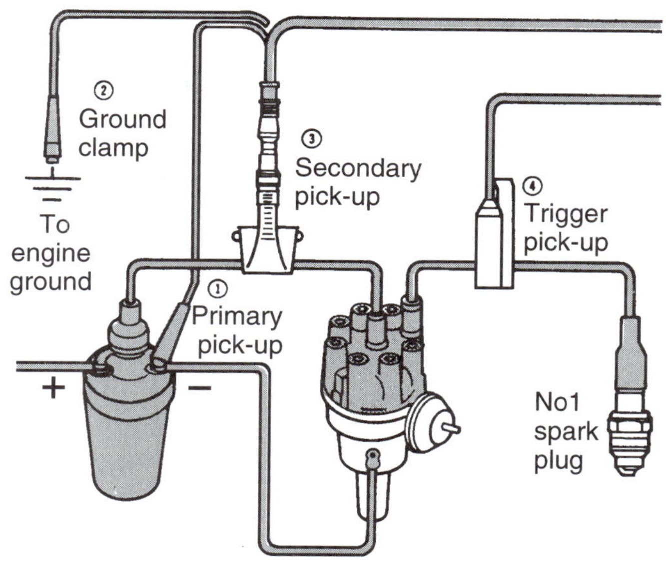

FIGURE 17–69 Typical hook-up of scope leads to an ignition system.

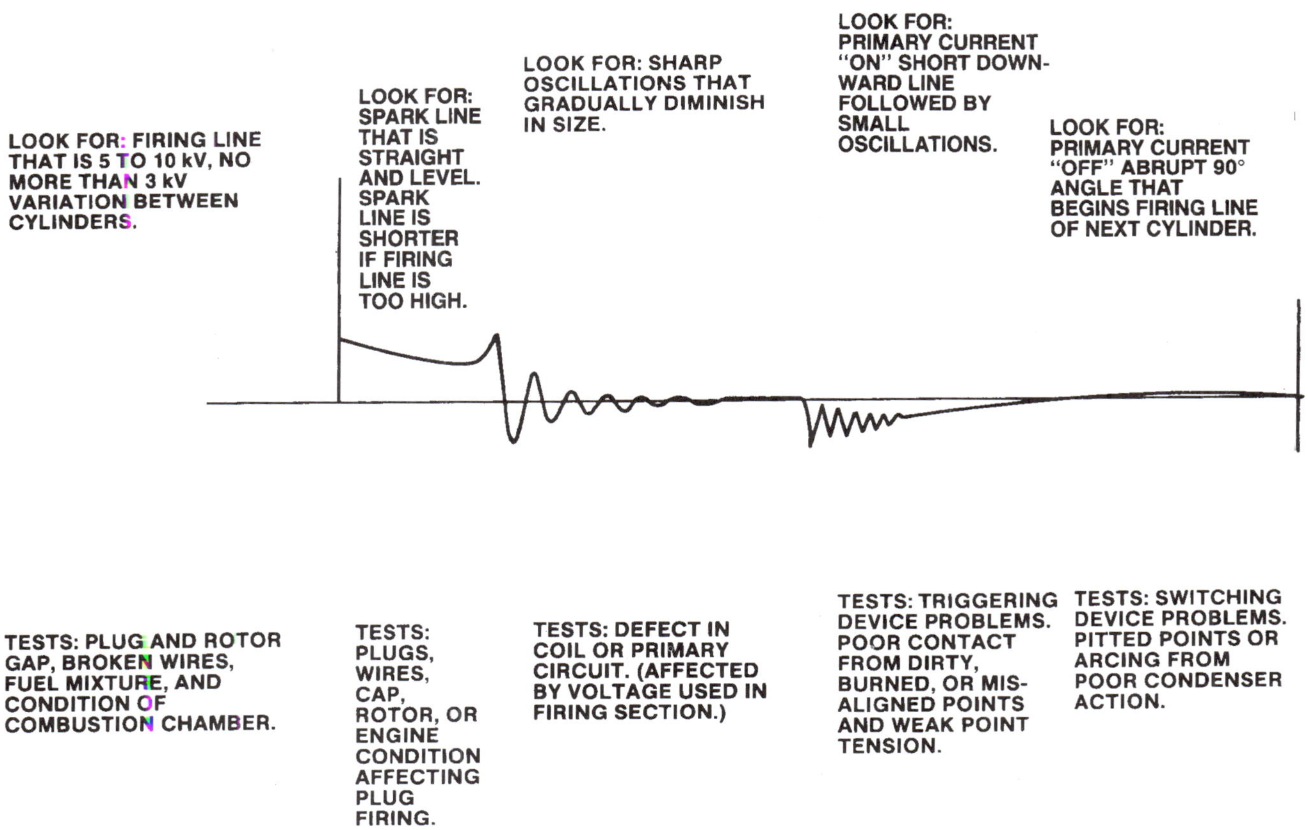

FIGURE 17–70 Typical secondary scope pattern.

(Courtesy of Snap-on Diagnostics)

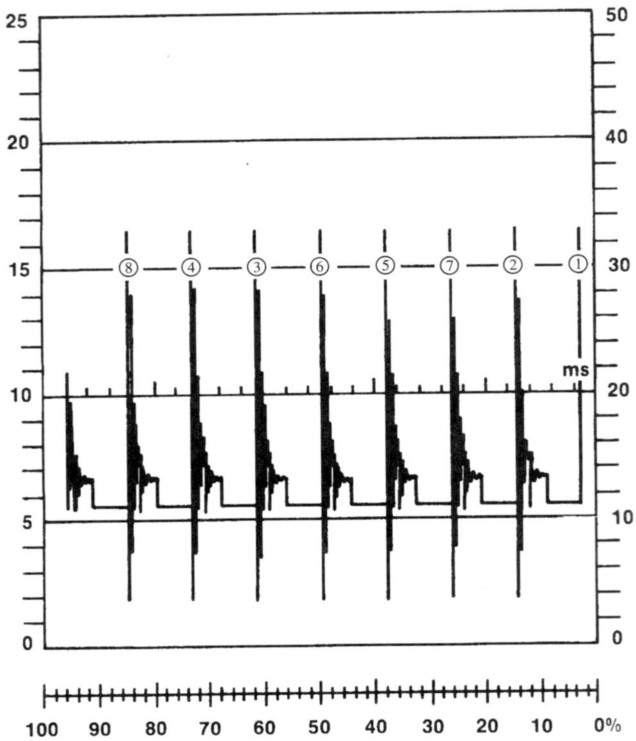

FIGURE 17–71 Display pattern.

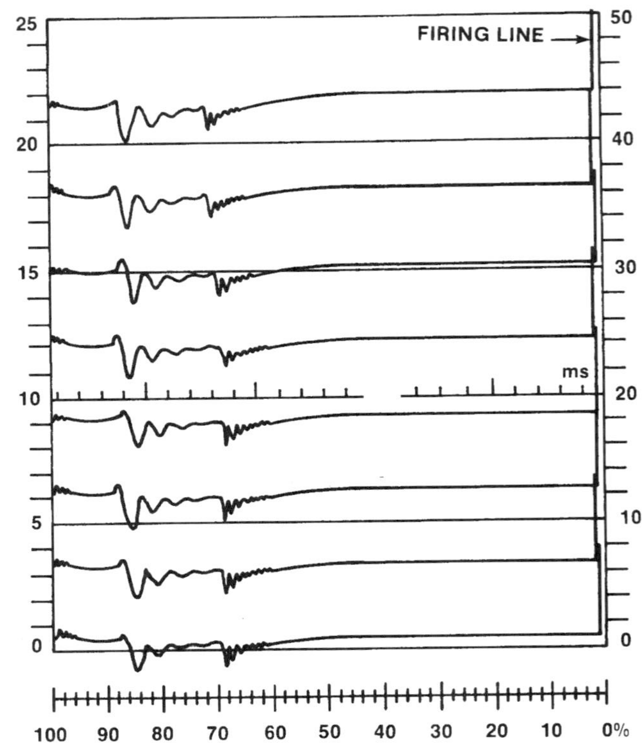

FIGURE 17–72 Raster pattern.

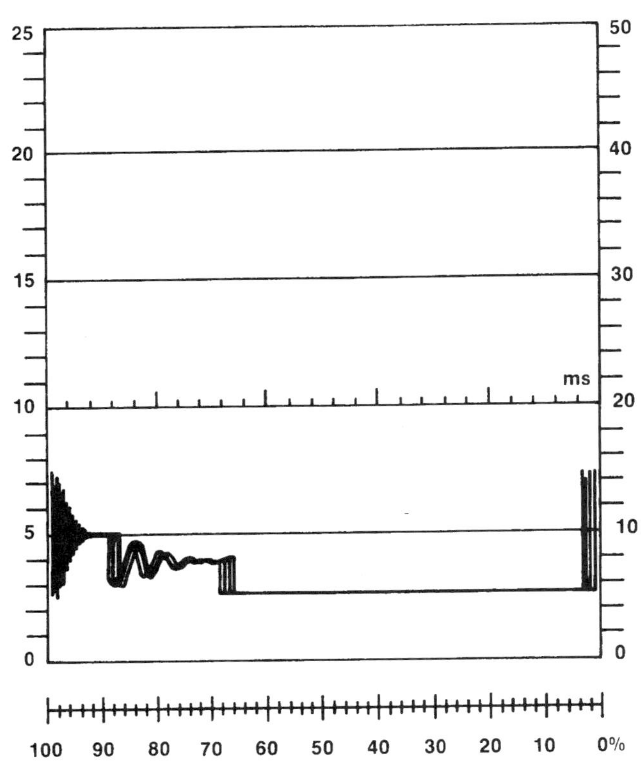

FIGURE 17–73 Superimposed pattern.

CASE STUDIES

CASE #1

The following case study is typical of an ignition related problem that technicians often need to diagnose. Knowledge of how ignition systems are constructed and of simple test procedures will help you perform the diagnosis quickly and efficiently.

Customer Complaint

A customer brought his vehicle into the shop with a possible ignition problem. The vehicle starts while cranking, but quits as soon as the ignition key is returned to the Run position. The customer changed the ignition coil but the problem remained.

Known Information

- ❑ Battery OCV is 12.66 volts
- ❑ Battery voltage while cranking is 11.24 volts
- ❑ Ignition coil has been replaced with a new one
- ❑ Spark plugs and wires are less than one year old
- ❑ Ignition switch has been tested, and is OK
- ❑ Simple schematic of ignition system is shown in Figure 17–74

Circuit Analysis

Answer the following questions on a *separate* sheet of paper.

1. With the known information above, what is the *most* likely cause of the ignition system problem that allows the vehicle to start while it is cranking only? ________________.
2. What preliminary checks should you perform before proceeding with more advanced tests on the ignition system components? ____.

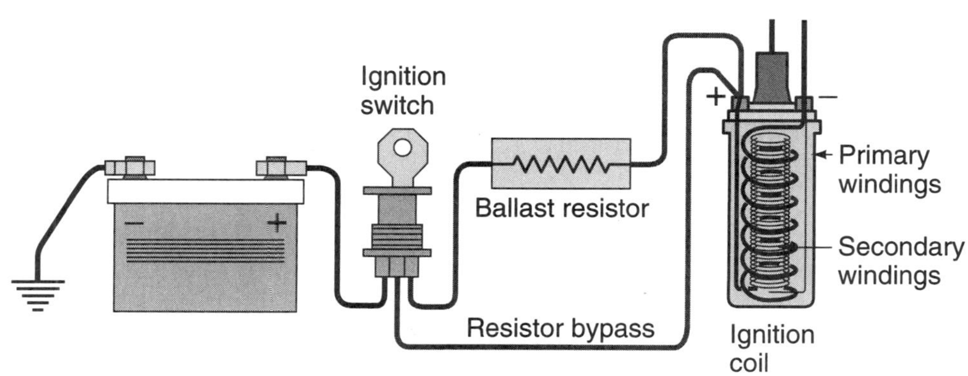

FIGURE 17–74 Simple schematic of ignition system.

(Courtesy of Snap-on Diagnostics)

3. What is the function of each of the components in Figure 17–74 as it relates to the ignition system.

Component	Function of the ignition system
1.	
2.	
3.	
4.	
5.	

4. What should the voltage be at the positive side of the coil with the key on, and the engine off? ________________. Why? ________________.

5. What should the voltage be at the positive side of the coil during engine cranking conditions? ________________. Why? ________________.

6. What is your analysis of the above case study? ________________________________.

Hands-On Vehicle Tests

The following seven hands-on vehicle checks are included in the **NATEF** (National Automotive Technicians Education Foundation) **Task List.** The task reference numbers are the same as the NATEF reference numbers. Some tasks are grouped together due to their close association with each other. Complete your answers to the below questions on a *separate* sheet of paper.

Performance Task #1 (NATEF #VIII-C-3)

Task Description
Inspect and test ignition primary circuit wiring and components; repair or replace as needed.

Performance Task #2 (NATEF #VIII-C-5)

Task Description
Inspect and test ignition system secondary circuit wiring and components; replace as needed.

Performance Task #3 (NATEF #VIII-C-8)Task Description

Task Description
Inspect and test ignition wiring harness and connectors; replace as needed

Task Objectives

- ❑ Obtain a vehicle that can be used for these tasks. What model and year of vehicle are you using for these tasks? ________________.
- ❑ What type of ignition system does this vehicle have? ________________.
- ❑ Describe the general appearance of the ignition system. ________________________________.

❑ Inspect and record the condition of the primary ignition system components.

Component	Condition	Test performed

❑ What tools and equipment would be useful in servicing the primary ignition system components?

____.

❑ Inspect and record the condition of the secondary ignition system components.

Component	Condition	Test performed

❑ What tools and equipment would be useful in servicing the secondary ignition system components?

____

____.

❑ What special service precautions or service factory service bulletins exist on this ignition system?

____

____.

Task Summary

After performing the above NATEF tasks, what can you determine that will be helpful in knowing how to check the ignition primary circuit and components, secondary circuit and components, and ignition wiring harness and connectors? ____

____

____

____.

Performance Task #4 (NATEF #VIII-C-4)

Task Description

Inspect and test distributor; service as needed.

Performance Task #5 (NATEF #VIII-C-6)

Task Description
Inspect and test ignition coil(s); replace as needed.

Performance Task #6 (NATEF #VIII-C-7)

Task Description
Check and adjust (where applicable) ignition system timing and timing advance/retard.

Performance Task #7 (NATEF #VIII-C-9)

Task Description
Inspect and test ignition pick-up sensor or triggering devices; replace as needed.

Task Objectives

- ❑ Obtain a vehicle that can be used for these tasks. What model and year of vehicle are you using for these tasks? ______________________________.
- ❑ What type of ignition system does this vehicle have? ______________________________.
- ❑ Describe the general appearance of the ignition system ______________________________
______________________________.
- ❑ What tests can be performed on the distributor in this vehicle? ______________________________
______________________________.
- ❑ What tools and equipment are being used to inspect and test the distributor? ______________________________
______________________________.
- ❑ List the steps used to check the primary and secondary windings in the ignition coil______________.
- ❑ What is the resistance reading of the coil primary windings? ______________________________.
- ❑ What is the resistance reading of the coil secondary windings?______________________________.
- ❑ List the tools and equipment used to check and adjust the ignition timing______________________________
______________________________.
- ❑ What are the recommended factory service manual procedures and precautions for adjusting the ignition timing on this vehicle? ______________________________
____
______________________________.
- ❑ What is the current ignition timing setting? ______________________________.
- ❑ What is the factory specification for ignition timing? ______________________________.
- ❑ What is the ignition timing setting after adjustment?______________________________.
- ❑ What type of primary pick-up or triggering mechanism does the ignition system have? ______________
______________________________.

- ❑ What are the service manual procedures and specifications for testing the primary triggering switch assembly? ____________________

  ____________________.

- ❑ What is the condition of the primary triggering switch assembly? ____________________

  ____________________.

After performing the above NATEF tasks, what can you determine will be helpful in knowing how to inspect and test the distributor, inspect and test the ignition coil, check and adjust the ignition timing, and inspect and test the primary triggering mechanism? ____________________

____

____

____________________.

Summary

- ❑ The combustion pressure is greatest when the piston is at ten degrees to twenty-three degrees ATDC.
- ❑ Engine detonation occurs when the normal combustion process is too slow, thus causing some unburned fuel/air mixture to ignite and create a colliding flame front.
- ❑ Pre-ignition is caused by an early ignition of the fuel/air mixture by a source other than the ignition system.
- ❑ The primary circuit in an ignition system is the low voltage circuit.
- ❑ The secondary circuit in an ignition system is the high voltage circuit.
- ❑ The ignition coil is a pulse transformer that generates a high voltage when a magnetic field moves across a wire.
- ❑ The ignition coil primary windings consist of a couple hundred turns of a twenty to twenty-four gauge copper wire coated with an insulating enamel.
- ❑ The ignition coil secondary windings consist of several thousand turns of a very fine copper wire coated with an insulating enamel.
- ❑ The soft iron laminated core in an ignition coil has a low inductive reluctance that freely expands or strengthens the magnetic field around it.
- ❑ Electromagnetism occurs in an ignition coil when current flowing through the primary windings create magnetic flux lines of force.
- ❑ Self induction in an ignition coil exists when the primary flux lines cross over onto themselves creating a counter electromotive force (CEMF).
- ❑ Mutual induction in an ignition coil exists when the primary flux lines cross over and create a current in the secondary windings.
- ❑ The capacitor (condenser) in an ignition system absorbs the surges of electrons induced by the collapse of the magnetic fields.

- ❑ The primary coil windings normally have a resistance of 2.0 Ω or less.
- ❑ The secondary windings normally have a resistance between 5,000 and 20,000 Ω.
- ❑ Ignition coil secondary tracking occurs when the windings short to the coil case.
- ❑ A ballast resistor or resistance wire limits the current flowing in the ignition primary circuit to three amps or less.
- ❑ An ignition-bypass system sends full battery voltage to the ignition coil anytime the engine is in the cranking mode.
- ❑ The breaker point ignition system consists of a mechanical set of points in the distributor that open and close to control the ignition primary circuit.
- ❑ Ignition dwell is the amount of time that the primary controlling switch is closed and conducting current.
- ❑ A solid state ignition system replaces the breaker points with an electronic triggering device and solid state ignition control module.
- ❑ The magnetic pulse generator, or PM generator, consists of a timing disc and pick-up coil assembly that operate on the basis of electromagnetism.
- ❑ The metal detection sensor does not have a permanent magnet, but a weak electromagnet.
- ❑ The Hall-effect sensor is based on the principle that if a current is allowed to flow through a thin conducting material, and when that material is exposed to a magnetic field, another voltage is produced.
- ❑ The heart of a Hall-effect system is a thin semiconductor gallium arsenate crystal.
- ❑ A schmitt trigger is a pulse shaping device that converts the Hall-effect signal to a clean square wave digital signal.
- ❑ A photoelectric sensor uses an LED that shines through a rotating slotted disc to a photo cell that converts the intermittent flashes of light into a voltage signal pulse.
- ❑ The distributor centrifugal advance mechanism advances the ignition timing in relation to engine speed.
- ❑ The distributor vacuum advance mechanism advances the ignition timing in relation to the engine vacuum conditions.
- ❑ Manifold vacuum is considered any vacuum below the throttle plates.
- ❑ Ported vacuum is considered any vacuum above the throttle plates.
- ❑ An electronic ignition (EI) system is a distributorless system consisting of a computer, ignition module, coil pack, position sensors, and associated wiring.
- ❑ A waste spark system fires one of a pair of spark plugs on the exhaust stroke while the mating spark plug fires towards the end of the compression stroke.
- ❑ The resistive element in a spark plug controls RFI by suppressing the AC voltage spike produced while the spark plug is firing.
- ❑ The display pattern on a scope screen displays all the cylinder patterns in a row from left to right.

- ❑ The raster pattern on a scope screen displays all the cylinder patterns above each other in a row.
- ❑ The superimposed pattern on a scope screen displays all the cylinder patterns stacked on top of one another.

Key Terms

after-top-dead-center (ATDC)
ballast resistor
before-top-dead-center (BTDC)
breaker point ignition
capacitor
centrifugal advance
coil-over-plug
detonation
dwell
E-coil
electronic ignition (EI)
firing order
hall effect sensor
inductive reluctance
magnetic pulse generator
magnetic suppression
manifold vacuum
metal detection sensor
mutual induction
photoelectric sensor
pick-up coil
PM generator
ported vacuum
pre-ignition
primary circuit
pulse transformer
radio frequency interference
rotor
saturation
schmitt trigger
secondary circuit
self induction
solid state ignition
vacuum advance
waste spark

Review Questions

Short Answer Essays

1. Describe the difference between engine detonation and pre-ignition.
2. Name the parts of the typical primary ignition system.
3. Name the parts of the typical secondary ignition system.
4. Describe how a coil transforms battery voltage to a voltage high enough to jump a spark plug gap.
5. Describe how a Hall effect sensor works as a triggering mechanism for an ignition system.

Fill-in-the-Blanks

1. In an engine cylinder, the combustion process must begin ________________________ on the ________________________ stroke.
2. On some ignition systems, to protect the primary switching circuit a ________________________ is wired in parallel to the circuit to absorb the surges of ________________________ induced by the collapse of the magnetic field.
3. The ________________________ in a distributor consists of a set of weights and springs that fly outward as engine rpm ________________________.

4. Vacuum that is above the throttle is considered ________________________ vacuum.
5. An electronic ignition system that consists of one coil for every two cylinders is called a ________________________ system.

ASE Style Review Questions

1. Technician A says the spark plug wire is part of the secondary ignition system. Technician B says the distributor cap is part of the primary ignition system. Who is correct?
   A. A only
   B. B only
   C. Both A and B
   D. Neither A nor B
2. Technician A says the vacuum advance unit in a distributor may be connected to manifold vacuum. Technician B says the vacuum advance unit in a distributor may be connected to ported vacuum. Who is correct?
   A. A only
   B. B only
   C. Both A and B
   D. Neither A nor B
3. Technician A says that engine detonation occurs when the fuel/air mixture ignites before ignition system spark. Technician B says a pre-ignition can occur from hot carbon deposits in the combustion chamber. Who is correct?
   A. A only
   B. B only
   C. Both A and B
   D. Neither A nor B
4. Technician A says the ignition coil is part of the primary system. Technician B says the primary triggering device is normally located after the coil in the circuit. Who is correct?
   A. A only
   B. B only
   C. Both A and B
   D. Neither A nor B

5. Technician A says the coil primary windings consist of a few thousand turn of a twenty to twenty four gauge wire. Technician B says the coil secondary windings consist of a few hundred turns of a very fine copper wire. Who is correct?
   A. A only
   B. B only
   C. Both A and B
   D. Neither A nor B

6. Technician A says a ballast resistor lowers the voltage to the coil positive terminal. Technician B says a resistance wire lowers the voltage to the coil positive terminal. Who is correct?
   A. A only
   B. B only
   C. Both A and B
   D. Neither A nor B

7. Technician A says the capacitor absorbs the surges of electrons in the coil. Technician B says the capacitor conducts electricity anytime the ignition primary system contact points are closed. Who is correct?
   A. A only
   B. B only
   C. Both A and B
   D. Neither A nor B

8. Technician A says the secondary windings in an ignition coil are tested by connecting the ohmmeter leads between the positive and negative terminals of the coil. Technician B says the coil primary resistance should be 2.0 Ω or less. Who is correct?
   A. A only
   B. B only
   C. Both A and B
   D. Neither A nor B

9. Technician A says the amount of time the contact points are closed in a point style distributor is the ignition dwell time. Technician B says the lower the ignition dwell, the wider the contact point gap. Who is correct?
   A. A only
   B. B only
   C. Both A and B
   D. Neither A nor B

10. A distributor with a PM generator is being discussed. Technician A says a permanent magnet is part of the switching assembly. Technician B says the increase and decrease of the voltage in the pick-up coil in relation to zero is what triggers the ignition module to turn the primary circuit on and off. Who is correct?
    A. A only
    B. B only
    C. Both A and B
    D. Neither A nor B

11. Technician A says a Hall-effect switch assembly produces an accurate square wave voltage signal. Technician B says the Hall-effect switch is a thin carbon and silicon crystal. Who is correct?
    A. A only
    B. B only
    C. Both A and B
    D. Neither A nor B

12. Technician A says that in a waste spark electronic ignition system one ignition coil is used for every two cylinders. Technician B says that in a waste spark electronic ignition system spark plugs also fire on the exhaust stroke. Who is correct?
    A. A only
    B. B only
    C. Both A and B
    D. Neither A nor B

13. Technician A says a Hall-effect switch requires its voltage to be amplified, inverted, and shaped into a clean square wave signal. Technician B says a metal detection sensor requires its voltage to be amplified, inverted, and shaped into a clean square wave signal. Who is correct?
    A. A only
    B. B only
    C. Both A and B
    D. Neither A nor B

14. Technician A says a magnetic suppression spark plug wire consists of a heavy copper core that is shielded with a metal jacket. Technician B says a flexible nylon core that is impregnated with carbon can suppress RFI in a spark plug wire. Who is correct?
    A. A only
    B. B only
    C. Both A and B
    D. Neither A nor B

15. Technician A says the firing order in a distributor is controlled by the placement of the secondary wires in the distributor cap. Technician B says the rotor assembly distributes the high voltage from the coil to the plug wires. Who is correct?
    A. A only
    B. B only
    C. Both A and B
    D. Neither A nor B

Appendix A

1. Technician A says the flow of free electrons in a circuit is called current flow. Technician B says that 6.28 billion billion electrons flowing past a point in one second is one amp. Who is correct?
   A. A only
   B. B only
   C. Both A and B
   D. Neither A nor B

2. Technician A says the base system of measurement for volts is the kilovolt. Technician B says the milliamp is the base unit of measurement for current. Who is correct?
   A. A only
   B. B only
   C. Both A and B
   D. Neither A nor B

3. Technician A says static electricity can occur whenever the technician slides across a vinyl seat in a car. Technician B says the best way to guard against static electricity is to use a static wrist strap. Who is correct?
   A. A only
   B. B only
   C. Both A and B
   D. Neither A nor B

4. Technician A says the diode acts like a one way check valve in an electrical circuit. Technician B says a diode allows current to flow in one direction only. Who is correct?
   A. A only
   B. B only
   C. Both A and B
   D. Neither A nor B

5. A circuit has the following four resistors wired in parallel: 10 Ω, 5 Ω, 10 Ω, and 2.5 Ω. The correct answer for the resulting total resistance is:
   A. 27.5 Ω
   B. 1.25 Ω
   C. 125 Ω
   D. None of the above

6. Technician A says using photovoltaic solar cells provide a way to capture the energy from the sun. Technician B says that a photon is the energy in sunlight that strikes the surface of a photovoltaic cell to create a voltage. Who is correct?
   A. A only
   B. B only
   C. Both A and B
   D. Neither A nor B

7. Technician A says an electrical circuit must be energized to measure the current flow through the circuit. Technician B says an ammeter is connected in series with the component or circuit being measured. Who is correct?
   A. A only
   B. B only
   C. Both A and B
   D. Neither A nor B

8. Two technicians are discussing a multifunction steering column switch. Technician A says the dimmer switch can be part of this style of switch. Technician B says the horn switch could be part of this type of switch. Who is correct?
   A. A only
   B. B only
   C. Both A and B
   D. Neither A nor B

9. Technician A says a diode is forward biased if the positive voltage is connected to the cathode side of the diode. Technician B says a diode is reversed biased if the positive voltage is connected to the anode side of the diode. Who is correct?
   A. A only
   B. B only
   C. Both A and B
   D. Neither A nor B

10. Technician A says a primary cell battery can be recharged to it's original condition. Technician B says a secondary cell battery cannot be recharged once it's original stored energy has been depleted. Who is correct?
    A. A only
    B. B only
    C. Both A and B
    D. Neither A nor B

11. Two technicians are discussing the results of a starter current draw test on a four-cylinder engine. Technician A says the minimum voltage should not drop below 10.5 volts. Technician B says the amperage draw of 125 amps is considered satisfactory. Who is correct?
    A. A only
    B. B only
    C. Both A and B
    D. Neither A nor B

12. A vehicle is being tested for a slow-crank condition. A starter current draw test indicates 450 amps. Technician A says this could be caused by a battery with a CCA rating too large for the system. Technician B says this could be caused by a pinion drive that slips when engaged with the flywheel ring gear. Who is correct?
    A. A only
    B. B only
    C. Both A and B
    D. Neither A nor B

13. Technician A says a delta wound winding consists of three windings connected together in a triangular shape with three output leads. Technician B says a delta wound winding provides a high amperage output for it's size. Who is correct?
    A. A only
    B. B only
    C. Both A and B
    D. Neither A nor B

14. Technician A says a thermocouple is installed in the exhaust system of a diesel engine to measure the exhaust temperature. Technician B says a thermocouple operates the same as a mercury glass thermometer. Who is correct?
    A. A only
    B. B only
    C. Both A and B
    D. Neither A nor B

15. Technician A says an IVR can be used on bimetallic gauges. Technician B says the IVR will open if the alternator overcharges the battery. Who is correct?
    A. A only
    B. B only
    C. Both A and B
    D. Neither A nor B

16. Technician A says a halogen headlamp uses a halogen filled inner bulb sealed in a glass housing. Technician B says in a composite halogen headlamp the entire headlamp assembly needs to be replace when burned out. Who is correct?
    A. A only
    B. B only
    C. Both A and B
    D. Neither A nor B

17. Technician A says the proper way to find the resistance of a "hot bulb" parallel circuit is to measure the amps and volts of the circuit, and then calculate the circuit resistance total. Technician B says the "hot bulb" resistance total of a parallel circuit can be determined by taking the "hot bulb" resistance of each circuit bulb and finding the reciprocal of the sum of each parallel branch. Who is correct?
    A. A only
    B. B only
    C. Both A and B
    D. Neither A nor B

18. *Voltage total* of a simple series/parallel circuit is being discussed. There are two series load devices with voltage drops of 6 volts each, and four load devices in parallel with a voltage reading across each device of 12 volts. What is the voltage total of this circuit?
    A. 12 volts
    B. 24 volts
    C. 60 volts
    D. None of the above

19. Technician A says a *short-to-voltage* has an unwanted copper-to-copper connection between two circuits. Technician B says a *short-to-voltage* may not blow the circuit fuse. Who is correct?
    A. A only
    B. B only
    C. Both A and B
    D. Neither A nor B

20. Technician A says a test light can be connected in series with a cycling circuit breaker to test for a short-to-ground. Technician B says a buzzer can be connected in series with a cycling circuit breaker to test for a short-to-ground. Who is correct?
    A. A only
    B. B only
    C. Both A and B
    D. Neither A nor B

21. Two technicians are discussing how a sneak circuit can affect circuits in an automotive electrical system. Technician A says current can back-feed through a circuit when two circuits share a common power feed. Technician B says current can back-feed through a circuit when two circuits share a common ground path. Who is correct?
    A. A only
    B. B only
    C. Both A and B
    D. Neither A nor B

22. A series circuit has three resistors of 20 Ω each. The voltage drop of each resistor is 4 volts. Technician A says the voltage total is 12 volts. Technician B says the resistance total is 60 Ω. Who is correct ?
    A. A only
    B. B only
    C. Both A and B
    D. Neither A nor B

23. Technician A says a knock sensor will produce a small voltage when there is an unusual ignition spark knock. Technician B says the knock sensor is a piezoelectric device that will produce a voltage under physical stress or vibration. Who is correct?
    A. A only
    B. B only
    C. Both A and B
    D. Neither A nor B

24. A circuit has four resistors wired in series: 10 Ω, 5 Ω, 4 Ω, and 2.5 Ω. What is the resistance total? The correct answer is:
    A. 21.5 Ω
    B. 1.05 Ω
    C. 20.5 Ω
    D. None of the above

25. Technician A says the pitch of a horn is controlled by the voltage going to the horn. Technician B says that in a dual-horn system, the pitch is normally the same for both horns. Who is correct?
    A. A only
    B. B only
    C. Both A and B
    D. Neither A nor B

26. The oil pressure light stays on when the engine is running. Technician A says the engine oil pressure could be low causing the circuit to operate as designed. Technician B says there could be a ground in the circuit between the indicator light and the pressure switch. Who is correct?
    A. A only
    B. B only
    C. Both A and B
    D. Neither A nor B

27. Two technicians are discussing the operation of NPN transistors. Technician A says the transistor is on when a voltage is applied to the base. Technician B says that when the transistor is on, current flows from the collector to the emitter. Who is correct?
    A. A only
    B. B only
    C. Both A and B
    D. Neither A nor B

28. Technician A says a potentiometer uses three wires—one movable wiper and two fixed ends. Technician B says a potentiometer can sense a variable voltage drop. Who is correct?
    A. A only
    B. B only
    C. Both A and B
    D. Neither A nor B

29. Technician A says that in a series circuit the voltage of each circuit load device is the same as the total circuit voltage. Technician B says the amperage of each series circuit branch is the same throughout the circuit. Who is correct?
    A. A only
    B. B only
    C. Both A and B
    D. Neither A nor B

30. Technician A says that adding resistive branches to a series circuit increases the total circuit amperage. Technician B says that if a series circuit is opened, the entire circuit will not operate. Who is correct?
    A. A only
    B. B only
    C. Both A and B
    D. Neither A nor B

31. Technician A says the battery stores the voltage pressure until the electrical system needs it. Technician B says the battery stores the energy in a chemical form. Who is correct?
    A. A only
    B. B only
    C. Both A and B
    D. Neither A nor B

32. Technician A says a CCA rating on a battery determines its ability to provide an amperage load for 30 seconds at 0° F with the terminal voltage dropping no lower than 7.2 volts. Technician B says the CCA rating can be divided be 5.25 to get the approximate A/H rating. Who is correct?
    A. A only
    B. B only
    C. Both A and B
    D. Neither A nor B

33. Technician A says battery temperature above 125° F helps the battery charge faster. Technician B says the battery negative cable should be disconnected before charging, to isolate the vehicle's electrical system from voltage conditions above 15.5 volts. Who is correct?
    A. A only
    B. B only
    C. Both A and B
    D. Neither A nor B

34. Technician A says a wye wound stator winding consists of three output connections and a common neutral junction. Technician B says a wye wound stator winding produces a high amperage output. Who is correct?
    A. A only
    B. B only
    C. Both A and B
    D. Neither A nor B

35. Technician A says the rule with the wire size AWG standard is the higher the number, the larger the wire conductor. Technician B says that in the metric system of determining wire size, the smaller the number, the smaller the wire conductor. Who is correct?
    A. A only
    B. B only
    C. Both A and B
    D. Neither A nor B

36. Technician A says a maxifuse is a fuse designed as a replacement for a fusible link in a modern vehicle. Technician B says a pacific fuse element is a fuse designed as a replacement for a fusible link in a modern vehicle. Who is correct?
    A. A only
    B. B only
    C. Both A and B
    D. Neither A nor B

37. Technician A says that analog meters use a pointer that moves across the face of a printed scale. Technician B says that analog meters are not polarity sensitive to the circuit being measured. Who is correct?
    A. A only
    B. B only
    C. Both A and B
    D. Neither A nor B

38. Technician A says the binary code assigned to a low digital signal is *zero.* Technician B says the binary code assigned to a high digital signal is *two.* Who is correct?
    A. A only
    B. B only
    C. Both A and B
    D. Neither A nor B

39. 150 milliamps total is flowing through a circuit that consists of three resistors in series, 30 Ω, 40 Ω, and 10 Ω. Technician A says the source voltage of the circuit is 120 volts. Technician B says the voltage is 1.2 volts. Who is correct?
    A. A only
    B. B only
    C. Both A and B
    D. Neither A nor B

40. A vehicle blows a taillight fuse when a trailer is hooked up to the circuit and energized. Technician A says there could be a short in the trailer taillight circuit. Technician B says there could be too much amperage for the vehicle taillight fuse when the trailer is added to the circuit. Who is correct?
    A. A only
    B. B only
    C. Both A and B
    D. Neither A nor B

41. *Voltage total* of a simple series/parallel circuit is being discussed. There are two series load devices with voltage drops of 6 volts each, and four load devices in parallel with a voltage reading across each device of 12 volts. What is the voltage total of this circuit?

    A. 12 volts

    B. 24 volts

    C. 60 volts

    D. None of the above

42. Technician A says a fused jumper wire can be used to bypass a switch to determine if a circuit is working. Technician B says a self-powered test light can be a useful tool for checking for circuit continuity. Who is correct?

    A. A only

    B. B only

    C. Both A and B

    D. Neither A nor B

43. Technician A says that back-feed in an electrical circuit occurs when the current in one circuit seeks an alternate path of flow because the original path is interrupted. Technician B says back-feed in an electrical circuit can *only* occur when the insulation breaks down between the wires of two separate circuits. Who is correct?

    A. A only

    B. B only

    C. Both A and B

    D. Neither A nor B

44. Two technicians are discussing the results of a three-minute charge test. Technician A says a reading of more than 15.5 volts on a conventional battery before three minutes time indicates a sulfated battery. Technician B says a load of 40 amps should be applied to the battery to get an accurate reading. Who is correct?

    A. A only

    B. B only

    C. Both A and B

    D. Neither A nor B

45. Technician A says a negative battery cable is always removed first when removing a battery. Technician B says a negative battery cable should always be connected first when replacing a battery. Who is correct?

    A. A only

    B. B only

    C. Both A and B

    D. Neither A nor B

46. Technician A says the starter solenoid moves a plunger to engage the overrunning clutch. Technician B says the two windings of the starter solenoid are called the pull-in and hold-in windings. Who is correct?
    A. A only
    B. B only
    C. Both A and B
    D. Neither A nor B

47. Technician A says the output winding is rotating in a DC generator. Technician B says the output winding is stationary in a AC generator. Who is correct?
    A. A only
    B. B only
    C. Both A and B
    D. Neither A nor B

48. Technician A says the positive-biased diodes in a rectifier bridge are attached directly to the alternator rear housing. Technician B says the negative-biased diodes in a rectifier bridge are mounted in a heat sink. Who is correct?
    A. A only
    B. B only
    C. Both A and B
    D. Neither A nor B

49. Technician A says that engine detonation occurs when the fuel/air mixture ignites before the ignition system spark. Technician B says a pre-ignition can occur from hot carbon deposits in the combustion chamber. Who is correct?
    A. A only
    B. B only
    C. Both A and B
    D. Neither A nor B

50. An ohmmeter is being used to test the windings in an ignition coil. Technician A says the secondary windings in an ignition coil are tested by connecting the ohmmeter leads between the positive and negative terminals of the coil. Technician B says the ignition coil primary resistance should be 2.0 W or less. Who is correct?
    A. A only
    B. B only
    C. Both A and B
    D. Neither A nor B

Appendix B

TYPICAL AUTOMOTIVE LIGHT BULBS (continued)

Trade Number	Design Volts	Design Amperes	Watts: P = I x E
192	13.0	0.33	4.3
194	14.0	0.27	3.8
194E-1	14.0	0.27	3.8
194NA	14.0	0.27	3.8
209	6.5	1.78	11.6
211-2	12.8	0.97	12.4
212-2	13.5	0.74	10.0
214-2	13.5	0.52	7.0
561	12.8	0.97	12.4
562	13.5	0.74	10.0
563	13.5	0.52	7.0
631	14.0	0.63	8.8
880	12.8	2.10	27.0
881	12.8	2.10	27.0
906	13.0	0.69	9.0
912	12.8	1.00	12.8
1003	12.8	0.94	12.0
1004	12.8	0.94	12.0
1034	12.8	1.80/0.59	23.0/7.6
1073	12.8	1.80	23.0
1076	12.8	1.80	23.0
1129	6.4	2.63	16.8
1133	6.2	3.91	24.2
1141	12.8	1.44	18.4
1142	12.8	1.44	18.4
1154	6.4	2.63/.75	16.8/4.5
1156	12.8	2.10	26.9
1157	12.8	2.10/0.59	26.9/7.6
1157A	12.8	2.10/0.59	26.9/7.6
1157NA	12.8	2.10/0.59	26.9/7.6
1176	12.8	1.34/0.59	17.2/7.6
1195	12.5	3.00	37.5
1196	12.5	3.00	37.5
1445	14.4	0.135	1.9
1816	13.0	0.33	4.3
1889	14.0	0.27	3.8
1891	14.0	0.24	3.4
1892	14.4	0.12	1.7
1893	14.0	0.33	4.6
1895	14.0	0.27	3.8
2057	12.8	2.10/0.48	26.9/6.1
2057NA	12.8	2.10/0.48	26.9/6.1
P25-1	13.5	1.86	25.1
P25-2	13.5	1.86	25.1
R19/5	13.5	0.37	
R19/10	13.5	0.74	
W10/3	13.5	0.25	

TYPICAL AUTOMOTIVE LIGHT BULBS

Trade Number	Design Volts	Design Amperes	Watts: P = I x E
37	14.0	0.09	1.3
37E	14.0	0.09	1.3
51	7.0	0.22	1.7
53	14.0	0.12	1.7
55	7.0	0.41	2.9
57	14.0	0.24	3.4
57X	14.0	0.24	3.4
63	7.0	0.63	4.4
67	13.5	0.59	8.0
68	13.5	0.59	8.0
70	14.0	0.15	2.1
73	14.0	0.08	1.1
74	14.0	0.10	1.4
81	6.5	1.02	6.6
88	13.0	0.58	7.5
89	13.0	0.58	7.5
90	13.0	0.58	7.5
93	12.8	1.04	13.3
94	12.8	1.04	13.3
158	14.0	0.24	3.4
161	14.0	0.19	2.7
168	14.0	0.35	4.9

Appendix C

SAE J1930 Revised SEP95

TABLE 1—CROSS REFERENCE AND LOOK UP

Existing Usage	Acceptable Usage	Acceptable Acronized Usage
A/C (Air Conditioning)	Air Conditioning	A/C
A/C Cycling Switch	Air Conditioning Cycling Switch	A/C Cycling Switch
A/T (Automatic Transaxle)	Automatic Transaxle[1]	A/T[1]
A/T (Automatic Transmission)	Automatic Transmission[1]	A/T[1]
AAT (Ambient Air Temperature)	**Ambient Air Temperature**	**AAT**
AC (Air Conditioning)	Air Conditioning	A/C
ACC (Air Conditioning Clutch)	Air Conditioning Clutch	A/C Clutch
Accelerator	Accelerator Pedal	AP
Accelerator Pedal Position	**Accelerator Pedal Position**[1]	**APP**[1]
ACCS (Air Conditioning Cyclic Switch)	Air Conditioning Cycling Switch	A/C Cycling Switch
ACH (Air Cleaner Housing)	Air Cleaner Housing[1]	ACL Housing1
ACL (Air Cleaner)	Air Cleaner[1]	ACL[1]
ACL (Air Cleaner) Element	Air Cleaner Element[1]	ACL Element[1]
ACL (Air Cleaner) Housing	Air Cleaner Housing[1]	ACL Housing[1]
ACL (Air Cleaner) Housing Cover	Air Cleaner Housing Cover[1]	ACL Housing Cover[1]
ACS (Air Conditioning System)	Air Conditioning System	A/C System
ACT (Air Charge Temperature)	Intake Air Temperature[1]	IAT[1]
Adaptive Fuel Strategy	Fuel Trim[1]	FT[1]
AFC (Air Flow Control)	Mass Air Flow	MAF
AFC (Air Flow Control(	Volume Air Flow	VAF
AFS (Air Flow Sensor)	Mass Air Flow Sensor	MAF Sensor
AFS (Air Flow Sensor)	Volume Air Flow Sensor	VAF Sensor
After Cooler	Charge Air Cooler[1]	CAC[1]
AI (Air Injection)	Secondary Air Injection[1]	AIR[1]
AIP (Air Injection Pump)	Secondary Air Injection Pump[1]	AIR Pump[1]
AIR (Air Injection Reactor)	Pulsed Secondary Air Injection[1]	PAIR[1]
AIR (Air Injection Reactor)	Secondary Air Injection[1]	AIR[1]
AIRB (Secondary Air Injection Bypass)	Secondary Air Injection Bypass[1]	AIR Bypass[1]
AIRD (Secondary Air Injection Diverter)	Secondary Air Injection Diverter[1]	AIR Diverter[1]
Air Cleaner	Air Cleaner[1]	ACL[1]
Air Cleaner Element	Air Cleaner Element[1]	ACL Element[1]
Air Cleaner Housing	Air Cleaner Housing[1]	ACL Housing[1]
Air Cleaner Housing Cover	Air Cleaner Housing Cover[1]	ACL Housing Cover[1]
Air Conditioning	Air Conditioning	A/C
Air Conditioning Sensor	Air Conditioning Sensor	A/C Sensor
Air Control Valve	Secondary Air Injection Control Valve[1]	AIR Control Valve[1]
Air Flow Meter	Mass Air Flow Sensor[1]	MAF Sensor[1]
Air Flow Meter	Volume Air Flow Sensor[1]	VAF Sensor[1]
Air Intake System	Intake Air System[1]	IA System[1]
Air Flow Sensor	Mass Air Flow Sensor[1]	MAF Sensor[1]
Air Management 1	Secondary Air Injection Bypass[1]	AIR Bypass[1]
Air Management 2	Secondary Air Injection Diverter[1]	AIR Diverter[1]
Air Temperature Sensor	Intake Air Temperature Sensor[1]	IAT Sensor[1]
Air Valve	Idle Air Control Valve[1]	IAC Valve[1]
AIV (Air Injection Valve)	Pulsed Secondary Air Injection[1]	PAIR[1]
ALCL (Assembly Line Communication Link)	Data Link Connector[1]	DLC[1]
Alcohol Concentration Sensor	Flexible Fuel Sensor[1]	FF Sensor[1]
ALDL (Assembly Line Diagnostic Link)	Data Link Connector[1]	DLC[1]

SAE J1930 Revised SEP95

TABLE 1—CROSS REFERENCE AND LOOK UP (CONTINUED)

Existing Usage	Acceptable Usage	Acceptable Acronized Usage
ALT (Alternator)	Generator	GEN
Alternator	Generator	GEN
Ambient Air Temperature	**Ambient Air Temperature**	**AAT**
AM1 (Air Management 1)	Secondary Air Injection Bypass[1]	AIR Bypass[1]
AM2 (Air Management 2)	Secondary Air Injection Diverter[1]	AIR Diverter[1]
APP (Accelerator Pedal Position)	**Accelerator Pedal Position[1]**	**APP[1]**
APS (Absolute Pressure Sensor)	Barometric Pressure Sensor[1]	BARO Sensor[1]
ATS (Air Temperature Sensor)	Intake Air Temperature Sensor[1]	IAT Sensor[1]
Automatic Transaxle	Automatic Transaxle[1]	A/T[1]
Automatic Transmission	Automatic Transmission[1]	A/T[1]
B+ (Battery Positive Voltage)	Battery Positive Voltage	B+
Backpressure Transducer	Exhaust Gas Recirculation Backpressure Transducer[1]	EGR Backpressure Transducer[1]
BARO (Barometric Pressure)	Barometric Pressure[1]	BARO[1]
Barometric Pressure Sensor	Barometric Pressure Sensor[1]	BARO Sensor[1]
Battery Positive Voltage	Battery Positive Voltage	B+
BLM (Block Learn Memory)	Long Term Fuel Trim[1]	Long Term FT[1]
BLM (Block Learn Multiplier)	Long Term Fuel Trim[1]	Long Term FT[1]
BLM (Block Learn Matrix)	Long Term Fuel Trim[1]	Long Term FT[1]
Block Learn Integrator	**Long Term Fuel Trim[1]**	**Long Term FT[1]**
Block Learn Matrix	Long Term Fuel Trim[1]	Long Term FT[1]
Block Learn Memory	Long Term Fuel Trim[1]	Long Term FT[1]
Block Learn Multiplier	Long Term Fuel Trim[1]	Long Term FT
BP (Barometric Pressure) Sensor	Barometric Pressure Sensor[1]	BARO Sensor[1]
BPP (Brake Pedal Position)	**Brake Pedal Position[1]**	**BPP[1]**
Brake Pressure	**Brake Pressure**	**Brake Pressure**
Brake Pedal Position	**Brake Pedal Position[1]**	**BPP[1]**
C3I (Computer Controlled Coil Ignition)	Electronic Ignition[1]	EI[1]
CAC (Charge Air Cooler)	Charge Air Cooler[1]	CAC[1]
Calculated Load Value	**Calculated Load Value**	**LOAD**
Camshaft Position	Camshaft Position[1]	CMP[1]
Camshaft Position Actuator	**Camshaft Position Actuator[1]**	**CMP Actuator[1]**
Camshaft Position Controller	**Camshaft Position Actuator[1]**	**CMP Actuator[1]**
Camshaft Position Sensor	Camshaft Position Sensor[1]	CMP Sensor[1]
Camshaft Sensor	Camshaft Position Sensor[1]	CMP Sensor[1]
Camshaft Timing Actuator	**Camshaft Position Actuator[1]**	**CMP Actuator[1]**
Canister	Canister[1]	Canister[1]
Canister	Evaporative Emission Canister[1]	EVAP Canister[1]
Canister Purge	**Evaporative Emission Canister Purge[1]**	**EVAP Canister Purge[1]**
Canister Purge Vacuum Switching Valve	Evaporative Emission Canister Purge Valve[1]	EVAP Canister Purge Valve[1]
Canister Purge Valve	Evaporative Emission Canister Purge Valve[1]	EVAP Canister Purge Valve[1]
Canister Purge VSV (Vacuum Switching Valve)	Evaporative Emission Canister Purge Valve[1]	EVAP Canister Purge Valve[1]
CANP (Canister Purge)	Evaporative Emission Canister Purge[1]	EVAP Canister Purge[1]
CARB (Carburetor)	Carburetor[1]	CARB[1]
Carburetor	Carburetor[1]	CARB[1]
Catalytic Converter Heater	**Catalytic Converter Heater**	**Catalytic Converter Heater**
CCC (Converter Clutch Control)	Torque Converter Clutch[1]	TCC[1]
CCO (Converter Clutch Override)	Torque Converter Clutch[1]	TCC[1]
CCS (Coast Clutch Solenoid)	**Coast Clutch Solenoid**	**CCS**

SAE J1930 Revised SEP95

TABLE 1—CROSS REFERENCE AND LOOK UP (CONTINUED)

Existing Usage	Acceptable Usage	Acceptable Acronized Usage
CCS (Coast Clutch Solenoid) Valve	**Coast Clutch Solenoid Valve**	**CCS Valve**
CCRM (Constant Control Relay Module)	**Constant Control RM**	**Constant Control RM**
CDI (Capacitive Discharge Ignition)	Distributor Ignition[1]	DI[1]
CDROM (Compact Disc Read Only Memory)	Compact Disc Read Only Memory[1]	CDROM[1]
CES (Clutch Engage Switch)	Clutch Pedal Position Switch	CPP Switch[1]
Central Multiport Fuel Injection	Central Multiport Fuel Injection[1]	Central MFI[1]
Central Sequential Multiport Fuel Injection	**Central Sequential Multiport Fuel Injection**	**Central SFI**
CFI (Continuous Fuel Injection)	Continuous Fuel Injection[1]	CFI[1]
CFI (Central Fuel Injection)	Throttle Body Fuel Injection[1]	TBI[1]
CFV	**Critical Flow Venturi**	**CFV**
Charcoal Canister	Evaporative Emission Canister	EVAP Canister[1]
Charge Air Cooler	Charge Air Cooler	CAC[1]
Check Engine	Service Reminder Indicator[1]	SRI[1]
Check Engine	Malfunction Indicator Lamp[1]	MIL[1]
CID (Cylinder Identification) Sensor	Camshaft Position Sensor	CMP Sensor[1]
CIS (Continuous Injection System)	Continuous Fuel Injection[1]	CFI[1]
CIS-E (Continuous Injection System Electronic)	Continuous Fuel Injection[1]	CFI[1]
CKP (Crankshaft Position)	Crankshaft Position[1]	CKP[1]
CKP (Crankshaft Position) Sensor	Crankshaft Position Sensor[1]	CKP Sensor[1]
CL (Closed Loop)	Closed Loop[1]	CL[1]
Closed Bowl Distributor	Distributor Ignition[1]	DI[1]
Closed Throttle Position	Closed Throttle Position[1]	CTP[1]
Closed Throttle Switch	Closed Throttle Position Switch[1]	CTP Switch[1]
CLS (Closed Loop System)	Closed Loop[1]	CL[1]
CLV	**Calculated Load Value**	**LOAD**
Clutch Engage Switch	Clutch Pedal Position Switch[1]	CPP Switch[1]
Clutch Pedal Position Switch	Clutch Pedal Position Switch[1]	CPP Switch[1]
Clutch Start Switch	Clutch Pedal Position Switch[1]	CPP Switch[1]
Clutch Switch	Clutch Pedal Position Switch[1]	CPP Switch[1]
CMFI (Central Multiport Fuel Injection)	Central Multiport Fuel Injection[1]	Central MFI[1]
CMP (Camshaft Position)	Camshaft Position[1]	CMP[1]
CMP (Camshaft Position) Sensor	Camshaft Position Sensor[1]	CMP Sensor[1]
COC (Continuous Oxidation Catalyst)	Oxidation Catalytic Converter[1]	OC[1]
Coast Clutch Solenoid	**Coast Clutch Solenoid**	**CCS**
Coast Clutch Solenoid Valve	**Coast Clutch Solenoid Valve**	**CCS Valve**
Condenser	Distributor Ignition Capacitor[1]	DI Capacitor[1]
Constant Control Relay Module	**Relay Module**	**RM**
Constant Volume Sampler	**Constant Volume Sampler**	**CVS**
Continuous Fuel Injection	Continuous Fuel Injection[1]	CFI[1]
Continuous Injection System	Continuous Fuel Injection System[1]	CFI System[1]
Continuous Injection System-E	Electronic Continuous Fuel Injection System[1]	Electronic CFI System[1]
Continuous Trap Oxidizer	Continuous Trap Oxidizer[1]	CTOX[1]
Coolant Temperature Sensor	Engine Coolant Temperature Sensor[1]	ECT Sensor[1]
CP (Crankshaft Position)	Crankshaft Position[1]	CKP[1]
CPP (Clutch Pedal Position)	Clutch Pedal Position[1]	CPP[1]
CPP (Clutch Pedal Position) Switch	Clutch Pedal Position Switch	CPP Switch[1]
CPS (Camshaft Position Sensor)	Camshaft Position Sensor[1]	CMP Sensor[1]
CPS (Crankshaft Position Sensor)	Crankshaft Position Sensor[1]	CKP Sensor[1]
Crank Angle Sensor	Crankshaft Position Sensor[1]	CKP Sensor[1]

SAE J1930 Revised SEP95

TABLE 1—CROSS REFERENCE AND LOOK UP (CONTINUED)

Existing Usage	Acceptable Usage	Acceptable Acronized Usage
Crankshaft Position	Crankshaft Position[1]	CKP[1]
Crankshaft Position Sensor	Crankshaft Position Sensor[1]	CKP Sensor[1]
Crankshaft Speed	Engine Speed[1]	RPM[1]
Crankshaft Speed Sensor	Engine Speed Sensor[1]	RPM Sensor[1]
Critical Flow Venturi	**Critical Flow Venturi**	**CFV**
CTO (Continuous Trap Oxidizer)	Continuous Trap Oxidizer[1]	CTOX[1]
CTOX (Continuous Trap Oxidizer)	Continuous Trap Oxidizer[1]	CTOX[1]
CTP (Closed Throttle Position)	Closed Throttle Position[1]	CTP[1]
CTS (Coolant Temperature Sensor)	Engine Coolant Temperature Sensor[1]	ECT Sensor[1]
CTS (Coolant Temperature Switch)	Engine Coolant Temperature Switch[1]	ECT Switch[1]
CVS	**Constant Volume Sampler**	**CVS**
Cylinder ID (Identification) Sensor	Camshaft Position Sensor[1]	CMP Sensor[1]
D-Jetronic	Multiport Fuel Injection[1]	MFI[1]
Data Link Connector	Data Link Connector[1]	DLC[1]
Detonation Sensor	Knock Sensor[1]	KS[1]
DFI (Direct Fuel Injection)	Direct Fuel Injection[1]	DFI[1]
DFI (Digital Fuel Injection)	Multiport Fuel Injection[1]	MFI[1]
DI (Direct Injection)	Direct Fuel Injection[1]	DFI[1]
DI (Distributor Ignition)	Distributor Ignition[1]	DI[1]
DI (Distributor Ignition) Capacitor	Distributor Ignition Capacitor[1]	DI Capacitor[1]
Diagnostic Test Mode	Diagnostic Test Mode[1]	DTM[1]
Diagnostic Trouble Code	Diagnostic Trouble Code[1]	DTC[1]
DID (Direct Injection - Diesel)	Direct Fuel Injection[1]	DFI[1]
Differential Pressure Feedback EGR (Exhaust Gas Recirculation) System	Differential Pressure Feedback Exhaust Gas Recirculation System[1]	Differential Pressure Feedback EGR System[1]
Digital EGR (Exhaust Gas Recirculation)	Exhaust Gas Recirculation[1]	EGR[1]
Direct Fuel Injection	Direct Fuel Injection[1]	DFI[1]
Direct Ignition System	Electronic Ignition System[1]	EI System[1]
DIS (Distributorless Ignition System)	Electronic Ignition System[1]	EI System[1]
DIS (Distributorless Ignition System) Module	Ignition Control Module[1]	ICM[1]
Distance Sensor	Vehicle Speed Sensor[1]	VSS[1]
Distributor Ignition	Distributor Ignition[1]	DI[1]
Distributorless Ignition	Electronic Ignition[1]	EI[1]
DLC (Data Link Connector)	Data Link Connector[1]	DLC[1]
DLI (Distributorless Ignition)	Electronic Ignition[1]	EI[1]
Driver	**Driver**	**Driver**
DS (Detonation Sensor)	Knock Sensor[1]	KS[1]
DTC (Diagnostic Trouble Code)	Diagnostic Trouble Code[1]	DTC[1]
DTM (Diagnostic Test Mode)	Diagnostic Test Mode[1]	DTM[1]
Dual Bed	Three Way + Oxidation Catalytic Converter[1]	TWC+OC[1]
Duty Solenoid for Purge Valve	Evaporative Emission Canister Purge Valve	EVAP Canister Purge Valve[1]
Dynamic Pressure Control	**Dynamic Pressure Control**	**Dynamic PC**
Dynamic Pressure Control Solenoid	**Dynamic Pressure Control Solenoid[1]**	**Dynamic PC Solenoid[1]**
Dynamic Pressure Control Solenoid Valve	**Dynamic Pressure Control Solenoid Valve[1]**	**Dynamic PC Solenoid Valve[1]**
E2PROM (Electrically Erasable Programmable Read Only Memory)	Electrically Erasable Programmable Read Only Memory[1]	EEPROM[1]
Early Fuel Evaporation	Early Fuel Evaporation[1]	EFE[1]
EATX (Electronic Automatic Transmission/ Transaxle)	Automatic Transmission[1]	A/T[1]
EC (Engine Control)	Engine Control[1]	EC[1]

SAE J1930 Revised SEP95

TABLE 1—CROSS REFERENCE AND LOOK UP (CONTINUED)

Existing Usage	Acceptable Usage	Acceptable Acronized Usage
ECA (Electronic Control Assembly)	Powertrain Control Module[1]	PCM[1]
ECL (Engine Coolant Level)	Engine Coolant Level	ECL
ECM (Engine Control Module)	Engine Control Module[1]	ECM[1]
ECT (Engine Coolant Temperature)	Engine Coolant Temperature[1]	ECT[1]
ECT (Engine Coolant Temperature) Sender	Engine Coolant Temperature Sensor[1]	ECT Sensor[1]
ECT (Engine Coolant Temperature) Sensor	Engine Coolant Temperature Sensor[1]	ECT Sensor[1]
ECT (Engine Coolant Temperature) Switch	Engine Coolant Temperature Switch[1]	ECT Switch[1]
ECU4 (Electronic Control Unit 4)	Powertrain Control Module[1]	PCM[1]
EDF (Electro-Drive Fan) Control	Fan Control	FC
EDIS (Electronic Distributor Ignition System)	Distributor Ignition System[1]	DI System[1]
EDIS (Electronic Distributorless Ignition System)	Electronic Ignition System[1]	EI System[1]
EDIS (Electronic Distributor Ignition System) Module	Distributor Ignition Control Module[1]	Distributor ICM[1]
EEC (Electronic Engine Control)	Engine Control[1]	EC[1]
EEC (Electronic Engine Control) Processor	Powertrain Control Module[1]	PCM[1]
EECS (Evaporative Emission Control System)	Evaporative Emission System[1]	EVAP System[1]
EEPROM (Electrically Erasable Programmable Read Only Memory)	Electrically Erasable Programmable Read Only Memory[1]	EEPROM[1]
EFE (Early Fuel Evaporation)	Early Fuel Evaporation[1]	EFE[1]
EFI (Electronic Fuel Injection)	Multiport Fuel Injection[1]	MFI[1]
EFI (Electronic Fuel Injection)	Throttle Body Fuel Injection[1]	TBI[1]
EGO (Exhaust Gas Oxygen) Sensor	Oxygen Sensor[1]	O2S[1]
EGOS (Exhaust Gas Oxygen Sensor)	Oxygen Sensor[1]	O2S[1]
EGR (Exhaust Gas Recirculation)	Exhaust Gas Recirculation[1]	EGR[1]
EGR (Exhaust Gas Recirculation) Diagnostic Valve	Exhaust Gas Recirculation Diagnostic Valve[1]	EGR Diagnostic Valve[1]
EGR (Exhaust Gas Recirculation) System	Exhaust Gas Recirculation System[1]	EGR System[1]
EGR (Exhaust Gas Recirculation) Thermal Vacuum Valve	Exhaust Gas Recirculation Thermal Vacuum Valve[1]	EGR TVV[1]
EGR (Exhaust Gas Recirculation) Valve	Exhaust Gas Recirculation Valve[1]	EGR Valve[1]
EGR TVV (Exhaust Gas Recirculation Thermal Vacuum Valve)	Exhaust Gas Recirculation Thermal Vacuum Valve[1]	EGR TVV[1]
EGRT (Exhaust Gas Recirculation Temperature)	Exhaust Gas Recirculation Temperature	EGRT[1]
EGRT (Exhaust Gas Recirculation Temperature) Sensor	Exhaust Gas Recirculation Temperature Sensor[1]	EGRT Sensor[1]
EGRV (Exhaust Gas Recirculation Valve)	Exhaust Gas Recirculation Valve[1]	EGR Valve[1]
EGRVC (Exhaust Gas Recirculation Valve Control)	Exhaust Gas Recirculation Valve Control[1]	EGR Valve Control[1]
EGS (Exhaust Gas Sensor)	Oxygen Sensor[1]	O2S[1]
EI (Electronic Ignition) (With Distributor)	Distributor Ignition[1]	DI[1]
EI (Electronic Ignition) (Without Distributor)	Electronic Ignition[1]	EI[1]
Electrically Erasable Programmable Read Only Memory	Electrically Erasable Programmable Read Only Memory[1]	EEPROM[1]
Electronic Engine Control	Electronic Engine Control[1]	Electronic EC[1]
Electronic Ignition	Electronic Ignition[1]	EI[1]
Electronic Spark Advance	Ignition Control[1]	IC[1]
Electronic Spark Timing	Ignition Control[1]	IC[1]
EM (Engine Modification)	Engine Modification[1]	EM[1]
EMR (Engine Maintenance Reminder)	Service Reminder Indicator[1]	SRI[1]
Engine Control	Engine Control[1]	EC[1]
Engine Coolant Fan Control	Fan Control	FC
Engine Coolant Level	Engine Coolant Level	ECL
Engine Coolant Level Indicator	Engine Coolant Level Indicator	ECL Indicator

SAE J1930 Revised SEP95

TABLE 1—CROSS REFERENCE AND LOOK UP (CONTINUED)

Existing Usage	Acceptable Usage	Acceptable Acronized Usage
Engine Coolant Temperature	<u>Engine Coolant Temperature</u>[1]	<u>ECT</u>[1]
Engine Coolant Temperature Sender	<u>Engine Coolant Temperature</u> Sensor[1]	<u>ECT</u> Sensor[1]
Engine Coolant Temperature Sensor	<u>Engine Coolant Temperature</u> Sensor[1]	<u>ECT</u> Sensor[1]
Engine Coolant Temperature Switch	<u>Engine Coolant Temperature</u> Switch[1]	<u>ECT</u> Switch[1]
Engine Modification	<u>Engine Modification</u>[1]	<u>EM</u>[1]
Engine Oil Pressure Sender	**<u>Engine Oil Pressure</u>** Sensor	**<u>EOP</u>** Sensor
Engine Oil Pressure Sensor	**<u>Engine Oil Pressure</u>** Sensor	**<u>EOP</u>** Sensor
Engine Oil Pressure Switch	**<u>Engine Oil Pressure</u>** Switch	**<u>EOP</u>** Switch
Engine Oil Temperature	**<u>Engine Oil Temperature</u>**	**<u>EOT</u>**
Engine Speed	<u>Engine Speed</u>[1]	<u>RPM</u>[1]
EOS (Exhaust Oxygen Sensor)	<u>Oxygen Sensor</u>[1]	<u>O2S</u>[1]
EOT (Engine Oil Temperature)	**<u>Engine Oil Temperature</u>**	**<u>EOT</u>**
EP (Exhaust Pressure)	**<u>Exhaust Pressure</u>**	**<u>EP</u>**
EPROM (Erasable Programmable Read Only Memory)	<u>Erasable Programmable Read Only Memory</u>[1]	<u>EPROM</u>[1]
Erasable Programmable Read Only Memory	<u>Erasable Programmable Read Only Memory</u>[1]	<u>EPROM</u>[1]
ESA (Electronic Spark Advance)	<u>Ignition Control</u>[1]	<u>IC</u>[1]
ESAC (Electronic Spark Advance Control)	<u>Distributor Ignition</u>[1]	<u>DI</u>[1]
EST (Electronic Spark Timing)	<u>Ignition Control</u>[1]	<u>IC</u>[1]
EVAP (Evaporate Emission) CANP (Canister Purge)	**<u>Evaporative Emission</u> Canister Purge**[1]	**<u>EVAP</u> Canister Purge**[1]
EVAP (Evaporative Emission)	<u>Evaporative Emission</u>[1]	<u>EVAP</u>[1]
EVAP (Evaporative Emission) Canister	<u>Evaporative Emission</u> Canister[1]	<u>EVAP</u> Canister[1]
EVAP (Evaporative Emission) Purge Valve	<u>Evaporative Emission</u> Canister Purge Valve[1]	<u>EVAP</u> Canister Purge Valve[1]
Evaporative Emission	<u>Evaporative Emission</u>[1]	<u>EVAP</u>[1]
Evaporative Emission Canister	<u>Evaporative Emission</u> Canister[1]	<u>EVAP</u> Canister[1]
EVP (Exhaust Gas Recirculation Valve Position) Sensor	<u>Exhaust Gas Recirculation</u> Valve Position Sensor[1]	<u>EGR</u> Valve Position Sensor[1]
EVR (Exhaust Gas Recirculation Vacuum Regulator) Solenoid	<u>Exhaust Gas Recirculation</u> Vacuum Regulator Solenoid[1]	<u>EGR</u> Vacuum Regulator Solenoid[1]
EVRV (Exhaust Gas Recirculation Vacuum Regulator Valve)	<u>Exhaust Gas Recirculation</u> Vacuum Regulator Valve[1]	<u>EGR</u> Vacuum Regulator Valve[1]
Exhaust Gas Recirculation	<u>Exhaust Gas Recirculation</u>[1]	<u>EGR</u>[1]
Exhaust Gas Recirculation Temperature	<u>Exhaust Gas Recirculation Temperature</u>[1]	<u>EGRT</u>[1]
Exhaust Gas Recirculation Temperature Sensor	<u>Exhaust Gas Recirculation Temperature</u> Sensor[1]	<u>EGRT</u> Sensor[1]
Exhaust Gas Recirculation Vacuum Solenoid Valve Regulator	**<u>Exhaust Gas Recirculation</u> Vacuum Regulator Solenoid Valve**[1]	**<u>EGR</u> Vacuum Regulator Solenoid Valve**[1]
Exhaust Gas Recirculation Vacuum Regulator Valve	**<u>Exhaust Gas Recirculation</u> Vacuum Regulator Valve**[1]	**<u>EGR</u> Vacuum Regulator Valve**[1]
Exhaust Gas Recirculation Valve	<u>Exhaust Gas Recirculation</u> Valve[1]	<u>EGR</u> Valve[1]
Exhaust Pressure	**<u>Exhaust Pressure</u>**	**<u>EP</u>**
4GR (Fourth Gear)	<u>Fourth Gear</u>	<u>4GR</u>
4WD (Four Wheel Drive)	**<u>Full Time Four Wheel Drive</u>**	**<u>F4WD</u>**
4WD (Four Wheel Drive)	**<u>Selectable Four Wheel Drive</u>**	**<u>S4WD</u>**
F4WD	**<u>Full Time Four Wheel Drive</u>**	**<u>F4WD</u>**
Fan Control	<u>Fan Control</u>	<u>FC</u>
Fan Control Module	<u>Fan Control</u> Module	<u>FC</u> Module
Fan Control Relay	<u>Fan Control</u> Relay	<u>FC</u> Relay
Fan Motor Control Relay	<u>Fan Control</u> Relay	<u>FC</u> Relay
Fast Idle Thermo Valve	<u>Idle Air Control</u> Thermal Valve[1]	<u>IAC</u> Thermal Valve[1]
FBC (Feed Back Carburetor)	<u>Carburetor</u>[1]	<u>CARB</u>[1]
FBC (Feed Back Control)	<u>Mixture Control</u>[1]	<u>MC</u>[1]

SAE J1930 Revised SEP95

TABLE 1—CROSS REFERENCE AND LOOK UP (CONTINUED)

Existing Usage	Acceptable Usage	Acceptable Acronized Usage
FC (Fan Control)	Fan Control	FC
FC (Fan Control) Relay	Fan Control Relay	FC Relay
FEEPROM (Flash Electrically Erasable Programmable Read Only Memory)	Flash Electrically Erasable Programmable Read Only Memory[1]	FEEPROM[1]
FEPROM (Flash Erasable Programmable Read Only Memory)	Flash Erasable Programmable Read Only Memory[1]	FEPROM[1]
FF (Flexible Fuel)	Flexible Fuel[1]	FF[1]
FI (Fuel Injection)	Central Multiport Fuel Injection[1]	Central MFI[1]
FI (Fuel Injection)	Continuous Fuel Injection[1]	CFI[1]
FI (Fuel Injection)	Direct Fuel Injection[1]	DFI[1]
FI (Fuel Injection)	Indirect Fuel Injection[1]	IFI[1]
FI (Fuel Injection)	Multiport Fuel Injection[1]	MFI[1]
FI (Fuel Injection)	Sequential Multiport Fuel Injection[1]	SFI[1]
FI (Fuel Injection)	Throttle Body Fuel Injection[1]	TBI[1]
Flame Ionization Detector	**Flame Ionization Detector**	**FID**
Flash EEPROM (Electrically Erasable Programmable Read Only Memory)	Flash Electrically Erasable Programmable Read Only Memory[1]	FEEPROM[1]
Flash EPROM (Erasable Programmable Read Only Memory)	Flash Erasable Programmable Read Only Memory[1]	FEPROM[1]
Flexible Fuel	Flexible Fuel[1]	FF[1]
Flexible Fuel Sensor	Flexible Fuel Sensor[1]	FF Sensor
Fourth Gear	Fourth Gear	4GR
FP (Fuel Pump)	Fuel Pump	FP
FP (Fuel Pump) Module	Fuel Pump Module	FP Module
Freeze Frame	**Freeze Frame**	**See Table 4**
Front Wheel Drive	**Front Wheel Drive**	**FWD**
FRZF (Freeze Frame)	**Freeze Frame**	**See Table 4**
FT (Fuel Trim)	Fuel Trim[1]	FT[1]
Fuel Charging Station	Throttle Body[1]	TB[1]
Fuel Concentration Sensor	Flexible Fuel Sensor[1]	FF Sensor[1]
Fuel Injection	Central Multiport Fuel Injection[1]	Central MFI[1]
Fuel Injection	Continuous Fuel Injection[1]	CFI[1]
Fuel Injection	Direct Fuel Injection[1]	DFI[1]
Fuel Injection	Indirect Fuel Injection[1]	IFI[1]
Fuel Injection	Multiport Fuel Injection[1]	MFI[1]
Fuel Injection	Sequential Multiport Fuel Injection[1]	SFI[1]
Fuel Injection	Throttle Body Fuel Injection[1]	TBI[1]
Fuel Level Sensor	Fuel Level Sensor	Fuel Level Sensor
Fuel Module	Fuel Pump Module	FP Module
Fuel Pressure	Fuel Pressure[1]	Fuel Pressure[1]
Fuel Pressure	**Fuel Pressure**	**See Table 4**
Fuel Pressure Regulator	Fuel Pressure Regulator[1]	Fuel Pressure Regulator[1]
Fuel Pump	Fuel Pump	FP
Fuel Pump Relay	Fuel Pump Relay	FP Relay
Fuel Quality Sensor	Flexible Fuel Sensor[1]	FF Sensor[1]
Fuel Regulator	Fuel Pressure Regulator[1]	Fuel Pressure Regulator[1]
Fuel Sender	Fuel Pump Module	FP Module
Fuel Sensor	Fuel Level Sensor	Fuel Level Sensor
Fuel System Status	**Fuel System Status**	**See Table 4**
FUEL SYS	**Fuel System Status**	**See Table 4**
Fuel Tank Unit	Fuel Pump Module	FP Module

SAE J1930 Revised SEP95

TABLE 1—CROSS REFERENCE AND LOOK UP (CONTINUED)

Existing Usage	Acceptable Usage	Acceptable Acronized Usage
Fuel Trim	Fuel Trim[1]	FT[1]
Full Time Four Wheel Drive	**Full Time Four Wheel Drive**	**F4WD**
Full Throttle	Wide Open Throttle[1]	WOT[1]
FWD	**Front Wheel Drive**	**FWD**
GCM (Governor Control Module)	Governor Control Module	GCM
GEM (Governor Electronic Module)	Governor Control Module	GCM
GEN (Generator)	Generator	GEN
Generator	Generator	GEN
Glow Plug	**Glow Plug**[1]	**Glow Plug**[1]
GND (Ground)	Ground	GND
Governor	Governor	Governor
Governor Control Module	Governor Control Module	GCM
Governor Electronic Module	Governor Control Module	GCM
Gram Per Mile	**Gram Per Mile**	**GPM**
GRD (Ground)	Ground	GND
Ground	Ground	GND
Heated Oxygen Sensor	Heated Oxygen Sensor[1]	HO2S[1]
HEDF (High Electro-Drive Fan) Control	Fan Control	FC
HEGO (Heated Exhaust Gas Oxygen) Sensor	Heated Oxygen Sensor[1]	HO2S[1]
HEI (High Energy Ignition)	Distributor Ignition[1]	DI[1]
High Speed FC (Fan Control) Switch	High Speed Fan Control Switch	High Speed FC Switch
HO2S (Heated Oxygen Sensor)	Heated Oxygen Sensor[1]	HO2S[1]
HOS (Heated Oxygen Sensor)	Heated Oxygen Sensor[1]	HO2S[1]
Hot Wire Anemometer	Mass Air Flow Sensor[1]	MAF Sensor[1]
IA (Intake Air)	Intake Air	IA
IA (Intake Air) Duct	Intake Air Duct	IA Duct
IAC (Idle Air Control)	Idle Air Control[1]	IAC[1]
IAC (Idle Air Control) Thermal Valve	Idle Air Control Thermal Valve[1]	IAC Thermal Valve[1]
IAC (Idle Air Control) Valve	Idle Air Control Valve[1]	IAC Valve[1]
IACV (Idle Air Control Valve)	Idle Air Control Valve[1]	IAC Valve[1]
IAT (Intake Air Temperature)	Intake Air Temperature[1]	IAT[1]
IAT (Intake Air Temperature) Sensor	Intake Air Temperature Sensor[1]	IAT Sensor[1]
IATS (Intake Air Temperature Sensor)	Intake Air Temperature Sensor[1]	IAT Sensor[1]
IC (Ignition Control)	Ignition Control[1]	IC[1]
ICM (Ignition Control Module)	Ignition Control Module[1]	ICM[1]
ICP (Injection Control Pressure)	**Injection Control Pressure**[1]	**ICP**[1]
IDFI (Indirect Fuel Injection)	Indirect Fuel Injection[1]	IFI[1]
IDI (Integrated Direct Ignition)	Electronic Ignition[1]	EI[1]
IDI (Indirect Diesel Injection)	Indirect Fuel Injection[1]	IFI[1]
Idle Air Bypass Control	Idle Air Control[1]	IAC[1]
Idle Air Control	Idle Air Control[1]	IAC[1]
Idle Air Control Valve	Idle Air Control Valve[1]	IAC Valve[1]
Idle Speed Control	Idle Air Control[1]	IAC[1]
Idle Speed Control	Idle Speed Control[1]	ISC[1]
Idle Speed Control Actuator	Idle Speed Control Actuator[1]	ISC Actuator[1]
IFI (Indirect Fuel Injection)	Indirect Fuel Injection[1]	IFI[1]
IFS (Inertia Fuel Shutoff)	Inertia Fuel Shutoff	IFS
Ignition Control	Ignition Control[1]	IC[1]
Ignition Control Module	Ignition Control Module[1]	ICM[1]
I/M (Inspection and Maintenance)	**Inspection and Maintenance**	**I/M**

SAE J1930 Revised SEP95

TABLE 1—CROSS REFERENCE AND LOOK UP (CONTINUED)

Existing Usage	Acceptable Usage	Acceptable Acronized Usage
IMRC (Intake Manifold Runner Control)	**Intake Manifold Runner Control**	**IMRC**
In Tank Module	Fuel Pump Module	FP Module
Indirect Fuel Injection	Indirect Fuel Injection[1]	IFI[1]
Inertia Fuel Shutoff	Inertia Fuel Shutoff	IFS
Inertia Fuel - Shutoff Switch	Inertia Fuel Shutoff Switch	IFS Switch
Inertia Switch	Inertia Fuel Shutoff Switch	IFS Switch
Injection Control Pressure	**Injection Control Pressure**[1]	**ICP**[1]
Input Shaft Speed	**Input Shaft Speed**	**ISS**
INT (Integrator)	Short Term Fuel Trim[1]	Short Term FT[1]
Inspection and Maintenance	**Inspection and Maintenance**	**I/M**
Intake Air	Intake Air	IA
Intake Air Duct	Intake Air Duct	IA Duct
Intake Air Temperature	Intake Air Temperature[1]	IAT[1]
Intake Air Temperature Sensor	Intake Air Temperature Sensor[1]	IAT Sensor[1]
Intake Manifold Absolute Pressure Sensor	Manifold Absolute Pressure Sensor[1]	MAP Sensor[1]
Intake Manifold Runner Control	**Intake Manifold Runner Control**	**IMRC**
Integrated Relay Module	Relay Module	RM
Integrator	Short Term Fuel Trim[1]	Short Term FT[1]
Inter Cooler	Charge Air Cooler[1]	CAC[1]
ISC (Idle Speed Control)	Idle Air Control[1]	IAC[1]
ISC (Idle Speed Control)	Idle Speed Control[1]	ISC[1]
ISC (Idle Speed Control) Actuator	Idle Speed Control Actuator[1]	ISC Actuator[1]
ISC BPA (Idle Speed Control By Pass Air)	Idle Air Control[1]	IAC
ISC (Idle Speed Control) Solenoid Vacuum Valve	Idle Speed Control Solenoid Vacuum Valve[1]	ISC Solenoid Vacuum Valve[1]
ISS (Input Shaft Speed)	**Input Shaft Speed**	**ISS**
K-Jetronic	Continuous Fuel Injection[1]	CFI[1]
KAM (Keep Alive Memory)	Non Volatile Random Access Memory[1]	NVRAM[1]
KAM (Keep Alive Memory)	Keep Alive Random Access Memory[1]	Keep Alive RAM[1]
KE-Jetronic	Continuous Fuel Injection[1]	CFI[1]
KE-Motronic	Continuous Fuel Injection[1]	CFI[1]
Knock Sensor	Knock Sensor[1]	KS[1]
KS (Knock Sensor)	Knock Sensor[1]	KS[1]
L-Jetronic	Multiport Fuel Injection[1]	MFI[1]
Lambda	Oxygen Sensor[1]	O2S[1]
LH-Jetronic	Multiport Fuel Injection[1]	MFI[1]
Light Off Catalyst	Warm Up Three Way Catalytic Converter[1]	WU-TWC[1]
Light Off Catalyst	Warm Up Oxidation Catalytic Converter[1]	WU-OC[1]
Line Pressure Control Solenoid Valve	**Line Pressure Control Solenoid Valve**	**Line PC Solenoid Valve**
LOAD (Calculated Load Value)	**Calculated Load Value**	**LOAD**
Lock Up Relay	Torque converter Clutch Relay[1]	TCC Relay[1]
Long Term FT (Fuel Trim)	Long Term Fuel Trim[1]	Long Term FT[1]
Long Term Fuel Trim	**Long Term FT**	**Long Term FT**
LONG FT	**Long Term Fuel Trim**	**See Table 4**
Low Speed FC (Fan Control) Switch	Low Speed Fan Control Switch	Low Speed FC Switch
LUS (Lock Up Solenoid) Valve	Torque Converter Clutch Solenoid Valve[1]	TCC Solenoid Valve[1]
M/C (Mixture Control)	Mixture Control[1]	MC[1]
MAF (Mass Air Flow)	Mass Air Flow[1]	MAF[1]
MAF (Mass Air Flow) Sensor	Mass Air Flow Sensor[1]	MAF Sensor[1]

SAE J1930 Revised SEP95

TABLE 1—CROSS REFERENCE AND LOOK UP (CONTINUED)

Existing Usage	Acceptable Usage	Acceptable Acronized Usage
Malfunction Indicator Lamp	Malfunction Indicator Lamp[1]	MIL[1]
Manifold Absolute Pressure	Manifold Absolute Pressure[1]	MAP[1]
Manifold Absolute Pressure Sensor	Manifold Absolute Pressure Sensor	MAP Sensor[1]
Manifold Differential Pressure	Manifold Differential Pressure[1]	MDP[1]
Manifold Surface Temperature	Manifold Surface Temperature[1]	MST[1]
Manifold Vacuum Zone	Manifold Vacuum Zone[1]	MVZ[1]
Manual Lever Position Sensor	Transmission Range Sensor[1]	TR Sensor[1]
MAP (Manifold Absolute Pressure)	Manifold Absolute Pressure[1]	MAP[1]
MAP (Manifold Absolute Pressure) Sensor	Manifold Absolute Pressure Sensor[1]	MAP Sensor[1]
MAPS (Manifold Absolute Pressure Sensor)	Manifold Absolute Pressure Sensor[1]	MAP Sensor[1]
Mass Air Flow	Mass Air Flow[1]	MAF[1]
Mass Air Flow Sensor	Mass Air Flow Sensor[1]	MAF Sensor[1]
MAT (Manifold Air Temperature)	Intake Air Temperature[1]	IAT[1]
MATS (Manifold Air Temperature Sensor)	Intake Air Temperature Sensor[1]	IAT Sensor[1]
MC (Mixture Control)	Mixture Control[1]	MC[1]
MCS (Mixture Control Solenoid)	Mixture Control Solenoid[1]	MC Solenoid[1]
MCU (Microprocessor Control Unit)	Powertrain Control Module[1]	PCM[1]
MDP (Manifold Differential Pressure)	Manifold Differential Pressure[1]	MDP[1]
MFI (Multiport Fuel Injection)	Multiport Fuel Injection[1]	MFI[1]
MIL (Malfunction Indicator Lamp)	Malfunction Indicator Lamp[1]	MIL[1]
Mixture Control	Mixture Control[1]	MC[1]
MLPS (Manual Lever Position Sensor)	**Transmission Range Sensor[1]**	**TR Sensor[1]**
Modes	Diagnostic Test Mode[1]	DTM[1]
Mono-Jetronic	**Throttle Body Injection[1]**	**TBI[1]**
Mono-Motronic	**Throttle Body Injection[1]**	**TBI[1]**
Monotronic	Throttle Body Fuel Injection[1]	TBI[1]
Motronic-Pressure	**Multiport Fuel Injection[1]**	**MFI[1]**
Motronic	Multiport Fuel Injection[1]	MFI[1]
MPI (Multipoint Injection)	Multiport Fuel Injection[1]	MFI[1]
MPI (Multiport Injection)	Multiport Fuel Injection[1]	MFI[1]
MRPS (Manual Range Position Switch)	Transmission Range Switch	TR Switch
MST (Manifold Surface Temperature)	Manifold Surface Temperature[1]	MST[1]
Multiport Fuel Injection	Multiport Fuel Injection[1]	MFI[1]
MVZ (Manifold Vacuum Zone)	Manifold Vacuum Zone[1]	MVZ[1]
NDS (Neutral Drive Switch)	Park/Neutral Position Switch[1]	PNP Switch[1]
Neutral Safety Switch	Park/Neutral Position Switch[1]	PNP Switch[1]
NGS (Neutral Gear Switch)	Park/Neutral Position Switch[1]	PNP Switch[1]
Non Dispersive Infrared	**Non Dispersive Infrared**	**NDIR**
Non Volatile Random Access Memory	Non Volatile Random Access Memory[1]	NVRAM[1]
NPS (Neutral Position Switch)	Park/Neutral Position Switch[1]	PNP Switch[1]
NVM (Non Volatile Memory)	Non Volatile Random Access Memory[1]	NVRAM[1]
NVRAM (Non Volatile Random Access Memory)	Non Volatile Random Access Memory[1]	NVRAM[1]
O2 (Oxygen) Sensor	Oxygen Sensor[1]	O2S[1]
O2S (Oxygen Sensor)	Oxygen Sensor[1]	O2S[1]
Oxygen Sensor Location	**Oxygen Sensor Location**	**See Table 4**
OBD (On Board Diagnostic)	On Board Diagnostic[1]	OBD[1]
OBD Status	**OBD Status**	**see Table 4**
OBD STAT	**OBD Status**	**see Table 4**
OC (Oxidation Catalyst)	Oxidation Catalytic Converter[1]	OC[1]

SAE J1930 Revised SEP95

TABLE 1—CROSS REFERENCE AND LOOK UP (CONTINUED)

Existing Usage	Acceptable Usage	Acceptable Acronized Usage
Oil Pressure Sender	Engine Oil Pressure Sensor	EOP Sensor
Oil Pressure Sensor	Engine Oil Pressure Sensor	EOP Sensor
Oil Pressure Switch	Engine Oil Pressure Switch	EOP Switch
OL (Open Loop)	Open Loop[1]	OL[1]
On Board Diagnostic	On Board Diagnostic[1]	OBD[1]
Open Loop	Open Loop[1]	OL[1]
OS (Oxygen Sensor)	Oxygen Sensor[1]	O2S[1]
OSS (Output Shaft Speed) Sensor	**Output Shaft Speed Sensor[1]**	**OSS Sensor[1]**
Output Driver	**Driver**	**Driver**
Output Shaft Speed Sensor	**Output Shaft Speed Sensor[1]**	**OSS Sensor[1]**
Oxidation Catalytic Converter	Oxidation Catalytic Converter[1]	OC[1]
OXS (Oxygen Sensor) Indicator	Service Reminder Indicator[1]	SRI[1]
Oxygen Sensor	Oxygen Sensor[1]	O2S[1]
P/N (Park/Neutral)	Park/Neutral Position[1]	PNP[1]
P/S (Power Steering) Pressure Switch	Power Steering Pressure Switch	PSP Switch
P- (Pressure) Sensor	Manifold Absolute Pressure Sensor[1]	MAP Sensor[1]
PAIR (Pulsed Secondary Air Injection)	Pulsed Secondary Air Injection[1]	PAIR[1]
Parameter Identification	**Parameter Identification**	**PID**
Parameter Identification Supported	**Parameter Identification Supported**	**See Table 4**
Park/Neutral Position	Park/Neutral Position[1]	PNP[1]
PC (Pressure Control) Solenoid Valve	**Pressure Control Solenoid Valve[1]**	**PC Solenoid Valve[1]**
PCM (Powertrain Control Module)	Powertrain Control Module[1]	PCM[1]
PCV (Positive Crankcase Ventilation)	Positive Crankcase Ventilation[1]	PCV[1]
PCV (Positive Crankcase Ventilation) Valve	Positive Crankcase Ventilation Valve[1]	PCV Valve[1]
Percent Alcohol Sensor	Flexible Fuel Sensor[1]	FF Sensor[1]
Periodic Trap Oxidizer	Periodic Trap Oxidizer[1]	PTOX[1]
PFE (Pressure Feedback Exhaust Gas Recirculation Sensor	**Feedback Pressure** Exhaust Gas Recirculation Sensor[1]	**Feedback Pressure** EGR Sensor[1]
PFI (Port Fuel Injection)	Multiport Fuel Injection[1]	MFI[1]
PG (Pulse Generator)	Vehicle Speed Sensor[1]	VSS[1]
PGM-FI (Programmed Fuel Injection)	Multiport Fuel Injection[1]	MFI[1]
PID (Parameter Identification)	**Parameter Identification**	**PID**
PID SUP	**Parameter Identification Supported**	**See Table 4**
PIP (Position Indicator Pulse)	Crankshaft Position[1]	CKP[1]
PNP (Park/Neutral Position)	Park/Neutral Position[1]	PNP[1]
Positive Crankcase Ventilation	Positive Crankcase Ventilation[1]	PCV[1]
Positive Crankcase Ventilation Valve	Positive Crankcase Ventilation Valve[1]	PCV Valve[1]
Power Steering Pressure	Power Steering Pressure	PSP
Power Steering Pressure Switch	Power Steering Pressure Switch	PSP Switch
Powertrain Control Module	Powertrain Control Module[1]	PCM[1]
Pressure Control Solenoid Valve	**Pressure Control Solenoid Valve[1]**	**PC Solenoid Valve[1]**
Pressure Feedback EGR (Exhaust Gas Recirculation)	Feedback Pressure Exhaust Gas Recirculation[1]	Feedback Pressure EGR[1]
Pressure Sensor	Manifold Absolute Pressure Sensor[1]	MAP Sensor[1]
Pressure Feedback EGR (Exhaust Gas **Recirculation**) **System**	**Feedback Pressure** Exhaust Gas Recirculation System[1]	**Feedback Pressure** EGR System[1]
Pressure Transducer EGR (Exhaust Gas Recirculation) System	Pressure Transducer Exhaust Gas Recirculation System[1]	Pressure Transducer EGR System[1]
PRNDL (Park- Reverse- Neutral- Drive- Low)	Transmission Range	TR
Programmable Read Only Memory	Programmable Read Only Memory[1]	PROM[1]

SAE J1930 Revised SEP95

TABLE 1—CROSS REFERENCE AND LOOK UP (CONTINUED)

Existing Usage	Acceptable Usage	Acceptable Acronized Usage
PROM (Programmable Read Only Memory)	Programmable Read Only Memory[1]	PROM[1]
PSP (Power Steering Pressure)	Power Steering Pressure	PSP
PSP (Power Steering Pressure) Switch	Power Steering Pressure Switch	PSP Switch
PSPS (Power Steering Pressure Switch)	Power Steering Pressure Switch	PSP Switch
PTOX (Periodic Trap Oxidizer)	Periodic Trap Oxidizer[1]	PTOX[1]
Pulsair	Pulsed Secondary Air Injection[1]	PAIR[1]
Pulsed Secondary Air Injection	Pulsed Secondary Air Injection[1]	PAIR[1]
Pulse Width Modulation	**Pulse Width Modulation**	**PWM**
PWM	**Pulse Width Modulation**	**PWM**
QDM (Quad Driver Module)	**Driver**	**Driver**
Quad Driver Module	**Driver**	**Driver**
Radiator Fan Control	Fan Control	FC
Radiator Fan Relay	Fan Control Relay	FC Relay
RAM (Random Access Memory)	Random Access Memory[1]	RAM[1]
Random Access Memory	Random Access Memory[1]	RAM[1]
Read Only Memory	Read Only Memory[1]	ROM[1]
Rear Wheel Drive	**Rear Wheel Drive**	**RWD**
Recirculated Exhaust Gas Temperature Sensor	Exhaust Gas Recirculation Temperature Sensor	EGRT Sensor[1]
Reed Valve	Pulsed Secondary Air Injection Valve[1]	PAIR Valve[1]
REGTS (Recirculated Exhaust Gas Temperature Sensor)	Exhaust Gas Recirculation Temperature Sensor[1]	EGRT Sensor[1]
Relay Module	Relay Module	RM
Remote Mount TFI (Thick Film Ignition)	Distributor Ignition[1]	DI[1]
Revolutions per Minute	Engine Speed[1]	RPM[1]
RM (Relay Module)	Relay Module	RM
ROM (Read Only Memory)	Read Only Memory[1]	ROM[1]
RPM (Revolutions per Minute)	Engine Speed[1]	RPM[1]
RWD	**Rear Wheel Drive**	**RWD**
S4WD	**Selectable Four Wheel Drive**	**S4WD**
SABV (Secondary Air Bypass Valve)	Secondary Air Injection Bypass Valve[1]	AIR Bypass Valve[1]
SACV (Secondary Air Check Valve)	Secondary Air Injection Control Valve[1]	AIR Control Valve[1]
SASV (Secondary Air Switching Valve)	Secondary Air Injection Switching Valve[1]	AIR Switching Valve[1]
SBEC (Single Board Engine Control)	Powertrain Control Module[1]	PCM[1]
SBS (Supercharger Bypass Solenoid)	Supercharger Bypass Solenoid[1]	SCB Solenoid[1]
SC (Supercharger)	Supercharger[1]	SC[1]
Scan Tool	Scan Tool[1]	ST[1]
SCB (Supercharger Bypass)	Supercharger Bypass[1]	SCB[1]
Secondary Air Bypass Valve	Secondary Air Injection Bypass Valve[1]	AIR Bypass Valve[1]
Secondary Air Check Valve	Secondary Air Injection Check Valve[1]	AIR Check Valve[1]
Secondary Air Injection	Secondary Air Injection[1]	AIR[1]
Secondary Air Injection Bypass	Secondary Air Injection Bypass[1]	AIR Bypass[1]
Secondary Air Injection Diverter	Secondary Air Injection Diverter[1]	AIR Diverter[1]
Secondary Air Switching Valve	Secondary Air Injection Switching Valve[1]	AIR Switching Valve[1]
Selectable Four Wheel Drive	**Selectable Four Wheel Drive**	**S4WD**
SEFI (Sequential Electronic Fuel Injection)	Sequential Multiport Fuel Injection[1]	SFI[1]
Self Test	On Board Diagnostic[1]	OBD[1]
Self Test Codes	Diagnostic Trouble Code[1]	DTC[1]
Self Test Connector	Data Link Connector[1]	DLC[1]
Sequential Multiport Fuel Injection	Sequential Multiport Fuel Injection[1]	SFI[1]

SAE J1930 Revised SEP95

TABLE 1—CROSS REFERENCE AND LOOK UP (CONTINUED)

Existing Usage	Acceptable Usage	Acceptable Acronized Usage
Service Engine Soon	Service Reminder Indicator[1]	SRI[1]
Service Engine Soon	Malfunction Indicator Lamp[1]	MIL[1]
Service Reminder Indicator	Service Reminder Indicator[1]	SRI[1]
SFI (Sequential Fuel Injection)	Sequential Multiport Fuel Injection[1]	SFI[1]
Shift Solenoid	**Shift Solenoid[1]**	**SS[1]**
Shift Solenoid Valve	**Shift Solenoid Valve[1]**	**SS Valve[1]**
Short Term FT (Fuel Trim)	Short Term Fuel Trim[1]	Short Term FT[1]
Short Term Fuel Trim	**Short Term Fuel Trim[1]**	**Short Term FT[1]**
SHRT FT	**Short Term Fuel Trim[1]**	**See Table 4**
SLP (Selection Lever Position)	Transmission Range	TR
SMEC (Single Module Engine Control)	Powertrain Control Module[1]	PCM[1]
Smoke Puff Limiter	Smoke Puff Limiter[1]	SPL[1]
SPARK ADV	**Spark Advance**	**See Table 4**
Spark Advance	**Spark Advance**	**See Table 4**
Spark Plug	**Spark Plug[1]**	**Spark Plug[1]**
SPI (Single Point Injection)	Throttle Body Fuel Injection[1]	TBI[1]
SPL (Smoke Puff Limiter)	Smoke Puff Limiter[1]	SPL[1]
SS (Shift Solenoid)	**Shift Solenoid[1]**	**SS[1]**
SRI (Service Reminder Indicator)	Service Reminder Indicator[1]	SRI[1]
SRT (System Readiness Test)	System Readiness Test[1]	SRT[1]
ST (Scan Tool)	Scan Tool[1]	ST[1]
Supercharger	Supercharger[1]	SC[1]
Supercharger Bypass	Supercharger Bypass[1]	SCB[1]
Sync Pickup	Camshaft Position[1]	CMP[1]
System Readiness Test	System Readiness Test[1]	SRT[1]
3-2TS (3-2 Timing Solenoid)	**3-2 Timing Solenoid**	**3-2TS**
3-2TS Valve (3-2 Timing Solenoid)Valve	**3-2 Timing Solenoid Valve**	**3-2TS Valve**
3-2 Timing Solenoid	**3-2 Timing Solenoid**	**3-2TS**
3-2 Timing Solenoid Valve	**3-2 Timing Solenoid Valve**	**3-2TS Valve**
3GR (Third Gear)	Third Gear	3GR
TAB (Thermactor Air Bypass)	Secondary Air Injection Bypass[1]	AIR Bypass[1]
TAC (Throttle Actuator Control)	**Throttle Actuator Control**	**TAC**
TAC (Throttle Actuator Control) Module	**Throttle Actuator Control Module[1]**	**TAC Module[1]**
TAD (Thermactor Air Diverter)	Secondary Air Injection Diverter[1]	AIR Diverter[1]
TB (Throttle Body)	Throttle Body[1]	TB[1]
TBI (Throttle Body Fuel Injection)	Throttle Body Fuel Injection[1]	TBI[1]
TBT (Throttle Body Temperature)	Intake Air Temperature[1]	IAT[1]
TC (Turbocharger)	Turbocharger[1]	TC[1]
TC (Turbocharger) Wastegate	**Turbocharger Wastegate[1]**	**TC Wastegate[1]**
TC (Turbocharger) Wastegate Regulating Valve	**Turbocharger Wastegate Regulating Valve[1]**	**TC Wastegate Regulating Valve[1]**
TCC (Torque Converter Clutch)	Torque Converter Clutch[1]	TCC[1]
TCC (Torque Converter Clutch) Relay	Torque Converter Clutch Relay[1]	TCC Relay[1]
TCC (Torque Converter Clutch) Solenoid	**Torque Converter Clutch Solenoid [1]**	**TCC Solenoid[1]**
TCC (Torque Converter Clutch) Solenoid Valve	**Torque Converter Clutch Solenoid Valve[1]**	**TCC Solenoid Valve[1]**
TCM (Transmission Control Module)	Transmission Control Module	TCM
TCCP (Torque Converter Clutch Pressure)	**Torque Converter Clutch Pressure**	**TCCP**
TFI (Thick Film Ignition)	Distributor Ignition[1]	DI[1]
TFI (Thick Film Ignition) Module	Ignition Control Module[1]	ICM[1]

SAE J1930 Revised SEP95

TABLE 1—CROSS REFERENCE AND LOOK UP (CONTINUED)

Existing Usage	Acceptable Usage	Acceptable Acronized Usage
TFP (Transmission Fluid Pressure)	**Transmission Fluid Pressure**	**TFP**
TFT (Transmission Fluid Temperature) Sensor	**Transmission Fluid Temperature Sensor**	**TFT Sensor**
Thermac	Secondary Air Injection[1]	AIR[1]
Thermac Air Cleaner	Air Cleaner[1]	ACL[1]
Thermactor	Secondary Air Injection[1]	AIR[1]
Thermactor Air Bypass	Secondary Air Injection Bypass[1]	AIR Bypass[1]
Thermactor Air Diverter	Secondary Air Injection Diverter[1]	AIR Diverter[1]
Thermactor II	Pulsed Secondary Air Injection[1]	PAIR[1]
Thermal Vacuum Switch	Thermal Vacuum Valve[1]	TVV[1]
Thermal Vacuum Valve	Thermal Vacuum Valve[1]	TVV[1]
Third Gear	Third Gear	3GR
Three Way + Oxidation Catalytic Converter	Three Way + Oxidation Catalytic Converter[1]	TWC+OC[1]
Three Way Catalytic Converter	Three Way Catalytic Converter[1]	TWC[1]
Throttle Actuator Control	**Throttle Actuator Control**	**TAC**
Throttle Actuator Control Module	**Throttle Actuator Control Module**	**TAC Module**
Throttle Body	Throttle Body[1]	TB[1]
Throttle Body Fuel Injection	Throttle Body Fuel Injection[1]	TBI[1]
Throttle Opener	Idle Speed Control[1]	ISC[1]
Throttle Opener Vacuum Switching Valve	Idle Speed Control Solenoid Vacuum Valve[1]	ISC Solenoid Vacuum Valve[1]
Throttle Opener VSV (Vacuum Switching Valve)	Idle Speed Control Solenoid Vacuum Valve[1]	ISC Solenoid Vacuum Valve[1]
Throttle Position	Throttle Position[1]	TP
Throttle Position Sensor	Throttle Position Sensor[1]	TP Sensor[1]
Throttle Position Switch	Throttle Position Switch[1]	TP Switch[1]
Throttle Potentiometer	Throttle Position Sensor[1]	TP Sensor[1]
TOC (Trap Oxidizer - Continuous)	Continuous Trap Oxidizer[1]	CTOX[1]
TOP (Trap Oxidizer - Periodic)	Periodic Trap Oxidizer[1]	PTOX[1]
Torque Converter Clutch	Torque Converter Clutch[1]	TCC[1]
Torque Converter Clutch Pressure	**Torque Converter Clutch Pressure**	**TCCP**
Torque Converter Clutch Relay	Torque Converter Clutch Relay[1]	TCC Relay[1]
Torque Converter Clutch Solenoid	**Torque Converter Clutch Solenoid**[1]	**TCC Solenoid**[1]
Torque Converter Clutch Solenoid Valve	**Torque Converter Clutch Solenoid Valve**[1]	**TCC Solenoid Valve**[1]
TP (Throttle Position)	Throttle Position[1]	TP[1]
TP (Throttle Position) Sensor	Throttle Position Sensor[1]	TP Sensor[1]
TP (Throttle Position) Switch	Throttle Position Switch[1]	TP Switch[1]
TPI (Tuned Port Injection)	Multiport Fuel Injection[1]	MFI[1]
TPNP (Transmission Park Neutral Position)	**Park/Neutral Position**[1]	**PNP**[1]
TPS (Throttle Position Sensor)	Throttle Position Sensor[1]	TP Sensor[1]
TPS (Throttle Position Switch)	Throttle Position Switch[1]	TP Switch[1]
TR (Transmission Range)	Transmission Range	TR
Track Road Load Horsepower	**Track Road Load Horsepower**	**TRLHP**
Transmission Control Module	Transmission Control Module	TCM
Transmission Fluid Pressure	**Transmission Fluid Pressure**	**TFP**
Transmission Fluid Temperature Sensor	**Transmission Fluid Temperature Sensor**	**TFT Sensor**
Transmission Park Neutral Position	**Park/Neutral Position**[1]	**PNP**[1]
Transmission Position Switch	Transmission Range Switch	TR Switch
Transmission Range Selection	Transmission Range	TR
Transmission Range Sensor	**Transmission Range Sensor**	**TR Sensor**
TRS (Transmission Range Selection)	Transmission Range	TR

SAE J1930 Revised SEP95

TABLE 1—CROSS REFERENCE AND LOOK UP (CONTINUED)

Existing Usage	Acceptable Usage	Acceptable Acronized Usage
TRSS (Transmission Range Selection Switch)	Transmission Range Switch	TR Switch
TSS (Turbine Shaft Speed) Sensor	**Turbine Shaft Speed Sensor[1]**	**TSS Sensor[1]**
Tuned Port Injection	Multiport Fuel Injection[1]	MFI[1]
Turbine Shaft Speed Sensor	**Turbine Shaft Speed Sensor[1]**	**TSS Sensor[1]**
Turbo (Turbocharger)	Turbocharger[1]	TC[1]
Turbocharger	Turbocharger[1]	TC[1]
Turbocharger Wastegate	**Turbocharger Wastegate[1]**	**TC Wastegate[1]**
Turbocharger Wastegate Regulating Valve	**Turbocharger Wastegate Regulating Valve[1]**	**TC Wastegate Regulating Valve[1]**
TVS (Thermal Vacuum Switch)	Thermal Vacuum Valve[1]	TVV[1]
TVV (Thermal Vacuum Valve)	Thermal Vacuum Valve[1]	TVV[1]
TWC (Three Way Catalytic Converter)	Three Way Catalytic Converter[1]	TWC[1]
TWC + OC (Three Way + Oxidation Catalytic Converter)	Three Way + Oxidation Catalytic Converter[1]	TWC+OC[1]
VAC (Vacuum) Sensor	Manifold Differential Pressure Sensor[1]	MDP Sensor[1]
Vacuum Switches	Manifold Vacuum Zone Switch	MVZ Switch[1]
VAF (Volume Air Flow)	Volume Air Flow[1]	VAF[1]
Valve Position EGR (Exhaust Gas Recirculation) System	**Valve Position Exhaust Gas Recirculation System[1]**	**Valve Position EGR System[1]**
Vane Air Flow	Volume Air Flow[1]	VAF[1]
Variable Control Relay Module	**Variable Control Relay Module**	**VCRM**
Variable Fuel Sensor	Flexible Fuel Sensor	FF Sensor[1]
VAT (Vane Air Temperature)	Intake Air Temperature[1]	IAT[1]
VCC (Viscous Converter Clutch)	Torque Converter Clutch[1]	TCC[1]
VCM	**Vehicle Control Module**	**VCM**
VCRM	**Variable Control Relay Module**	**VCRM**
Vehicle Control Module	**Vehicle Control Module**	**VCM**
Vehicle Identification Number	**Vehicle Identification Number**	**VIN**
Vehicle Speed Sensor	Vehicle Speed Sensor[1]	VSS[1]
VIN (Vehicle Identification Number)	**Vehicle Identification Number**	**VIN**
VIP (Vehicle In Process) Connector	Data Link Connector[1]	DLC[1]
Viscous Converter Clutch	Torque Converter Clutch[1]	TCC[1]
Voltage Regulator	Voltage Regulator	VR
Volume Air Flow	Volume Air Flow[1]	VAF[1]
VR (Voltage Regulator)	Voltage Regulator	VR
VSS (Vehicle Speed Sensor)	Vehicle Speed Sensor[1]	VSS[1]
VSV (Vacuum Solenoid Valve) (Canister)	Evaporative Emission Canister Purge Valve[1]	EVAP Canister Purge Valve[1]
VSV (Vacuum Solenoid Valve) (EVAP)	Evaporative Emission Canister Purge Valve[1]	EVAP Canister Purge Valve[1]
VSV (Vacuum Solenoid Valve) (Throttle)	Idle Speed Control Solenoid Vacuum Valve[1]	ISC Solenoid Vacuum Valve[1]
Warm Up Oxidation Catalytic Converter	Warm Up Oxidation Catalytic Converter[1]	WU-OC[1]
Warm Up Three Way Catalytic Converter	Warm Up Three Way Catalytic Converter[1]	WU-OC[1]
Wide Open Throttle	Wide Open Throttle[1]	WOT[1]
WOT (Wide Open Throttle)	Wide Open Throttle[1]	WOT[1]
WOTS (Wide Open Throttle Switch)	Wide Open Throttle Switch[1]	WOT Switch[1]
WU-OC (Warm Up Oxidation Catalytic Converter)	Warm Up Oxidation Catalytic Converter[1]	WU-OC[1]
WU-TWC (Warm Up Three Way Catalytic Converter)	Warm Up Three Way Catalytic Converter[1]	WU-TWC[1]

Recommended Terms and Recommended Acronyms See Table 2

[1] Emission-Related Term

Bold indicates new/revised entry

Appendix D

Metric Conversions

	to convert these	to these,	multiply by:
TEMPERATURE	Centigrade Degrees	Fahrenheit Degrees	1.8 then +32
	Fahrenheit Degress	Centigrade Degrees	0.556 after - 32
LENGTH	Millimeters	Inches	0.03937
	Inches	Millimeters	25.4
	Meters	Feet	3.28084
	Feet	Meters	0.3048
	Kilometers	Miles	0.62137
	Miles	Kilometers	1.60935
AREA	Square Centimeters	Square Inches	0.155
	Square Inches	Square Centimeters	6.45159
VOLUME	Cubic Centimeters	Cubic Inches	0.06103
	Cubic Inches	Cubic Centimeters	16.38703
	Cubic Centimeters	Liters	0.001
	Liters	Cubic Centimeters	1000
	Liters	Cubic Inches	61.025
	Cubic Inches	Liters	0.01639
	Liters	Quarts	1.05672
	Quarts	Liters	0.94633
	Liters	Pints	2.11344
	Pints	Liters	0.47317
	Liters	Ounces	33.81497
	Ounces	Liters	0.02957
WEIGHT	Grams	Ounces	0.03527
	Ounces	Grams	28.34953
	Kilograms	Pounds	2.20462
	Pounds	Kilograms	0.45359
WORK	Centimeter Kilograms	Inch Pounds	0.8676
	Inch Pounds	Centimeter Kilograms	1.15262
	Meter Kilograms	Foot Pounds	7.23301
	Foot Pounds	Newton Meters	1.3558
PRESSURE	Kilograms/Sq. Cm	Pounds/Sq. Inch	14.22334
	Pounds/Sq. Inch	Kilograms/Sq. Cm	0.07031
	Bar	Pounds/Sq. Inch	14.504
	Pounds/Sq. Inch	Bar	0.06895

Appendix E

DECIMAL AND METRIC EQUIVALENTS

DECIMAL AND METRIC EQUIVALENTS

Fractions	Decimal (in.)	Metric (mm)	Fractions	Decimal (in.)	Metric (mm)
1/64	0.015625	0.397	33/64	0.515625	13.097
1/32	0.03125	0.794	17/32	0.53125	13.494
3/64	0.046875	1.191	35/64	0.546875	13.891
1/16	0.0625	1.588	9/16	0.5625	14.288
5/64	0.078125	1.984	37/64	0.578125	14.684
3/32	0.09375	2.381	19/32	0.59375	15.081
7/64	0.109375	2.778	39/64	0.609375	15.478
1/8	0.125	3.175	5/8	0.625	15.875
9/64	0.140625	3.572	41/64	0.640625	16.272
5/32	0.15625	3.969	21/32	0.65625	16.669
11/64	0.171875	4.366	43/64	0.671875	17.066
3/16	0.1875	4.763	11/16	0.6875	17.463
13/64	0.203125	5.159	45/64	0.703125	17.859
7/32	0.21875	5.556	23/32	0.71875	18.256
15/64	0.234275	5.953	47/64	0.734375	18.653
1/4	0.250	6.35	3/4	0.750	19.05
17/64	0.265625	6.747	49/64	0.765625	19.447
9/32	0.28125	7.144	25/32	0.78125	19.844
19/64	0.296875	7.54	51/64	0.796875	20.241
5/16	0.3125	7.938	13/16	0.8125	20.638
21/64	0.328125	8.334	53/64	0.828125	21.034
11/32	0.34375	8.731	27/32	0.84375	21.431
23/64	0.359375	9.128	55/64	0.859375	21.828
3/8	0.375	9.525	7/8	0.875	22.225
25/64	0.390625	9.922	57/64	0.890625	22.622
13/32	0.40625	10.319	29/32	0.90625	23.019
27/64	0.421875	10.716	59/64	0.921875	23.416
7/16	0.4375	11.113	15/16	0.9375	23.813
29/64	0.453125	11.509	61/64	0.953125	24.209
15/32	0.46875	11.906	31/32	0.96875	24.606
31/64	0.484375	12.303	63/64	0.984375	25.003
1/2	0.500	12.7	1	1.00	25.4

Glossary

A-circuit: A generator regulator curcuit that uses an external grounded field circuit. In the A-circuit, the regulator is on the ground side of the field coil.

Alassandro Volta: Italian physicist, 1745–1827. Invented the electrophorus which was a device for storing an electrical charge.

Alternating current (AC): Electrical current that changes directing from positive to negative.

American wire gauge (AWG): Size standard assigned to wire diameter. The higher the number, the smaller the wire conductor.

Ampere: Measurement of the movement of free electrons called current flow. One ampere is equal to 6.28 billion billion free electrons flowing past a point in one second.

Amp-hour rating (A/H): Battery rating that is the amount of current a fully charged battery can produce over a twenty hour period at 80(F before the terminal voltage reaches a minimum voltage of 10.5 volts.

Amplitude: Voltage measurement represented on an oscilloscope screen by the vertical Y axis.

Analog meter: An electrical meter operated by a D'Arsonval movement that uses a pointer needle that moves across the face of a printed scale.

Analog-to-digital (A/D) converter: A device to convert analog signals to digital signals used by the computer.

Analog voltage signal: A signal produced by an input sensor that is continuously variable within a certain voltage range.

André Marie Ampére: French scientist and mathematician, 1775–1836. Established the importance of the relationship between electricity and magnetism.

Anode: The positive charged electrode in a voltage cell.

Armature: The rotating component of an electric motor that consists of a conductor wound around a laminated iron core to create a magnetic field.

Atom: Smallest part of an element, consisting of protons, neutrons, and electrons.

Autofuse: A flat blade type fuse used for circuit protection in modern vehicles.

B-circuit: A generator regulator circuit that is internally grounded. In the B circuit, the voltage regulator controls the power side of the field circuit.

Back-feed: When the current in one circuit seeks an alternate path of flow when the original path is interrupted.

Ballast resistor: A resistive element enclosed in a ceramic block. Used to reduce the voltage to the ignition coil.

Base: The center layer of a bipolar transistor that activates the switching function of the transistor.

Bendix inertia drive: A starter drive pinion that depends on the inertia movement of the starter armature spinning to mesh the drive pinion forward into the ring gear teeth.

Bimetallic strip: Two dissimilar metals joined together, and when heated will expand at different rates and cause the strip to bend momentarily until cooled off again.

Binary code: A series of numbers represented by 1's and 0's. Any number or word can be translated into a combination of binary 1's and 0's.

Bit: A computer term for a binary digit.

Bound electron: Inner orbit(s) of electrons rotating around the nucleus of the atom.

Byte: A computer term for eight bits of information.

Cathode: The negative charged electrode of a voltage cell.

Capacitor: A device used to store electrical charges from a circuit. Also called a condenser.

Center high mounted stop light (CHMSL): A brake light mounted in the high center rear of the vehicle . Required on all vehicles sold in the United States since 1986.

Chemical energy: Electricity produced from a battery when two dissimilar metals and an electrolyte are subjected to a completed circuit.

Circuit breaker: An electrical circuit load control device. May be designed to be self-resetting, or manual-resetting.

Circuit protection device: An in-line fuse or circuit breaker that protects an electrical circuit from excess amperage.

Clamping diode: A diode that is connected in parallel with a coil to prevent voltage spikes from the coil from reaching other components in the circuit.

Cold cranking ampere rating (CCA): The ability of the battery to provide an amperage load for thirty seconds at 0° F without the terminal voltage dropping below 7.2 volts.

Collector: The portion of the bipolar transistor that receives the majority of the current carriers.

Commutator: A series of conducting segments located around one end of the armature.

Condenser: See capacitor.

Continuity light: See self-powered test light.

Conventional battery: A lead acid battery that uses a lead antimony mixture in the cell plates.

Conventional flow theory: An electrical theory that states that current flows from a positive potential to a negative potential.

Current flow: The flow of free electrons from positive to negative. Measured in amperes.

Daylight running lights (DRL): Headlights that operate automatically during daylight hours at a reduced voltage.

Deep cycle: A lead acid battery that uses approximately 2.75 percent antimony on the positive plate and a calcium alloy on the negative plate.

Delta wound: A stator winding in an AC generator that consists of three connections connected together in a triangular shape and with three output leads.

Depletion zone: See PN junction.

Depressed park: A wiper circuit that parks the wiper blades off the glass past the end of their normal stroke, usually below the edge of the lower windshield molding.

Detonation: A combustion chamber condition that occurs when the normal combustion process is too slow, causing some unburned fuel/air mixture to ignite and create a secondary flame front that collides with the normal flame front.

Dielectric: An insulator material.

Digital multimeter (DMM): Also known as digital volt ohm meter (DVOM). Uses a digital readout display that may also automatically record a measured value.

Digital storage oscilloscope (DSO): Also known as a scope meter. Converts the input voltage signal into digital information. For screen display or transfer to other diagnostic equipment.

Digital voltage signal: Also known as a square wave signal. A signal that produces a square wave from the rapid on and off cycling of the circuit.

Diode: An electrical one-way check valve that allows current to flow in one direction only.

Diode trio: Rectifies current from the stator to create the magnetic field excitation in the rotor field coil.

Direct current (DC): Steady-state electrical potential that remains at a specific voltage in one direction.

Double-filament bulb: An incandescent bulb with two coiled tungsten filaments that serve more than one electrical function.

Duty cycle: The relationship between the amount of on time and the off time in a digital signal.

Dwell: The amount of time that the primary ignition system points are closed. Measured in degrees.

Electrolysis: Also called gassing. When the battery electrolyte is converted into hydrogen and oxygen gas.

Electronic erasable programmable read only memory (EEPROM): Computer memory chip that can only be programmed with special test equipment.

Electromagnet: A magnet formed when current flows through a coil of wire wrapped around a soft iron core.

Electromagnetism: Magnetic field that occurs when a current flows through a conductor.

Electromotive force (EMF): See voltage

Electron: Negatively charged particle of an atom.

Electron flow theory: An electrical theory that states that current flows from a negative potential to a positive potential.

Electrolyte: A chemical compound of alkali, salt, or acid solution that is capable of conducting electrical energy.

Electrostatic discharge (ESD): The process by which charged objects transfer their charge to any neighboring object.

Electrostatic field: The field that is between two oppositely charged plates.

Element: A cluster of one type of atom.

Emitter: The outer layer of the transistor that supplies the majority of the current carriers.

Farad: The standard unit of measurement for a capacitor.

Field coils: Heavy copper wire wrapped around an iron core called a pole shoe.

Field current: The current going to the field windings of a motor or generator.

Fixed resistor: A resistive element normally made of carbon/graphite-based material mixed with a resin bonding agent.

Flux density: The concentration of the magnetic lines of force.

Forward bias: A positive voltage that is applied to the p-type material or anode of a diode.

Free electrons: Valence electrons that become dislodged from their orbit and move to a neighboring atom.

Frequency: The number of cycles within a period of time.

Full-wave rectification: The conversion of a complete AC voltage signal to a DC voltage.

Fusible link: An in-line wire in the circuit that is made of meltable material with a special heat resistant insulation. The link will melt when the designed circuit current flow is exceeded.

George Simon Ohm: German physicist, 1787-1854. Identified the relationship between resistance, current, and voltage in an electrical circuit known as Ohm's Law.

Grid growth: A condition that shorts out the plates in the battery when little metallic fingers grow through the separators to the plates.

Grounded circuit: Also referred to as a short-to-ground. An unwanted path for current to flow directly to ground.

Half-wave rectification: Rectification of one half of an AC voltage.

Hall-effect switch: A sensor that operates on the principle that if a current is allowed to flow through thin conducting material being exposed to a magnetic field, another voltage is produced.

Halogen headlamp: A type of headlamp that uses halogen gas in an inner bulb to allow the tungsten filament to burn brighter.

Hard shell connector: An electrical connector that has one to a dozen wires molded separately into the connector.

Hydrometer: A testing instrument that measures the specific gravity content of the electrolyte to determine a battery's state-of-charge condition.

Incandescent lamp: A device that produces a light as a result of current flowing through a coiled tungsten wire filament inside a glass envelope.

Inductive pick-up: An adapter available for a meter to measure the strength the magnetic field, generated by the current flow in a circuit.

Instrument voltage regulator (IVR): Maintains a constant voltage to the gauge under all battery load and charging conditions.

Integrated circuit (IC) chip: A complex circuit consisting of thousands of transistors, diodes, resistors, capacitors, and other electronic devices that are formed on a small silicon chip.

Interval wiper: A wiper system that operates in time intervals depending on the delay setting requested by the driver.

James Watt: Scottish inventor, 1736–1819. Developed a comparison to what his steam engines could produce in place of a normal size horse, now known as Watt's Law.

Joule: Base unit of energy measurement.

Keep alive memory (KAM): A volatile RAM memory that retains information as long as it is connected to a battery power source.

Kirchoff's voltage law: A circuit law for voltage in a series circuit, where the sum of the individual voltage drops of each resistive load equals the source voltage.

Kirchoff's current law: A circuit law for current in a parallel circuit, where the sum of the current in the individual branch circuits equals the total circuit amperage.

Knock sensor: A piezoelectric device that senses combustion chamber detonation to control spark knock.

Lead antimony: A battery grid plate that contains up to a ten percent mixture of lead and lead alloys to strengthen the plate grid frame.

Lead peroxide: A reddish brown paste material pressed into the battery positive plate.

Light emitting diode (LED): A gallium arsenide diode that converts the energy developed when holes and electrons collide during normal diode operation into light.

Logic gates: Electronic circuits that act as gates to output voltage signals depending on different combinations of input signals.

Magnetism: A natural invisible force resulting from atoms aligning within certain materials that can then repel or attract other materials.

Maintenance free: A lead acid battery that uses a lead calcium mixture in the cell plates.

Matter: Any substance that occupies a space and contains a mass.

Maxifuse: A large flat blade-type fuse used as a replacement design for fusible links in electrical circuits.

Mercury switch: A vial that contains mercury that conducts electricity when electrical contacts come into contact with the mercury.

Metri-pack connector: Similar to a weather-pack connector, except they do not have a seal on the cover half of the connector.

Microprocessor: The decision making chip in the computer that interprets input information, makes calculations, and makes decisions.

Minifuse: A small blade-type fuse designed to save space and allow more electrical circuits to have individual fuses.

Molecule: A group of elements.

MPMT switch: An electrical switch that contains multiple wipers that work in unison together. Also known as a ganged switch.

Multifunction switch: A combination switch on the steering column consisting of a turn signal switch, headlight switch, hazard switch, horn switch, and flash-to-pass switch.

NATEF: The abbreviation for the National Automotive Technician Education Foundation. Dedicated to developing national standards for training automotive technicians.

Neutral safety switch: A safety switch that prevents a starter from cranking unless the automatic transmission is in park or neutral, or the clutch is pushed in for a manual transmission.

Neutrons: Particle of the nucleus of the atom that carries no charge.

Non-depressed park: A wiper circuit that parks the wiper blades at the end of their normal stroke, at the edge of the lower windshield molding.

Non-powered test light: Simple test light tool that connects one end to ground, and the other end to a voltage source to light the bulb.

Normally closed (N/C) switch: A switch denoting that the contacts are normally closed until opened by an outside force.

Normally open (N/O) switch: A switch denoting that the contacts are normally open until closed by an outside force.

NPN Transistor: A transistor that contains two N-type materials, and one P-type material.

Nucleus: The core of the atom that contains positively charged protons, and neutrons with no charge.

Ohm's law: The current in an electrical circuit is inversely proportional to the resistance of the circuit, and directly proportional to the voltage.

Open circuit: A break in the current flow of a complete circuit.

Open circuit voltage: the voltage of a battery in a no-load condition. Normally used on a maintenance free battery to check the state-of-charge condition.

Overrunning clutch drive: A starter drive pinion that wedges spring-loaded rollers between an inner and outer housing to engage the pinion teeth into the flywheel teeth.

Pacific fuse element: A quick replacement fuse developed to replace the fusible link.

Parallel circuit: A circuit that provides two or more paths for current to flow.

Parasitic drain: Unwanted current drain still present in the electrical system after the ignition switch is in the off position.

Photons: Pure energy that contains no mass. When light energy strikes the surface of a photovoltaic cell, the energy in the photon is given to a free electron.

Piezoelectricity: Electricity produced from pressure being applied to quartz or barium titanate crystals.

PN junction: The area where the anode and cathode material join together in a diode. Also called the depletion zone.

PNP transistor: A transistor that contains two P-type materials, and one N-type material.

Pole shoe: The soft iron core inside the field coil that helps concentrate and direct the magnetic lines of force in the field coil.

Potentiometer: A variable resistor, with three wire connections, that acts as a circuit divider to provide accurate voltage drop readings proportional to movement.

Power of ten: Base unit of measurement for the metric system.

Pre-ignition: a combustion chamber condition that occurs by an early ignition of the fuel/air mixture, and by a source other than the spark plug and ignition system.

Primary wire: Term used for a wiring circuit that carries a low voltage.

Primary cell: A cell that cannot be recharged to it's original voltage condition.

Programmable read only memory (PROM): Memory chip that contains specific data that pertains to the exact vehicle in which a computer is installed.

Proton: Positive charged particle of an atom. Protons and neutrons form together to form the nucleus.

Pull-to-seat connector: An electrical terminal that is pulled into the connector housing to seat it.

Pulse width: The amount of time the voltage is on or off in a digital circuit.

Pulse width modulation: The on and off cycling of a component. The pulse width varies, but its frequency remains constant.

Push-to-seat connector: An electrical terminal that is pushed into the connector housing to seat it.

Random access memory (RAM): Stores temporary information that can be read from or written to by the CPU. RAM can be designated as volatile or nonvolatile.

Read only memory: Computer memory chip that stores permanent information used to instruct the computer on what to do in response to input data.

Recombination battery: A completely sealed lead acid battery that uses electrolyte in a gel form.

Rectification: The converting of an AC current to a DC current.

Relay: A switch that uses low current in a circuit coil to magnetically control the high current in another circuit through a set of mechanical contact points.

Reserve capacity (R/C): A battery rating that is the amount of time in minutes a fully charged battery can be discharged at 25 amperes while maintaining a minimum battery voltage of 10.5 volts at 80° F in the event of a charging system failure.

Resin core: See rosin core.

Resistance: Any force or substance that restricts or opposes the flow of current in a circuit.

Reverse bias: A positive voltage is applied to the N-type material, or cathode, of a diode.

Rheostat: A variable resistor with two terminals. One connected to a fixed end of the resistor. The other end connected to a movable wiper contact.

Rosin core: A type of flux cleaning agent that is used with solder to repair electrical connections. Also known as resin core.

Rotary switch: An electrical switch that allows for multiple functions to be made from the same control knob. Normally used on an analog meter.

Rotor: The component of an AC generator that rotates and creates the magnetic field.

Scan tool: A diagnostic tool used to communicate with the vehicle's computer and access stored diagnostic trouble codes.

Schmitt trigger: An electronic circuit used to convert an analog input signal to a clean square wave digital signal.

Sealed-beam headlamp: A light assembly consisting of a glass lens fused to a parabolic reflector that is sprayed with vaporized aluminum. The inside of the lamp is filled with argon gas to prevent the filament from oxidizing.

Secondary wire: Term used for a wiring circuit that carries a high voltage, such as spark plug wires.

Secondary cell: A cell that can be recharged once it's original stored energy has been depleted.

Seebeak effect: Also known as thermoelectricity. The heating of two dissimilar metals joined at one end will produce a voltage at the open ends.

Self-powered test light: Also known as a continuity light. A simple test light that contains an internal battery that will light the lamp bulb when there is a complete circuit.

Semiconductor: A stable element with exactly four electrons in it's structure.

Separator: A micro-porous material used to prevent contact between each positive and negative plate.

Series circuit: An electrical circuit that provides a single path for current flow from the electrical source through all the circuit's components, and back to the source.

Series/parallel circuit: An electrical circuit that has some loads in series and some in parallel.

Short circuit: An unwanted or accidental bypass of the current flow in a circuit.

Short-to-ground: See grounded circuit.

Short-to-voltage: An unwanted copper-to-copper connection between two separate circuits.

Sine Wave: A scope pattern that shows the positive and negative voltage generated during one revolution.

Single-filament bulb: An incandescent bulb with one coiled tungsten filament.

Slip rings: The copper rings at one end of the rotor in an AC generator that provide a current path connection from the carbon brushes to the rotor coil.

Solenoid: An electromagnetic device that uses movement of a plunger to exert a pulling or holding force.

Specific gravity: The weight of a given volume of liquid in comparison to the weight of the same volume of water with a specific gravity of 1.000.

Solar light energy: Electricity generated by using light energy from the Sun and capturing it in a photovoltaic solar cell.

Solderless connector: Electrical connector that is crimped to the wire with special pliers.

SPDT switch: Single-pole double-throw switch that has one input wire and two output wires

SPST switch: Single-pole single-throw switch that has one input wire and one output wire. Switch may be on/off controlled, or momentary.

Square wave: See digital voltage signal.

Static electricity: Electricity that is without motion or at rest, and will produce an electrical charge whenever two dissimilar nonmetallic materials are rubbed lightly together.

Stator: The stationary coil windings in an AC generator in which electricity is produced.

Stepped resistor: Typical resistor with two or more fixed resistive values. Normally used to control the current in a heater blower motor circuit.

Sulfuric acid: A conductive and reactive mixture of H_2SO_4 consisting of sixty four percent water and thirty six percent sulfuric acid, by weight.

Tap splice: A type of solderless connector that joins a new wire to an existing circuit.

Thermistor: A variable resistor made from a semiconductor material that changes resistance in relation to temperature changes.

Thermocouple: A small device that gives off a voltage when two dissimilar metals are heated. Normally used to monitor exhaust temperatures.

Thermoelectricity: Electricity produced when two dissimilar metals are heated to generate an electrical voltage.

Three-phase: The resulting overlap of three-phase stator windings produces a three-phase voltage output.

Transistor: A three-layer semiconductor used as a very fast switching device.

Unintentional open circuit: An unwanted break in the current flow in a electrical circuit.

Valence electron: Electrons in the outer most orbit of the atom.

Variable resistor: A resistor that provides for an infinite number of resistance values within a set range.

Voltage: The electrical pressure that causes the free electrons to move through a circuit. Also referred to as EMF or electromotive force.

Voltage drop: A resistance in an electrical circuit that reduces the electrical pressure or voltage available after the resistance. The voltage drop can be the result of resistance in the load component, the conductors, connection points, or unwanted resistance.

Voltage limiter: Automatically determines what the field circuit voltage is depending on the required amount of charging.

Voltage regulator: Used to control the output voltage of an AC generator based on charging system demand.

Voltaic cell: Two unlike metals and an alkali, salt, or acid solution combined to produce a voltage.

Voltaic pile: A series of small round copper and zinc plates separated by cardboard soaked in saltwater.

Watt's law: The amount of work done in one second by one volt moving a charge of one amp through a resistance of one ohm is equal to one watt. 1 horsepower = 746 watts.

Waveform: The picture produced on an oscilloscope screen. Also called a scope trace.

Weather-pack connector: Electrical connectors that have a rubber seal on the terminal ends and on the covers of the connector halves. Used on electrical circuits for moisture and corrosion protection.

Wye wound: Stator windings that consist of a common junction and three output connections that branch out in a "Y" pattern.

Zener diode: A diode that allows reverse current to flow above a designed-in voltage limit.

Index

A

C

E

H

I

J

K

L

M

S

T

U

V

W

X

Z